Nonconventional Yeasts in Biotechnology

Springer
Berlin
Heidelberg
New York
Barcelona
Budapest
Hong Kong
London
Milan
Paris
Santa Clara
Singapore
Tokyo

Klaus Wolf

Nonconventional Yeasts in Biotechnology

A Handbook

With 94 Figures and 45 Tables

 Springer

Prof. Dr. Klaus Wolf
RWTH Aachen
Institut für Biologie IV (Mikrobiologie)
Worringer Weg
52056 Aachen, Germany

Cover Illustration: Kluyveromyces lactis. A scanning electron micrograph of *K. lactis* strain CBS 2359. Courtesy of Professor Masako Osumi, Department of Biology, Japan Women's University, Tokyo (see Chap. 5, Fig. 1)

ISBN-13: 978-3-642-79858-0   e-ISBN-13: 978-3-642-79856-6
DOI: 10.1007/978-3-642-79856-6

Library of Congress Cataloging-in-Publication Data. Nonconventional yeasts in biotechnology: a handbook/ Klaus Wolf. p. cm. Includes bibliographical references and index.   1.
Yeast fungi—Biotechnology—Handbooks, manuals, etc.   I. Wolf. K. (Klaus), 1994– .   TP248.27.Y43N66   1996
660'.62—dc20   96-4935

© Springer-Verlag Berlin Heidelberg 1996
Softcover reprint of the hardcover 1st edition 1996

Cover design: Springer-Verlag, Erich Kirchner

Typesetting: Best-set Typesetter Ltd., Hong Kong

SPIN: 10037326   39/3137/SPS – 5 4 3 2 1 0 – Printed on acid-free paper

Preface

It is a difficult task to define a yeast and, even more so, a nonconventional yeast. According to the yeast taxonomists Guilliermond and Lodder, yeasts are unicellular fungi which reproduce by budding or fission. In this sense, only unicellular organisms belong to the yeast group, but many yeast species are dimorphic and produce yeast and hyphal forms. Similarly, many filamentous fungi have yeast stages. Therefore the reader will miss one or the other organism in this volume.

In this book, all yeasts except *Saccharomyces* and *Schizosaccharomyces* are classified as nonconventional. This may change over the years, as is evident from the wealth of information collected for some of these yeasts, as well as their applications in many fields. Therefore some of the nonconventional yeasts of today will be the conventional yeasts of tomorrow.

The reader is always referred to the excellent guides to genetics and molecular biology of *Saccharomyces* by Guthrie and Fink (1991)[1] and of *Schizosaccharomyces* by Alfa, Fantes, Hyams, McLeod and Warbrick (1993).[2]

In contrast to these books, this volume is aimed to provide much more background information on the organisms and their biotechnological use.

Since many techniques are used in all yeast systems, there will be some overlap. This may, however, be helpful for the reader, since the experimental variants are adaptations of a technique for the special yeast.

This book will be a good introduction into the special properties of yeasts for beginners and also for current practitioners. Molecular biology is progressing so fast, that some of the techniques will certainly have been altered by the time of appearance of this book. Nevertheless, the detailed descriptions by the authors should help in avoiding experimental pitfalls.

The volume is preceded by a chapter on principles and methods of yeast classification in order to provide a framework for the nontaxonomically trained user of yeasts and to give an impression of the wealth of potentially useful organisms. Identifying and classifying organisms is of great importance to many areas of science, including agriculture, medicine, biology, biotechnology and the food industry, and also for assessment of industrial property rights.

---

[1] Guide to yeast genetics and molecular biology. Methods in enzymology, vol 194. Academic Press, New York

[2] Experiments with fission yeast. A laboratory course manual. Cold Spring Harbor Laboratory Press, Cold Spring Harbor

Two more chapters are devoted to an introduction into the basic techniques of protoplast fusion and karyotyping. Variants of these techniques will also be found for some of the yeast species.

I thank all the authors for their care in preparing their chapters, and all my colleagues for their help in editing this book.

Aachen   Klaus Wolf
Summer 1995

Contents

Chapter 2

Protoplast Fusion of Yeasts
Martin Zimmermann and Matthias Sipiczki 83

Chapter 3

Electrophoretic Karyotyping of Yeasts
Martin Zimmermann and Philippe Fournier 101

Chapter 6

Pichia pastoris
Koti Sreekrishna and Keith E. Kropp 203

Chapter 9

Hansenula polymorpha (Pichia angusta)
Hans Hansen and Cornelis P. Hollenberg............................. 293

Chapter 10

Yarrowia lipolytica
Gerold Barth and Claude Gaillardin 313

Chapter 13

Trichosporon
Jakob Reiser, Urs A. Ochsner, Markus Kälin, Virpi Glumoff,
and Armin Fiechter 581

List of Contributors

(Their addresses can be found at the beginning of their respective chapters.)

Barth, G.   313
Boekhout, T.   1
Breunig, K.D.   139
Dohmen, R.J.   117
Fiechter, A.   581
Fournier, P.   101
Fukuhara, H.   139
Gaillardin, C.   313
Glumoff, V.   581
Hansen, H.   293
Hollenberg, C.P.   117, 293
Kälin, M.   581
Kropp, K.E.   203
Kunze, G.   389

Kunze, I.   389
Kurtzman, C.P.   1
Mauersberger, S.   411
Ochsner, U.A.   581
Ohkuma, M.   411
Reiser, J.   581
Schunck, W.H.   411
Sibirny, A.A.   225, 277
Sipiczki, M.   83
Sreekrishna, K.   203
Takagi, M.   411
Wésolowski-Louvel, M.   139
Zimmermann, M.   83, 101

Principles and Methods Used in Yeast Classification, and an Overview of Currently Accepted Yeast Genera

Teun Boekhout[1] and Cletus P. Kurtzman[2]

1
Introduction

Yeasts are of benefit to mankind because they are widely used for production of foods, wine, beer, and a variety of biochemicals. Yeasts also cause spoilage of foods and beverages, and are of medical importance. At present, approximately 700 yeast species are recognized, but only a few are commonly known. Relatively few natural habitats have been thoroughly investigated for yeast species; consequently, we can assume that many more species await discovery. Because yeasts are widely used in traditional and modern biotechnology, the exploration for new species should lead to additional novel technologies.

Several definitions have been used to describe the yeast domain. According to Guilliermond (1912) and Lodder (1970), yeasts are unicellular fungi which reproduce by budding or fission. In this sense, only true unicellular fungi are regarded as yeasts, but, in reality, many yeast species are dimorphic and produce pseudohyphae and hyphae in addition to unicellular growth. Similarly, many hyphal fungi are dimorphic and are usually referred to as yeast-like. Because of the overlap in morphological appearance, some authors regard yeasts merely as fungi that produce unicellular growth, but that otherwise are not different from filamentous fungi (Flegel 1977), or as unicellular fungal growth forms which have resulted as a response to a commonly encountered set of environmental pressures (Kendrick 1987). Oberwinkler (1987) placed the yeasts in a phylogenetic framework and defined them as unicellular, ontogenetic stadia of either asco- or basidiomycetes (see also van der Walt 1987). In summary, yeasts are ascomycetous or basidiomycetous fungi that reproduce vegetatively by budding or fission, and that form sexual states which are not enclosed in a fruiting body. Molecular comparisons show the ascomycetous yeasts to be phylogenetically distinct from the filamentous species. In this chapter, we present an introduction to the principles, trends, and methods of yeast systematics with the aim of providing a framework for the non-taxonomically trained user of yeasts.

[1] Centraalbureau voor Schimmelcultures, Yeast Division, Julianalaan 67, 2628 BC Delft, The Netherlands
[2] National Center for Agricultural Utilization Research, USDA, ARS, 1815 North University Street, Peoria, Illinois 61604, USA

2
Some Principles of Yeast Taxonomy

Identifying, naming, and placing organisms in their proper evolutionary frame-work is of importance to many areas of science that include agriculture, medicine, the biological sciences, biotechnology, food industry, and for assessment of indus-trial property rights. As evolutionary uniqueness can be expressed at all levels between gene and the whole organism, comparative investigations also need to be performed at various levels of biological development and diversification. Areas of comparison include morphology, physiology, genetics, biochemistry, ecology, and molecular genetics. Taxonomic concepts change as the result of developments in science and philosophy. As a consequence, several different species concepts have been applied in yeast systematics.

The phenetic species concept is based on discontinuities of phenotypic charac-teristics. In the past, delimitation of yeast taxa was mainly based on morphological and physiological differences between strains or groups of strains. Kreger-van Rij (1984b), for instance, defined yeast species as an assemblage of clonal populations. The reliability of the phenetic approach largely depends on the quality and number of characters investigated. Interpretation of physiological data is complicated, because many of the carbon sources used in discriminatory growth tests can be metabolized by common pathways (Barnett 1977; Golubev 1989). Furthermore, the metabolism of many mono-, di- and trisaccharides is controlled by only one or a few genes (Winge and Roberts 1949; Barnett 1968), and physiological charac-teristics are not always genetically stable and reproducible (Scheda 1966; Scheda and Yarrow 1966). Extension of the series of carbon and nitrogen compounds used in growth tests may increase the significance of this approach to yeast classi-fication. Phaff (1989) suggested that compounds with a complicated metabolic route will be most useful in this respect. The weakness of the phenetic approach has been stressed by several authors (Phaff 1981, 1989; Kurtzman et al. 1983; Kurtzman and Phaff 1987; van der Walt 1987). However, from a practical point of view, fermentation and assimilation reactions are still widely used for identification.

The biological species concept assumes the existence of arrays of Mendelian populations which are reproductively isolated from other population arrays (e.g., Dobzhansky 1976). The occurrence of a perfect (sexual) state following mating of complementary strains is commonly interpreted to indicate conspecificity of the mated strains. However, unless viability of the F_1 and F_2 generations is verified, the presumption of conspecificity may be incorrect because some closely related spe-cies may mate but the progeny are not viable (for examples see Kurtzman 1987). Among basidiomycetous yeasts, gene exchange has only been documented in a limited number of heterothallic species such as *Rhodosporidium toruloides* (Banno 1967), and *Filobasidiella neoformans* (Kwon-Chung 1980). By definition, the bio-logical species concept cannot be applied to asexual (anamorphic) yeast species such as the genus *Candida*, but measurements of DNA relatedness can lead to an approximation.

A third species concept is the evolutionary or phylogenetic species concept (Wiley 1981), which regards a species as a single phylogenetically derived lineage. The phylogenetic species concept cannot be applied directly to the yeast domain because of the lack of genealogical evidence (e.g., lack of fossils), and because of morphological and physiological plasticity. However, the increasing number of molecular evolutionary studies of yeasts, e.g., those using sequence analysis of ribosomal DNA, should result in a clearer understanding of the applicability of the phylogenetic species concept to yeasts. Presently, many taxa are best interpreted as genetically uncertain entities whose definition needs to be tested by the analysis of independent phylogenetically derived character sets. Ideally, these approaches, together with a critical evaluation of phenetic and genetic data, should lead to a stable species concept.

3
Trends in the Systematics of Yeasts

The first period of yeast systematics (until approximately 1960) is characterized by a thorough study of morphology, comparative nutritional physiology, and conventional genetics. Important workers in this period were M. Reess (morphology), E.C. Hansen (application of pure cultures and physiology), A.J. Kluyver (physiology), L.J. Wickerham (physiology, genetics, ecology), and A. Guilliermond, O. Winge and C.C. Lindegren (genetics). Comparative taxonomic studies performed at the CBS Yeast Division (e.g., Stelling-Dekker 1930; Lodder 1934; Diddens and Lodder 1942), resulted in a series of monographs, which made the so-called Delft School well known.

Initially, responses on only a limited number of carbon and nitrate compounds were used for taxonomic purposes. Wickerham (1951) extended this series, and today approximately 60 tests are being performed routinely, including fermentation and assimilation of carbon compounds, assimilation of nitrogen compounds, vitamin requirements, resistance to cycloheximide, temperature requirements, etc. (see Sect. 5, Methods).

Genetic studies revealed the presence of different sexual strategies. Sexual cycles of ascomycetous yeasts may be haplontic, diplontic, or diplohaplontic. Yeast species may be homothallic, heterothallic, or a combination of these. Monokaryotic fruiting is tentatively assumed to occur in species presently placed in the genus *Mrakia* (Fell and Statzell Tallman 1984a), and apomixis has been observed in the heterobasidiomycete genus *Itersonilia* (Boekhout 1991a). Incompatibility systems of basidiomycetous yeasts are bipolar, tetrapolar, or modified tetrapolar, and mating factors can be biallelic or multiallelic (Fell 1984; Fell and Statzell Tallman 1984a,b; Kwon-Chung and Fell 1984; Wong 1987).

The second period of yeast systematics (1960 until present) is characterized by an extension of morphological characteristics because of the introduction of the electron microscope, the application of biochemical criteria, and the introduction of molecular genetic studies. Transmission electron microscopy has revealed differences between ascomycetous and basidiomycetous yeasts. Ascomycetous yeasts have electron-transparent cell walls, mainly made up of β-glucans, and a thin electron-dense outer layer which consists of α-mannans. Basidiomycetous yeasts have lamellate and electron-dense cell walls which are mainly built up of β-glucans (Fig. 1; Kreger-van Rij and Veenhuis 1971). Bud formation is also different in these two groups of yeasts. Ascomycetous yeasts show holoblastic budding, i.e., the entire cell wall seems to be involved in the formation of the newly formed wall of the bud, while basidiomycetous yeasts have enteroblastic budding in which only the inner cell wall layer is involved in this process (Figs. 1, 2). Other ultrastructural differences between these two groups of yeasts are found in the mitotic apparatus (Heath 1978; Heath et al. 1987), and can be summarized as follows. Ascomycetous

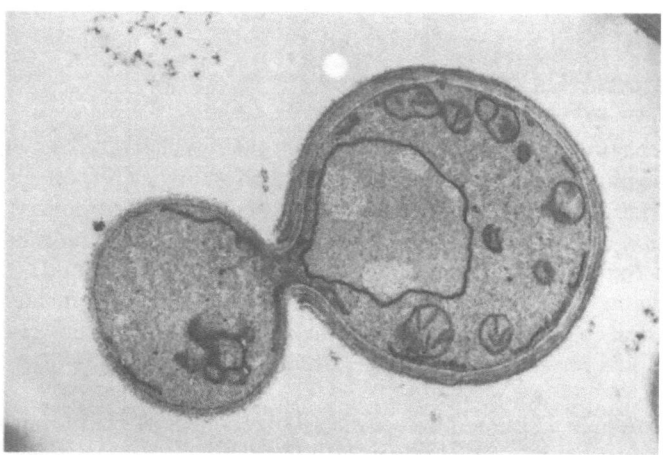

Fig. 1. Multilayered cell wall and enteroblastic budding in *Rhodotorula acuta* (CBS 7053, ×13 500)

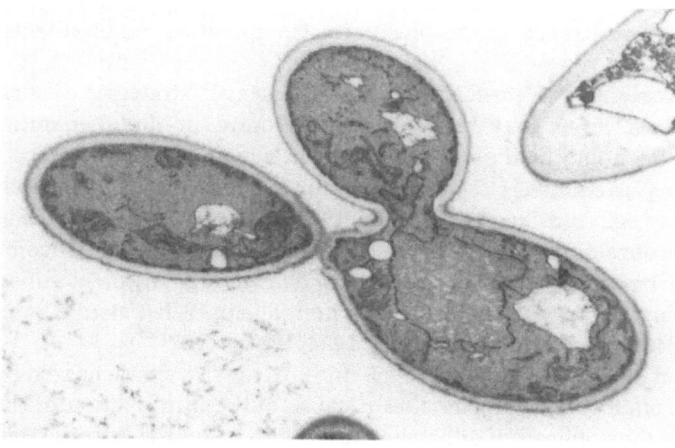

Fig. 2. Bilayered cell wall and holoblastic budding in *Saccharomyces cerevisiae* (CBS 1171, ×8400)

yeasts have a layered spindle pole body (SPB) which is closely associated with the nuclear envelope, and spindle formation takes place inside the mother cell. Elongation of the mitotic spindle is towards the bud. The SPB of basidiomycetous yeasts is not layered, is initially positioned in the cytoplasm, and spindle formation takes place inside the bud. Elongation of the spindle is towards the mother cell.

Septal ultrastructure shows important differences between the two classes of yeasts. Septa of many ascomycetous yeasts have one or several micropores. These are very thin electron-dense connections between two adjacent cells (Fig. 3). Additionally, diaphragma-like pores occur as well, and Woronin bodies may be present. Pores of *Ambrosiozyma* species are swollen around the pore, thus having some resemblance with the dolipores of basidiomycetes (Fig. 4). Heterobasidiomycetous yeasts show a greater variation in septal ultrastructure. In the cytoplasm, a structure made up of modified endoplasmic reticulum, the parenthesome, may be present. The parenthesome can have a different morphology. Tremellaceous yeasts usually have dolipores in which the septum is swollen around a central pore (Figs. 5, 6). *Filobasidiella* and *Bulleromyces* have a parenthesome made up of U-shaped vesicles (Tremellales type). Other basidiomycetous yeasts (e.g., *Itersonilia*) lack a parenthesome. Another group of basidiomycetous yeasts has diaphragma-like pores reminiscent of those found in the higher ascomycetes, but without Woronin bodies. These yeasts may be related to the smuts (Ustilaginales). A third group of basidiomycetous yeasts has micropore-like structures (Boekhout et al. 1992). The fine structure of septa seems to reflect the affinity to higher taxonomic categories.

Biochemical characteristics, such as carbohydrate composition of cell walls and capsules (e.g., Weijman and Rodrigues de Miranda 1983; Weijman and Golubev 1987; Suzuki and Nakase 1988; Prillinger et al. 1993), proton magnetic resonance spectra of cell walls (e.g., Spencer and Gorin 1969, 1970), number of isoprene units of the coenzyme Q (e.g., Yamada and Kondo 1973; Yamada et al. 1973, 1987), cytochromes (e.g., Claisse et al. 1970; Fiol and Claisse 1987; Montrocher and

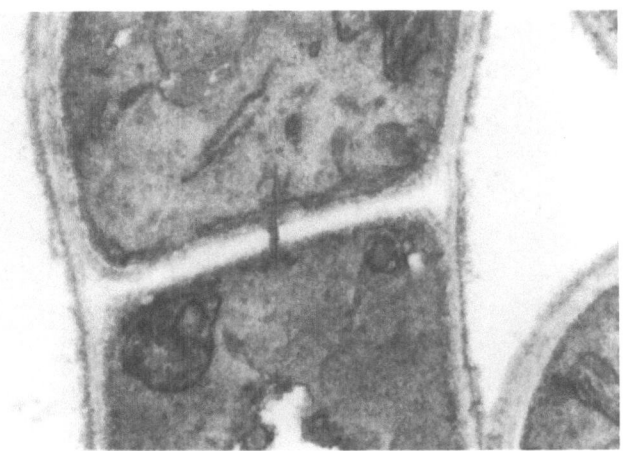

Fig. 3. Micropore in *Sporothrix guttuliformis* (CBS 437.76, ×27 000)

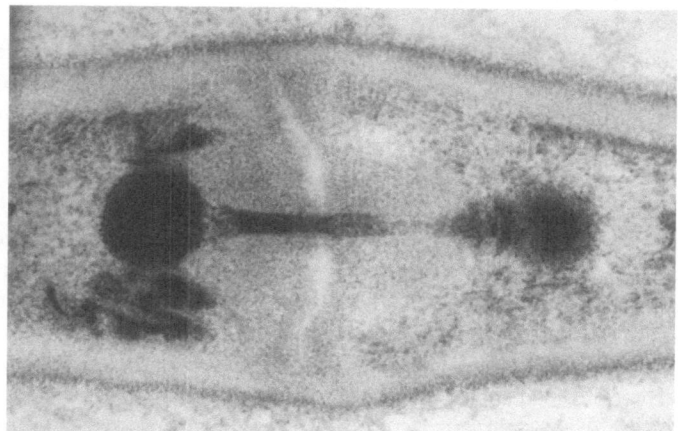

Fig. 4. Dolipore-like structure in *Ambrosiozyma platypodis* (CBS 4111, ×36 000)

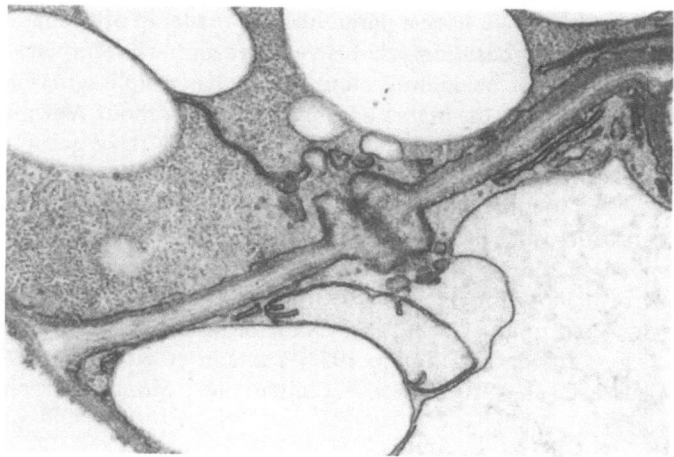

Fig. 5. Dolipore in *Trichosporon beigelii* (CBS 5791, ×25 000)

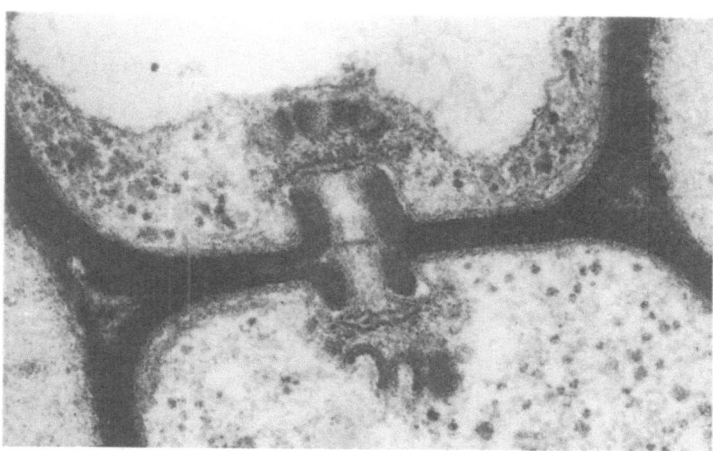

Fig. 6. Dolipore with parenthesome in *Bulleromyces albus* (CBS 500 × CBS 7440, ×60 000)

Claisse 1988), fatty acid composition (e.g., Cottrell et al. 1986; Viljoen et al. 1986; Westhuizen et al. 1991), and isozyme patterns (e.g., Yamazaki et al. 1983, 1985) have been used for taxonomic distinctions.

The introduction of DNA studies provides, in principle, an objective parameter for estimating evolutionary distances among taxa. Different methods offer resolution at different taxonomic levels. The taxonomic value of nucleic acid base composition (mol% G+C) is mainly exclusionary (see Sect. 5, Methods). Phenotypically similar strains differing more than ca. 2–3% in their base composition are usually regarded as different species (Kurtzman 1985; Kurtzman and Phaff 1987; Phaff 1989), while strains with the same base composition do not necessarily represent one and the same species. The range of nucleic acid base compositions differs for ascomycetous and basidiomycetous yeasts. Most ascomycetous yeasts have a mol% G+C lower than 50, whereas most basidiomycetous yeasts have a mol% G+C above 50 (Table 1).

DNA hybridization studies are used to determine DNA similarity between species. Commonly used methods (for reviews see Jahnke 1987; Kurtzman 1993) include spectrophotometric analysis of heteroduplex formation, membrane-bound reassociation techniques using isotopes or fluorochromes and methods in which the reassociated heteroduplex is bound on hydroxylapatite columns (Britten et al. 1974). Recently, fluorometric and colorimetric methods using microtiter plates have been developed (Ezaki et al. 1989; Hara et al. 1991) and are being applied in yeast taxonomy (Kaneko and Banno 1991). DNA-binding percentages above 65–70% are interpreted to indicate conspecificity (Phaff 1981; Kurtzman 1987; Kurtzman et al. 1980). A proportional relationship has been suggested between the occurrence of gene flow and high values of DNA similarity (Lachance 1985). A positive correlation seems to exist between DNA complementarity and interfertility (Phaff 1981; Kurtzman and Phaff 1987; Kurtzmen et al. 1980). However, low values of DNA similarity, even up to ca. 25%, do not necessarily exclude gene exchange (Kurtzman et al. 1980; Aulakh et al. 1981; Vaughan Martini and Kurtzman 1985; Kurtzman and Phaff 1987). Consequently, a rigid application of any lower limit up to ca. 25% DNA similarity as the sole criterion for species delimitation does not seem justified. Intermediate values of DNA similarity (40–70%) sometimes are interpreted to indicate the presence of infraspecific taxa (Kurtzman et al. 1980; Phaff et al. 1987).

Pulsed-field electrophoretic techniques are a promising tool for yeast systematics, as they provide the possibility for study of individual chromosomes (Table 2; see Chap. 3, this Vol.). Many species show a considerable variation in number and size of chromosomal DNAs (see Boekhout et al. 1993b, and references therein). Chromosomal length polymorphisms are apparent in most species studied so far. In contrast, all strains investigated of the medical yeast *Malassezia pachydermatis* have similar karyotypes. In combination with approaches like gene assessment, macrorestriction analysis, densitometry, chromosome-based hybridization experiments, and random amplification of polymorphic DNA (RAPD), electrophoretic karyotyping may provide answers on the role of chromosomal diversification in speciation processes.

Table 1. Distribution of mol% G+C among ascomycetous and basidiomycetous yeast species. (Data from Barnett et al. 1990)

Mol% G + C	Percentage of ascomycetous yeasts	Percentage of basidiomycetous yeasts
25–29	0.75	
30–34	14.5	
35–39	28.0	1.0
40–44	32.5	1.0
45–49	16.0	10.0
50–54	5.5	35.0
55–59	2.0	31.0
60–64	0.75	18.0
65–69		4.0

Molecular approaches applied at or below the species level are restriction analysis of mtDNA (e.g., McArthur and Clark-Walker 1983) and rDNA (Laaser et al. 1987; Vilgalys and Hester 1990; Molina et al. 1993a,b; Shen and Lachance 1993), and random amplification of polymorphic DNA (RAPD) (for review on PCR see Foster et al. 1993). RAPD is a promising tool in biotyping and epidemiology.

4
Phylogeny

Most present-day yeast taxonomists follow the opinion that a taxonomic scheme should represent the phylogeny of the group of organisms concerned. One has to assume an orthologous relationship between the characters studied and the actual, but unknown, evolutionary relationship of the group of organisms concerned (Ragan 1988). Phylogenetic reconstructions based on sequence analysis of ribosomal RNA (rRNA) and ribosomal DNA (rDNA) recently received much attention (for reviews see Bruns et al. 1991; Kurtzman 1992, and references therein). These molecules are considered to be chronometers because of their universal occurrence, functional constraints, and the presence of both variable and less variable regions (Woese 1987). Application of PCR and universal primers (White et al. 1990) makes it easy to compare different species. The most frequently used numerical methods for sequence comparison are parsimony, distance methods, and maximum likelihood methods (Felsenstein 1988). Resulting phylogenetic trees need to be statistically tested to set confidence limits for the branching order, e.g., by bootstrap or jackknife analysis (Felsenstein 1988).

Nucleotide sequences of 5S rRNA (ca. 120 nucleotides) are highly conserved, and were found to correlate well with septal ultrastructure within the basidiomycetous yeasts (Templeton 1983). However, the Uredinales (rust fungi), which show diaphragma-like pores, were found to cluster with doliporous species (Gottschalk and Blanz 1985). However, this unexpected result needs verification, as the use of contaminant strains cannot be ruled out.

Entire 18S rRNA or rDNA of a number of yeasts species have been sequenced (Hendriks et al. 1992a,b; Van de Peer et al. 1992). Some genera are found to be phylogenetically heterogeneous, e.g., *Pichia* and *Candida*. Most phylogenetic studies on yeasts performed thus far have been based on partial sequences of the 18S and/or 25-28S rRNA or rDNA. This approach appears justified, as the phylogenetic resolution of partial sequence comparisons was found to be similar to that from entire sequences (McCarroll et al. 1983; Lane et al. 1985). Several yeast genera have been studied, e.g., *Debaryomyces*, *Saccharomyces*, and *Schizosaccharomyces* (Kurtzman and Robnett 1991), *Lipomyces* and *Myxozyma* (Kurtzman and Liu 1990), *Metschnikowia* (Mendoca-Hagler et al. 1993), *Torulaspora* and *Zygosaccharomyces* (Yamada et al. 1991a), *Debaryomyces* (Kurtzman and Robnett 1991; Yamada et al. 1991b), and several basidiomycetous yeasts (Guého et al. 1990; Yamada et al. 1989a,b; 1990a,b, Yamada and Kawasaki 1989a; Yamada and Nakagawa 1992; Fell et al. 1992).

Resolution of close genetic relationships from rRNA sequence divergence was examined from comparisons of sibling species. The sibling pairs *Pichia mississippiensis/P. amylophila*, *P. americana/P. bimundalis*, *Issatchenkia scutulata* var. *scutulata/var. exigua*, *Saccharomyces cerevisiae/S. bayanus/S. pastorianus* showed identical sequences in four areas of the 18S rRNA, whereas differentiation was found to occur in the most variable region of the 25S rRNA (Peterson and Kurtzman 1991).

Within the basidiomycetous yeasts, a correlation was observed between clusters based on partial sequences of 18S and/or 28S rRNA or rDNA, carbohydrate composition of cell walls, and septal ultrastructure. Clusters 1, 2, and 3 of Fell et al. (1992) contain species which lack a dolipore and have no xylose in their cell walls or capsules. Clusters 4 and 5 contain doliporous species which contain xylose in their cell walls. However, cluster 6, which includes species lacking xylose, seems closely related to clusters 4 and 5. Unfortunately, the septal ultrastructure of the species belonging to this cluster is unknown. Some of these species clusters have a uniform coenzyme Q composition, whereas others are found to be heterogeneous in this respect. To a certain extent, the clustering of species based on partial 28S rRNA sequences correlates with the capacity to assimilate D-glucuronic acid and *myo*-inositol (Boekhout et al. 1993a).

5
Methods

5.1
Morphology

Morphological characteristics are still of great taxonomic importance. Generic differentiation of yeasts is often based on morphology, e.g., conidiogenesis, presence or absence of hyphae, pigmentation, characteristics of ascus formation, and morphology of asci, ascospores, teliospores, basidia, etc. Some morphological characteristics indicate whether imperfect (anamorphic) yeasts belong to the asco- or to the basidiomycetes, e.g., mode of conidiogenesis (enteroblastic budding versus holoblastic budding), presence of ballistoconidia or clamp connections, and ultrastructure of cell walls and pores. Due to the pleomorphic character of many yeasts and yeast-like organisms, the use of standardized experimental conditions for the investigation of morphological features is strongly recommended.

5.1.1
Vegetative Morphology

Yeasts show different modes of vegetative reproduction (sometimes referred to as conidiogenesis).

Blastoconidia (Bud Formation)

Young cells, formed by cell wall expansion at a limited locus, gradually increase in size during maturation. The succession of conidia is schizolytic. Depending on the positioning of the bud formation, three types of blastoconidiogenesis are distinguished: (1) monopolar budding in which buds originate at only one pole of the cell (example: *Malassezia*, Fig. 7); (2) bipolar budding where buds originate with a broad base at both poles of the cell (examples: *Hanseniaspora*, *Nadsonia*, Fig. 8). (3) multipolar (multilateral) budding where buds originate over a wider range of the cell surface (examples: *Saccharomyces* (Fig. 9), *Kluyveromyces*, *Pachysolen*, *Pichia*).

In monopolar and bipolar conidiogenesis, formation of successive conidia may occur at the same locus (percurrent conidiogenesis), finally resulting in the forma-

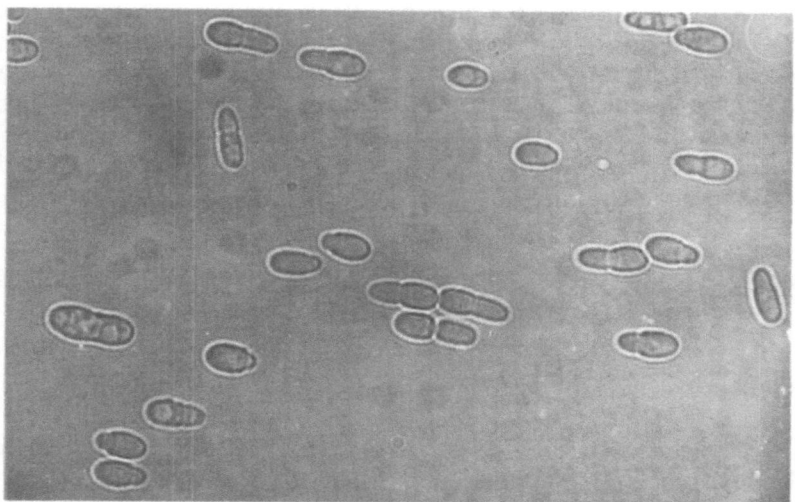

Fig. 7. Monopolar budding in *Malassezia pachydermatis* (CBS 1879, 3d YPGA)

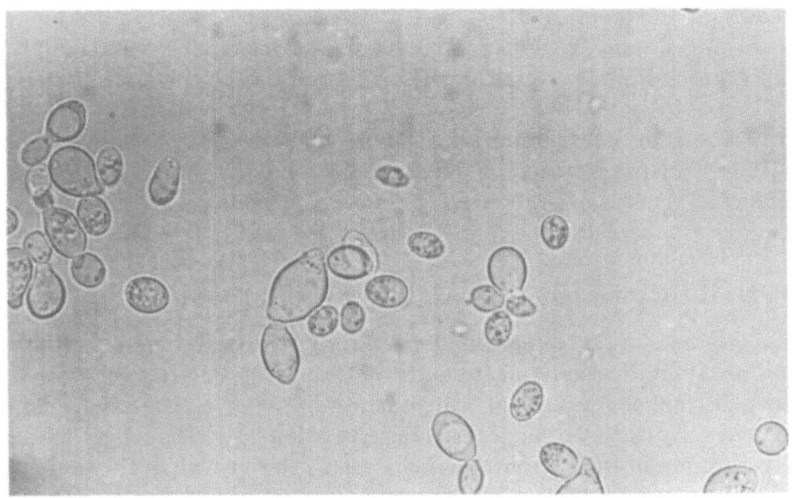

Fig. 8. Bipolar budding in *Nadsonia fulvescens* (CBS 2596, 10d YMA)

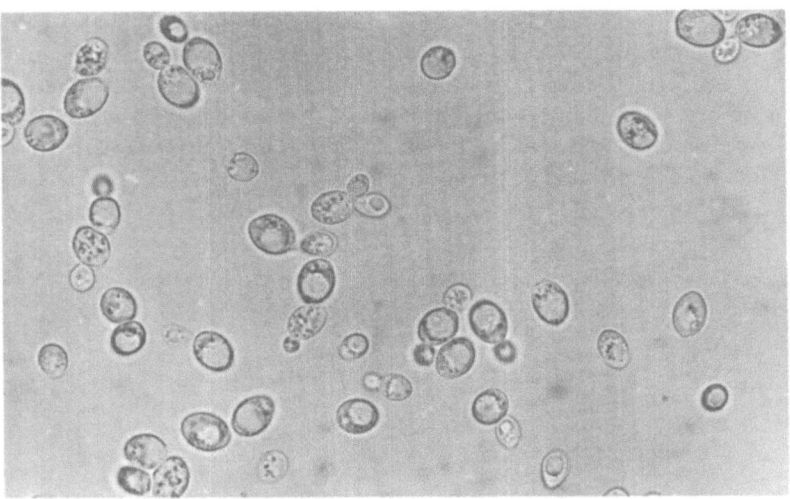

Fig. 9. Multipolar budding in *Saccharomyces cerevisiae* (CBS 1171, 3d YPGA)

tion of annellidic scars (annellations). In multipolar budding, young cells usually originate at different loci. For many basidiomycetous yeasts, formation of conidia is limited to an area around the poles of the cells. Contrary to bipolar budding, these buds have a relatively narrow base, and frequently proliferate sympodially (examples: *Cryptococcus, Rhodotorula*).

A special type of blastoconidiogenesis is the formation of ballistoconidia (examples: *Bullera, Sporobolomyces*), which are actively discharged and originate at tapering outgrowths (sterigmata) on the cell. The formation of ballistoconidia may be stimulated by growth on malt extract agar, cornmeal agar, or yeast morphology agar. Ballistoconidia may be distinctly bilaterally symmetrical (example: *Sporobolomyces*) or more or less rotationally symmetrical (example: most species of *Bullera*).

Buds formed on short denticles or long stalks frequently occur among basidiomycetous yeasts. The presence of conidia formed on stalks is characteristic for a limited number of genera (examples: *Sterigmatomyces*, *Fellomyces*, *Tsuchiyaea*, *Kurtzmanomyces*, *Kockovaella*).

Arthroconidia (Thalloconidia)

Young cells originate after disarticulation of differentiated hyphae. Young arthroconidia are cylindrical and finally round off (examples: *Geotrichum*, *Arxula*, *Trichosporon*). Fission, as it occurs in *Schizosaccharomyces* (Fig. 10), can be considered as percurrent arthroconidiogenesis. Here, two daughter cells separate after schizolytic dissolution of the septum, leaving a scar. After growth at the poles of the new daughter cells, annellidic scars are left.

Hyphae, Pseudohyphae, Chlamydospores, and Endospores

Many yeasts have the capability to form hyphae or pseudohyphae. Hyphae are not constricted at their septa, whereas pseudohyphae show distinct constrictions. Pseudohyphae are formed when more or less elongate budded yeast cells adhere in branched or unbranched chains. Different types of pseudohyphae have been distinguished (see Diddens and Lodder 1942). Proliferation occurs acropetally, so that the youngest cell is formed at the apex of the chain of cells.

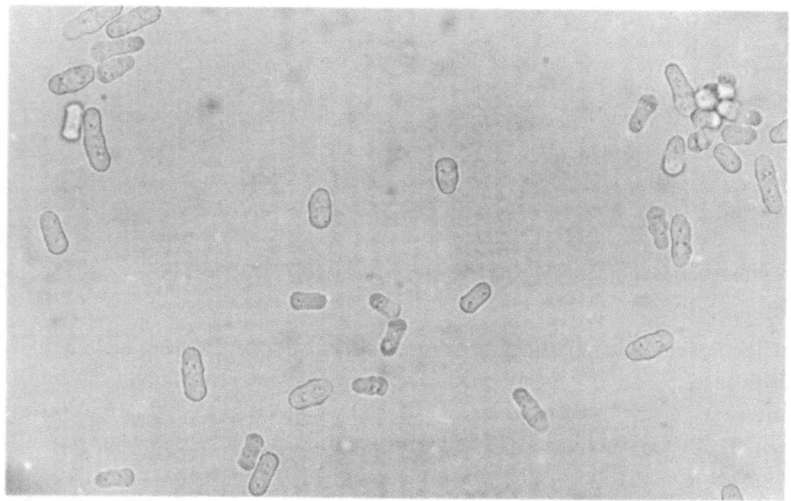

Fig. 10. Fission in *Schizosaccharomyces pombe* (CBS 356, 3d YPGA)

Dimorphism, the alternate occurrence of unicellular and hyphal and/or pseudohyphal phases occurs in many yeasts (example: *Candida albicans*, Fig. 11). Life cycles of many basidiomycetous yeasts are dimorphic as well. Vegetative monokaryotic yeast cells alternate with dikaryotic hyphae on which the sexual form of sporulation may be formed.

Hyphae and pseudohyphae are best studied in slide cultures, as their formation is stimulated by reduced oxygen levels. It is also useful to investigate microscopically the margins of long-standing colonies on culture plates at low magnification (objective 4× or 10×) for the possible occurrence of hyphae and pseudohyphae.

Chlamydospores are thick-walled vegetative resting spores which occur in some yeasts. The formation of chlamydospores is regarded as a diagnostic test for *Candida albicans* (Fig. 12). Slide cultures with rice agar give good results with this species. It may be difficult to differentiate chlamydospores from teliospores morphologically (see below).

Endospores also occur in some yeasts such as *Candida, Cryptococcus, Trichosporon, Cystofilobasidium,* and *Leucosporidium.* They are vegetative cells formed endogenously inside other cells and mostly occur in long-standing cultures.

5.1.2
Generative Morphology

Yeast species may differ widely in their strategies for sexual reproduction. Life cycles may be haplontic, diplontic, or haplo-diplontic. Homothallic yeast species may form asci or basidia without mating, whereas heterothallic species require strains of opposite mating type. For detailed morphogenetic information on the mating processes of ascomycetous and basidiomycetous yeasts, the reader is re-

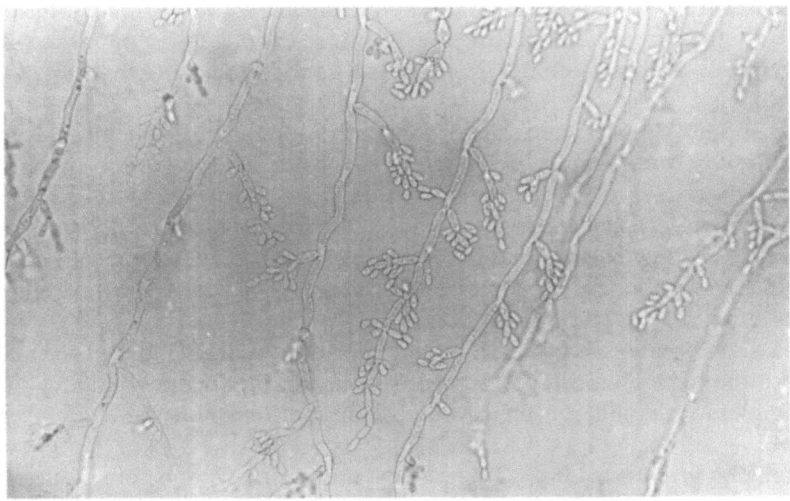

Fig. 11. Pseudomycelium in *Candida albicans* (CBS 5982, 3d Dalmau plate on rice agar)

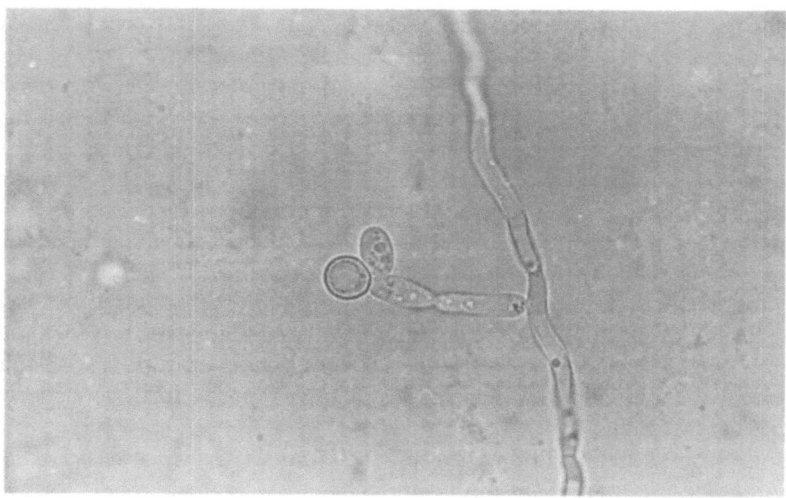

Fig. 12. Chlamydospore in *Candida albicans* (CBS 5982, 3d Dalmau plate on rice agar)

ferred to specialized references (e.g., Flegel 1981; Wong 1987; Casselton et al. 1989; Kurjan 1991; Lipke and Kurjan 1992).

Ascomycetes

Homothallic ascomycetous yeasts can form asci in different ways. In parent cell-bud conjugation, two haploid nuclei, one each from the parent cell and the bud, fuse and give rise to the diploid phase. Subsequent meiosis results in the formation of haploid ascospores (example: *Debaryomyces*, Fig. 13). In *Nadsonia*, a comparable mechanism seems to occur. Here the diploid nucleus, newly formed by fusion of nuclei from the parent cell and bud, migrates from the bud to the opposite pole of the parent cell, where it becomes enclosed in a newly formed bud that serves as an ascus (Miller and Phaff 1984).

In other species, short protuberances (gametangia) frequently develop adjacent to a septum, fuse, and form a diploid proascus. After meiosis, one to four, or sometimes more, ascospores are formed (examples: *Saccharomycopsis*, *Galactomyces* (Fig. 14), *Dipodascus*). Representatives of the Lipomycetaceae form multispored asci directly on apparent vegetative cells, or after fusion of short protuberances that originate either on a single cell or on two adjacent cells (van der Walt, pers. comm.).

Besides the preceding strategies, autodiploidization of the haploid phase may occur. Both the haploid and diploid yeast cells may reproduce by budding, but only the diploid cells undergo meiosis and form asci (example: *Saccharomyces cerevisiae*).

Heterothallic species require pairing of opposite mating types before induction of plasmogamy and karyogamy. When mixed on an appropriate medium, complementary cells each form conjugation tubes that grow toward one another, fuse, and allow passage of the haploid nuclei, which then diploidize. After meiosis,

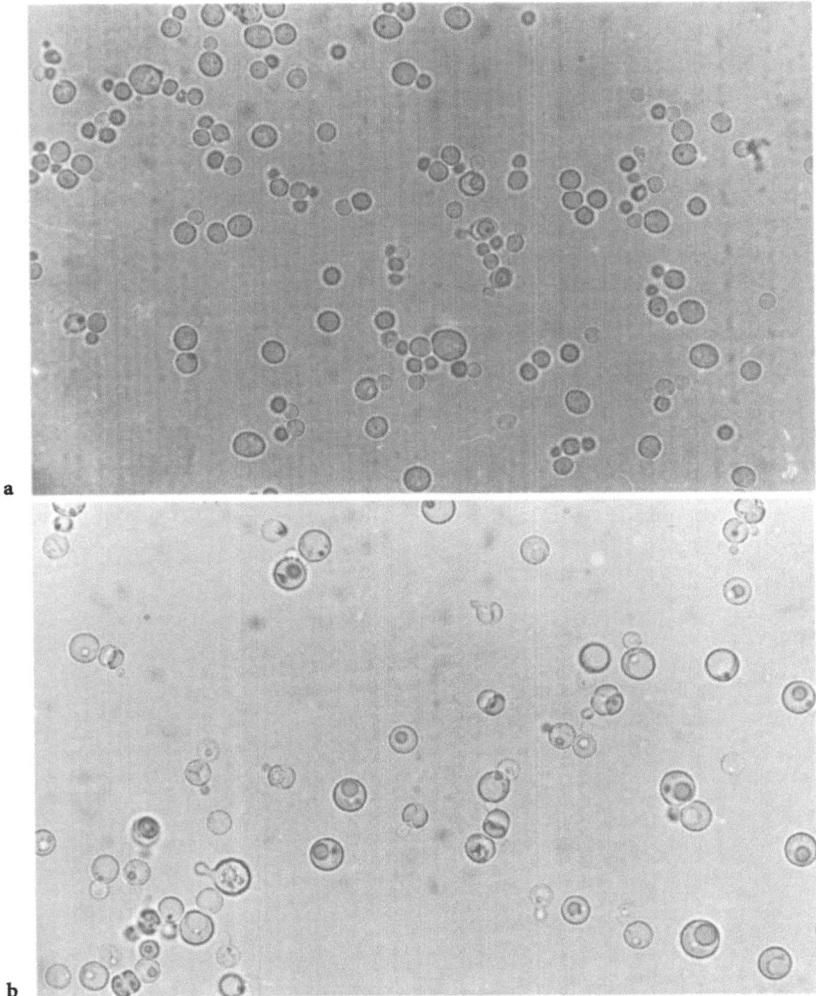

Fig. 13. a Asci with ascospore in *Debaryomyces hansenii* (CBS 767, 7d V8-agar). b Asci with ascospores in *Wingea robertsii* (now transferred to *Debaryomyces*, CBS 5637, 5d YPGA)

haploid ascospores are formed, either one or both of the conjugating vegetative cells serving as the ascus (examples: *Pichia*, and *Zygosaccharomyces* Fig. 15). Conjugation also can occur between hyphae (example: *Zygoascus*). In many species, haploid, diploid, polyploid, or aneuploid cells may propagate by budding, resulting in rather complicated life cycles. Diploid cells do not require prior conjugation to undergo ascosporulation. In the case of *Saccharomycodes ludwigii*, ascospores of the opposite mating type conjugate while still in the ascus and produce diploid cells capable of forming new ascospores. Asci can be persistent or evanescent.

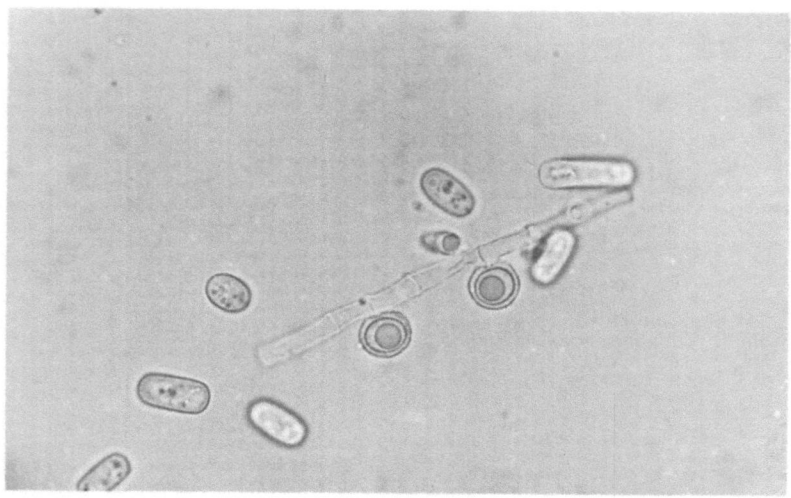

Fig. 14. Asci with ascospore and arthroconidia in *Galactomyces geotrichum* (CBS 772.71, 3d YPGA)

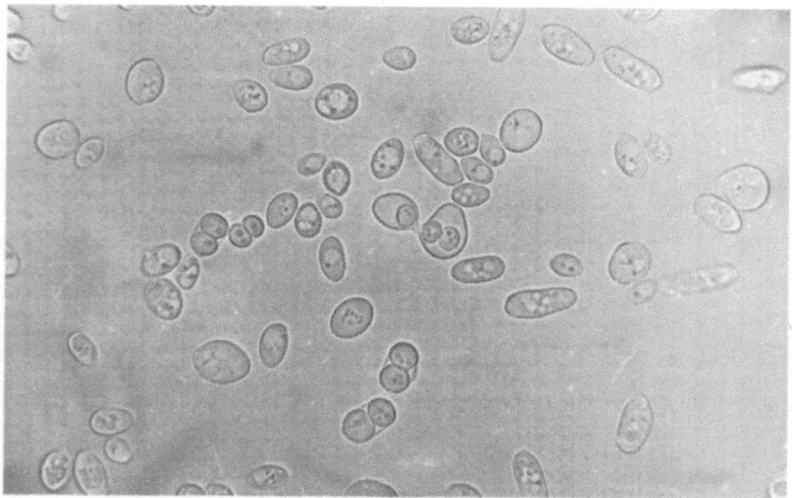

Fig. 15. Ascus with ascospores in *Zygosaccharomyces bailii* (CBS 1097, 7d YPGA)

Ascospore morphology is often used for genus delimitation. For most yeasts, the number of ascospores varies from one to four (or to eight). However, multispored asci occur in several genera (examples: *Ascoidea*, *Lipomyces*, *Dipodascus*). Ascospores are usually hyaline, but occasionally pigmented (e.g., *Lipomyces*), and can be globose, ellipsoidal, hat-shaped, saturn-shaped, arrow-like, or elongate-fusiform with thread-like appendages. The spore surface may be smooth, verrucose, or ridged.

Basidiomycetes

Many basidiomycetous yeast species have dimorphic life cycles in which monokaryotic yeast phases alternate with dikaryotic hyphal phases. Clamp connections are frequently present. The incompatibility system can be bipolar, tetrapolar, or modified tetrapolar (Bandoni 1963). The presence of dissimilar mating factors results in completion of conjugation, plasmogamy, and karyogamy. Many species form teliospores which are thick-walled probasidia (examples: *Sporidiobolus*, *Rhodosporidium* (Fig. 16), *Leucosporidium*, *Cystofilobasidium*). They can only be differentiated from vegetative chlamydospores by karyology (karyogamy and meiosis), and typically germinate with basidia. Teliospores may be intercalary or terminal, single or in small clusters, (sub)globose or angular, hyaline or pigmented, and are usually smooth. However, teliospores of *Tilletiaria anomala* are covered with warts. The teliospores germinate, often enhanced by being soaked in water for several weeks (Fell et al. 1969), by transversely septate or one-celled basidia on which basidiospores are formed. Monokaryotic fruiting, i.e., formation of apparent sexual structures on monokaryotic hyphae without an obvious sexual mechanism, has been observed to occur in several taxa (example: *Mrakia*).

Some species do form basidia directly at the dikaryotic hyphae (examples: *Filobasidiella*, *Filobasidium*, *Bulleromyces*). The basidia of *Filobasidiella* and *Filobasidium* are one-celled, clavate to capitate (Filobasidiales type), whereas in

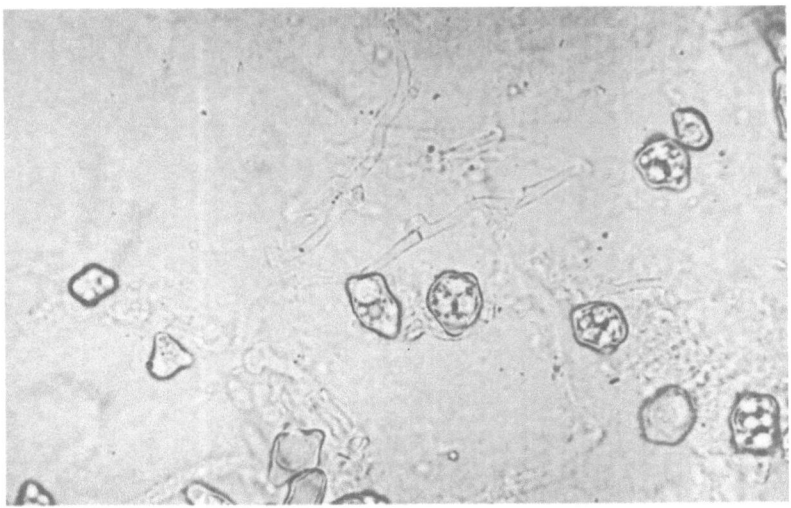

Fig. 16. Teliospores and hyphae with clamp connections in *Rhodosporidium toruloides* (CBS 14 × CBS 349, 7d PDA)

Bulleromyces they are longitudinally or obliquely septate (Tremellales type, Fig. 17). Basidiospores in *Filobasidiella* are formed basipetally, viz. the youngest spore originates at the base of a chain of spores.

5.2
Physiological Characterization of Yeasts

Physiological characterization is still important for the identification of unknown yeasts strains. Two methods are commonly used, namely the auxanographic technique and growth in liquid medium. Recently, several commercial kits have been introduced that use a limited series of compounds (e.g., API 32C). Anaerobic utilization of sugars is generally tested by measuring the amount of CO_2 that is trapped in Durham tubes. It has been argued that this method is not very accurate for detecting fermentation in slowly fermenting yeast species because the CO_2 may not evolve rapidly enough to be collected as a gas bubble. Because of this, many yeast species long considered to be non-fermentative proved capable of producing ethanol (van Dijken et al. 1986). However, for identification purposes, the use of Durham tubes is recommended because of easy preparation and scoring. An initial fermentation test using only glucose is advisable. If no fermentation is detected, other sugars need not be tested. The sugars tested for identification purposes usually include glucose, galactose, maltose, sucrose, lactose, raffinose, trehalose, and xylose.

5.2.1
Fermentation Tests

Sugars are tested at a concentration of 2% (w/v) (or 4% for raffinose) in a basal medium of 2% (w/v) yeast extract, yeast infusion, or 0.3% yeast extract-0.5% peptone (Wickerham 1951). Depending on the size of the test tubes used, the

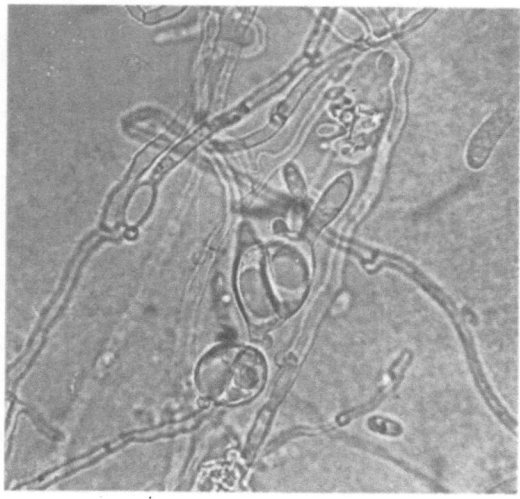

Fig. 17. Phragmobasidia and hyphae with clamp connections in *Bulleromyces albus* (CBS 6302 × CBS 7441, 14d CMA)

volume of medium ranges from 3.0 to 4.5 ml. Each tube contains a small inverted insert tube to collect CO_2 that may be formed.

Sugars are filter-sterilized at a 3× concentration and added aseptically to the autoclaved basal medium concentrate, which often contains bromthymol blue as pH indicator.

Fermentation tubes are inoculated with 0.1–0.2 ml of a heavy aqueous cell suspension prepared from an actively growing slant culture. Tubes are incubated at 25–28 °C and observed every few days for up to 4 weeks. Psychrophilic species require 12 or 17 °C.

5.2.2
Assimilation Tests

Assimilation of carbon compounds can be tested by the auxanographic method or in test tubes with liquid medium. Auxanograms can be read after 2 to 3 days, whereas test tubes need up to 4 weeks. However, the latter method is more sensitive. For identification purposes, growth reactions on the following compounds are usually determined:

• Hexoses: (D-glucose, D-galactose, L-rhamnose, L-sorbose).

• Pentoses: (D-xylose, D-ribose, L-arabinose, D-arabinose).

• Disaccharides: (sucrose, maltose, cellobiose, trehalose, lactose, melibiose).

• Trisaccharides: (raffinose, melezitose).

• Polysaccharides: (soluble starch, inulin).

• Alcohols: (erythritol, ribitol (adonitol), D-mannitol, m-inositol, methanol, ethanol, glycerol, glycol, propane 1,2 diol, butane 2,3 diol, galactitol (dulcitol), D-glucitol (sorbitol)).

• Organic acids: (succinate, citrate, DL-lactate, D-gluconate, D-glucuronate, D-galacturonate, 2-keto-D-gluconate, 5-keto-D-gluconate).

• Glycosides: (a-methyl-D-glucoside, arbutin, salicin).

• Other compounds: (glucono-δ-lactone, D-glucosamine-HCl, N-acetyl glucosamine, decane, hexadecane).

Carbon Assimilation by Auxanogram

Nitrogen-containing agar: 0.5% w/v $(NH_4)_2SO_4$, 0.1% KH_2PO_4, 0.05% $MgSO_4$, 2% agar (Difco), 0.1% v/v vitamin solution (according to Wickerham 1951) in distilled water. Commercially available Yeast Nitrogen Base with agar is a convenient alternative.

An inoculum of the strain to be tested is suspended in 5 ml sterile distilled water, poured in a petri dish (diam. ca. 9–11 cm), and thoroughly mixed with 10 ml of cooled, molten nitrogen-containing agar.

After solidification, crystals of the compounds to be tested are placed on the surface of the agar. Glucose is used as a positive control. Growth is examined after 2–3 days.

Assimilation of Carbon Compounds in Liquid Medium

Stock 10× solutions of Yeast Nitrogen Base (Difco), each with the carbon source to be tested, are sterilized by filtration. Final concentrations of the carbon sources are 0.5%, except for raffinose, which is 1%.

The pH of the 10× stock needs to be readjusted to 5.6 for those carbon compounds that are not neutral, such as organic acids. Water blanks consisting of test tubes with 4.5 ml of distilled water are autoclaved, and each then receives 0.5 ml of the appropriate 10× stock solution.

Soluble starch and inulin are exceptions to this treatment. Each is made as a 0.5% aqueous solution. For soluble starch, 4.5-ml amounts are placed in test tubes and autoclaved for 15 min. Because inulin is easily degraded by heat, the solution is filter-sterilized and 4.5 ml is added to sterile test tubes.

The soluble starch and inulin tubes each receive 0.5 ml of 10× Yeast Nitrogen Base. Assimilation tubes are inoculated with an aqueous suspension of starved yeast cells as described by Wickerham (1951). Turbidity, used as a measurement of growth, is scored after 7, 14, and 21, or 28 days at 25 °C.

Growth is faster and more consistent if tubes are shaken during incubation on a rocking shaker (30 × min⁻¹), a reciprocal shaker, or a rotary shaker. Psychrophilic species require lower temperatures, e.g., 12 or 17 °C.

Nitrogen Assimilation by Auxanogram

The following compounds are tested for routine identifications: nitrate, ethylamine, L-lysine, and cadaverine. Additional compounds are nitrite, creatine, creatinine, and imidazole.

Carbon-containing agar: 2% w/v glucose, 0.1% w/v KH_2PO_4, 0.05% w/v $MgSO_4 \cdot 7H_2O$, 2% w/v agar (Difco), 0.1% v/v vitamin solution (according to Wickerham 1951) in distilled water. Commercially available Yeast Carbon Base with agar is a convenient alternative.

An inoculum of the strain of interest is made in 5 ml sterile distilled water, poured into a petri dish (diam. ca. 9–11 cm), and thoroughly mixed with ca. 10 ml molten, cooled agar. After solidification, some crystals of the compounds to be tested are placed on the surface of the agar. Peptone or $(NH_4)_2SO_4$ are used as positive controls. Growth is recorded after 2–3 days.

Alternatively, assimilation of nitrogen compounds can be tested in liquid medium as described by Wickerham (1951).

5.2.3
Vitamin Requirements

Vitamin requirements are tested by slightly inoculating a test tube containing 5 ml vitamin-free medium (Difco). If growth occurs, a second tube is inoculated from the first to rule out carryover of vitamins.

5.2.4
Other Tests

A number of other tests may facilitate identification of strains: resistance to cyclo-heximide (actidion), growth on 50% glucose, production of acetic acid, urease activity, staining reaction with Diazonium Blue B, starch production, and growth at 17, 25, and 37 °C.

Resistance to Cycloheximide

In 10× concentrated Yeast Nitrogen Base (Difco) 0.5 ml of a filter-sterilized 0.1% or 1.0% cycloheximide is added to 4.5 ml 0.5% w/v glucose in distilled water, giving final concentrations of 0.01% (100 ppm) or 0.1% (1000 ppm) cycloheximide, respectively. After inoculation with the strain of interest, growth is scored.

Growth on 50% Glucose

A small amount of inoculum is streaked on slants or plates containing yeast extract agar with 50% w/v glucose. Growth is scored, usually at 1 and 2 weeks.

Production of Acetic Acid

A small amount of inoculum streaked on slants or plates of YPGA with 0.5% w/v $CaCO_3$. Production of acetic acid results in a clear zone around the culture.

Urease Activity

Difco Bacto Urea R broth (0.5 ml) is dispensed into tubes and stored in a freezer. After inoculation, tubes are placed at 37 °C, or the maximum growth temperature for the strain if lower than 37 °C. Change of color of the suspension to deep red after ca. 4–20 h indicates urease activity.

Diazonium Blue B Reaction

This test is used to differentiate between ascomycetous and basidiomycetous yeasts. The latter organisms show a color reaction in this test.

Directly before use, 0.1% w/v Diazonium Blue B (DBB) salt is dissolved in ice-cold 0.1 M Tris-HCl, pH 7.0. Strains are inoculated on YMA plates or slants for 10 days to 3 weeks, and a drop of the freshly prepared ice-cold DBB reagent is placed on the culture. Immediate change of color of the colony to dark red indicates a positive DBB reaction.

Extracellular Starch Production

Iodine (0.33%) and potassium iodide (0.66%) are dissolved in distilled water (Lugol's solution). A drop of reagent is added to the culture in an assimilation tube containing D-glucose. Dark blue coloration indicates presence of extracellar starch-like compounds. It is often helpful to have duplicate glucose assimilation tubes in order to test for starch-like compounds at 1 and 2 weeks after inoculation.

5.3
Mating

5.3.1
Ascomycetes

Strains can be grow for 2 days on YMA and paired by mixing a loopful of growth from each strain on fresh YMA. Alternatively, up to eight strains may be mixed in one test, with the strategy that if mating is detected, strains will then be paired in all combinations. Mixtures should be examined under the microscope at daily intervals, for at least a week, although mating usually occurs within the first day or two after mixing. Mated cells are joined by fused conjugation tubes that are usually long enough to be easily discerned. Occasionally, conjugation tubes are quite short and require careful observation for detection. Conjugating cells may be of equal size or one of the conjugants can be quite small and misinterpreted as a bud if the conjugation tube is short. If ascosporulation does not follow mating, the pairings can be repeated on other media that may be more conducive to ascospore formation such as those listed below. Lack of ascosporulation following mating may also indicate incompatibility, which could be due to ploidy differences or phylogenetic divergence sufficiently great to prevent karyogamy (Kurtzman et al. 1980).

5.3.2
Basidiomycetes

The strains of interest are mixed in pairs near the margin of an agar plate (e.g., modified Flegel's conjugation medium, PDA, MEA, or YMPA). With a flattened needle, three to five streaks are made across the petri dish. When many strains need to be investigated, the procedure described above may be followed. Mating reactions frequently occur along these streaks.

The first indication of the occurrence of a mating reaction is the formation of conjugation tubes, followed by the development of dikaryotic hyphae, which frequently grow submerged. Formation of clamp connections, dikaryotic cells, teliospores, and/or basidia can indicate a sexual reaction.

Commonly used media include corn meal agar, modified Flegel's conjugation medium, V8 juice agar, 5% malt extract agar, potato-dextrose agar, and 1.5% malt extract-0.05% yeast extract-0.25% peptone agar. For mating of *Filobasidiella neoformans*, the following media have been suggested: corn meal agar, hay infusion agar, V8 juice agar, pigeon manure agar, or sunflower seed agar (see Appendix).

Germination of teliospores is sometimes difficult, and it can be enhanced by a temporary increase of the temperature to 50–55 °C for 5 or 10 min (Bandoni et al. 1971), or soaking in sterile distilled water for 2–10 weeks at 12 °C (Fell et al. 1969) prior to transfer to appropriate media for germination, e.g., corn meal agar, or 2% water agar.

5.4
Nuclear Staining

The number of nuclei and their behavior play an important role in life cycles of yeasts. A number of rapid and reliable fluorescent nuclear staining techniques have been developed. Nuclear staining of thick-walled and/or pigmented cells, such as teliospores, is sometimes difficult. Some fluorochromes permit quantification of nuclear DNA, such as para-rosanaline (Feulgen), propidium iodide, and 4′-6′-diamino-2-phenolindole (DAPI). For nonfluorescent staining with Giemsa is convenient. Cells can be fixed and stained using 1.5-ml microfuge tubes, or adhered to a cover glass using egg albumin as an adhesive.

5.4.1
Staining Nuclei Using DAPI (Coleman et al. 1981)

Stock solution: 1 mg DAPI ml^{-1} distilled water, store at 4 °C in the dark. Working solution: 0.5 μg DAPI ml^{-1} McIlvaine's buffer, pH 4.4 (44.1 ml 0.2 M Na$_2$HPO$_4$ · 2H$_2$O and 55.9 ml 0.1 M citric acid).

Fix cells in 70% ethanol for 60 min in a 1.5-ml centrifuge tube; centrifuge.

Rinse 5 min in McIlvaine's buffer, pH 4.4; centrifuge.

Resuspend cells in DAPI working solution for ca. 3 h, or overnight; centrifuge.

Mount cells in 90% v/v glycerol in McIlvaine's buffer and seal coverglass with nail polish. Preparations made in this way may last longer than a year when stored in the dark.

For rapid analysis of the number of nuclei, yeast cells may be suspended directly in DAPI working solution. An additional advantage of DAPI is that mitochondria stain as well. The emission of DAPI depends on the mol% G+C, and therefore the method has to be applied with caution for quantitative fluorescence microscopy.

5.4.2
Staining Nuclei with Propidium Iodide
(After Uno et al. 1984; Eilan et al. 1992)

Propidium iodide (PI) stock solution: 150 μg PI ml^{-1} NS buffer (20 mM tris-HCl pH 7.6, 0.25 M sucrose, 1 mM MgCl$_2$, 0.1 mM ZnSO$_4$, 0.1 mM CaCl$_2$, 0.8 mM phenylmethylsulphonylfluoride (PMSF, toxic), 0.05% b-mercaptoethanol).

Propidium iodide working solution: 1.5–10 μg PI ml^{-1} NS buffer, and containing 50–500 μg RNase ml^{-1}.

Rnase stock solution: 10 mg RNase ml^{-1} 0.1 M Na-acetate, pH 5.2. Heat 15 min at 100 °C, cool down slowly, and adjust pH to 7.4 by adding 0.1 volumes 0.1 M Tris-HCl (pH 7.5).

Fix cells with 50% ethanol for 10 min, and subsequently with 70% ethanol for 12 h. Stain with PI on a microscope slide for 30 min to 16 h (under a coverglass), remove excess dye with tissue, and seal with nail polish.

Staining of RNA by PI may cause high background fluorescence. Preparations sometimes improve after a longer incubation time (e.g., cells of *Galactomyces geotrichum* need up to ca. 4 days). Propidium iodide can be used for cytometric quantification of DNA as the emission is independent of the mol% G+C.

5.4.3
Staining Nuclei with Mithramycin and Ethidium Bromide
(After Barlogie et al. 1976)

Mithramycin working solution: 100 μg mithramycin ml^{-1} 15 mM MgCl$_2$ + 30% ethanol.

Ethidium bromide working solution: 25 μg ethidium bromide ml^{-1} 0.1 M Tris + 0.6% NaCl, pH 7.4 (Cole 1983).

Mix equal volumes of mithramycin and ethidium bromide working solutions, and suspend cells in this mixture for ca. 30 min. Preparations can be viewed directly. Thick-walled cells may give better results after gentle heating.

Acriflavine has been used to stain meiotic chromosomes in fungi (*Neurospora crassa*) (Raju 1986), and may be useful for yeasts as well.

5.4.4
Staining Nuclei with Giemsa (After Bauer 1987)

Giemsa stock solution: 0.76 g Giemsa powder ml^{-1} 50% (v/v) glycerol in absolute methanol (Gurr 1965).

Phosphate buffer (pH 7.0): mix 60.8 ml 0.15 M Na$_2$HPO$_4 \cdot$ 2H$_2$O with 39.8 ml 0.15 M KH$_2$PO$_4$.

Cells are dried for 20 min at room temperature, fixed for 30 min in a 3:1 mixture of 92% ethanol and acetic acid, repeatedly rinsed with water, and hydrolyzed in 1N HCl for 7 min at 60 °C.

After rinsing once with water and five times with phosphate buffer (pH 7.0), the cells are stained with Giemsa working solution (one volume Giemsa stock solution and nine volumes phosphate buffer, pH 7.0) for 2 h.

After rinsing in phosphate buffer (pH 7.0), and dipping in water, the preparations are dried and ready for use.

5.5
DNA

5.5.1
Isolation

Methods used for DNA isolation for taxonomic studies need to be rapid and reliable. Several methods are in use by yeast systematists. Two commonly used large-scale DNA isolation methods are described below. The first method is based on adsorption of DNA on hydroxylapatite columns as described by Britten et al. (1970). The second is based on extraction with various reagents and subsequent precipitation with ethanol or iso-amylalcohol. Small-scale methods for DNA isolation (minipreps) are widely used for PCR amplification. We describe a miniprep method which we have successfully applied to PCR amplification of ribosomal DNA. Further DNA isolation protocols can be found in Cryer et al. (1975), Holm et al. (1986), Taylor and Natvig (1987) Johnston (1988), Mann and Jeffery (1989), and Treco (1989).

DNA Isolation Using Hydroxylapatite (Britten et al. 1970)

DNA isolated by this method has been used extensively for spectrophotometric analysis of base composition, and DNA reassociation experiments.

Phosphate buffer 5.1 M: 2.4 M $Na_2HPO_4 \cdot 2H_2O$ + 2.7 M $NaH_2PO_4 \cdot 1H_2O$ in distilled water.

Grow cells in ca. 400 ml YM, YPG, or YPM broth, until the late logarithmic or early stationary phase.

Harvest by centrifugation or, in the case of extensive hyphal growth, by filtration through a Buchner funnel. Wash with tap water and with saline EDTA (0.1 M NaCl, 0.15 M EDTA). Cells can be stored at −20 °C until use.

Usually ca. 5 g wet-packed cells is sufficient. Add an equal volume of lysing buffer [900 ml 10 M urea, 56.2 ml 5.1 M phosphate buffer, 43.8 ml 20% sodium dodecylsulfate (SDS)], break cells at least three times with, e.g., a French Press, and check microscopically (the majority of cells should be broken). Repeat if necessary.

Centrifuge broken cell debris for 10 min at 10 000 rpm. The supernatant contains the DNA.

Suspend ca. 5 g hydroxylapatite in washing buffer [800 ml 10 M urea, 50 ml 5.1 M phosphate buffer, 150 ml distilled water], and pour column.

Pour supernatant containing the DNA on the hydroxylapatite column and let DNA adsorb. Wash with ca. 200 ml washing buffer until A_{260} = 0.0.

Remove excess urea with ca. 30 ml low phosphate buffer [1 ml 5.1 M phosphate buffer and 480 ml distilled water], elute DNA with eluting buffer (30 ml 5.1 M phosphate buffer and 450 ml distilled water), and collect DNA fractions with a fraction collector at A_{260} (threshold value: $A_{260} = 0.25$, which equals ca. 12.5 μg DNA ml^{-1}).

Pool fractions and measure A_{260}.

Dialyze overnight against $0.1 \times$ SSC ($1 \times$ SSC: 0.15 M NaCl, 15 mM trisodium citrate. $2H_2O$), repeat once, and measure absorbance at 260, 230, and 280 nm. Ratios for purified DNA are $A_{230/260} = 0.5$ and $A_{260/280} = 1.85$.

For spectrophotometric DNA reassociations, the A_{260} has to be at least 1.5. If necessary, concentrate the sample as follows.

Add ca. 2.5 vol of ice-cold ethanol, let DNA precipitate for 24 h, centrifuge 1 h at 17 000 rpm, remove supernatant and, add 0.1 SSc to a final A_{260} of 3.0. The samples can be stored at $-20\,^{\circ}$C until further use.

DNA Isolation by a Modified Marmur Method

Isolation of DNA from yeasts by a modification of Marmur's (1961) method has been used successfully in many laboratories, and the procedure given here is similar to that reported by Price et al. (1987).

Strains are usually grown for 3 days at $25\,^{\circ}$C on a rotary shaker (200 rpm) in Fernbach flasks containing 1500 ml of YM broth and harvested by centrifugation. Two flasks are usually prepared for each strain.

Following harvesting, the cells are suspended in $2\times$ SSE buffer (Timberlake 1978) and either broken in a Braun cell homogenizer (B. Braun Biotech, Allentown, PA) with 0.5–mm glass beads or by enzymatic digestion (see Chap. 2, this Vol.).

Following cell breakage, sodium perchlorate and sodium sarcosine are added to the suspension to give concentrations of 1 M and 1%, respectively, and the mixture is then emulsified by swirling with an equal volume of chloroform: isoamyl alcohol (CIA) (24:1, v/v).

The emulsion is maintanined for 3 h on a rotary shaker and then separated by centrifugation. The upper DNA-containing aqueous layer is removed by a wide-mouth pipette, and the DNA is precipitated by addition of 1.3 vol. of cold ($-20\,^{\circ}$C) ethanol.

The precipitate is collected by centrifugation, dissolved in 20 ml of $1 \times$SSC containing 2 mg α-amylase and 2 mg of pancreatic RNase, and incubated overnight at room temperature on a rotary shaker. One mg of pronase is then added, and the solution is incubated for an additional 4 h.

The preparation is placed in a 300-ml Erlenmeyer flask with an equal volume of CIA, emulsified for 30 min on a rotary shaker, and then centrifuged. The upper layer is removed with a wide-mouth pipette and placed in a beaker where the DNA is spooled following addition of 1.3 volumes of cold ethanol.

The spooled DNA can be dissolved in 20 ml of 0.001 M sodium phosphate buffer for later use.

At this point, the DNA may be further purified by cesium chloride gradient ultracentrifugation or by hydroxylapatite chromatography.

If the latter option is chosen, the DNA spool is dissolved in 20 ml of 0.001 M sodium phosphate buffer and treated with a mixture of enzymes containing 2 mg a-amylase, 2 mg pancreatic RNase, and 400 units T1 RNase.

The DNA-enzyme solution is dialyzed against 0.001 M sodium phosphate buffer overnight at room temperature. The solution is treated with CIA as before and adjusted to 0.2 M with sodium phosphate buffer if the G+C content of the DNA is under 55% or else to 0.15 M. The solution is now passed through a hydroxylapatite column and the DNA eluted with 0.5 M sodium phosphate buffer.

Miniprep Method for Isolation of DNA for PCR Amplification
(After Raeder and Broda 1985)

Cells are grown in 25 ml YM broth on a 200 rpm shaker at 25 °C for ca. 48 h, and harvested by centrifugation. After being washed with tap water, the cells are lyophilized overnight.

The lyophilized cell mass is pulverized with a pipette tip, and shaken with glass beads (diameter 0.5 mm) for ca. 15 min using a wrist-action shaker.

Following breaking, 1000 μl extraction buffer (200 mM Tris-HCl, pH 8.4; 200 mM NaCl; 25 mM EDTA; 0.5% SDS) is added. After pelleting by centrifugation, 600 μl of the supernatant is transferred to a new tube, to which 420 μl phenol is added.

After vortexing, 180 μl chloroform is added. Centrifuge for 10 min at 14 000 rpm; 550 μl of the (upper) aqueous phase is transferred to a new tube without disturbing the interphase. Add 1 vol chloroform, and vortex briefly.

Centrifuge 5 min at 14 000 rpm, and transfer 400 μl of the (upper) aqueous phase to a new tube. Add 0.54 vol isopropanol, and shake briefly.

Centrifuge the precipitated DNA for 10–60 s at 14 000 rpm. Discard the supernatant, and wash the pellet once with 70% ethanol.

Centrifuge for 3 min at 14 000 rpm, and decant the ethanol carefully. Add 100 μl TE buffer (10 mM Tris-HCl; 1 mM EDTA, pH 8.0), loosen the pellet, and solubilize the DNA for 1 h (or overnight) at 55°C. The DNA can be stored at −20 °C.

5.5.2
Analysis of Base Composition

Spectrophotomeric Determination of Mol% G+C

Dilute the DNA with 0.1 × SSC in quartz cuvettes until an A_{260} of 0.3–0.4. *Candida parapsilosis* CBS 604 (mol% G+C = 40.8) is used as a reference strain, and 0.1× SSC is used as a blank. Increase of temperature is 0.5 °C min^{-1}, and the A_{260} is recorded using a microprocessor-controlled spectrophotometer. T_m values from melting curves can be calculated graphically, or from the first or second derivative (Jahnke 1987). Mol% G+C can be calculated as follows (in 0.1 × SSC): mol% G + C = T_m × 2.08 − 106.4. Melting curves of basidiomycetous yeasts frequently show one or two shoulders, which probably represent mtDNA and/or rDNA.

Determinations of mol% G+C obtained with HPLC are frequently lower than those obtained from spectrophotometric analyses. This may be due to the presence of mtDNA in the fractions. A short protocol for analyzing the mol% G + C using the high performance liquid chromatography (HPLC) method can be found in Hamamoto et al. (1986) and Nakase et al. (1989).

Determination of Mol% G+C Content from Buoyant Density

When a cesium chloride solution is spun in an ultracentrifuge, a density gradient is formed. The position of DNA in the gradient is determined by its G+C content, which can be calculated from the relative position of a second DNA of known density. Determinations are generally made in an analytical ultracentrifuge, but a preparative ultracentrifuge will also serve the purpose.

Begin by making a stock cesium chloride solution in the following manner. Add 130 g CsCl to 75 ml of 10 mM Tris buffer, pH 8.5. Dissolve and treat for 20 min with 2 g activated charcoal to remove any material absorbing at 260 nm. Remove the charcoal by filtration through Whatman No. 1 filter paper. Determine the exact concentration of CsCl from the refractive index. Solutions are usually approximately 1.87 g ml^{-1}, and about 405 μl is used in a total volume of 500 μl.

Combine the required amount of CsCl stock solution, 1 μg of undetermined DNA, 1 μg of *Micrococcus luteus* reference DNA (buoyant density 1.7311 g ml^{-1}), and bring to a final volume of 500 μl with distilled water.

Load centrifuge cells. Centrifugation is generally for 20 h at 44 000 rpm.

Determine distances of peak center points for unknown and reference DNAs and determine the G+C content using the following equations (Schildkraut et al. 1962).

$$p = p_o + 4.2w^2\left(r^2 + r_o^2\right) \times 10^{-10} \text{g/cm}^3, \tag{1}$$

where p is the density of unknown DNA, p_o the density of known DNA, w the radians/s (2π radians/revolution), r the distance of unknown from the center of rotation, and r_o the distance of standard from the center of rotation.

$$\text{Mol\% G+C} = \left(p - 1.66\right)/0.098 \times 100. \tag{2}$$

5.5.3
Hybridization of Nuclear DNA

Hybridization or reassociation of nuclear DNA is a reliable means for estimating the extent of genetic relatedness between strains (Price et al. 1978; Kurtzman and

Phaff 1987). DNAs that have approximately 80% or more nucleotide similarity can form a duplex under appropriate conditions of incubation. Because duplex formation requires considerable nucleotide similarity, genetic resolution from measurements of reassociation extends only to the distance of sibling species.

Protocols commonly used to measure DNA reassociation have been compiled by Kurtzman (1993), and the following account represents only a brief summary of these methods. Methods for measuring DNA relatedness fall into two general categories: (1) the free-solution technique, in which all of the reactants are solubilized, and (2) the filter-binding technique, in which the DNA of the strain is immobilized on nitrocellulose or other filter materials and DNA from the other strain is solubilized in the buffer surrounding the membrane. Each method has its strengths and, when properly done, each provides the same measure of relatedness (Seidler and Mandel 1971; Kurtzman et al. 1980).

In order to satisfy reaction kinetics, free-solution hybridizations require fragmentation of the genomic DNA in 400–500 base pairs. The DNA can be sheared to this length by double passage through a French pressure cell at 10 000 psi or greater. Any fragments that escape shearing can be removed by passage of the solution of sheared DNA through a 0.45-μm membrane filter. The DNA bound to filters for filter hybridization is not sheared, but the probe is of sheared DNA.

Free-Solution Hybridization

Spectrophotometric Method. The spectrophotometric (optical) method is a convenient non-isotopic procedure that has the advantage of simultaneously providing data for estimates of genome size. The rationale for this method is based on the observation that DNA reassociation is a concentration-dependent, second-order reaction. As a result, if 50 μg/ml of DNA reassociates at a certain rate, 25 μg/ml will take twice as long. Consequently, if a mixture of two DNAs reassociates at the same rate as an equivalent concentration of unmixed DNA, the organisms providing this DNA belong to the same species. If the reaction time of the mixture is the sum of that of the two unmixed DNAs, the organisms are different species. Because the genome sizes of yeasts are relatively small, the midpoint of the reaction, which is often used as a reference, is reached within about 2–12 h, depending on reaction conditions and actual genome sizes.

Hydroxylapatite Method. Hydroxylapatite (HA) is a form of calcium phosphate that preferentially binds double-stranded DNA when in an appropriate concentration of phosphate buffer. This specificity provides a means for separating renatured DNA duplexes from free-solution reassociation reactions. The technique requires use of radiolabeled probe DNA, but has the potential for relative ease in processing large numbers of samples.

S1 Nuclease Method. DNA hybridization mixtures are freed of single-stranded, unhybridized DNA by hydrolysis with S1 endonuclease. The double-stranded hybrids are then removed from the reaction mix by precipitation with trichloroacetic acid or collected on Whatman DE-81 filters. The probe DNA is radiolabeled.

Filter Hybridization

Unsheared, single-stranded DNA is attached to nitrocellulose or nylon membrane filters. The filters are incubated in a liquid medium containing a sheared, single-stranded probe that is either labeled with a radioisotope or has another means for detection. Following incubation, the filters are washed and assayed for extent of binding by the probe. The filter method is amenable to easily performing large numbers of comparisons, but leaching of unsheared target DNA from filters during incubation can be a problem.

Interpretation of DNA Hybridization Data

Data from DNA reassociation experiments require interpretation in the context of the species concept. Dobzhansky (1976) has championed the idea that species can be described in terms of genetics and that, among sexually reproducing and outbreeding organisms, species can be defined as mendelian populations or arrays of populations that are reproductively isolated from other population arrays. The concept seems apparent for mammals, but its application to yeasts is less straightforward because not all taxa are known to have sexual cycles, and some that do may show little outbreeding. Further, species formation does not usually leave a clear-cut separation between groups comprising the new species, and geneticaly intermediate populations may initially survive. Consequently, in order to use DNA reassociation data to define species, there must be some knowledge of the extent of DNA divergence to be found among members of a yeast species.

On the bases of comparisons between the extent of nuclear DNA relatedness and fertility of progeny arising from conventional genetic crosses, conspecific strains generally exhibit DNA relatedness in excess of 70%. Varietal designations can be accorded to those strains showing 40–70% DNA relatedness unless genetic crosses demonstrate the absence of interfertility. An exception is the varieties of *Issatchenkia scutulata*, which show only 25% DNA relatedness, but exhibit some intervarietal fertility (Kurtzman 1987; Kurtzman and Phaff 1987).

5.5.4
Amplification of Yeast DNA Using Polymerase Chain Reaction (PCR)

DNA isolated by different methods may be used for enzymatic amplification. We obtained good results with DNA isolated according to the miniprep method as described earlier. Either DNA in TE buffer or diluted DNA in TE/10 buffer (add 4 µl of the DNA solution in TE buffer to 1 ml TE/10 buffer), can be used as template in the PCR reactions.

TE/10: 10 mM Tris-HCl pH 8.0, 0.1 mM EDTA pH 8.0.

Amplification of Double-stranded DNA (dsDNA) by Symmetrical PCR Amplification

Prepare a mastermix containing the following (for one strain): $8.25\,\mu l$ double-distilled (dd) H_2O, $5\,\mu l$ $10\times$ buffer ($25\,ml$ $0.5\,M$ KCl; $5\,ml$ $0.1\,M$ Tris-HCl pH 8.4; $1.25\,ml$ $25\,mM$ $MgCl_2$; $50\,mg$ gelatin; $18.75\,ml$ ddH_2O); $9\,\mu l$ nucleotide mix [$0.25\,ml$ of each dNTP (Li-salt, Boehringer) $100\,mmol\,l^{-1}$], $1.25\,\mu l$ each of forward (F) and reverse (R) primers ($10\,pmol/\mu l$).

After mixing (vortex mix and centrifuge) $0.25\,\mu l$ *Taq* polymerase is added, followed by $25\,\mu l$ of DNA in TE/10. After carefully mixing by hand and short centrifugation, two droplets of sterile mineral oil are added, after which thermocycling is performed.

Typical cycling parameters for dsPCR are the following: initial denaturation for $60\,s$ at $94\,°C$, annealing for $60\,s$ at 50–$60\,°C$, extension for $90\,s$ at $72\,°C$; number of cycles 20–30, and hold at $4\,°C$. After amplification, $100\,\mu l$ chloroform is added, mix well (vortex), and centrifuge for $5\,min$ at $10\,000\,rpm$. Remove the lower, DNA-containing, layer to a new centrifuge tube. Store in freezer until further use.

Visualization of the amplified products is performed by electrophoresis in 1% agarose minigels in $1\times$ TBE ($0.045\,M$ Tris borate, $0.001\,M$ EDTA pH 8.0), using agarose dye mix (2% bromophenol blue; 2% xylene cyanol; $1\,ml$ ddH_2O, $0.2\,g$ sucrose) and a DNA ladder [$10\,\mu l$ DNA ladder stock ($1\,\mu g\,ml^{-1}$, Gibco); $10\,\mu l$ TE buffer; $50\,\mu l$ ddH_2O; $100\,\mu l$ agarose dye mix (see above)].

Staining is performed with ethidium bromide ($8 \times 10^{-5}\,\mu g\,\mu l^{-1}$) and visualized on a UV transilluminator.

Amplification of Single-stranded DNA (ssDNA) by Asymmetrical Amplification

We successfully applied the method of Kaltenboeck et al. (1992). Prepare a mastermix containing the following (for one reaction): $67.5\,\mu l$ ddH_2O; $10.0\,\mu l$ single-stranded (ss) buffer ($12.5\,ml$ $1\,M$ KCl; $2.5\,ml$ $1\,M$ Tris-HCl pH 8.4; $625\,\mu l$ $1\,M$ $MgCl_2$; $9.37\,ml$ H_2O; $2.5\,\mu l$ Tween 20; $2.5\,\mu l$ Nonidet P-40); $4.0\,\mu l$ F or R primer ($10\,pmol\,\mu l^{-1}$); after mixing (vortex) and short centrifugation ($10\,000\,rpm$), $0.5\,\mu l$ *Taq* polymerase is added.

After careful mixing (hand) and short centrifugation, $10\,\mu l$ dsDNA from the symmetric amplification is added. Mix carefully (hand, centrifuge) and add two droplets of sterile mineral oil.

Typical parameters for asymmetric amplification of DNA are the following: initial denaturation for $30\,s$ at $92\,°C$, annealing for $60\,s$ at 50–$60\,°C$, extension for $120\,s$ at $72\,°C$; repeat 20–30 times, followed by a final extension for $4\,min$ at $72\,°C$, and store at $4\,°C$.

After thermocycling is completed, add 100 μl chloroform, mix well (vortex) and centrifuge 5 min at 10 000 rpm. Transfer DNA-containing layer to a new centrifuge tube.

Check asymmetric PCR products on a 1% agarose/2% Nusieve agarose minigel in 1× TPE (0.09 M Tris-phosphate, 0.02 M EDTA pH 8.0). Depending on the primers used and size of the DNA to be amplified, the cycling parameters used may need modification.

Single-stranded DNA has to be cleaned (e.g., Geneclean II, Bio 101, La Jolla, CA; or by washing in Millipore Ultrafree-MC NMWL cellulose filter-microfuge tubes) to remove unincorporated nucleotides. The ssDNAs can be used directly in sequencing reactions, e.g., using Sequenase T7 DNA polymerase (USB, Cleveland, OH). In most cases, use of 5.0 μl ssDNA solution yielded satisfactory sequencing results. For protocols on manual and/or automated DNA sequencing procedures, the reader should consult specialized references.

Universal primers for amplification of the nuclear small-subunit rDNA, internally transcribed spacers (ITS 1 and 2), and mitochondrially encoded small and large subunit rDNA are given by White et al. (1990). For amplification of ca. 600 nucleotides near the 5′ end of the large subunit rDNA, we successfully used the primers NL 1, NL 3 (both forward), and NL 2 and NL 4 (both reverse; O'Donnell 1993). Nishida and Sugiyama (1993) used the following primers to amplify the nuclear small subunit rDNA of some ascomycetous yeast-like fungi: 5′-ATCTGGTTGATCCTGCCAGT-3′ and 5′-GATCCTTCCGCAGGTTCACC-3′. For basidiomycetous yeasts slightly modified internal primers have to be used (J.W. Fell, pers. comm.).

5.5.5
Electrophoretic Karyotyping

Electrophoretic patterns of chromosomal DNAs (electrophoretic karyotypes) provide data of potential use in systematic studies of yeasts (Table 2, see also Chap. 3, this Vol.). The analysis of karyotypes by pulsed-field techniques has received wide application in yeast systematics and genetics. Number and size of individual chromosomes, estimates of total genome sizes, chromosomal rearrangements, and gene assessment are among the characters studied. Taxonomically, these data are mainly used at or below the species level, e.g. to differentiate species, strains, and/or populations.

Due to the presence of chromosomal length polymorphisms, aneuploidy, and comigrating bands, a straightforward taxonomic interpretation of electrophoretic karyotypes is sometimes complicated. Ribosomal DNA-containing chromosomes frequently show considerable size variation (Maleszka and Clark-Walker 1989; Rustchenko-Bulgac 1991). Many species studied show rather variable karyotypes, e.g., *Candida albicans* (Iwaguchi et al. 1990; Rustchenko-Bulgac 1991). In contrast, other species show remarkably stable karyotypes, e.g., the basidiomycetous yeast

Table 2. Electrophoretic parameters successfully applied to separate chromosomes in some selected yeasts by pulsed-field electrophoresis

Species	Run time (h)	Pulse time	Agarose (%)	System	Reference
Saccharomyces cerevisiae	18	50 s	1.5	OFAGE	Carle and Olson (1985)
	23 7	60 s 100 s	1.0	CHEF	Bakalinsky and Snow (1990)
	18 6	60 s 35 s	1.0	TAFE	Vezinhet et al. (1990)
Schizosacchamyces pombe	130	60 min	0.6	CHEF	Vollrath and Davis (1987)
	162	75 min	1.0	PULSAPHORE	Smith et al. (1987)
	150	60 min	1.0	PULSAPHORE	Smith et al. (1987)
Kluyveromyces marxianus and *K. lactis*	25 18 25 48	30–300 s 300 s 600 s 600–1200s	1.5	OFAGE	Sor and Fukuhara (1989)
	32 20 40	20–120 s 120 s 120–600 s	1.5	CHEF	Sor and Fukuhara (1989)
	65 18	300 s 65 s	1.0	OFAGE	Steensma et al. (1988)
Pachysolen tannophilus	70 24 24 48	60–120 s 300 s 500 s 600–800 s	1.2	PULSAPHOR	Maleszka and Skrzypek (1990)
Filobasidiella neoformans	22 20 18	20 min 5 min 120 s	1.0	CHEF	Perfect et al. (1989)
	16 32	50–130 s 170–300 s	0.6	CHEF	Kwon-Chung et al. (1992)

Table 2 (*contd.*)

Species	Run time (h)	Pulse time	Agarose (%)	System	Reference
Candida albicans	48 24 48	4 min 9 min 14 min	1.0	TAFE	McEachern and Hicks (1991)
	14 18 18 18	2 min 4 min 7 min 10 min	0.6	TAFE	McEachern and Hicks (1991)
Malassezia furfur	36 36	300 s 300–600 s	1.0	CHEF	Boekhout and Bosboom (1994)
Malassezia pachydermatis	48	50–300 s	1.0	CHEF	Boekhout and Bosboom (1994)

Malassezia pachydermatis. All strains of this species show a slight length variation of only the smallest chromosomal DNA (Boekhout and Bosboom 1994). In *Saccharomyces cerevisiae*, a correlation has been demonstrated between similarities of DNA as observed in DNA reassociation experiments, and similarities of karyotypes (Vaughan Martini et al. 1993).

At present, it is possible to separate chromosomal DNAs of up to ca. 6 Mb (Gunderson and Chu 1991), but even 10 Mb has been reported (Zhang et al. 1991). Separation of chromosomal DNAs larger then ca. 6 Mb is still difficult, and requires long pulse and run times. Chromosomal DNAs of *Saccharomyces cerevisiae*, *Pichia canadensis* (*Hansenula wingei*), and *Schizosaccharomyces pombe* are commercially available as size standards.

For each yeast species with an unknown karyotype, the parameters used during electrophoresis have to be optimized (Table 2). As a rule of thumb, the size of the chromosomal DNAs is inversely proportional to the agarose concentration of the gels and the field strength to be used. The pulse time is directly proportional to the chromosomal size.

Several modifications of the methods used have been described. Mathaba et al. (1993) used microtiter trays for culturing and preparing agarose blocks of *Candida albicans*. Cook (1984) and Dear and Cook (1991) described a method for the isolation of nuclear DNA using cells encapsulated in agarose minibeads. McCluskey et al. (1990) and Gardner et al. (1993) analyzed electrophoretic karyotypes of a variety of plant pathogenic fungi and *Saccharomyces cerevisiae*, respectively, without protoplasting the cells. Other modifications are the use of macrorestriction enzymes (see Dixon and Kinghorn 1990; Gardiner 1991), hybridization with chromosome-specific probes (Török et al. 1992), and densitometric analysis (Mahrous et al. 1990).

Protoplasts of ascomycetous yeasts usually are prepared using zymolyase (de Jonge et al. 1986). However, this enzyme is not suitable for the preparation of protoplasts of basidiomycetous yeasts. Therefore, we prefer to use Novozym 234 (Novo Biolabs) for protoplasting both asco- and basidiomycetous yeasts. Further protocols can be found in Anand and Southern (1988), Sambrook et al. (1989) and Johnston (1988) (see also Chap. 3, this Vol.). We have used the following protocol successfully for the preparation of agar-embedded protoplasts of many basidiomycetous yeasts (according to de Jonge et al. 1986). Alternatively, protoplasts can be made before embedding in low-melting point agarose.

Preparation of Agar-Embedded Protoplasts Using Novozym 234

Rotary-shaken (ca. 200 rpm) cultures are grown until late logarithmic phase in 50 ml 1% yeast extract-0.5% peptone-4% glucose (YPG) broth at 25 °C. For basidiomycetous yeasts, 0.05% yeast extract-0.5% peptone-0.7% malt extract (YPM) broth is preferred. Psychrophilic species require lower temperatures, e.g., 12 or 17 °C.

Approximately 0.5×10^9 cells are harvested by centrifugation, and washed in 2 ml 0.05 M EDTA, pH 7.5.

The pellet is resuspended in ca. 2 ml buffer made up of 0.05 M EDTA, 10 mM Tris-HCL, and 10 mM dithiothreitol (DTT), final pH 7.5.

After centrifugation, the pellet is washed with 2 ml CPE buffer [100 ml 40 mM citric acid, 120 mM Na_2PO_4 (pH 6.0) and 4 ml 0.5 M EDTA, pH 7.5]. Centrifuge and resuspend in 0.3 ml CPES buffer (CPE buffer containing 1.2 M sorbitol and 5 mM DTT).

Dissolve 2 mg Novozym 234 in 0.5 ml 1% w/v low-melting agarose in CPE buffer at 38 °C. Mix equal volumes of Novozym 234/agarose solution and the cell suspension (when preparing protoplasts of hyphal species or large-sized cells, the volumes have to be adjusted).

Pipette the cell/Novozym/agarose mixture in a precooled matrix and allow to gel on ice. The agarose/cell blocks are incubated for 1 h at 30 °C in ca. 5 ml CPE buffer.

The agarose blocks are rinsed with ca. 5 ml NDS (0.5 M EDTA, pH 7.5, 10 mM Tris-HCl, pH 7.5), 1% v/v sodium N-lauroylsarcosinate, and subsequently placed for 16 h at 50 °C in 2 ml NDS buffer to which 0.4 mg proteinase K and 1% v/v sodium N-lauroylsarcosinate have been added.

After lysis, the agarose blocks are rinsed in NDS buffer containing 1% v/v sodium N-lauroylsarcosinate and stored in NDS buffer containing 1% v/v sodium N-lauroylsarcosinate and 0.2 mg ml^{-1} proteinase K. Sodium N-lauroylsarcosinate and proteinase K have to be added just prior to use. The agarose/cell blocks can be stored for several months at 4 °C.

Alternatively, protoplasts of cells can be prepared before embedding in low-melting agarose as follows. Add 2 mg Novozym 234 in 0.5 ml CPE to the resuspended cells in 0.3 ml CPES (see above), and allow degradation of the cell wall for 2–3 h at 30 °C.

Check protoplast formation microscopically, and in the case of sufficient cell wall degradation, centrifuge for 3 min at 13 000 rpm. Resuspend the protoplasts in 0.3 ml CPES, after which an equal volume 1% low-melting agarose in CPE buffer is added.

Pipette the protoplast/agarose mixture in a precooled matrix and allow to gel on ice. Rinse the agarose blocks in 2 ml NDS, 1% v/v lauroylsarcosinate, and incubate for 16 h at 50 °C in NDS to which 1% v/v lauroylsarcosinate and 0.02% Proteinase K are added. Proceed according to the above protocol.

Prior to electrophoresis, the agarose blocks are rinsed in 0.5× TBE (1× TBE: 90 mM Tris base, 90 mM boric acid, 2.5 mM EDTA disodium salt, final pH 8.2) for 1 h. Depending on the size of the chromosomal DNAs to be separated, 0.5–1.5% agarose gels (chromosomal grade) in 0.25–0.5× TBE are prepared.

The positioning of the agarose/cell blocks in the gel inserts is facilitated by initially flooding the inserts with 0.5× TBE buffer. 0.25–0.5× TBE is used as electrophoretic buffer, and the temperature is kept between 12 and 15 °C. Some electrophoretic parameters, taken from the literature, are presented in Table 2.

After completion, the gels are stained in $0.5 \mu g ml^{-1}$ ethidium bromide for 30–120 min, destained with demineralized water, and the DNA visualized with an UV transilluminator. If the DNA banding is obscured by a smear, an RNase treatment is suggested (de Jonge et al. 1986) as follows.

Incubate the gel for 2 h at 37 °C with gentle shaking in a sealed bag containing 30 ml 0.5× TBE and 1.5 ml RNase solution ($500 \mu g ml^{-1}$ pancreatic RNase + 100 units ml^{-1} T1 RNase in 10 mM Tris-HCl, pH 7.5, 15 mM NaCl, heated for 10 min at 100 °C and slowly cooled to room temperature). Rinse twice with electrophoresis buffer for 60 min.

6
Overview of Yeast Genera

6.1
Teleomorphic Ascomycetous Genera

Ambrosiozyma van der Walt, Mycopath. Mycol. Appl. 46, 305 (1972).
Two species: *A. cicatricosa* (Scott et van der Walt) van der Walt, *A. monospora* (Saito) van der Walt.
Vegetative reproduction is by multipolar budding, pseudohyphae, and septate hyphae. The septa are dolipore-like. Asci are spheroidal to ovoidal, generally formed on hyphae, and persistent or deliquescent. Ascospores are hat-shaped with 1–4 per ascus.

The species are often isolated from insects or insect tunnels in woody plants. Species are slow or weak fermenters of sugars. *A. monospora* forms CoQ-7. Kurtzman and Robnett (1995) demonstrated from rDNA sequence analysis that *Ambrosizyma and Hormoaxus* are congeneric and transferred the three *Mormoascus* species to *Ambrosiozyma*.

Arthroascus **von Arx**, Antonie van Leeuwenhoek J. Microbiol. 38, 289 (1972).
Three species: *A. javanensis* (Klöcker) von Arx, *A. schoenii* (Nadson et Krasilnikov) Bab'eva et al., *A. fermentans* Lee et al.
Vegetative reproduction is by multipolar budding, pseudohyphae, and septate hyphae. The septa have a single micropore. Asci are swollen hyphal cells that deliquesce at maturity. Asci form up to four spores which are spheroidal, subspheroidal with a circumfluent ledge, or occasionally hat-like. Some spores may be warty. Habitats are soil and plant materials.
Two of the species are nonfermentative. *A. javanensis* forms CoQ-8.

Species of *Arthroascus* were recently transferred to *Saccharomycopsis* because phylogenetic analysis of rDNA sequences showed species of both genera to be members of a single clade (Kurtzman and Robnett 1995). The species assigned to *Arthroascus* were characterized by asci that are swollen hyphal cells, but this type of ascus formation may occasionally be seen among other members of *Saccharomycopsis*.

Arxiozyma **van der Walt and Yarrow**, S. African J. Bot. 3, 340 (1984).
One species: *A. telluris* (van der Walt) van der Walt et Yarrow.
Anamorph: *Candida pintolopesii* (van Uden) S.A. Meyer et Yarrow.
Vegetative reproduction is by multipolar budding and pseudohyphae. The persistent asci arise from the conversion of vegetative cells and are spheroidal to ovoidal. Ascospores are warty and spheroidal to ovoidal with 1–2 per ascus.

The species ferments sugars and is isolated primarily from poultry and other birds. CoQ-6 is formed.

Ascoidea **Brefeld and Lindau**, Unters. Gesammtgeb. Mykol. 9, 91 (1891).
The six species are listed by Gams and Grinbergs (1970).
Vegetative reproduction is predominantly by true hyphae but budding yeast cells and pseudohyphae are produced by some species. Asci are formed on hyphae and are ellipsoidal or acicular. Ascospores are released through a terminal opening in the ascus. New asci develop within the remains of previous asci. Ascospores are generally hat-shaped and may number from 16 to several hundred per ascus.

The species are isolated from tree bark, bark beetles, or the slime flux of trees. Fermentation of sugars is absent or weak.

Ashbya **Guilliermond**, Rev. Gén. Bot. 40, 328 (1928).
One species: *Ashbya gossypii* (Ashby et Nowell) Guilliermond.
Vegetative reproduction is by septate hyphae. Asci are elongate to fusiform, deliquescent at maturity, and produce 8 or more two-celled, needle-like ascospores that have long, whip-like appendages. The species can often be found on cotton where it forms a yellow discoloration of the bolls.

Sugars are not fermented. *A. gossypii* is a commercial source of riboflavin and in culture the presence of riboflavin turns the medium yellow.

Phylogenetic analysis of rDNA sequences from *Ashbya*, *Eremothecium*, *Holleya* and *Nematospora* showed species of all four genera to be closely related members of a single clade. Consequently, Kurtzman (1995) placed all species in *Eremothecium*, the genus of taxonomic priority.

Botryoascus von Arx, Antonie van Leeuwenhoek J. Microbiol. 38, 289 (1972).

One species: *B. synnaedendrus* (Scott et van der Walt) von Arx.

Vegetative reproduction is by pseudohyphae and septate hyphae.

Budding yeast cells were reported not to be present. The septa have plasmodesmata. Asci are spheroidal to elongate, deliquescent at maturity, and contain 1–4 hat-shaped spores.

The species is non-fermentative and found in the tunnels of insects. CoQ-8 is formed. *Botryoascus* is identical to *Saccharomycopsis* except that von Arx (1972) reported budding cells to be absent. Budding cells were detected in two strains of *B. synnaedendrus*, and rDNA analysis places it among species of *Saccharomycopsis*, the genus to which it was recently returned (Kurtzman and Robnett 1995).

Cephaloascus Hanawa, Jpn. J. Dermatol. Urol. 20, 103 (1920).

Two species: *C. albidus* Kurtzman, *C. fragans* Hanawa.

Vegetative growth is by budding cells, pseudohyphae, and true hyphae. Hyphal septa have a single, central micropore. Asci are deliquescent and borne in clusters on erect branched ascophores. Ascophores may be smooth and hyaline, or roughened and with a brown pigmentation. Asci produce 1–4 hat-shaped spores. One of the species is homothallic while the other is heterothallic. *C. fragrans* is common to various types of wood, although one strain was from a human ear. *C. albidus* was isolated from cranberry pumace. Fermentation of sugars is weak or absent.

Citeromyces Santa María, Inst. Nac. Invest. Agron., Sec. Bioquímica, Cuaderno No. 258, 269 (1957).

One species: *C. matritensis* (Santa María) Santa María.

Vegetative reproduction is by multilateral budding. Pseudohyphae and true hyphae are not produced. Asci are spheroidal, heavy-walled and persistent. Each forms 1–2 spheroidal ascospores that have roughened surfaces. The species is heterothallic.

The species is a strong fermenter of sugars and is usually isolated from sugar concentrates and slime fluxes. CoQ-8 is formed.

Clavispora Rodrigues de Miranda, Antonie van Leeuwenhoek J. Microbiol. 45, 479 (1979).

Two species: *C. lusitaniae* Rodrigues de Miranda, *C. opuntiae* Phaff et al.

Vegetative reproduction is by multipolar budding and pseudohyphae. True hyphae are not formed. Asci are spheroidal to elongate and develop following conjugation of complementary mating types. Ascospores are clavate, finely warted, and released at maturity. There are 1–2 spores per ascus. *C. lusitaniae* is known from human and animal sources, while *C. opuntiae* has only been isolated from rotting *Opuntia* cactus. The species ferment sugars and both produce CoQ-8.

Coccidiascus **Chatton**, C. R. Soc. Biol. 75, 117 (1913).
One Species: *C. legeri* Chatton.
Vegetative reproduction is by multipolar budding. Pseudohyphae and true hyphae do not occur. Asci are elongate and curved and form up to eight spindle-shaped ascospores. The species is known only from the tissue of *Drosophila* spp.

Cyniclomyces **van der Walt and Scott**, Mycopath. Mycol. Appl. 43, 279 (1971).
One species: *C. guttulatus* (Robin) van der Walt & D.B. Scott.
Vegetative reproduction occurs by multipolar budding, occurring predominantly near the poles of the cell. Pseudohyphae occur, but true hyphae are not formed. Asci are elongate and are seldom deliquescent. Ascospores are generally elongate, 1–4 per ascus.

The species weakly ferments sugars. *C. guttulatus* is restricted to the feces and stomach contents of rabbits. Laboratory cultivation requires a high CO_2 atmosphere.

Debaryomyces **Lodder and Kreger-van Rij**, Taxon 27, 306 (1978).
The presently accepted 14 species are described by Kurtzman and Robnett (1991) and Yamada et al. (1991a).
Vegetative reproduction is by multipolar budding and pseudohyphae. The species are homothallic, and asci show parent cell-bud conjugation or, infrequently, cell-cell conjugation. Asci form 1–4 spores. For most species, asci are persistent. Ascospores are usually spheroidal and often roughened, but certain species may form roughened spores with equatorial ledges, spores with spiral ridges, or smooth, lenticular spores. All produce CoQ-9, and are commonly found in soil, plant products, foods, and in clinical specimens.

Comparisons of ribosomal RNA sequence similarities resulted in the transfer to *Debaryomyces* of several species from other genera, namely *Schwanniomyces occidentalis* (Kurtzman and Robnett 1991), *Wingea robertsii* (Kurtzman and Robnett 1994), and *Pichia carsonii* and *P. etchellsii* (Yamada et al. 1991). The species give a weak to strong fermentation of sugars. The species formerly classified as *Schwanniomyces*, *S. occidentalis*, shows a high amylase activity (De Mot et al. 1984a) and a transformation system has been developed (Klein and Favreau 1988; see Chap. 4, this Vol.).

Dekkera **van der Walt**, Antonie van Leeuwenhoek J. Microbiol. 30, 273 (1964).
Two species: *D. anomala* Smith et van Grinsven, *D. bruxellensis* van der Walt.
Anamorph: *Brettanomyces*.
Vegetative reproduction is by multipolar budding and pseudohyphae. Some of the yeast cells have an ogival shape at one end. Asci are spheroidal to elongate, form 1–4 hat-shaped spores, and become deliquescent at maturity. All produce CoQ-9. The species are usually isolated from beer, wine and soft drinks.

The species may show a variable fermentation of sugars, which is stimulated by the presence of oxygen (Custers effect). Species are noted for a vigorous production of acetic acid and this causes early cell death in cultures.

Dipodascopsis **Batra and P. Millner,** Proc. Int. Symp. Taxon. Fungi, Madras 1973, 187 (1978).
Two species: *D. uninucleata* (Biggs) Batra et Millner, *D. tothii* Zsolt.
Vegetative growth is occasionally by budding, but predominantly by true hyphae. Hyphal septa have a single, central pore. Asci are initiated by the fusion of gametangia from adjacent hyphae or adjacent hyphal cells. The asci become elongated and produce from 30 to over 100 ellipsoidal to reniform ascospores which are released when the ascus tip deliquesces. The species are homothallic. Isolates have been obtained from tree wounds and insects. Sugars are not fermented.

Dipodascus **Lagerheim,** Jb. Wiss. Bot. 24, 549 (1892).
The 13 species have been reported by de Hoog et al. (1986).
Anamorph: *Geotrichum*.
Vegetative reproduction is by hyphae and arthroconidia. Hyphal septa have micropores. Asci are initiated from the fusion of gametangia and are spheroidal or, more frequently, elongate. Ascospores are hyaline, ellipsoidal and, depending on the species, from 4 to over 100 in each ascus. The tips of asci deliquesce for spore release. Some of the species are heterothallic. Strains are obtained from plant materials and insects, usually those associated with trees. Most species are non-fermentative.

Endomyces **Reess,** Botanische Untersuchungen über die Alcoholgärungspilze, p. 77. A. Felix, Leipzig (1870).
Three species: *E. cortinarii* Redhead et Malloch, *E. parasiticus* Fayod, *E. polyporicola* (Schumacher et Ryvarden) de Hoog, M.Th. Smith & Guého.
Vegetative reproduction is by septate hyphae that form small chains of blastoconidia. Asci form on hyphae and are ellipsoidal to elongate. The asci produce 2–12 helmet-shaped spores that are released by deliquescence of the ascus apex. Species of *Endomyces* are parasitic on mushrooms.

Eremothecium **Borzi,** Nuovo Giorn. Bot. Ital. 20, 452–456 (1888).
Five species: *E. cymbalariae* Borzi, *E. ashbyi* (Guilliermond ex Routien) Batra, *E. coryli* (Peglion) Kurtzman, *E. gossypii* (Ashby et Nowell) Kurtzman, *E. sinecaudum* (Holley) Kurtzman.
Vegetative reproduction is by pseudohyphae and true hyphae. *E. coryli* and *E. sinecaudum* also show budding whereas the other three species do not. Asci, which are elongate or spindle-shaped, are produced on hyphae and may be terminal or intercalary. There are generally at least 8 ascospores per ascus, and they are released at maturity by deliquescence of the ascus wall. Ascospores are elongate, straight or curved, and sharply tapered on one end. The ascospores of *E. coryli* have a whip-like appendage. The species are homothallic; they have been isolated from cotton bolls, citrus, several other plant species and an insect.

Sugars are not fermented. Cultures of *E. ashbyi* and *E. gossypii* are often yellow-orange in color from formation of riboflavin, and are used for commercial production of this vitamin.

Species assigned to *Ashbya*, *Eremothecium*, *Holleya*, and *Nematospora* show considerable phenotypic similarity, but had been assigned to their respective genera because of the presence or absence of budding cells and differences in ascospore shape and the number of isoprene units on their coenzyme Q molecules. Phylogenetic analysis of rDNA sequences showed the species to be closely related members of a single clade, Kurtzman (1995) placed all species in *Eremothecium*, the genus of taxonomic priority.

Galactomyces **Redhead et Malloch,** Can. J. Bot. 55, 1701 (1977).
Two species: *G. geotrichum* (E. E. Bulter et L. J. Petersen) Redhead et Malloch, *G. reessii* (van der Walt) Redhead et Malloch.
Anamorph: *Geotrichum.*
Vegetative reproduction is by hyphae and arthroconidia. Hyphal septa have micropores. Asci are spheroidal to ellipsoidal and arise from the fusion of hyphal gametangia. Ascospores are spheroidal, pale brown, and roughened. Each ascus forms 1 or, rarely, 2 spores that are released at maturity. *G. geotrichum* has both heterothallic and homothallic strains, whereas *G. reessii* appears to be homothallic. The species are common in soil, plant material, food (dairy products), and clinical specimens. The species may weakly ferment sugars.

Guilliermondella **Nadson and Krassilnikov,** C. Séanc. hebd. Acad. Sci. 187, 307 (1928).
One species: *G. selenospora* Nadson et Krassilnikov.
Vegetative reproduction is by multipolar budding, mainly occurring near the poles of the parent cell. Pseudohyphae and true hyphae are also formed. The septa of true hyphae have plasmodesmata. Asci are formed following conjugation of adjacent hyphal cells and form 1–4 spores that may be spheroidal, reniform or sickle-shaped. Ascospores are released at maturity. The species is homothallic. Isolates have been obtained from tanning fluid and the slime fluxes of trees. Of the commonly tested sugars, only glucose is fermented. The species produces CoQ-8.

Kurtzman and Robnett (1995) showed from similarity of rDNA sequences that *G. selenospora* is a member of the *Saccharomycopsis* clade. On the basis of this analys is, they transferred *G. selenospora* to *Saccharomycopsis*.

Hanseniaspora **Zikes,** Zentbl. Bakteriol. Parasitenk, Abt. 2, 30, 145 (1911).
The six species are discussed by Meyer et al. (1978).
Anamorph: *Kloeckera.*
Members of this genus are noted for producing apiculate or lemon-shaped cells. Vegetative reproduction is by bipolar budding and occasionally by formation of rudimentary pseudohyphae. Asci are generally elongate, forming 1–4 spores. Ascospores may be hat-shaped or spheroidal, and with or without an equatorial ring. Species forming hat-shaped ascospores have deliquescent asci. The species are most frequently isolated from soil, fruits, and plant exudates.

The species ferment glucose and occasionally other sugars. All member of the genus produce CoQ-6

Holleya Yamada, J. Gen. Appl. Microbiol. 32, 447 (1986).
One species: *H. sinecauda* (Holley) Yamada.
Vegetative reproduction is by multipolar budding, pseudohyphae, and true hyphae. Asci are elongate and deliquesce at maturity. Each ascus produces 4–8 needle-shaped spores that show a central globose swelling at germination.

The species is non-fermentative, forms CoQ-9, and has been isolated from the seeds of oriental and yellow mustard. On the basis of rDNA comparisons, Kurtzman (1995) transferred *H. sinecauda* to the genus *Eremothecium*.

Hormoascus von Arx, Antonie van Leeuwenhoek J. Microbiol. 38, 289 (1972).
Three species: *H. ambrosiae* (van der Walt et D.B. Scott) van der Walt et von Arx, *H. philentomus* (van der Walt et al.) van der Walt et von Arx, *H. platypodis* (Baker et Kreger-van Rij) von Arx.
Vegetative reproduction is by multipolar budding, pseudohyphae, and true hyphae. The hyphae produce dolipore-like septa. Asci are spheroidal to elongate and may show cell-to-cell conjugation. The asci produce 1–4 hat-shaped ascospores that are released at maturity. All three species have been isolated from insect tunnels under bark.

Species often show a weak fermentation of sugars and are characterized by CoQ-7.

Members of *Hormoascus* assimilate nitrate as a sole source of nitrogen whereas species of *Ambrosiozyma* do not. Von Arx (1972) used this growth reaction to separate the two genera. Kurtzman and Robnett (1995) demonstrated from rDNA sequence analysis that *Hormoascus* and *Ambrosiozyma* are congeneric and transferred species of the former genus to *Ambrosiozyma*.

Issatchenkia Kudriavzev, Die Systematik der Hefen, p. 161. Akademie Verlag, Berlin (1960).
Four species: *I. occidentalis* (Phaff, Miller et Miranda) Kurtzman, Smiley et Johnson, I. orientalis Kudriavzev, I. scutulata (Phaff, Miller et Miranda) Kurtzman, Smiley et Johnson, *I. terricola* (van der Walt) Kurtzman, Smiley et Johnson.
Vegetative reproduction is by multipolar budding and by pseudohyphae. Asci are spheroidal to elongate, persistent, and form 1–4 roughened, spheroidal spores. The species are heterothallic. Common habitats are soil, fruits, flowers, and tree exudates. I. orientalis and its anamorph C. krusei are also isolated from human and animal sources.

All species ferment glucose but not other commonly tested sugars. Species are characterized by CoQ-7.

Kluyveromyces van der Walt emend. van der Walt, Antonie van Leeuwenhoek J. Microbiol. 31, 314 (1965).
The presently accepted 13 species are described by Fuson et al. (1987).
Vegetative reproduction is by multipolar budding and pseudohyphae. Asci are spheroidal to elongate and generally unconjugated. A genus characteristic is deli-

quescence of the ascus wall at maturity. Ascospores are smooth and spheroidal to reniform. Most species form 1–4 ascospores, but *K. africanus* may produce up to 16–20, and *K. polysporus* can produce in excess of 100. The species appear to be homothallic except for K. lactis. Species of *Kluyveromyces* are isolated from soil, water, fruit and other plant materials, tree fluxes, dairy products, *Drosophila* and occasionally from clinical specimens.

All species ferment glucose and usually at least one other sugar. The species form CoQ-6. Because of their ability to ferment lactose, *K. lactis* (anamorph *Candida sphaerica*) and *K. marxianus* (anamorph *Candida kefyr*) have been used industrially to produce ethanol from waste dairy products such as whey. Lactose utilization from whey has also been reported for *K. marxianus* (as *K. fragilis*; Jenq et al. 1989). *Kluyveromyces marxianus* shows a high inulinase activity (Vandamme and Derycke 1983). *Kluyveromyces lactis* is used for heterologous gene expression (Reiser et al. 1990; van den Berg et al. 1990; Fleer et al. 1991; Bergkamp et al. 1992; Gellisen et al. 1992; Romanos et al. 1992; see Chap. 5, this Vol.).

Lipomyces **Lodder and Kreger-van Rij,** The yeasts, a taxonomic study, p. 669. North-Holland Publ. Co., Amsterdam (1952).
The presently accepted five species are described by Kurtzman and Liu (1990).
Vegetative reproduction is by multipolar budding and pseudohyphae. Asci are ellipsoidal to elongate, frequently conjugated, and deliquescent at maturity. Ascospores are spheroidal to elongate and covered with warts and ridges. The spores are pigmented, amber to brown and up to 30 may be produced in a single ascus. The species are isolated from soil.

Lipomyces is non-fermentative. Most species form CoQ-9, but *L. lipofer* (synonym *Waltomyces lipofer*) produces CoQ-10. In culture, the species are quite mucoid because of the copious production of an extracellular polysaccharide. Lipomyces kononenkoae was found to have a high extracellular amylase activity (De Mot et al. 1984a). *Lipomyces lipofer* and *L. starkey* are oleaginous (Ratledge 1986, 1993; Rezanka 1991).

Lodderomyces **van der Walt,** Antonie van Leeuwenhoek J. Microbiol. 32, 1 (1966).
One species: *L. elongisporus* (Recca et Mrak) van der Walt.
Vegetative reproduction is by multipolar budding and pseudohyphae. Asci are spheroidal to ellipsoidal, unconjugated, and persistent. Each ascus forms 1–2 spores, and the spores are elongate, slightly tapered, and may have somewhat blunt ends. Isolates have come from soil and orange juice. The species ferments sugars, utilizes higher alkanes, and forms CoQ-9.

Metschnikowia **Kamienski,** Trav. Soc. Imp. Natural. St. Petersbourg 30, 363 (1899).
The presently accepted eight species are listed by Mendonça-Hagler et al. (1993).
Vegetative reproduction is by multipolar budding and pseudohyphae. Asci are unusual in shape and range from clavate to sphero- or ellipsopedunculate. They are nonconjugated, persistent, and for some species, arise from the morphogenesis of chlamydospores. Ascospores are unusual in shape, elongated and arrow-like (Fig. 18). Generally, there are 1–2 spores per ascus. Some of the species are

Fig. 18. Ascus with needle-shaped ascospores in *Metschnikowia zobellii* (6d McClary's agar)

heterothallic. The species are isolated from seawater, freshwater, flowers, invertebrates, and *Drosophila*.

The species ferment sugars and form CoQ-9. Some strains of *M. pulcherrima* produce pulcherrimin, a reddish pigment that may diffuse into the growth medium.

Nadsonia **Sydow,** Ann. Mycol. 10, 347 (1912).
Three species: *Nadsonia commutata* Golubev, *Nadsonia elongata* Konokotina, *Nadsonia fulvescens* (Nadson et Konokotina) Sydow.
Vegetative reproduction is by bipolar budding. Asci are persistent, elongate, and formed by a parent-bud conjugation. Depending on the species, either the parent cell or an opposite bud becomes the ascus. Ascospores are spheroidal and roughened with 1–2 per ascus. The species are commonly found in soil and the slime fluxes of trees.

Two of the species ferment sugars, and CoQ-6 is present in those species that have been examined.

Nematospora **Peglion,** Atti R. Accad. Naz. Lincei, Ser. 5, 6, 216 (1897).
One species: *Nematospora coryli* Peglion.
Vegetative reproduction is predominantly by multilateral budding, but also by pseudohyphae and true hyphae. The yeast cells often have unusual shapes. Asci are relatively large, elongate, and the walls are deliquescent at maturity. Asci generally form 8 spores that are quite elongated and have a whip-like terminal appendage.

The species ferments sugars and has CoQ-5 or 6. *Nematospora* shows many similarities with *Holleya*, *Ashbya*, and *Eremothecium*, and, as discussed earlier, *N.*

coryli was transferred to the genus *Eremothecium* (Kurtzman 1995). The species causes diseases in hazelnuts, cotton bolls, and various beans.

Pachysolen Boidin and Adzet, Bull. Trimest. Soc. Mycol. Fr. 73, 331 (1957).
One species: *P. tannophilus* Boidin et Adzet.
Vegetative reproduction is by multipolar budding and pseudohyphae. Ascus formation by this genus is unique. A diploid cell, which may show an attached conjugant, gives rise to a heavy-walled, refractile tube that flares outward slightly on the end. The ascus develops within the distal flared end and forms 4 hat-shaped ascospores that are released at maturity. The species is homothallic. It forms CoQ-8. Isolates of *P. tannophilus* are from tanning liquors and leather.

The species ferments a variety of sugars. Of interest to biotechnology is the ethanolic fermentation of D-xylose (Schneider et al. 1981; Dekker 1982; Kurtzman et al. 1982), D-galactose and glycerol (Maleszka et al. 1982), and the bioconversion of wheat straw (Detroy et al. 1982). *P. tannophilus* is also noted for production of an extracellular polysaccharide.

Pachytichospora van der Walt, Bothalia 12, 563 (1978).
One species: *P. transvaalensis* (van der Walt) van der Walt.
Vegetative reproduction is by multipolar budding and poorly developed pseudohyphae. Asci are ellipsoidal, unconjugated, and persistent. The feature that distinguishes this genus from *Saccharomyces* is its ascospores, which are spheroidal to ellipsoidal and have notably thickened walls. There are usually 1–2 spores per ascus. The species is heterothallic. Strains have been isolated from soil and an animal cecum. The species is fermentative and forms CoQ-6.

Pichia E.C. Hansen, Zentbl. Bakteriol. Parasitenk., Abt. 2, 12, 529 (1904) emend. Kurtzman, Antonie van Leeuwenhoek J. Microbiol. 50, 209 (1984).
The presently accepted ca. 100 species are described by Kurtzman (1984) and Barnett et al. (1990).
Vegetative reproduction is by multipolar budding and pseudohyphae and/or true hyphae are sometimes present. The septa of species with true hyphae are characterized by a single, central micropore. Asci may be spheroidal to ellipsoidal, conjugated or unconjugated, and persistent or deliquescent. Species generally form 4 ascospores per ascus, but rarely there may be as many as 8. Most of the species have hat-shaped ascospores, but a few are known for spheroidal spores that may be roughened or have ledges. Members of the genus are common in soil, water, trees, fruits, insects, and clinical specimens.

Species are fermentative and nonfermentative, and have CoQ-7, -8, or -9. *Pichia* appears to be extremely heterogeneous. Some of the species have, somewhat prematurely, been moved to the genus *Yamadazyma* Billon-Grand. Before additional changes are made, all of the species need careful comparison by molecular methods. The genus nearly doubled in size with the transfer of *Hansenula* species to *Pichia* (Kurtzman 1984). Saturn-spored species have been reassigned to *Williopsis* and *Saturnispora*, and those with smooth spheroidal spores to *Debaryomyces*.

Besides *Yamadazyma*, the genus *Hyphopichia* von Arx & van der Walt is presently considered a synonym of *Pichia*.

Pichia stipitis (anamorph *Candida shehatae*) is of biotechnological importance, because of its fermentation of xylose (Dellweg et al. 1984; Björling and Lindman 1989; Prior et al. 1989) a major component of plant biomass. For *Pichia pastoris* and *P. angusta* (synonym *Hansenula polymorpha*) heterologous gene expression systems have been developed (e.g., Reiser et al. 1990; Buckholz and Gleeson 1991; Clare et al. 1991; Gellissen et al. 1992; Romanos et al. 1992; Cregg et al. 1993). *Pichia nakazawae* was found to have high amylase activity (De Mot et al. 1984a; see also Chap. 6 and 9, this Vol.). *Pichia guilliermondi* is a producer of riboflavin (see Chap. 7, this Vol.). *Pichia methanolica* mutants defective in acetyl-CoA synthetase can be used in a biosensor (see Chap. 8, this Vol.).

Saccharomyces **Meyen ex Reess,** Sber. Phys. Med. Soz. Erlangen 9, 190 (1877).
The presently accepted ten species have been defined by Vaughan Martini and Martini (1987) and Kurtzman and Robnett (1991).
Vegetative reproduction is by multipolar budding and occasionally by pseudohyphae. Asci are ellipsoidal to elongate, unconjugated, and persistent. Ascospores are smooth, spheroidal to short ellipsoidal, and 1–4 are usually formed in an ascus. Species are commonly isolated from soil, fruits, foods, beverages, and, rarely, from clinical specimens.

The species are fermentative and characterized by CoQ-6. Four of the species in this genus, which form the *S. cerevisiae* sibling species complex, are widely used for bread-making, production of beer, wine, distilled beverages, and fuel alcohol. *Saccharomyces cerevisae* is used for heterologous protein production (e.g., Sleep et al. 1991; Gellissen et al. 1992; Romanos et al. 1992).

Saccharomycodes **Hansen,** Zentlbl. Bakteriol. Parasitenk., Abt. 2, 12, 529 (1904).
One species: *S. ludwigii* Hansen.
Yeast cells are apiculate and vegetative reproduction is by bipolar budding and by poorly developed pseudohyphae. Asci are spheroidal to ellipsoidal, unconjugated, and persistent. Asci produce 1–4 ascospores that are spheroidal and have a fine subequatorial ledge.

The species is fermentative, forms CoQ-6, and has been isolated from slime fluxes of trees.

Saccharomycopsis **Schiönning,** C. R. Trav. Lab. Carlsberg 6, 101 (1903).
The ten species were described by Kurtzman and Robnett (1995). Vegetative reproduction is by multipolar budding, arthrospores, pseudohyphae, and true hyphae. Hyphal septa of most species are characterized by plasmodesmata. Asci are spheroidal to ellipsoidal, conjugated or unconjugated, and either free or attached to hyphae. The asci may be either deliquescent or persistent. There are usually 1–4 ascospores per ascus, but ascospore morphology varies markedly among species. Some of the species form hat-shaped spores, while others have reniform, spheroidal or ellipsoidal spores, two of which have circumfluent ledges. Strains have been isolated from trees, pollen, fruit, insects, and various starchy foods.

The species give a slow or weak fermentation of sugars and form CoQ-8. *Saccharomycopsis fibuligera* is noted for the production of amylase. Some of the species were previously assigned to *Arthroascus, Botryoascus, Guilliermondella, Endomyces,* and *Endomycopsella.* Kurtzman and Robnett (1995) demonstrated from rDNA sequence analysis that the ten species currently accepted are members of the same clade and therefore appear congeneric. With the apparent exception of *Endomyces,* the preceding genera are considered to be synonyms of *Saccharomycopsis.*

Saturnispora **Liu and Kurtzman,** Antonie van Leeuwenhoek J. Microbiol. 60: 21 (1991).

Four species: *S. ahearnii* Kurtzman, *S. dispora* (Dekker) Liu et Kurtzman, *S. saitoi* (Kodama, Kyono et Kodama) Liu et Kurtzman, *S. zaruensis* (Nakase et Komagata) Liu et Kurtzman.

Vegetative reproduction is by multipolar budding and occasionally pseudohyphae. Asci are spheroidal to elongate and may be unconjugated or show conjugation between either independent cells or a parent cell and a bud. Asci may be persistent or deliquescent. Asci contain 1–4, or rarely 8, saturn-shaped ascospores. All four species are homothallic. Common habitats are soil and the exudates of trees.

The species ferment sugars and have CoQ-7. Species of *Saturnispora* can be readily separated from members of the genus *Williopsis,* which also form saturn-shaped ascospores, because they do not assimilate D-xylose or glycerol.

Schizosaccharomyces **Lindner,** Wochenschr. Brau. 10, 1298 (1893).

Three species: *S. japonicus* Yukawa et Maki, *S. octosporus* Beyerinck, *S. pombe* Lindner.

Vegetative reproduction occurs by fission which makes this genus unique. Occasionally, true hyphae are also produced. Asci are elongated, conjugated or unconjugated, and either persistent or deliquescent. Ascospores are smooth, spheroidal to ellipsoidal, and generally 4–8 per ascus (Fig. 19). One of the species is heterothallic. The species have CoQ-9 or -10. The species are isolated from fruits and fruit juices, wines, and high-sugar substrates.

The recently segregated genera *Hasegawaea* and *Octosporomyces,* are in this treatment considered to be synonyms of *Schizosaccharomyces.* The species are strong fermenters of sugars and have been used for the production of ethanol. Systems for heterologous protein secretion have been developed (e.g., Reiser et al. 1990; Toyama and Okayama 1990; Buckholz and Gleeson 1991; Romanos et al. 1992).

Sporopachydermia **Rodr. de Miranda,** Antonie van Leeuwenhoek J. Microbiol. 44, 439 (1978).

Three species: *S. cereana* Rodr. de Miranda, *S. lactativora* Rodr. de Miranda, *S. quercuum* Lachance.

Vegetative reproduction is by multipolar budding and occasionally by rudimentary pseudohyphae. Asci are spheroidal to elongate, conjugated or unconjugated,

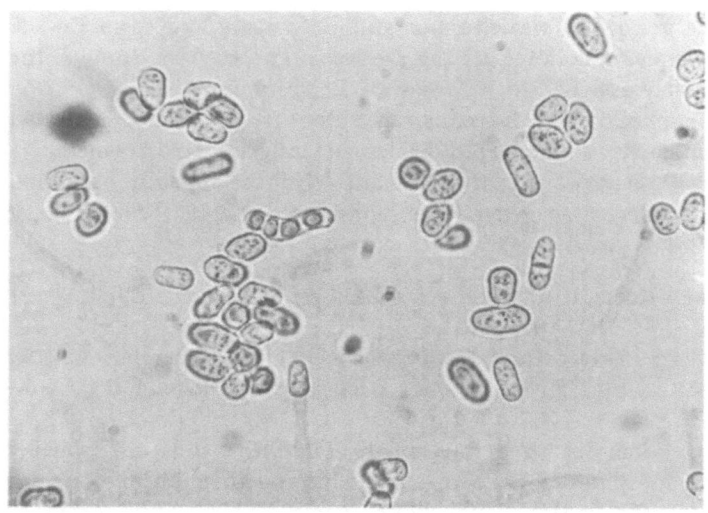

Fig. 19. Ascus with ascospores in *Schizosaccharomyces pombe* (CBS 356, 7d YPGA)

and deliquescent. Ascospores are smooth, spheroidal to ellipsoidal, and have very thick walls. Asci produce 1–4 spores. The species are heterothallic. The species have been isolated from seawater, an asphalt waste lagoon, rotting cacti, and fluxes from oak trees.

Two of the species do not ferment sugars. The genus is characterized by thick-walled ascospores, an offensive odor when grown on certain media, and the utilization of *m*-inositol as a carbon source.

***Stephanoascus* M.Th. Smith et al.,** Antonie van Leeuwenhoek J. Microbiol. 42, 119 (1976).
Two species: *S. ciferrii* M.Th. Smith et al., *S. farinosus* de Hoog et al.
Anamorph: *Blastobotrys*.
Vegetative reproduction is by multipolar budding yeast cells, pseudohyphae, and true hyphae. Hyphal septa are characterized by plasmodesmata. The spheroidal asci form on hyphae and have a distinctive apical cap cell. Asci are persistent and produce 1–4 ascospores which have a hemispherical shape. One of the species is heterothallic. Isolates are from soil, fruiting bodies of fungi, and from human and animal sources.

Fermentation is either slow or absent. *S. farinosus* has a maximum growth temperature of 25 °C, whereas *S. ciferrii* can grow at 40 °C.

***Torulaspora* P. Lindner,** Jb. Versuchs.-Lehranst. Brau. Berl. 7, 448 (1904).
Three species: *T. delbrueckii* (Lindner) Lindner, *T. globosa* (Klöcker) van der Walt & E. Johannsen, *T. pretoriensis* (van der Walt & Tscheuschner) van der Walt & E. Johannsen.
Vegetative reproduction is by multipolar budding and rudimentary pseudohyphae. Asci are spheroidal to ellipsoidal, persistent, and often have a tapered

protuberance. Asci produce 1–4 spores, which may be either rough or smooth, depending on the species. Two of the species are heterothallic. Strains are frequently isolated from soil, fruits, fruit juices, and other plant products, and occasionally from human and animal sources.

The species strongly ferment sugars and are characterized by CoQ-6. Separation of *Torulaspora*, *Saccharomyces*, and *Zygosaccharomyces* has been problematic. The tapered protuberances formed on asci of *Torulaspora* are not usually produced by *Saccharomyces* or *Zygosaccharomyces*.

Wickerhamia **Soneda**, Nagaoa, Mycol. J. Nagao Inst. 7, 9 (1960).
One species: *W. fluorescens* Soneda.
Vegetative reproduction is by bipolar budding of the apiculate yeast cells. Pseudohyphae and hyphae are not produced. Asci are ellipsoidal, unconjugated, and deliquescent. Up to 16 ascospores may be produced in an ascus, but 1 or 2 are usually observed. The ascospores somewhat resemble a baseball cap.

The species is fermentative and forms CoQ-9. The single known isolate was obtained from squirrel dung.

Among the ascomycetous yeasts are four genera that exhibit bipolar budding. These are *Wickerhamia*, *Saccharomycodes*, *Nadsonia*, and *Hanseniaspora*. The main morphological distinction among the genera is based on ascospore shape.

Wickerhamiella **van der Walt**, Antonie van Leeuwenhoek J. Microbiol. 39, 121 (1973).
One species: *W. domercqii* van der Walt.
Vegetative reproduction is by multipolar budding. The cells are among the smallest produced by yeasts. The sexual state of this species is unusual. Asci are conjugated and ellipsoidal. A single oblong ascospore forms, and at maturity the end of the ascus deliquesces and the spore appears to be forcibly ejected. The empty ascus is somewhat elongated and wrinkled. The two known isolates originated from a wine vat and the effluent of a sugar cane factory. However, the species is nonfermentative.

Williopsis **Zender**, Bull. Soc. Bot. Genève 17, 272 (1925).
The five species are described by Kurtzman (1991) and Liu and Kurtzman (1991). Vegetative reproduction is by multipolar budding and, in some species, pseudohyphae. Asci are spheroidal to ellipsoidal, persistent or deliquescent, and may be unconjugated or show parent-bud or cell-to-cell conjugations. Ascospores are saturn-shaped with 1–4 per ascus. Common habitats include soil, water, tree fluxes and other plant sources, and human and animal feces.

The species are fermentative and synthesize CoQ-7. The similarities between *Williopsis* and *Saturnispora* were noted earlier.

Yarrowia **van der Walt and von Arx**, Antonie van Leeuwenhoek J. Microbiol. 46, 517 (1980).
One species: *Y. lipolytica* (Wickerham et al.) van der Walt et von Arx.
Anamorph: *Candida lipolytica* (F.C. Harrison) Diddens et Lodder.
Vegetative reproduction is by multipolar budding, arthrospores, pseudohyphae, and true hyphae. Hyphal septa are characterized by a single central micropore.

Asci are spheroidal to ellipsoidal and usually attached to hyphae. Asci form 1–4 ascospores, which are released at maturity. The morphology of ascospores ranges from spheroidal and roughened, to hat-shaped, crescentiform, or saucer-like. The species is heterothallic and ascospore shape is influenced by the mating strains that are paired. CoQ-9 is formed. Isolates are from soil, agricultural and industrial processing wastes, lipidic and proteinaceous materials, and animal and human clinical specimens.

The species is non-fermentative. *Y. lipolytica* is an important industrial yeast (see Chap. 10, this Vol.). It is markedly proteolytic and lipolytic, and because of its ability to grow on hydrocarbons, has been used to produce single-cell protein from petroleum. More important is its capability to produce high yields of citric acid (Furukawa et al. 1982). For a review on the molecular genetics and biotechnological aspects of the species the reader is referred to Heslot (1990), and for heterologous protein production to Franke et al. (1988), Buckholz and Gleeson (1991), Reiser et al. (1990), and Romanos et al. (1992).

Zygoascus **M.Th. Smith**, Antonie van Leeuwenhoek J. Microbiol. 52, 25 (1986). One species: *Z. hellenicus* Smith.

Vegetative reproduction is by multipolar budding, pseudohyphae and true hyphae. Hyphal septa have a single central pore. Asci are spheroidal to ellipsoidal and form on the bridge between two conjugating hyphal cells of opposite mating type, thus giving a distinct, diagnostic morphology. Asci are persistent and form 1–4 hemispheroidal to hat-shaped ascospores.

Strains have been isolated from grape must, organic industrial waste, and mastitic cows. The species is fermentative.

Zygosaccharomyces **Barker**, Phil. Trans. R. Soc., Ser. B 194, 467 (1901). The currently accepted nine species are listed by Kurtzman (1990).

Vegetative reproduction is by multipolar budding and sometimes by pseudohyphae. Asci are persistent, and usually consist of two conjugating cells, each of which contains one or more ascospores. Ascospores are smooth, spheroidal to ellipsoidal, and 1–4 are formed per ascus.

The species is fermentative and forms CoQ-6. Isolates are commonly obtained from wine, various foods, fruit, trees, and *Drosophila*. Species of *Zygosaccharomyces* often cause spoilage of acidified foods and condiments. The distinction between *Zygosaccharomyces*, *Saccharomyces*, and *Torulaspora* has been noted in the discussion of the latter genus.

Zygozyma **van der Walt et al.**, Syst. Appl. Microbiol. 9, 115 (1987). Four species: *Z. arxii* van der Walt et al., *Z. oligophaga* van der Walt et von Arx, *Z. smithiae* van der Walt et al., *Z. suomiensis* Smith et al.

Vegetative reproduction is by multipolar budding. Asci are ellipsoidal to elongate, deliquescent and form 1–4 smooth ovoidal to elongate spores.

The species are non-fermentative and form CoQ-8. In culture, colonies are mucoid because of the production of large amounts of extracellular polysaccharide.

6.2
Anamorphic Ascomycetous Genera

Aciculiconidium **D.S. King and S.C. Jong,** Mycotaxon 3, 401 (1976).
One species: *A. aculeatum* (Phaff et al.) D.S. King & S.C. Jong. Vegetative growth is by budding yeast cells that form a multitude of shapes and by true hyphae. In addition, needle-shaped cells with rounded bases are formed on hyphae and as buds on yeast cells. These unusual and unique cells characterize the genus. The two known strains have been isolated from Drosophila spp.
The species weakly ferments sugars and forms CoQ-9.

Arxula **van der Walt et al.** Antonie van Leeuwenhoek J. Microbiol. 57, 59 (1990).
Two species: *A. adeninivorans* (Middelhoven et al.) van der Walt et al., *A. terrestre* (van der Walt et Johannsen) van der Walt et al.
Vegetative reproduction is by budding, arthroconidia and true hyphae. Hyphal septa are multiperforate.
The species are non-fermentative, xerotolerant, and form CoQ-9. The genus was defined because of its xerotolerance, budding, formation of arthroconidia, and production of narrow, slow-growing hyphae. However, comparisons of rDNA sequences show the *Arxula* spp. to be members of a clade composed of *Blastobotrys* spp., *Sympodiomyces parvus*, *Stephanoascus* spp., and several species of *Candida* (Kurtzman and Robnett 1995). Several aromatic compounds are assimilated by *A. adeninivorans* (Middelhoven et al. 1991), and two extracellular acid phosphatases have been characterized from this species (Büttner et al. 1991; see also Chap. 11, this Vol.).

Brettanomyces **Kufferath and van Laer,** Bull. Soc. Chim. Belge 30, 270 (1921).
The five species are listed by Smith et al. (1990).
Teleomorph: *Dekkera*.
Vegetative reproduction is by multipolar budding and pseudomycelium. Some of the yeast cells have an ogival shape at one end. The species may show variable fermentation of sugars, which is stimulated by the presence of oxygen (Custers effect). Species are noted for vigorous production of acetic acid and this causes early cell death in cultures. All species have CoQ-9. The species have been isolated from beer, wine and soft drinks.
Eeniella nana M.Th Smith et al. (1981) has been reclassified as *Brettanomyces* based on partial 28S ribosomal DNA sequences (Boekhout et al. 1994). Contrary to a series of papers (Jong and Lee 1986; Lee and Jong 1986), we consider *B. custersianus* and *B. naardenensis* as asexual species.

Candida **Berkhout,** De schimmelgeslachten *Monilia, Oidium, Oospora* en *Torula,* p. 41. Thesis, Utrecht (1923).
The most recent summary of the ca. 160 *Candida* species is given by Barnett et al. (1990).
Candida is a taxonomic form-genus to which the ascomycetous species that exhibit multipolar budding and have no known sexual state are assigned. The genus

Torulopsis Berlese is now considered a synonym of *Candida*, and consequently, *Candida* includes those species that form pseudohyphae and hyphae as well as those that do not. Because *Candida* is so broadly defined, phylogenetic affinities may be found with any teleomorphic ascomycetous genus that exhibits multilateral budding.

Several *Candida* species are of medical importance, e.g., *C. albicans*, *C. glabrata*, *C. parapsilosis*, and *C. tropicalis*.

Several *Candida* species have been investigated for biotechnological aspects, e.g., *Candida utilis* (Pasari et al. 1989) and *C. maltosa* (see Chap. 12, this Vol.) are used for biomass production from carbohydrate substrates, *Candida blankii* may be a good candidate for single-cell protein production using pentoses as substrate (Meyer et al. 1992), and *C. tropicalis* has been found to utilize hydrocarbons (Cook et al. 1973).

Kloeckera Janke, Zentbl. Bakteriol. Parasitenk., Abt. 2, 76, 161 (1928).
This genus represents the anamorphic state of *Hanseniaspora* and only one species, *K. lindneri* (Klöcker) Janke, remains for which the ascosporic state has not yet been found. Vegetative reproduction is by bipolar budding. The species ferments glucose and has CoQ-6. It has been isolated from soil in Java.

Myxozyma van der Walt et al., Sydowia 34, 191 (1981).
The five species are described by Barnett et al. (1990).
Vegetative reproduction is by multipolar budding. Pseudohyphae and true hyphae are not produced. Growth in culture is mucoid.

The species are non-fermentative and form either CoQ-8 or-9. Isolates have been obtained from soil, bark beetles, rotting cacti, and a corticolous lichen. Ribosomal RNA sequence comparisons indicate a close relationship to *Lipomyces*, though *Myxozyma* probably is not the anamorph of *Lipomyces* (Kurtzman and Liu 1990).

Oosporidium Stautz, Phytopath. Z. 3, 163 (1931).
One species: *O. margaritiferum* Stautz.
Vegetative reproduction is by multipolar budding, pseudohyphae, and asexual endospores. Cultures often show a light pink to light orange color. Analysis of rDNA sequences shows this taxon to be a dimorphic member of the euascomycetes (Kurtzman and Robnett 1995).

The species is non-fermentative. Strains have been obtained from slime fluxes of trees.

Saitoella Goto et al., J. Gen. Appl. Microbiol. 33, 75 (1987).
One species: *S. complicata* Goto et al.
Vegetative reproduction is by enteroblastic budding. Pseudohyphae and hyphae are not formed.

The species is non-fermentative and forms CoQ-10. Isolates have been obtained from soil. Colonies in culture are orange and butyrous and somewhat similar to *Rhodotorula*. However, the Diazonium Blue B reaction is negative, indicating *Saitoella* to be an ascomycete. It may be phylogenetically related to *Taphrina* (Nishida and Sugiyama 1993).

Schizoblastosporion **Ciferri,** Arch. Protistenk. 71, 405 (1930).
Two species: *S. starkeyi-henricii* Ciferri, *S. chiloense* Ramírez et González.
Vegetative growth is by budding on a broad base, almost exclusively at the poles of
the cells. *S. starkeyi-henricii* sometimes produces a primitive pseudomycelium. *S.
chiloense* forms pseudohyphae and true hyphae.

The species are non-fermentative and have been isolated from soil, rotten
wood, and stomach contents.

Sympodiomyces **Fell and Statzell,** Antonie van Leeuwenhoek J. Microbiol. 37, 359
(1971).
One species: *S. parvus* Fell et Statzell.
Vegetative growth is by budding and true hyphae. Conidiophores arise from some
cells and produce terminal conidia by sympodial proliferation. Phylogenetic placement of this genus is discussed under *Arxula*.

The species is non-fermentative. All isolates are from seawater of the Indian and
Pacific Oceans.

Trigonopsis **Schachner,** Z. Ges. Brauwesen 52, 137 (1929).
One species: *T. variabilis* Schachner.
Vegetative growth is by multilateral budding. Pseudohyphae and hyphae are absent. This species is unique because of the triangular shape (actually tetrahedral) of
the yeast cells. The proportion of triangular to ellipsoidal cells is strain-dependent
and is also influenced by the composition of the culture medium. Budding on the
triangular cells is from the cell apices.

The species is non-fermentative. Isolates are from beer and grape must.

6.3
Teleomorphic Heterobasidiomycetous Genera

Bulleromyces **Boekhout and A. Fonseca,** Antonie van Leeuwenhoek J. Microbiol.
59, 90 (1991).
One species: *B. albus* Boekhout et A. Fonseca.
Anamorph: *Bullera*.
Heterothallic, but occasionally self-sporulating. Hyphae producing clamp connections, haustorioid branches, and transversely, obliquely or longitudinally
septate phragmobasidia, which germinate with yeast cells, hyphae, or
ballistoconidia. Teliospores are absent. Vegetative reproduction is by polar
budding, and rotationally symmetrical ballistoconidia. The species contain xylose
in the cell walls and have CoQ-10. Hyphal septa have dolipores with a
parenthesome made up of U-shaped vesicles. Commonly isolated from such habitats as air and leaves.

Chionosphaera **Cox,** Mycologia 68, 503 (1976).
One species: *C. apobasidialis* Cox.
Heterothallic. Basidiocarps stilboid, made up of hyphae without clamp connec-
tions. Basidia clavate. Vegetative reproduction may occur by budding.

Ballistoconidia are absent (Cox 1976). No data on cell wall composition and CoQ are available. Isolated from dead corticated branches of tree.

***Cystofilobasidum* Oberwinkler and Bandoni,** Syst. Appl. Microbiol. 4, 116 (1983). Four species: *C. bisporidii* (Fell et al.) Oberwinkler & Bandoni, *C. capitatum* (Fell et al.) Oberwinkler et Bandoni, *C. infirmominiatum* (Fell et al.) Hamamoto et al., *C. lari-marini* (Saëz et Nguyen) Fell et Tallman.
Anamorph: *Cryptococcus*.
Heterothallic or homothallic. Hyphae with or without clamp connections, and lacking haustorioid branches. Teliospores are present, germinating with capitate or pyriform holobasidia. Vegetative reproduction usually is by polar budding, but endospores may be present. Ballistoconidia are absent. The species contain xylose in the cell wall and have CoQ-8. Hyphal septa have dolipores without a parenthesome. Isolated from various substrates, like seawater, plankton, soil, trees, food, and birds.

***Erythrobasidium* Hamamoto et al.,** J. Gen. Appl. Microbiol. 34, 285 (1988).
One species: *E. hasegawianum* Hamamoto et al.
Anamorph: *Rhodotorula*.
Homothallic. Hyphae have clamp connections and holobasidia, but teliospores and haustorioid branches are absent. Vegetative reproduction is usually by polar budding. Ballistoconidia are absent. The species has no xylose in the cell walls, and has CoQ-10(H_2). Hyphae have a simple diaphragma-like pore (Suh et al. 1993). Isolated from spent brewers's yeast.

***Filobasidiella* Kwon-Chung,** Mycologia 67, 1198 (1975).
One species: *F. neoformans* Kwon-Chung with two varieties, var. *neoformans* and var. *bacillispora*.
Anamorph: *Cryptococcus*.
Heterothallic, but occasionally self-sporulating. Hyphae have clamp connections, haustorioid branches, and clavate holobasidia, on which chains of basidiospores are formed basipetally. Teliospores are absent. Vegetative reproduction is by polar budding. Ballistoconidia are absent. The species has xylose in the cell walls and has CoQ-10. Septa have dolipores without parenthesomes (Kwon-Chung and Popkin 1976). Isolated from man, pigeon droppings, *Eucalyptus* species and soil, but also from fruit juice.

Filobasidiella neoformans is of medical importance, and can cause meningitis. The natural habitat of the variety *neoformans* is known to be pigeon droppings, and variety *bacillispora* is associated with *Eucalyptus* trees (Ellis and Pfeiffer 1990, 1992).

***Filobasidium* Olive,** J. Elisha Mitchell Scient. Soc. 84, 261 (1968).
Five species: *F. floriforme* Olive, *F. capsuligenum* Rodrigues de Miranda, *F. uniguttulatus* Kwon-Chung, *F. elegans* Bandoni & Oberwinkler, *F. globisporum* Bandoni et Oberwinkler.
Anamorphs: *Cryptococcus*.

Heterothallic, but apparently also homothallic. Hyphae have clamp connections, haustorioid branches, and form cylindrically clavate holobasidia. Teliospores are absent. Vegetative reproduction is by polar budding. Ballistoconidia are absent. The species contain xylose in the cell walls and have CoQ-9 or -10. Septa have dolipores with or without parenthesomes. Parenthesomes, if present, are made up of U-shaped vesicles (Kreger-van Rij and Veenhuis 1971; Moore and Kreger-van Rij 1972). *Filobasidium floriforme, F. elegans,* and *F. globisporum* are known from plant material, *F. uniguttulatus* is known from clinical specimens, and *F. capsuligenum* is isolated from soil and wine-making products.

Filobasidium capsuligenum shows extracellular amylolytic activity because of production of *a*-amylase and glucoamylases (De Mot et al. 1984a; De Mot and Verachtert 1985).

Leucosporidium **Fell et al.,** Antonie van Leeuwenhoek J. Microbiol. 35, 438 (1969). Three species: *L. antarcticum* Fell et al., *L. fellii* Giménez-Jurado, *L. scottii* Fell et al. Heterothallic and homothallic. Hyphae may or may not have clamp connections. Teliospores are present and germinate by septate or aseptate metabasidia. Haustorioid branches are absent. Vegetative reproduction is by polar budding. Ballistoconidia are absent. The species lack xylose in the cell walls and have CoQ-7,-9 or -10 (Yamada and Nakagawa 1990). Hyphae have diaphragma-like septa with a central pore (Kreger-van Rij and Veenhuis 1971). Isolated from soil, fresh water, seawater, seaweeds, food, and trees.

The reported heterogeneity in the coenzyme Q may indicate taxonomic heterogeneity, which needs to be examined carefully. The degradation of L(+)-tartaric acid by *L. fellii* may be of interest to the wine industry (Giménes-Jurado and van Uden 1989). Aromatic compounds are assimilated by *L. scottii* (Middelhoven et al. 1992).

Kondoa **Yamada et al.,** J. Gen. Appl. Microbiol. 35, 383 (1989). One species: *K. malvinella* (Fell et Hunter) Yamada et al. Heterothallic. Hyphae have clamp connections. Teliospores are present and germinate with two-celled metabasidia (Fell 1970). Haustorioid branches are absent. Vegetative reproduction is by polar budding. Ballistoconidia are absent. The species lacks xylose in the cell wall, and has CoQ-9 (Yamada et al. 1989a). Isolated from seawater.

Kondoa is considered a genus distinct from *Rhodosporidium*, because of dissimilar partial nucleotide sequences of the 18S and 26S ribosomal RNA, stalked teliospores, and two-celled metabasidia (Yamada et al. 1989a).

Mrakia **Yamada and Komagata,** J. Gen. Appl. Microbiol. 33, 456 (1987). Four species: *M. frigida* (Fell et al.) Yamada et Komagata, *M. gelidum* (Fell et al.) Yamada & Komagata, *M. nivalis* (Fell et al.) Yamada et Komagata, *M. stokesii* (Fell et al.) Yamada et Komagata. Self-sporulating. Hyphae do not form clamp connections and haustorioid branches. Teliospores are present and germinate with one- to three-celled metabasidia (Fell and Statzell Tallman 1984a). Vegetative reproduction is by polar budding. Ballistoconidia are absent. The species contain xylose in the cell wall, and have CoQ-8 (Yamada and Komagata 1987). Isolated from antarctic soil and snow.

Isozyme and whole-cell protein patterns, and partial 18S and 26S ribosomal RNA sequences of all species are similar or identical (Yamada and Matsumoto 1988, Yamada and Kawasaki 1989a; Fell et al. 1992; Vancanneyt et al. 1992). Therefore, it is expected that the number of species can be reduced.

Rhodosporidium Banno, J. Gen. Appl. Microbiol. 13, 192 (1967).

As several species have been described recently, we present a list of the eight species presently known: *R. dacryoideum* Fell et al., *R. diobovatum* Newell & Hunter, *R. fluviale* Fell et al., *R. kratochvilovae* Hamamoto et al., *R. lusitaniae* Fonseca et Sampaio, *R. paludigenum* Fell et Statzell Tallman, *R. sphaerocarpum* Newell et Fell, *R. toruloides* Banno.

Anamorphs: *Rhodotorula*.

Homothallic or heterothallic. Hyphae with or without clamp connections. Teliospores are present and germinate with transversely septate metabasidia. Haustorioid branches are absent. Vegetative reproduction usually is by polar budding. Ballistoconidia are absent. The species lack xylose in the cell wall and have CoQ-9 or -10. Hyphae have diaphragma-like septa with a central pore (Kreger-van Rij and Veenhuis 1971; Johnson-Reid and Moore 1972). Isolated from various substrates such as soil, fresh water, seawater, mangrove swamps, air, and wood pulp.

Rhodosporidium kratochvilovae, R. diobovatum, R. paludigenum, R. toruloides, and *R. sphaerocarpum* form one species cluster based on partial 28s rRNA sequences, whereas *R. dacryoideum* is in a separate cluster (Fell et al. 1992).

Rhodosporidium lusitaniae has been reported to degrade phenolic compounds (Fonseca and Sampaio 1992). *Rhodosporidium toruloides* can accumulate large amounts of lipids (Ratledge 1978, 1982, 1986; Ratledge and Evans 1989). This process was found to be largely influenced by the nitrogen sources used (e.g., Evans and Ratledge 1984a,b).

Sporidiobolus Nyland, Mycologia 41, 686 (1949).

Four species: *S. johnsonii* Nyland, *S. pararoseus* Fell et Tallman, *S. ruineniae* Holzschu et al., *S. salmonicolor* Fell et Tallman.

Anamorph: *Sporobolomyces*.

Homothallic or heterothallic. Hyphae have clamp connections. Teliospores are present and germinate with phragmo- or holometabasidia. Haustorioid branches are absent. Vegetative reproduction usually is by polar budding. Bilaterally symmetrical ballistoconidia are present. The species lack xylose in the cell wall and have CoQ-10. Hyphae have diaphragma-like septa with a central pore (Kreger-van Rij and Veenhuis 1971; Boekhout et al. 1992). The species occur mainly on leaves, but have also been isolated from air, fruit, skin, fodder, seawater, wood chips, and oil.

Species of *Sporidiobolus* belong to one species cluster based on partial nucleotide sequences of 28S ribosomal RNA (Fell et al. 1992). Because of rather high DNA relatedness, *S. salmonicolor* is sometimes considered synonymous with *S.*

johnsonii (Boekhout 1991b). Fell et al. (1992) observed a number of differences in partial nucleotide sequences of the 28S rRNA (Fell et al. 1992) between the two species, which may justify their distinction. Torularhodine is found to be the main carotenoid pigment, but β-carotene, torulene, and γ-carotene occur as well (Valadon 1976).

Sterigmatosporidium **Kraepelin and U. Schulze,** Antonie van Leeuwenhoek J. Microbiol 48, 479 (1982).
One species: *S. polymorphum* Kraepelin et Schulze.
Heterothallic, but apparently homothallic strains occur as well. Hyphae have clamp connections and haustorioid branches. Teliospore-like cells are present. The life cycle has not been fully elucidated, because the sites of karyogamy and meiosis could not be identified with certainty. Vegetative reproduction is by multipolar budding on short denticles. Ballistoconidia are absent. The species contains xylose in the cell wall and has CoQ-10 (Yamada et al. 1989b). It has been isolated from waterlogged planks in an old mine.

Tilletiaria **Bandoni and Johri,** Can. J. Bot. 50, 39 (1972).
One species: *T. anomala* Bandoni et Johri.
Homothallic. Hyphae lack clamp connections and haustorioid branches. Teliospores are present and germinate with phragmobasidia (Bandoni and Johri 1972). Yeast cells are absent, but ballistoconidia are present. The species lacks xylose in the cell wall and has CoQ-10 (Boekhout 1991b). Hyphal septa have dolipore-like structures (Boekhout et al. 1992). It has been isolated from decaying wood.

Tremellales with Yeast Phases
Yeast stages have been isolated from a limited number of species belonging to the genera *Tremella* Persoon and *Sirobasidium* Lagerheim & Patouillard, (e.g., Bandoni 1984). Most species are heterothallic, but homothallic species occur as well. Hyphae have clamp connections, haustorioid branches, and form obliquely or longitudinally septate basidia. Teliospores are absent. Vegetative reproduction usually is by polar budding. Ballistoconidia may be formed by yeast stages of *Sirobasidium*. These genera are characterized by the presence of xylose in the cell wall, and have CoQ-9 or -10 (Kuraishi et al. 1985; Yamada et al. 1987). Septa have a dolipore with a parenthesome made up of U-shaped vesicles (Tremellales type; Patrignani et al. 1984), or lack a parenthesome. Most species occur on decaying wood, or are mycoparasitic.

Partial nucleotide sequences of 28S rRNA of some *Tremella* species and *Sirobasidium magnum* suggest a close phylogenetic relationship with certain species of *Cryptococcus* (Fell et al. 1992). This is also supported by a number of physiological and biochemical characteristics (e.g., assimilation of *m*-inositol and extracellular carbohydrate composition).

Xanthophyllomyces **Golubev,** Yeast 11, 105 (1995).
One species: *X. dendrorhous* Golubev.
Anamorph: *Phaffia rhodozyma* Miller et al. (= *Rhodomyces dendrorhous*).

Homothallic, hyphae and haustorioid branches are absent, but chlamydospores may be present. Mother cell-bud conjugation, induced by polyols (Golubev 1995), results in cylindrical holobasidia on which basidiospores are formed. The cell wall contains xylose and rhamnose (Golubev 1995) and CoQ-10 is present (Yamada and Kawasaki 1989b).

The species is mainly isolated from exuded sap from trees (e.g., *Betula* ssp.), especially from arctic and alpine habitats.

6.4
Anamorphic Heterobasidiomycetous Genera

Bensingtonia Ingold, Trans. Br. Mycol. Soc. 86, 325 (1986).
Nine species are listed in Boekhout (1991b).
Vegetative reproduction is by polar budding, bilaterally symmetrical ballistoconidia, hyphae and pseudohyphae. The species lack xylose in the cell wall and have CoQ-9. Hyphae have diaphragma-like septa with a central pore (Boekhout et al. 1992). The species have been isolated mainly from leaves, but also from fungi. Partial nucleotide sequences of the 28S rRNA indicate that the genus in its present circumscription may be polyphyletic (Fell et al. 1992).

Bullera Derx, Annls Mycol. 28, 11 (1930).
Thirteen species are listed in Boekhout (1991b).
Teleomorph: *Bulleromyces.*
Vegetative reproduction is by polar budding, rotationally or bilaterally symmetrical ballistoconidia, hyphae, and pseudohyphae. The species contain xylose in the cell wall and have CoQ-10. The dikaryophase of *B. variabilis* has a dolipore with a parenthesome made up of U-shaped vesicles (Tremellales type; Boekhout et al., 1991). The species have been isolated mainly from leaves, but also from fruit, plants, larvae of Buprestid beetles, wood, frozen salmon, air, etc.

Mating, followed by formation of dikaryotic hyphae with clamp connections and probasidium-like structures has been observed in *B. variabilis* (Boekhout et al. 1991). *Bullera alba* seems phylogenetically close to *Cryptococcus* (Fell et al. 1992). Recently, species with bilaterally symmetrical ballistoconidia have been transferred to *Udeniomyces* (Nakase and Takematsu 1992).

Cryptococcus Kützing, Linnaea, 371 (1833).
Thirty-eight species are listed in Barnett et al. (1990).
Recently additional species have been added (Chernov and Bab'eva 1988; Fonseca and Van Uden 1991; Vaughan-Martini 1991; Vishniac and Kurtzman 1992).
Vegetative reproduction usually is by polar budding, and, more rarely by pseudohyphae and hyphae. Ballistoconidia are absent. The species contain xylose in the cell wall and have CoQ-8, -9 or-10. Septa of *C. laurentii* have dolipores without parenthesomes (Rhodes et al. 1981). The species have been isolated from diverse substrates such as soil, fruit, water, man, animals, wine, leaves, fungi.

Apiotrichum Stautz is considered a synonym of *Cryptococcus* (Golubev 1981; Weijman et al. 1988; Barnett et al. 1990). Mating, followed by formation of hyphae and chlamydospore-like structures has been observed to occur in *C. laurentii* (Kurtzman 1973). The genus seems phylogenetically heterogeneous based on partial 28S rRNA sequences (Fell et al. 1992).

Cryptococcus flavus and *C. tsukubaensis* are amylolytic (de Mot et al. 1984b). *Cryptococcus laurentii* (Ratledge and Evans 1989) and *Cryptococcus curvatus* have been reported to accumulate large amounts of lipids (Ratledge 1978, 1982, 1986; Park et al. 1990), using whey as a substrate (*C. curvatus*; Ykema et al. 1988, 1989a). Mutants of this latter species able to produce cocoa butter equivalents have been described (Ykema et al. 1989b). It has been suggested that strain IFO 1322 (= CBS 4567) of *C. albidus* (= *C. terricolus*) may be useful for single-cell oil production (Boulton and Ratledge 1984).

Degradation of phenol by an immobilized mixed culture of *C. elinovii* and *Pseudomonas putida* has been reported (Zache and Rehm 1989). Utilization of a variety of aromatic compounds was reported for *C. diffluens* (as *C. albidus* var. *diffluens*), *C. terreus* (Mills et al. 1971), *C. humicolus*, and *C. laurentii* (Middelhoven et al. 1992). Several species, e.g., *C. flavus*, *C. laurentii* and *C. luteolus* form β-carotene as the main pigment, but lycopene and γ-carotene are formed as well (Valadon 1976).

Fellomyces **Yamada and Banno**, J. Gen. Appl. Microbiol. 30, 524 (1984).
Four species: *F. fuzhouensis* (Yue) Yamada et Banno, *F. horovitziae* Spaaij et al., *F. penicillatus* (Rodrigues de Miranda) Yamada et Banno, *F. polyborus* (D.B. Scott et van der Walt) Yamada et Banno.
Vegetative reproduction is by budding and by conidia formed on stalks. Ballistoconidia are absent, but pseudohyphae may be present. The species contain xylose in the cell wall (Yamada et al. 1988) and have CoQ-10. Isolated from diverse substrates, such as food, flowers, tree, and fungi.

Based on partial ribosomal RNA sequences, a considerable heterogeneity has been observed among heterobasidiomycetous yeasts forming conidia on stalks (Yamada et al. 1989a; Guého et al. 1990). Mainly according to biochemical characteristics (composition of extracellular carbohydrates, CoQ) five genera are currently distinguished that form stalked conidia, namely *Fellomyces*, *Kurtzmanomyces*, *Sterigmatomyces*, *Tsuchiyaea*, and *Kockovaella*.

Itersonilia **Derx**, Bull. Bot. Gdn Buitenzorg, Ser. 3, 17, 471 (1948).
One species: *I. Perplexans* Derx.
Vegetative reproduction of the yeast stage usually is by polar budding, and more rarely by ballistoconidia. Dikaryotic hyphae have clamp connections, chlamydospores, teliospore-like cells, and form ballistoconidia. Monokaryotic hyphae form incomplete clamp connections. The species contains xylose in the cell wall and has CoQ-9. Commonly isolated from leaves. *Itersonilia perplexans* is reported to be a plant pathogen (Channon 1956, 1963; Gandy 1966).

Kockovaella **Nakase et al.**, J. Gen. Appl. Microbiol. 37, 178 (1991).
Two species: *K. imperatae* Nakase et al., *K. thailandica* Nakase et al.

Vegetative reproduction is by formation of conidia on stalks, ballistoconidia and, more rarely, by budding. Hyphae and pseudohyphae are absent. The species contain xylose in the cell wall and have CoQ-10 (Nakase et al.1991). The species have been isolated from leaves.

The genus *Kockovaella* forms conidia on stalks, and ballistoconidia. Because of the presence of xylose, an affilation is inferred with the Tremellales (Boekhout et al. 1993a).

Kurtzmanomyces Yamada et al., J. Gen. Appl. Microbiol. 34, 505 (1988).
Two species: *K. nectairii* (Rodrigues de Miranda) Yamada et al., *K. tardus* Giménez-Jurado et van Uden.
Vegetative reproduction is by conidium formation on stalks. Ballistoconidia and pseudohyphae are absent. The species lack xylose in the cell wall and have CoQ-10 (Yamada et al. 1988). The species has been isolated from cheese and water.

Malassezia Baillon, Traité Bot. Méd. Cryptog., P. 243, O. Doin, Paris (1889).
Three species: *M. furfur* (Robin) Baillon, *M. pachydermatis* (Weidman) Dodge, *M. sympodialis* R.B. Simmons et Guého.
Vegetative reproduction is by monopolar and more rarely by sympodial budding (*M. sympodialis*). Ballistoconidia, hyphae and pseudohyphae are absent. The species are characterized by lack of xylose in the cell wall and have CoQ-9 (Hamamoto et al. 1992). The species have been isolated mainly from medical sources, dandruff, ears of dogs etc.

Species of *Malassezia* are of medical importance. *Malassezia furfur* and *M. sympodialis* require the addition of olive oil or long-chain fatty acids in the culture medium.

Phaffia Miller et al., Int. J. Syst. Bact. 26, 287 (1976).
One species: *P. rhodozyma* Miller et al.
Vegetative reproduction is by polar budding. Ballistoconidia and hyphae are absent, but spherical chlamydospores may be present. The species contains xylose in the cell wall (Sugiyama et al. 1985) and has CoQ-10 (Yamada and Kawasaki 1989b).

Phaffia rhodozyma ferments D-glucose (Miller et al. 1976). Partial 18S rRNA nucleotide sequences suggest that the genus is phylogenetically distinct from *Cryptococcus* (Yamada and Kawasaki 1989b). The species produces the carotenoid pigment astaxanthin, and is used as a dietary pigment source of salmonids, crustaceans, and chickens (Johnson et al. 1977, 1980). Recently, mutants with an increase in pigment production have been isolated (An et al. 1989; Lewis et al. 1990).

Rhodotorula F.C. Harrison, Trans. R. Soc. Can., Sect. 3, 21, 349 (1927).
Thirty-three species are listed by Barnett et al. (1990). Vegetative reproduction usually is by polar budding. Ballistoconidia are absent, but hyphae and pseudohyphae may be present. The species lack xylose in the cell wall, but fucose and rhamnose may be present. They have CoQ-9,-10 or -10 (H_2). The species have been isolated from a wide range of substrates.

Rhodotorula in its present circumscription seems phylogenetically heteroge-neous, as can be concluded from the species distribution among the dendrograms based on partial rRNA sequences (Fell et al. 1992).

Many *Rhodotorula* species form torulene or torularhodine as the main pigment, but γ-carotene and β-carotene are usually present as well and some species also form neurosporene and/or lycopene (Valadon 1976).

Degradation of phenol is reported for *R. glutinis* var. *glutinis* (Nei 1971a,b) and *R. rubra* (Katayama-Hirayama et al. 1991). *Rhodotorula glutinis, R. minuta, R. rubra,* and *R. aurantiaca* are reported to utilize a variety of aromatic compounds (Mills et al. 1971; Middelhoven et al. 1992). *Rhodotorula graminis, R. glutinis* var. *glutinis, R. gracilis* and *R. mucilaginosa* are oleaginous (Nilsson et al. 1943; Ratledge 1978, 1982, 1986; Ratledge and Hall 1979; Ratledge and Evans 1989; Rolph et al. 1989). Two extracellular lipases are secreted by *R. rubra* (as *R. pilimanae*; Muderhwa et al. 1986).

Sporobolomyces **Kluyver et van Niel**, Centbl. Bakt. Parasitenk., Abt. 2, 63, 19 (1924).
Eighteen species are listed by Boekhout (1991b).
Vegetative reproduction is by polar budding and ballistoconidia, but hyphae and pseudohyphae may be present as well. The species lack xylose in the cell wall, but fucose may be present. They have CoQ-10 or CoQ-10 (H_2). The species occur mainly in the phyllosphere, but have been isolated from a wide range of other substrates.

Partial rRNA nucleotide sequences suggest phylogenetic heterogeneity of the genus in its present delimitation (Fell et al. 1992). *Sporobolomyces roseus* contains torularhodine as its main pigment, but torulene and *β*-carotene occur as well (Valadon 1976).

Sterigmatomyces **Fell**, Antonie van Leeuwenhoek J. Microbiol. 32, 101 (1966).
Two species: *S. elviae* Sonck et Yarrow, *S. halophilus* Fell.
Vegetative reproduction is by budding on stalks (Fig. 20). Ballistoconidia, hyphae, and pseudohyphae are absent. The species lack xylose in the cell walls, and have CoQ-9. The species have been isolated from diverse substrates, e.g., man, pastry, and soil.

Sympodiomycopsis **J. Sugiyama et al.**, Antonie van Leeuwenhoek J. Microbiol. 59, 99 (1991).
One species: *S. paphiopedili* J. Sugiyama et al.
Vegetative reproduction is by polar, annellidic, and sympodial budding. Ballistoconidia are absent. Hyphae are present and have diaphragma-like septa with a central pore (Suh et al. 1993). The species contains trace amounts of xylose in the cell wall and has CoQ-10 (Sugiyama et al. 1991). The species has been isolated from orchid nectar.

Tilletiopsis **Derx**, Annls Mycol. 28, 3 (1930).
Six species are listed by Boekhout (1991b).

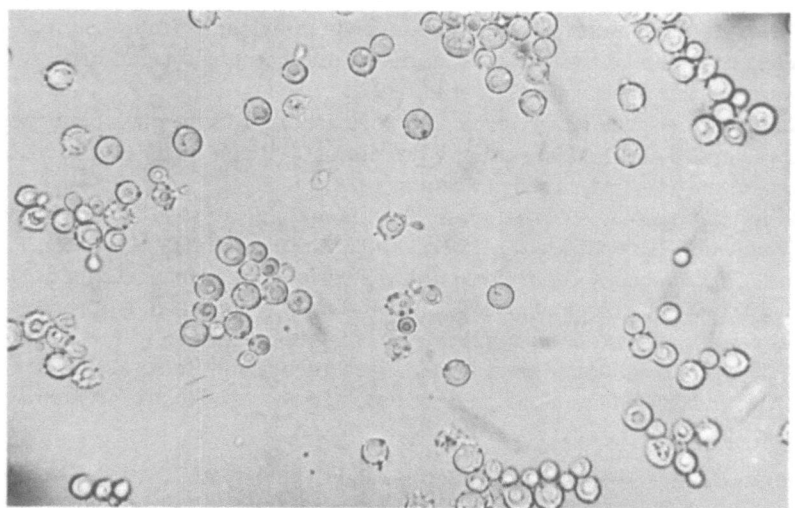

Fig. 20. Vegetative reproduction on stalks in *Sterigmatomyces halophilus* (CBS 5632, 7d YPGA)

Vegetative reproduction of the yeast stage usually is by polar budding. Hyphae and ballistoconidia are present. The species lack xylose in the cell wall and have CoQ-10. The species mainly occur on leaves, but have been isolated from other sources as well, e.g., fungi, man, etc. *Tilletiopsis pallescens* is being tested as an antagonist for use in biological control of powdery mildews.

Trichosporon **Behrend**, Berl. Klin. Wschr. 21, 467 (1890).
Nineteen species are listed by Guého et al. (1992).
Vegetative reproduction is by budding and arthroconidia. Ballistoconidia are absent, but hyphae are present. Septa have dolipores or micropore-like structures. Parenthesomes, if present, are tubular (Guého et al. 1992). The species contain xylose in the cell wall (Weijman 1979) and have CoQ-9 or -10 (Guého et al. 1992).

Because of recent changes in species concepts the observations reported below may refer to different taxa. *Sarcinosporon* King & Jong is considered a synonym of *Trichosporon* Behrend (Guého et al. 1992). Several species of *Trichosporon* are of medical importance.

Trichosporon pullulans degrades starch and pullulan due to the production of α-amylase and glucoamylase(s) (De Mot et al. 1984b; De Mot and Verachtert 1986). Two lactose hydrolases have been characterized from *T. cutaneum* (West et al. 1990), and degradation of phenol is reported for this species (Shoda and Udaka 1980; Neujahr and Varga 1970; Mills et al. 1971). *Trichosporon cutaneum*, *T. moniliiforme*, and *T. dulcitum* are found to assimilate a wide variety of aromatic compounds (Reiser et al. 1990; Middelhoven et al. 1992). *Trichosporon beigelii* is able to utilize cheese whey as a carbon and energy source for biomass production.

Trichosporon cutaneum and *T. pullulans* are able to accumulate substantial amounts of lipid (Ratledge 1978, 1982, 1986; Ratledge and Evans 1989). Addition of sodium chloride was found to increase lipid production in *T. beigelii*, which makes this species a potential candidate for single-cell oil production (Tahoun et al. 1987). A transformation system has been developed for *T. cutaneum* (Reiser et al. 1990, see Chap. 13, this Vol.).

Tsuchiyaea **Yamada et al.,** J. Gen. Appl. Microbiol. 34, 509 (1988).
One species: *T. wingfieldii* (van der Walt et al.) Yamada et al.
Vegetative reproduction is by budding and conidium formation on stalks. Ballistoconidia and pseudohyphae are absent. The species contains xylose in the cell wall and has CoQ-9 (Yamada et al. 1988). It has been isolated from frass of a Scolytid beetle.

Some Notes on *Vanrija* and Anamorphs of Smuts

Vanrija **R.T. Moore,** Bot. Mar. 23, 367 (1980).
This genus has been introduced to accommodate certain basidiomycetous *Candida* species (Moore, 1980, 1987). The genus in its present circumscription seems heterogeneous as it comprises species lacking xylose (e.g., *Vanrija antarctica*), and species containing xylose in their cell walls (e.g., *Vanrija curvata* and *V. nivalis*). Consequently, several of the species are better maintained in genera like *Cryptococcus*, *Mrakia*, and *Leucosporidium*.

Anamorphs of Ustilaginales (Smuts)

Several basidiomycetous yeast-like organisms, classified as diversely as *Candida* (*Sporobolomyces, Vanrija*) antarctica (Goto et al.) Kurtzman et al. *Candida* 107, *Sterigmatomyces aphidis* Henniger & Windisch, *Trichosporon oryzae* Ito et al., and *Pseudozyma prolifica* Bandoni, most likely represent anamorphs of Ustilaginales (Boekhout 1987). The taxonomy of this group of organisms urgently needs revision. Partial 28S ribosomal RNA sequences of *Candida* (*Sporobolomyces*) *antarctica* suggest a position rather close to *Malassezia* (Fell et al. 1992).

Candida 107 (NCYC 911), which appparently is closely related to or identical with *C. antarctica*, accumulates large amounts of fatty acids (ca. 41 % of the dry weight, and ca. 40 % of the total fatty acid content as saturated triglycerides; Gill et al. 1977). *Candida* (*Vanrija, Sporobolomyces*) *antarctica* is reported to produce extracellular mannosylerythritol lipids when grown on soybean oil as a carbon source (Kitamoto et al. 1990a,b). The industrial synthesis of glucoside esters by B lipase from *C. antarctica* is in development (Björkling et al. 1991).

7
Appendix

7.1
Media

Many of these media are commercially available. (* Recipe given below)

General media

- Malt extract agar (MEA, Difco) *
- Potato-dextrose agar (PDA, Difco) *
- Yeast extract-peptone-glucose agar (YPGA) *
- Yeast extract-malt extract-agar (YMA, Difco) *
- Yeast morphology agar (YMoA, Difco)

Media for ballistoconidium formation

- Malt extract agar (MEA, Difco) *
- Corn meal agar (CMA, Difco)
- Yeast morphology agar (YMoA, Difco)

Sporulation media commonly used for ascomycetous yeasts

- Corn meal agar (CMA, Difco)
- Gorodkowa agar *
- 2% or 5% Malt extract-3% agar (MEA, Difco) *
- McClary's acetate agar *
- V8 juice agar and diluted V8 juice agar *
- YM agar and diluted YM agar *

Media used for mating of basidiomycetous yeasts

- Conjugation medium according to Flegel (modified) *
- Corn meal agar (CMA, Difco)
- 5% Malt extract agar (MEA, Difco) *
- Potato-dextrose agar (PDA, Difco) *
- V8 juice agar *
- Malt extract-yeast extract-peptone agar *

Filobasidiella (Cryptococcus) neoformans requires specific media for induction of the teleomorph

- Hay infusion agar (Know-Chung and Bennett 1978) *
- Pigeon manure agar (Staib 1981) *
- Sunflower seed agar (Pal et al. 1991) *

Media for endospore formation

- Corn meal agar (CMA, Difco)
- Yeast extract-malt extract agar (YMA, Difco) *
- Malt extract agar (MEA, Difco) *
- Potato-dextrose agar (PDA, Difco) *

Medium for chlamydospore formation in *Candida albicans*

- Rice agar (Difco) *

7.2
Recipes of Some Media

Conjugation medium after Flegel (modified)

0.5% w/v glucose, 0.1% w/v $(NH_4)_2SO_4$, 0.02% w/v $MgSO_4.7H_2O$, 0.01% w/v $CaCl_2$, 1.5% w/v agar in distilled water. After sterilization, 0.01 v/v of a sterile vitamin solution is added.

Gorodkowa agar

0.5% w/v NaCl, 0.1% w/v glucose, 1% w/v peptone, 3% w/v agar in distilled water.

Hay infusion agar

50.0 g of partially decomposed hay is added to 1000 ml distilled water. After autoclaving and filtering, the volume is adjusted to 1000 ml, after which 0.2% w/v K_2HPO_4 and 1.5% w/v agar is added.

Malassezia furfur and *M. sympodialis* agar (Leeming and Notman medium)

1% w/v peptone, 0.5% w/v glucose, 0.01% w/v yeast extract, 0.4% w/v desiccated oxbile, 0.1% v/v glycerol, 0.05% w/v glycerol monostearate, 0.05% v/v Tween 60, 1% v/v high fat cow's milk, 1.5% w/v agar in distilled water.

Malt extract agar (MEA)

2 to 5% w/v malt extract, 3% w/v agar in distilled water

Malt extract-yeast extract-pepton agar

1.5% w/v malt extract, 0.05% w/v yeast extract, 0.25% w/v pepton, 3% w/v agar in distilled water.

McClary's acetate agar

0.1% w/v glucose, 0.12% w/v NaCl, 0.98% w/v KC_2HO_2, 0.07% w/v $MgSO_2.7H_2O$, 0.25 w/v yeast extract, 2% w/v agar in distilled water.

Pigeon manure agar (Staib 1981)

100 g of pigeon manure is added to 900 ml distilled water, and magnetically stirred for 1 h. After several filtrations, the filtrate is passed through a membrane filter (pore size 0.2 μm). 40 U ml^{-1} penicillin G and 80 U ml^{-1} streptomycin are added. Equal volumes of pigeon manure filtrate (ca. 40 °C) and 4% nutrient-free agar of ca. 60 °C are mixed, after which the pH is adjusted to 6.2.

Potato-dextrose agar (PDA)

100 g scrubbed and diced potatoes are added to 300 ml distilled water, and placed overnight at 4 °C. After filtration through filter paper, the filtrate is sterilized

(120 °C, 20 min). 230 ml potato extract, 20 g glucose and 20 g agar are added to 770 ml distilled water, and sterilized 20 min at 110 °C.

Sunflower seed agar (Pal et al. 1991)

22 g of ground sunflower seed are boiled in 400 ml distilled water for 1 h. After filtration through cheese cloth the volume is adjusted to 1000 ml, to which 0.01% w/v KH_2PO_4, 0.005% w/v $MgSO_4$, and 2% w/v agar are added.

Rice agar

2 g of peeled rice are boiled in 100 ml distilled water for 45 min. After filtration through cheese cloth, the volume is adjusted to 100 ml, and 2% w/v agar is added.

V8 juice agar

Add 40 g bakers yeast to 100 ml V8 juice (Campbell Soup Co.; pH 6.8). After steaming (10 min) and filtration, adjust the volume and pH to 6.8. Melt 2% w/v agar in 100 ml distilled water and add to V8 juice.

Vitamin solution (Wickerham 1951)

20 μg Biotin, 2 μg folic acid, 2 mg calcium pantothenate, 10 mg m-inositol, 400 mg niacin, 200 μg p-aminobenzoic acid, 440 μg pyridoxine-HCl, 200 μg riboflavine, and 400 μg thiamine HCl are added to 1000 ml distilled water and filter-sterilized. For easy preparation, a stock solution is 1000× concentrated, and subsequently diluted.

Yeast extract-peptone-glucose agar (YPGA)

4% w/v glucose, 0.5% w/v peptone, 2% w/v agar and 1% w/v yeast extract in distilled water (or 50% yeast infusion with 50% distilled water).

Yeast extract-malt extract agar (YMA)

0.05% w/v yeast extract-0.7% w/v malt extract-0.1% w/v peptone-3% w/v agar in distilled water.

Yeast infusion

Suspend 1 kg baker's yeast in 5 l demineralized water. Add egg white of two eggs and shake well, sterilize at 120 °C, filter directly, and dispense in bottles. Sterilize again and filter the yeast infusion just prior to use.

7.3
Culture Collections

Yeast strains can be obtained from a number of culture collections. The most important are listed below:

- ATCC, American Type Culture Collection, 12301 Parklawn Drive, Rockville, Maryland 20852-1776, USA.

- CBS, Yeast Division, Julianalaan 67, 2628 BC Delft, The Netherlands.

- DSM, Deutsche Sammlung von Mikroorganismen und Zellkulturen, Mascheroder Weg 1B, Braunschweig, Germany.

- IFO, Institute of Fermentation, 17-85, Juso-honmachi 2-chome, Yodogawa-ku, Osaka 532, Japan.

- JCM, Japan Collection of Microorganisms, Riken, Wako-shi, Saitama 351-01, Japan.

- MUCL, Mycothèque de l'Université Catholique de Louvain, Place Croix du Sud 3, 1348 Louvain-la-Neuve, Belgium.

- NCYC, National Collection of Yeast Cultures, AFRC Institute of Food Research, Colney Lane, Norwich NR4 7UA, UK.

- NRRL, ARS Culture Collection, National Center for Agricultural Utilization Research, USDA, ARS, 1815 North University Street, Peoria, Illinois 61604, USA.

- PYC, Portuguese Yeast Collection, Centro de Biología, Instituto Gulbenkian de Ciencia, Rua da Quinta Grande, Apartado 14, 2781 Oeiras, Portugal.

- VKM, All-Union Collection of Microorganisms, Institute of Biochemistry and Physiology of Microorganisms, Pushchino on the Oka, Russia.

- Yeast Genetics Stock Center, Department of Biophysics and Medical Physics, University of California, Berkeley, California 94720, USA.

Acknowledgments. Wilma Batenburg-van der Vegte and Annette Beijn kindly assisted in the preparation of the plates, for which they are thanked.

References

An G-H, Schuman DB, Johnson EA (1989) Isolation of *Phaffia rhodozyma* mutants with increased astaxanthin content. Appl Environ Microbiol 55: 116-124

Anand R, Southern EM (1988) Pulsed field electrophoresis. In: Rickwood D, Hames BD (eds) Gel electrophoresis of nucleic acids. A practical approach. IRL Press, Oxford, pp 101-123

Aulakh HS, Straus SE, Kwon-Chung KJ (1981) Genetic relatedness of *Filobasidiella neoformans* (*Cryptococcus neoformans*) and *Filobasidiella bacillispora* (*Cryptococcus bacillisporus*) as determined by deoxyribonucleic acid composition and sequence homology studies. Int J Syst Bacteriol 31: 97-103

Bakalinsky AT, Snow R (1990) The chromosome constitution of wine strains of *Saccharomyces cerevisiae*. Yeast 6: 367-382

Bandoni RJ (1963) Conjugation in *Tremella mesenterica*. Can J Bot 41: 467-474

Bandoni RJ (1984) Tremellales with a yeast phase (Sirobasidiaceae and Tremellaceae) In: Kreger-van Rij NJW (ed) The yeasts, a taxonomic study, 3rd edn. Elsevier, Amsterdam, pp 541-544

Bandoni RJ, Johri BN (1972) *Tilletiaria*: a new genus in the Ustilaginales. Can J Bot 50: 39–43

Bandoni RJ, Lobo KJ, Brezdan SA (1971) Conjugation and chlamydospores in *Sporobolomyces odorus*. Can J Bot 49: 683–686

Banno I (1967) Studies on the sexuality of *Rhodotorula*. J Gen Appl Microbiol 13: 167–196

Barlogie B, Spitzer G, Hart JS, Johnston DA, Buchner T, Schumann J, Drewinko B (1976) DNA histogram analysis of human hemopoietic cells. Blood 48: 245–258

Barnett JA (1968) Biochemical differentiation of taxa with special reference to the yeasts. In: Ainsworth GC, Sussman AS (eds) The fungi, vol 3. Academic Press, London, pp 557–595

Barnett JA (1977) The nutritional tests in yeast systematics. J Gen Microbiol 99: 183–190

Barnett JA, Payne RW, Yarrow D (1990) Yeasts: characteristics and identification. Cambridge University Press, Cambridge

Batra LR (1973) Nematosporaceae (Hemiascomycetidae): taxonomy, pathogenicity, distribution, and vector relations. Tech Bull US Dep Agric 1469: 1–71

Bauer R (1987) Uredinales – germination of basidiospores and pycnospores. Stud Mycol 30: 111–125

Bergkamp RJM, Kool IM, Geerse RH, Planta RJ (1992) Multiple-copy integration of the α-galactosidase gene from *Cyamopsis tetragonoloba* into the ribosomal DNA of *Kluyveromyces lactis*. Curr Genet 21: 365–370

Björkling F, Godtfredsen SE, Kirk O (1991) The future impact of industrial lipases. Tibtech 9: 360–363

Björling T, Lindman B (1989) Evaluation of xylose-fermenting yeasts for ethanol production from spent sulfite liquor. Enzym Microb Technol 11: 240–246

Boekhout T (1987) Systematics of anamorphs of Ustilaginales (smut fungi) – a preliminary survey. Stud Mycol 30: 137–149

Boekhout T (1991a) Systematics of *Itersonilia*: a comparative phenetic study. Mycol Res 95: 135–146

Boekhout T (1991b) A revision of ballistoconidia-forming yeasts and fungi. Stud Mycol 33: 1–194

Boekhout T, Bosboom RW (1994) Karyotyping of *Malassezia* yeasts: taxonomic and epidemiological implications. Syst Appl Microbiol 17: 146–153

Boekhout T, Fonseca A, Batenburg-van der Vegte WH (1991) *Bulleromyces* genus novum (Tremellales), a teleomorph for *Bullera alba*, and the occurrence of mating in *Bullera variabilis*. Antonie van Leeuwenhoek J Microbiol 59: 81–93

Boekhout T, Kurtzman CP, O'Donnell K, Smith MT (1994) Phylogeny of the yeast genera *Mansemiaspura* (anamorph *Klueckera*), *Dekkera* (anamorph *Brettanomyces*), and *Eeniella* as inferred from partial 26S ribosomal DNA nucleotide sequences. Int J Syst Bact 44: 781–786

Boekhout T, Yamada Y, Weijman ACM, Roeijmans HJ, Batenburg-van der Vegte WH (1992) The significance of coenzyme Q, carbohydrate composition and ultrastructure for the taxonomy of ballistoconidia-forming yeasts and fungi. Syst Appl Microbiol 15: 1–10

Boekhout T, Fonseca A, Sampaio J-P, Golubev WI (1993a) Classification of heterobasidiomycetous yeasts: characteristics and affiliation of genera to higher taxa of heterobasidiomycetes. Can J Microbiol 39: 276–290

Boekhout T, Renting M, Scheffers WA, Bosboom R (1993b) The use of karyotyping in the systematics of yeasts. Antonie van Leeuwenhoek J Microbiol 63: 157–163

Boulton CA, Ratledge C (1984) *Cryptococcus terricolus*, an oleaginous yeast re-appraised. Appl Microbiol Biotechnol 20: 72–76

Britten RJ, Pavich M, Smith J (1970) A new method for DNA purification. Yearb Carnegie Inst 68: 400–402

Britten RJ, Graham DE, Neufeld BR (1974) Analysis of repeating DNA sequences by reassociation. In: Grossman L, Moldave K (eds) Methods in enzymology, vol 29, Nucleic acids and protein synthesis. Academic Press, New York, pp 363–418

Bruns TD, White TJ, Taylor JW (1991) Fungal molecular systematics. Annu Rev Ecol Syst 22: 525–564

Bucknolz RG, Gleeson MAG (1991) Yeast systems for the commercial production of heterologous proteins. Bio/technology 9: 1067–1072

Büttner R, Bode R, Birnbaum D (1991) Characterization of extracellular acid phosphatases from the yeast *Arxula adeninivorans*. Zentralbl Mikrobiol 146: 399–406

Carle GF, Olson MV (1985) A karyotype for yeast. Proc Natl Acad Sci USA 82: 3756–3760

Casselton LA, Mutasa ES, Tymon A, Mellon FM, Little PFR, Taylor S, Benhagen J, Stratman R (1989) The molecular analysis of basidiomycete mating type genes. In: Nevalainen H, Penttilä M (eds) Proc EMBO-Alko Workshop Molec. Biol Filamentous Fungi. Found Biotech Ind Ferm Res 6: 139–148

Channon AG (1956) Association of a species of *Itersonilia* with parsnip canker in Great Britain. Nature (Lond) 178: 217

Channon AG (1963) Studies on parsnip canker I. The causes of the diseases. Ann Appl Biol 51: 1–15

Chernov IY, Bab'eva IP (1988) New species *Cryptococcus* yeast from tundra soil. Mikrobiologiya 57: 1031–1034

Claisse ML, Péré-Aubert GA, Clavillier LP, Slonimski P (1970) Méthodes d'estimation de la concentration des cytochromes dans les cellules entières de levures. Eur J Biochem 16: 430–438

Clare JJ, Rayment FB, Ballantine SP, Sreekrishna K, Romanos MA (1991) High-level expression of tetanus toxin fragment C in *Pichia pastoris* strains containing multiple tandem integrations of the gene. Bio/technology 9: 455–460

Cole GT (1983) *Graphiola phoenicis*: a taxonomic enigma. Mycologia 75: 93–116

Coleman AW, Maguire MJ, Coleman JR (1981) Mithramycin- and 4'-6'diamidino-2-phenylindole (DAPI)-DNA staining for fluorescence microspectrophotometric measurement of DNA in nuclei, plastids, and virus particles. J Histochem Cytochem 29: 959–968

Cook PR (1984) A general method for preparing intact nuclear DNA. EMBO J 3: 1837–1842

Cook WL, Massey JK, Ahearn DG (1973) Degradation of crude oil by yeasts and its effects on *Lesbistes reticulatus*. In: Ahearn DG, Meyers SP (eds) The microbial degradation of oil. Louisiana State University, Baton Rouge, pp 279–282

Cottrell M, Kock JLF, Lategan PM, Britz TJ (1986) Long-chain fatty acid composition as an aid in the classification of the genus *Saccharomyces*. Syst Appl Microbiol 8: 166–168

Cox DE (1976) A new homobasidiomycete with anomalous basidia. Mycologia 68: 481–510

Cregg JM, Vedvick TS, Raschke WC (1993) Recent advances in the expression of foreign genes in *Pichia pastoris*. Bio/technology 11: 905–910

Cryer DR, Eccleshall R, Marmur J (1975) Isolation of DNA. In: Prescott DM (ed) Methods in cell biology, vol 12. Academic Press, New York, pp 39–44

Dear PH, Cook PR (1991) Cellular gels, purifying and mapping long DNA molecules. Biochem J 273: 695–699

de Hoog GS, Rantio-Lehtimäki AH, Smith MTh (1985) *Blastobotrys*, *Sporothrix* and *Trichosporiella*: generic delimitation, new species, and a Stephanoascus teleomorph. Antonie van Leeuwenhoek J Microbiol 51: 79–109

de Hoog GS, Smith MTh, Guého E (1986) A revision of the genus *Geotrichum* and its teleomorphs. Stud Mycol 29: 1–131

de Jonge P, de Jongh FCM, Meijers R, Steensma HY, Scheffers WA (1986) Orthogonal-field-alternation gel electrophoresis banding patterns of DNA from yeasts. Yeast 2: 193–204

Dekker RFH (1982) Ethanol production from D-xylose and other sugars by the yeast *Pachysolen tannophilus*. Biotechnol Lett 4: 411–416

Dellweg H, Rizzi M, Methner H, Debus D (1984) Xylose fermentation by yeasts 3. Comparison of *Pachysolen tannophilus* and *Pichia stipitis*. Biotechnol Lett 6: 395–400

De Mot R, Verachtert H (1985) Purification and characterization of extracellular amylolytic enzymes from the yeast *Filobasidium capsuligenum*. Appl Environ Microbiol 50: 1474–1482

De Mot R, Verachtert H (1986) Secretion of *a*-amylase and multiple forms of glucoamylase by the yeast *Trichosporon pullulans*. Can J Microbiol 32: 47–51

De Mot R, Andries K, Verachtert H (1984a) Comparative study of starch degradation and amylase production by ascomycetous yeast species. Syst Appl Microbiol 5: 106–118

De Mot R, Demeersman M, Verachtert H (1984b) Comparative study of starch degradation and amylase production by non-ascomycetous, yeast species. Syst Appl Microbiol 5: 421–432

Detroy RW, Cunningham RL, Bothast RJ, Bagby MO, Herman A (1982) Bioconversion of wheat straw cellulose/hemicellulose to ethanol by *Saccharomyces uvarum* and *Pachysolen tannophilus*. Biotech Bioeng 24: 1105–1113

Diddens HA, Lodder J (1942) Die anaskosporogenen Hefen, zweite Hälfte. Noord-Hollandse Uitgevers Maatschappij, Amsterdam

Dixon R, Kinghorn JR (1990) Separation of large DNA molecules by pulsed-field gel electrophoresis. Soc gen Microbiol Quart 17: 86–88

Dobzhansky T (1976) Organismic and molecular aspects of species formation. In: Ayala FJ (ed) Molecular evolution. Sinauer, Sunderland, pp 95–105

Eilam T, Bushnell WR, Anikster Y, McLaughlin DJ (1992) Nuclear DNA content of basidiospores of selected rust fungi as estimated from fluorescence of propidium iodide-stained nuclei. Phytopathology 82: 705–712

Ellis DH, Pfeiffer TJ (1990) Natural habitat of *Cryptococcus neoformans* var. *gattii*. J Clin Microbiol 28: 430–431

Ellis DH, Pfeiffer T (1992) The ecology of *Cryptococcus neoformans*. Eur J Epidemiol 8: 321–325

Evans CT, Ratledge C (1984a) Effect of nitrogen source on lipid accumulation in oleaginous yeasts. J Gen Microbiol 130: 1693–1704

Evans CT, Ratledge C (1984b) Influence of nitrogen metabolism on lipid accumulation by *Rhodosporidium toruloides* CBS 14. J Gen Microbiol 130: 1705–1710

Ezaki T, Hashimoto Y, Yabuuchi E (1989) Fluorometric deoxyribonucleic acid-deoxyribonucleic acid hydridization in microdilution wells as an alternative to membrane filter hybridization in which radioisotopes are used to determine genetic relatedness among bacterial strains. Int J Syst Bacteriol 39: 224–229

Fell JW (1984) Teliospore-forming yeasts. In: Kreger-van Rij NJW (ed) The yeasts, a taxonomic study, 3rd edn. Elsevier, Amsterdam, pp 491–495

Fell JW (1970) Yeasts with heterobasidiomyctous life cycles. Chap 3. In: Ahearn DG (ed) Recent trends in yeast research. School of arts and sciences, Georgia State University, Atlanta, pp 49–66

Fell JW, Statzell Tallman A (1989a) *Leucosporidium* Fell, Statzell, Hunter & Phaff. In: Kreger-van Rij NJW (ed) The yeasts, a taxonomic study, 3rd edn. Elsevier, Amsterdam, pp 496–508

Fell JW, Statzell Tallman A (1984b) *Rhodosporidium* Banno. In: Kreger-van Rij NJW (ed) The yeasts, a taxonomic study, 3rd edn. Elsevier, Amsterdam, pp 509–531

Fell JW, Statzell AC, Hunter IL, Phaff HJ (1969) *Leucosporidium* gen. n., the heterobasidiomycetous stage of several yeasts of the genus *Candida*. Antonie van Leeuwenhoek J Microbiol 35: 433–462

Fell JW, Statzell-Tallman A, Lutz MJ, Kurtzman CP (1992) Partial rRNA sequences in marine yeasts: a model for identification of marine eukaryotes. Mol Mar Biol Biotechnol 1: 175–186

Felsenstein J (1988) Phylogenies from molecular sequences: inference and reliability. Annu Rev Genet 22: 521–565

Fiol JB, Claisse M (1987) Les cytochromes des *Kluyveromyces*: détermination et implications systématiques. Mycopathologia 78: 177–184

Fleer R, Yeh P, Amellal N, Maury I, Fournier A, Bacchetta F, Baduel P, Jung G, L'Hote H, Becquart J, Fukuhara H, Mayaux JF (1991) Stable multicopy vectors for high-level secretion of recombinant human serum albumin by *Kluyveromyces* yeasts. Bio/technology 9: 968–975

Flegel TW (1977) Let's call a yeast a yeast. Can J Microbiol 23: 945–946

Flegel TW (1981) The pheromonal control of mating in yeasts and its phylogenetic implication: a review. Can J Microbiol 27: 373–389

Fonseca A, Sampaio JP (1992) *Rhodosporidium lusitaniae* sp. nov., a novel homothallic basidiomycetous yeast species from Portugal that degrades phenolic compounds. Syst Appl Microbiol 15: 47–51

Fonseca A, van Uden N (1991) *Cryptococcus yarrowii* sp. nov., a novel yeast species from Portugal. Antonie van Leeuwenhoek J Microbiol 59: 177–181

Foster LM, Kozak KR, Loftus MG, Stevens JJ, Ross IK (1993) The polymerase chain reaction and its application to filamentous fungi. Mycol Res 97: 769–781

Franke AE, Kaczmarek FS, Eisenhard ME, Geoghegan KF, Danley DE, De Zeeuw JR, O'Donnell MM, Gollaher MG, Davidow LS (1988) Expression and secretion of bovine prochymosin in *Yarrowia lipolytica*. Dev Ind Microbiol 29: 43–57

Furukawa T, Ogino T, Matsuyoshi T (1982) Fermentative production of citric acid from n-paraffins by *Saccharomycopsis Lipolytica*. J Ferment Technol 60: 281–286

Fuson GB, Presley HL, Phaff HJ (1987) Deoxyribonucleic acid base sequence relatedness among members of the yeast genus *Kluyveromyces*. Int J Syst Bacteriol 37: 371–379

Gams W, Grinbergs J (1970) *Ascoidea corymbosa* n. spec., ein hefeähnlicher Pilz im Bast von *Araucaria araucana*. Acta Bot Neerl 19: 794–798

Gandy DG (1966) *Itersonilia perplexans* on chrysanthemums: alternative hosts and ways of overwintering. Trans Br Mycol Soc 49: 499–507

Gardiner K (1991) Pulsed field gel electrophoresis. Anal Chem 63: 658–665

Gardner DCJ, Heale SM, Stateva LI, Oliver SG (1993) Treatment of yeast cells with wall lytic enzymes is not required to prepare chromosomes for pulsed-field gel analysis. Yeast 9: 1053–1055

Gellissen G, Melber K, Janowicz ZA, Dahlems UM, Weydemann U, Piontek M, Strasser AWM, Hollenberg CP (1992) Heterologous protein production in yeast. Antonie van Leeuwenhoek J Microbiol 62: 79–93

Gill CO, Hall MJ, Ratledge C (1977) Lipid accumulation in an oleaginous yeast (*Candida* 107) growing on glucose in single-stage continuous culture. Appl Environ Microbiol 33: 231–239

Giménez-Jurado G, van Uden N (1989) *Leucosporidium fellii* spec. nov., a basidiomycetous yeast that degrades L(+)-tartaric acid. Antonie van Leeuwenhoek 55: 133–141

Golubev WI (1981) New combinations in *Cryptococcus* yeasts. Mikol Fitopatol 15: 467–468

Golubev WI (1995) Perfect state of *Rhodomyces dendrorhous* (*Phaffia rhodozyma*). Yeast 11: 101–110

Golubev WI (1989) Catabolism of *m*-inositol and taxonomic value of D-glucuronate assimilation in yeasts. Mikrobiologiya 58: 276–283

Goto S, Sugiyama J, Hamamoto M, Komagata K (1987) *Saitoella*, a new anamorph genus in the Cryptococcaceae to accommodate two Himalayan yeasts isolates formerly identified as *Rhodotorula glutinis*. J Gen Appl Microbiol 33: 75–85

Gottschalk M, Blanz P (1985) Untersuchungen an 5S ribosomalen Ribonukleinsäuren als Beitrag zur Klärung von Systematik und Phylogeny der Basidiomyzeten. Z Mykol 51: 205–243

Guého E, Kurtzman CP, Peterson SW (1990) Phylogenetic relationship among species of *Sterigmatomyces* and *Fellomyces* as determined from partial rRNA sequences. Int J Syst Bacteriol 40: 60–65

Guého E, Smith MTh, de Hoog GS, Billon-Grand G, Christen R, Batenburg-van der Vegte WH (1992) Contributions to a revision of the genus *Trichosporon*. Antonie van Leeuwenhoek J Microbiol 61: 289–316

Guilliermond A (1912) Les levures. Doin, Paris

Gunderson K, Chu G (1991) Pulsed-field electrophoresis of megabase-sized DNA. Mol Cell Biol 11: 3348–3354

Gurr E (1965) The rational use of dyes in biology. Leonard Hill, London

Hamamoto M, Sugiyama J, Komagata K (1986) DNA base composition of strains in the genera *Rhodosporidium*, *Cystofilobasidium*, and *Rhodotorula* determined by reverse-phase high-performance liquid chromatography. J Gen Appl Microbiol 32: 215–223

Hamamoto M, Uchida K, Yamaguchi H (1992) Ubiquinone system and DNA base composition of strains in the genus *Malassezia* determined by high-performance liquid chromatography. J Gen Appl Microbiol 38: 79–82

Hara T, Shimoda T, Nonaka K, Ogata S (1991) Colorimetric detection of DNA-DNA hybridization in microdilution wells for taxonomic application on bacterial strains. J Ferment Bioengin 72: 122–124

Heath IB (1978) Experimental studies of fungal mitotic systems. A review. In: Heath IB (ed) Nuclear division in the fungi. Academic Press, New York, pp 89–176

Heath IB, Ashton ML, Kaminskyi SGW (1987) Mitosis as a phylogenetic marker among the yeasts – review and observations on novel mitotic systems in freeze substituted cells of the Taphrinales. Stud Mycol 30: 279–297

Hendriks L, Goris A, Van de Peer Y, Neefs J-M, Vancanneyt M, Kersters K, Hennebert G, De Wachter R (1992a) Phylogenetic analysis of five medically important *Candida* species as deduced on the basis of small ribosomal subunit RNA sequences. J Gen Microbiol 137: 1223–1230

Hendriks L, Goris A, Van de Peer Y, Neefs J-M, Vancanneyt M, Kersters K, Berny J-F, Hennebert G (1992b) Phylogenetic relationships among Ascomycetes and Ascomycete-like yeasts as deduced from small ribosomal subunit RNA sequences. Syst Appl Microbiol 15: 98–104

Heslot H (1990) Genetics and genetic engineering of the industrial yeast *Yarrowia lipolytica*. In: Fiechter A (ed) Advances in biochemical engineering/biotechnology, vol 43. Springer, Berlin Heidelberg New York, pp 43–73

Holm C, Meeks-Wagner DW, Fangman WL, Botstein D (1986) A rapid, efficient method for isolating DNA from yeast. Gene 42: 169–173

Iwaguchi S, Homma M, Tanaka K (1990) Variations in the electrophoretic karyotype analysed by the assignment of DNA probes in *Candida albicans*. J Gen Microbiol 136: 2433–2442

Jahnke K-D (1987) Assessing natural relationships by DNA analysis: techniques and applications. Stud Mycol 30: 227–246

Jenq WJ, Speckman RA, Crang RE, Steinberg MP (1989) Enhanced conversion of lactose to glycerol by *Kluyveromyces fragilis* utilizing whey permeate as a substrate. Appl Environ Microbiol 55: 573–578

Johnson EA, Conklin DE, Lewis MJ (1977) The yeast *Phaffia rhodozyma* as a dietary source for salmonids and crustaceans. J Fish Res Board Can 34: 2417–2421

Johnson EA, Lewis MJ, Grau CR (1980) Pigmentation of egg yolks with astaxanthin from the yeast *Phaffia rhodozyma*. Poult Sci 59: 1777–1782

Johnson-Reid JA, Moore RT (1972) Some ultrastructural features of *Rhodosporidium toruloides* Banno. Antonie van Leeuwenhoek J Microbiol 38: 417–435

Johnston JR (1988) Yeast genetics, molecular aspects. In: Campbell I, Duffus JH (eds) Yeast, a practical approach. IRL Press, Oxford, pp 107–123

Jong S-C, Lee F-L (1986) The new species *Dekkera naardenensis*, teleomorph of *Brettanomyces naardenensis*. Mycotaxon 25: 147–152

Kaltenboeck B, Spatafora JW, Zhang X, Kousalas KG, Blackwell M, Storz J (1992) Efficient production of single-stranded DNA as long as 2 kb for sequencing of PCR-amplified DNA. Biotechniques 12: 164–171

Kaneko Y, Banno I (1991) Reexamination of *Saccharomyces bayanus* strains by DNA-DNA hybridization and electrophoretic karyotyping. IFO Res Comm 15: 30–41

Katayama-Hirayama K, Tobita S, Hirayama K (1991) Degradation of phenol by yeast *Rhodotorula*. J Gen Appl Microbiol 37: 147–156

Kendrick B (1987) Yeasts and yeast-like fungi – new concepts and new techniques. Stud Mycol 30: 479–486

Kitamoto D, Akiba S, Hioki C, Tabuchi T (1990a) Extracellular accumulation of mannosylerythritol lipids by a strain of *Candida antarctica*. Agric Biol Chem 54: 31–36

Kitamoto D, Haneishi K, Nakahara T, Tabuchi I (1990b) Production of mannosylerythritol lipids by *Candida antarctica* from vegetable oils. Agric Biol Chem 54: 37–40

Klein RD, Favreau MA (1988) Transformation of *Schwanniomyces occidentalis* with an ADE2 gene cloned from S. occidentalis. J Bacteriol 170: 5572–5578

Kreger-van Rij NJW (1984a) *Saccharomycopsis* Schiönning. In: Kreger-van Rij NJW (ed) The yeasts, a taxonomic study, 3rd edn. Elsevier, Amsterdam, pp 399–413

Kreger-van Rij NJW (1984b) The species. In: Kreger-van Rij NJW (ed) The yeasts, a taxonomic study, 3rd edn. Elsevier, Amsterdam, p 16

Kreger-van Rij NJW, Veenhuis M (1971) A comparative study of the cell wall structure of basidiomycetous and related yeasts. J Gen Microbiol 68: 87–95

Kuraishi H, Katayama-Fujimura Y, Sugiyama J, Yokoyama T (1985) Ubiquinone systems in fungi I. Distribution of ubiquinones in the major families of ascomycetes, basidiomycetes, and deuteromycetes, and their taxonomic implications. Trans Mycol Soc Jpn 26: 383–395

Kurjan J (1991) Cell-cell interactions involved in yeast mating. In: Dworkin M (ed) Microbial cell-cell interactions. American Society for Microbiology, Washington, pp 113–144

Kurtzman CP (1973) Formation of hyphae and chlamydospores by *Cryptococcus laurentii*. Mycologia 65: 388–395

Kurtzman CP (1984) Synonymy of the yeast genera *Hansenula* and *Pichia* demonstrated through comparisons of deoxyribonucleic acid relatedness. Antonie van Leeuwenhoek J Microbiol 50: 209–217

Kurtzman CP (1985) Molecular taxonomy of fungi. In: Bennett JW, Lasure LL (eds) Gene manipulation in fungi. Academic Press, New York, pp 35–56

Kurtzman CP (1987) Prediction of biological relatedness among yeasts from comparison of nuclear DNA complementarity. Stud Mycol 30: 459–468

Kurtzman CP (1990) DNA relatedness among species of the genus *Zygosaccharomyces*. Yeast 6: 213–219

Kurtzman CP (1991) DNA relatedness among saturn-spored yeasts assigned to the genera *Williopsis* and *Pichia*. Antonie van Leeuwenhoek J Microbiol 60: 13–19

Kurtzman CP (1992) rRNA sequence comparisons for assessing phylogenetic relationships among yeasts. Int J Syst Bacteriol 42: 1–6

Kurtzman CP (1995) Relationships among the genera *Ashbya*, *Eremothecium*, *Holleya* and *Nematospora* determined from rDNA sequence divergence. J Indust Microbiol 14: 523–530

Kurtzman CP (1993) DNA-DNA hybridization approaches to species identification in small genome organisms. Methods Enzymol 224: 335–348

Kurtzman CP, Liu Z (1990) Evolutionary affinities of species assigned to *Lipomyces* and *Myxozyma* estimated from ribosomal RNA sequence divergence. Curr Microbiol 21: 387–393

Kurtzman CP, Phaff HJ (1987) Molecular taxonomy. In: Rose AH, Harrison JS (eds) The yeasts, 2nd edn, vol 1. Academic Press, London, pp 63–94

Kurtzman CP, Robnett CJ (1991) Phylogenetic relationships among species of *Saccharomyces*, *Schizosaccharomyces*, *Debaryomyces* and *Schwanniomyces* determined from partial ribosomal RNA sequences. Yeast 7: 61–72

Kurtzman CP, Robnett CJ (1994) Synonymy of the yeast genera *Wingea* and *Debaryomyces*. Antonie van Leeuwenhoek J Microbiol 66: 337–342

Kurtzman CP, Robnett CJ (1995) Molecular relationships among hyphal ascomycetous yeasts and yeastlike tara. Can J Bot 73: S824–S830

Kurtzman CP, Smiley MJ, Johnson CJ, Wickerham LJ, Fuson GB (1980) Two new and closely related heterothallic species. *Pichia amylophila* and *Pichia mississippiensis*: characterization by hybridization and deoxyribonucleic acid reassociation. Int J Syst Bacteriol 30: 208–216

Kurtzman CP, Bothast RJ, VanCauwenberge JE (1982) Conversion of D-xylose to ethanol by the yeast *Pachysolen tannophilus*. US Patent 4, 359, 534

Kurtzman CP, Phaff HJ, Meyer SA (1983) Nucleic acid relatedness among yeasts. In: Smith ARW, Spencer JFT, Spencer DM (eds) Yeast genetics. Fundamental and applied aspects. Springer, Berlin Heidelberg New York, pp 139–166

Kwon-Chung KJ (1980) Nuclear genotypes of spore chains in *Filobasidiella neoformans* (*Cryptococcus neoformans*). Mycologia 72: 418–422

Kwon-Chung KJ, Bennett JE (1978) Distribution of a and a mating types of *Cryptococcus neoformans* among natural and clinical isolates. Am J Epidemiol 108: 337–340

Kwon-Chung KJ, Fell JW (1984) The genus *Filobasidiella* Kwon-Chung. In: Kreger-van Rij NJW (ed) The yeasts, a taxonomic study, 3rd edn. Elsevier, Amsterdam, pp 472–482

Kwon-Chung KJ, Popkin TJ (1976) Ultrastructure of septal complex in *Filobasidiella neoformans* (*Cryptococcus neoformans*). J Bacteriol 126: 524–528

Kwon-Chung KJ, Wickes BL, Stockman L, Roberts GD, Ellis D, Howard DH (1992) Virulence, serotype, and molecular characteristics of environmental strains of *Cryptococcus neoformans* var. *gattii*. Infect Immun 60: 1869–1874

Laaser G, Möller E, Jahnke K-D, Prillinger H, Prell H (1987) Ribosomal restriction analysis as a taxonomic tool in separating physiologically similar basidiomycetous yeasts. Syst Appl Microbiol 11: 170–175

Lachance M-A (1985) Current views on the yeast species. Microbiol Sci 2: 122–126

Lane DJ, Pace B, Olsen GJ, Stahl DA, Sogin ML, Pace NR (1985) Rapid determination of 16S ribosomal RNA sequences for phylogenetic analysis. Proc Natl Acad Sci USA 82: 6955–6959

Lee F-L, Jong S-C (1986) New species of *Dekkera custersiana* and *D. lambica*, teleomorphs of *Brettanomyces*. Mycotaxon 25: 455–460

Lewis MJ, Ragot N, Berlant MC, Miranda M (1990) Selection of astaxanthin-overproducing mutants of *Phaffia rhodozyma* with β-ionone. Appl Environ Microbiol 56: 2944–2945

Lipke PN, Kurjan J (1992) Sexual agglutination in budding yeasts: structure, function, and regulation of adhesion glycoproteins. Microbiol Rev 56: 180–194

Liu Z, Kurtzman CP (1991) Phylogenetic relationships among species of *Williopsis* and *Saturnospora* gen. nov. as determined from partial rRNA sequences. Antonie van Leeuwenhoek J Microbiol 60: 21–30

Lodder J (1934) Die anaskosporogenen Hefen, erste Hälfte. Noord-Hollandsche Uitgevers Maatschappij, Amsterdam

Lodder J (ed) (1970) The yeasts, a taxonomic study, 2nd edn. North-Holland Publ, Amsterdam

Mahrous M, Lott TJ, Meyer SA, Sawant AD, Ahearn DG (1990) Electrophoretic karyotyping of typical and atypical *Candida albicans*. J Clin Microbiol 28: 876–881

Maleszka R, Clark-Walker GD (1989) A petite positive strain of *Kluyveromyces lactis* has a 300 kb deletion in the rDNA cluster. Curr Genet 16: 429–432

Maleszka R, Skrzypek M (1990) Assignment of cloned genes to electrophoretically separated chromosomes of the yeast *Pachysolen tannophilus*. FEMS Microbiol Lett 69: 79–82

Maleszka R, Wang PY, Schneider H (1982) Ethanol production from D-galactose and glycerol by *Pachysolen tannophilus*. Enzyme Microb Technol 4: 349–352

Mann W, Jeffery J (1989) Isolation of DNA from yeasts. Anal Biochem 178: 82–87

Marmur J (1961) A procedure for the isolation of DNA from microorganisms. J Mol Biol 3: 208–218

Mathaba LT, Franklyn KM, Warmington JR (1993) A rapid technique for the isolation of DNA from clinical isolates of *Candida albicans*. J Microbiol Meth 17: 17–25

McArthur CR, Clark-Walker GD (1983) Mitochondrial DNA size diversity in the *Dekkera/Brettanomyces* yeasts. Curr Genet 7: 29–35

McCarroll R, Olsen GJ, Stahl YD, Woese CR, Sogin ML (1983) Nucleotide sequence of the *Dictyostelium discoideum* small-subunit ribosomal acid inferred from the gene sequence: evolutionary implications. Biochemistry 22: 5858–5868

McCluskey K, Russell BW, Mills D (1990) Electrophoretic karyotyping without the need for generating protoplasts. Curr Genet 18: 385–386

McEachern MJ, Hicks JB (1991) Dosage of the smallest chromosome affects both the yeast-hyphal transition and the white-opaque transition of *Candida albicans* WO-1. J Bacteriol 173: 7436–7442

Mendonça-Hagler LC, Hagler AN, Kurtzman CP (1993) Phylogeny of *Metschnikowia* species estimated from partial rRNA sequences. Int J Syst Bacteriol 43: 368–373

Meyer PS, Du Preez JC, Kilian SG (1992) Isolation and evaluation of yeasts for biomass production from bagasse hemicellulose hydrolysate. Syst Appl Microbiol 15: 161–165

Meyer SA, Smith MT, Simione FP Jr (1978) Systematics of *Hanseniaspora* Zikes and *Kloeckera* Janke. Antonie van Leeuwenhoek J Microbiol 44: 79–96

Middelhoven WJ, de Jong IM, de Winter M (1991) *Arxula adeninivorans*, a yeast assimilating many nitrogen and aromatic compounds. Antonie van Leeuwenhoek 59: 129–137

Middelhoven WJ, Koorevaar M, Schuur GW (1992) Degradation of benzene compounds by yeasts in acidic soils. Plant Soil 145: 37–43

Miller MW, Phaff HJ (1984) *Nadsonia* Sydow. In: Kreger-van Rij NJW (ed) The yeasts, a taxonomic study, 3rd edn. Elsevier, Amsterdam, pp 279–285

Miller MW, Yoneyama M, Soneda M (1976) *Phaffia*, a new yeast genus in the Deuteromycotina (Blastomycetes). Int J Syst Bacteriol 26: 286–291

Mills SC, Child JJ, Spencer JFT (1971) The utilization of aromatic compounds by yeasts. Antonie van Leeuwenhoek J Microbiol 37: 281–287

Molina FI, Jong S-C, Huffman JL (1993a) PCR amplification of the 3′ external transcribed and intergenic spacers of the ribosomal DNA repeat unit in three species of *Saccharomyces*. FEMS Microbiol Lett 108: 259–264

Molina FI, Shen P, Jong S-C (1993b) Validation of the species concept in the genus *Dekkera* by restriction analysis of genes coding for rRNA. Int J Syst Bacteriol 43: 32–35

Montrocher R, Claisse M (1988) Spectrophotometric analysis of yeasts: cytochrome spectra of some *Candida* and related taxa. J Gen Appl Microbiol 34: 221–232

Moore RT (1980) Taxonomic proposals for the classification of marine yeasts and other yeast-like fungi including the smuts. Bot Mar 23: 361–373

Moore RT (1987) Additions to the genus *Vanrija*. Bibl Mycol 108: 167–173

Moore RT, Kreger-van Rij NJW (1972) Ultrastructure of *Filobasidium olive*. Can J Microbiol 18: 1949–1951

Muderhwa JM, Ratomahenina R, Pina M, Graille UJ, Galzy P (1986) Purification and properties of the lipases from *Rhodotorula pilimanae* Hedrick & Burke. Appl Microbiol Biotechnol 23: 348–354

Nakase T, Takematsu A (1992) *Udeniomyces*, a new ballistosporous anamorphic yeast genus in the Cryptococcaceae proposed for three *Bullera* species which produce large bilaterally symmetrical ballistospores. FEMS Microbiol Lett 100: 497–502

Nakase T, Okada G, Sugiyama J, Itoh M, Suzuki M (1989) *Ballistosporomyces*, a new ballistospore-forming anamorphic yeast genus. J Gen Appl Microbiol 35: 289–309

Nakase T, Itoh M, Takematsu A, Mikata K, Banno I, Yamada Y (1991) *Kockovaella*, a new ballistospore-forming anamorphic yeast genus. J Gen Appl Microbiol 37: 175–197

Nei N (1971a) Microbiological decomposition of phenol I. Isolation and identification of phenol metabolizing yeasts. J Ferment Technol 49: 655–660

Nei N (1971b) Microbiological decomposition of phenol II. Decomposition of phenol by *Rhodotorula glutinis*. J Ferment Technol 49: 852–860

Neujahr HY, Varga JM (1970) Degradation of phenols by intact cells and cell-free preparations of *Trichosporon cutaneum*. Eur J Biochem 13: 37–44

Nilsson R, Enebo L, Lundin H, Myrbäck K (1943) Mikrobielle Fettsynthese unter Verwendung von *Rhodotorula glutinis* nach dem Lufthefeverfahren. Sven Kem Tidskr 55: 41–51

Nishida H, Sugiyama J (1993) Phylogenetic relationships among *Taphrina*, *Saitoella*, and other higher fungi. Mol Biol Evol 10: 431–436

Oberwinkler F (1987) Heterobasidiomycetes with ontogenetic yeast-stages – systematic and phylogenetic aspects. Stud Mycol 30: 61–74

Oberwinkler F, Bandoni R, Blanz P, Kisimova-Horovitz L (1983) *Cystofilobasidium*: a new genus in the Filobasidiaceae. Syst Appl Microbiol 4: 114–122

O'Donnell K (1993) *Fusarium* and near relatives. In: Reynolds DP, Taylor JW (eds) The fungal holomorph: mitotic, meiotic and pleomorphic speciation in fungal systematics CAB, Wallingford, pp 225–233

Pal M, Onda C, Hasegawa A (1991) Sexual compatibility and environmental isolates of *Cryptococcus neoformans*. Jpn J Med Mycol 32: 101–106

Park WS, Murphy PA, Glatz BA (1990) Lipid metabolism and cell composition of the oleaginous yeast *Apiotrichum curvatum* grown at different carbon to nitrogen ratios. Can J Microbiol 36: 318–326

Pasari AB, Korus RA, Heimsch RC (1989) A model for continuous fermentations with amylolytic yeasts. Biotechnol Bioeng 33: 338–343

Patrignani G, Pellegrini S, Gerola FM (1984) Difference in septal pore apparatus ultrastructure of *Tremella mesenterica*. Caryologia 37: 77–86

Perfect JR, Magee BB, Magee PT (1989) Separation of chromosomes of *Cryptococcus neoformans* by pulsed field gel electropheresis. Infect Immunol 57: 2624–2627

Peterson SW, Kurtzman CP (1991) Ribosomal RNA sequence divergence among sibling species of yeasts. Syst Appl Microbiol 14: 124–129

Phaff HJ (1981) The species concept in yeasts: physiologic, morphologic, genetic, and ecological parameters. In: Stewart GG, Russell I (eds) Current developments in yeast research. Pergamon Press, Oxford, pp 635–643

Phaff HJ (1989) Trends in yeast research. Yeast (Spec Issue) 5: 341–349

Phaff HJ, Starmer WT, Tredick-Kline J (1987) *Pichia kluyveri* sensu lato – a proposal for two new varieties and a new anamorph. Stud Mycol 30: 403–414

Price CW, Fuson GB, Phaff HJ (1978) Genome comparison in yeast systematics: delimitation of species within the genera *Schwanniomyces*, *Saccharomyces*, *Debaryomyces*, and *Pichia*. Microbiol Rev 42: 161–193

Prillinger H, Oberwinkler F, Umile C, Tlachac K, Bauer R, Dörfler C, Taufratzhofer E (1993) Analysis of cell wall carbohydrates (neutral sugars) from ascomycetous and basidiomycetous yeasts with and without derivatization. J Gen Appl Microbiol 39: 1–34

Prior BA, Kilian SG, Du Preez JC (1989) Fermentation of D-xylose by the yeasts *Candida shehatae* and *Pichia stipitis*. Prospects and problems. Proc Biochem 24: 21–32

Raeder U, Broda P (1985) Rapid preparation of DNA from filamentous fungi. Lett Appl Microbiol 1: 17–20

Ragan MA (1988) Ribosomal RNA and the major lines of evolution: a perspective. Biosystems 21: 177–188

Raju NB (1986) A simple fluorescent staining method for meiotic chromosomes of *Neurospora*. Mycologia 78: 901–906

Ratledge C (1978) Lipids and fatty acids. In: Rose AH (ed) Economic microbiology, vol 2, Primary products of metabolism. Academic Press, London, pp 263–302

Ratledge C (1982) Microbial oils and fats: an assessment of their commercial potential. Prog Ind Microbiol 16: 119–206

Ratledge C (1986) Lipids. In: Pape H, Rehm H-J (eds) Biotechnology, vol 4, 1st edn., Microbial products II. VCH, Weinheim, pp 185–213

Ratledge C (1993) Single cell oils – have they a biotechnological future? Tibtech 11: 278–284

Ratledge C, Evans CT (1989) Lipids and their metabolism. In: Rose AH, Harrison JS (eds) The yeasts, vol 3, 2nd edn. Academic Press, London, pp 367–455

Ratledge C, Hall MJ (1979) Accumulation of lipid by *Rhodotorula glutinis* in continuous culture. Biotechnol Lett 1: 115–120

Reiser J, Glumoff V, Kälin M, Ochsner U (1990) Transfer and expression of heterologous genes in yeasts other than *Saccharomyces cerevisiae*. In: Fiechter A (ed) Advances in biochemical engineering/biotechnology vol 43. Springer, Berlin Heidelberg New York, pp 75–102

Rezanka T (1991) Overproduction of microbial lipids and lipases. Folia Microbiol 36: 211–224

Rhodes JC, Kwon-Chung KJ, Popkins TJ (1981) Ultrastructure of the septal pore complex in hyphae of *Cryptococcus laurentii*. J Bacteriol 145: 1410–1412

Rolph CE, Moreton RS, Harwood JL (1989) Acyl lipid metabolism in the oleaginous yeast *Rhodotorula gracilis* (CBS 3043). Lipids 24: 715–720

Romanos MA, Scorer CA, Clare JJ (1992) Foreign gene expression in yeast: a review. Yeast 8: 423–488

Rustchenko-Bulgac EP (1991) Variations of *Candida albicans* electrophoretic karyotypes. J Bacteriol 173: 6586–6596

Sambrook J, Fritsch EF, Maniatis T (1989) Molecular cloning, a laboratory manual, 2nd edn. CSH Press, New York, pp 6.50–6.59

Scheda R (1966) Merkmalsveränderungen bei Hefen der Gattung *Saccharomyces*. Monatsschr Brau 19: 256–258

Scheda R, Yarrow D (1966) The instability of physiological properties used as criteria in the taxonomy of yeasts. Arch Mikrobiol 55: 209–225

Schildkraut CL, Marmur J, Doty P (1962) Determination of the base composition of deoxyribonucleic acid from its buoyant density in CsCl. J Mol Biol 4: 430–433

Schneider H, Wang PY, Chan YK, Maleszka R (1981) Conversion of D-xylose into ethanol by the yeast *Pachysolen tannophilus*. Biotechnol Lett 3: 89–92

Seidler RJ, Mandel M (1971) Quantitative aspects of DNA renaturation: DNA base composition, state of chromosome replication, and polynucleotide homologies. J Bacteriol 106: 608–614

Shen R, Lachance M-A (1993) Phylogenetic study of ribosomal DNA of cactophilic *Pichia* species by restriction mapping. Yeast 9: 315–330

Shoda M, Udaka S (1980) Preferential utilization of phenol rather than glucose by *Trichosporon cutaneum* possessing a partially constitutive catechol 1,2-oxygenase. Appl Environ Microbiol 39: 1129–1133

Sleep D, Belfield GP, Ballance DJ, Steven J, Jones S, Evans LR, Moir PD, Goodey AR (1991) *Saccharomyces cerevisiae* strains that overexpress heterologous proteins. Bio/technology 9: 183–187

Smith CL, Matsumoto T, Niwa O, Klco S, Fan J-B, Yanagida M, Cantor CR (1987) An electrophoretic karyotype for *Schizosaccharomyces pombe* by pulsed field electrophoresis. Nucleic Acids Res 15: 4481–4489

Smith MTh, van der Walt JP, Johannsen E (1976) The genus *Stephanoascus* gen. nov. (Ascoideaceae). Antonie van Leeuwenhoek J Microbiol 42: 119–127

Smith MTh, Batenburg-van der Vegte WH, Scheffers WA (1981) *Eeniella*, a new yeast genus of the Torulopsidales. Int J Syst Bacteriol 31: 196–203

Smith MTh, Yamazaki M, Poot GA (1990) *Dekkera, Brettanomyces* and *Eeniella*: electrophoretic comparison of enzymes and DNA-DNA homology. Yeast 6: 299–310

Sor F, Fukuhara H (1989) Analysis of chromosomal DNA patterns of the genus *Kluyveromyces*. Yeast 5: 1–10

Spencer JFT, Gorin PAJ (1969) Systematics of the genus *Candida* Berkhout: proton magnetic resonance spectra of the mannans and mannose-containing polysaccharides as an aid in classification. Antonie van Leeuwenhoek J Microb 35: 33–44

Spencer JFT, Gorin PAJ (1970) Systematics of the genus *Torulopsis*: proton magnetic resonance spectra of the mannose-containing polysaccharides as an aid in classification. Antonie van Leeuwenhoek J Microbiol 36: 509–524

Staib F (1981) The perfect state of *Cryptococcus neoformans*, *Filobasidiella neoformans*, on pigeon manure agar. Zentralbl Bakteriol, Parasitenkd, Abt 1, 248: 575–578

Steensma HY, de Jongh FCM, Linnekamp M (1988) The use of electrophoretic karyotypes in the classification of yeasts: *Kluyveromyces marxianus* and *K. lactis*. Curr Genet 14: 311–317

Stelling-Dekker NM (1930) Die sporogenen Hefen. Verh K Akad Wet Sect 2, pp 1–547

Sugiyama J, Kukagawa M, Chiu SW, Komagata K (1985) Cellular carbohydraubiq composition, DNA base composition, ubiquinone systems, and diazoneum blue B color test in the genera *Rhodosporidium*, *Leucosporidium*, *Rhodotorula* and related basidiomycetous yeasts. J Gen Appl Microbiol 31: 519–550

Sugiyama J, Tokuoka K, Suh S-O, Hirata A, Komagata K (1991) *Sympodiomycopsis*: a new yeast-like anamorph genus with basidiomycetous nature from orchid nectar. Antonie van Leeuwenhoek J Microbiol 59: 95–108

Suh S-O, Hirata A, Sugiyama J, Komagata K (1993) Septal ultrastructure of basidiomycetous yeasts and their taxonomic implications with observations on the ultrastructure of *Erythrobasidium hasegawianum* and *Sympodiomycopsis paphiopedili*. Mycologia 85: 30–37

Suzuki M, Nakase T (1988) The distribution of xylose in the cells of ballistosporous yeasts: application of high performance liquid chromatography without derivatization to the analysis of xylose in whole cell hydrolysates. J Gen Appl Microbiol 34: 95–103

Tahoun MK, El-Merleb Z, Salam A, Youssef A (1987) Biomass and lipids from lactose or whey by *Trichosporon beigelii*. Biotechnol Bioeng 24: 358–360

Taylor JW, Natvig DO (1987) Isolation of fungal DNA. In: Fuller MS, Jaworski A (eds) Zoosporic fungi in teaching and research. Southeastern Publ Corp, Athens, pp 252–258

Templeton AR (1983) Systematics of basidiomycetes based on 5S rRNA sequences and other data. Nature (Lond) 303: 731

Timberlake WE (1978) Low repetitive DNA content in *Aspergillus nidulans*. Science 202: 973–975

Török T, Royer C, Rockhold D, King AD (1992) Electrophoretic karyotyping of yeasts, and southern blotting using whole chromosomes as templates for the probe preparation. J Gen Appl Microbiol 38: 313–325

Toyama R, Okayama H (1990) Human chorionic gonadotropin α and human cytomegalovirus promoters are extremely active in the fission yeast *Schizosaccharomyces pombe*. FEBS Lett 268: 271–221

Treco DA (1989) Preparation of yeast DNA. In: Ausubel FM (ed) Current protocols in molecular biology Wiley, New York, pp 13.11.1–13.11.5

Uno I, Matsumoto K, Adachi K, Ishihana T (1984) Characterization of cyclic AMP-requiring yeast mutants altered in the catalytic subunit of proteinkinase. J Biol Chem 259: 12508–12513

Valadon LRG (1976) Carotenoids as additional taxonomic characters in fungi: a review. Trans Br Mycol Soc 67: 1–15

Vancanneyt M, Van Lerberge E, Berny J-F, Hennebert GL, Kersters K (1992) The application of whole-cell protein electrophoresis for the classification and identification of basidiomycetous yeast species. Antonie van Leeuwenhoek J Microbiol 61: 69–78

Vandamme EJ, Derycke DG (1983) Microbial inulases: fermentation process, properties, and applications. Adv Appl Microbiol 29: 139–175

van den Berg JA, van der Laken KJ, van Ooyen AJJ, Renniers TCHM, Rietveld K, Schaap A, Brake AJ, Bishop RJ, Schultz K, Moyer D, Richman M, Shuster JR (1990) *Kluyveromyces* as a host for heterologous gene expression: expression and secretion of prochymosin. Bio/technology 8: 135–139

Van de Peer Y, Hendriks L, Goris A, Neefs J-M, Vancanneyt M, Kersters K, Berny J-F, Hennebert G, De Wachter R (1992) Evolution of basidiomycetous yeasts as deduced from small ribosomal subunit RNA sequences. Syst Appl Microbiol 15: 250–258

van der Walt JP (1987) The yeasts – a conspectus. Stud Mycol 30: 19–31

van der Westhuizen JPJ, Kock JLF, Smith EJ (1991) The potential use of cellular long-chain fatty acid composition in the taxonomy of the carotenoid pigment producing genera *Rhodosporidium* Banno and *Rhodotorula* Harrison. Syst Appl Microbiol 14: 282–290

Vaughan Martini A (1991) Intraspecific discontinuity within the yeast species *Cryptococcus albidus* as revealed by nDNA/nDNA reassociations. Exp Mycol 15: 140–145

Vaughan Martini A, Kurtzman CP (1985) Deoxyribonucleic acid relatedness among species of the genus *Saccharomyces* sensu stricto. Int J Syst Bacteriol 35: 508–511

Vaughan Martini A, Martini A (1987) Three newly delimited species of *Saccharomyces* sensu stricto. Antonie van Leeuwenhoek J Microbiol 53: 77–84

Vaughan Martini A, Martini A, Cardinale G (1993) Electrophoretic karyotyping as a taxonomic tool in the genus *Saccharomyces*. Antonie van Leeuwenhoek 63: 145–156

Vezinhet F, Blondin B, Hallet J-N (1990) Chromosomal DNA patterns and mitochondrial DNA polymorphism as tools for identification of enological strains of *Saccharomyces cerevisiae*. Appl Microbiol Biotechnol 32: 568–571

Vilgalys R, Hester M (1990) Rapid genetic identification and mapping of enzymatically amplified ribosomal DNA from several *Cryptococcus* species. J Bacteriol 172: 4238–4246

Viljoen BC, Kock JLF, Lategan PM (1986) Long-chain fatty acid composition of selected genera of yeasts belonging to the Endomycetales. Antonie van Leeuwenhoek J Microbiol 52: 45–51

Vishniac HS, Kurtzman CP (1992) *Cryptococcus antarcticus* sp. nov. and *Cryptococcus albidosimilis* sp. nov., basidioblastomycetes from Antarctic soils. Int J Syst Bacteriol 42: 547–553

Vollrath D, Davis RW (1987) Resolution of DNA molecules greater than 5 megabases by contour-clamped homogeneous electric fields. Nucleic Acids Res 15: 7865–7876

Weijman ACM (1979) Carbohydrate composition and taxonomy of *Geotrichum*, *Trichosporon* and allied genera. Antonie van Leeuwenhoek J Microbiol 45: 119–127

Weijman ACM, Golubev WI (1987) Carbohydrate patterns and taxonomy of yeasts and yeast-like fungi. Stud Mycol 30: 361–371

Weijman ACM, Rodrigues de Miranda L (1983) Xylose distribution within and taxonomy of the genera Bullera and *Sporobolomyces*. Antonie van Leeuwenhoek J Microbiol 49: 559–562

Weijman ACM, Rodrigues de Miranda L, van der Walt JP (1988) Redefinition of *Candida* Berkhout and the consequent emendation of *Cryptococcus* Kützing and *Rhodotorula* Harrison. Antonie van Leeuwenhoek J Microbiol 54: 545–553

West M, Emerson GW, Sullivan PA (1990) Purification and properties of two lactose hydrolases from *Trichosporon cutaneum*. J Gen Microbiol 136: 1483–1490

White TJ, Bruns T, Lee S, Taylor J (1990) Amplification and direct sequencing of fungal ribosomal RNA genes for phylogenetics. In: Innis MA, Gelfand DH, Sninsky JJ, White TJ (eds) PCR protocols: a guide to methods and applications. Academic Press, New York, pp 315–322

Wickerham LJ (1951) Taxonomy of yeasts. Tech Bull US Dept Agric 1029: 1–56

Wiley EO (1981) Phylogenetics. The theory and practice of phylogenetics systematics. Wiley, New York

Winge O, Roberts C (1949) Inheritance of enzymatic characters in yeasts, and the phenomenon of long-term adaptation. CR Trav Carlsberg Sér Physiol 24: 263–315

Woese CR (1987) Bacterial evolution. Microbiol Rev 51: 221–271

Wong GJ (1987) A comparison of the mating system of *Tremella mesenterica* and other modified bifactorial species. Stud Mycol 30: 431–441

Yamada Y, Kawasaki H (1989a) The molecular phylogeny of the Q_8-equipped basidiomyc-etous yeast genera *Mrakia* Yamada et Komagata and *Cystofilobasidium* Oberwinkler et Bandoni based on the partial sequences of 18S and 26S ribosomal ribonucleic acids. J Gen Appl Microbiol 35: 173-183

Yamada Y, Kawasaki H (1989b) The genus *Phaffia* is phylogeneticlly separate from the genus *Cryptococcus* (Cryptococcaceae). Agric Biol Chem 53: 2845-2846

Yamada Y, Komagata K (1987) *Mrakia* gen. nov., a heterobasidiomycetous yeast genus for the Q_8-equipped, selfsporulating organisms which produce a unicellular metabasidium, formerly classified in the genus *Leucosporidium*. J Gen Appl Microbiol 33: 455-457

Yamada Y, Kondo K (1973) Coenzyme Q system in the classification of the yeast genera *Rhodotorula* and *Cryptococcus*, and the yeast-like genera *Sporobolomyces* and *Rhodosporidium*. J Gen Appl Microbiol 19: 59-77

Yamada Y, Matsumoto A (1988) An electrophoretic comparison of enzymes in strains of species in the genus *Mrakia* Yamada et Komagata (Filobasidiaceae). J Gen Appl Microbiol 34: 201-208

Yamada Y, Nakagawa (1990) The molecular phylogeny of the basidiomycetous yeast species *Leucosporidium scottii* based on the partial sequences of 18S and 26S ribosomal ribo-nucleic acids. J Gen Appl Microbiol 36: 63-68

Yamada Y, Nakagawa Y (1992) The phylogenetic relationships of some heterobasidiomycetous yeast species based on the partial sequences of 18S and 26S ribosomal RNAs. J Gen Appl Microbiol 38: 559-565

Yamada Y, Ohishi T, Kondo I (1973) Coenzyme Q system in the classification of the ascosporogenous yeast genera *Hansenula* and *Pichia*. J Gen Appl Microbiol 19: 189-208

Yamada Y, Banno I, von Arx JA, van der Walt JP (1987) Taxonomic significance of the coenzyme Q system in yeasts and yeast-like fungi. Stud Mycol 30: 299-308

Yamada Y, Kawasaki H, Itoh M, Banno I, Nakase T (1988) *Tsuchiyaea* gen. nov., an anamor-phic yeast genus for the Q_9-equipped organisms whose reproduction is either by enteroblastic budding or by the formation of conidia which are disjointed at a septum in the mid-region of the sterigma and whose cells contain xylose. J Gen Appl Microbiol 34: 507-510

Yamada Y, Nakagawa Y, Banno I (1989a) The phylogenetic relationship of the Q_9-equipped species of the heterobasidiomycetous yeast genera *Rhodosporidium* and *Leucosporidium* based on the partial sequences of 18S and 26S ribosomal ribonucleic acids: the proposal of a new genus *Kondoa*. J Gen Appl Microbiol 35: 377-385

Yamada Y, Kawasaki H, Nakase T, Banno I (1989b) The phylogenetic relationship of the conidium-forming anamorphic yeast genera *Sterigmatomyces*, *Kurtzmanomyces*, *Tsuchiyaea*, and *Fellomyces*, and the teleomorphic yeast genus *Sterigmatosporidium* on the basis of the partial sequences of 18S and 26S ribosomal ribonucleic acids. Agric Biol Chem 53: 2993-3001

Yamada Y, Nakagawa Y, Banno I (1990a) The molecular phylogeny of the Q_{10}-equipped species of the heterobasidiomycetous yeast genus *Rhodosporidium* Banno based on the partial sequences of 18S and 26S ribosomal ribonucleic acids. J Gen Appl Microbiol 36: 435-444

Yamada Y, Nagahama T, Kawasaki H, Banno I (1990b) The phylogenetic relationships of the genera *Phaffia* Miller, Yoneyama et Soneda and *Cryptococcus* Kützing emend. Phaff et Spencer (Cryptococcaceae) based on the partial sequences of 18S and 26S ribosomal ribonucleic acids. J Gen Appl Microbiol 36: 403-414

Yamada Y, Nagahama T, Banno I (1991a) The molecular phylogeny of the Q_9-equipped ascomycetous teleomorphic yeast genus *Debaryomyces* Lodder et Kreger-van Rij based on the partial sequences of 18S and 26S ribosomal ribonucleic acids. J Gen Appl Microbiol 37: 277-288

Yamada Y, Nagahama T, Banno I, Giménez-Jurado G, van Uden N (1991b) The phylogenetic relationship of *Kurtzmanomyces tardus* Giménez-Jurado et van Uden

(Cryptococcaceaea) based on the partial sequences of 18S and 26S ribosomal RNA's. J Gen Appl Microbiol 37: 321–324

Yamazaki M, Goto S, Komagata K (1983) An electrophoretic comparison of the enzymes of *Saccharomyces* yeasts. J Gen Appl Microbiol 29: 305–318

Yamazaki M, Goto S, Komagata K (1985) Taxonomic studies of the genus *Tilletiopsis* on physiological properties and electrophoretic comparison of enzymes. Trans Mycol Soc Jpn 26: 13–22

Ykema A, Verbree EC, Kater MM, Smith H (1988) Optimization of lipid production in the oleaginous yeast *Apiotrichum curvatum* in whey permeate. Appl Microbiol Biotechnol 29: 211–218

Ykema A, Kater MM, Smit H (1989a) Lipid production in whey permeate by an unsaturated fatty acid mutant of the oleaginous yeast *Apiotrichum curvatum*. Biotechnol Lett 11: 477–482

Ykema A, Verbree EC, Nijkamp HJH, Smit H (1989b) Isolation and characterization of fatty acid autotrophs from the oleaginous yeast *Apiotrichum curvatum*. Appl Microbiol Biotechnol 32: 76–84

Zache G, Rehm H-J (1989) Degradation of phenol by a coimmobilized entrapped mixed culture. Appl Microbiol Biotechnol 30: 426–432

Zhang TY, Smith CL, Cantor CR (1991) Secondary pulsed field gel electrophoresis: a new method for faster separation of larger DNA molecules. Nucleic Acids Res 19: 1291–1296

CHAPTER 2

Protoplast Fusion of Yeasts

Martin Zimmermann[1] and Matthias Sipiczki[2]

1
Introduction

This chapter deals with the possible use of protoplast fusion for hybridization and genetic analysis of yeasts. For a more general overview the reader is referred to a review by Ferenczy (1981) on microbial protoplast fusion.

Protoplast fusion has been used in yeast genetics since 1976, when Sipiczki reported on the first fusion of *Schizosaccharomyces pombe* strains at a conference in Schliersee, Germany. The first papers on the use of protoplast fusion in yeasts date from 1977 (Ferenczy and Maraz 1977; Fournier et al. 1977; Sipiczki and Ferenczy 1977a,b; van Solingen and van der Plaat 1977). At that time, protoplast fusion was used mostly in fungi and plants. Since that time, it has become an important tool in the genetics of nonconventional (i.e., non-*Saccharomyces*) and industrial yeasts.

The general rationale of protoplast fusion is that the cell walls are removed enzymatically, so that protoplasts are formed. These protoplasts are kept in a medium that is stabilized osmotically. Under the influence of a fusogenic agent such as polyethylene glycol or a strong electric field (Weber et al. 1981), protoplasts will aggregate and subsequently fuse. A mixture of calcium ions and PEG has been used as fusogenic agent since the first reports on protoplast fusion in yeast. This method had found wide use in plants and fungi before. The process of protoplast fusion will bring together the cytoplasms, organelles, and nuclei of the different strains, even in cases where a sexual hybridization is impossible. If a suitable selection is available, fusants or hybrids can be isolated.

Strains from one species, from the same genus, or from different genera have been fused, and viable fusants have been obtained. The resulting fusants are then classified as intraspecific, interspecific, or intergeneric, respectively. As yeast taxonomy is still subject of intense discussion (see Chap. 1, this Vol.), this classification should be considered with some caution.

The fusants can be analyzed with regard to their morphology, physiology, and genetics. A genetic analysis is possible when the fusant can undergo meiosis or

[1] Institut für Biologie IV (Mikrobiologie) der Rheinisch-Westfälischen Technischen Hochschule Aachen, Worringer Weg, 52056 Aachen, Germany
[2] Department of Genetics, L.K. University, P.O. Box 56, 4010 Debrecen, Hungary

when mitotic segregation, either spontaneous or induced by haploidizing agents, can be observed. Haploidizing agents are for instance benomyl (an agent disturbing the assembly of the microtubules), p-fluorophenylalanine (Sipiczki and Ferenczy 1977a,b; van de Broock et al. 1983) or UV light (Whelan et al. 1980). Sometimes, the fusant can be mated to another strain and the resulting progeny can be analyzed.

Protoplast fusion has been used for genetical research concerning nuclear and cytoplasmatic genes and for strain improvement. In the following, some results will be highlighted.

2
Transfer of Cytoplasmic Genes

Fusing *Saccharomyces cerevisiae* strains, Ferenczy and Maraz (1977) demonstrated that mitochondrial genes can be transferred irrespective of the mating type. Maraz and Ferenczy (1980) were able to obtain protoplasts from *Saccharomyces cerevisiae* which contained only mitochondria, but no nuclei. They used a centrifugation step to remove the nuclei containing protoplasts from those without nuclei. When fusing these small protoplasts with protoplasts containing nuclei, a selective transfer of mitochondrial genomes was accomplished. Working on *Schizosaccharomyces pombe*, Lückemann et al. (1979) showed that the average frequency of transmission and recombination of mitochondrial genes was the same after protoplast fusion as after normal mating. In individual clones, however, the frequencies were very different; a strong influence of the regeneration time, not of the mating process itself, was demonstrated. These findings were supported by Maraz and Subik (1981), who showed for *Saccharomyces cerevisiae* that the behavior of mitochondrial genes was identical in sexual and somatic hybridization experiments. These results opened the way for genetic analysis of mitochondrial genes in cases where sexual mating was impossible. Allmark et al. (1978) could differentiate mitochondrially and nuclearily encoded respiratory deficiencies in *Kluyveromyces lactis*. After protoplast fusion of *Saccharomycopsis lipolytica* (*Yarrowia lipolytica*) strains, Matsuoka et al. (1982) obtained only fusants where the cytoplasms had fused (called cybrids). They could observe the transfer of resistance to oligomycin; karyogamy did not take place.

DNA killer plasmids could be transferred from *Kluyveromyces lactis* to *Saccharomyces cerevisiae* where they were stable and expressed (Gunge and Sakaguchi 1981) and to *Candida pseudotropicalis* (Sugisaki et al. 1985). Goodey and Bevan (1983) obtained fusion products of *Saccharomyces cerevisiae* which contained the nuclei of one parent and the combined cytoplasms of both parents. They could demonstrate the transfer of the dsRNA (double-stranded) killer plasmid as well as of the 2-μ plasmid. DsRNA killer plasmids could also be transferred by protoplast fusion into brewing yeasts and used as selection markers (Vondrejs et al. 1982; Bendova et al. 1983; Röcken 1984).

3
Production of Polyploid Strains

It was expected that technical yeasts might be improved by polyploidization. Johannsen et al. (1985) tried to improve the ability of *Candida shehatae* to ferment xylose to ethanol by producing polyploid strains. Using mutant strains as fusion partners, they produced diploid, triploid, and tetraploid fusants. They observed karyogamy and recombination of the parental markers among the progeny of the fusants. The ethanol production was slightly improved. Gupthar (1987) tried to improve *Pichia stipitis* by a similar approach. The resulting fusants were aneuploid, meiotic and mitotic segregation was possible, and proved that karyogamy had taken place; the ethanol fermentation, however, was not improved.

A *Candida blankii* strain was used to produce single cell protein from bagasse. As its cells were rather small, Gericke and van Zyl (1992) tried successfully to facilitate the cell harvest by producing polyploid strains.

The carotinoid production of *Phaffia rhodozyma* was improved by Chun et al. (1992) after polyploidization. The fusants were analyzed by mitotic segregation, proving that karyogamy had taken place.

From the haplontic yeast *Schizosaccharomyces pombe* polyploid hybrids were constructed by repeated protoplast fusion. The ploidy could be increased up to tetraploidy; the fusants, however, were not very stable and often mitotic chromosome losses occurred (Molnar and Sipiczki 1993).

4
Fusion of Strains with Identical Mating Type

Strains from heterothallic yeasts can mate only with a partner having the opposite mating type. Protoplast fusion can overcome that barrier. Stable diploid *Schizosaccharomyces pombe* strains with h⁻ mating type were obtained by Sipiczky and Ferenczy (1977b). The strains could be mated with h⁺ strains and showed recombination of parental markers. In *Rhodosporidium*, Sipiczky and Ferenczy (1977a) induced hybridization and recombination without a heterokaryotic phase when they fused two "a" strains. Stahl (1978) showed that the mating types of *Saccharomycopsis* (*Yarrowia*) *lipolytica* control only the first reaction of cell recognition, and not karyogamy or sporulation. He obtained +/+ diplonts which sporulated and showed recombination of parental genes. The results are different from those observed in *Saccharomyces cerevisiae* or *Schizosaccharomyces pombe*.

Maraz et al. (1978) could surmount the mating barrier in *Saccharomyces cerevisiae* fusing strains of identical mating type. The resulting diploid had the same mating type as the parents and did not sporulate, but could copulate with a strain of the opposite mating type. Arima and Takano (1979) found that multiple fusions took place, resulting in diploid, triploid, and tetraploid *Saccharomyces cerevisiae* strains from the same experiment. They could demonstrate that in the fusants karyogamy had occurred. To this end, they mated the fusants to strains of opposite mating type and analyzed the resulting spores.

By using electrophoretic karyotyping, Hoffmann et al. (1987) extended the analysis to chromosomes for which no marker was present. They found diploid to tetraploid *Saccharomyces cerevisiae* hybrids, and the banding patterns correlated with the genetical analysis.

5
Establishment of Parasexual Genetic Systems

Yeast species for which a sexual mating is missing or unknown can be hybridized by protoplast fusion. The resulting fusants can undergo mitotic segregation and the data from this segregation can be used to establish genetic maps. Fournier et al. (1977) obtained hybrids from *Candida tropicalis* and observed recombination events. Whelan et al. (1980) introduced UV light as an agent to induce mitotic segregation in *Candida albicans*. Poulter et al. (1981) established a parasexual cycle in *Candida albicans*, using four markers and demonstrating mitotic crossing-over. Kakar and Magee (1982) demonstrated complementation groups in auxotrophic mutants of *Candida albicans* by protoplast fusion.

Sarachek and Rhoads (1981) determined the ability of fusants to regenerate as a function of the regenerative capacity of both parents. Fusants could regenerate, even if one partner could hardly do so. They showed that in the primary event of fusion, many protoplasts interact, a finding that was supported by the production of polyploid *Saccharomyces cerevisiae* fusants. After protoplast fusion in *Candida albicans*, more or less stable heterokaryons are formed. The transfer of genetic material between the nuclei is temperature-dependent (Sarachek and Weber 1984) and mostly partial hybrids originate. The heterokaryotic cells can produce segregant-defective variants (Sarachek and Weber 1986), indicating significant genetic instabilities.

A review of Scherer and Magee (1990) shows how far genetics of *Candida albicans* has progressed. Similar experiments have been carried out in other imperfect yeasts. Klinner et al. (1984) have demonstrated linkage groups of auxotrophic markers in *Candida maltosa*. When using mitotic segregation induced by UV, Klinner and Boettcher (1985) observed chromosomal rearrangements after fusion, i.e., different fusant clones gave reproducibly different cosegregation frequencies for the markers. This finding paralleled the observations of Boettcher et al. (1979), who after protoplast fusion found cosegregation of markers that were not linked as determined by sexual analysis in *Pichia guilliermondii*. Fusants in *Pichia guilliermondii* behaved differently from hybrids obtained by mating. Further work on *Pichia guilliermondii* showed that besides homokaryotic fusants heterokaryotic cells originated (Klinner and Boettcher 1984a,b). The results depended on the strains used; the fusants seemed aneuploid and recombination of parental markers occurred. This strain specificity also occurred in fusions within *Rhodosporidium toruloides* (Becher and Boettcher 1979). The stability of the fusants depended on the strains used.

Parasexual hybridizations have also been done in *Torulopsis glabrata* (Whelan and Kwong-Chung 1987) and *Candida boidinii* (Lakhchev et al. 1992). Corner and Poulter (1989) have used protoplast fusion of *Candida tropicalis* strains, auxotrophic for adenine with *Candida albicans* strains with known *ade1* and *ade2* genes to establish complementation groups for the adenine defects in Candida tropicalis. Parasexual genetics by protoplast fusion have allowed the analysis of imperfect yeasts; the results, however, show that one must take care when genetic maps are established. The results can be specific for a strain rather than a species and, even within one strain, different hybrid clones can yield different results.

6
Fusion of Strains Belonging to Different Species or Genera

As it became obvious that sexual barriers could be overcome by protoplast fusion, many groups tried to hybridize yeasts from different species or genera. One important objective of these experiments was to transfer useful traits such as the ability to ferment lactose or xylose into industrial strains of *Saccharomyces cerevisiae*. When fusing auxotrophic strains from *Kluyveromyces lactis* and *Saccharomyces cerevisiae*, Galeotti and Clark-Walker (1983) obtained hybrids which had most of their genomes from the *Kluyveromyces* parent. They showed that a fragment of chromosome XV from *Saccharomyces cerevisiae* must have been integrated into the genome of the hybrid.

Upon fusion of a *Saccharomyces cerevisiae* brewing yeast with *Kluyveromyces lactis* (Taya et al. 1984), an intermediate hybrid was obtained. It showed complementation of natural markers, i.e., auxotrophy for nicotinic acid, pantothenate, and a good fermentation of lactose. Farahnak et al. (1986) fused *Saccharomyces cerevisiae* with *Kluyveromyces fragilis* and obtained hybrids with good fermentation of lactose and high ethanol tolerance. Fusants of *Saccharomyces cerevisiae* and *Zygosaccharomyces fermentati* showed combined traits from both parents, as they were able to grow on lactate and cellobiose (Pina et al. 1986). Fusants of *Saccharomyces cerevisiae* with *Candida shehatae* or *Pichia stipitis* were not stable, and the ethanol production was not improved (Gupthar 1992). Hybrids between *Candida utilis* and *Saccharomyces cerevisiae* were produced and analyzed by Perez et al. (1984). During mitotic segregation, the chromosomes from *Saccharomyces cerevisiae* were lost rapidly. Genes from *Saccharomyces cerevisiae* segregated together in the hybrid, though they are not linked in *Saccharomyces cerevisiae*. One hybrid could be mated to a *Saccharomyces cerevisiae* strain (Perez and Benitez 1986) and the resulting mating product produced ascospores. Some of these ascospores showed markers from both parents of the fusion experiment. The spores, however, could no longer be mated, since they were impaired in karyogamy. De Richard et al. (1984) fused respiratory deficient strains from *Candida utilis* with auxotrophic strains from *Saccharomyces cerevisiae*. They found respiratory competent fusants which strongly resembled the *Candida utilis* parent. They concluded that only mitochondrial genes had been transferred.

As it is often difficult to introduce auxotrophic markers into industrial *Saccharomyces cerevisiae* yeasts, Spencer et al. (1980) used respiratory-deficient mutants of *Saccharomyces cerevisiae* as partners for fusion experiments. They could restore mitochondrial functions in these yeasts by fusion with other yeasts from different species (Spencer and Spencer 1981). A fusion of *Saccharomyces diastaticus* with *Hansenula capsulata* yielded sporulating hybrids. Hybrids between *Saccharomyces diastaticus* and yeasts from other species or genera (*Hansenula capsulata, Hansenula wingei, Torulopsis glabrata, Candida pseudotropicalis, Saccharomyces rouxii, Saccharomyces kluyveri, Saccharomyces bayanus*) were analyzed by DNA-DNA reassociation. It was found that the hybrids had high homology to the *Saccharomyces diastaticus* parent and only very little homology to the other parent. There was one exception: a hybrid between *Torulopsis glabrata* and *Saccharomyces diastaticus* had only little homology to both parents. The authors concluded that only few chromosomes or parts of chromosomes had been transferred into the dominant partner.

Fusion of *Saccharomyces cerevisiae* with *Schwanniomyces castellii* (Tamaki 1986) yielded sporulating hybrids. It was postulated that both strains were closely related as the spores showed perfect disomic segregation of the auxotrophic markers from the parents. Hintz (1987) did not obtain fusion products between these species, though she could show by labeling of the protoplasts that fusion had taken place. Two days after the fusion experiment, she observed that the nuclei did not fuse. When both parental strains were used for intraspecific fusions, the nuclei fused much earlier.

The dominance of one fusion partner had already been demonstrated by Sipiczki (1979) in a fusion of *Schizosaccharomyces pombe* with *Schizosaccharomyces octosporus*. The hybrids were unstable and resembled the *Schizosaccharomyces octosporus* parent.

The same phenomenon was observed by Groves and Oliver (1984) after the fusion of *Kluyveromyces lactis* and *Yarrowia lipolytica*. The hybrids had a DNA content almost equal to the sum of both parents, but the genome was almost exclusively derived from the *Yarrowia* parent, as shown by analytical density-gradient centrifugation of the DNA and two-dimensional electrophoresis of the cell proteins.

Hybrids between *Kluyveromyces lactis* and *Kluyveromyces fragilis* often had a DNA content lower than the sum of the two parents, and the mitochondrial DNA came exclusively from the *Kluyveromyces lactis* parent (Whittaker and Leach 1978). Johannsen et al. (1984) fused several strains from the genus *Kluyveromyces* and found only hybrids that were unstable on rich medium.

After fusion of *Pichia stipitis* with *Candida shehatae*, Gupthar and Garnett (1987) observed stable fusants. They could induce segregation and recombination of parental markers. Selebano et al. (1993) obtained hybrids between these species which were very similar to the *Pichia stipitis* parent. By means of electrophoretic karyotyping, they showed that only minor portions of DNA from the *Candida shehatae* parent had been transferred.

The transfer of single chromosomes could be demonstrated in fusion products of *Kluyveromyces marxianus* with its anamorph *Candida macedoniensis* by Zimmermann et al. (1988) using electrophoretic karyotyping. Fusion products between *Saccharomyces cerevisiae* and *Kluyveromyces marxianus* were very similar to the *Kluyveromyces marxianus* parent (Witte et al. 1989). Attempts to determine the *Saccharomyces cerevisiae* part in the fusants' genomes had failed. Hybrids were obtained by illegitimate mating of *Kluyveromyces marxianus* and *Kluyveromyces thermotolerans*. The analysis by means of physiological analysis, isoenzyme patterns, DNA reassociation, and RFLPs of rDNA showed that only incomplete addition of the parental genomes had taken place (Vaughan-Martini et al. 1987). The hybridization process was described as some sort of transformation rather than copulation. Provost et al. (1978) fused *Candida tropicalis* and *Saccharomycopsis lipolytica* and found unstable hybrids whose genomes were obviously composed mainly of the genome of one parent and only minor portions of the other one. Hybrids between *Candida boidinii* and *Candida tropicalis* had electrophoretic karyotypes resembling that of the *Candida boidinii* parent (Kobori et al. 1991).

Skala et al. (1988) fused *Saccharomyces cerevisiae* with *Saccharomyces fermentati*. Two types of hybrids were recovered. One type seemed to contain the total *Saccharomyces fermentati* genome plus one additional *Saccharomyces cerevisiae* chromosome. The other type was thought to have the complete *Saccharomyces cerevisiae* genome plus one chromosome from *Saccharomyces fermentati*.

All these results indicate that interspecific and intergeneric hybridization is possible in yeasts. Complete addition of genomes or integration of intact chromosomes, however, seems to be a rare event. Integration of minor parts of DNA from one parent into the chromosomes of the other parent seems to be the more likely consequence.

7
Practical Recommendations

In the following section, practical hints for the preparation, treatment, and fusion of protoplasts are given. These hints can serve only as general guideline, as the experimental conditions must be optimized for each strain system. The first point is the choice of suitable fusion partners. As the formation of a fusant may be a rare event, the parental strains must bear stable markers to allow the selection of a fusant. Markers frequently used are different complementing auxotrophies, respiratory deficiencies, or differences in the utilization of carbon sources. The efficiency of a selective system should be checked before a fusion experiment is done.

A next point is the protoplasting of the cells. Many strains from most genera can be protoplasted by either Novozym G234 or by Zymolyase together with reducing agents. Novozym shows good results with strains containing a high amount of α-1,3-glucan, while Zymolyase works best on strains with β-1,3-glucan. Other protoplasting enzymes are for instance Lyticase, Glusulase, or Helicase. A mixture of different enzymes can help if a single preparation is ineffective. The susceptibility of a strain towards protoplasting enzymes can be used as a trait in yeast taxonomy (Bastide et al. 1975). When using Zymolyase, it is often advantageous to cleave the disulfide bonds of the proteins in the cell wall. To this end, reducing agents such as mercaptoethanol or dithiothreitol can be used. Mercaptoethanol is volatile and highly toxic; it might also injure the cell membrane (Sipiczki et al. 1985). The negative effects can be avoided if it is only used in a pretreatment and omitted in the enzyme treatment. The physiological status of the cells often has a tremendous effect on the protoplasting efficiency. Cells from the early to middle logarithmic phase are usually more easily protoplasted, so the cultivation of the cells can be optimized.

The synthesis of the cell wall can be impaired by the addition of 2-deoxy-D-glucose (DOG) to the medium prior to protoplasting. This will make the cells more susceptible to protoplasting enzymes. The treatment with 2-DOG should be kept short (Foury and Goffeau 1973; Sipiczki and Ferenczy 1978). If no conditions can be found which lead to the formation of protoplasts, a screening for producers of protoplasting enzymes can serve as a last resort (Kaul et al. 1993). The protoplasting of the cells can be monitored microscopically by adding distilled water to the protoplast suspension on the slide. Protoplasts will then burst. Cell walls can be stained by UVITEX 2B. Further information can be found in the papers by Torres-Bauza and Riggsby (1980), Stephen and Nasim (1981), Sipiczki et al. (1985), and Lin and Levin (1990). The stabilizer which creates the osmotic pressure necessary to keep the protoplasts from bursting can have some effect on the regeneration of the protoplasts. The osmotic pressure of the medium should be equal to or slightly higher than the pressure in the cytoplasm. Often 0.7–0.9 M sorbitol or 0.6–0.7 M KCl are used. The osmotic value of the buffer must also be considered. Too high a concentration of stabilizer will cause the protoplasts to shrink and impair fusion. Novozym will be greatly reduced in activity by concentrations of sorbitol exceeding 0.7 M KCl or 0.9 M sorbitol. Other substances, such as mannitol, can be tried, either alone or in a mixture with sorbitol. When trying to protoplast osmotolerant yeasts, higher concentrations of stabilizer might be needed (Spencer et al. 1988). For further information read Sipiczki et al. (1985).

A good protoplasting scheme should convert more than 99% of the cells into protoplasts within 1 h. If the cells tend to form clumps during protoplasting, one should try to increase the agitation.

When optimizing the conditions for protoplasting, the following parameters should be monitored:

• Initial number of colony-forming units (cfu) on an unstabilized medium.

• Cell number as counted in a hemocytometer as a control for the lysis of cells during protoplasting.

• Colony-forming units on an unstabilized medium after protoplasting, giving the amount of cells which were not protoplasted.

- Colony-forming units on a stabilized medium, giving, compared to step 3, the regeneration rate of the protoplasts.

- Regeneration of protoplasts on the selective medium.

- Regeneration of protoplasts on the selective medium supplemented according to the markers, giving some impression of the stress that the selective medium imposes on the regenerating cells.

The aim of the optimization can be defined as follows:

- The cell number should remain constant during protoplasting.

- The cfu on unstabilized medium should decrease by some two orders of magnitude.

- The cfu on stabilized rich medium should equal the cfu prior to protoplasting.

- There should be no regeneration on selective medium.

- The regeneration on the selective medium supplemented for the markers should be as good as on rich medium.

To reach this aim, treat the protoplasts as carefully as possible, i.e., keep the protoplasting step as short as possible with as little enzyme as possible, do not vortex the protoplasts, pipette them slowly through wide-bore pipettes, centrifuge them as gently as possible (1000 rpm in a standard laboratory centrifuge), and do not subject them to elevated temperatures. To induce protoplast fusion, protoplasts from both strains are mixed in similar numbers, centrifuged, and suspended in a solution of 25–30% PEG 6000 and 10–100 mM Ca^{2+}. It is well established that PEG and Ca^{2+} are both necessary to induce fusion. Work from Kobori et al. (1991) indicates that filter-sterilized PEG leads to much higher fusion rates than PEG that was autoclaved. Kavanagh et al. (1991) found that the use of calcium propionate can greatly improve hybrid recovery.

After the PEG treatment, the protoplasts are transferred to regeneration medium. Usually, it is better to embed the protoplasts in regeneration agar than to use liquid media or plate the protoplasts onto the surface of a plate (Svoboda 1978). The protoplast suspension should be diluted in regeneration medium to reduce the PEG concentration below the inhibitory concentration of PEG.

As the regeneration of the protoplasts in selective medium might impose too much pressure onto the cells, alternative procedures have been developed. These involve the encapsulation of the protoplasts in alginate (Vidoli et al. 1982; Hansen et al. 1990) or agar-gelatine beads (Spencer et al. 1989). The beads can be incubated in rich medium, allowing for a better regeneration. After regeneration, the cells are transferred into selective medium. These procedures, however, allow no precise determination of the fusion rate. It is advisable to separate the beads into many flasks in order to obtain fusion products which have originated independently.

It is possible to stain the cells prior to protoplasting with different fluorescent dyes that do not impair the regeneration of the cells. After the fusion experiment, protoplasts which have fused can be detected, as they show the mixed color. In the same experiment, the nuclei can also be stained with DAPI, so their development can also be followed (Hintz 1987). The protoplasts might take several days or even weeks to regenerate, therefore the plates should be incubated in a humid atmosphere in order to keep them from drying up.

Most authors stabilize the fusion products by subculturing them several times on selective medium. During this step, one should look for sectored colonies, which indicate segregation processes. Though one normally wants to have stable fusion products, these segregants should not be discarded, as they can serve as additional proof for a fusion. The stability of the fusants can be checked by culturing on rich medium for some passages, plating on rich medium, and subsequently plating on selective media. When segregants are obtained, one should look for colonies that bear the parental markers and search for recombinants.

8
Analysis of the Fusants

Stable fusants can be analyzed further by investigating traits of the parental strains, which were not selected for. Such traits are for instance morphological features, DNA content, or isoenzyme patterns. Strain-specific gene probes, DNA-DNA reassociation, RFLP markers, RAPD markers, or electrophoretic karyotyping can be used to assess the contribution of each parent to the fusant's genome.

As an example, a protoplast fusion protocol is given that is used in a genetical course in our laboratory. It works with strains of *Saccharomyces cerevisiae Kluyveromyces marxianus, Schwanniomyces occidentalis* and *Candida macedoniensis*. Figure 1 gives the suitable controls.

8.1
Preparation of Protoplasts

1. Preculture: 24 h in YEPD, 50 ml.

2. Main culture: 16 h in YEPD, 500 ml, inoculated with 5 ml preculture.

3. Centrifuge cells from 200 ml culture and resuspend cells in 8 ml EDTA solution, 100 mM pH 7.5.

4. Wash twice with EDTA solution.

5. Wash once with SORTRISCA (1.2 M sorbitol, 10 mM $CaCl_2$, 100 mM TRIS/HCl, pH 7.5).

6. Resuspend cells in 8 ml SORTRISCA.

7. Add 20 μl mercaptoethanol and 1 ml Zymolyase, 1 mg/ml in SORTRISCA, filter-sterilized CAUTION: mercaptoethanol is highly toxic!

8. Incubate at 35 °C with mild agitation.

9. Check for protoplasts by microscopy.

10. Wash protoplasts five times with SORTRISCA.

11. Regenerate protoplasts by:

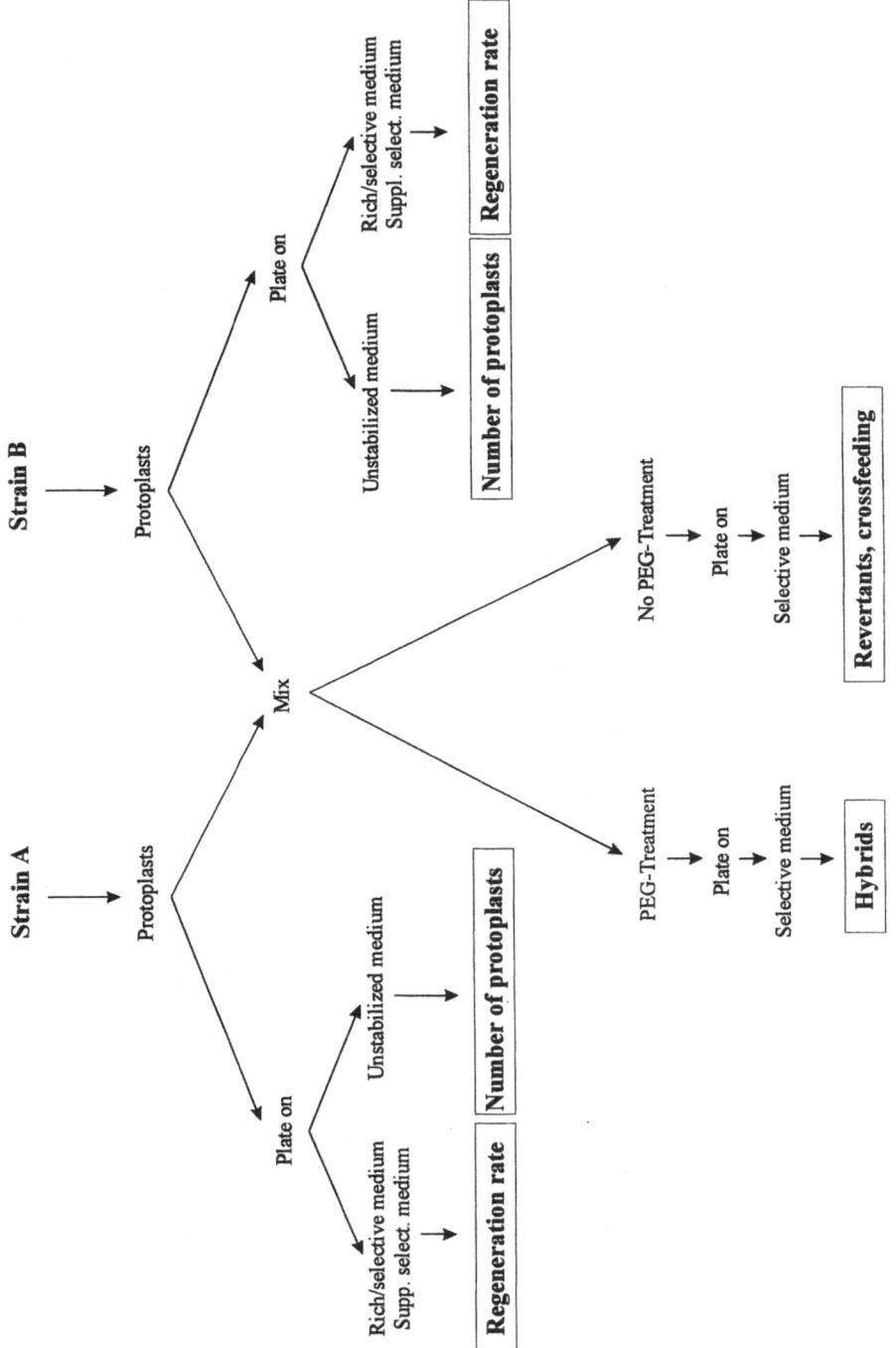

Fig. 1

a) plating on YEPD without sorbitol,

b) mixing of 100 μl suspension/dilution with 4 ml liquified YEPD agar stabilized with 1.2 M sorbitol, warmed to 50 °C, and plating on stabilized medium.

12. Incubate 5–7 days, calculate regeneration rate.

8.2
Fusion of Protoplasts

1. Prepare protoplasts from two strains.

2. Mix 1×10^8 protplasts from either strain.

3. Centrifuge the protoplasts.

4. Resuspend pellet in 5 ml PEG-solution (25–30% PEG 4000 or 6000 in 10–100 mM calcium).

5. Incubate 30 min at 30 °C.

6. Dilute PEG by adding five times 1 ml stabilized medium at intervals of 3 min.

7. Add 100 μl of suspension to 4 ml liquified, stabilized selective agar and plate onto selective agar.

Controls

• Regeneration on stabilized rich medium.

• Regeneration of a mixture of protoplasts without PEG treatment.

• Regeneration of protoplasts from either strain alone after PEG treatment on rich and on selective medium.

9
Additional Protocols

9.1
Alginate Encapsulation of Protoplasts
(Vidoli et al. 1982; Hansen et al. 1990)

1. Mix equal volumes of protoplast suspension and 5% alginate solution.

2. Pour the mixture dropwise into 100 mM $CaCl_2$, the drops will form beads.

3. Transfer the beads into several batches of rich, stabilized medium and incubate until the protoplasts have regenerated into cells.

4. Wash the beads twice with 1.2 M sorbitol.

5. Transfer the beads into selective medium and incubate several days.

6. Dissolve the beads in a solution of 2.5% sodium-hexa-meta-phosphate and plate the cells onto selective medium.

9.2
Induction of Haploidization or Mitotic Segregation by p-Fluoro-Phenylalanine (Wilson et al. 1983)

1. Inoculate 20 ml YEPD with the fusant, incubate 24 h.

2. Inoculate 50 ml YEPD containing 100 μg/ml p-fluoro-phenylalanine with 0.5 ml preculture, incubate 48 h.

3. Dilute the culture appropriately and plate onto YEPD agar, incubate some days.

4. Replica plate onto selective medium, identify auxotrophic colonies.

9.3
Staining of Cells Prior to Protoplasting (Hintz 1987)

Suitable dyes are fluorescein isothiocyanate (FITC) and rhodamin isothiocyanate (RITC), both from Sigma.

1. The dyes are dissolved in ethanol (FITC) or water (RITC) at a concentration of 2.5 mg/ml.

2. Add 3 to 4 ml of the solution to 500 ml of culture 5 h prior to protoplasting.

3. Observe the cells by fluorescent microscopy using an excitation wavelength of 410 nm. The staining remains visible for almost 40 h (Hintz 1987).

A vital stain of the nuclei is possible by adding 1 μg/ml DAPI to the culture medium.

10
Concluding Remarks

The survey of the literature has shown that possible consequences of protoplast fusion are:

- Exchange of cytoplasm and cytoplasmatic genes
- Polyploidization due to fusion of several protoplasts
- Aneuploidy due to segregation of polyploids or diploids
- Complete addition of two or more parental genomes
- Exchange of single chromosomes or addition of single chromosomes from one parent to the genome of the other parent
- Integration of small pieces of DNA from one parent into the genome of the other.

It becomes clear from this list of possible events (which is probably not complete) that the analysis of a fusant can be a formidable task. Nevertheless, protoplast fusion is a valuable tool for the hybridization and genetic analysis of yeasts which cannot be crossed by conventional methods.

References

Allmark BM, Morgan AJ, Whittaker PA (1978) The use of protoplast fusion in demonstrating chromosomal and mitochondrial inheritance of respiratory-deficiency in *Kluyveromyces lactis*, a petite-negative yeast. Mol Gen Genet 159: 297–299

Arima K, Takano I (1979) Multiple fusion of protoplasts in *Saccharomyces* yeasts. Mol Gen Genet 173: 271–277

Bastide M, Trave P, Bastide JM (1975) L'hydrolyse enzymatique de la paroi appliqué à la classification des lévures. Ann Microbiol (Inst Pasteur) 126 A: 275–294

Becher D, Boettcher F (1979) Hybridization of *Rhodosporidium toruloides* by protoplast fusion. In: Advances in protoplast research. Publishing House of the Hungarian Academy of Sciences, Budapest, pp 105–111

Bendova O, Kupcova L, Janderova B, Vondrejs V, Vernerova J (1983) Ein Beitrag zur Brauereihefehybridisierung. Monatsschr Brauwiss 36: 167–171

Boettcher F, Becher D, Klinner U, Samsonova IA, Schilova B (1979) Genetic structure of yeast hybrids constructed by protoplast fusion. In: Advances in protoplast research. Publishing House of the Hungarian Academy of Sciences, Budapest, pp 99–104

Chun SB, Chin JE, Bae S, An GH (1992) Strain improvement of *Phaffia rhodozyma* by protoplast fusion. FEMS Microbiol Lett 93: 221–226

Corner BE, Poulter RTM (1989) Interspecific complementation analysis by protoplast fusion of *Candida tropicalis* and *Candida albicans* adenine auxotrophs. J Bacteriol 171: 3586–3589

de Richard MS, van de Broock MR (1984) Protoplast fusion between a petite strain of *Candida utilis* and *Saccharomyces cerevisiae* respiratory-competent cells. Curr Microbiol 10: 117–120

Farahnak F, Seki T, Ryu DDY, Ogrydziak D (1986) Construction of lactose-assimilating and high-ethanol-producing yeasts by protoplast fusion. Appl Environ Microbiol 51: 362–367

Ferenczy L (1981) Microbial protoplast fusion. In: Glover SW, Hopwood DA (eds) Genetics as a tool in microbiology. 31st Symp Soc Gen Microbiol, University Press, Cambridge, pp 1–34

Ferenczy L, Maraz A (1977) Transfer of mitochondria by protoplast fusion in *Saccharomyces cerevisiae*. Nature 268: 524–525

Fournier P, Provost A, Bourguignon C, Heslot H (1977) Recombination after protoplast fusion in the yeast *Candida tropicalis*. Arch Microbiol 115: 143–149

Foury F, Goffeau A (1973) Combination of 2-deoxyglucose and snail-gut enzyme treatments for preparing sphaeroplasts of *Schizosaccharomyces pombe*. J Gen Microbiol 75: 227–229

Galeotti CL, Clark-Walker GD (1983) Changes of gene expression in fusion products between *Saccharomyces cerevisiae* and *Kluyveromyces lactis*. In: Nagley P, Linnane AW, Peacock WJ, Payeman JA (eds) Manipulation and expression of genes in eukaryotes. Academic Press Australia, Melbourne, pp 159–166

Gericke M, Van-Zyl WH (1992) Improvement of the cell volume of *Candida blankii* through protoplast fusion. J Ind Microbiol 10: 117–122

Goodey AR, Bevan EA (1983) Production and genetic analysis of yeast hybrids. Curr Genet 7: 69–72

Groves DP, Oliver SG (1984) Formation of intergeneric hybrids of yeast by protoplast fusion of *Yarrowia* and *Kluyveromyces* species. Curr Genet 8: 49–55

Gunge N, Sakaguchi K (1981) Intergeneric transfer of desoxyribonucleic acid killer plasmids, *pGKI1* and *pGKI2*, from *Kluyveromyces lactis* into *Saccharomyces cerevisiae* by cell fusion. J Bacteriol 147: 155–160

Gupthar AS (1987) Construction of a series of *Pichia* strains with increased DNA content. Curr Genet 12: 605–610

Gupthar AS (1992) Segregation of altered parental properties in fusions between *Saccharomyces cerevisiae* and the D-xylose fermenting yeasts *Candida shehatae* and *Pichia stipitis*. Can J Microbiol 38: 1233–1237

Gupthar AS, Garnett HM (1987) Hybridization of *Pichia stipitis* with its presumptive imperfect partner *Candida shehatae*. Curr Genet 12: 199–120

Hansen M, Roecken W, Emeis CC (1990) Construction of yeast strains for the production of low-carbohydrate beer. J Inst Brew 96: 125–129

Hintz M (1987) Versuche zur somatischen Hybridisierung von *Schwanniomyces* und *Saccharomyces* mit dem Ziel der Ethanolgewinnung aus Stärke. Dissertation, RWTH Aachen

Hoffmann M, Zimmermann M, Emeis CC (1987) Orthogonal field alternation gel electrophoresis as a means for the analysis of somatic hybrids obtained by protoplast fusion of different *Saccharomyces* strains. Curr Genet 11: 599–603

Johannsen E, Halland L, Opperman A (1984) Protoplast fusion within the genus *Kluyveromyces* van der Walt emend. van der Walt. Can J Microbiol 30: 540–552

Johannsen E, Eagle L, Bredenhann G (1985) Protoplast fusion used for the construction of presumptive polyploids of the D-xylose fermenting yeast *Candida shehatae*. Curr Genet 9: 313–319

Kakar SN, Magee PT (1982) Genetic analysis of *Candida albicans*: identification of different isoleucine-valine, methionine, and arginine alleles by complementation. J Bacteriol 151: 1247–1252

Kaul W, Rossow U, Emeis CC (1993) Screening for microorganisms with cell wall lytic activity to produce protoplast-forming enzymes. Appl Microbiol Biotechnol 39: 574–576

Kavanagh K, Walsh M, Whittaker PA (1991) Enhanced intraspecific protoplast fusion in yeast. FEMS Microbiol Lett 65: 283–266

Klinner U, Boettcher F (1984a) Hybridization of yeasts by protoplast fusion: early events after fusion in *Pichia guilliermondii*. Z Allg Mikrobiol 24: 539–544

Klinner U, Boettcher F (1984b) Hybridization of yeasts by protoplast fusion: ploidy levels of hybrids resulting from fusions in haploid strains of *Pichia guilliermondii*. Z Allg Mikrobiol 24: 533–537

Klinner U, Boettcher F (1985) Chromosomal rearrangements after protoplast fusion in the yeast *Candida maltosa*. Curr Genet 9: 619–621

Klinner U, Samsonova IA, Boettcher F (1984) Genetic analysis of the yeast *Candida maltosa* by means of induced parasexual processes. Curr Microbiol 11: 241–246

Kobori H, Takata Y, Osumi M (1991) Interspecific protoplast fusion between *Candida tropicalis* and *Candida boidinii*: characterization of the fusants. J Ferment Bioeng 72: 439–444

Lakhchev K, Penkova R, Ivanova V, Tuneva D (1992) Genetic analysis of methylotrophic yeast *Candida boidinii* PLD1. Antonie Leeuwenhoek J Microbiol 61: 185–194

Lin LF, Levin RE (1990) Relative effectiveness of yeast cell wall digesting enzymes on *Yarrowia lipolytica*. Microbios 63: 109–115

Lückemann G, Sipiczki M, Wolf K (1979) Transmission, segregation, and recombination of mitochondrial genomes in zygote clones and protoplast fusion clones of yeast. Mol Gen Genet 177: 185–187

Maraz A, Ferenczy L (1980) Selective transfer of fungal cytoplasmatic genetic elements by protoplast fusion. Curr Microbiol 4: 343–345

Maraz A, Subik J (1981) Transmission and recombination of mitochondrial genes in *Saccharomyces cerevisiae* after protoplast fusion. Mol Gen Genet 181: 131–133

Maraz A, Kiss M, Ferenczy L (1978) Protoplast fusion in *Saccharomyces cerevisiae* strains of identical and opposite mating types. FEMS Microbiol Lett 3: 319–322

Matsuoka M, Uchida K, Aiba S (1982) Cytoplasmic transfer of oligomycin resistance during protoplast fusion of *Saccharomycopsis lipolytica*. J Bacteriol 152: 530–533

Molnar M, Sipiczki M (1993) Polyploidy in the haplontic yeast *Schizosaccharomyces pombe*: construction and analysis of strains. Curr Genet 24: 45–52

Perez C, Benitez J (1986) Defective karyogamy in meiotic segregants of a *Candida utilis-Saccharomyces cerevisiae* hybrid. Curr Genet 10: 639–642

Perez C, Vallin C, Benitez J (1984) Hybridization of *Saccharomyces cerevisiae* with *Candida utilis* through protoplast fusion. Curr Genet 8: 575–580

Pina A, Calderon IL, Benitez T (1986) Intergeneric hybrids of *Saccharomyces cerevisiae* and *Zygosaccharomyces fermentati* obtained by protoplast fusion. Appl Environ Microbiol 51: 995–1003

Poulter R, Jeffery K, Hubbard MJ, Shepherd MG, Sullivan PA (1981) Parasexual genetic analysis of *Candida albicans* by spheroplast fusion. J Bacteriol 146: 833–840

Provost A, Bourguignon C, Fournier P, Heslot H (1978) Intergeneric hybridization in yeasts through protoplast fusion. FEMS Microbiol Lett 3: 309–312

Röcken W (1984) Übertragung des Killerplasmids von einer Killerhefe auf eine untergärige Bierhefe durch Protoplastenfusion. Monatsschr Brauwiss 9: 384–389

Sarachek A, Rhoads DD (1981) Production of heterokaryons of *Candida albicans* by protoplast fusion: effects of differences in proportions and regenerative abilities of fusion partners. Curr Genet 4: 221–222

Sarachek A, Weber DA (1984) Temperature-dependent internuclear transfer of genetic material in heterokaryons of *Candida albicans*. Curr Genet 8: 181–187

Sarachek A, Weber DA (1986) Segregant-defective heterokaryons of *Candida albicans*. Curr Genet 10: 685–693

Scherer S, Magee PT (1990) Genetics of *Candida albicans*. Microbiol Rev 54: 226–241

Selebano ET, Govinden R, Pillay D, Pillay B, Gupthar AS (1993) Genomic comparisons among parental and fusant strains of *Candida shehatae* and *Pichia stipitis*. Curr Genet 23: 468–471

Sipiczki M (1979) Interspecific protoplast fusion in fission yeast. Curr Microbiol 3: 37–40

Sipiczki M, Ferenczy L (1977a) Fusion of *Rhodosporidium* (*Rhodotorula*) protoplasts. FEMS Microbiol Lett 2: 203–205

Sipiczki M, Ferenczy L (1977b) Protoplast fusion of *Schizosaccharomyces pombe* auxotrophic mutants of identical mating-type. Mol Gen Genet 151: 77–81

Sipiczki M, Ferenczy L (1978) Enzymatic methods for enrichment of fungal mutants. I. Enrichment of *Schizosaccharomyces pombe* mutants. Mutat Res 50: 163–173

Sipiczki M, Heyer WD, Kohli J (1985) Preparation and regeneration of protoplasts and spheroplasts for fusion and transformation of *Schizosaccharomyces pombe*. Curr Microbiol 12: 169–174

Skala J, Luty J, Kotylak Z (1988) Interspecific protoplast fusion between the yeasts *Saccharomyces cerevisiae* and *Saccharomyces fermentati*. Curr Genet 13: 101–104

Spencer JFT, Spencer DM (1981) The use of mitochondrial mutants in hybridizaton of industrial yeasts III. Restoration of mitochondrial function in petites of industrial yeast strains by fusion with respiratory-competent protoplasts of other yeast species. Curr Genet 4: 177–180

Spencer JFT, Laud P, Spencer DM (1980) The use of mitochondrial mutants in the isolation of hybrids involving industrial yeast strains II. Use in isolation of hybrids obtained by protoplast fusion. Mol Gen Genet 178: 651–654

Spencer JFT, Spencer DM, Whittington-Vaughan P, Miller R (1983) Use of mitochondrial mutants in the isolation of hybrids involving industrial yeast strains IV. Characterization of an intergeneric hybrid, *Saccharomyces diastaticus x Hansenula capsulata*, obtained by protoplast fusion. Curr Genet 7: 159–164

Spencer JFT, Spencer DM, Bizeau C, Vaughan-Martini A, Martini A (1985) The use of mitochondrial mutants in hybridization of industrial yeast strains V. Relative parental contribution to the genomes of interspecific and intergeneric yeast hybrids obtained by protoplast fusion, as determined by DNA reassociation. Curr Genet 9: 623–625

Spencer JFT, Spencer DM, Reynolds N (1988) Genetical manipulation of non-conventional yeasts by conventional and non-conventional methods. J Basic Microbiol 28: 321–333

Spencer JFT, Spencer DM, Schiappacasse MC, Heluane H, Reynolds N, de Figueroa LI (1989) Two new methods for recovery and genetic analysis of hybrids after fusion of yeast protoplasts. Curr Microbiol 18: 285–287

Stahl U (1978) Zygote formation and recombination between like mating types in the yeast *Saccharomycopsis lipolytica* by protoplast fusion. Mol Gen Genet 160: 111–113

Stephen ER, Nasim A (1981) Production of protoplasts in different yeasts by mutanase. Can J Microbiol 27: 550–553

Sugisaki Y, Gunge N, Sakaguchi K, Yamasaki M, Tamura G (1985) Transfer of DNA killer plasmids from *Kluyveromyces lactis* to *Kluyveromyces* and *Candida pseudotropicalis*. J Bacteriol 164: 1373–1375

Svoboda A (1978) Fusion of yeast protoplasts induced by polyethylene glycol. J Gen Microbiol 109: 169–175

Tamaki H (1986) Genetic analysis of intergeneric hybrids obtained by protoplast fusion in yeast. Curr Genet 10: 491–494

Taya M, Honda H, Kobayashi T (1984) Lactose-utilizing hybrid strain derived from *Saccharomyces cerevisiae* and *Kluyveromyces lactis* by protoplast fusion. Agric Biol Chem 48: 2239–2243

Torres-Bauza LJ, Riggsby WS (1980) Protoplasts from yeast and mycelial forms of *Candida albicans*. J Gen Microbiol 119: 341–349

van de Broock MR, Sierra M, de Figueroa LC (1983) Ploidy reduction using p-fluorophenylalanine of fusion products of *Saccharomyces cerevisiae*. Curr Microbiol 8: 13–16

van Solingen P, van der Plaat JB (1977) Fusion of yeast spheroplasts. J Bacteriol 130: 946–947

Vaughan-Martini A, Sidenberg DG, Lachance MA (1987) Analysis of a hybrid between *Kluyveromyces marxianus* and *Kluyveromyces thermotolerans* by physiological profile comparison, isoenzyme electrophoresis, DNA reassociation, and restriction mapping of ribosomal RNA. Can J Microbiol 33: 971–978

Vidoli R, Yamazaki H, Nasim A, Veliky IA (1982) A novel procedure for the recovery of hybrid products from protoplast fusion. Biotechnol Lett 4: 781–784

Vondrejs V, Psenicka L, Kupcova L, Dostalova R, Janderova B, Bendova O (1982) The use of a killer factor in the selection of hybrid yeast strains. Folia Biol (Praha) 29: 372–384

Weber H, Foerster W, Berg H, Jacob HE (1981) Parasexual hybridization of yeasts by electric field stimulated fusion of protoplasts. Curr Genet 4: 165–166

Whelan WL, Kwom-Chung KJ (1987) Parasexual genetics of *Torulopsis glabrata*. J Bacteriol 169: 4991–4994

Whelan WL, Partridge RM, Magee PT (1980) Heterozygosity and segregation in *Candida albicans*. Mol Gen Genet 180: 107–113

Whittaker PA, Leach SM (1978) Interspecific hybrid production between the yeasts *Kluyveromyces lactis* and *Kluyveromyces fragilis* by protoplast fusion. FEMS Microbiol Lett 4: 31–34

Wilson JJ, Khachatourians GG, Ingledew WM (1982) Protoplast fusion in the yeast *Schwanniomyces alluvius*. Mol Gen Genet 186: 95–100

Witte W, Grossmann B, Emeis CC (1989) Molecular probes for the detection of *Kluyveromyces marxianus* chromosomal DNA in electrophoretic karyotypes of intergeneric protoplast fusion products. Arch Microbiol 152: 441–446

Zimmermann M, Hoffmann-Hintz M, Kolvenbach M, Emeis CC (1988) OFAGE banding patterns of different yeast genera and of intergeneric hybrids. J Basic Microbiol 28: 241–247

CHAPTER 3

Electrophoretic Karyotyping of Yeasts

Martin Zimmermann[1] and Philippe Fournier[2]

1
Introduction and Theory

Electrophoretic karyotyping means the separation of intact chromosomal DNA according to its size on an agarose gel. Depending on the number and size of the chromosomes present in a strain, a specific banding pattern will be obtained. In order to reach this goal, two demands must be met. First, it is important to prepare the DNA without degradation by mechanical stress or by DNAses. Second, a method for the electrophoretic separation of the extremely large molecules must be developed. Conventional DNA electrophoresis is able to separate molecules of up to 50 kilobases. Yeast chromosomes range from several hundred to several thousand kilobases.

Since the mid-1980s, these problems have been solved by a number of groups. The preparation of the DNA is usually carried out by lysing cells or protoplasts which have been encapsulated in agarose, thus preventing mechanical stress. DNAses are inhibited by a high concentration of EDTA.

For electrophoretic separation, several pulsed-field devices have been developed. Since the theory of pulsed-field gel electrophoresis (PFGE) is rather complex, a short, simplified explanation will be given. For a detailed description of PFGE techniques and theory, see Gemmill (1991) and Chu (1991). The DNA is electrophoresed through an agarose gel, the driving forces are electrical fields which are turned on alternatively. These fields are orientated towards each other under an angle of more than 90°, so the DNA will be forced to reorientate after each change of the field. Longer molecules will take more time to reorientate and therefore move were slowly than shorter ones (Fig. 1). Schwartz and Koval (1989) were able to observe the molecular dynamics of the DNA in PFGE after labeling the DNA with fluorescent dyes.

The separation process is subject to many parameters. These include the type and concentration of agarose, the temperature, the switching interval (also called

[1] Institut für Biologie IV (Mikrobiologie) der Rheinisch-Westfälischen Technischen Hochschule Aachen, Worringer Weg, 52056 Aachen, Germany
[2] Institut National de la Recherche Agronomique, Centre de Biotechnologies Agro-Industrielles, Laboratoire de Génétique Moléculaire et Cellulaire, 78850 Thiverval-Grignon, France

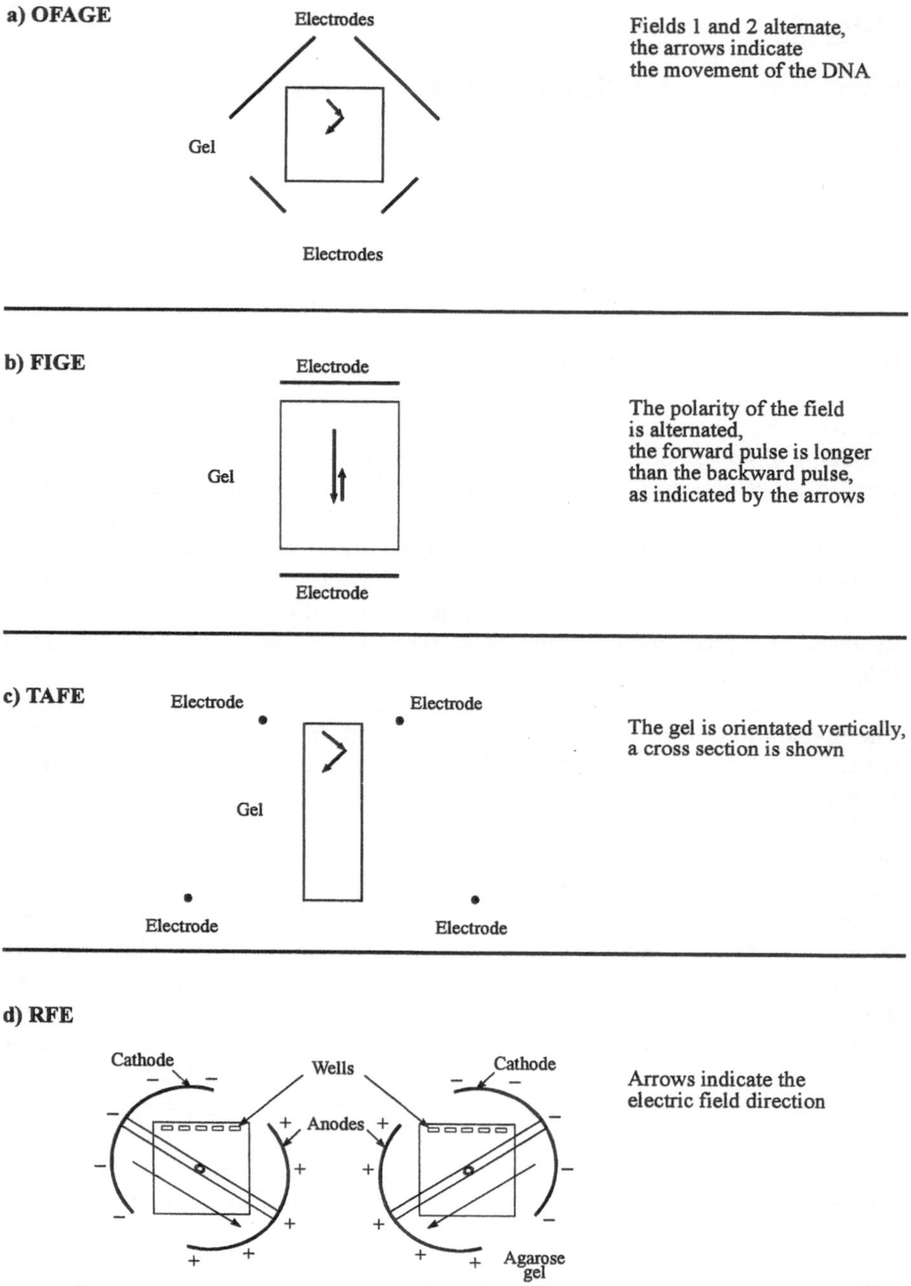

a) OFAGE

Electrodes

Gel

Electrodes

Fields 1 and 2 alternate,
the arrows indicate
the movement of the DNA

b) FIGE

Electrode

Gel

Electrode

The polarity of the field
is alternated,
the forward pulse is longer
than the backward pulse,
as indicated by the arrows

c) TAFE

Electrode   Electrode

Gel

Electrode   Electrode

The gel is orientated vertically,
a cross section is shown

d) RFE

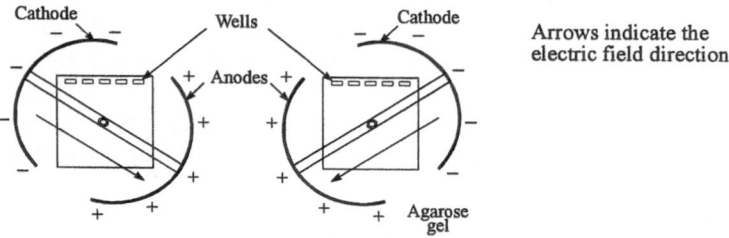

Cathode   Wells   Cathode

Anodes

Agarose
gel

Arrows indicate the
electric field direction

Fig. 1a–d. Placement of electrodes in different PFGE devices. *OFAGE* Orthogonal Field Alternation Gel Electrophoresis; *FIGE* Field-Inversion Gel Electrophoresis; *TAFE* Transverse Alternating Field Electrophoresis; *RFE* Rotating Field Electrophoresis pulse time), the field strength, the field geometry, the DNA concentration, and the DNA topology. The influence of these parameters has been investigated systematically by Cantor et al. (1988), Mathew et al. (1988a,b,c) and Doggett et al. (1992).

During the past years, several different PFGE techniques have been published, like Field-Inversion Gel Electrophoresis (FIGE; Carle et al. 1986); Contour-Clamped Homogeneous Electric Field (CHEF; Chu et al. 1986); Transverse Alternating Field Electrophoresis (TAFE; Gardiner and Patterson 1988); Rotating System (RGE or RFE; for a review see Eby 1990); Secondary Pulsed Field Gel Electrophoresis (Zhang et al. 1991) and Zero-Integrated-Field Electrophoresis (ZIFE; Noolandi and Tunnel 1992). All these techniques have proven useful for electrophoretic karyotyping of yeasts. They differ primarily in the geometry of the electrodes and the mechanisms by which the change of the fields is accomplished (see Fig. 1).

Some systems, i.e., CHEF or RGE, allow for a change of the angle between the fields during the experiment. This can result in a faster separation (Chu and Gunderson 1991).

The first electrophoretic karyotypes were obtained from *Saccharomyces cerevisiae* (Carle and Olson 1984, 1985; Schwartz and Cantor 1984). The chromosomes were separated into 11 bands whereas with variations of the initial techniques all 16 chromosomes can be visualized (Chu and Gunderson 1991). These karyotypes agreed well with the data obtained from genetic investigations. It could be shown that the DNA bands corresponded to intact chromosomes, and that different chromosomes were separated into different bands. When using as probes cloned genes whose map position was known, each band could be identified. Smith et al. (1987), as well as Vollrath and Davies (1987), published karyotypes of *Schizosaccharomyces pombe*, which proved that the separation range of PFGE could be extended to molecules of more than 5 megabases.

2
Fields of Application

2.1
Yeast Taxonomy

Sor and Fukuhara (1989) investigated the genus *Kluyveromyces*. Data from electrophoretic karyotypes were correlated to restriction patterns of mitochondrial DNA. The species *K. lactis*, *K. fragilis*, and *K. marxianus* could be differentiated. Within *K. marxianus* two groups were found. A close relationship between *K. marxianus* and its presumptive anamorph *Candida macedoniensis* was demonstrated. Here, too, significant chromosome polymorphisms were detected. Steensma et al. (1988) also showed that *K. marxianus* and *K. lactis* are different species, though they can be hybridized by mating. The banding patterns of the parental *K. marxianus* and *K. lactis* strains differed; the banding pattern of the hybrids was the sum of the parental banding patterns. An exchange of genetic material during the meiosis of the hybrids could not be demonstrated. Some degree of species-specific banding patterns was also observed in the medically important species of *Candida tropicalis, Candida krusei,* and *Candida guilliermondii* (Suzuki et al. 1988). Other taxonomical studies involved species of the genera *Saccharomyces, Hortaea, Filobasidiella* and *Malazessia* (Vaughan-Martini et al. 1993; Boekhout et al. 1993).

2.2
Study of Chromosome Polymorphisms

Strains belonging to the same species often show different banding patterns due to chromosome length polymorphism. Examples for such polymorphism are *Saccharomyces cerevisiae* (Hoffmann et al. 1987), some *Candida* yeasts (Suzuki et al. 1988; Lott et al. 1993), *Yarrowia lipolytica* (Naumova et al. 1993), *Hansenula polymorpha* (Marri et al. 1993) and *Torulaspora pretoriensis* (Oda and Tonomura 1995). The clinically important yeast *Candida albicans* was the subject of many studies, such as Magee and Magee (1987), Merz et al. (1988), Rustchenko-Bulgac et al. (1990), Rustchenko-Bulgac (1991), and Asakura et al. (1991). Here again, polymorphisms were found which could be correlated with morphological mutations and used for epidemiological studies. These polymorphisms could be correlated with alterations in the size of the rDNA cistrons as, for instance, in chromosome XII of *Saccharomyces cerevisiae* or chromosome III of *Schizosaccharomyces pombe* (Pasero and Marilley 1993). Another reason might be chromosomal translocation events that were observed in *Candida albicans* (Thrash-Bingham and Gorman 1992).

2.3
Typing of Yeast Strains

The chromosomal banding pattern of a strain is often specific enough to be used to recognize a strain in fermentation processes or in epidemiological studies. Examples are investigations concerning enological *Saccharomyces* strains (Bakalinsky and Snow 1990; Vezinhet et al. 1990; Bidenne et al. 1992; Querol et al. 1992) and clinical *Candida albicans* strains (Merz et al. 1988). Several papers come to the conclusion that the electrophoretic karyotype is superior in resolution to RFLP or isozyme pattern, that is to say that strains could be differentiated by their karyotype but not by their RFLP pattern (Merz et al. 1988; Carruba et al. 1991; Vazquez et al. 1991, 1993; Magee et al. 1992; Lehmann et al. 1992).

2.4
Genome Mapping

After a gene has been cloned, it can be assigned to electrophoretically separated chromosomes by Southern blotting and hybridization, even if a genetic map is not available. This was done, for instance, with genes from *Candida albicans* (Magee et al. 1988), DNA fragments containing potential centromeres from *Kluyveromyces lactis* (Heus et al. 1990), or several genes from *Yarrowia lipolytica* (Naumova et al. 1993).

Physical mapping of the whole genome has been done for *Saccharomyces cerevisiae* (Link and Olson 1991; Thierry and Dujon 1992) and for *Schizosaccharomyces pombe* (Fan et al. 1991). These references give protocols for the digestion of whole DNA preparations and of individual chromosomes with rare cutting enzymes such as *Not* I and *Sfi* I. The genetic data were correlated to data obtained by separation of intact chromosomes and of large restriction fragments and subsequent Southern hybridization.

Chromosome fragmentation has been used for mapping and evidenced by PFGE techniques (Vollrath et al. 1988), and this was used to show the presence of centromeres on cloned DNA fragments (Fournier et al. 1993).

Chromosomal rearrangement (transposition, recombination) were also studied with karyotyping (Chibana et al. 1994; Rustchenko et al. 1994; Suzuki et al. 1994), as well as in DNA repair (Dardalhon and Averbeck, 1995; Contopoulou et al. 1987).

2.5
Characterization of Hybrids

Electrophoretic karyotypes have also been used to analyze yeast hybrids obtained either by sexual or by somatic hybridization. In the case of *Saccharomyces* hybrids, data from the electrophoretic karyotype could be compared with data from genetic analysis. Both methods led to similar results (Hoffmann et al. 1987; Schillberg et al. 1991). Selebano et al. (1993) analyzed hybrids between *Candida shehatae* and *Pichia stipitis*, and Kobori et al. (1991) characterized fusants between *Candida boidinii* and *Candida tropicalis*. Intergeneric hybrids between *Saccharomycopsis fibuligera* and *Yarrowia lipolytica* were analyzed by Nga et al. (1992). Hybrids between *Kluyveromyces marxianus* and *Candida macedoniensis*, which are regarded as anamorphs (see above), and intergeneric hybrids between *Saccharomyces cerevisiae* and *Kluyveromyces marxianus* were analyzed by Zimmermann et al. (1988). These studies indicated that only small portions of DNA from one parent were integrated into the other parents' DNA when strains from different species or genera were fused.

2.6
Probe Preparation and Transformation

Chromosome bands can be isolated and used to prepare chromosome-specific probes for Southern blots (Török et al. 1992). Chromosome III from *Saccharomyces cerevisiae* has been isolated by PFGE and used for transformation experiments (Goto et al. 1990).

2.7
Miscellaneous

Many groups have investigated different yeasts by means of PFGE. In the following, some yeast species and genera are listed for which electrophoretic karyotypes have been determined.

De Jonge et al. (1986) used the OFAGE technique to determine the banding patterns of yeasts from the genera *Brettanomyces, Candida, Cryptococcus, Filobasidiella, Geotrichum, Hansenula, Kluyveromyces, Pachysolen, Pichia, Rhodosporidium, Rhodotorula, Saccharomyces, Saccharomycodes, Saccharomycopsis, Schizosaccharomyces,* and *Zygosaccharomyces.* Using conditions which separated chromosomes of up to 1600 kb, they noticed considerable variations in numbers and sizes of chromosomes, even within strains from the same species. They found that *Saccharomyces cerevisiae* was quite exceptional for its high number of rather small chromosomes (smaller than 500 kb).

Candida maltosa was investigated by Tanaka et al. (1987), who found eight chromosome bands. Johnston and Mortimer (1986) and Johnston et al. (1988) analyzed *Saccharomyces, Candida albicans, Candida utilis, Kluyveromyces lactis, Pichia canadensis, Rhodosporidium toruloides, Saccharomycopsis fibuligera, Schwanniomyces occidentalis,* and *Schizosaccharomyces pombe.* Other yeasts for which electrophoretic karyotypes have been determined include *Candida parapsilosis* (Lott et al. 1993); *Candida tropicalis* and *Candida boidinii* (Kobori et al. 1991); *Candida utilis* (Stoltenburg et al. 1992); *Phaffia rhodozyma* (Adrio et al. 1995; Nagy et al. 1994); *Pichia stipitis* and *Candida shehatae* (Passoth et al. 1992; Selebano et al. 1993); *Yarrowia lipolytica;* and related yeasts from the genera *Saccharomycopsis, Endomycopsella,* and *Endomyces* (Naumova et al. 1993).

This short survey of the literature shows that electrophoretic karyotyping is a powerful means for the analysis of yeast genomes. Possible pitfalls of the method, however, should not be neglected. One should bear in mind that the number of bands may not be equal to the number of chromosomes. Different chromosomes of similar size can comigrate. Chromosomes which are too large to be separated by the experimental conditions choosen might escape detection. Size determinations by PFGE are problematic as long as independent calibration methods are missing. PFGE should therefore be used in conjunction with other methods.

3
Practical Recommendations

An excellent review has been written as a practical guide by Birren and Lai (1993), so that only general remarks are provided here.

3.1
Sample Preparation

Many different procedures have been published concerning the preparation of DNA suitable for PFGE. The following procedures have consistently worked well in our laboratories for the last 6 years. They are derived from the procedures given by Schwarz and Cantor (1984) and Carle and Olson (1985). In these procedures the DNA is protected against mechanical stress by embedding the cells or protoplasts in agarose, and high concentrations of EDTA are used to inhibit DNAses. In both procedures the cells are protoplasted. In most cases, cells from the early log phase can be protoplasted more easily than cells from the late log phase or from the stationary phase. Sometimes the medium can be optimized in order to facilitate the protoplasting step. The medium described by Svilha et al. (1961) was used to improve the preparation of *Yarrowia lipolytica* plugs. It consists of 3% glucose, 0.5% yeast extract, 0.5% peptone, 0.0745% L-methionine, 0.0675% DL-homocysteine thiolactone, 0.1% DL-methionine methyl sulfonium chloride. The amount of DNA remaining in the well could be reduced significantly if the cells were grown on this medium.

3.1.1
Procedure A: Protoplast Formation by Zymolyase

Many ascomyceteous yeast like *Saccharomyes*, *Kluyveromyces*, or some *Candida* strains can be prepared by this procedure.

1. Grow the cells in rich medium such as YEPD to late logarithmic phase.

2. Centrifuge the cells from a 20-ml culture and wash the cells twice in 50 mM EDTA, pH 7.5.

3. Adjust the cell number to 2×10^9/ml and mix 1.5 ml suspension with 0.5 ml Zymolyase (2 mg/ml), 50 μl mercaptoethanol and 2.5 ml low-melting agarose (1% in 0.125 mM EDTA, pH 7.5).

4. Pour the mixture into small petri dishes or in insert (plug) moulds and let it soldify.

5. Cut out the plugs according to the size of the wells and incubate them for 24 h at 37 °C in 5 ml LET-buffer (0.5 M EDTA, 0.01 M Tris, pH 7.5). Sometimes 2% mercaptoethanol must be added to the buffer.

6. Replace the LET buffer with NDS solution (0.5 M EDTA, 0.01 M Tris, 1% N-lauroylsarcosinate, pH 9.5 with 0.5 mg/ml Proteinase K) and incubate the plugs for 24 h at 50 °C. In many cases, the plugs will become clear after this treatment.

7. Rinse the plugs several times with LET buffer and incubate them in LET buffer containing 500 U *T1*-RNAse at 37 °C for 24 h. The plugs can be stored in this solution.

8. Prior to electrophoresis, incubate the plugs to be used in electrophoresis buffer to equilibrate for some time.

The Zymolyase treatment can be shortened by starting the protoplasting in liquid medium for 10 min in osmotically stabilized medium and subsequently producing agarose plugs. The plugs are then incubated in stabilized medium containing 1 mg/ml Zymolyase for 2 h at 37 °C. The stabilization can be accomplished by sucrose, sorbitol, mannitol, or salts like KCl, MgCl, or MgSO$_4$, depending on the yeast strain. The stabilization should be checked microscopically for protoplast-like forms and for the absence of lysis.

3.1.2
Procedure B: Protoplast Formation by Novozym

Most strains that cannot be protoplasted by Zymolyase can be protoplasted by Novozym, as is the case with many basidiomyceteous yeasts, but also with *Schizosaccharomyces pombe*.

1. Prepare protoplasts according to your optimized procedure or grow cells to late log phase.
2. Wash cells twice in 0.1 M citrate-phosphate buffer, pH 5.8 containing 0.6 M potassium chloride.
3. Protoplast cells with Novozym G 234, 20 mg/10^{10} cells in stabilized buffer (see step (1), check for protoplasts by microscopy. This step will take about 1 h.
4. Centrifuge the protoplasts and wash them in stabilized buffer.
5. Mix 1.5 ml of the suspension containing 3 to 5 × 10^9 protoplasts gently but thoroughly with 2.5 ml LM-agarose and let the mixture solidify.
6. Cut out the plugs and proceed as in procedure A, step (6).

Recently, several groups have found that chromosome plugs can be produced without the use of protoplasting enzymes (McClusky et al. 1990; Kwan et al. 1991; Gardner et al. 1993; Ibeas and Jimenez 1993). As the electrophoretic behaviour of the DNA depends on the DNA concentration in the plugs, it is important to use similar amounts if banding patterns are to be compared.

3.1.3
Markers

Suitable size markers are bacteriophage *lambda* concatemeres, *Saccharomyces cerevisiae*, *Schizosaccharomyces pombe*, and *Hansenula wingei* chromosomes, which are all commercially available as agarose plugs or beads.

Recently, Zhang et al. (1993) found out that several other yeasts could serve as size markers for the range from 0.7 to 3.3 megabases. However, it should be noted that Pasero and Marilley (1993) detected significant variations in the sizes of chromosomes bearing rDNA clusters, i.e., chromosome XII from *Saccharomyces cerevisiae* and chromosome III from *Schizosaccharomyces pombe*. Isolates belonging to one strain can differ by several hundred kilobases with regard to these chromosomes. Several physical parameters influence the migration (Mathew et al. 1988a,b). Therefore care should be taken when chromosome or genome sizes are to be deduced from PFGE data.

3.2
Electrophoresis Apparatus

Several manufacturers sell PFGE equipment. All these devices have a good record for their performance, so no recommendation for a specific vendor will be given.

Many high-quality data are still obtained by groups who use home-made devices. Carle and Olson (1984) and Schwartz et al. (1989) have published information on the construction of PFGE units.

Some groups sterilize their electrophoresis unit before use. We did not find this to be necessary, but it cannot be wrong to do it. However, before chemicals are used to sterilize the unit, one must make sure that the material is resistent to the chemicals. Most plastic ware is not resistent to ethanol!

3.3
Electrophoresis Conditions

The electrophoresis should not be carried out in the presence of ethidium bromide, as this will influence the migration of the DNA. When unknown samples are to be analyzed, it is advisable to start with conditions that will separate smaller chromosomes, such as the ones of *Saccharomyces cerevisiae*. Such conditions are:

Buffer	0.4 × TBE
Gel	1% agarose (GTG agarose)
Temperature	13 °C
Voltage	200 V, or 10 V per centimeter
Pulse time	100 s
Run time	30 h

These conditions will separate molecules from 50 to 1600 kb.

Figure 2a shows the separation of *Saccharomyces cerevisiae* chromosomes as an example. Chromosomes which are significantly larger than the ones from *Saccharomyces cerevisiae* should migrate as a single band at the top of the gel. If the samples are not degraded, there should be no smear of DNA.

In order to separate larger chromosomes, the voltage should be reduced and the pulse time and the run time should be increased.

We have observed that improved separation can be obtained by increasing agarose concentration. This does not affect the migration, but must be compensated by a proportional increase of the run time.

A next run could be as follows:

Buffer	0.4 × TBE
Gel	1% agarose
Temperature	13 °C
Voltage	100 V or 5 V cm^{-1}
Pulse time	2, 5, 10, 20 min for 24 h each

These conditions should separate chromosomes from 1 to 3.5 megabases. Even larger chromosomes can be separated as follows:

Buffer	0.4 × TBE
Gel	0.8% agarose
Temperature	13 °C
Voltage	40 V or 2 V cm^{-1}

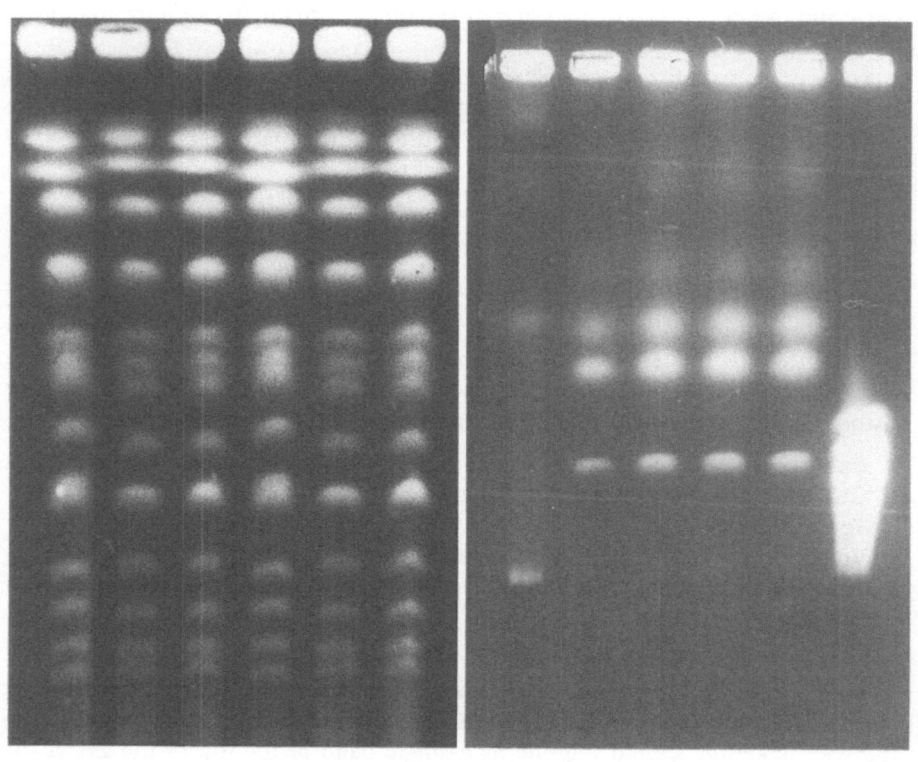

Fig. 2. Separation of yeast chromosomes by PFGE. **a** Chromosomes of *S. cerevisiae*: size range from 260 to 1600 kb; **b** Chromosomes of *Schizosaccharomyces pombe* (lane *1*), *Trichosporon beigelii* (lanes *2–5*), and *Saccharomyces cerevisiae* (lane *6*): size range from 500 to 6000 k*b*

Pulse time	60 min
Run time	6 days

Now the chromosomes from *Schizosaccharomyces pombe* should be separated, giving a range from 3.5 to 5–6 megabases. Figure 2b shows the separation of chromosomes from *Schizosaccharomyces pombe*, *Trichosporon beigelii*, and *Saccharomyces cerevisiae*.

The control of the temperature is very important if the results are to be reproducible.

The conditions given here are rather simple examples which make no use of the possibilities some sophisticated equipments offer, such as the generation of different field geometries or of pulse-time ramps. For a fine tuning of the separation these possibilities may be very helpful.

Starting with conditions separating smaller molecules will save time and effort, especially if it is not clear whether the samples are good. The resolution of the chromosomes can be further improved when different conditions are used successively (Sor 1988; Tunnel and Lalande 1988; Brody and Carbon 1989). When the voltage is increased, the pulse time should be decreased according to the formula:

$$W = E^{1.4} \times Tp,$$

where W is a window function affecting the size range that can be separated, E is the field strength, and Tp the pulse time (Gunderson and Chu 1991).

One example is the separation of *Yarrowia lipolytica* chromosomes which could be improved by the following scheme:

- 47.7 h at 40 V with 55-min pulse time

- 70 h at 43 V with 50-min pulse time

- 48 h at 50 V with 40-min pulse time

in a CHEF DR II system
or

- 90 h at 20 mA with 70-min pulse time

- 75 h at 25 mA with 50-min pulse time

in a TAFE system.

Some vendors sell agaroses specially designed for PFGE. They claim that separations can be accomplished much faster. The choice of the agarose can be dictated by the size of the chromosomes to be separated. The staining and photography of the gels can be done by standard procedures, often a destaining in distilled water improves the result.

3.4
Blotting of the Gels

Most groups use standard procedures for the blotting of the gels; the following procedure from Nguyen et al. (1988) has proven to give good results reproducibly:

1. Stain, destain, and take a picture of the gel.

2. Soak the gel in 0.25 M HCl for 15 min, not longer, as that might lead to a bad transfer. Alternatively, the gel can be exposed to UV-light, 5 s, 3.5 mW/cm², 254 nm.

3. Rinse the gel with water and than soak it in 0.5 M NaOH, 1.5 M NaCl for 20 min. Repeat this step once.

4. Do capillary transfer overnight using the alkaline solution and a GeneScreen membrane (NEN). The membrane should be prepared by first immersing it in water, then in the NaOH solution.

5. After the transfer, rinse the membrane twice for 15 min in 50 mM NaPi, pH 6.5.

6. Expose the membrane to 0.2 mW/cm² for 8 min at 254 nm UV light in order to achieve crosslinking of the DNA to the membrane.

7. Strip the membrane in 2 mM EDTA, 0.1% SDS at 85 °C before the first and all successive hybridizations.

8. Prehybridize the membrane in buffer for at least 2 h. The buffer is: 4 ml 20 × SSPE, 2 ml 50 × Denhardt, 1 ml 10% SDS, 250 μl salmon sperm DNA at 12 mg/ ml, water up to 25 ml.

9. Hybridize in the same buffer plus 10% Dextran sulfate at 65 °C overnight.

10. The probe is labeled by random priming, boiled with the salmon sperm DNA, and added to the hybridization mixture.

11. After hybridization, wash four times for 5 min at 65 °C in buffer A, then three times in buffer B.

   Buffer A 40 mM NaPi, pH: 7.2, 1 mM EDTA, 5% SDS.

   Buffer B same as buffer A, but with 1% SDS.

12. Overnight exposure should be enough to detect a gene.

13. Stripping of the membrane is as described above.

14. The blots are kept in stripping buffer and should never be dried. We have successfully rehybridized the same membrane ten times without significant loss of signal.

Acknowledgments. We thank S. Casaregola for critical reading of the manuscript.

References

Adrio J, Lopez M, Casqueiro J, Fernandez C, Veiga M (1995) Electrophoretic karyotype of the astaxanthin-producing yeast *Phaffia rhodozyma*. Curr Genet 27: 447–450

Asakura K, Iwaguchi SI, Homma M, Sukai T, Higashide K, Tanaka K (1991) Electrophoretic karyotypes of clinically isolated yeasts of *Candida albicans* and *Candida glabrata*. J Gen Microbiol 137: 2531–2538

Bakalinsky AT, Snow R (1990) The chromosomal constitution of wine strains of *Saccharomyces cerevisiae*. Yeast 6: 367–382

Bidenne C, Blondin B, Dequin S, Vezinhet F (1992) Analysis of the chromosomal DNA polymorphism of wine strains of *Saccharomyces cerevisiae*. Curr Genet 22: 1–7

Birren B, Lai E (1993) Pulsed field gel electrophoresis: a practical guide. Academic Press, London

Boekhout T, Renting M, Scheffers WA, Bosboom R (1993) The use of karyotyping in the systematics of yeasts. Antonie Leeuwenhoek J Microbiol 63: 157–163

Brody H, Carbon J (1989) Electrophoretic karyotype of *Aspergillus nidlans*. Proc Natl Acad Sci USA 86: 6260–6263

Cantor CR, Gaal A, Smith CL (1988) High-resolution separation and accurate size determination in pulsed-field gel electrophoresis of DNA. 3. Effect of electrical field shape. Biochemistry 27: 9216–9221

Carle GF, Olson MV (1984) Separation of chromosomal DNA molecules from yeast by orthogonal field alternation gel electrophoresis. Nucleic Acids Res 12: 5647–5664

Carle GF, Olson MV (1985) An electrophoretic karyotype for yeast. Proc Natl Acad Sci USA 82: 3756–3760

Carle GF, Frank M, Olson MV (1986) Electrophoretic separation of large DNA molecules by periodic inversion of the electric field. Science 232: 65–68

Carruba G, Pontieri E, De Bernardis F, Martino P, Cassone A (1991) DNA fingerprinting and electrophoretic karyotype of environmental and clinical isolates of *Candida parapsilosis*. J Clin Microbiol 29: 916–922

Chibana H, Iwaguchi SI, Homma M, Chindamporn A, Nakagawa Y, Tanaka K (1994) Diversity of tandemly repetitive sequences due to short periodic repetitions in the chromosomes of *Candida albicans*. J Bacteriol 176: 3851–3858

Chu G (1991) Bag model for DNA migration during pulsed field electrophoresis. Proc Natl Acad Sci USA 88: 11071–11075

Chu G, Gunderson K (1991) Separation of large DNA by a variable angle CHEF apparatus. Anal Biochem 194: 439–446

Chu G, Vollrath D, Davies RW (1986) Separation of large DNA molecules by contour-clamped homogeneous electric field. Science 234: 1582–1585

Contopoulou CR, Cook VE, Mortimer RK (1987) Analysis of DNA double strand breakage and repair using orthogonal field alternation gel electrophoresis. Yeast 3: 71–76

Dardalhon M, Averbeck D (1995) Pulsed-field gel electrophoresis analysis of the repair of psoralen plus UVA induced DNA photoadducts in *Saccharomyces cerevisiae*. Mutat Res 336: 49–60

De Jonge P, De Jonge FCM, Meijers R, Steensma HY, Scheffers WA (1986) Orthogonal field alternation gel electrophoresis banding patterns of DNA from yeasts. Yeast 2: 193–204

Dogget NA, Smith CL, Cantor CR (1992) The effect of DNA concentration on mobility in pulsed field gel electrophoresis. Nucleic Acids Res 20: 859–864

Eby MJ (1990) Pulsed field separations: continued evolution. Biotechnology 8: 243–245

Fan JB, Grothues D, Smith C (1991) Alignement of *Sfi* I Sites with *Not* I restriction map of *Schizosaccharomyces pombe* genome. Nucleic Acids Res 19: 6289–6294

Fournier P, Abbas A, Chasles M, Kudla B, Ogrydziak DM, Yaver D, Xuan JW, Peito A, Ribet A-M, Feynerol C He F, Gaillardin C (1993) Colocalization of centromeric and replicative functions on autonomously replicating sequences isolated from the yeast *Yarrowia lipolytica*. Proc Natl Acad Sci USA 90: 4912–4916

Gardiner K, Patterson D (1988) Transverse alternating electrophoresis. Nature 331: 371–372

Gardner DCJ, Heale SM, Stateva LI, Oliver SG (1993) Treatment of yeast cells with wall lytic enzymes is not required to prepare chromosomes for pulsed-field gel analysis. Yeast 9: 1053–1055

Gemmill RM (1991) Pulsed field gel electrophoresis. In: Chrambach A, Dunn MJ, Radola BJ (eds) Advances in electrophoresis 4. Verlag Chemie, Weinheim

Goto K, Motoyoshi T, Tamura G, Obata T, Hara S (1990) Chromosomal transformation in *Saccharomyces cerevisiae* with DNA isolated by pulsed field gel electrophoresis. Agric Biol Chem 5: 1499–1504

Gunderson K, Chu G (1991) Pulsed field electrophoresis of megabase-sized DNA Mol Cell Biol 11: 3348–3354

Heus JJ, Zonneveld BJM, Steensma HY, van den Berg JA (1990) Centromeric DNA of *Kluyveromyces lactis*. Curr Genet 18: 517–522

Hoffmann M, Zimmermann M, Emeis C-C (1987) Orthogonal field alternation gel electrophoresis (OFAGE) as a means for the analysis of somatic hybrids obtained by protoplast fusion of different *Saccharomyces strains*. Curr Genet 11: 599–603

Ibeas JI, Jimenez J (1993) Electrophoretic karyotype of budding yeasts with intact cell wall Nucleic Acids Res 21: 3902

Johnston JR, Mortimer RK (1986) Electrophoretic karyotyping of laboratory and commercial strains of *Saccharomyces* and other yeasts. Int J Syst Bacteriol 36: 569–572

Johnston JR, Contopoulou CR, Mortimer RK (1988) Karyotyping of yeast strains of several genera by field inversion gel electrophoresis. Yeast 4: 191–198

Kobori H, Takata Y, Osumi M (1991) Interspecific protoplast fusion between *Candida tropicalis* and *Candida boidinii*: characterization of the fusants. J Ferment Bioeng 72: 439–444

Kwan HS, Li CC, Chiu SW, Cheng SC (1991) A simple method to prepare intact yeast chromosomal DNA for pulsed field gel electrophoresis. Nucleic Acids Res 19: 1347

Lehmann PF, Khazan H, Wu LC, Wickes RI, Kwon-Chung KJ (1992) Karyotype and isozyme profiles do not correlate in *Kluyveromyces marxianus* var. *marxianus*. Mycol Res 96: 637–642

Link AJ, Olson MV (1991) Physical map of the *Saccharomyces cerevisiae* genome at 110-kilobase resolution. Genetics 127: 681–698

Lott TJ, Kuykendall RJ, Welbel SF, Pramanik A, Lasker BA (1993) Genomic heterogeneity in the yeast *Candida parapsilosis*. Curr Genet 23: 463–467

Magee BB, Magee PT (1987) Electrophoretic karyotypes and chromosome numbers in *Candida* species. J Gen Microbiol 133: 425–430

Magee BB, Koltin Y, Gorman JA, Magee PT (1988) Assignement of cloned genes to the seven electrophoretically separated *Candida albicans* chromosomes. Mol Cell Biol 8: 4721–4726

Magee PT, Bowdin L, Staudinger J (1992) Comparison of molecular typing methods for *Candida albicans*. J Clin Microbiol 30: 2674–2679

Marri L, Rossolini GM, Satta G (1993) Chromosome polymorphisms among strains of *Hansenula polymorpha* (syn. *Pichia angusta*). Appl Environ Microbiol 59: 939–941

Mathew KM, Smith CL, Cantor CR (1988a) High-resolution separation and accurate size determination in pulsed-field gel electrophoresis of DNA. 1. DNA size standards and the effect of agarose and temperature. Biochemistry 27: 9204–9210

Mathew KM, Smith CL, Cantor CR (1988b) High-resolution separation and accurate size determination in pulsed-field gel electrophoresis of DNA. 2. Effect of pulse time and electric field strength and implications for models of the separation process. Biochemistry 27: 9210–9216

Mathew KM, Hui CF, Smith CL, Cantor CR (1988c) High-resolution separation and accurate size determination in pulsed-field gel electrophoresis of DNA. 4. Influence of DNA Topology. Biochemistry 27: 9222–9226

McClusky K, Russel BW, Mills D (1990) Electrophoretic karyotyping without the need for generating protoplasts. Curr Genet 18: 385–386

Merz WG, Connelly C, Hieter P (1988) Variation of electrophoretic karyotypes among clinical isolates of *Candida albicans*. J Clin Microbiol 26: 842–845

Nagy A, Garamszegi N, Vágvölgyi C, Ferenczy L (1994) Electrophoretic karyotypes of *Phaffia rhodozyma* strains. FEMS Microbiol Lett 123: 315–318

Naumova E, Naumov G, Fournier P, Nguyen HV, Gaillardin C (1993) Chromosomal polymorphism of the yeast *Yarrowia lipolytica* and related species: electrophoretic karyotyping and hybridization with cloned genes. Curr Genet 23: 450–454

Nga BH, Abu Baker FD, Loh GH, Chiu LL, Harashima S, Oshima Y, Heslot H (1992) Intergeneric hybrids between *Saccharomycopsis fibuligera* and *Yarrowia lipolytica*. J Gen Microbiol 138: 223–227

Nguyen C, Roux D, Mattei MG, Delapeyriere O, Goldfarb M, Birnbaum D, Jordan BR (1988) The FGF-related oncogenes *hst* and *int.2*, and the *bcl.1* locus are contained within one megabase in band q13 of chromosome 11, while the *fgf.5* oncogene maps to 4q21. Oncogene 3: 703–708

Noolandi J, Tunnel C (1992) Preparation, manipulation and pulse strategy for one-dimensional pulsed-field gel electrophoresis (ODPFGE) Chap. 7. In: Burmeiter M, Ulanovsky L (eds) Methods in molecular biology 12. Humana Press, Totowa

Oda Y, Tonomura K (1995) Molecular genetic properties of the yeast *Torulaspora pretoriensis*: characterization of chromosomal DNA and genetic transformation by Saccharomyces cerevisiae – based plasmids. Curr Genet 27: 131–134

Pasero P, Marilley M (1993) Size variation of rDNA clusters in the yeasts *Saccharomyces cerevisiae* and *Schizosaccharomyces pombe*. Mol Gen Genet 236: 448–452

Passoth V, Hansen M, Klinner U, Emeis CC (1992) The electrophoretic banding patterns of the chromosomes of *Pichia stipitis* and *Candida shehatae*. Curr Genet 22: 429–431

Querol A, Barrio E, Ramon D (1992) A comparative study of yeast strain characterization. Syst Appl Microbiol 15: 439–446

Rustchenko-Bulgac EP (1991) Variation of *Candida albicans* electrophoretic karyotypes. J Bacteriol 173: 6586–6596

Rustchenko-Bulgac EP, Sherman F, Hicks JB (1990) Chromosomal rearrangements associated with morphological mutants provide a means for genetic variation of *Candida albicans*. J Bacteriol 172: 1276–1283

Rustchenko EP, Howard DH, Sherman F (1994) Chromosomal alternations of *Candida albicans* are associated with the gain and loss of assimilating functions. J Bacteriol 176: 3231–3241

Schillberg S, Zimmermann M, Emeis CC (1991) Analysis of hybrids obtained by rare-mating of *Saccharomyces* strains. Appl Microbiol Biotechnol 35: 242–246

Schwartz DC, Cantor CR (1984) Separation of yeast chromosome-sized DNA by pulsed field gradient gel electrophoresis. Cell 37: 67–75

Schwartz DC, Koval M (1989) Conformational dynamics of individual DNA molecules during gel electrophoresis. Nature 338: 520–522

Schwartz DC, Smith CL, Baker M, Hsu M (1989) Pulsed electrophoresis instrument. Nature 342: 575–576

Selebano ET, Govinden R, Pillay D, Pillay B, Gupthar AS (1993) Genomic comparisons among parental and fusant strains of *Candida shehatae* and *Pichia stipitis*. Curr Genet 23: 468–471

Smith CL, Matsumoto T, Niwa O, Klco S, Fan J, Yanagida M, Cantor CR (1987) An electrophoretic karyotype for *Schizosaccharomyces pombe* by pulsed field gel electrophoresis. Nucleic Acids Res 15: 4481–4489

Sor F (1988) A computer program allows the separation of a wide range of chromosome sizes by pulsed field gel electrophoresis. Nucleic Acids Res 16: 4853–4863

Sor F, Fukuhara H (1989) Analysis of chromosomal DNA patterns of the genus *Kluyveromyces*. Yeast 5: 1–10

Steensma HY, De Jongh FCM, Linnekamp M (1988) The use of electrophoretic karyotypes in the classification of yeasts: *Kluyveromyces marxianus* and *K. lactis*. Curr Genet 14: 311–317

Stoltenburg R, Klinner U, Ritzerfeld P, Zimmermann M, Emeis CC (1992) Genetic diversity of the yeast *Candida utilis*. Curr Genet 22: 441–446

Suzuki T, Kobayashi I, Mizuguchi I, Banno T, Tanaka K (1988) Electrophoretic karyotypes in medically important *Candida* species. J Gen Appl Microbiol 34: 409–416

Suzuki T, Hitomi A, Magee PT, Sakaguchi S (1994) Correlation between polyploidy and auxotrophic segregation in the imperfect yeast *Candida albicans*. J Bacteriol 176: 3345–3353

Svilha G, Schlenk F, Dainko JL (1961) Spheroplasts of *Candida utilis*. J Bacteriol 82: 808–814

Tanaka H, Takagi M, Yano K (1987) Separation of chromosomal DNA of *Candida maltosa* on agarose gels using the OFAGE Technique. Agric Biol Chem 51: 3161–3163

Thierry A, Dujon B (1992) Nested chromosomal fragmentation in yeast using the meganuclease *I-Sce I*: a new method for physical mapping of eukaryotic genomes. Nucleic Acids Res 20: 5625–5631

Thrash-Bingham C, Gorman JA (1992) DNA translocations contribute to chromosome length polymorphisms in *Candida albicans*. Curr Genet 22: 93–100

Török T, Royer C, Rockhold D, King D (1992) Electrophoretic karyotyping of yeasts and Southern blotting using whole chromosomes as templates for the probe preparation. J Gen Appl Microbiol 38: 313–325

Tunnel C, Lalande M (1988) Resolution of *Schizosaccharomyces pombe* chromosomes by field inversion gel electrophoresis. Nucleic Acids Res 16: 4727

Vaughan-Martini A, Martini A, Cardinali G (1993) Electrophoretic karyotyping as a taxonomic tool in the genus *Saccharomyces*. Antonie Leeuwenhoek J Microbiol 63: 157–163

Vazquez JA, Beckley A, Sobel JD, Zervos MJ (1991) Comparison of restriction enzyme analysis and pulsed-field gradient gel electrophoresis as typing systems for *Candida albicans*. J Clin Microbiol 29: 962–996

Vazquez JA, Beckley A, Donabedian S, Sobel J, Zervos MJ (1993) Comparison of restriction enzyme analysis versus pulsed-field gradient gel electrophoresis as a typing system for *Torulopsis glabrata* and *Candida* species other than *C. albicans*. J Clin Microbiol 31: 2021–2030

Vezinhet F, Blondin B, Hallet JN (1990) Chromosomal DNA patterns and mitochondrial DNA polymorphisms as tools for identification of oenological strains of *Saccharomyces cerevisiae*. Appl Microbiol Biotechnol 32: 568–571

Vollrath D, Davies RW (1987) Resolution of greater than 5 mega-basepair DNA by contour-clamped homogenous electric fields. Nucleic Acids Res 15: 7865–7876

Vollrath D, Davies RW, Connelly C, Hieter P (1988) Physical mapping of large DNA by chromosome fragmentation. Proc Natl Acad Sci USA 85: 6027–6032

Zhang TY, Smith CL, Cantor CR (1991) Secondary pulsed field gel electrophoresis: a new method for faster separation of larger DNA molecules. Nucleic Acids Res 19: 1291–1295

Zhang TY, Fan JB, Ringquist S, Smith CL, Cantor CR (1993) The 0.7 to 3.5 megabase chromosomes from *Candida*, *Kluyveromyces* and *Pichia* provide accurate size standards for pulsed field gel electrophoresis. Electrophoresis 14: 290–295

Zimmermann M, Hoffmann-Hintz M, Kolvenbach M, Emeis C-C (1988) Ofage banding patterns of different yeast genera and of intergeneric hybrids. J Basic Microbiol 28: 241–249

CHAPTER 4

Schwanniomyces occidentalis

R. Jürgen Dohmen and Cornelis P. Hollenberg

1
History of *Schwanniomyces occidentalis* Research

The ascomycetous yeast *Schwanniomyces occidentalis* (*Schw. occidentalis*) was first described by Kloecker in 1909, who isolated it from soil of the island of St. Thomas in the West Indies, hence the species name. Subsequently, a variety of other species, namely *Schw. castellii*, *Schw. alluvius*, an *Schw. persoonii* were accepted under the same genus name (Phaff 1970). These were distinguished from *Schw. occidentalis* by their different fermentation properties. On the basis of molecular analysis involving DNA reassociation experiments, Price et al. (1978) placed all four species under the same species name, *Schw. occidentalis*. The former species *Schw. occidentalis*, *Schw. castellii*, and *Schw. alluvius* showed more than 97% sequence homology and were therefore renamed *Schw. occidentalis* var. *occidentalis*. The former species *Schw. persoonii*, however, had only about 80% sequence complementarity with *Schw. occidentalis* and for that reason was named *Schw. occidentalis* var. *persoonii*. Recent analysis of the long-chain fatty acid compositions (Cottrell et al. 1986) and of the nucleic acid sequences of the cytochrome c gene, as well as of ribosomal RNAs, supports this view, the latter placing *Schwanniomyces* in close proximity to *Candida* and especially to *Debaryomyces* (Amegadzie et al. 1990; Kurtzmann and Robnett 1991). The high degree of identity of the rRNA sequences prompted Kurtzman and Robnett to suggest the transfer of *Schw. occidentalis* to the genus *Debaryomyces* (see Chap. 1, this Vol.). In this chapter we retain the name *Schw. occidentalis*, since the high degree of noncomplementarity of the DNA of *Schwanniomyces* and *Debaryomyces* species found in reassociation experiments (Price et al. 1978) argues strongly in favor of a discrimination of these two species.

This genus *Schwanniomyces* is a member of the family of Saccharomycetaceae (order Endomycetalis). It belongs to the same subfamily (Saccharomycoideae) as well-studied genera such as *Saccharomyces*, *Kluyveromyces*, and *Hansenula* (Lodder 1984).

An important criterion in the diagnosis of the genus *Schwanniomyces* is the

Institut für Mikrobiologie, Heinrich-Heine-Universität Düsseldorf, Universitätsstr. 1, Gebäude 26.12, 40225 Düsseldorf, Germany unique shape of its ascospores, which distinguishes it from other ascomycetous yeasts (Phaff and Miller 1984; see also Sect. 5.1).

The number of research groups studying the properties of the yeast *Schw. occidentalis* has increased significantly over the last 15 years as it has become clear that this organism is equipped with an extremely powerful amylolytic system. The degradation of starch is the basis for a variety of important industrial processes like beer brewing and the production of potable spirits. Because starch-containing raw materials are inexpensive and widely available, investigators have increased their efforts to characterize the amylolytic system of *Schw. occidentalis*, as well as the yeast itself. In this chapter we summarize the current knowledge on *Schw. occidentalis* with emphasis on the available methodology for its genetic manipulation and its application in biotechnological processes.

2
Physiology

Schw. occidentalis is capable of utilizing a wide variety of organic compounds as carbon sources. These include, glucose, fructose, galactose, D-xylose (the latter two cannot be assimilated by var. *persoonii*), sucrose, raffinose, cellobiose, trehalose, lactose (by some strains), succinate, citrate, ethanol, n-alkanes (with the exception of var. persoonii), as well as maltose, *iso*-maltose, pullulan, dextrin, and soluble and raw starch. Melibiose cannot be utilized. This yeast is unable to grow on vitamin-free media or to assimilate nitrate (McCann and Barnett 1984; Phaff and Miller 1984). Phytate (*myo*-inositol hexakiphosphate) can be utilized as a sole phosphate source by *Schw. occidentalis*. This abundant plant seed constituent is hydrolyzed into inorganic phosphate and inositol mono- to pentaphosphate by the enzyme phytase. *Schw. occidentalis* showed the highest phytase activity of the yeast species assayed (Lambrechts et al. 1992).

Schw. occidentalis is Crabtree-negative and shows a strong Pasteur effect, meaning that oxidative metabolism is not repressed by high sugar concentration and that there is little fermentation under aerobic conditions (Ingledew 1987; Poinsot et al. 1987). Poinsot et al. (1987) isolated *Schwanniomyces* petite mutants, one of which appeared to be deficient in cytochrome b. This mutant displayed a greatly decreased Pasteur effect and a significant Crabtree effect, resulting in increased fermentation rates under both aerobic and anaerobic conditions. The analysis of these cytochrome-deficient mutants also resulted in the discovery of alternative respiratory pathways in *Schw. occidentalis* (Dubreucq et al. 1990a,b).

The most remarkable property of *Schw. occidentalis* is its ability to efficiently degrade starch as the result of the combined action of secreted α-amylase and glucoamylase. The expression of both enzymes is inducible and repressed by glucose (see Sect. 10).

3
Media

Schw. occidentalis can be grown on the same complete (YP) and synthetic minimal (SD, if dextrose is the carbon source) media as *S. cerevisiae* (Sherman 1993). YP media contain 10 g/l Bacto-yeast extract and 20 g/l Bacto-peptone (both from Difco, Detroit, USA) plus usually 2% of one of the carbon sources listed above (see Sect. 2). Synthetic minimal media contain 6.7 g/l yeast nitrogen base without the amino acids (Difco) plus required amino acids and carbon source. Synthetic minimal media can be buffered with 0.2 M NaPO4 (pH 6.3) to maintain enzymatic activity of the amylases (see Sect. 10).

Various media have been described for the induction of sporulation. Ferreira and Phaff (1959) used yeast autolysate/glucose agar with incubation for 5 days at room temperature. James and Zahab (1984) employed media containing 1.3% yeast extract, 0.5% peptone, and 1.5% glucose (originally described by Kreger van Rij 1977) with pH adjusted to 4.5. The plates were incubated at 19 °C for 20 days. Kurtzman et al. (1972) sporulated on yeast-malt extract agar. Wilson et al. (1982) obtained asci on agar plates containing 0.5% yeast extract and 5% glucose.

4
Available Strains

The *Schw. occidentalis* type strains are available from the American type culture collection (ATCC) in Rockville, MD, USA, or from the Centraalbureau voor Schimmelcultures in Delft, Holland. A large number of auxotrophic mutants have been produced in several laboratories (Johannsen and van der Walt 1980; Wilson et al. 1982; James and Zahab 1984). Until now, only a few auxotrophic mutations have been defined, mainly by their complementation with the respective genes. Klein and Favreau (1988) isolated an *ade2* mutant that could be complemented by transformation with plasmids containing the *Schw. occidentalis ADE2* gene. Dohmen et al. (1989) described a strain carrying a mutation (*trp5*) in the tryptophan synthetase gene which was complemented by the *S. cerevisiae TRP5* gene. The availability of an established transformation system (see Sect. 5) now enables the generation of defined auxotrophic mutations via homologous recombination using the sequences of cloned genes (see Sect. 6).

Mutants with defects in the carbon catabolite repression of amylase expression have been isolated in a number of laboratories (McCann and Barnett 1984; Sills et al. 1984b; Ingledew 1987; Boze et al. 1989b). Respiration deficient mutants have been mentioned above (see Sect. 2).

5
Genetic Techniques

5.1
Description and Life Cycle

Schwanniomyces cells are ovoidal or egg-shaped ($3.5-8.0 \times 4.5-9.0\,\mu m$) and reproduce vegetatively by multilateral budding. Pseudomycelia are not formed or are at
least not characteristic (Phaff and Miller 1984). Cells are haploid and monotypic,
since intraspecies or interspecies crosses are not observed. Hybridization studies
carried out with a variety of auxotrophic mutants of different *Schwanniomyces*
strains failed to detect any recombinants, suggesting that mating between free-
occurring cells does not occur to any genetically detectable extent (<1 in 10^8 mated
cells; Johannsen and van der Walt 1980). However, fusion between a mother and a
daughter cell does occur and precedes the formation of ascospores (Kreger-van Rij
1977). Since there is no detectable formation of recombinants in crosses, it is
assumed that this type of conjugation takes place before the detachment of the
daughter cell (Johannsen and van der Walt 1980). Electron microscopical studies
by Kreger-van Rij (1977) suggest that after fusion of the mother cell with a small
bud, from which it was first completely separated by a wall, and nuclear fusion,
meiosis takes place in the mother cell. Fused mother cells could be distinguished
from cells with a young bud by the presence of a dark ring in the wall at the site of
the fusion, and occasionally by the presence of remnants of the cross wall. Usually
one, or more rarely, two walnut-like ascospores are formed, which are characterized by a wide, flat equatorial ring and long protuberances (Kurtzman et al. 1972).
The spores from two different *Schw. occidentalis* var. *persoonii* strains differed
from those of var. *occidentalis* strains, as well as from each other, by the shape of
the equatorial ring and the presence or form of the protuberances (Kurtzman et al.
1972). Thus, based on the current knowledge, *Schwanniomyces* strains appear to be
genetically isolated (Johannsen and van der Walt 1980; Ingledew 1987). In the life
cycle, diploidization results from the fusion of sister nuclei such that meiosis does
not serve as a means for mixing of genetic information. Johannsen and van der
Walt (1980) pointed out that this rather rudimentary sexual cycle still entails the
advantage of resulting in the formation of ascospores, specialized cells which
enable the survival under unfavorable conditions.

5.2
Strain Construction

As described above, attempts to obtain intrageneric recombinants from crosses
have unfortunately been unsuccessful (Johannson and van der Walt 1980). Wilson
et al. (1982), on the other hand, were able to generate prototrophic recombinants
by polyethylene glycol-induced protoplast fusion of auxotrophic mutants. In some
cases, these recombinants were unstable heterokaryons, in others, they were stable
diploids (or aneuploids) which gave rise to mitotic segregants with new combinations of auxotrophies. The mitotic segregation frequency was increased by the

addition of p-fluorophenylalanine. The fusion products showed an increased frequency of asci with two spores (11–18%) compared to the parental strains (0–4%). Work by James and Zahab (1984), who studied hybrids generated by protoplast fusion by diad analysis of two-spored asci, revealed that the spore clones retained the ploidy of the hybrid. (The spores were freed for isolation by treating the asci with Zymolyase.) These investigators concluded that the increased ploidy of the hybrids is not recognized by the cells during the meiotic cycle. In summary, the fusion of *Schwanniomyces* protoplasts appears to be a useful method for the construction of polysomic or polyploid strains and could be used for the characterization of mutants in complementation tests and even for a genetic analysis by meiotic segregation, the latter, however, being very much more complicated than e.g., for *S. cerevisiae*. Protoplast fusion, however, seems not to provide a means to generate haploid recombinants with new marker combinations.

Protoplasts of *Schw. occidentalis* cells can be generated with Zymolyase, Helicase, Glusulase, β-Glucuronidase or Mutanase (Stephen and Nasim 1981; Wilson et al. 1982; James and Zahab 1984; Klein and Favreau 1988).

5.3
Mutagenesis

Standard mutagenesis protocols using UV irradiation, treatment with EMS or N-methyl-N'-nitro-N-nitrosoguanidine have been employed (Johannsen and van der Walt 1980; McCann and Barnett 1984; Klein and Favreau 1988; Boze et al. 1989b). Since *Schw. occidentalis* cells are generally haploid, the generation and identification of recessive mutations is comparably easy. Cytochrome-deficient mutants were induced by treatment with ethidium bromide, manganese, or acriflavine (Poinsot et al. 1987; Dubreucq et al. 1990a).

5.4
Transformation

Two different chemical protocols have been independently developed for the transformation of *Schw. occidentalis* cells. Klein and Favreau (1988) used a modification of the procedure described by Beggs (1978), which employs the generation of spheroplasts. Dohmen et al. (1989, 1991) developed an easier method, that is based on the polyethylene glycol-induced transformation of whole cells described by Klebe et al. (1983). The modified method (given below) enables the long-term storage of competent cells and yields up to 1000 transformants per μg plasmid DNA. With slight modifications, this protocol has also proven useful in the transformation of other yeast species such as *S. cerevisiae*, *Schizosaccharomyces pombe*, *Kluyveromyces lactis* and *Hansenula polymorpha* (Dohmen et al. 1991).

Protocol for the transformation of *Schw. occidentalis*:

Cells are grown in YPD (20 ml per transformation) to an OD_{600} of 0.6, washed in 0.5 vol of 1.25 M KCl/30 mM/10 mM Bicin-NaOH (pH 8.35)/3% ethylene glycol/5%

DMSO, and resuspended in 0.01 vol of the same solution. Aliquots of 0.2 ml of the cell suspension are then frozen in dry-ice/acetone or by placing them in a <−70 °C freezer, where they can be stored for months before transformation.

Transformation of such competent cells is achieved by adding the DNA to the frozen cells (transformation with highly purified DNAs can be increased by the addition of sonicated and heat-denatured calf thymus DNA; Schiestl and Gietz 1989), followed by rapid agitation in an Eppendorf mixer at 37 °C (mixing at room temperature, which is often more convenient, results in slightly lower yields).

After addition of 1.4 ml of 40% polyethylene glycol 1000 (Roth, Karlsruhe, Germany; available in the US through Atomergic Chemetals Corp., Farmingdale, NY)/ 0.2 M Bicin-NaOH (pH 8.3) the samples are mixed by gentle inversion and incubated for 1 h at 30 °C.

The cells are then pelleted by centrifugation at 3000 g for 5 min, washed once with 1.6 ml of 0.15 M NaCl/10 mM Bicin-NaOH (pH 8.35), resuspended in an appropriate volume of the same buffer, and plated onto selective media. Transformant colonies are generally obtained after 2–3 days incubation at 30 °C.

More recently, a high-efficiency electrotransformation protocol for *Schw. occidentalis* has been described by Costaglioli et al. (1994).

5.5
Gene Disruptions and Deletions

The availability of a transformation system enables the generation of null alleles of cloned genes as well as the introduction of mutants alleles into the genome via homologous recombination (for review of the strategies that have been worked out in *S. cerevisiae*, see Stearns et al. 1990). These approaches are hampered in some yeast species such as *S. pombe* and *H. polymorpha* by the frequent occurrence of nonhomologous recombination (Moreno et al. 1991; Hansen and Hollenberg, Chap. 9, this Vol.). Although no thorough quantitation has been carried out, the results obtained thus far suggest that in *Schw. occidentalis*, as in *S. cerevisiae*, homologous recombination appears to predominate. Dohmen et al. (1990) generated null alleles of the *Schw. occidentalis* glucoamylase (*GAM1*) gene by two different approaches. In the first strategy (Shortle et al. 1982), the targeted integration of a plasmid that contained an internal fragment of the glucoamylase into the wild-type *GAM1* gene resulted in the generation of two inactive copies of the *GAM1* gene, one being truncated at the 5' end, the other at the 3' end. In these experiments about 70% of several hundred transformants displayed a *gam1* phenotype, indicating that correct integration of the plasmid via homologous recombination had taken place, which was confirmed by Southern analysis. The second approach was a one-step gene transplacement (Rothstein 1983). In this case, the entire sequence of the *GAM1* gene was stably replaced by the marker gene *TRP5* (see Sect. 7). Four out of 30 stable Trp⁺ colonies displayed the desired *gam1* phenotype. Similar results were obtained by Amegadzie et al. (1990), who generated an insertion allele of the *Schw. occidentalis CYC1* gene by the one-step gene replacement method. Three out of 23 analyzed transformants displayed the *cyc1* phenotype. Recently, we have generated a null mutant of the *Schw. occidentalis LEU2* gene using the two-step gene replacement strategy (K. Schuchart, R.J.D., A.W.M. Strasser, and C.P.H., unpubl. results; Rothstein 1983). In conclusion, these techniques now established for *Schw. occidentalis* enable both the generation of useful laboratory strains with multiple stable marker mutations (see also Sect. 7) and the functional analysis of cloned genes and their mutant alleles as well as the construction of strains with improved properties for industrial processes (see Sect. 11).

6
Chromosomal DNA

Schw. occidentalis is a haploid yeast (Ferreira and Phaff 1959) with a DNA content of $5-6 \times 10^{-14}$ g/cell, which is about twice as much as present in *S. cerevisiae* cells (Wilson et al. 1982; Leusch et al. 1985). This corresponds to a genome size of about 3×10^7 base pairs. The G + C content is 35.2–35.4% (Price et al. 1978). Johnston et al. (1988) resolved at least five chromosomes by field inversion gel electrophoresis. Del Pozo et al. (1993) detected a minimum of seven chromosomes by pulse field gel electrophoresis (see also Chap. 3, this Vol.).

7
Genes and Genetic Markers

The cloning of 14 different *Schw. occidentalis* genes has been reported. Sequence information is available for 8 of them (Table 1). With the exception of the glucoamylase (*GAM1*), all the cloned genes for which such assays have been pub-

Table 1. Cloned *Schw. occidentalis* genes

Gene	Function	Size (aa)	Expression in S. cerevisiae	Reference
ADE1	Adenine synthesis	n.d.	n.d.	Prakash and Seligy (1988)
ADE2	Adenine synthesis	n.d.	Yes	Klein and Favreau (1988)
ODC1	Uracil synthesis	n.d.	Yes	Klein and Roof (1988)
HIS4	Histidine synthesis	n.d.	Yes	Dohmen et al. (1989)
EG1	Unknown	158	n.d.	Prakash and Seligy (1988)
INV1	Invertase	533	n.d.	Klein et al. (1989)
AMY1	α-Amylase	512	Yes	Strasser et al. (1989)
GAM1	Glucoamylase	958	No	Dohmen et al. (1990)
CYC1	Cytochrome c	110	Yes	Amegadzie et al. (1990)
GDH1	Glutamate dehydrogenase	459	Yes	De Zoysa et al. (1991)
SWA2	α-Amylase	507	Yes	Claros et al. (1993)
SCR1	Cycloheximide resistance	n.d.	Yes	Del Pozo et al. (1993)
SCR2	Ribosomal protein	106	Yes	Del Pozo et al. (1993)
LEU2	Leucine synthesis	n.d.	n.d.	Schuchart et al. (unpubl.)
AMG1	Subunit NADH: Q6 Oxidoreductase	487	Yes	Fabry et al. (unpubl.)

lished were found to be functionally expressed in *S. cerevisiae* (Table 1), suggesting that the transcription, translation, and secretion signals are well conserved between these two yeasts. TATA-like elements were found in all sequenced promoter regions of *Schw. occidentalis* genes.

Interestingly, some genes, when introduced into *S. cerevisiae*, are regulated in a way similar to that in *Schw. occidentalis*, while others are not. For example, both the *AMY1* and the *SWA2* genes are glucose-repressed in *Schw. occidentalis*. However, only the SWA2 gene retains this regulation upon transformation into *S. cerevisiae* (Dohmen et al. 1989; Abarca et al. 1991). Even more striking is the following observation made by Amegadzie et al. (1990). The *Schw. occidentalis* CYC1 gene is induced by oxygen but, unlike its counterpart in *S. cerevisiae*, it is not subject to glucose repression. However, when introduced into *S. cerevisiae*, the *Schw. occidentalis* gene is not only oxygen-induced but also glucose-repressed. These observations demonstrate a strong conservation of both the *cis-* and *trans-*acting elements of this regulation. An appealing model, when comparing these two yeasts, is that the glucose-regulated form of the *CYC1* gene was the evolutionary predecessor of the glucose-independent form of this gene in the Crabtree-negative *Schw. occidentalis*. The *cis*-acting element in the *CYC1* promoter would have been retained in *Schw. occidentalis* upon the loss of the Crabtree effect, whereas the specific interaction of this element with *trans*-acting binding factors has been lost. Alternatively, the *trans*-acting factors in *S. cerevisiae* could have evolved to recognize preexisisting sequences present in the promoters of the *CYC1* gene in both yeasts.

The conservation of expression signals between genes of *S. cerevisiae* and *Schw. occidentalis* has simplified the isolation of a number of *Schw. occidentalis* genes through expression cloning or by complementation of well-defined *S. cerevisiae* mutants. More such clonings are to be expected in the near future. Particularly useful in the generation of systems for the genetic manipulation of *Schw. occidentalis* has been the cloning of genes involved in amino acid and nucleotide biosynthesis like *ADE1*, *ADE2*, *ODC1* (the *Schw. occidentalis* equivalent of the *S. cerevisiae URA3* gene), *HIS4*, and *LEU2* (Table 1). The *ADE2* gene has been employed as a selectable marker in *Schw. occidentalis* transformations (Klein and Favreau 1988). *Schw. occidentalis ade2* mutants transformed with plasmids containing the *ADE2* gene were either integrated or maintained as extrachromosomal elements without detectable mitotic loss. The latter are subject to rearrangements yielding a variety of plasmids, including some of high molecular weight, which led the authors to conclude that the fragment containing the *ADE2* gene also contains an autonomous replication sequence (Klein and Favreau 1988). Autonomous replication sequences, named *SwARS1* and *SwARS2*, were also identified, respectively, on a 1.5-kb fragment isolated from a clone containing the *Schw. occidentalis HIS4* gene (Dohmen et al. 1989) and in the 3′ flanking region of the *GAM1* gene (Dohmen et al. 1990). *SwARS* plasmids carrying the *S. cerevisiae TRP5* gene as a selectable marker (Fig. 1) replicate autonomously in both *S. cerevisiae* and *Schw. occidentalis* and thus are convenient shuttle vectors in the analysis of gene expression and function in these two yeast species. Unlike the *ADE2* plasmids described

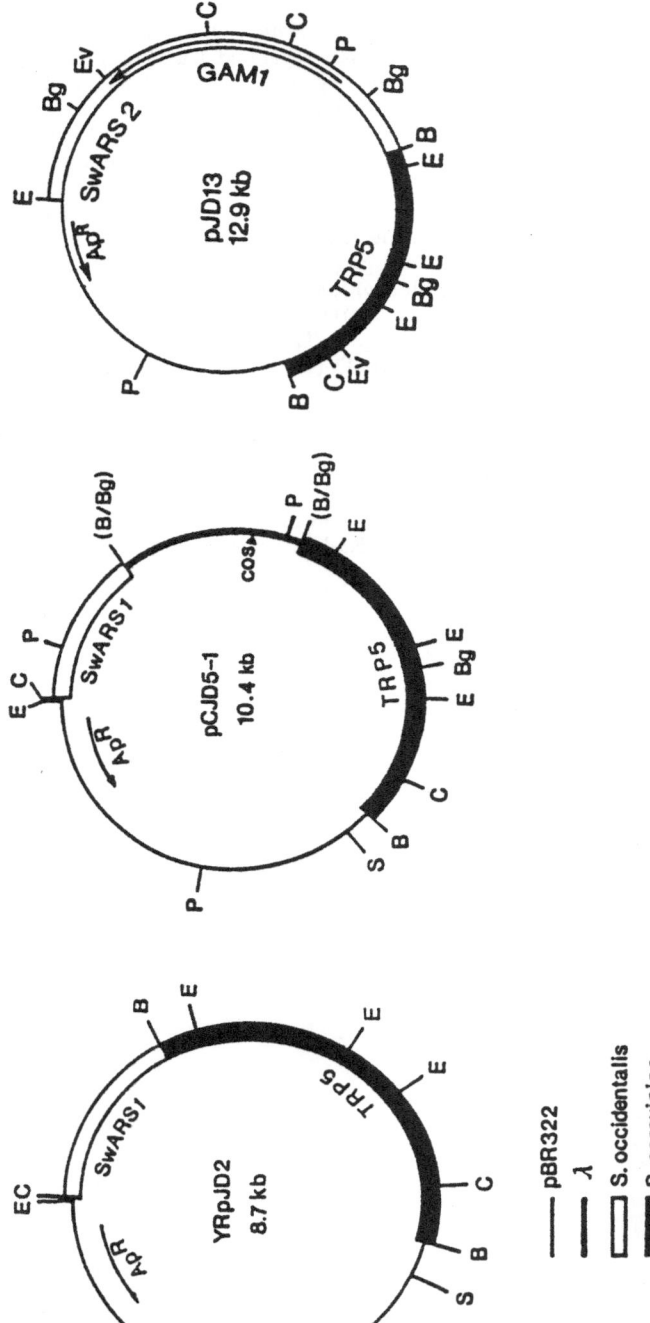

Fig. 1. Shuttle vectors for the transformation of *Schw. occidentalis* and *S. cerevisiae*. Plasmid YRpJD2 is pBR322-based and contains *SwARS1*, an autonomous replication sequence that originated from the *Schw. occidentalis* HIS4 gene (Dohmen et al. 1989) and the *S. cerevisiae* TRP5 gene contained in a 3.2 kb *BamHI* fragment from pYe(trp5)1-53 (Walz et al. 1953). This vector was the basis for a plasmid that lead to a fivefold overexpression of α-amylase (Amy1) in *Schw. occidentalis* (Dohmen et al. 1989). The cosmid pCJD5-1 was obtained by inserting a 1.7-kb *BglII* fragment isolated from pHC79 (Hohn and Collins 1980) into YRpJD2. This fragment containes the cos element of bacteriophage λ, which allows the in vitro packaging of such cosmids carrying inserts of about 40 kb into λ particles (Hohn 1980). Such a cosmid pool of genomic *Schw. occidentalis* DNA for the cloning of *Schw. occidentalis* genes in *Schw. occidentalis* through the complementation of appropriate mutants has been constructed (A.W.M. Strasser, R.J. Dohmen, and C.P. Hollenberg, unpubl. results). As a first such example, it was successfully used in the cloning of a gene, the expression of which is required for the expression of glucoamylase in *Schw. occidentalis* (P. Fabry, A.W.M Strasser, R.J. Dohmen, and C.P. Hollenberg, unpubl. results). Plasmid pJD13 was derived from YRpJD2 by replacing its *SwARS1*-containing *EcoRI/BamHI* fragment with a fragment containing the *GAM1* gene and in its 3' flanking region an autonomous replication sequence named *SwARS2*

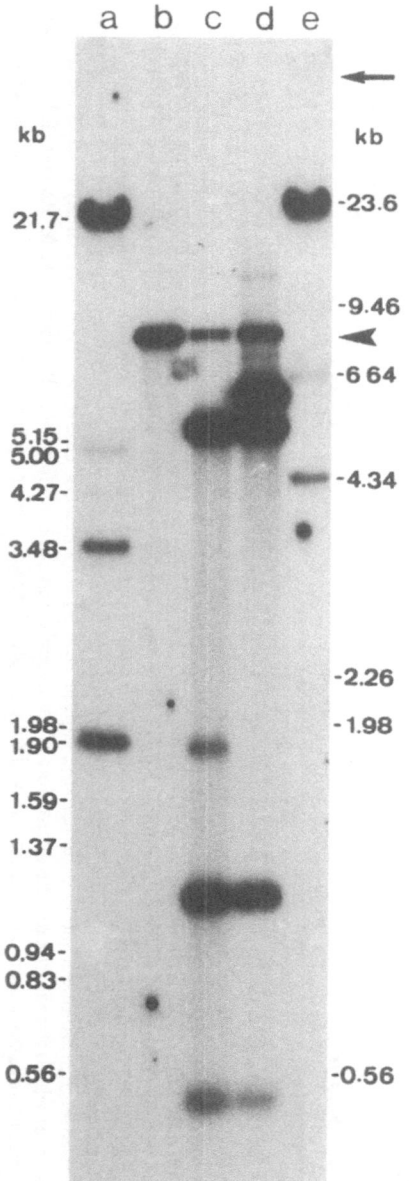

Fig. 2. Comparison of average copy numbers of *SwARS* plasmids in *Schw. occidentalis* transformants. DNAs were separated in a 0.8% agarose gel, blotted onto Hybond N (Amersham), and hybridized with radio-labeled pJD13 DNA. *Lanes a and e* λ DNA cut with, respectively, *Eco*RI and *Hind*III, or *Hind*III. (Some bands are cross-hybridizing with the probe used. The *positions* of the marker bands *indicated at the sides* were deduced from the ethidium bromide stained gel) *Lanes b-d* DNA prepared from *Schw. occidentalis* and cut with *Eco*RI; *lane b* untransformed strain RJD11; *lane c* RJD11 transformed with plasmid YRpJD2 (see Fig. 1); *lane d* RJD11 transformed with plasmid pJD13 (see Fig. 1). The position of the wells of the gel is shown by an *arrow*. An *arrowhead* indicates the position of an 8.3-kb fragment of the chromosomal *GAM1* locus. The average copy number of the plasmids per cell was estimated to be 5–10 by comparing the intensity of the band of the chromosomal 8.3-kb fragment with those of the 5.3-kb fragments of the two plasmids

by Klein and Favreau (1988), *SwARS* plasmids are mitotically unstable, with an average of five to ten copies per cell (Fig. 2), resembling plasmids containing *S. cerevisiae ARS* elements in *S. cerevisiae* transformants. In contrast to the versatility of *SwARS* vectors, *ARS* plasmids are extremely unstable when introduced into

Schw. occidentalis, yielding an average of only one copy per cell under selective conditions (Dohmen et al. 1989).

8
Vector Systems

Klein and Favreau (1988, 1991) have described vectors for the transformation of *Schw. occidentalis* which were based on the *ADE2* marker and a putative replication sequence (see also Sect. 7). The reason for their stable maintenance and their engagement in recombination events is not understood.

We developed a series of vectors, which employ the *S. cerevisiae TRP5* gene as a selectable marker (Dohmen et al. 1989; R.J.D., A.W.M. Strasser, and C.P.H., unpubl. results). This set includes a plasmid for integrative transformation (pAT153-TRP5, not shown), the plasmid YRpJD2 and a cosmid (pCJD5-1); these latter two are maintained as extrachromosomal plasmids in *Schw. occidentalis* and *S. cerevisiae* due to the presence of the *SwARS1* element (Fig. 1). They enable the generation of gene deletions (Dohmen et al. 1990), the introduction of mutated genes or of gene fusions, as well as of extra copies resulting in increased expression of a gene of interest (Dohmen et al. 1989). Moreover, they provide the possibility of cloning *Schw. occidentalis* genes through complementation of the respective recessive mutations.

No *CEN* plasmids are as yet available for *Schw. occidentalis*.

9
Heterologous Gene Expression

The vector systems discussed above provide a basis for the utilization of *Schw. occidentalis* as a host for heterologous gene expression. The different *Schw. occidentalis* genes that have been cloned (Table 1) provide a source for promoters, terminators, as well as for sequences encoding signal sequences for the secretion of proteins in *Schw. occidentalis*. Of particular interest are the promoters of genes like *INV1*, *AMY1*, *SWA2*, and *GAM1*, which are tightly regulated. These promoters should even allow the expression of putatively growth-inhibitory genes. *Schw. occidentalis* cells containing such genes, e.g., expressed from the tightly glucose-repressed *GAM1* promoter, could be grown to high densities before the expression of these genes is induced. It has been suggested that *Schw. occidentalis* might be particularly attractive as a host for the expression of secretory proteins since, unlike the situation in *S. cerevisiae*, hyperglycosylation of proteins was not observed for certain proteins (Klein and Favreau 1991). However, there is as yet no evidence available that the same protein would be hyperglycosylated in *S. cerevisiae* and not in *Schw. occidentalis*. In fact, the only such studies available demonstrate that the Amy1 and the Gam1 proteins are modified similarly in *S. cerevisiae* and *Schw. occidentalis* (Strasser et al. 1989; Dohmen et al. 1990). Reasons that make *Schw. occidentalis* an attractive alternative eukaryotic expression system will be discussed in Sect. 11.

10
The Amylolytic System

The main reason that *Schw. occidentalis* has gained such high attention in recent years is its outstanding amylolytic system. It had been a goal for a number of years to generate *S. cerevisiae* strains that secrete α-amylase and glucoamylase with debranching activity (see below) since they would combine high fermentation rates and ethanol tolerance with the ability to degrade starch completely (Sills and Stewart 1982).

Because of their close relationship to *S. cerevisiae*, other yeast species are favorable candidates as donors for genes encoding amylolytic enzymes. A large number of yeast strains have been described that are able to grow on starch (Lodder 1984). However, only a very few degrade starch with high efficiency as a result of the combined action of α-amylase (EC 3.2.11), glucoamylase (EC 3.2.13) and debranching activity, i.e., hydrolysis of α-1,6 glucosidic bonds (Spencer-Martins and van Uden 1979; Clementi et al. 1980; Sills and Stewart 1982; Touzi et al. 1982; De Mot et al. 1984).

Of these strains, *Schw. occidentalis* expresses the most significant debranching activity (Sills and Stewart 1982), which is a property of the glucoamylase enzyme of this yeast (Wilson and Ingledew 1982; Sills et al. 1984a). Both the α-amylase and the glucoamylase secreted by *Schw. occidentalis* moreover have the advantage of being inactivated during the pasteurization conditions usually employed in brewing (Sills et al. 1987). The expression of these enzymes and their biochemistry has been studied extensively (Clementi et al. 1980; Oteng-Gyang et al. 1981; Moranelli et al. 1987; Wilson et al. 1982; Calleja et al. 1984; Sills et al. 1984a; Simòes-Mendes 1984; Clementi and Rossi 1986; Boze et al. 1987; Dowhanick et al. 1990; Deibel et al. 1988; Boze et al. 1989a; Dowhanick et al. 1990).

α-Amylase is an endoenzyme that hydrolyzes α-1,4 glucosidic bonds in starch releasing maltose, maltotriose, and higher oligosaccharides, and with very low efficiency glucose (Ingledew 1987).

Glucoamylase is an exoenzyme releasing glucose from the nonreducing end of a starch molecule by hydrolyzing α-1,4 glucosidic bonds. The *Schw. occidentalis* glucoamylase, moreover, has the ability to bypass the α-1,6 branch points by virtue of its debranching activity. This property of glucoamylase enzyme enables *Schw. occidentalis* to utilize pullulan as a sole carbon source (Wilson and Ingledew 1982). This polysaccharide is composed of maltotriose units linked through α-1,6 glucosidic bonds.

The synergistic action of α-amylase and glucoamylase secreted by *Schw. occidentalis* results in efficient and complete degradation of starch (Fuji and Kawamura 1985; Boze et al. 1989a). Kinetic studies on the degradation of starch by the two amylases revealed that α-amylase activity is especially important when complex substrates like glycogen, potato, or barley starch are used (Boze et al. 1989a). Release of glucose from these substrates by glucoamylase is rather slow. The role of the endoenzyme α-amylase is to generate free nonreducing ends, which are the substrates for the exoenzyme glucoamylase. The synergistic effect of these two enzymes was also reflected by growth properties of *S. cerevisiae* cells expressing either or both of the two enzymes (Dohmen et al. 1990; Strasser et al. 1990, 1991). Growth of such strains secreting both enzymes on media containing crude starch as a sole carbon source was significantly better than that of those which secreted either of the enzymes (R.J.D., P.G. Seeboth, A.W.M. Strasser, and C.P.H., unpubl. results).

Contradictory reports have described the biochemical identification of one or two glucoamylases and of one or two α-amylases produced by different *Schw. occidentalis* strains (reviewed by Ingledew 1987). At present, it is unclear whether some of the identified polypeptides represent different forms of the respective proteins expressed from the same gene or whether they are products of separate genes. The way to clearly resolve these issues will be to analyze the different genes involved and to use them for the generation of null mutations in the different strains.

Two different *Schw. occidentalis* α-amylase genes (*AMY1* and *SWA2*) have been cloned and sequenced (Strasser et al. 1989; Abarca et al. 1991; Wu et al. 1991; Claros et al. 1993). The two proteins display 66% identity and 77% similarity (Claros et al. 1993). Both proteins share highly conserved regions found in all α-amylases that have been analyzed (Strasser et al. 1989; Claros et al. 1993). Both are about 54 kDa in size and contain two potential N-glycosylation sites (Asn × Ser/ Thr), only one of which appears to be modified by a core glycosylation (Strasser et al. 1989). Gene deletions that would help to understand the roles of the two α-amylase genes in *Schw. occidentalis* have not yet been described.

A single glucoamylase gene (*GAM1*) from *Schw. occidentalis* has been cloned and sequenced (Dohmen et al. 1990). It encodes a 958-amino acid protein (including signal sequence) which, without further posttranslational modification, has a calculated molecular weight of 106.5 kDa. Signal sequence processing after Ala^{22}, without any further modifications, would yield a protein of 104 kDa that contains ten potential N-glycosylation sites. Treatment of the native 145-kDa protein purified from medium of a *Schw. occidentalis* culture with endoglycosidase H revealed that about 23 kDa of its molecular weight is contributed by N-linked sugars. Some, if not all, of the other modifications that account for the difference between the apparent and the calculated molecular weights are presumably due to O-glycosylations. A *Schw. occidentalis* strain that was completely deleted for the sequence of the *GAM1* coding region was generated (Dohmen et al. 1990). Glucoamylase in the culture supernatant or on the surface of intact cells of such *gam1Δ* strains was in the range of the detection limit (<1% of the activity found in cultures of congenic *GAM1* strains; Fig. 3B, and data not shown). The *gam1Δ* strains were still able, though very poorly, to grow on media containing starch or pullulan as a sole carbon source (Dohmen et al. 1990). This phenomenon could be explained by either the presence of a second very weakly expressed glucoamylase gene or by the ability of α-amylase together with a putative α-glucosidase to release a small amount of glucose from these substrates. In any case, the *GAM1*

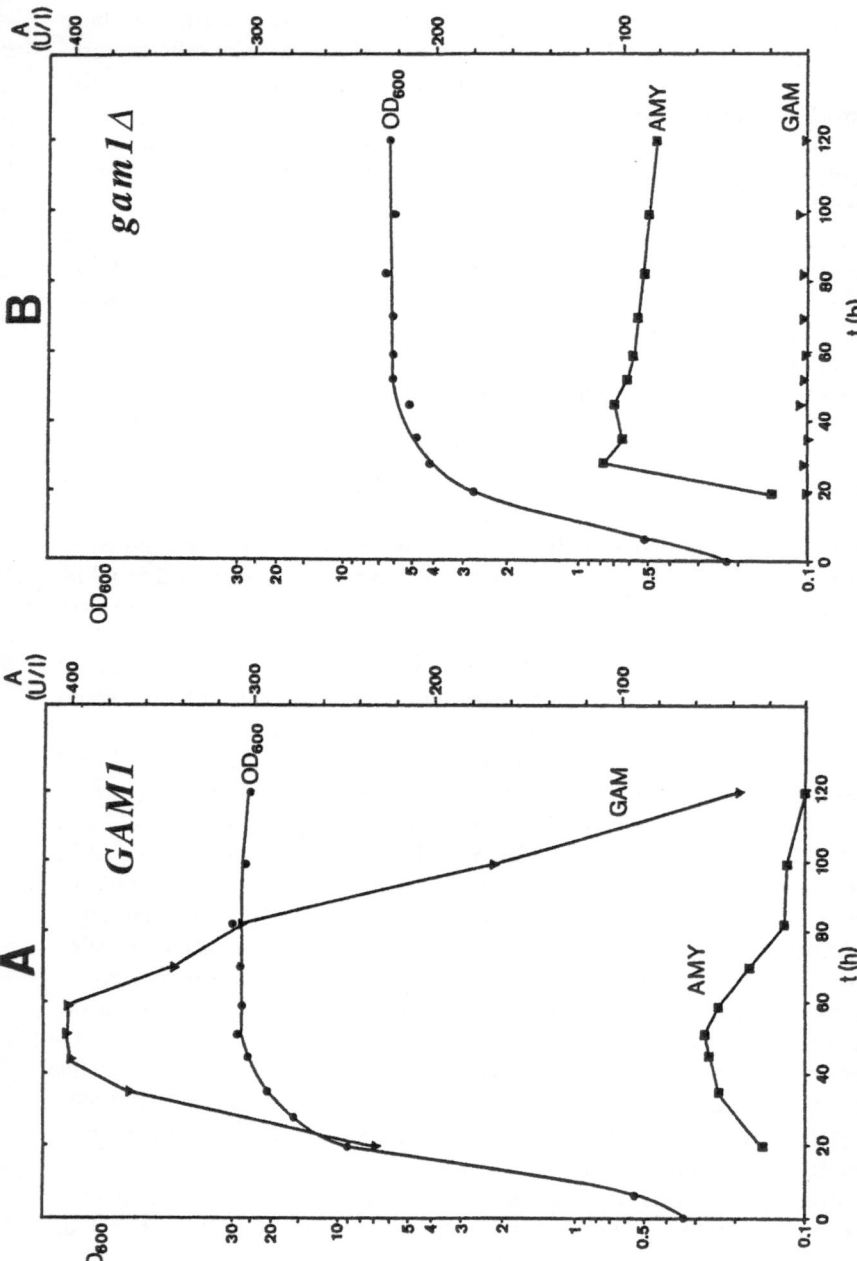

Fig. 3A,B. Amylase activities secreted by *Schw. occidentalis* GAM1 (strain RJD11) and an isogenic *gam1Δ* strain. The cells were grown in YP media with 2% soluble starch at 30 °C. α-Amylase (AMY, *squares*) and glucoamylase (GAM, *triangles*) activity in the medium was determined as described in Dohmen et al. (1990). The values are units per liter (U/l) culture supernatant. The optical density of the culture at 600_{nm} (*circles*) was plotted logarithmically. A "wild-type" strain RJD11. B *gam1Δ::TRP5* strain derived from RJD11. (Dohmen et al. 1990)

gene is clearly the main contributor to the glucoamylase activity of the strains tested.

Expression of amylase genes is tightly regulated in *Schw. occidentalis*. The α-amylase genes *AMY1* and *SWA2*, as well as the *GAM1* gene, are subject to glucose repression (Clementi et al. 1980; McCann and Barnett 1984; Ingledew 1987; Dohmen et al. 1989, 1990; Dowhanick et al. 1990; Abarca et al. 1991). The expression of α-amylase and glucoamylase is induced (in the absence of glucose) by maltose, soluble starch, and melizitose (Clementi and Rossi 1986) and is repressed at elevated temperatures (37 °C; Calleja et al. 1984). Both enzymes are inactivated at pH values below 3.5 and by temperatures above 55 °C.

Schw. occidentalis displays a Kluyver effect on maltose and starch media. Both compounds cannot be metabolized in anaerobiosis due to repression of amylase and α-glucosidase expression (Calleja et al. 1982; De Mot et al. 1985; Boze et al. 1987; Violla et al. 1992). Some time ago, mutants that were unable to grow on starch were isolated (A. Nasim, pers. comm.). Recently, one of the complementing genes, *AMG1*, was cloned and identified to be the gene encoding the 51-kDa subunit of the NADH dehydrogenase (complex I) of the repiratory chain (Fabry 1993; Fabry and Hollenberg, in prep.) previously sequenced from *Neurospora crassa* (Weiss et al. 1991). The *amg1* mutant was in addition unable to grow with maltose as sole carbon source. Apparently the deficiency in the respiratory chain leads to inability to express active glucoamylase and to inefficient maltose takeup (Fabry 1993).

Although *Schw. occidentalis* expresses maltose permease and α-glucosidase, glucoamylase activity, at least in the strains tested, appears to be rate-limiting in the utilization of maltose, as *gam1Δ* strains show significant reduction in growth on maltose media (Dohmen et al. 1990).

Glucose, the product of glucoamylase activity, is at the same time a repressor of its expression. This feedback regulation of glucoamylase expression is apparently the reason why the highest expression of this enzyme is frequently observed in the stationary phase of growth (e.g., Clementi et al. 1980, 1986). When the concentration of the glucoamylase substrate is sufficiently low (usually in the stationary phase) such that substrate hydrolysis does not generate repressing concentrations of glucose, glucoamylase expression can proceed uninhibited. This effect is even more obvious when the copy number of the *GAM1* gene is increased (Fig. 4A). The induction of glucoamylase from multiple gene copies, in a culture with 2% soluble starch, results in a rapid increase in the glucose concentration in the medium. This then results in a decrease of Gam1 activity until the glucose concentration has dropped sufficiently low to allow another round of induction. A similar but less striking up and down of glucose concentration and Gam1 activity is observed in the culture without additional copies of the *GAM1* gene (Fig. 4B). As a result of this feedback regulation, the increase in glucoamylase activity due to the additional five to ten copies of the *GAM1* gene is not all that high. This is in contrast to a similar experiment, in which additional copies of the *AMY1* gene were introduced into *Schw. occidentalis* on a related *SwARS* plasmid yielding a fivefold overproduction of α-amylase (Dohmen et al. 1989).

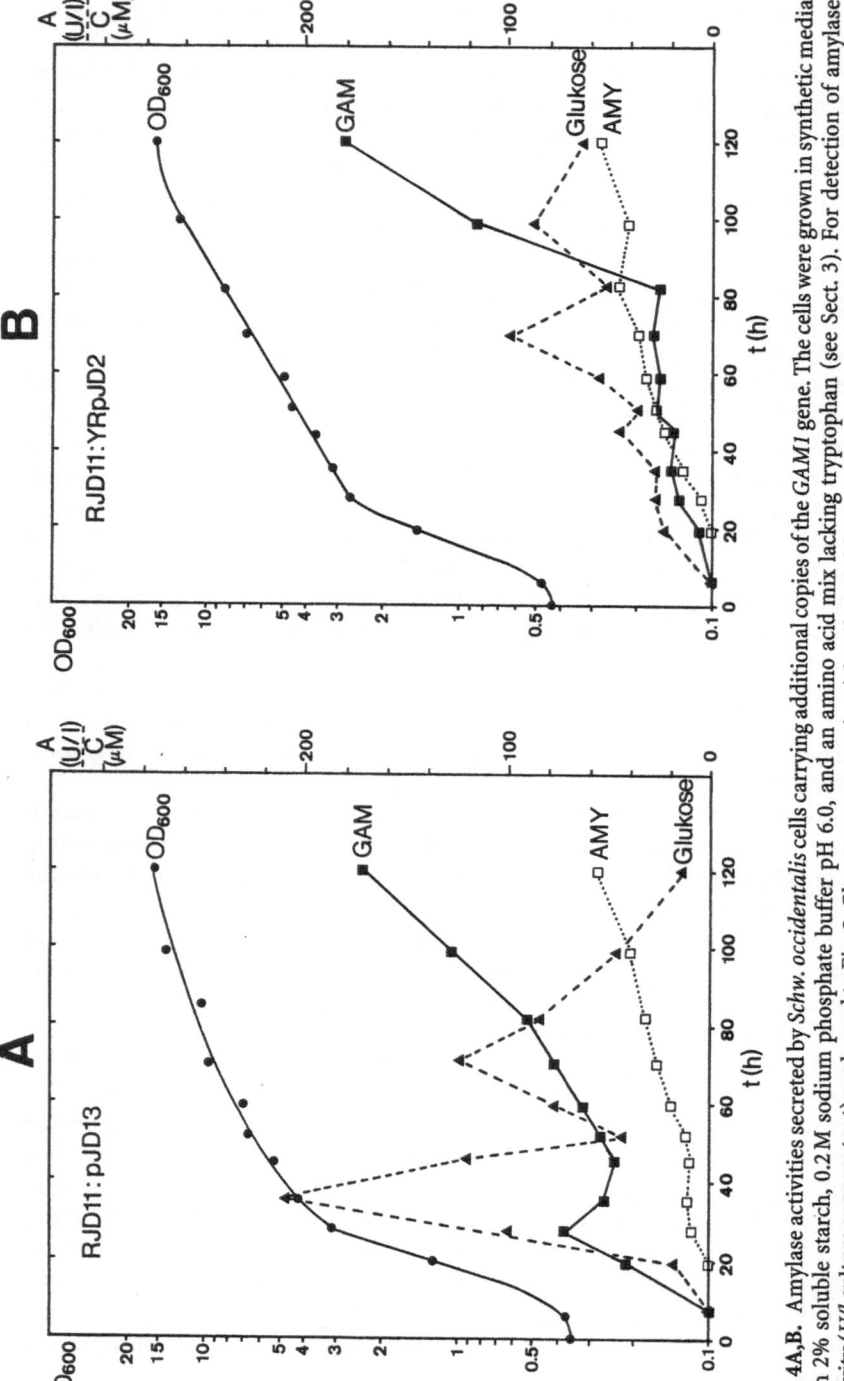

Fig. 4A,B. Amylase activities secreted by *Schw. occidentalis* cells carrying additional copies of the *GAM1* gene. The cells were grown in synthetic media with 2% soluble starch, 0.2 M sodium phosphate buffer pH 6.0, and an amino acid mix lacking tryptophan (see Sect. 3). For detection of amylase activity (*U/l* culture supernatant) see legend to Fig. 3. Glucose concentration (given in *μmol/l*, *triangles*) was determined with glucose dehydrogenase/mutarotase (Merck) and NAD. The optical density of the culture at 600$_{nm}$ (*circles*) was plotted logarithmically. **A** Strain RJD11 transformed with plasmid pJD13 containing the *GAM1* gene (see Figs. 1 and 2). **B** Strain RJD11 transformed with plasmid YRpJD2 (see Figs. 1 and 2). Note the difference in amylase secretion when compared to cells grown in rich media (Fig. 3)

11
Industrial Applications

Schw. occidentalis has become a target for investigators concerned with biotechnological processes because of its ability to degrade starch completely. Starch is one of the most abundant renewable carbon sources (~1 × 10⁹ tons/year) on planet earth (Hollenberg and Wilhelm 1987). Starch conversion is by mass the most important biotechnological process. The enzymes required for starch breakdown, α-amylase, and glucoamylase, are among the biotechnologically most highly produced enzymes worldwide (Stewart 1987). Starch is used as a raw material in biotechnological processes leading to high dextrose and high fructose corn syrups, the latter being used in the food and soft drink industry, and in ethanol fermentation. The latter process includes the production of ethanol, which is used as an octane booster in lead-free gasoline, and of alcoholic beverages (for review see Finn 1987).

Commercial utilization of starch from corn or barley grain (for review see Tubb 1986) requires, after milling of the raw material, "gelatinization" (by cooking) and "liquefaction" (partial hydrolysis with heat-stable bacterial α-amylases). Subsequently, starch "saccharification" results from treatment with α-amylase and glucoamylase. For ethanol production, the resulting sugars are then usually fermented by strains of *S. cerevisiae*. In processes like the brewing of beer or the production of malt whiskey, saccharification by the amylolytic enzymes present in barley malt is incomplete such that 25% of the starch material remains as unfermentable dextrin (Tubb 1986). These dextrins are, besides ethanol, the main contributors of calories in beer. In order to reduce the dextrin content of the product, a variety of efforts involving *Schw. occidentalis* have been made to introduce amylase activity, in particular glucoamylase with debranching activity, into the fermentation step. These include the addition of amylases obtained from *Schw. occidentalis* cultures (Sills et al. 1983) and the introduction of amylase genes into *S. cerevisiae*. As discussed above (Sect. 10), the latter attempts have yielded *S. cerevisiae* strains that are equipped with the *Schw. occidentalis* genes encoding α-amylase and glucoamylase. Such strains are able to degrade starch completely (Strasser et al. 1990). The future will show whether such strains will gain relevance in industrial processes.

Another possible application for *Schw. occidentalis* strains is their direct utilization in SCP production from inexpensive starchy raw materials (Touzi et al. 1982), which could replace the more expensive molasses, or in ethanol fermentation (Calleja et al. 1982). The latter process will require further strain improvement, since the ethanol tolerance of those studied was not sufficiently high (Ingledew 1987; De Mot et al. 1985). Other problems, the resolution of which has already begun (see Sect. 4), include the glucose repression of amylase expression and the absence of a Crabtree effect, combined with the presence of a Kluyver effect. Calleja et al. (1982) bypassed these problems by preceding the anaerobic fermentation with a period of aerobic growth enabling amylase secretion. Now that the respective genes have been cloned, strains with altered regulation and strength of amylase gene expression can also be constructed by approaches employing gene technology (e.g., Dohmen et al. 1989).

The ability of *Schw. occidentalis* to grow in inexpensive media might also contribute to make it a choice as a host for the production of heterologous proteins. The availability of vectors for its genetic manipulation and of tightly regulated promoters (see Sects. 7, 8) make *Schw. occidentalis* an attractive system for this purpose, as well as for more general studies.

Acknowledgments. We are grateful to Christopher Byrd for his comments on the manuscript.

References

Abarca D, Fernández-Lobato M, del Pozo L, Jiménez A (1991) Isolation of a new gene (SWA2) encoding an α-amylase from *Schwanniomyces occidentalis* and its expression in *Saccharomyces cerevisiae*. FEBS Lett 279: 41–44

Amegadzie BY, Zitomer RS, Hollenberg CP (1990) Characterization of the cytochrome c gene from the starch-fermenting yeast *Schwanniomyces occidentalis* and its expression in baker's yeast. Yeast 6: 429–440

Beggs JD (1978) Transformation of yeast by a replicating hybrid plasmid. Nature 275: 104–108

Boze H, Moulin G, Galzy P (1987) Influence of culture conditions on the yield and amylase biosynthesis in continuous culture by *Schwanniomyces castellii*. Arch Microbiol 148: 162–166

Boze H, Guyot JB, Moulin G, Galzy P (1989a) Kinetics if the amyloglucosidase of *Schwanniomyces castellii*. In: Martini A, Vaughan Martini A (eds) Yeast as a main protagonist of biotechnology. Yeast 2: 117–121

Boze H, Moulin G, Galzy P (1989b) Isolation and characterization of a hexokinase mutant of *Schwanniomyces castellii*: consequences on cell production in continous culture. Yeast 5: 469–476

Calleja GB, Levy-Rick S, Lusena CV, Moranelli F, Nasim A (1982) Direct and quantitative conversion of starch to ethanol by the yeast *Schwanniomyces alluvius*. Biotechnol Lett 4: 543–547

Calleja GB, Levy-Rick S, Moranelli F, Nasim A (1984) Thermosensitive export of amylases in the yeast *Schwanniomyces alluvius*. Plant Cell Physiol 25: 757–761

Claros MG, Abarca D, Fernández-Lobato M, Jiménez A (1993) Molecular structure of the SWA2 gene encoding an AMY1-related α-amylase from *Schwanniomyces occidentalis*. Curr Genet 24: 75–83

Clementi F, Rossi J (1986) α-Amylase and glucoamylase production by *Schwanniomyces castellii*. Antonie Leeuwenhoek J Microbiol 52: 343–352

Clementi F, Rossi J, Costamagna L, Rosi J (1980) Production of amylases by *Schwanniomyces castellii* and *Endomycopsis fibuligera*. Antonie Leeuwenhoek 46: 399–406

Costaglioli P, Meilhoc E, Masson JM (1994) High-efficiency electrotransformation of the yeast *Schwanniomyces occidentalis*. Curr Genet 27: 26–30

Cottrell M, Viljoen BC, Kock JLF, Lategan PM (1986) The longchain fatty acid compositions of species representing the genera *Saccharomyces, Schwanniomyces and Lipomyces*. J Gen Microbiol 132: 2401–2403

Deibel MR, Hiebsch RR, Klein RD (1988) Secreted amylolytic enzymes from *Schwanniomyces occidentalis*: purification by fast protein liquid chromatography (FPLC) and preliminary characterization. Prep Biochem 18: 77–120

Del Pozo L, Abarca D, Hoenicka J, Jimenez A (1993) Two different genes from *Schwanniomyces occidentalis* determine ribosomal resistance to cycloheximide. Eur J Biochem 213: 849–857

De Mot R, Andries K, Verachtert H (1984) Comparative study of starch degradation and amylase production by ascomycetous yeast species. Syst Appl Microbiol 5: 106–118

De Mot R, van Dijck K, Donkers A, Verachtert H (1985) Potentialities ad limitations of direct alcoholic fermentations of starchy material with amylolytic yeast. Appl Microbiol Biotechnol 22: 222–226

De Zoysa PA, Connerton IF, Watson DC, Johnston JR (1991) Cloning, sequencing and expression of the *Schwanniomyces occidentalis* NAPD-dependent glutamate dehydrogenase gene. Curr Genet 20: 219–224

Dohmen RJ, Strasser AWM, Zitomer RS, Hollenberg CP (1989) Regulated overproduction of α-amylase by transformation of the amylolytic yeast *Schwanniomyces occidentalis*. Curr Genet 15: 319–325

Dohmen RJ, Strasser AWM, Dahlems UM, Hollenberg CP (1990) Cloning of the *Schwanniomyces occidentalis* glucoamylase gene (*GAM1*) and its expression in *Saccharomyces cerevisiae*. Gene 95: 111–121

Dohmen RJ, Strasser AWM, Höner CB, Hollenberg CP (1991) An efficient transformation procedure enabling long-term storage of competent cells of various yeast genera. Yeast 7: 691–692

Dowhanick TM, Russell I, Scherer SW, Stewart GG, Seligy VL (1990) Expression and regulation of glucoamylase from the yeast *Schwanniomyces castellii*. J Bacteriol 172: 2360–2366

Dubreucq E, Boze H, Fouilhé M, Moulin G, Galzy P (1990a) Alternative respiration pathways in *Schwanniomyces castellii*. I. Isolation and characterization of cytochrome-deficient mutants. Antonie Leeuwenhoek J Microbiol 57: 123–130

Dubreucq E, Boze H, Moulin G, Galzy P (1990b) Alternative respiration pathways in *Schwanniomyces castellii*. II. Characterization of oxidation pathways. Antonie Leeuwenhoek J Microbiol 57: 130–137

Fabry P (1993) Analyse der Stärkeverwertung der Hefe *Schwanniomyces occidentalis*. PhD Thesis, Heinrich-Heine-Universität, Düsseldorf

Ferreira JD, Phaff HJ (1959) Life cycle and nuclear behavior of a species of the yeast genus *Schwanniomyces*. J Bacteriol 78: 352–361

Finn RK (1987) Conversion of starch to liquid sugar and ethanol. In: Biotec I: Microbial genetic engineering and enzyme technology. Gustav Fischer, Stuttgart, pp 101–107

Fuji M, Kawamura Y (1985) Synergistic action of α-amylase and glucoamylase on hydrolysis of starch. Biotechnol Bioeng 27: 260–265

Hohn B (1980) In vitro packaging of λ and cosmids. In: Wu R (ed) Recombinant DNA. Methods Enzymol 68: 299–309

Hohn B, Collins J (1980) A small cosmid for efficient cloning of large DNA fragments. Gene 11: 291–298

Hollenberg CP, Wilhelm M (1987) New substrates for old organisms. In: Biotec I: Microbial genetic engineering and enzyme technology. Gustav Fischer, Stuttgart, pp 21–31

Ingledew WM (1987) *Schwanniomyces*: a potential superyeast? CRC Crit Rev Biotechnol 5: 159–176

James AP, Zahab DM (1984) Genetic system of *Schwanniomyces alluvius* determined by diad analysis of fusion products. J Bacteriol 160: 1105–1108

Johannsen E, van der Walt JP (1980) Hybridization studies within the genus *Schwanniomyces* Klöcker. Can J Microbiol 26: 1199–1203

Johnston JR, Contopoulou CR, Mortimer RK (1988) Karyotyping of yeasts strains of several genera by field inversion gel electrophoresis. Yeast 4: 191–198

Klebe RJ, Harriss JV, Sharp D, Douglas MG (1983) A general method for polyethylene-glycol-induced transformation of bacteria and yeast. Gene 25: 333–341

Klein RD, Favreau MA (1988) Transformation of *Schwanniomyces occidentalis* with an *ADE2* gene cloned from *S. occidentalis*. J Bacteriol 170: 5572–5578

Klein RD, Favreau MA (1991) A DNA fragment containing the *ADE2* gene from *Schwanniomyces occidentalis* can be maintained as an extrachromosomal element. Gene 97: 183–189

Klein RD, Roof LL (1988) Cloning of the orotidine 5'-phosphate decarboxylase (ODC) gene of *Schwanniomyces occidentalis* by complementation of the *ura3* mutation of *S. cerevisiae*. Curr Genet 13: 29–35

Klein RD, Poorman RA, Favreau MA, Shea MH, Hatzenbuhler NT, Nulf SC (1989) Cloning and sequence analysis of the gene encoding invertase from the yeast *Schwanniomyces occidentalis*. Curr Genet 16: 145–152

Klöcker A (1909) Deux nouveaux genres de la famille des Saccharomycètes. CR Trav Lab Carlsberg 7: 273–278

Kreger-van Rij NJW (1977) Electron microscopy of sporulation in *Schwanniomyces alluvius*. Antonie Leeuwenhoek J Microbiol 43: 55–64

Kurtzman CP, Robnett CJ (1991) Phylogentic relationships among species of *Saccharomyces, Schizosaccharomyces, Debaryomyces* and *Schwanniomyces* determined from partial ribosomal RNA sequences. Yeast 7: 61–72

Kurtzmann CP, Smiley MJ, Baker FL (1972) Scanning electron microscopy of ascospores of *Schwanniomyces*. J Bacteriol 112: 1380–1382

Lambrechts C, Boze H, Moulin G, Galzy P (1992) Utilization of phytate by some yeasts. Biotechnol Lett 14: 61–66

Leusch HG, Hoffmann M, Emeis CC (1985) Fluorometric determination of the total DNA content of different yeast species using 4',6-diamidino-2-phenyl-indol-dihydrochloride. Can J Microbiol 31: 1164–1166

Lodder J (1984) The yeasts: a taxonomic study. 3rd edn. North Holland Publishing, Amsterdam

McCann AK, Barnett JA (1984) Starch utilization by yeasts: mutants resistant of carbon catabolite repression. Curr Genet 8: 525–530

Moranelli F, Yaguchi M, Calleja GB, Nasim A (1987) Purification and characterization of the extracellular α-amylase activity of the yeast *Schwanniomyces alluvius*. Biochem Cell Biol 65: 899–908

Moreno S, Klar A, Nurse P (1991) Molecular genetic analysis of fission yeast *Schizosaccharomyces pombe*. In: Guthrie C, Fink GR (eds) Guide to yeast genetics and molecular biology. Methods Enzymol 194: 795–823

Oteng-Gyang K, Moulin G, Galzy P (1981) A study of the amylolytic system of *Schwanniomyces castellii*. Z Allg Mikrobiol 21: 537–544

Phaff HJ (1970) *Schwanniomyces* Klöcker. In: Lodder J (ed) The yeasts. A taxonomic study. North-Holland Publishing, Amsterdam, pp 756–766

Phaff HJ, Miller MW (1984) *Schwanniomyces* Klöcker. In: Lodder J (ed) The yeasts: a taxonomic study. 3rd edition. North Holland Publishing, Amsterdam, pp 423–426

Poinsot C, Moulin G, Claisse M, Galzy P (1987) Isolation and characterization of a mutant of *Schwanniomyces castellii* with altered respiration. Antonie Leeuwenhoek J Microbiol 53: 65–75

Prakash K, Seligy VL (1988) A temporally expressed gene from *Schwanniomyces alluvius* and detection of homologous sequences in other yeasts. Gene 73: 131–140

Price CW, Fuson GB, Phaff HJ (1978) Genome comparison in yeast systematics: delimitation of species within the genera *Schwannionyces, Saccharomyces, Debaryomyces* and *Pichia*. Microbiol Rev 42: 161–193

Rothstein RJ (1983) One-step gene disruption in yeast. In: Wu R, Grossman L, Moldave K (eds) Methods Enzymol 101: 202–211

Schiestl RH, Gietz RD (1989) High efficiency transformation of intact yeast cells using single stranded nucleic acids as a carrier. Curr Genet 16: 339–346

Sherman F (1983) Getting started with yeast. In: Guthrie C, Fink GR (eds) Guide to yeast genetics and molecular biology. Methods Enzymol 194: 3–21

Shortle D, Haber JE, Botstein D (1982) Lethal disruption of the yeast actin gene by integrative DNA transformation. Science 217: 371–373

Sills AM, Stewart GG (1982) Production of amylolytic enzymes by several yeast species. J Inst Brew 88: 313–316

Sills AM, Panchal CJ, Stewart GG (1983) The production and use of yeast amylases in the brewing of low carbohydrate beer. Proceedings of the 19th European Brewery Courention Congress, London, IRL Press, Oxford, pp 377–384

Sills AM, Sauder ME, Stewart GG (1984a) Isolation and characterization of the amylolytic system of *Schwanniomyces castellii*. J Inst Brew 90: 311–314

Sills AM, Zygora PSJ, Stewart GG (1984b) Characterization of *Schwanniomyces castellii* mutants with increased productivity of amylases. Appl Microbiol Biotechnol 20: 124–128

Sills AM, Panchal CJ, Russel I, Stewart GG (1987) Production of amylolytic enzymes by yeasts and their utilization in brewing. CRC Crit Rev Biotechnol 5: 105–115

Simoes-Mendes B (1984) Purification and characterization of the extracellular amylases of the yeast *Schwanniomyces alluvius*. Can J Microbiol 30: 1163–1170

Spencer-Martins I, van Uden N (1979) Extracellular amylolytic system of the yeast *Lipomyces kononenkoae*. Eur J Appl Microbiol Biotechnol 6: 241–250

Stearns T, Ma H, Botstein D (1990) Manipulating yeast genome using plasmid vectors. Methods Enzymol 185: 280–297

Stephen ER, Nasim A (1981) Production of protoplasts in different yeasts by mutanase. Can J Microbiol 27: 550–553

Stewart GG (1987) The biotechnological relavance of starch-degrading enzymes. CRC Crit Rev Biotechnol 5: 89–93

Strasser AWM, Selk R, Dohmen RJ, Niermann T, Bielefeld M, Seeboth P, Tu G, Hollenberg CP (1989) Analysis of the α-amylase gene of *Schwanniomyces occidentalis* and secretion of its gene product in transformants of different yeast genera. Eur J Biochem 184: 699–706

Strasser AWM, Janowicz ZA, Dohmen RJ, Roggenkamp RO, Hollenberg CP (1990) Prospects of yeasts in biotechnology. Agro-Industry Hi-Tech 1: 21–24

Strasser AWM, Janowicz ZA, Roggenkamp RO, Dahlems U, Weydemann U, Merckelbach A, Gellissen G, Dohmen RJ, Piontek M, Melber K, Hollenberg CP (1991) Applications of genetically manipulated yeasts. In: Peberdy JF, Caten CE, Ogden JE, Bennett JW (eds) Applied molecular genetics of fungi. Cambridge University Press, Cambridge, pp 161–169

Touzi A, Prebois JP, Moulin G, Deschamps F, Galzy P (1982) Production of food yeast from starchy substrates. Eur J Appl Microbiol Biotechnol 15: 232–236

Tubb RS (1986) Amylolytic yeasts for commercial applications. Trends Biotechnol 4: 98–104

Twigg AJ, Sherrat D (1980) Trans-complementable copy number mutants of plasmid ColE1. Nature 283: 216–218

Violla P, Boze H, Moulin G, Galzy P (1992) Transport and hydrolysis of maltose by *Schwanniomyces castellii*. J Basic Microbiol 32: 57–63

Walz A, Ratzkin B, Carbon J (1978) Control of expression of a cloned yeast (*Saccharomyces cerevisiae*) gene (*TRP5*) by a bacterial insertion element (IS2) Proc Natl Acad Sci USA 75: 6172–6176

Weiss H, Friedrich T, Hofhaus G, Preiss D (1991) The respiratory chain NADH dehydrogenase (complex I) of mitochondria. Eur J Biochem 197: 563–576

Wilson JJ, Ingledew WM (1982) Isolation and characterization of *Schwanniomyces alluvius* amylolytic enzymes. Appl Environ Microbiol 44: 301–307

Wilson JJ, Khachatourians GG, Ingledew WM (1982) Protoplast fusion in the yeast *Schwanniomyces alluvius*. Mol Gen Genet 186: 95–100

Wu FM, Wang TT, Hsu WH (1991) The nucleotide sequence of *Schwanniomyces occidentalis* α-amylase gene. FEMS Micobiol Lett 82: 313–318

CHAPTER 5

Kluyveromyces lactis

Micheline Wésolowski-Louvel[1], Karin D. Breunig[2], and Hiroshi Fukuhara[1]

1
History of *Kluyveromyces lactis* Research

Genetic studies of *Kluyveromyces lactis* (Dombrowski) van der Walt began in the early 1960s. *Saccharomyces lactis* was the name of the yeast at that time. Since then, the genus *Kluyveromyces* has been the object of intensive taxonomical studies. The position of *Kluyveromyces lactis* with respect to other members of the genus has been discussed in detail by a number of authors (van der Walt 1970; Johannsen and van der Walt 1978; Sidenberg and Lachance 1986; Lachance 1989; Fuson et al. 1987; Vaughan-Martini and Martini 1987; see also Kurtzman and Phaff 1989, for molecular taxonomy of yeasts). *K. lactis* had been thought to be closely related to the so-called *K. fragilis* (*K. marxianus*), another lactose-assimilating yeast well known in industry. On the basis of cellular hybridization studies, Johannsen (1980) and van der Walt and Johannsen (1979) considered *K. lactis* as a variety of *K. marxianus* (Hansen) van der Walt, but later taxonomic studies cited above showed that they are distinct species, as judged by various molecular criteria. DNA complementarity is less than 15–20% according to Fuson et al. (1987) and Vaughan-Martini and Martini (1987); the electrophoretic karyotype and the mitochondrial DNA restriction pattern are also completely different between the two yeasts (see below). At present, *K. lactis* is considered as a separate species from *K. marxianus*. *K. lactis* now incorporates *K. drosophilarum* and *K. vanudenii*. To our knowledge, the latest review on *Kluyveromyces* systematics is that of Lachance (1993).

Since the pioneer work by Halvorson and collaborators on the genetics of "*Saccharomyces lactis*" (Herman and Halvorson 1963a,b; Tingle et al. 1968; Tingle 1967; Herman 1963), several laboratories have developed genetically labeled strains which formed the genetic background of the strains we are now using in various genetic and biochemical studies. Stimulated by the recent biotechnological interest, the research on *K. lactis* gained increasing attention. Because of its dis-

[1] Institut Curie, Section de Biologie, Bat. 110, Centre Universitaire Paris XI, Orsay 91405, France
[2] Institut für Mikrobiologie, Heinrich-Heine-Universität Düsseldorf, 40225 Düsseldorf, Germany

tinctive physiological properties, *K. lactis* has become an important alternative to the classical *Saccharomyces cerevisiae*. This chapter will consider, as often as possible, the question how *K. lactis* is similar to, or different from *S. cerevisiae*.

2 Physiology

K. lactis cells are spherical to oval, somewhat smaller than *S. cerevisiae* cells (Fig. 1). The natural habitat of this species is diverse, but many strains were originally isolated from milk-derived products in which the major carbon source is lactose. These strains indeed grow well on lactose, while *S. cerevisiae* does not. Also in contrast to *S. cerevisiae*, in which the fermentative metabolism dominates, the

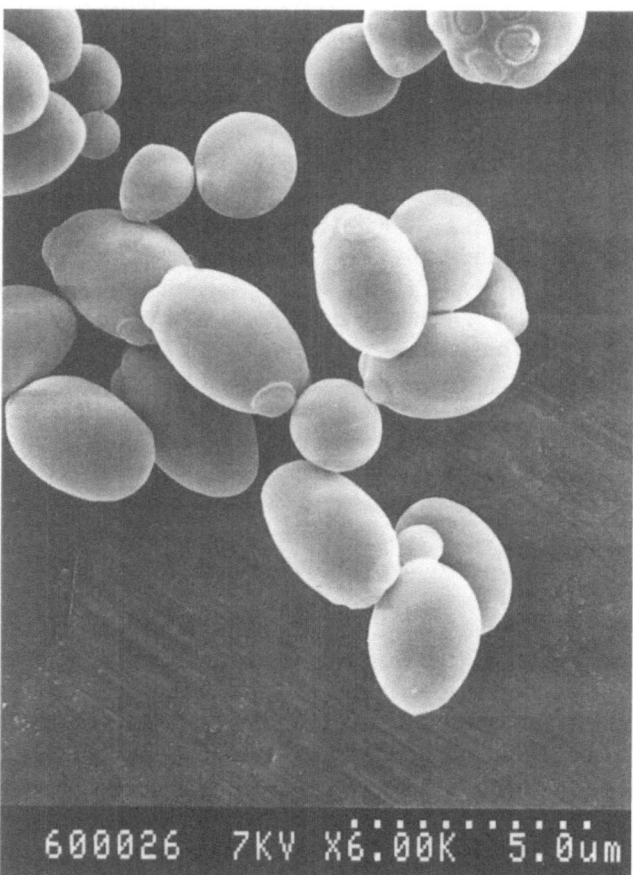

Fig. 1. *Kluyveromyces lactis*. A scanning electron micrograph of *K. lactis* strain CBS 2359. (Courtesy of Professor Masako Osumi, Department of Biology, Japan Women's University, Tokyo)

majority of yeast species, including *K. lactis*, seem to be essentially aerobic organisms. *K. lactis* cultures, which smell fruity, suggesting the presence of organic acid esters, are easily distinguishable from the familiar *S. cerevisiae* cultures. Under strict anaerobic conditions, most *K. lactis* strains cannot grow even with added sterols and fatty acids, although there is a report on anaerobic growth on glucose and fructose (Entian and Barnett 1983). In aerobic conditions, *K. lactis* ferments glucose, producing ethanol, but the respiratory system also functions throughout the growth phase. Apparently, biogenesis of the respiratory system in *K. lactis* is not sensitive to glucose repression. It has been described that many petite-negative species, which are generally thought to be obligate aerobes, are not subject to glucose repression (Bulder 1964a; de Deken 1966) with the known exception of *Schizosaccharomyces pombe* (Heslot et al. 1970; Goffeau et al. 1975). *K. lactis* is one of such petite-negative species. Indeed, it seems to be relatively insensitive to glucose repression. In contrast, *S. cerevisiae* grows on glucose essentially by alcoholic fermentation even in aerated conditions, and the respiratory pathway is induced only when the glucose level approaches exhaustion. The so-called glucose effect constitutes a major control mechanism of many metabolic processes in *S. cerevisiae*, whilst the response of *K. lactis* to glucose is clearly different (see Sect. 9).

3
Growth Media

K. lactis assimilates a wide variety of carbon sources. Table 1 lists selected substrates that distinguish *K. lactis* from *S. cerevisiae* and *Schizosaccharomyces pombe*. Compared to them, *K. lactis* can grow on a much wider range of substrates. Table 2 shows the growth rates of *K. lactis* in liquid cultures with the carbon sources commonly used in the laboratory. The growth rate on lactose is quite high, and the respiratory substrates are excellent carbon sources for growth. Comprehensive reviews are available on the utilization of sugars by various yeasts (Barnett 1976, 1992).

For routine laboratory studies, the basic culture conditions are the same as those generally used for *S. cerevisiae*. The complete medium contains 1% yeast extract, 1% (sometimes 2%) peptone, supplemented with appropriate carbon sources (2% glucose in the routine "complete glucose medium"). The basic minimal medium is made of 0.67% Yeast Nitrogen Base without amino acids, Difco, supplemented with carbon sources. As respiratory substrates, glycerol, DL-lactate, ethanol or succinate may be used. A specific medium is ME agar (5% malt extract Difco, with 3% agar; Wickerham 1951) which is used for mating and sporulation (see below). Various brands of microbiological grade agar were found satisfactory for culture, but for plate assays of the killer toxin, Oxoid no.1 agar is preferred to Difco Bacto agar, as the response is more sensitive, possibly due to reduced contents of unkown inhibitors of the toxin.

K. lactis is usually grown at a temperature of 25–28 °C. The upper limit is approximately 40 °C. At this critical temperature, *S. cerevisiae* appears to be more

Table 1. Growth of *K. lactis* on various carbon sources. (Barnett et al. 1990)

Substrate	*K. lactis*	*S. cerevisiae*	*Sch. pombe*
D-Glucose	+	+	+
D-Galactose	+	+, −	−, D
L-Sorbose	+	−	−
Sucrose	+	+	+
Maltose	+	+	+
Cellobiose	+	−	−
Salicin	+	−	−
Arbutin	+	−	−
Melibiose	−	+, −	+, −
Lactose	+, −	−	−
Raffinose	+	+, −	+
Melezitose	+	+, −	−
D-Mannitol	+	−, D	−
Ethanol	+	+, −	−
Glycerol	+	+, −	−
Xylitol	+, D	−	−
Butane-2,3-diol	+	−	−
Citrate	+, −	−	−
DL-Lactate	+	D	−
Succinate	+	−, D	−
Ethylamine	+	−	+, −
L-Lysine	+	−	+, −
Cadaverine	+	−	+, W
No niacin	−	+	−
+0.1% cycloheximide	+	−	−

D: delayed growth; W: weak positive growth.

Table 2. Growth rates of *K. lactis* on common carbon sources

Carbon source	Lactose	Glucose	Glycerol	Galactose	Ethanol
Doubling time, min	78	84	96	108	110
Final cell density	8.4	10.6	7.6	10.2	6.4

Strain 2359/152 (a methionine-requiring clone of CBS 2359) was grown at 28 °C in a 5-ml medium containing yeast extract and peptone, 1% each, supplemented with the indicated carbon source at 2%. The final cell density, expressed as cell number per ml ($\times 10^8$), was taken after 48 h of shake culture.

resistant than *K. lactis*. A heat shock of 50 °C, 10 min, kills a large fraction of *K. lactis* cells (therefore, caution should be taken when plating with molten agar).

4
Available Strains

When we trace the origins of the current laboratory strains, we find that only a few independent isolates had been used in early genetic crosses and mutant isolation.

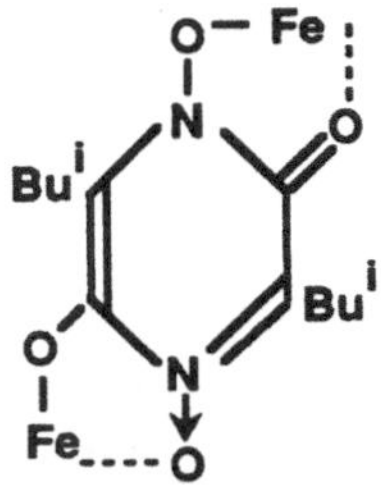

Pulcherrimin

Fig. 2. Pulcherrimin. This red pigment is an iron complex of 2:5 diisobutyl-1:4-dihydroxy-3:6-dioxopiperazine (Cook and Slater 1956). The structure can be viewed as a leucine dimer. It is produced by several species of yeast, including *Kluyveromyces lactis*, and is reported to have antibiotic activity

Most *K. lactis* researchers took advantage of the known mating types of the strains used by Halvorson and collaborators. Thus, the major genetic background seems to come from three NRRL strains:

- NRRL Y-1140, mating type **a** (CBS 2359, ATCC 8585),

- NRRL Y-1118, mating type α (CBS 6315, ATCC 8563),

- NRRL Y-1205, mating type α (CBS 2360, ATCC 8651).

- NRRL Y-1140 (CBS 2359) is probably the most widely used strain. Several investigators have proposed that this strain be the reference material for genetic and molecular studies (Annual Workshop Biology of *Kluyveromyces* II, Rome, 1989), although the type strain of the species is NRRL Y-8279 (CBS 683). NRRL Y-1140 has the particularity that it carries the killer plasmids pGKL1 and 2 (see later sections).

The strain NRRL Y-1205 (CBS 2360) also has its characteristic traits. It is one of the natural variant strains that have a recessive *rag1* allele, deficient in low-affinity glucose transport (Goffrini et al. 1990). A similar allele was also found in several other strains (CBS 141, CBS 5618, and CBS 8043) among the 30 independent isolates examined (CBS collection). As the *rag1* strains show particular responses to mitochondrial inhibitors and possibly to glucose repression (see below), care should be taken as to the allelic status of the *RAG1* locus in certain areas of research. Under normal conditions of culture, the *rag1* strains do not show any obvious deficiency for growth.

Liquid cultures of *K. lactis* may be conserved for months in the refrigerator. Old, large colonies on agar plates are sometimes surrounded by a pink-colored ring due to the pigment pulcherrimin (Fig. 2). For long-term conservation, strains are best maintained frozen at −70 to −80 °C in complete liquid medium containing 30–40% glycerol.

5
Genetic Techniques

5.1
Life Cycle

K. lactis strains are usually heterothallic. The complementary mating types are designated **a** and α. A gene coding for a protein highly homologous to the mating factor α of *S. cerevisiae* has been identified (Brake et al. 1988). The life cycle of *K. lactis* is essentially similar to that of *S. cerevisiae*. However, the diploid phase of *K. lactis* is transitory for the majority of cells (Herman and Halvorson 1963a; Herman and Roman 1966). Cultures of diploid cells tend to sporulate spontaneously. This particular feature is taken into account in the protocols of genetic experiments with *K. lactis*.

5.2
Sexual Crosses and Tetrad Analysis

The following procedure of genetic cross stems from the description given by Herman and Halvorson (1963a).

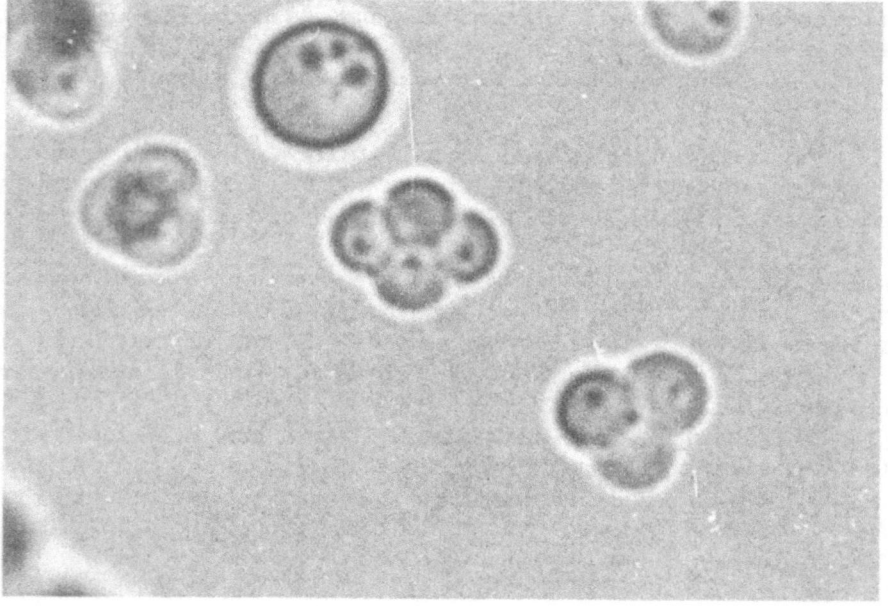

Fig. 3. Asci of *Kluyveromyces lactis*. Optical microscopy of asci from a cross of two laboratory strains

Two strains of complementary genotypes are grown overnight in a liquid complete medium. On a ME agar plate, a drop of one culture is superposed with a drop of the partner culture. The plate is incubated for 2 days at 28 °C. The diploid colonies arising from the zygotes are selected by replica plating onto a minimal plate. After 2 days of incubation at 28 °C, the diploid colonies are again replicated onto a ME plate and left overnight at 28 °C. The four-spored asci (Fig. 3) are easily recognizable under the microscope. Massive sporulation can usually be obtained in 2–3 days, but some crosses may need a much longer time, in which case the plates are kept in the refrigerator for several days, then examined for the presence of asci. For dissection of asci, a loopful mass of the asci-containing population is mixed with 90 μl of complete glucose medium and 15 μl of Glusulase (Endo Laboratories). The mixture is kept at room temperature for 6 min, then placed on ice to stop digestion. The asci can be dissected by any type of micromanipulator. The spores of *K. lactis* are somewhat smaller than those of *S. cerevisiae*, but can be micromanipulated in the same way. These asci are evanescent, that is, they tend to release spores. Therefore, the asci should not be overdigested before dissection. Generally, *K. lactis* strains are very sensitive to the cell wall-digesting enzymes as compared to *S. cerevisiae*. For asci digestion or spheroplasting, the amount of Glusulase or Zymolyase required may be roughly 10–20% of that used for *S. cerevisiae* manipulation.

5.3
Mutagenesis

Chemical and radiation mutagenesis of *K. lactis* can be performed by the standard methods used for *S. cerevisiae* (for example, Sherman et al. 1974). Under ultraviolet light, 50% survival was observed around the dose of 25–30 J/m^2 (250–300 erg/cm^2) with three strains tested. During test plating after mutagenesis, cells suspended in water can be stored at 4 °C for several days without loss of viability, but die massively at room temperature.

6
Chromosomal DNA

6.1
Chromosomal DNAs and Genome Size

Pulsed field gel electrophoresis distinguishes six DNA bands (CBS 2359, CBS 2360, CBS 683, CBS 141) (Steensma et al. 1988; Sor and Fukuhara 1989). They are numbered I through VI, from the smallest of about 1 million base pairs (Mbp) to the largest (about 3 Mbp). The bands IV and V often separate poorly, depending on the strain. In addition, the band IV may appear sometimes diffuse due to the presence of the repetitive ribosomal DNA. The sum of the molecular weights suggests a genome size of about 12 Mbp, which may not be significantly different from the 15-Mbp genome size of *S. cerevisiae*, considering the low accuracy of the measurement. While the latter yeast has a completely different karyotype with 16 chromosomes of much smaller size, there are, however, some indications that the local order of genes can often be similar in both species. Such a possibility may be worth consideration in the search for genes in *K. lactis* chromosomes. Comparative data are still very limited on this interesting point.

6.2
Chromosome Separation by Pulsed Field Gel Electrophoresis
(see also Chap. 3, this Vol.)

Early experiments (De Jonge et al. 1986; Johnston et al. 1988) have shown that *K. lactis* chromosomes contained DNA molecules of relatively large sizes. Separation of this size range of DNA can be achieved by the techniques recently developed by several laboratories (Sor 1988; Steensma et al. 1988; Sor and Fukuhara 1989; Viovy et al. 1992; Sor 1992). Figure 4 shows the electrophoretic karyotypes of various *K. lactis* strains of independent origins. The patterns are remarkably homogeneous, and some strains of *K. drosophilarum* show a similar pattern (Sor and Fukuhara 1989). For electrophoretic separation of chromosomal DNA, as shown in Fig. 4, an experimental protocol was provided by Dr. F. Sor (Institut Curie, Orsay):

Yeast is grown overnight in 5 ml liquid medium (standard glucose complete medium) to early stationary phase (about 2×10^8 cells/ml). Cells are collected by centrifugation (5000 rpm for 5 min) and suspended in 50 mM EDTA containing 50 μg/ml Zymolyase 100T (Seikakagu Corp., Tokyo) at a density of 2×10^9 cells/ml.

An equal volume of 1% low gelling temperature agarose (SeaPlaque, FMC Corp., or equivalent, dissolved in 125 mM EDTA and held at 50 °C) is added. The mixture is poured in 100-μl molds (Pharmacia, Uppsala, or equivalent, depending on the electrophoresis apparatus used) and allowed to harden at 4 °C.

The agarose blocks are removed from the moulds and placed in 5 ml (for 10–12 blocks) of a solution containing 500 mM EDTA, pH 8–9, 7.5% mercaptoethanol-2, 50 mM Tris, pH8. These blocks are incubated at 37 °C for at least 4 h, usually overnight.

Then the solution is withdrawn and replaced by 5 ml of a solution containing 500 mM EDTA, pH 8–9, 1% Sarkosyl, and Proteinase K (1 mg/ml). The incubation is continued for 6–24 h at 50 °C, after which the blocks are stored at 4 °C in 500 mM EDTA/1% Sarkosyl. They are stable for years.

The CHEF electrophoresis is performed on a 15-cm gel (Gene Navigator, Pharmacia) using a standard low M_r agarose (Biorad, ref. 162-0102 or equivalent) at gel concentrations depending on chromosome size. Usually 1% gels allow a wide spread of the chromosomes on the gel (Fig. 4A). This gives a good resolution for Southern blot analysis. To improve the separation between the chromosomes IV and V, which migrate together in certain strains, 1.5% gel (Fig. 4B) may be used.

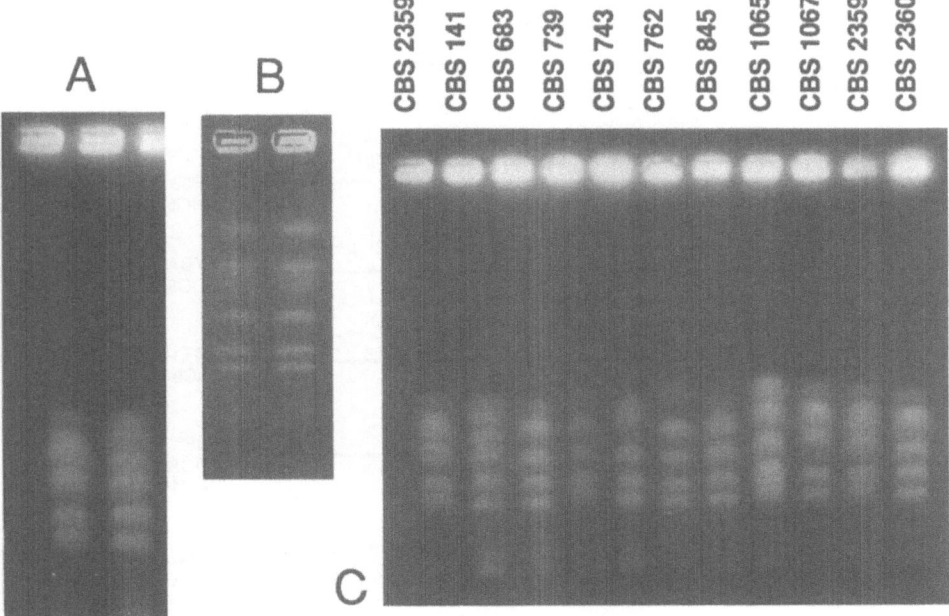

Fig. 4A–C. Pulsed field gel electrophoresis of *K. lactis* DNA. CHEF migration of *K. lactis* strains. A 1% agarose gel: *left lane* CBS 2359; *right lane* CBS 141. B 1.5% gel; *both lanes* CBS 2359. C Various *K. lactis* strains: *from left to right* CBS 2359, CBS 141, CBS 683, CBS 739, CBS 743, CBS 762, CBS 845, CBS 1065, CBS 1067, CBS 2359 (2×), CBS 2360. Electrophoresis was run at 50 V for 190 h, and the pulse time varies from 1200 to 2500 s with a logarithmic ramp

In a current protocol, electrophoresis is run at 130 V for 73 h. The temperature of the buffer is maintained at 9–10 °C. For the first 16 h the pulse time is 100 s, then it varies from 180 to 360 s with a linear ramp during the following 51 h, to be of 360 s for the last 6 h. The conditions should be adjusted as a function of the equipment used, because the separation is very sensitive to the configuration of gels and electrodes.

7
Genes and Genetic Markers

Auxotrophic mutations for various amino acids have become available from several laboratories. Most of the strains in which these mutations were obtained seem to have, at least in part, NRRL Y-1140-related background. To describe the genetic loci, the basic rules used for *S. cerevisiae* have been spontaneously adopted by most *K. lactis* workers: a locus is written by three letters followed by a locus number, all in italics. In practice, however, the published names of mutations are often degenerate and redundant, because they have been isolated independently by different

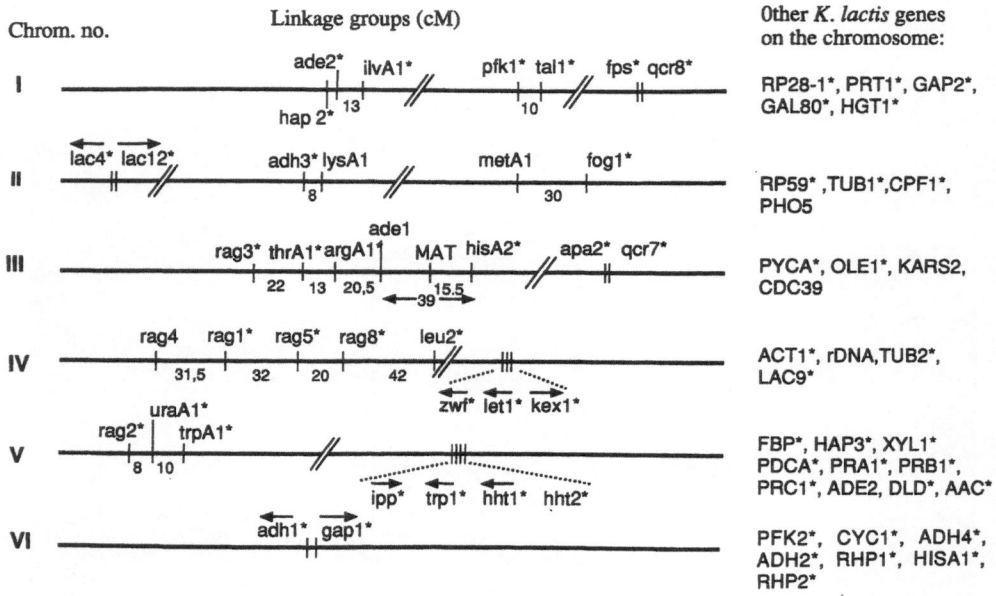

Fig. 5. A linkage map of chromosomal genes. This map is a compilation of data assembled from heterogeneous sources (see text). Chromosome sizes are not in proportion. The *numbers* are the distances in centimorgans calculated from genetic crosses and the position with respect to the centromeres is not known. *Asterisks* indicate that these genes have been cloned and localized by hybridization with chromosome blots. The *trp1* marker on chromosome V is homologous to the *trp1* mutation of *S. cerevisiae*, and the *ade1* mutation, which is homologous to the *ade1* mutation of *S. cerevisiae* (Zonneveld, unpubl. data) is allelic to the *ade1* mutation of the strains WM52 and W600B (Brunner et al. 1973; Tingle et al. 1968). The *ADE2* gene on chromosome V is homologous to the *ADE2* gene of *S. cerevisiae* (Zonneveld, unpubl. data) and corresponds to the *ade1-1* mutation from Dickson's strains (Sheetz and Dickson 1980), and WM12 strain (Tingle et al. 1968). The *ade2* marker on chromosome I is from W600B strain and is different from the former. The *uraA1* mutation is allelic to the cloned *KlURA3*. The *LEU2* gene is homologous to the *LEU2* gene of *S. cerevisiae*. Many of the *K. lactis* gene probes used for chromosomal hybridization in the authors' laboratory were kindly provided by the colleagues who cloned them (as indicated here in *parentheses*): *KlADH1* (Saliola et al. 1990); *PRA, PRB, PRC* (Shuster 1990), and *KlURA3* (Shuster et al. 1987); *LAC4, LAC9,* and *KlGAL80* (Breunig et al. 1984; Salmeron and Johnston 1986; Zenke et al. 1993); *KlACT1* and *RP59* (Deshler et al. 1989; Larson and Rossi 1991); *KlCYC1* (Clark-Walker 1991; Freire Picos et al. 1993); *KlTRP1* (Stark and Milner 1989); *KlPFK1* and *KlPFK2* (Heinisch et al. 1993); *KlQCR8, KlQCR7, KlCPF1* (Mulder et al. 1994a,c,d) and *KlHAP3* (Mulder et al. 1994b); *KlLEU2* (Zhang et al. 1992); *RP28-1, PRT1, GAP2, HGT1, PYCA, OLE1, XYL1,* and *PDCA* (Ménart, unpubl. data); *KLHAP2* (Nguyen et al. 1995); *KlFBP* (Zaror et al. 1993); *KlPHO5* (Ferminan and Dominguez, unpubl. data). Some of the data are personal communications from the authors: *FOG1*, and its genetic linkage to the *metA1* marker (Goffrini and Ficarelli, unpubl. data); *KlAAC* (Viola and Galeotti, unpubl. data); *KlDLD* (Lodi et al. 1994, and Lodi and Goffrini, pers. comm.), *KlTUB1, KlTUB2, KARS2, KlCDC39, ade1,* and its genetic linkage to the *argA1* marker (Zonneveld, unpubl. data)

laboratories. For example, there are two *ade1* corresponding to different functions. To avoid confusion with already described loci, some laboratories add a provisional identification letter before the locus number (for example, *lysA1*). Another problem concerns the genes whose functional equivalent is known in *S. cerevisiae*. A *trp1* of *K. lactis* may or may not correspond to *trp1* of *S. cerevisiae*, depending on the laboratory who named it. There is some consensus that the loci of known function be written, for example, *KlTRP1* or *KlURA3* to show their equivalence to *S. cerevisiae TRP1* or *URA3*, respectively. The genes cloned by functional complementation of known *S. cerevisiae* mutations tend to adopt this expression, although, strictly speaking, they do not mean genetic loci. A unified nomenclature in *K. lactis* genetics remains to be established. A compilation of early auxotrophic mutations is in preparation.

The *ade* and *leu* markers from one of the classical strains, W600B, have been in use in many laboratories. Unfortunately, the origin of this strain is not clear. Despite our search, helped by Dr. C.P. Kurtzman and Dr. A. Herman (NRRL, Peoria), the original record of this strain could not be recovered. However, W600B and some of its meiotic progeny show a few characteristic traits: deletion and duplication of specific chromosomal segments as well as its high sensitivity to glucose repression. These traits are all found in the strain NRRL Y-1118, which had been used by the Wisconsin group. It is therefore likely that W600B is a close relative of Y-1118.

Compared to the impressive genetic map of *S. cerevisiae*, describing more than 800 loci, *K. lactis* map is still a desert. Only several genetic linkage groups have been reported (Tingle et al. 1968; Sheetz and Dickson 1980; Wésolowski-Louvel et al. 1992b) and some were physically mapped to the individual chromosomes of *K. lactis*. Five of the six centromeres have recently been cloned (Heus et al. 1990). Genes tightly linked to the centromeres of known chromosomes have not been reported. Chromosomal telomere sequences have been reported by McEachern and Blackburn (1994). These sequences show a complex pattern distinct from that of *S. cerevisiae*, but conserve the TG-rich motif. Figure 5 is a sample of linkage data including unpublished results contributed by different laboratories.

8
K. lactis Genes vs. *S. cerevisiae* Genes

8.1
Sequence Homology of Gene Products

The question in most minds is the phylogenetic distance between *K. lactis* and *S. cerevisiae*. The GC content of nuclear DNA is fairly similar in *S. cerevisiae* (39%) and *K. lactis* (40%). Comparison of ribosomal DNA sequences (Verbeet et al. 1984; Barns et al. 1991; Hendriks et al. 1992) indicates that *K. lactis* is much closer to *S. cerevisiae* (0.0217 changes per nucleotide on average) and *Torulopsis glabrata* (0.0155) than to *Hansenula polymorpha* (0.0340), *Candida parapsilosis* (0.0368), or *Candida krusei* (0.0655), but that the nontranscribed spacer region is completely divergent in all these species. We may obtain some idea of relatedness by comparing the amino acid sequences of equivalent proteins from the two species. Some of the available data are shown in Table 3.

We can note that abundantly produced proteins, such as those of glycolytic enzymes, generally show a high degree of amino acid identity with *S. cerevisiae* products. Also ribosomal protein sequences are particularly well conserved. Regulatory proteins sometimes have diverged considerably except for interspersed conserved blocks (Salmeron and Johnston 1986; Jakobsen and Pelham 1991; Oberyé et al. 1993; Zenke et al. 1993). The cloning of *K. lactis* homologues is therefore being used by an increasing number of yeast researchers to facilitate assignment of functional domains.

The use of known *S. cerevisiae* gene probes to clone, by sequence homology, the equivalent genes in *K. lactis*, or vice versa, is an obvious approach, but many *S. cerevisiae* probes give rather weak hybridization signals on the Southern blots of *K. lactis* genomic digests, because of dispersed homology of nucleotide sequences.

Table 3. Amino acid sequence identity of proteins between *K. lactis* and *S. cerevisiae*

Gene/product K. lactis	S. cerev.	Function	Amino and identity, % global/local[a]	Reference
KlACT1	ACT1	Actin	97	(31)
KlS33	S33	Ribosomal protein	95	(28)
KlL25	L25	Ribosomal protein	93	(29)
KlL32	L32	Ribosomal protein	91	(30)
PR59	CRY1	Ribosomal protein	95	(1)
KlADH1	ADH1	Alcohol dehydrogenase	85	(35)
KlADH2	ADH1	Alcohol dehydrogenase	81	(2, 35)
KlADH3	ADH3	Alcohol dehydrogenase	80	(35)
KlADH4		Alcohol dehydrogenase		(34)
KlCYC1	CYC1	Cytochrome c	89	(3, 33)
KlAPA2	APA2	Bis 5(nucleosidyl)tetraphospatase	51	(38)
KlQCR7	QCR7	Respiratory complex III subunit 7	69	(38)
KlQCR8	QCR8	Respiratory complex III subunit 8	90	(4)
KlAAC2	AAC2	ADP/ATP translocator	89	(5)
KlLEU2	LEU2	IPM dehydrogenase	85	(6)
KlURA3	URA3	OMP decarboxylase	81	(7)
KlTUB1	TUB1	Tubulin	78	(8)
KlTUB2	TUB2	Tubulin	77	(8)
KlPGK	PGK	Phosphoglycerate kinase	83	(9)
KlFPS	FPS	Farnesyl biphosphate synthase	74	(38)
KlGAP	GAP	Glyceraldehyde P dehydrogenase	82	(10)
KlIPP	IPP	Inorganic pyrophosphatase	83	(11)
KlPFK1	PFK1	Phosphofructokinase subunit	73	(12)
KlPFK2	PFK2	Phosphofructokinase subunit	70	(12)
KlTAL1	TAL1	Transaldolase	76	(41)
KlZWF	ZWF	Glucose-6-phosphate dehydrogenase	70	(40)
KlDLD	DLD	Dlactate-cytochrome c reductase	80	(13)
RAG2	PGI	Phosphoglucose isomerase	86	(14)
RAG3	PDC2	Pyruvate decarboxylase regulator	50/70	(39)

Table 3 (*contd.*)

Gene/product *K. lactis*	*S. cerev.*	Function	Amino and identity, % global/local[a]	Reference
RAG5	HXK2	Hexokinase	73	(15)
RAG1	LGT1	Low-affinity glucose permease	68	(16)
KlMOL2	MOL2	Thiamine synthesis	60–65	(17)
KlHSP12	HSP12	Heat shock protein	59	(17)
KlTRP1	TRP1	PRA isomerase	53	(11)
KEX1	KEX2	Processing protease	73	(18, 32)
BiP	KAR2	HSP family protein	77	(19)
KlHAP3	HAP3	Transcription factor	88	(21)
KlTFIIB	TFIIB	Transcription factor	67	(37)
KlGAL80	GAL80	Transcription factor	60/90	(25)
KlREB1	REB1	DNA binding protein	40/82	(27)
KlGAL11	GAL11	Transcription factor	37	(26)
KlABF1	ABF1	Transcription factor	28/75	(23)
LAC9	GAL4	Transcription factor	24/41–88	(22)
KlHAP2	HAP2	Transcription factor	19/94	(20)
KlSTE12	STE12	Transcription factor	10–20/78	(24)

[a] Conserved functional domain(s).

References: (1) Larson and Rossi 1991; (2) Shain et al. 1992; (3) Clark-Walker 1991; (4) Mulder et al. 1994c; (5) Viola et al. 1992; (6) Zhang et al. 1992; (7) Shuster et al. 1987; (8) B. Zonneveld, pers. comm.; (9) Fournier et al. 1990; (10) Shuster 1990; (11) Stark and Milner 1989; (12) Heinisch et al. 1993; (13) Lodi et al. 1994; (14) Wésolowski-Louvel et al. 1988a; (15) Prior et al. 1993b; (16) Goffrini et al. 1990 and Prior et al. 1993a; (17) D. Walsh and P. Meacock, pers. comm.; (18) Tanguy-Rougeau et al. 1988; (19) Lewis and Pelham 1990; (20) Nguyen et al. 1995; (21) Mulder et al. 1994b; (22) Salmeron and Johnston 1986; (23) Gonçalves et al. 1992; (24) Yuan et al. 1992; (25) Zenke et al. 1993; (26) Mylin et al. 1991; (27) Morrow et al. 1993; (28) Hoekstra et al. 1992; (29) Bergkamp-Steffens et al. 1992; (30) Eng and Warner 1991; (31) Deshler et al. 1989; (32) Tanguy-Rougeau 1991; (33) Freire Picos et al. 1993; (34) Saliola et al. 1991; (35) Saliola et al. 1990; (36) Chang and Dickson 1988; (37) Na and Hampsey 1993; (38) Mulder et al. 1994a; (39) Prior 1993; (40) Wésolowski-Louvel, unpubl. data; (41) Jacoby et al. 1993.

Such genes are expected to be better identified by colony hybridization of genomic libraries in which specific DNA segments are amplified in individual colonies.

Alternatively, isolation of *K. lactis* homologues may be accomplished by complementation of *S. cerevisiae* mutants. This strategy was successful even in cases where interaction between homologous and heterologous protein components was required for function (Eng and Warner 1991; Heinisch et al. 1993; Na and Hampsey 1993). Several genes encoding transcriptional regulators also could be cloned by complementation: *GAL4* (Salmeron and Johnston 1986); *REB1* (Morrow et al. 1993); *CPF1* (Mulder et al. 1994a); *ABF1* (Gonçalves et al. 1992; Oberyé et al. 1993); *HSF* (Jakobsen and Pelham 1991); *GAL11* (Mylin et al. 1991); *HAP2* (Nguyen et al. 1995); *HAP3* (Mulder et al. 1994b).

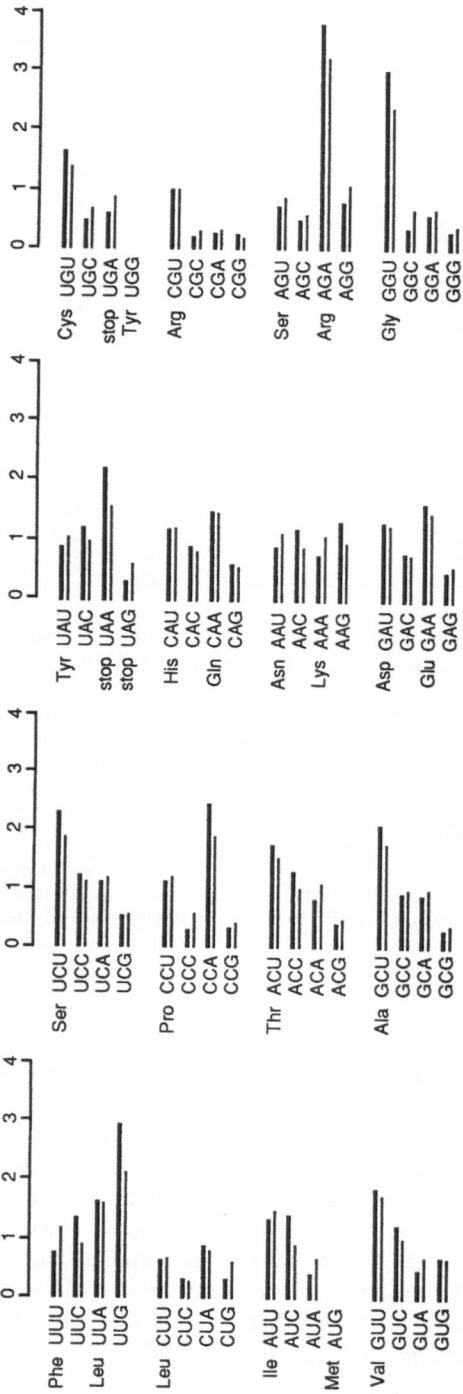

Fig. 6. Relative synonymous codon usage (RSCU) in *K. lactis* and *S. cerevisiae*. Adapted from the data by Lloyd and Sharp (1993). *K. lactis* and *S. cerevisiae* are very similar in codon selection. The RSCU scale is shown at the *top*. *Thick bars* represents *K. lactis* and *thin bars* S. *cerevisiae*

Table 4. Codon adaptation index (CAI) of *K. lactis* genes. (Based on data from Lloyd and Sharp 1993)

	Gene	Product	CAI
Ribosome	RPL32	Ribosomal protein L32	0.90
	RPL30	Ribosomal protein L30	0.87
	RPL25	Ribosomal protein L25	0.87
	RPL41	Ribosomal protein L41	0.83
	CPY59	Ribosomal protein 59	0.78
	RPS33	Ribosomal protein S33	0.63
Glycoslysis	GAP1	Glyceraldehyde-phosphate dehydrogenase	0.77
	ADH1	Alcohol dehydrogenase	0.74
	ADH3	Alcohol dehydrogenase	0.73
	PGK1	Phosphoglycerate kinase	0.73
	ADH4	Alcohol dehydrogenase	0.72
	ADH2	Alcohol dehydrogenase	0.59
	RAG2	Phosphoglucose isomerase	0.54
	RAG1	Glucose transporter	0.42
	PFK1	Phosphofructokinase subunit α	0.33
	FBP1	Fructose-1,6-bisphosphatase	0.32
	PFK2	Phosphofructokinase subunit beta	0.30
Respiration	CYC1	Cytochrome c	0.85
	TAL1	Transaldolase	0.69
	ZWF	Glucose-6-phosphate dehydrogenase	0.22
GAL/LAC genes	GAL1	ATP-galactose-1-phosphotransferase	0.30
	GAL10	UDP-galactose-4-epimerase	0.29
	LAC12	Lactose permease	0.26
	GAL7	Glactose-1-phosphate uridyltransferase	0.25
	LAC4	beta-galactosidase	0.23
Regulation	REB1	DNA binding protein	0.21
	SUA7	Transcription factor	0.17
	GAL11	Transcription Transcription factor	0.16
	HSF	Heat shock transcription factor	0.16
	ABF1	ARS binding factor	0.15
	RAG3	Pyruvate decarboxylase regulator	0.14
	GAL80	Negative regulator of LAC9	0.11
	LAC9	Positive regulator of LAC4/GAL genes	0.11
Others	ACT1	Actin	0.79
	IPP	Inorganic phosphatase	0.70
	HHT1	Histone H3	0.69
	LEU2	beta-isopropylmalate dehydrogenase	0.47
	KAR2	BiP/HSP70 homologue	0.45
	URA3	Orotidine-5-phosphate decarboxylase	0.23
	ERD2	ER protein retention receptor	0.22
	TRP1	Anthranylate isomerase	0.19
	KEX1	Serine protease	0.13
	LET1	Hydrogenase-like protein	0.10

Noncoding regions have widely diverged between *S. cerevisiae* and *K. lactis*; only protein-binding sites seem to be conserved. At least there are potential binding sites for *S. cerevisiae* regulatory factors in a number of *K. lactis* genes: *LEU2* (Zhang et al. 1992); *CYC1* (Freire Picos et al. 1993); *GAL1-GAL10* (Webster and Dickson 1988); *GAL80* (Zenke et al. 1993); *QCR7* and *QCR8* (Mulder et al. 1994a,c); *PFK2* (Heinisch et al. 1993, and pers. comm.). In most cases, however, their functional significance remains to be established.

The evidence that equivalent activators exist in *K. lactis* is supported by the fact that most *S. cerevisiae* promoters that have been tested so far (see Table 6) also proved to function in *K. lactis*. Among them are strong constitutive as well as regulated ones, confirming the conservation of regulatory sequences. A case where regulation of a *S. cerevisiae* gene was not maintained after transfer to *K. lactis* has also been reported. The metal-inducible *CUP1* gene of *S. cerevisiae* is constitutively expressed in *K. lactis* (Macreadie et al. 1991).

8.2
Codon Usage

Codon selection in *K. lactis* chromosomal genes has been reviewed by Lloyd and Sharp (1993) on the basis of 47 available gene sequences. It appears that, in all essential features, codon usage in *K. lactis* is very similar to that in *S. cerevisiae* (Fig. 6). These authors concluded that mutational biases and abundances of particular tRNAs have not diverged between the two species. Highly expressed genes have high codon usage bias (Table 4). We note that the single cytochrome c gene of *K. lactis* has a very high codon usage bias (0.85), which might be related with the aerobic nature of this species. In *S. cerevisiae*, the two cytochrome c (*iso*-1 and *iso*-2) genes have a CAI of 0.37 and 0.18 respectively (*iso*-1 is usually about ten times more abundant than *iso*-2).

9
Regulation of Carbon Metabolism

9.1
Lactose and Galactose Metabolism

One of the properties that distinguishes *K. lactis* from *S. cerevisiae* is the ability to use lactose as a sole source of carbon and energy. Lactose degradation by cells or purified enzymes has found a number of applications in the dairy industry. Large-scale fermentation and enzyme preparation from *K. lactis* are therefore well-established fields of activity.

Lactose metabolism is controlled by two linked genes, *LAC4* and *LAC12*, encoding β-galactosidase (Dickson and Markin 1978; Dickson 1980; Poch et al. 1992) and lactose permease (Chang and Dickson 1988), respectively. These genes, when transferred to *S. cerevisiae*, were sufficient to allow growth on lactose (Sreekrishna and Dickson 1985).

The lactose permease, which enables the cells to take up the chromogenic glycoside 5-bromo-4-chloro-3-indolyl-α-D-glucopyranoside (X-gal), allows an estimation of *lacZ* or *LAC4* gene expression by plating on X-gal containing media. Use of the *E. coli lacZ* reporter gene in gene expression studies, however, requires *lac4* mutant strains. *lac4* point mutations, as well as disruption mutants, have been reported (Sheetz and Dickson 1981; Das and Hollenberg 1982; Gödecke et al. 1991).

The first genetic studies on lactose metabolism (Herman 1963; Herman and Halvorson 1963a,b) identified two unlinked loci, *LAC1* and *LAC2*, each of which can exist as a recessive mutation in the natural population (Herman and Halvorson 1963b). For example, NRRL Y-1140 is *lac1 LAC2* and Y-1118 is *LAC1 lac2*. Mating between such strains yielded meiotic *lac1 lac2* segregants showing reduced β-galactosidase activity (Boze et al. 1987). It is not known to date how these genes relate to the *LAC* genes that have been studied at the molecular level.

Major control of lactose metabolism occurs at the level of transcription of the many genes involved in the process (Lacy and Dickson 1981). These genes include *LAC4* (Sheetz and Dickson 1981), *LAC12* (Sreekrishna and Dickson 1985), *LAC5*, *LAC8*, *LAC9* (Sheetz and Dickson 1980), *LAC10* (Dickson et al. 1981), and *LAC11* (Riley and Dickson 1984). *LAC5*, *LAC8*, and *LAC11* encode the enzymes of the Leloir pathway of galactose metabolism and have been renamed *GAL7*, *GAL10*, and *GAL1*, respectively, in accordance with the homologous *S. cerevisiae* genes (Riley and Dickson 1984). These genes are coregulated with *LAC4* and *LAC12* at the transcriptional level by the regulatory genes *LAC9* and *KlGAL80* (Webster and Dickson 1988; Lacy and Dickson 1981; Wray et al. 1987; Zenke et al. 1993). *LAC9* or *KlGAL4* encodes a transcriptional activator, the *K. lactis* equivalent of Gal4p (Witte and Dickson 1990; Salmeron and Johnston 1986; Wray et al. 1987; Pan et al. 1990). Lac9p function is regulated by the negative regulator Gal80p in essentially the same way as Gal4p (Salmeron et al. 1989; Dickson et al. 1990; Zenke et al. 1993). Mutant phenotypes suggested that *KlGAL80* could be identical to the regulatory gene *LAC10* (Dickson et al. 1981).

Lac9p-controlled promoters can be activated by Gal4p and vice versa (Salmeron and Johnston 1986; Riley et al. 1987a), and both activators recognize the same consensus sequence (Breunig and Kuger 1987). Thus, the widely used *GAL1–GAL10* promoter of *S. cerevisiae* is efficiently expressed and regulated in *K. lactis*. Interestingly, the *K. lactis* genes *GAL7*, *GAL10*, and *GAL1* are not only similarly arranged, but the number and positioning of Lac9p binding sites in the *GAL1–GAL10* divergent promoter have also been conserved (Webster and Dickson 1988).

The regulon is induced by either lactose or galactose in the medium, resulting in a 50- to 250-fold activation of transcription (Dickson and Markin 1980; Dickson and Barr 1983; Das et al. 1985; Webster and Dickson 1988; Breunig 1989b). Induction by lactose requires β-galactosidase activity, indicating that activation of Lac9p function requires galactose (Dickson and Barr 1983; Breunig, unpub.). Both sugars can be taken up by the *LAC12* gene product. Galactose, but not lactose, can also enter the cells by (an)other carrier(s), but induction under these conditions is less efficient (Riley et al. 1987b; Zachariae and Breunig, submitted). To inactivate the

Gal80p repressing function, the *GAL1* gene product, galactokinase, is required (Meyer et al. 1991; Meyer 1993; Zachariae and Breunig, submitted). By mutation, a regulatory function is separable from the galactose phosphorylating activity and some *gal1* mutants are still inducible (Riley and Dickson 1984; Meyer et al. 1991; Meyer 1993). In *S. cerevisiae*, the same regulatory function seems to be redundantly contained within the *GAL1* and *GAL3* gene products (Meyer et al. 1991; Bhat and Hopper 1992).

In contrast to *S. cerevisiae*, in which induction of the *GAL1–GAL10* promoter takes many hours when cells are shifted from glucose to galactose medium, induction in *K. lactis* is rapid, since glucose repression is not as tight (see below). In addition, the promoter can be converted into a constitutive one, highly active even on glucose, by disrupting the *KlGAL80* gene (Zenke et al. 1993). In contrast, in a *S. cerevisiae gal80* mutant, this promoter is still largely repressed (Johnston and Hopper 1982; Nehlin et al. 1991). There is some evidence that this difference is at least partly due to the activator concentration, which, in glucose-grown cells of *K. lactis*, seems to be higher than in *S. cerevisiae* (see below and discussion in Zenke et al. 1993).

The *LAC4–LAC12* divergent promoter is regulated in a similar way as the *GAL1–GAL10* promoter but its size of 2.6 kb pairs (Chang and Dickson 1988; Gödecke et al. 1991) makes it less suitable in cloning experiments. Truncated versions have been used, but gave variable results possibly due to influences of flanking sequences or genetic background (Das et al. 1985; Ruzzi et al. 1987; Fleer et al. 1991b). In its chromosomal context, however, it was adopted for heterologous gene expression and gave high-level production strains (Van den Berg et al. 1990). By manipulating the expression of the *LAC9* gene, the *LAC4* promoter strength can be varied (Kuger et al. 1990; Gödecke et al. 1991; Zachariae and Breunig 1993; Zachariae et al. 1993). This offers the possibility to adjust the rate of transcription of heterologous genes according to their influence on cellular growth, alleviating the need to change the promoter or growth conditions.

9.2
Glucose Repression of Lactose/Galactose Metabolism

Glucose repression on the lactose/galactose system in *K. lactis* has already been much discussed (for example, Tingle and Halvorson 1972; Ferrero et al. 1978). The simultaneous presence of glucose and galactose in the medium can inhibit galactose induction, but the preference for glucose over galactose utilization under these conditions varies between strains (Breunig 1989b). In *S. cerevisiae*, this phenomenon of glucose repression, which controls many metabolic processes, is particularly pronounced and has been studied extensively. Experiments addressing the molecular basis of glucose repression of *GAL* gene expression in *S. cerevisiae* suggested a complex regulation involving overlapping Gal4p-mediated and Gal4p-independent mechanisms (Finley et al. 1990; Flick and Johnston 1990; Lamphier and Ptashne 1992). The Gal4-dependent pathway seems to consist primarily of a fivefold repression of *GAL4* gene expression by glucose, mediated by

MIG1 (Griggs and Johnston 1991; Nehlin et al. 1991). In *K. lactis LAC9* gene expression is controlled by autoregulation (Zachariae and Breunig 1993) and glucose repression (Kuzhandaivelu et al. 1992; Zachariae et al. 1993), and thus Lac9p-dependent and -independent mechanisms overlap already at the level of *LAC9* gene expression.

Galactose induction and glucose repression are two counteracting processes, and the efficiency of glucose repression crucially depends on the activity of Lac9p. Any manipulations that increase Lac9p activity (like the *gal80* mutation; Zenke et al. 1993) or rate of synthesis (Zachariae et al. 1993) allow induction to overcome the inhibitory effect of glucose. One cause for the differences in glucose repression of *LAC/GAL* genes between strains is an allelic variation in the *LAC9* promoter which affects the rate of *LAC9* expresson (Kuzhandaivelu et al. 1992; Zachariae et al. 1993). When the stronger promoter variant was introduced into a glucose-repressible strain, repression was alleviated (Kuzhandaivelu et al. 1992; Zachariae et al. 1993).

However, even in the repressible strains, glucose repression is not very pronounced. Genes that are tightly repressed by glucose in *S. cerevisiae*, like those encoding invertase or respiratory functions, are only moderately affected in *K. lactis*. A correlation between glucose sensitivity and glucose consumption exists between different *K. lactis* strains and the rate of glucose consumption is generally lower than in *S. cerevisiae* (Weirich 1992; Weirich et al. submitted), indicating that the regulation of glucose uptake may be a crucial determinant of glucose repression.

9.3
Regulation of Fermentation and Respiration

K. lactis is unable to grow anaerobically even on fermentable substrates. This phenomenon has been attributed to a combination of different factors that result in lowering of glycolytic flux under anaerobic conditions (or when respiration is blocked by inhibitors; Barnett 1992). A major determinant that limits the glycolytic flux seems to be the rate of sugar uptake. Some strains show a dependence on respiration even for growth on glucose as indicated by the so-called Rag⁻ phenotype (Rag means restistance to Antimycin A on glucose; Ferrero et al. 1978; Goffrini et al. 1989). A high variability with respect to glucose utilisation and ethanol production rates between strains has been observed, and a correlation exists between low glucose consumption rates and the Rag⁻ phenotype (Weirich 1992).

Natural Rag⁻ variants have been identified as mutant alleles of an inducible hexose transporter gene (*RAG1*) (Goffrini et al. 1990; Chen et al. 1992b; Wésolowsi-Louvel et al. 1992a). Some 13 other complementation groups have been identified by using the Rag⁻ phenotype in screening for glycolysis mutants (Wésolowski-Louvel et al. 1992b). One of these, *RAG2*, has been identified as the phosphoglucose isomerase gene (Wésolowski-Louvel et al. 1988a; Goffrini et al. 1991). *RAG5* encodes a hexokinase which in *K. lactis* seems to be a unique enzyme that phosphorylates glucose and fructose (Prior et al. 1993b). Some of these *RAG* genes interact with each other. For example, the regulation of the expression of the *RAG1* gene occurs at the level of transcription, and this transcription is under the control of *RAG4*, *RAG5*, and *RAG8* (Chen et al. 1992b; Wésolowski-Louvel et al. 1992b).

In contrast to *S. cerevisiae*, in which glycolysis mutants in general appear out of condition, *K. lactis* mutants appear in much better shape, probably because the cells can efficiently metabolize glucose through the pentose-phosphate pathway (Heinisch et al. 1993). In line with this, growth on glucose of mutants lacking phosphofructokinase activity was abolished when the transaldolase or the transketolase genes were mutated (Jacoby et al. 1993). Most of the known glycolysis mutants grow normally on respiratory substrates.

In contrast, respiratory-deficient mutants, isolated as small colonies on 2% glycerol/0.1% glucose agar plates (Herman and Griffin 1968; Del Guidice and Puglisi 1974) or obtained by gene disruption (Mulder et al. 1994c,d) often showed reduced growth rates even on glucose media.

These findings emphasize the role of oxidative metabolism which in *K. lactis* plays a more important role in glucose/fructose assimilation than in *S. cerevisiae*. The molecular basis for this difference is not yet clear. However, the variability between *K. lactis* strains with respect to this phenotype, which ranges from largely fermentative to largely oxidative glucose metabolism, offers an experimental approach to address this point.

It was recently shown that JA6, a fermentative strain with high glucose consumption and ethanol production rates, contains a hexose transporter gene, *KHT2*, not present in other strains (Weirich et al., submitted). Like the low-affinity transporter gene *RAG1*, it belongs to the family of *HXT* genes of *S. cerevisiae*. (Bisson et al. 1993) and is located right downstream of *RAG1*. Southern analysis of different strains using *RAG1*- and *KHT2*-specific probes revealed a high heterogeneity at that locus, but the rate of glucose consumption was not strictly correlated with the absence or presence of the *KHT2* gene (K. Breunig, unpubl. data).

Deletion of either one or both of these transporter genes leads to only a marginal reduction in the growth rate, indicating that at least one additional glucose carrier exists. A gene for a potential high-affinity glucose carrier has been isolated (Ménart and M. Wésolowski-Louvel, unpubl.). However, the hexose transporter mutant phenotypes in *K. lactis* indicate that there is a much lower redundancy in transporter genes than in *S. cerevisiae* (reviewed by Bisson et al. 1993).

Interestingly, mutants with reduced sugar uptake have been obtained as suppressors of a *Klggs1* mutant (Luyten et al. 1993). The *GGS1* gene of *S. cerevisiae* encodes a subunit of the trehalose-6-phosphate synthase/phosphatase complex (Bell et al. 1992; Van Aelst et al. 1993). Its mutant phenotype is similar to that of *S. cerevisiae*, showing growth inhibition by glucose and fructose and absence of all glucose-induced effects (Van Aelst et al. 1993). Thus, in both yeasts, trehalose metabolism seems to be involved in the control of glucose effects.

As mentioned above, the respiratory system in *K. lactis* does not seem to be glucose-repressed, at least not to the same extent as in *S. cerevisiae*, in line with its role during growth on fermentable carbon sources. The same is true for alcohol dehydrogenase. Four *ADH* genes (*KlADH1* to *KlADH4*; Saliola et al. 1990, 1991;

Shain et al. 1992) have been isolated, none of which was glucose-repressed to an extent similar to the *ADH2* gene of *S. cerevisiae*. Two of the enzymes, the products of *KlADH3* and *KlADH4* genes, are probably associated with mitochondria. The *KlADH4* promoter has a new feature not yet found for *S. cerevisiae* genes: it is induced by ethanol.

In *S. cerevisiae*, many enzymes of the respiratory system are under the control of the HAP2/HAP3/HAP4 regulatory complex. *K. lactis* possesses all the components of the complex which are capable of complementing the corresponding *hap* mutations of *S. cerevisiae*. Nevertheless, mutations of these genes do not seem to seriously affect the respiratory system of normal *K. lactis* strains, illustrating the difference for major energetic metabolism between these two species (Mulder et al. 1994b; Nguyen et al. 1995) in regulatory circuitry.

The interaction between respiratory and fermentative pathways is perhaps one of the key features that distinguishes *K. lactis* from *S. cerevisiae*. An explanation for this at the molecular level is an interesting challenge.

10
Mitochondria

10.1
Mitochondrial Mutations

The well-known cytoplasmic *petite colony* mutants of *S. cerevisiae* are respiratory-deficient. These mutants (also called *rho⁻* mutants) arise spontaneously, at an extremely high frequency, as a result of large deletion or complete loss of mitochondrial DNA. It is not known why *S. cerevisiae* has such an unstable mitochondrial genome. Known as a "petite-negative yeast" (Bulder 1963, 1964a,b), *K. lactis* does not produce *rho⁻* type deletions of mitochondrial DNA. Only recently, strains producing *rho⁻* mutants have been isolated from *K. lactis*. These mutants were obtained from strains which happened to contain a specific nuclear mutation called *mgi* (Hardy et al. 1989; Maleszka and Clark-Walker 1989; Chen and Clark-Walker 1993). It therefore appears that the petite negativeness is associated with the allelic status of a few nuclear genes. A similar situation had been reported for *Schizosacharomyces pombe*, another petite-negative yeast (Haffter and Fox 1992; Massardo et al. 1994). The nature of these nuclear genes is not known at present.

K. lactis is sensitive to various mitochondria-specific inhibitors, typically, erythromycin, chloramphenicol, oligomycin, and antimycin A. These antibiotics efficiently inhibit the growth of *K. lactis* on respiratory substrates (Zennaro et al. 1977). Response to chloramphenicol seems to be somewhat variable, depending on the strain, and tends to show clonal variation of sensitivity. *K. lactis* has been reported to be much more resistant to antimycin A than is *S. cerevisiae* (Brunner et al. 1987), but a concentration of $5\text{--}10\,\mu M$ is still sufficient to block mitochondrial complex III function. Several mitochondrial markers have been described which confer resistance to erythromycin (Algeri et al. 1977; Del Giudice and Brunner

1977), oligomycin (Brunner et al. 1977; Brunner and Tuena de Cobos 1980) and antimycin A (Brunner et al. 1987; Coria et al. 1989). Spontaneous clones resistant to erythromycin (4 mg/ml) or oligomycin (3 μg/ml) can be readily obtained after prolonged culture on the drug-containing plates, but many of them are highly unstable and revert when transferred to drug-free media. In a case of oligomycin resistance, Ragnini and Fukuhara (1989) found that the unstable resistance was due to the amplification of a specific segment of mitochondrial DNA, and that the amplified segment can be maintained only in the presence of the drug.

Cycloheximide is a specific inhibitor of cytoplasmic protein synthesis in *S. cerevisiae*, whilst erythromycin specifically blocks mitochondrial protein synthesis. *K. lactis*, like many petite-negative yeasts, is resistant to cycloheximide (van der Walt 1970). Thus there is a need for a specific inhibitor of cytoplasmic protein synthesis when one wishes to distinguish mitochondrial from cytoplasmic protein synthesis. Trichodermin (Leo Pharmaceuticals Co. Ltd.; reviewed by Vasquez 1979) has been reported to be an efficient inhibitor of cytoplasmic protein synthesis in this yeast (Marmiroli et al. 1979). Unfortunately, this drug is not commercially available. Trichodermin-resistant mutants are not known in *K. lactis*.

10.2
Mitochondrial DNA

Mitochondrial DNA amounts to almost 10% of the total cellular DNA, suggesting an average copy number of about 30 per cell. The molecule is circular (Sanders et al. 1974), about 39 kbp long (Wésolowski et al. 1981). Restriction sites and the map positions of major genes are known (Fig. 7; Hardy et al. 1989; Wilson et al. 1989; Skelly et al. 1991). Some polymorphic rearrangements of sequences have been noted, including a large translocation of a segment in one strain (Brunner and Coria 1989; Skelly et al. 1991).

The gene sequence data from *K. lactis* mitochondrial DNA are still fragmentary. Genes for cytochrome oxidase subunit 1 (Hardy et al. 1989), subunit 2 (Hardy and Clark-Walker 1990), and apocytochrome b gene (Brunner and Coria 1989) have been sequenced. According to these results and the tRNA gene analysis (Hardy and Clark-Walker 1989; Wilson et al. 1989), the mitochondrial genetic code in *K. lactis* appears to be very similar to that of *S. cerevisiae*. Neither the CUN codon nor its corresponding tRNA gene (*thr2* of *S. cerevisiae*) were found so far. CGN codons were found only in the introns of the cytochrome oxidase subunit 1 gene, but their corresponding tRNA gene (*arg2* of *S. cerevisiae*) has not been identified. ATA (supposed methionine codon) are rarely used, but, in the cytochrome oxidase subunit 1 introns, it is much more frequent than ATG. Twenty-two tRNA genes out of the probable 24 have been mapped and sequenced. On average, there are about ten nucleotide changes per tRNA with respect to *S. cerevisiae* counterparts.

Mitochondrial coding sequences are generally well conserved in the yeast species so far examined. The high adenine-thymine content (more than 80%) in these mitochondrial DNAs is reflected by the exceedingly high frequency of A and T at the third position of codons. Hybridization probes made of *S. cerevisiae* and *K. lactis* mitochondrial genes have been conveniently used in the mapping studies of

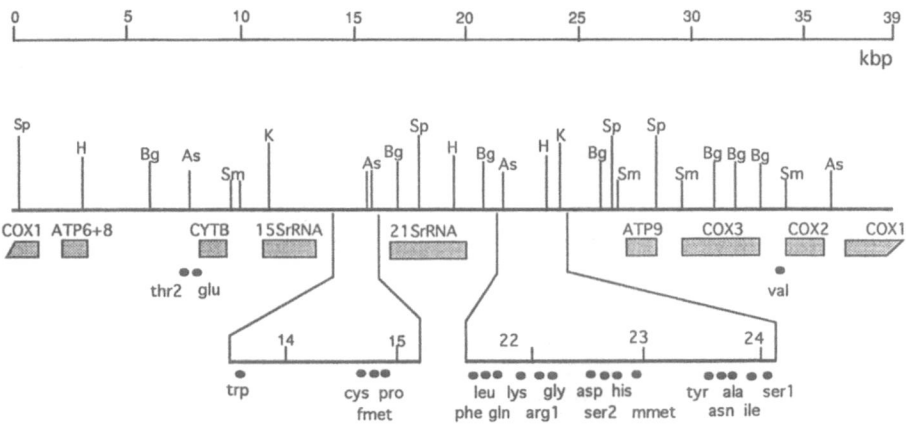

Fig. 7. Genetic and physical map of *K. lactis* mitochondrial DNA. The strain is 2360/7 (*a lysA1* clone of CBS 2360). The map is circular, 39 kbp long. Restriction sites are Sp (*Spe*I), H (*Hind*III), B (*Bgl*II), As (*Asu*III), Sm (*Sma*I), K (*Kpn*I). The *hatched boxes* indicate the limits of the gene location. *COX1, 2,* and *3* are the genes for cytochrome oxidase subunits 1, 2, and 3, respectively. *ATP 6, 8,* and *9* code for the subunits 6, 8, and 9 of the mitochondrial ATPase. *CYTB* is cytochrome b gene. tRNA genes (*dots*) are represented by corresponding single-letter amino acid names. The major transfer RNA gene regions are *enlarged* to show the detail. This map is based on the data by Ragnini and Fukuhara (1988), Wilson et al. (1989), and Coria et al. (1990). The map of WM37 (isomitochondrial to NRRL Y-1140; Coria et al. 1990) is very similar to NRRL Y-1205, except for the presence of an additional *Ava*I site. An insertion of about 1250 bp to the right of the *ATP9* gene was found in the strain W600B (Coria et al. 1990). The map from WM27 (isomitochondrial to NRRL Y-1118) appears to deviate considerably from the above map, with a translocation of *ATP9* gene (Coria et al. 1990). All the genes so far known are transcribed *from left to right*

many other yeasts, including the species of *Pichia, Williopsis, Kluyveromyces, Yarrowia,* and *Candida,* generally giving unambiguous hybridization signals (Fukuhara et al. 1993), but some *Candida* species show considerable divergence of gene sequences. Introns have been found in the large ribosomal subunit RNA gene (homologous to the ω sequence of *S. cerevisiae*; Jacquier and Dujon 1983; Wilson and Fukuhara 1991) and in the cytochrome oxidase subunit 1 gene (Hardy and Clark-Walker 1991), but not in the apocytochrome b gene (Brunner and Coria 1989). The ribosomal ω intron of *K. lactis* does not carry the endonuclease gene present in the *S. cerevisiae* ω element.

A nonanucleotide sequence, (A/T) TATAAGTA, or its variants, is present at the head of the transcription units in *S. cerevisiae* mitochondrial DNA. This motif is recognized by the unique RNA polymerase of the mitochondria. The last A of the motif is the starting A of the primary transcript. This signal appears to be also used by *K. lactis* mitochondrial transcription system (Osinga et al. 1982), as well as by the mitochondria of several species from *Williopsis* and *Pichia* (Dinouël et al. 1993). Numerous GC-rich clusters that are potential recombinational hot spots are characteristically present in both *K. lactis* and *S. cerevisiae* mitochondrial DNAs

(Ragnini and Fukuhara 1988), but not in the mitochondrial DNA of *K. marxianus* strains.

11
A Few Notes on Biochemical Procedures

11.1
Cell Mass Determination

The yeast cell density is most conveniently determined by absorbance in a spectrophotometer. However, this "absorbance", including light dispersion, can vary severalfold, depending on the internal geometry of each photometer. For a given absorbance value, *K. lactis* represents about twice as many cells as *S. cerevisiae*, reflecting the difference of cell size. As an indicative value, an absorbance of 1 at 600 nm, in a Zeiss M4QIII spectrophotometer in the standard 4-ml cuvette with 1-cm optical path, corresponds to a cell density of about 1.4×10^7 per ml. There exist floculating strains, which are unpractical for density determination or quantitative plating, but they are infrequent, as far as we know.

11.2
Cell Extracts for Preparation of Nucleic Acids

11.2.1
Nucleic Acids Prepared from Spheroplasts

Spheroplasting is the preferred procedure to prepare nucleic acids. As mentioned above, *K. lactis* cells are more sensitive to Zymolyase or Glusulase than *S. cerevisiae* cells. The amount of enzyme and the digestion time required for spheroplasting vary considerably, depending on the strain, but care should be taken to avoid spontaneous lysis due to overdigestion. Sorbitol is most often used as an osmotic stabilizer at a concentration of 1.2 M. Some investigators prefer 1.5 M (Sreekrishna et al. 1984). KCl at 0.6 M has also been used (Gunge et al. 1981).

Basically, the spheroplasting techniques are similar to those used for *S. cerevisiae*. Because of the fragility of spheroplasts, addition of 0.1% bovin serum albumin in the buffers is recommended to protect against traces of detergents. A very dense suspension of aerobic yeast cells often quickly turns acidic during spheroplasting and blocks the cell wall digestion. It is important to maintain the initial pH with occasional addition of NaOH.

11.2.2
Nucleic Acids Prepared by Mechanical Extraction

Rapid mechanical procedures are sometimes convenient, for example, for the preparation of RNA. Mechanical disruption of *K. lactis* cells raises no particular problems with respect to *S. cerevisiae*. A Braun homogenizer (Melsungen,

Germany) at full setting releases about the same fraction (say 50%) of cellular proteins under the conditions applied for S. cerevisiae. An example of mechanical disruption, on a relatively large scale, is given below.

About 10 g (wet weight) of a washed cell pellet (about 1 l of early stationary phase culture) are suspended in an appropriate buffer to a final volume of 40 ml in a Braun bottle; 20 g of glass beads (diameter 0.5 mm, Braun) are added, and the mixture is cooled to ice-water temperature. Cells are disrupted for 30 s at the top setting, and the bottle is immediately returned to the ice-water bath. The extract is separated from beads by decantation. Examples of suggested disruption buffers are (1) 50 mM Tris-HCl, pH 8, 5 mM EDTA for DNA preparation, (2) 50 mM Tris-HCl pH7.2/150 mM NaCl/10 mM EDTA, optionally including 1% sodium thioglycolate and 1/10 volume of boiled 1% macaloid suspension (Faye et al. 1974) for RNA preparation.

11.3
Small-Scale Preparation of DNA

The procedure is based on that of Sherman et al. (1982) for S. cerevisiae.

1. Grow cells in 2 ml glucose complete medium. Collect cells by centrifugation and suspend them in 0.25 ml of 1 M sorbitol plus 0.05 ml of 20 mM EDTA, pH 7.5.

2. Transfer the cells to a 1.5-ml microfuge tube, add 0.01 ml of Zymolyase 100 T (stock 2.5 mg/ml) and incubate at 37 °C for 60 min.

3. Spin the tube in a microfuge for 1 min, resuspend cells in 0.25 ml of 50 mM Tris pH 7.4 /20 mM EDTA. Add 0.025 ml of 10% sodium dodecylsulfate, mix well. Incubate the mixture at 65 °C for 30 min.

4. Add 0.1 ml of 5 M potassium acetate and place the tube in ice for 30 min. Spin for 5 min. Transfer the supernatant to a fresh tube and add 1 vol isopropanol at room temperature. Mix and leave at room temperature for 5 min.

5. Spin 10 s, pour off the supernatant, air dry the pellet.

6. Dissolve the pellet in 0.15 ml of TE (10 mM Tris, pH 8, 1 mM EDTA). Add 0.015 ml of 3 M sodium acetate, mix and precipitate DNA with 0.1 ml of isopropanol. Spin briefly to pellet DNA. Pour off the supernatant, dry and dissolve DNA in 40 μl of TE, pH 8. Use 5 μl of the preparation for a restriction digest.

11.4
Large-Scale Preparation of Nuclear and Mitochondrial DNA

The guanine-cytosine contents of nuclear and mitochondrial DNAs are close to those of S.cerevisiae. These DNAs can be easily separated by density gradient centrifugation in CsCl in the presence of intercalating dyes according to the various protocols described for S. cerevisiae. The following is an example.

1. Phenol-treated, ethanol-precipitated total DNA (obtained either by protoplasting or by mechanical disruption from 1 l culture) is dissolved in 9 ml of
Tris-EDTA buffer (above) and mixed with 9.0 g of CsCl. 0.1 ml of bisbenzimide
(Hoechst 33258, stock solution 10 mg/ml water) is added. This fills just two
sealable tubes for the VTi65 vertical rotor of the Beckman centrifuge.

2. Spin overnight at 55 000 rpm, the upper band (mitochondrial DNA) and the
lower band (nuclear DNA), detected by lateral ultraviolet illumination,
are separately collected and diluted with a fresh CsCl-dye solution and recentri-
fuged.

3. The collected DNA is concentrated by dialysis against polyethyleneglycol powder (Merck, PEG 6000). DNA is then dialyzed against 1 mM Tris-HCl pH 8/
0.1 mM EDTA. This dilute buffer allows reconcentration of the DNA solution by
evaporation in a vacuum centrifuge.

At this stage, each DNA is considered to be pure enough for restriction enzyme
analysis and for most hybridization studies, except that polysaccharides are often
visibly present. For DNA reassociation studies involving optical measurements,
DNA should be purified by published specific procedures involving hydroxyapa-
tite chromatography (Price et al. 1978; Kurtzman et al. 1980) which give more
fragmented, but optically purer material.

11.5
Cell Disruption for Enzyme Assays

11.5.1
Disruption by Braun Homogenizer

Cells from 100 ml culture (about 10^8 cells/ml, A_{600} of 6–10 in Zeiss photometer) are
pelleted and washed once with water and once with 0.1 M Tris-HCl pH 7.5. The
pellet is resuspended in 1 vol. of the same buffer and 1 vol. of glass beads.

Cells are disrupted for 1.5 min at top speed in a cooled Braun homogenizer
equipped with an adapter for 1.5-ml Eppendorf tubes. The supernatant is collected
by decantation, and the glass pellet is rinsed twice with 1/4 of the initial volume of
buffer.

The supernatant and washes are mixed and centrifuged at 3000 rpm for 10 min at
4 °C. The pellet is washed with 1/2 volume of buffer.

The pooled supernatants give a protein concentration of about 30 mg/ml.

11.5.2
Disruption by Vortexing

Alternatively, the cells, mixed with glass beads as above, can be disrupted by
vortexing four times 30 s at top speed, with periods of 1 min on ice in between.
Extraction efficiency appears to be comparable to the first method.

11.5.3
Permeabilized Cells

Some enzymes such as β-galactosidase or invertase can be directly assayed on whole cells permeabilized with toluene. The case of β-galactosidase measurement is given as an example.

β-galactosidase assay is performed according to the method described by Miller (1972). The colorless substrate o-nitrophenyl-β-D-galactoside (ONPG) is hydrolyzed into galactose and yellow O-nitrophenol absorbing at 420 nm.

1. Cells from 1 ml of culture are collected by centrifugation (5 min, 4000 rpm) at room temperature. The pellet is suspended in 1 ml of Z buffer and 30 μl of toluene.

2. Vortex at top speed for 15 s. Then incubate the tubes, with the top open, for 30 min at 37 °C with shaking to allow toluene evaporation; 20–200 μl of permeabilized cells are added to Z buffer (final volume of 1 ml).

3. After 5 min at 28 °C to equilibrate temperature, the reaction is started by the addition of 200 μl of ONPG. The reaction is stopped by adding 0.5 ml of 1 M Na_2CO_3 solution. Record the time of the reaction.

4. Remove cells by a 5-min centrifugation. The absorbance at 420 nm is determined on the yellow supernatant. The enzyme activity may be expressed as nanomoles of o-nitrophenol produced per min per A_{600} unit of cells (roughly 10^8 cells). One nmol/ml of o-nitrophenol gives A_{420} of 0.0045 with a 1 cm light-path cuvette.

In this β-galactosidase assay, the use of chloroform (50 μl) and sodium dodecylsulfate (0.0025%), instead of toluene, avoids the evaporation step.

Reagents.

Z buffer: 0.06 M Na_2HPO_4/ 0.04 M NaH_2PO_4/ 0.01 M KCl/ 0.001 M $MgSO_4$/ 0.05 M mercaptoethanol-2, pH 7.0.

ONPG stock solution: 4 mg/ml in 0.1 M Na phosphate buffer, pH 7.0.

11.6
Gene Fusions

To study the transcriptional regulation of a gene, the *E.coli* β-galactosidase gene *lacZ* has been frequently used in *S. cerevisiae*. The upstream regulatory sequence of a gene is fused with the *lacZ* coding sequence. This constitutes a convenient system to monitor the response of the regulatory element(s) to various experimental conditions. Since *K. lactis* has its own β-galactosidase, the *lacZ* system can only be used in the β-galactosidase-deficient mutant hosts such as the *lac4* mutants described before. As an alternative, another gene fusion system has been developed that can be applied to *K. lactis* and *S. cerevisiae*. The kanamycin resistance gene Km^R from the bacterial transposon Tn903 confers resistance to the antibiotic G418 pSK-kan401 (phase 1)

EcoRI   SacI   KpnI   BamHI   (SalI/XhoI)   8th codon
GAATTCGAGCTCGGTACCCGGGGATCC TCTAGAGTCGAGGCCGCGATTAAATTCCAAC ATG GAT ...Km'

pSK-kan 1105 (phase 2)

EcoRI   SacI   KpnI   BamHI   (SalI/XhoI)   8th
GAATTCGAGCTCGGTACCCGGGGATCCTCGAGCGATCCTCTAGAGTCGAGGCCGCGATTAAATTCCAAC ATG ...

pSK-kan 1238 (phase 3)

EcoRI   SacI   KpnI   BamHI   (SalI/XhoI)   8th
GAATTCGAGCTCGGTACCCGGGGATCCGCGATCCTCTAGAGTCGAGGCGCGATTAAATTCCAAC ATG ...

Fig. 8. Km^R gene fusion as a reporter system. 3-'aminoglycoside phosphotransferase of the transposon Tn*903* confers resistance to Kanamycin in bacteria and resistance to G418 on yeast. N-terminal amino acids of this enzyme are nonessential for activity, and the second ATG, at 8th codon, can be used as initiator codon. Thus the gene can easily be fused to a promoter-carrying sequence. Such a fusion system with dominant selection is of particular interest for the gene studies in nonconventional species which often have no genetic markers. The three vectors allow fusing, in all three reading frames, of the promoter-less Km^R gene with a promoter-containing sequence of interest. The unique cloning sites before the Km^R sequence are *Bam*HI, *Sac*I, and *Kpn*I. The plasmids contain the *URA3* marker, as well as both pKD1 (S11 fragment, 0.9 kbp) and 2-μ origins of replication to allow transformation of *K. lactis* and *S. cerevisiae*. The sequences in *bold type* are common to all three vectors

on a wide range of yeast cells. Since several codons at the N-terminus of the Km^R gene are dispensable for the 3'-aminoglycoside phosphotransferase activity, this part is deleted and replaced by cloning sites. Various transcriptional regulatory elements can be introduced at these sites. Vectors carrying such cloning sites have been constructed which allow Km^R fusion in three reading frames (Fig. 8). Practical information on the use of the Km^R system is given at the end of Sect. 14.

11.7
DNA-Binding Studies

For the analysis of *K. lactis* DNA-binding proteins, protocols established for *S. cerevisiae* are generally applicable. The following method (Breunig and Kuger 1987; Zachariae et al. 1993) worked well for the detection of Lac9 and other DNA-binding factors.

1. 75 A$_{600}$ units of cells were harvested at 4 °C, washed once with ice-cold TMEGA buffer (0.2 M Tris/HCl pH7.8, 0.3 M (NH$_4$)$_2$SO$_4$, 10 mM EDTA, 1 mM DTT, 10% glycerol), resuspended in 0.4 ml of the same buffer containing 1 mM PMSF, 4 μM Pepstatin, 4 μM Leupeptin, and 14 μg/ml Aprotinin, and broken with glass beads as described above.

2. After centrifugation, 200 μl of the supernatant were centrifuged for 1 h at 100 000 g in a Beckman TLA 45 rotor.

3. The supernatant (S100) with a protein concentration of 25 to 15 mg/ml was frozen in aliquots in liquid nitrogen and can be stored for a year or more.

4. The S100 can either be used directly in electophoretic mobility shift assays or subjected to heparin-Sepharose affinity chromatography according to standard protocols.

The low-abundant transcription factor Lac9 was detectable in the S100 when the ratio of protein to nonspecific competitor DNA (sonicated calfthymus DNA worked well for Lac9, but poly[dIdC] may be preferable for other DNA-binding proteins) was carefully titrated for each extract. Protease activity is certainly a problem in these extracts as in *S. cerevisiae*. A protease A gene has been cloned from *K. lactis* by Chiron Corp. but the difference in protease activity between a wild-type and a disruption mutant was not as pronounced as for the *S. cerevisiae* *pep4* mutant.

12
Plasmids

12.1
Circular Plasmids

Vectors derived from the 2-μ plasmid of *S. cerevisiae* can transform *K. lactis* strains, but are very unstable, requiring a selective pressure to be maintained. A replicating vector system specific for *K. lactis* has become available. This system

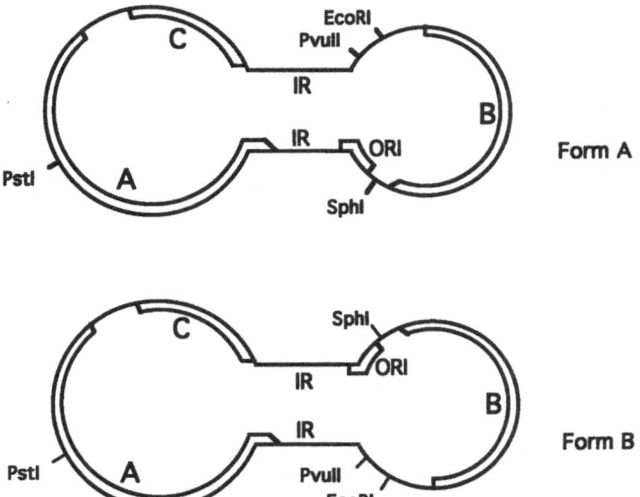

Fig. 9. Isomeric forms of the plasmid pKD1. pKD1, 4757 bp long, has three open reading frames. *A* is a recombinase gene analogous to the FLP gene of 2-μ, mediating the inversion of DNA segments between the two IRs. *B* and *C* are believed to be required for proper partition of the plasmid. *IRs* are inverted repeats 346 base pairs long. *ORI* is the region containing the autonomous replication function. A locus, *cis*-acting on ORI, and required for stable maintenance of the plasmid is probably located near the *Eco*RI site. By convention, the nucleotide numbering begins at the *Eco*RI site and goes *clockwise* on the B form

(Chen et al. 1988; Wésolowski-Louvel at al. 1990) is based on the plasmid pKD1 isolated from the strain UCD 51-130 (CBS 2105); (Chen et al. 1986; Falcone et al. 1986). The latter strain was formerly called *K. drosophilarum* (hence the name pKD1), but is now classified as *K. lactis*. pKD1 (Fig. 9) belongs to the 2-μ family plasmids sharing the same type of gene organization, although there is little homology of nucleotide sequence. The host range is also quite different. By genetic crosses, pKD1 plasmid has been propagated to other strains of *K. lactis*, thus creating a series of cir+ transformation host strains. Analysis of individual functional elements of pKD1 has been described by Bianchi et al. (1987, 1989, 1991) and Bianchi (1992). The only other circular plasmid known among the strains of the genus *Kluyveromyces* is pKW1 (Chen et al. 1992a) found in *Kluyveromyces waltii* CBS 6430. This plasmid does not replicate in *K. lactis*.

12.2
Linear DNA Plasmids and the Killer System

This subject has been reviewed by Gunge (1986), Gunge and Kitada (1988), Volkert et al. (1989), and Stark et al. (1990).

The linear DNA plasmids pGKL1 and pGKL2 were first found in *K. lactis* IFO 1267 (NRRL Y-1140, CBS 2359; Gunge et al. 1981). The two plasmids are present as a pair in the cell and confer the killer phenotype. Very similar, if not identical, plasmids were also present in four other *K. lactis* isolates (CBS 1065, pKL2A, and 2B; CBS 5618, pKL3A, and 3B; CBS 8043, pKL4A, and 4B; NRRL Y-1115, pKL5A, and 5B). The gene organization of the pGKL plasmids is shown in Fig. 10.

pGKL1 is a 8.8-kbp double-stranded DNA carrying the genes coding for the killer toxin subunit proteins. The toxin kills sensitive cells of various yeast species. When the killer strain is cured of the pGKL1 plasmid, it loses the killer character and becomes sensitive to the toxin. Most *K. lactis* strains are sensitive to the toxin, but a few strains are resistant, due to the presence of Mendelian determinants. The toxin, secreted into the medium, is a large protein made up of three subunits. During the secretion process, the precursor proteins are shortened by endopeptidases into mature proteins. One of the processing proteases has been identified as Kex1 (Wésolowski-Louvel et al. 1988b; Tanguy-Rougeau et al. 1988), whose known target is a lysine-arginine dipeptide sequence. Kex1 is an equivalent of *S. cerevisiae* Kex2 peptidase, which is required for the processing of the dsRNA-dependent killer toxin and of the mating factor α. However, *K. lactis kex1* mutants of mating type α are not defective for mating.

The N-terminal signal peptides from pGKL1 ORF2 and ORF4 have been used to direct the secretion of foreign proteins from *K. lactis* (Baldari et al. 1987; Tokunaga et al. 1988).

pGKL2 has been sequenced by Tommasino et al. (1988). This plasmid, 13.4 kbp long, is necessary for the maintenance of pGKL1. Nonkiller mutants are available in which pGKL1 only or both plasmids had been lost. Both pGKL1 and pGKL2 have inverted terminal repeats (Sor et al. 1983) whose 5' ends have a covalently attached

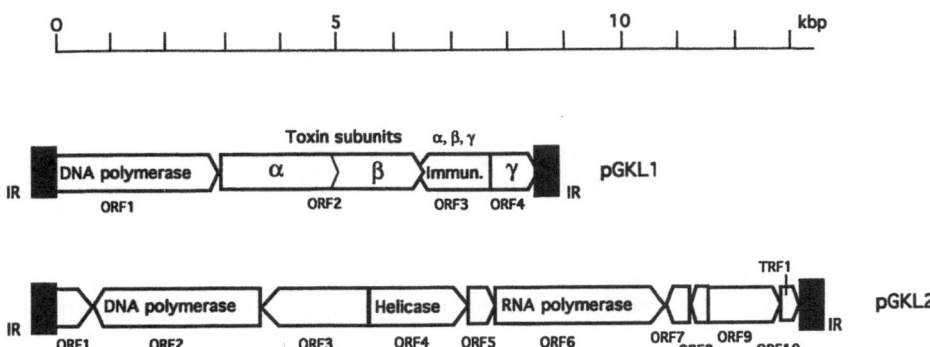

Fig. 10. Structure of the killer plasmids pGKL1 and pGKL2. This schematic representation is based on the following sources: Wésolowski et al. 1982b; Stark et al. 1984; Sugisaki et al. 1985; Hishinuma et al. 1984, 1986; Stark and Boyd 1986; Sor and Fukuhara 1985; Jung et al. 1987; Fukuhara 1987; Tokunaga et al. 1987, 1989; Tommasino et al. 1988; Stark 1988; Wilson and Meacock 1988; Gorbalenya et al. 1989; Tommasino 1991; McNeel and Tamaoki 1991; Schaffrath et al. 1992; Soond and Meacock, pers. comm.; Wésolowski-Louvel et al. 1988b. Known or putative functions of the plasmid elements are the following: pGKL1 genes. *ORF1* DNA polymerase; *ORF2* α (chitinase) and β subunits of toxin; *ORF3* determinant of immunity to toxin; *ORF4* toxin subunit γ (G1-arrest toxin); *IR* left and right terminal inverted repeats, 202 bp long; *TP* terminal protein, about 28 kDa. pGKL2 genes. *ORF1* Dispensable for plasmid maintenance; *ORF2* DNA polymerase; *ORF4* DNA-dependent ATPase/helicase; *ORF6* RNA polymerase; *ORF10* terminal region recognition factor (TRF1); *IR* left and right terminal inverted repeats, 184 bp long; *TP* terminal protein, about 36 kDa

protein (Kikuchi et al. 1984; Stam et al. 1986). Replication is probably initiated by these terminal proteins, very much as in the case of the bacteriophage φ29 (Kikuchi et al. 1985; Kitada and Gunge 1988; Wésolowski-Louvel and Fukuhara 1990). This cytoplasmic killer system appears to function with its own transcription machinery to express plasmid genes (Romanos and Boyd 1988). One of the pGKL2-borne genes (ORF6) is thought to encode an RNA polymerase. Interestingly, host genes cannot be expressed on these plasmids unless their promoters are replaced by the plasmid-specific promoter signals. The killing mechanism is not known, but appears to be distinct from that of the dsRNA killer system of *S. cerevisiae*. A few hypotheses have been proposed for the nature of the toxin (Sugisaki et al. 1983, 1984; White et al. 1989; Bradshaw 1990; Butler et al. 1991). The killer character does not seem to be a barrier to mating with sensitive partners. In such crosses, practically all the progeny of the cross will have the killer plasmids, showing a typical pattern of cytoplasmic inheritance.

Two point mutations are known which occurred in the ORF2 of pGKL1 plasmid. Existence of recombination between pGKL1 DNAs has been demonstrated by the use of these mutations (Wésolowski et al. 1982c). Two important questions, that is, copy number control and mitotic partition mechanism of the linear plasmids, still remain entirely open.

pGKL plasmids have been transferred to *S. cerevisiae*, in which they replicated stably (Gunge and Sakaguchi 1981; Gunge et al. 1982; Fujimura et al. 1987; Gunge et al. 1990). For unknown reasons, only the *rho⁻* strains of *S. cerevisiae* were found to be compatible with this replication (Gunge and Yamane 1984). pGKL plasmids have also been stably transferred to *Kluyveromyces marxianus* strains (previously classified as *K. fragilis* and *Candida pseudotropicalis*; Sugisaki et al. 1985).

A DNA segment isolated from the killer plasmid was shown to contain an ARS activity (Thompson and Oliver 1986; Trueman et al. 1990). This activity might be fortuitous, but explains why the first *K. lactis* transformants obtained by pGKL1 derivatives (de Louvencourt et al. 1983) contained only circularized killer plasmids. Probably their replication relied on this ARS activity (see Sect. 14, Vector Systems).

12.3
RNA Plasmids

RNA plasmids have not been found so far in *K. lactis* strains. Among 150 strains examined in the *Kluyveromyces* genus, only one strain of *K. aestuarii* and two strains of *K. waltii* contained what appears to be a dsRNA plasmid (Fukuhara, unpubl.). Killer phenotype was not associated with these plasmids when tested against *S. cerevisiae*, *Candida glabrata*, and *K. lactis*.

12.4
Killer Assay

Killer activity can be tested against a few tester strains. Besides *S. cerevisiae* and *K. lactis* 2360/7 (Wésolowski et al. 1982a), *Candida glabrata* CBS 138 may also be used, because this strain is highly sensitive to a wide range of killer toxin types. A practical procedure is as follows.

1. Ten μl of a fresh culture of the sensitive tester strain are placed in the middle of a plastic petri dish.

2. Ten ml of molten glucose complete agar medium (1.5% Oxoid agar no. 1, kept at 45 °C) are poured and quickly mixed by swirling the dish.

3. On the solidified agar plate, colonies of the killer strain are streaked or replica-plated and incubated at 20 °C for 3 days. Killer activity is evaluated from the diameter of the growth inhibition zone (Fig. 11).

Two percent galactose complete medium adjusted to pH 4.5 with 50 mM Na-phosphate buffer was found to be optimal for killer tests on plates, but this medium cannot be used for galactose-negative mutants or species such as *K. waltii* and *K. delphensis*. The test can be done on glucose plates, but the sensitivity appears to be lower (see also Gunge et al. 1981, for an alternative protocol).

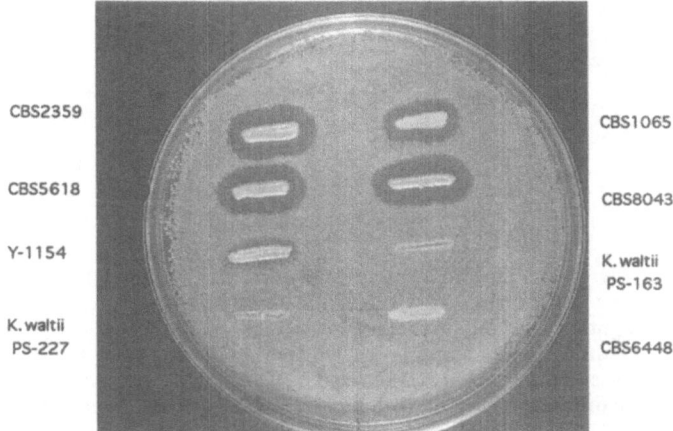

CBS2359

CBS5618

Y-1154

K. waltii
PS-227

CBS1065

CBS8043

K. waltii
PS-163

CBS6448

Fig. 11. Plate assay of killer phenotype. The agar lawn contains *Candida glabrata* as a killer-sensitive tester. The indicated killer and nonkiller strains were streaked on the agar surface and incubated for 3 days. The tester cells cannot grow around the *K. lactis* colonies secreting the killer toxin. *K. waltii* strains (provided by Dr. M.-A. Lachance, University of Western Ontario) contained nonkiller RNA plasmids. *CBS6448* is a strain of *Endomycopsella crataegensis*, which contained also RNA plasmids

12.5
Detection of Plasmids in Colony Lysates

Circular and linear plasmids and their derivatives can be quickly detected in many individual colonies by electrophoresis of minilysates, as follows.

Solutions: $5 \times$ TE (50 mM Tris-HCl/5 mM EDTA, pH 7.5). Zymolyase 20 T (Kirin Breweries, Tokyo; stock solution at 1 mg/ml, stored frozen in small fractions), 5% SDS (sodium dodecylsulfate), proteinase K (Boehringer, 2 mg/ml, stored in small fractions).

1. Take a 1–2-mm-sized colony with a sterile toothpick, suspend in 20 μl of $5 \times$ TE in an Eppendorf tube.

2. Add 3 μl of Zymolyase, vortex 1 s and incubate for 1 h at 37 °C.

3. During incubation, prepare 0.6% agarose gel for electrophoresis, with as many sample wells as necessary.

4. Add 2 μl of SDS, vortex, 5 μl of Proteinase K, vortex, and incubate at 60–65 °C for 1 h.

5. Add 5 μl of blue sauce (the standard mix for nucleic acid electrophoresis, Maniatis et al. 1982) and microfuge for 1 min.

6. Take 10–20 μl supernatant, avoiding the viscous pellet if any, and run the electrophoresis until the blue dye moves a few cm. Stain the gel with ethidium bromide (0.5 μg/ml water) for observation under UV. Figure 12 shows examples

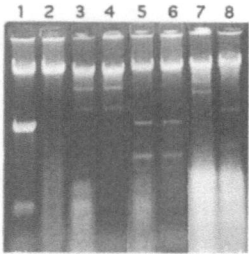

Fig. 12. Analysis of colony lysates to detect plasmids. Mini-lysates of individual colonies were electrophoresed on agarose gel. DNA was stained with ethidium bromide and photographed under ultraviolet light. pGKL1 and pGKL2 migrate as sharp bands separated from the mass of cellular DNA. **Lane 1** *Candida vacciniae* CBS 7318 (contains RNA plasmids); **lane 2** id. with pancreatic RNase; **lane 3** *Pichia etchellsii* CBS 2011 (contains two linear DNA plasmids pPE1A and pPE1B; unpubl.); **lane 4** id. with RNase; **lane 5** *Pichia scaptonizae* CBS 8167 (contains uncharacterized plasmids); **lane 6** id. with RNase; **lane 7** *Kluyveromyces lactis* CBS 2359 (contains the linear plasmids pGKL1 and 2; **lane 8** *Kluyveromyces lactis* MD2/1 (contains the circular plasmid pKD1)

of this analysis. Optionally, RNA (and RNA plasmids) may be removed by adding $2\,\mu l$ of pancreatic RNase $(200\,\mu g/ml)$ at the end of the Zymolyase step. Compared with the circular pKD1-derived plasmids, the linear DNA plasmids give characteristically sharp thin bands.

12.6
Preparation of Killer Plasmid DNAs

The isolation procedure for the linear plasmids must include a Proteinase K (or pronase) treatment to remove the terminally attached proteins. Earlier procedures have been improved by Stam et al. (1986) such that the plasmid DNAs, extremely rich in AT bases, can be separated from other cellular DNAs by CsCl-bisbenzimide centrifugation (see Sect. 12.2). The killer DNAs form a sharp band on top of the mitochondrial DNA band. Separation between pGKL1 and pGKL2 requires a preparative agarose (0.5–0.6%) gel electrophoresis. pGKL1-less mutants (Niwa et al. 1981; Wésolowski et al. 1982a) can be conveniently used for the isolation of pGKL2 DNA.

13
Vector Systems

13.1
Transformation Markers

The standard *S. cerevisiae* markers such as *TRP1*, *URA3*, and *LEU2* genes have been conveniently used in *K. lactis*, because corresponding mutations in *K. lactis* can be complemented by these genes under their native promoter. *K. lactis* genes corresponding to these genes have now been cloned and sequenced (*KlURA3*, Shuster et al. 1987; *KlTRP1*, Stark and Milner 1989; *KlLEU2*, Zhang et al. 1992). Besides auxotrophic markers, G418 resistance can also be used (see below).

13.2
pKD1 Plasmid-Derived Vectors

Since pKD1 has a 2-μ-type organization, recombinant vectors carrying the plasmid replication origin can be constructed in the way similar to the 2-μ-derived vectors. KEp6 is one of the representative vectors (Fig. 13). It is composed of the integrative vector YIp5 (carrying *URA3*, and pBR322 sequence) and pKD1 origin of replication. The maintenance of these vectors requires a cir+ host (carrying resident pKD1) and the stability of this simple form of vectors is high enough to be used for construction of DNA libraries. However, a more stable form of vectors can be obtained using the totality of the pKD1 sequence. 2-μ-family plasmids have a few intergenic regions in which foreign sequences can be introduced without drastically affecting the plasmid stability (Chen et al. 1989; Chinery and Hinchcliffe 1989). pE1 is an example of this type of stable autonomous vectors (Fig. 13B). Such full-sequence vectors, which do not need cir+ hosts, have been successfully used in the production of heterologous proteins (Fleer et al. 1991a,b; Ogawa et al. 1990).

The host range of pKD1-derived vectors extends to some other species of *Kluyveromyces* including *K. dobzhanskii*, *K. aestuarii*, *K. waltii*, *K. thermotolerans*, *K. wickerhami*, and some strains of *K. marxianus* (Chen et al. 1989). The host range could only be tested by the use of a "full-sequence" vector, since the KEp6 type plasmids require cir+ hosts for maintenance. In addition, the Km^R marker has to be used in such tests, because wild-type yeasts have no auxotrophic markers. Some yeast species are G418-resistant and cannot be tested by this marker (see below).

13.3
ARS Vectors

Several autonomously replicating sequences (ARS) of chromosomal origin have been isolated (Das and Hollenberg 1982; Sreekrishna et al. 1984; Fabiani et al. 1990). KARS is the name used by Hollenberg's group. A KARS whose sequence is known was shown to contain an undecanucleotide closely related to the *S. cerevisiae* ARS consensus core (Broach et al. 1982). Some KARSs can replicate in both *K. lactis* and *S. cerevisiae* (Fabiani et al. 1990). KARS-based vectors have been used in gene expression studies reported by several authors (Das et al. 1985; Leonardo et al. 1987), although these vectors are usually much less stable than the plasmid-derived vectors. High amounts of DNA seem to be required for transformation with KARS-based vectors.

13.4
Centromeric Vectors

As mentioned before, most of the *K. lactis* centromere sequences have been cloned. When associated with these DNAs, the KEp6 type vectors can form highly stable

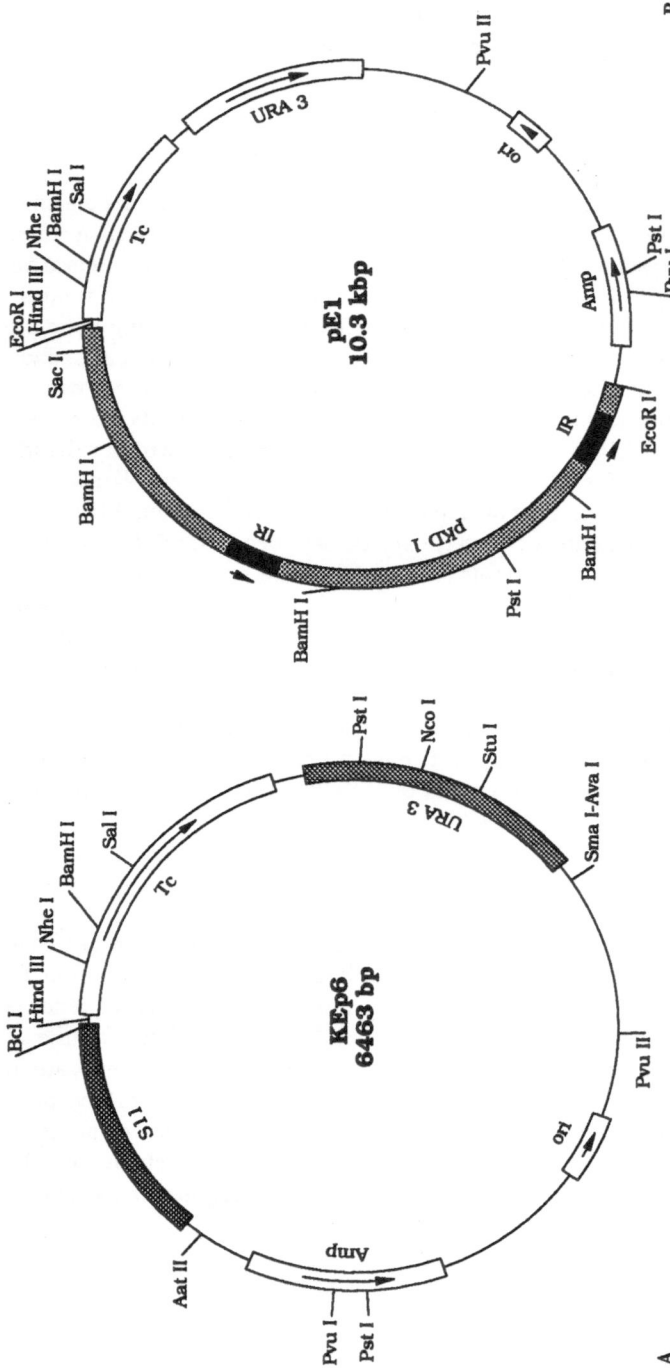

Fig. 13A,B. Two typical pKD1-derived vectors. *Left* KEp6 is a standard multicopy vector obtained by introducing the replication origin of pKD1 (a 914-bp fragment, S11) into the *Eco*RI site of the pBR322-derived plasmid YIp5. This vector requires cir⁺ hosts. Transformation marker is *URA3* for *K. lactis uraA* hosts. *Right* pE1 contains the total sequence of pKD1 which was opened at the unique *Eco*RI site where YIp5 was introduced. This is an autonomous vector stable in cir⁰ hosts. A similar full-sequence vector (pCXJ-kan1) is available with a Km^R marker for transformant selection. The unique *Sph*I site is another insertion point to construct a highly stable full-sequence vector analogous to pE1 (Bianchi 1992).

low copy number plasmids (1–2 copies per cell). KCp491 is one such vector (Fig. 14A). Centromeric vectors have been used by Prior et al. (1993b) and Zachariae et al. (1993).

13.5
K. lactis /S. cerevisiae Shuttle Vectors and Shuttle Libraries

pKD1-derived plasmids can replicate in *S. cerevisiae*, but the transformants are extremely unstable. 2-μ-based vectors are also highly unstable in *K. lactis*. By introducing the 2 μ origin of replication into pKD1-derived vectors, one obtains recombinant plasmids that can replicate stably in the two species. There appears to be no incompatibility between the two origins. The shuttle vectors allow the exploitation of many mutations identified in *S. cerevisiae* for *K. lactis* studies and vice versa. pSK1 (Fig. 14B) is typical of such shuttle vectors on which *K. lactis* and *S. cerevisiae* genomic libraries can be constructed (Prior et al. 1993a).

13.6
Expression and Secretion Vectors

pUC19- or pBR322-derived plasmids carrying the pKD1 origin of replication have been constructed in which *S. cerevisiae* promoters and terminators (from *ADH1*, *PHO5*, or *PGK*) were included together with a cloning site between them. Some of these vectors have been modified so that a DNA sequence coding for a secretion signal peptide (from a *K. lactis* killer toxin gene ORF2) was placed in front of the cloning site. These constructions have been used to produce heterologous proteins from *K. lactis* (Fleer et al. 1991a,b). Figure 15 shows a few examples of expression and secretion vectors constructed by X.J. Chen (1987) in the authors' laboratory.

13.7
Killer Plasmid DNAs as a Possible Vector

The high copy number killer plasmids are obviously a possible source of gene vectors. Particularly, it was expected that the killer toxin genes of pGKL1 may be replaced by heterologous DNA sequences. After an early attempt to explore this possibility (de Louvencourt et al. 1983), two major problems have become apparent. First, the fact that these DNAs are linear molecules with terminally attached proteins makes it difficult to amplify them in conventional *E. coli* systems. Second, the expression of genes on the linear plasmids requires the specific transcription signals of the plasmids. A solution to the first problem was the direct introduction of foreign DNA sequences into the plasmids by in vivo targeting through homologous recombination (Kämper et al. 1989a,b; Tanguy-Rougeau et al. 1990; Chen et al. 1991).

To solve the second obstacle, there were two possibilities: (1) expression of the foreign gene by its fusion with the dispensable toxin subunit gene, so that the foreign gene is expressed under the toxin gene promoter (Kämper et al. 1989a), or

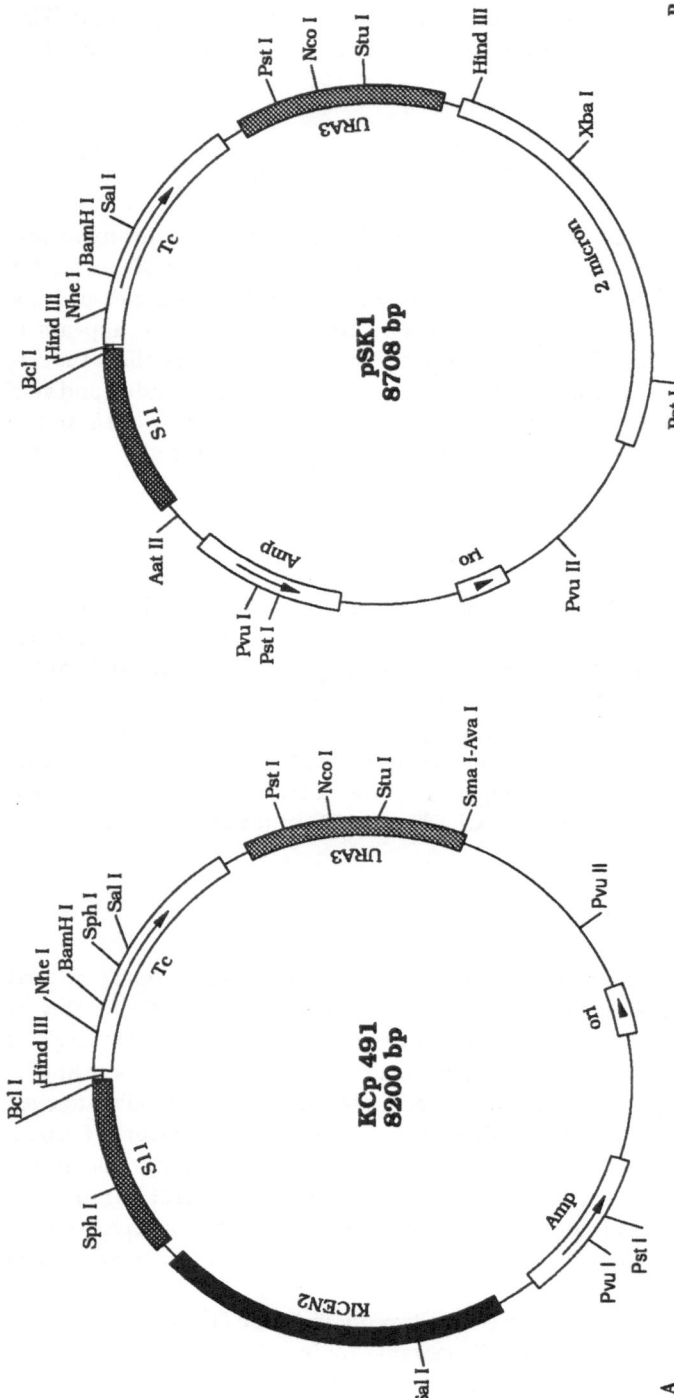

Fig. 14A,B. Centromeric vector and *K. lactis-S. cerevisiae* shuttle vector. *Left* Centromeric vector KCp 491. The centromere fragment (*KlCEN2*) comes from chromosome II (provided by Dr. B. Zonneveld, Clusius Laboratory, University of Leiden). The plasmid is otherwise identical to KEp6. *Right* Shuttle vector pSK1. A 2-μ origin of replication was added to KEp6 plasmid. pSK1 can replicate in *K. lactis* and in *S. cerevisiae* cir⁺ hosts which are *uraA* and *ura3*, respectively

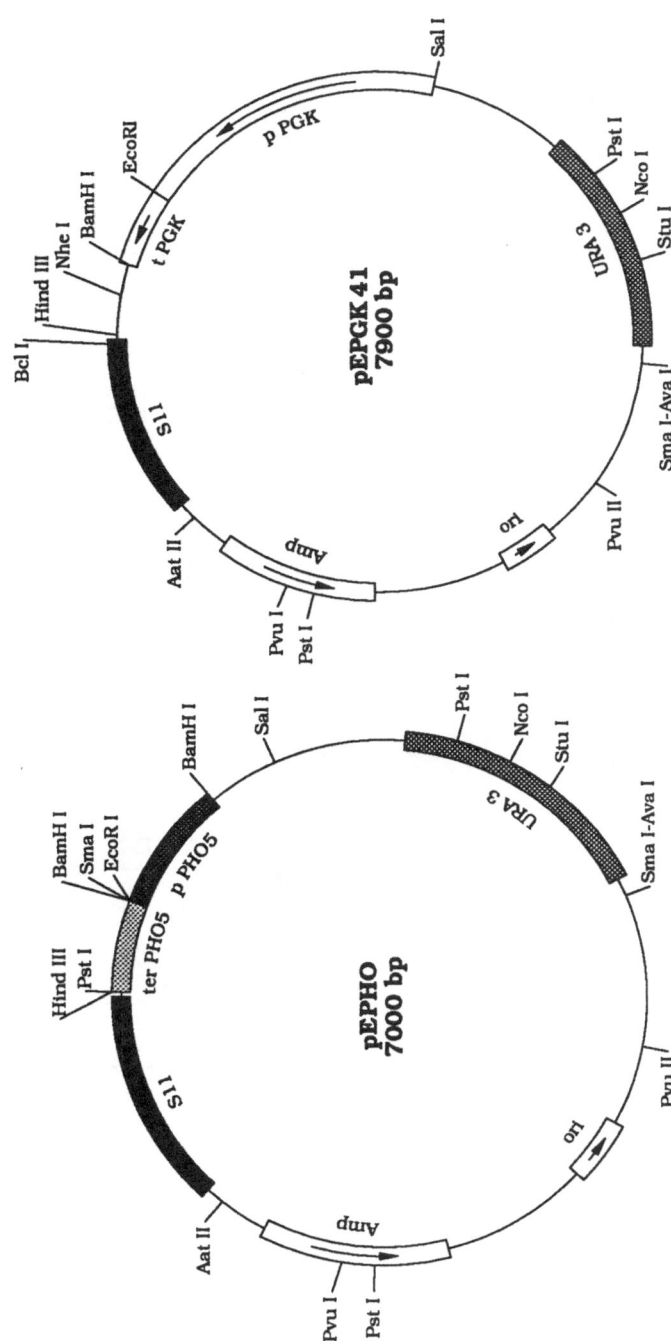

Fig. 15A–C. Expression and secretion vectors. **A** Expression vectors pEPHO, 7000 bp and pEPGK41, 7900 bp. PHO5 and PGK promoters from *S. cerevisiae* are used. *PHO5* is repressible by inorganic phosphate. PGK is a strong constitutive promoter. **B** Secretion vectors pSPHO4, 7500 bp and pSPGK1, 7900 bp. Nucleotide sequence coding for the N-terminal signal peptide of pGKL1 ORF2 was chemically synthesized and inserted behind the promoter sequence. The inserted sequence is followed by an unique cloning site to allow fusion of a coding sequence. **C** *Eco*RI synthetic fragment encoding the killer toxin secretion signal

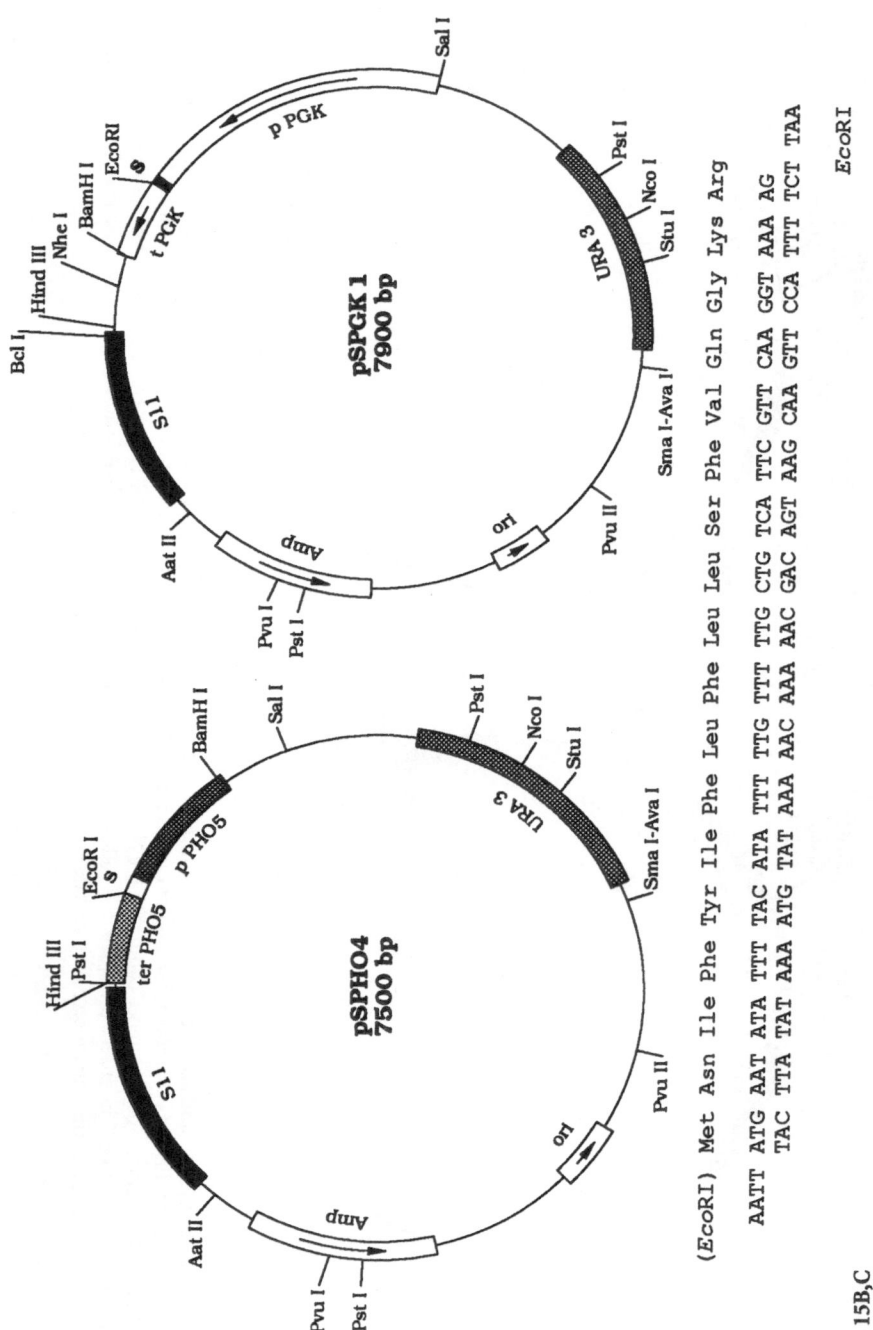

B

(EcoRI) Met Asn Ile Phe Tyr Ile Phe Leu Phe Leu Leu Ser Phe Val Gln Gly Lys Arg

AATT ATG AAT ATA TTT TAC ATA TTT TTG TTT CTG TCA TTC GTT CAA GGT AAA AG   EcoRI
     TAC TTA TAT AAA ATG TAT AAA AAC AAA AAC AGT GAC AGT AAG CAA GTT CCA TTT TCT TAA

C

Fig. 15B,C

(2) integration of the foreign gene which had been linked in vitro to an isolated plasmid promoter (Tanguy-Rougeau et al. 1990). Both approaches have worked as expected, showing that the linear plasmids can be modified to allow expression of a desired gene. Some of the modified linear plasmids have been shown to be highly stable. However, transcription under the control of the plasmid promoters does not seem to be strong enough to support high level production of a protein. The level of human interleukin 1β produced from the killer plasmids was low compared to the results obtained with circular vectors (Cong 1994). In order to exploit the linear killer plasmids as efficient expression vectors, it will be necessary to develop strong promoters that would function on these cytoplasmic plasmids. A mutant promoter of this kind has been described (Cong et al. 1994).

14
Transformation Procedures

14.1
Various Methods of Transformation

To introduce DNA into *K. lactis* cells, the three methods (spheroplasting, lithium salt treatment, or electroporation) that are used for *S. cerevisiae* can be applied with minor modifications. Although the transformation frequency of spheroplasts is high, the overall yield of transformants is somewhat inferior to that of *S. cerevisiae*, because of a lower regeneration rate of *K. lactis* spheroplasts (in the order of 1%).

The efficiency of the lithium salt method seems to vary considerably with strains. Use of lower concentrations of the Li salt improves the transformation frequency in some cases (see below).

Electroporation generally gives high frequencies of transformation. When linearized DNA is to be integrated into chromosomes, electroporation is suspected to give increased proportions of nonhomologous integration as compared to the spheroplast method. Still, gene replacement at the homologous sites can be obtained, but at lower frequencies as compared to *S. cerevisiae*. A useful transformation procedure has been reported by Dohmen et al. (1991), which is an adaptation to yeast of the general method originally described by Klebe et al. (1983). The method does not require spheroplast formation and allows long-term storage of competent cells at −70 °C. This method is convenient for routine replicative transformation as well as for integrative transformation. The electroporation method is still preferred when high-frequency transformation is required, for example, gene cloning by in vivo complementation using a genomic library, because lower amounts of DNA are required.

In *S. cerevisiae*, disruption of an essential gene resulting in recessive lethals can be performed on diploid hosts, so that the meiotic products prove to segregate 2:0. In *K. lactis*, this procedure is slightly complicated by the fact that the diploid state is transitory. The diploids have to be maintained under prototrophic or other selective pressure during the integration experiment.

Some protocols of transformation procedures are given below.

14.2
Transformation by Spheroplasting

The protocol is based on the procedure described by Bianchi et al. (1987) for pKD1 -derived vectors, with a few modifications. All operations are at room temperature if not indicated otherwise.

1. Grow yeast in 100 ml glucose complete medium from 2–3 μl of fresh overnight culture. Cells are harvested at the exponential phase of culture. The cell density should be less than 10^7 cells/ml.

2. Centrifuge cells (5 min, 5000 rpm), wash once in 10 ml of 0.1% BSA (or water).

3. Suspend cells in 10 ml SEM, add 10 μl of mercaptoethanol-2 (final 0.014 M), incubate at 30 °C for 10 min. Wash cells twice with 10 ml of 1.2 M sorbitol.

4. Suspend cells in 10 ml of SEC. Check absorbance by diluting 100–150 μl of suspension in 3 ml of water.

5. Add 0.1 ml of cytohelicase. Incubate at 30 °C for not more than 20 min. Check absorbance at 5-min intervals. Ideally, the absorbance should decrease to 10–20% of the initial value.

6. Centrifuge spheroplasts at 2500 rpm for 5 min, wash twice with 1.2 M of sorbitol, once with SCa.

7. Suspend spheroplasts in 1 ml SCa. At this stage, the suspension may be stored at 4 °C for a week.

8. 100-μl portions of suspension are placed in 10-ml sterile plastic tubes. Each tube corresponds to one transformation test.

9. Add DNA (0.1–0.5 μg, plus 10 μg of salmon sperm DNA) and let stand for 15 min at 22 °C. DNA volume should not exceed 10 μl. Two tubes without DNA are kept for no DNA control and for regeneration test.

10. Add 1 ml of PEG and incubate for 20 min at 22 °C. Centrifuge for 5 min at 2500 rpm. Discard the supernatant completely by decantation and suction with a Pasteur pipette.

11. Add 150 μl of SOS and incubate at 30 °C for 1 hour.

12. Add 5 ml of 0.6 M KCl and centrifuge at 5000 rpm for 5 min. Discard the supernatant completely. Suspend spheroplasts in 5 ml of KCl.

13. 50 μl of suspension are mixed with 5 ml of molten top agar (48 °C) and immediately poured onto a selective agar plate prewarmed at 37 °C. Incubate the plates for 5–7 days.

14. Regeneration test is done by plating the spheroplast suspension diluted 10^4–10^5 times with addition of all the requirements. Regeneration rate, which is

highly dependent on the strain, can vary between 0.5 and 5%. A variable fraction of spheroplasts undergoes lysis during PEG treatment. Just proceed ahead.

With CsCl-purified DNA, a transformation frequency of about 3×10^3 per μg DNA can be obtained. Plating too many protoplasts on a plate may have adverse effects on the regeneration rate.

Solutions: Cytohelicase (Industrie Biologique Française, Clichy, France), 100 mg/ml (roughly equivalent to Glusulase), filter-sterilized; BSA, bovine serum albumin 1 mg/ml, filter-sterilized; 1.2 M sorbitol, sorbitol 21.6 g/100 ml water; SEM, 1.2 M sorbitol/ 25 mM EDTA 14 mM mercaptoethanol-2; SEC, 1.2 M sorbitol/ 10 mM EDTA/ 0.1 M Na citrate pH 5.8; SCa, 1.2 M sorbitol/ 10 mM $CaCl_2$; PEG, 20% w/v polyethylene glycol 4000 (Merck) in 10 mM Tris-HCl pH 7.4/ 10 mM $CaCl_2$; filter-sterilized; SOS, 1.2 M sorbitol/7 mM $CaCl_2$/ 0.4 volume of glucose complete medium, filter-sterilized; 0.6 M KCl, 4.5 g/100 ml water; sonicated salmon sperm DNA 10 mg/ml. Top agar, 0.6 M KCl/ 0.8% Bacto agar/0.67% Yeast Nitrogen Base w/o amino acids Difco/ 2% glucose supplemented with auxotrophic requirements as necessary; agar plates, the same as the top agar except that concentration is 2%.

14.3
Transformation by Electroporation

The protocols derived from the procedure described by Meilhoc et al. (1990) seem to have been adopted by many laboratories.

14.3.1
Transformation by the Electropulsateur

The apparatus used for electroporation in the authors' laboratory is the Electropulsateur from S.A. Jouan (Rue Bobby Sands, C.P.3203, 44 805 SaintHerblain, France; model TRX GHT 1287 ref. 51700100, electrode set ref. 51700150). A procedure for this apparatus has been communicated to us by Dr. R. Fleer (Department of Biotechnology CRVA, Rhone-Poulenc Rorer, Vitry-sur-Seine, France). A slightly modified protocol, as practised in the authors laboratory, is given below.

1. Grow cells as in the preceding protocol.

2. Harvest cells by centrifugation and wash once with EP buffer.

3. Suspend cells in 1 ml of DTT medium and incubate at 28 °C for 30 min.

4. Centrifuge and wash cells in EP; suspend cells in 1 ml EP (about 10^9 cells/ml).

5. Take 100 μl of suspension (about 10^8 cell) in a sterile Eppendorf tube, mix with 2–4 μl of transforming DNA. Cool on ice.

6. Place the suspension between the electrodes (1 cm wide, 4 mm apart, alcohol-washed). The drop is supported by the electrodes on the surface of a plastic

petri- dish which can be placed on ice. (This setting was found more convenient than the use of a cuvette, especially when many samples are to be processed.)

7. Apply 1 kV for 20 ms.

8. Transfer the drop into 1 ml of glucose complete medium, and incubate for 1 h at 30 °C to allow regeneration.

9. Plate 0.1 to 0.3 ml on appropriate selective media, and incubate for 2–3 days.

Solutions EP buffer: 10 mMTris-HCl, pH 7.5/270 mM sucrose/ 1 mM MgCl$_2$, autoclaved.

DTT medium: mix 100 ml of glucose complete medium, 2.5 ml of 0.1 M dithiothreitol and 2.0 ml of 1 M Hepes buffer, pH 8, sterilized by filtration.

In a typical case, 30 ng of DNA gave 1000 transformants; higher amounts of DNA may reduce the transformation frequency. Cell viability is around 30–50%, which is about 100 times higher than the spheroplast method.

14.3.2
Transformation by the Gene Pulser

The Gene Pulser/Pulse controller system from Biorad Laboratories (ref. 165-2077) is also in use in different laboratories. Biorad and Jouan systems differ by the shape of the electric pulse ("exponential" versus "square"). The following procedure has been communicated to us by Dr. A. Dominguez (University of Salamanca, Spain).

A culture grown overnight is used as the inoculum. Grow cells overnight in 200 ml of YED (1% yeast extract Difco, 1% glucose) until early to mid-exponential phase (a density of A$_{600}$ 0.8–1.4; 1.4–3 × 10^7 cells/ml). Centrifuge cells and wash with water.

Cells are suspended in 20 ml of pretreatment buffer (1% glucose, 1% yeast extract, 25 mM dithiothreitol, 20 mM HEPES, pH 8).

After incubation for 30 min at 30 °C with shaking, cells are collected by centrifugation and resuspended in EB buffer (10 mM TrisHCl, pH 7.5/270 mM sucrose/ 1 mM lithium acetate) at a concentration of 2–3 × 10^9 cells/ml.

50-μl aliquots are mixed with DNA in an Eppendorf tube and kept on ice for 15 min. The final volume is 50–55 μl.

An electroporation cuvette of 2 mm path (ref. 165–2086) is used. Voltage 1000 V, capacitance 25 μF, and resistance 400 ohm. After the pulse, 1 ml of cold YED is added to the suspension, mixed thoroughly, transferred to a sterile Eppendorf tube, kept 15 min on ice and 60 min at 30 °C, then plated on appropriate media.

Various parameters in this protocol have been extensively studied by Sanchez et al. (1993). They reported a transformation efficiency of 10^6–10^7 transformants per μg DNA in using pKD1-derived circular vectors.

14.4
Transformation by a LiCl Method

The protocol was contributed by Dr. S. Ménart and Dr. M. Bolotin-Fukuhara (University of Paris XI, Orsay).

1. Grow cells in 10 ml of complete medium to A_{600} of 0.6–1.0. Collect cells by centrifugation and suspend them in 400 μl TE plus 400 μl 20 mM LiCl. Incubate at 28 °C with shaking.

2. For transformation, mix 100 μl of the above suspension with 0.2–1.0 μg of transforming DNA in a volume not larger than 10 μl. Incubate for 30 min at 28 °C.

3. Add 100 μl of 70% polyethyleneglycol (PEG 4000) and incubate for 1 h. Heat shock at 42 °C for 5 min. Collect cells by 1 min centrifugation, rinse three times with 100 μl of water, suspend them in 100 μl of water before plating on appropriate agar plates.

14.5
Transformation of Frozen Competent Cells

This protocol was adapted from Dohmen et al. (1991)'s procedure above. It employs polyethlyene glycol and dimethylsulfoxide (DMSO).

1. Cells grown in glucose complete medium (100 ml) to a density of about 10^7/ml (A_{6000} about 0.6), are harvested and washed with water, then with a half-volume of bicine buffer. The pellet is suspended in 1/50 volume of the same buffer; 200-μl portions of the suspension are distributed in sterile Eppendorf tubes and stored frozen at −70 °C until use.

2. Transformation of the frozen cells is carried out at a temperature above 15 °C. To the top of the frozen competent cells, add 0.1–5 μg of transforming DNA plus 50 μg of sonicated salmon sperm DNA. The volume of DNA must not exceed 20 μl.

3. Allow cells to thaw with vigorous agitation at 37 °C for 5 min. Add slowly 1.4 ml of 40% PEG/ 0.2 M Bicine solution. Mix by gentle inversion of the tube, before incubation for 1 h at 30 °C.

4. Pellet the cells at 3000 g for 5 min (20 °C). Wash the pellet with 1.5 ml of 0.15 M NaCl/10 mM bicine. Finally, resuspend the cells in the same buffer before plating on appropriate selection medium.

Reagents: bicine buffer, 1 M Sorbitol/10 mM Bicine-NaOH (pH 8.35)/3% PEG 1000/ 5% DMSO, filter-sterilized and stored at 4 °C.

40% PEG 1000/ 0.2 M Bicine-NaOH (pH 8.35), filter-sterilized and stored at room temperature.

14.6
Release of Plasmids from K. lactis Transformants

Total yeast DNA is prepared from individual transformants and used to transform competent *E. coli* cells to recover the yeast plasmid for further characterization.

1. Grow a 5-ml culture in selective medium to saturation at 28 °C.

2. Collect the cells by centrifugation at 2500 rpm for 5 min in a refrigerated centrifuge (4 °C).

3. Remove the supernatant and resuspend the cells in 200 μl of 50 mM Tris-HCl (pH 7.5), 60 mM EDTA, 0.4% Triton X-100, 2.5 M LiCl. Transfer to a microfuge tube (1.5 ml).

4. Add 0.3 g of acid-washed glass beads and 0.2 ml of phenol/chloroform: isoamylalcohol (25 : 24 : 1, volume ratio).

5. Vortex for 5 min.

6. Centrifuge in a microfuge for 2 min (room temperature).

7. Transfer the aqueous layer to a fresh tube and add 1 ml of 96% ethanol. Mix by inversion.

8. Centrifuge in a microfuge for 2 min. Discard the supernatant. Resuspend the pellet in 50 μl of 1 × TE pH 8.

9. Use 2 μl of the plasmid preparation to transform 100 μl of competent *E. coli* cells.

14.7
Use of G418 Resistance Marker in Transformation

Study of nonconventional yeasts often suffers from the absence of available auxotrophic mutants as transformation hosts. Sensitivity of many yeasts, including *K. lactis*, to the antibiotic G418 allows the use of the kanamycin resistance gene (coding for a 3' aminoglycoside-phosphotransferase) as a transformation marker. A few comments on the use of this marker may be appropriate. G418 is a ribosomal inhibitor in many eukaryotic cells. It is equivalent to the antibacterial agents kanamycin or neomycin. The kanamycin resistance gene, Km[R], isolated from the bacterial transposon Tn*903* can be expressed in *S. cerevisiae* from its natural promoter (Jimenez and Davies 1980) and this also applies to *K. lactis* (Sreekrishna et al. 1984; Chen and Fukuhara 1988). *K. lactis* strains are clearly more sensitive to G418 than is *S. cerevisiae* (Fig. 16A). However, when *K. lactis* cells are maintained at low concentrations of G418 (up to about 50 μg/ml), resistant colonies tend to appear spontaneously at relatively high frequencies (10^{-5}). At 200 μg/ml, their frequency is extremely low. This concentration of G418 (a stock solution of Geneticin, Sigma ref. G5013, at 100 mg/ml water, filter-sterilized and frozen) was

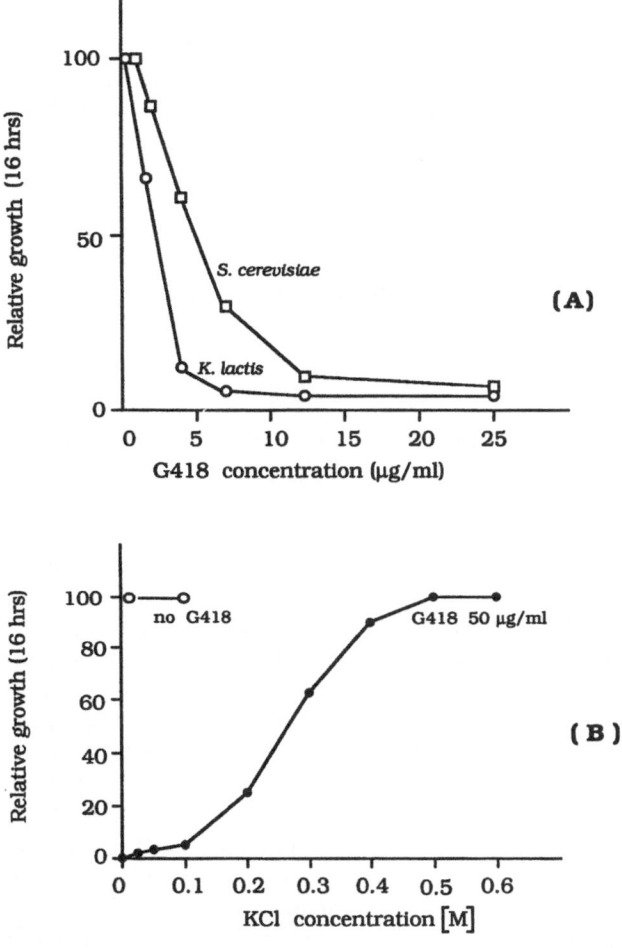

Fig. 16A,B. Effect of G418 on the growth of *K. lactis* and *S. cerevisiae*. A Range of effective G418 concentration. B Effect of salf concentration on the sensitivity to G418

adopted in the authors' laboratory for routine selection of G418 resistant transformants. The Km^R gene introduced into *K. lactis* by a multicopy vector can confer resistance to G418 at the level of more than 2 mg/ml. In practice, G418 should not be used in high salt media (such as the standard Yeast Nitrogen Base plates), because the yeast becomes insensitive to the drug at the salt concentrations exceeding 0.1 M NaCl or KCl (see Fig. 16B). Sorbitol at high concentration does not interfere with G418 action. The standard complete media (1% yeast extract/ 1% peptone/ 2% glucose) can be used as such in combination with G418. It should be noted that many yeast species are not sensitive to G418 (Table 5). Among the

Table 5. Examples of yeast species whose strains are generally resistant to G418. (200 μg/ml in glucose complete agar medium, 20 °C)

Candida albicans	Pichia anomala	Williopsis saturnus
Candida glabrata	Pichia farinosa	Yarrowia lipolytica
Candida parapsilosis	Pichia guilliermondii	
Candida saitoana	Pichia jadinii	
Candida santamaria	Pichia kluyveri	
Candida shehatae	Pichia membranaefaciens	
Candida tropicalis	Pichia stipis	

Many yeast strains show a variable degree of resistance to the drug as a strain-specific and not species-specific character. This list includes only those species in which all the strains tested gave homogeneously a resistant phenotype.

sensitive species, the sensitivity may vary greatly from strain to strain. The response of various strains of *K. lactis* to G418 appears to be quite homogeneous.

When G418 resistance is the selection phenotype, the spheroplast transformation protocol described above should be modified because of the salt effects. The following is a modification of the procedure described by Sreekrishna et al. (1984) from which KCl was omitted.

After the 1 h incubation of spheroplasts in SOS (see the spheroplast transformation protocol described above), add 3 ml of 1.2 M sorbitol. Centrifuge the suspension at 4000 rpm for 5 min. Suspend the pellet in 0.8 ml of 1.2 M sorbitol (in 10 ml for the regeneration test); 50–100, and 200-μl portions were diluted in 5-ml regeneration top agar (melted and kept at 46 °C). This modified top agar contains glucose complete medium/1.2 M sorbitol/2% agar. The mix is poured on a prewarmed base plate of the same composition as this top agar. When solidified, the plates are incubated at 28 °C for 16 h. A second 5 ml of top agar containing 0.16 ml of G418 stock solution (50 mg/ml) is spread on the surface, and incubation is continued for a few days to allow colony development. If one omits the first 16-h incubation in the absence of G418, the appearance of transformants is delayed and their number is much lower.

In the case of transformation by electroporation, the transformants are first plated on glucose complete medium, and incubated at 30 °C for 15 h, then replica-plated onto G418 medium.

15
K. lactis for Industrial Application

K. lactis is known as a source of β-galactosidase, but attempts at large-scale use of *K. lactis* for protein production are relatively recent. As this yeast is present in various milk products, it is accepted as "generally recognized as safe (GRAS)" for industrial use. It can be grown to a high cell mass in a fed-batch fermenter up to a yield of near 100 g dry mass per liter in corn steep/glucose media (Fleer et al. 1991b). The recent interest in *K. lactis* seems to come especially from its ability to secrete high molecular weight proteins. The fact that the linear plasmid-encoded toxin is secreted into culture media suggested that *K. lactis* was capable of excreting very large proteins, even if such a capacity is not an exceptional characteristic of this species. In order to achieve high production levels of secreted heterologous proteins, such as human proteins, industrial firms have been trying to use yeasts in place of bacteria (recently reviewed by Shuster 1991; Fleer 1992; Romanos et al. 1992). Although many such attempts had been made with the familiar *S. cerevisiae*, *K. lactis* has become one of hopeful alternatives to achieve production of secreted proteins (van den Berg et al. 1990; Fleer et al. 1991a,b). Table 6 shows a few examples of heterologous proteins produced from *S. cerevisiae* and *K. lactis*.

It has been shown that the secretion of foreign proteins can be directed not only by the *K. lactis* killer toxin signal peptide (Baldari et al. 1987; see Fig. 15c), but also by various signal peptides from heterologous sources including the mating factor α of *S. cerevisiae* or even human serum albumin preprosequence. To our knowledge, proteases that could interfere with secreted protein production have not been detected in *K. lactis* culture supernatants.

For high-level production of heterologous proteins, two types of strategies are considered: one is based on the expression of chromosomally integrated foreign protein genes (which are highly stable through mitosis), and the other relies on multicopy plasmid vectors carrying the foreign gene (maintained at a high gene dosage). Both approaches have given successful examples: prochymosin was produced from chromosomally integrated genes (van den Berg et al. 1990), and human serumalbumin from pKD1 plasmid-based vectors (Fleer et al. 1991b; Yeh et al.

Table 6. Some examples of heterologous proteins produced by *K. lactis* and *S. cerevisiae*

Protein produced	Vector	Promoter	Secretion signal	Approx. yield/l flask/fermenter
S. cerevisiae systems				
Human GRF	2μ	MFα	MFα	30 mg/–
Human apoE	2μ	PHO5	MFα	40 mg/–
Human IL-6	2μ	GAL1	MFα	30 mg/–
Human IL-1β	2μ	GPD	Amylase	20 mg/–
Human PTH	2μ	MFα	MFα	100 mg/–
Human serumalbumin	2μ	Diverse	Diverse	40–150 mg/–
Murine amylase	2μ	GAL7	–	60 mg[a]/370 mg[a]
Tick anticoagulant	2μ	GAL10	MFα	250 mg/–
Tetanos toxin C	2μ	GAL/ADH	–	90 mg[a]/>1 g[a]
K. lactis systems				
Human IL-1β	pKD1	PHO5/PGK	Killer	80 mg
Human serumalbumin	pKD1	LAC4/PGK	Native	400 mg/a few grams
HBsAg	Integ.	LAC4	–	12 mg[a]/–
Bovin prochymosin	Integ.	LAC4	MFα	–/a few grams

[a] Intracellular accumulation. The information is mostly based on the data compiled and communicated to us by Dr. R. Fleer (Rhône-Poulenc Rorer; see review by Fleer 1992).

1992), both with a yield in the order of grams per liter. Multiple integration of a plant α-galactosidase gene into ribosomal DNA produced also a high level of secreted production of the enzyme (Bergkamp et al. 1992). Although strong transcription of heterologous genes can be often obtained, many posttranscriptional and posttranslational steps remain out of control, such as messenger RNA stability, codon choice and glycosylation, and so on. As mentioned before, the codon choice in *K. lactis* is very similar to that of *S. cerevisiae*, and differs much from the mammalian genes. *K. lactis* can use the consensus N-glycosylation sites (Asn-X-Tyr/Ser) which sometimes remain silent in human sequences. Recent reviews by Romanos et al. (1992) and Fleer (1992) may serve as a starting point of literature search on these subjects. Whatever the strategy, high level production requires not only an efficient expression system, but also optimization of fermentation processes which will only be achieved through a better knowledge of general and specific physiological regulation in *K. lactis*.

Acknowledgments. Our colleagues, too many to enumerate, kindly contributed unpublished information. We also thank Dr. Masako Osumi (Japan Women's University, Tokyo), who kindly provided the electron micrograph of *K. lactis* cells. The authors' own work was supported in part by the Commission of the European Communities (BIOT-CT91-0267).

References

Algeri AA, Marmiroli N, Viola AM, Puglisi PP (1977) Dependence of cytoplasmic on mitochondrial protein synthesis in *K. lactis* CBS 2360. II. Genetic studies. Mol Gen Genet 150: 141–145

Baldari C, Murray JAH, Ghiara P, Cesareni G, Galeotti CL (1987) A novel leader peptide which allows efficient secretion of a fragment of human interleukin 1β in *Saccharomyces cerevisiae*. EMBO J 6: 229–234

Barnett JA (1976) The utilization of sugars by yeasts. Adv Carbohydr Chem Biochem 32: 125–234

Barnett JA (1992) Some controls on oligosaccharide utilization by yeasts: the physiological basis of the Kluyver effect. FEMS Microbiol Lett 100: 371–378

Barnett JA, Payne RW, Yarrow D (1990) Yeast: characteristics and identification. 2nd edn. Cambridge University Press, Cambridge

Barns SM, Lane DJ, Sogin ML, Bibeau C, Weisburg WG (1991) Evolutionary relationship among pathogenic *Candida* species and relatives. J Bacteriol 173: 2250–2255

Bell W, Klaassen P, Ohnacker M, Boller T, Herwijer M, Schoppink P, van der Zee P, Wiemken A (1992) Characterization of the 56-kDa subunit of trehalose-6-phosphate synthase and cloning of its gene reveals identity with the product of *CIF1*, a regulator of carbon catabolite inactivation. Eur J Biochem 209: 951–959

Bergkamp RJM, Kool IM, Geerse RH, Planta RJ (1992) Multiple-copy integration of the α-galactosidase gene from *Cyamopsis tetragonoloba* into the ribosomal DNA of *Kluyveromyces lactis*. Curr Genet 21: 365–370

Bergkamp-Steffens GK, Hoekstra R, Planta RJ (1992) Structural and putative regulatory sequences of *Kluyveromyces* ribosomal protein genes. Yeast 8: 903–922

Bhat PJ, Hopper JE (1992) Overproduction of the GAL1 or GAL3 protein causes galactose-independent activation of the GAL4 protein: evidence for a new model of induction for the yeast GAL/MEL regulon. Mol Cell Biol 12: 2701–2707

Bianchi MM (1992) Site-specific recombination of the circular $2\,\mu$m-like plasmid pKD1 requires integrity of the recombinase A and of the partitioning genes B and C. J Bacteriol 174: 6703–6706

Bianchi MM, Falcone C, Chen XJ, Wésolowski-Louvel M, Frontali L, Fukuhara H (1987) Transformation of the yeast *Kluyveromyces lactis* by new vectors derived from the 1.6-μm circular plasmid pKD1. Curr Genet 12: 185–192

Bianchi MM, Frontali L, Fukuhara H (1989) Active recombination of pKD1-derived vectors with resident pKD1 in *Kluyveromyces lactis* transformation. Curr Genet 15:253–260

Bianchi MM, Santarelli R, Frontali L (1991) Plasmid functions involved in the stable propagation of pKD1 circular plasmid in *Kluyveromyces lactis*. Curr Genet 19: 155–161

Bisson LF, Coons DM, Kruckeberg AL, Lewis DA (1993) Yeast sugar transporters. CRC Crit Rev Biochem Mol Biol 28: 259–308

Boze H, Nicol D, Moulin G, Galzy P (1987) The role of genes *LAC1* and *LAC2* in the biosynthesis of lactose metabolism enzymes by *Kluyveromyces lactis*. Acta Microbiol Hung 34: 73–83

Bradshaw HD (1990) Killer toxins. Nature 345: 299

Brake A, Irvine B, Masiarz F, Shultz K (1988) Structure of genes encoding precursors of two *Kluyveromyces lactis* transported proteins. Yeast 4: S436

Breunig KD (1989a) Multicopy plasmids containing the gene of the transcriptional activator LAC9 are not tolerated by *K. lactis* cells. Curr Genet 15: 143–148

Breunig KD (1989b) Glucose repression of *LAC* gene expression in yeast is mediated by the transcriptional activator LAC9. Mol Gen Genet 216: 422–427

Breunig KD, Kuger P (1987) Functional homology between the yeast regulatory proteins GAL4 and LAC9: LAC9-mediated transcriptional activation in *Kluyveromyces lactis* involves protein binding to a regulatory sequence homologous to the GAL4 protein-binding site. Mol Cell Biol 7: 4400–4406

Breunig KD, Dahlems U, Das S, Hollenberg CP (1984) Analysis of a eukaryotic β-galactosidase gene: the N-terminal end of the yeast *K. lactis* protein shows homology to the *E. coli* LacZ gene product. Nucleic Acids Res 12: 2327–2342

Broach JR, Li YY, Feldman J, Jarayam M, Abraham J, Nasmyth KA, Hicks JB (1982) Localization and sequence analysis of yeast origin of replication. Cold Spring Harbor Symp Quant Biol 47: 1165–1173

Brunner A, Coria R (1989) Cloning and sequencing of the gene for apocytochrome b of the yeast *Kluyveromyces lactis* strain WM27 (NRRL Y-17066) and WM37 (NRRL Y-1140). Yeast 5: 209–218

Brunner A, Tuena de Cobos A (1980) Extrachromosomal oligomycin-resistant mutants of the petite-negative yeast *Kluyveromyces lactis*. Properties of mitochondrial ATPase and cross-resistance to inhibitors of phosphoryl transfer reaction. Mol Gen Genet 178: 351–355

Brunner AL, Mas J, Celis E, Matoon JR (1973) Cytoplasmic and nuclear inheritance of resistance to alkylguanidines and ethidium bromide in a petite-negative yeast. Biochem Biophys Res Commun 53: 638–644

Brunner A, Tuena de Cobos A, Griffith DE (1977) The isolation and genetic characterization of extrachromosomal chloramphenicol- and oligomycin-resistant mutants from the petite-negative yeast *Kluyveromyces lactis*. Mol Gen Genet 152: 183–191

Brunner AL, Mendoza VR, Tuena de Cobos A (1987) Extrachromosomal genetics in the yeast *Kluyveromyces lactis*. Isolation and characterization of antimycin-resistant mutants. Curr Genet 11: 475–482

Bulder CJAE (1963) On respiratory deficiency in yeasts. PhD Thesis, Technische Hoge school, Delft

Bulder CJAE (1964a) Induction of petite mutations and inhibition of synthesis of respiratory enzymes in various yeasts. Antonie Leeuwenhoek J Microbiol 30: 1–9

Bulder CJAE (1964b) Lethality of the petite mutation in petite-negative yeasts. Antonie Leeuwenhoek J. Microbiol 30: 442

Butler AR, O'Donnel RW, Martin VJ, Gooday GW, Stark MJR (1991) *Kluyveromyces lactis* toxin has an essential chitinase activity. Eur J Biochem 199: 483–488

Chang X-D, Dickson RC (1988) Primary structure of the lactose permease gene from the yeast *Kluyveromyces lactis*. Presence of an unusual transcript structure. J Biol Chem 263: 16696–16703

Chen X-J (1987) Etude du plasmide pKD1 et développement de systèmes d'expression de gènes chez la levure *Kluyveromyces lactis*. PhD Thesis, University of Paris XI, Paris

Chen X-J, Clark-Walker GD (1993) Mutations in *MGI* genes convert *Kluyveromyces lactis* into a petite-positive yeast. Genetics 133: 517–525

Chen X-J, Fukuhara H (1988) A gene fusion system using the aminoglycoside 3'-phosphotransferase gene of the kanamycin-resistance transposon Tn*903*: use in the yeast *Kluyveromyces lactis* and *Saccharomyces cerevisiae*. Gene 69: 181–192

Chen X-J, Saliola M, Falcone C, Bianchi MM, Fukuhara H (1986) Sequence organization of the circular plasmid pKD1 from the yeast *Kluyveromyces drosophilarum*. Nucleic Acids Res 14: 4471–4481

Chen X-J, Wésolowski-Louvel M, Tanguy-Rougeau C, Fukuhara H, Bianchi MM, Fabiani L, Saliola M, Falcone C, Frontali L (1988) A gene cloning system for *Kluyveromyces lactis* and isolation of a chromosomal gene required for killer toxin production. J Basic Microbiol 28: 211–220

Chen X-J, Bianchi MM, Suda K, Fukuhara H (1989) Host range of the pKD1 derived plasmids in yeasts. Curr Genet 16: 95–98

Chen X-J, Wésolowski-Louvel M, Tanguy-Rougeau C, Fukuhara H (1991) Promoter activity associated with the left inverted terminal repeat of the killer plasmid k1 from yeast. Biochimie 73: 1195–1203

Chen X-J, Cong YS, Li YY, Wésolowski-Louvel M, Fukuhara H (1992a) Characterization of a circular plasmid from the yeast *Kluyveromyces waltii*. J Gen Microbiol 138: 337–345

Chen X-J, Wésolowski-Louvel M, Fukuhara H (1992b) Glucose transport in the yeast *Kluyveromyces lactis*. II. Transcriptional regulation of the glucose transporter gene RAG1. Mol Gen Genet 33: 97–105

Chinery SA, Hinchcliffe E (1989) A novel class of vector for yeast transformation. Curr Genet 16: 21–25

Clark-Walker GD (1991) Contrasting mutation rates in mitochondrial and nuclear genes of yeasts versus mammals. Curr Genet 20: 195–198

Cong YS (1994) Etude des plasmides circulaires et linéaires de levures. PhD Thesis, University of Paris XI, Paris

Cong YS, Wésolowski-Louvel M, Fukuhara H (1994) Creation of a functional promoter by rearrangement in a *Kluyveromyces lactis* linear plasmid. Gene 147: 125–129

Cook AH, Slater CA (1956) The structure of pulcherrimin. J Chem Soc Part 1956: 4133–4135

Coria RO, Garcia M, Brunner A (1989) Mitochondrial cytochrome b genes with a six-nucleotide deletion or single-nucleotide substitutions confer resistance to antimycin A in the yeast *Kluyveromyces lactis*. Mol Microbiol 3: 1590–1604

Coria RO, Zalce ME, Mendoza VR, Brunner A (1990) Restriction site variation, length polymorphism and changes in gene order in the mitochondrial DNA of the yeast *Kluveromyces lactis*. Antonie Leeuvenhoek J Microbiol 58: 227–234

Das S, Hollenberg CP (1982) A high frequency transformation system for the yeast *Kluyveromyces lactis*. Curr Genet 6: 123–128

Das S, Breunig KD, Hollenberg CP (1985) A positive regulatory element is involved in the induction of the β-galactosidase gene from *Kluyveromyces lactis*. EMBO J 4: 793–798

De Deken RH (1966) The Crabtree effect and its relation to the petite mutation in petite-negative yeasts. J Gen Microbiol 44: 157–165

De Jonge P, de Jongh FCM, Meijer SR, Steensma HY, Scheffers NA (1986) Orthogonal-field-alteration gel electrophoresis banding patterns of DNA from yeasts. Yeast 2: 193–204

Del Giudice L, Brunner A (1977) Chromosomal and extrachromosomal inheritance of erythromycin-resistance in the "petite-negative" yeast *Kluyveromyces lactis*. Mol Gen Genet 152: 325–329

Del Giudice L, Puglisi PP (1974) Induction of RD mutants in *K. lactis* by nitrosoguanidine. Biochem Biophys Res Commun 59: 565–571

De Louvencourt L, Fukuhara H, Heslot H, Wésolowski M (1983) Transformation of *Kluyveromyces lactis* by killer plasmid DNA. J Bacteriol 154: 737–742

Deshler JO, Larson GP, Rossi JJ (1989) *Kluyveromyces lactis* maintains *Saccharomyces cerevisiae* intron-encoded splicing signals. Mol Cell Biol 9: 2208–2213

Dickson RC (1980) Expression of a foreign eukaryotic gene in *Saccharomyces cerevisiae*: β-galactosidase from *Kluyveromyces lactis*. Gene 10: 347–356

Dickson RC, Barr K (1983) Characterization of lactose transport in *Kluyveromyces lactis*. J Bacteriol 154: 1245–1251

Dickson RC, Markin JS (1978) Molecular cloning and expression in *E. coli* of a yeast gene coding for β-galactosidase. Cell 15: 123–130

Dickson RC, Markin JS (1980) Physiological studies of β-galactosidase induction in *Kluyveromyces lactis*. J Bacteriol 142: 777–785

Dickson RC, Sheetz RM, Lacy LR (1981) Genetic regulation: yeast mutants constitutive for β-galactosidase activity have an increased level of β-galactosidase messenger ribonucleic acid. Mol Cell Biol 1: 1048–1056

Dickson RC, Gerardot CJ, Martin AK (1990) Genetic evidence for similar negative regulatory domains in the yeast transcription activators GAL4 and LAC9. Nucleic Acids Res 18: 5213–5217

Dinouël N, Drissi R, Miyakawa I, Sor F, Rousset S, Fukuhara H (1993) Linear mitochondrial DNAs of yeasts. Closed loop structure of the termini and possible linear-circular conversion mechanisms. Mol Cell Biol 13: 2315–2323

Dohmen RJ, Strasser AWM, Höner CB, Hollenberg CP (1991) An efficient transformation procedure enabling long-term storage of competent cells of various yeast genera. Yeast 7: 691–692

Eng FJ, Warner JR (1991) Structural basis for the regulation of splicing of a yeast messenger RNA. Cell 65: 797–804

Entian K-D, Barnett JA (1983) Some genetical and biochemical attempts to elucidate the energetics of sugar uptake and explain the Kluyver effect in the yeast *Kluyveromyces lactis*. Curr Genet 7: 323–325

Fabiani L, Aragona M, Frontali L (1990) Isolation and sequence analysis of a *K. lactis* chromosomal DNA element able to autonomously replicate in *S. cerevisiae* and *K. lactis*. Yeast 6: 69–76

Falcone C, Saliola M, Chen X-J, Bianchi MM, Frontali L, Fukuhara H (1986) Analysis of a 1.6-µm circular plasmid from the yeast *Kluyveromyces drosophilarum*. Structure and molecular dimorphism. Plasmid 15: 248–252

Faye G, Kujawa C, Fukuhara H (1974) Physical and genetic organization of petite and grande yeast mitochondrial DNA. IV. In vivo transcription products of mitochondrial DNA and localization of 23S ribosomal RNA gene in petite mutants of *Saccharomyces cerevisiae*. J Mol Biol 88: 185–203

Ferrero I, Rossi C, Landini MP, Puglisi PP (1978) Role of mitochondrial protein synthesis in catabolite repression of the petite-negative yeast *Kluyveromyces lactis*. Biochem Biophys Res Commun 80: 340–348

Finley RL Jr, Chen S, Ma J, Byrne P, West RW Jr (1990) Opposing regulatory functions of positive and negative elements in USA$_G$ control transcription of the yeast *GAL* genes. Mol Cell Biol 10: 5663–5670

Fleer R (1992) Engineering yeast for high level expression. Curr Opinion Biotechnol 3: 486–496

Fleer R, Chen XJ, Amellal N, Yeh P, Gault N, Faucher D, Folliard F, Fukuhara H, Mayaux J-F (1991a) High-level secretion of correctly processed recombinant human interleukin-1β in *Kluyveromyces lactis*. Gene 107: 285–295

Fleer R, Yeh P, Amellal N, Fournier A, Bacchetta F, Baduel P, Jung G, L'Hôte H, Becquart J, Fukuhara H, Mayaux J-F (1991b) Stable multicopy vectors for high-level secretion of recombinant human serumalbumin in *Kluyveromyces* yeasts. Bio/Technology 9: 968–975

Flick JS, Johnston M (1990) Two systems of glucose repression of the GAL1 promoter in *Saccharomyces cerevisiae*. Mol Cell Biol 10: 4757–4769

Fournier A, Fleer R, Yeh P, Mayaux J-F (1990) The primary structure of the 3-phosphoglycerate kinase (PGK) gene from *Kluyveromyces lactis*. Nucleic Acids Res 18: 365

Freire Picos MA, Rodriguez Torres AM, Ramil E, Cerdan ME, Breunig KD, Hollenberg CP, Zitomer RS (1993) Sequence of a cytochrome c gene from *Kluyveromyces lactis* and its upstream region. Yeast 9: 201–204

Fujimura H, Hishinuma F, Gunge N (1987) Terminal segment of *Kluyveromyces lactis* linear DNA plasmid pGKL2 supports autonomous replication of hybrid plasmids in *Saccharomyces cerevisiae*. Curr Genet 12: 99–104

Fukuhara H (1987) The RF1 gene of the killer DNA of yeast may encode a DNA polymerase. Nucleic Acids Res 15: 10046

Fukuhara H, Sor F, Drissi R, Dinouel N, Miyakawa I, Rousset S, Viola AM (1993) Linear mitochondrial DNAs of yeasts. Frequency of occurrence and general features. Mol Cell Biol 13: 2309–2314

Fuson GB, Presley HL, Phaff HJ (1987) Deoxyribonucleic acid base sequence relatedness among members of the yeast genus *Kluyveromyces*. Int J Syst Bacteriol 37: 371–379

Gödecke A, Zachariae W, Arvanitidis A, Breunig KD (1991) Coregulation of the *Kluyveromyces lactis* lactose permease and β-galactosidase genes is achieved by interaction of multiple LAC9 binding sites in a 2.6 kbp divergent promoter. Nucleic Acids Res 19: 5351–5358

Goffeau A, Briquet M, Colson AM, Delhez J, Foury F, Labaille F, Landry Y, Mohar O, Moena E (1975) Stable pleiotropic respiratory-deficient mutations of a "petite-negative" yeast. In: Tzagoloff A (ed) Membrane biogenesis. Chap 3. Plenum Press, New York

Goffrini P, Algeri AA, Donnini C, Wésolowski-Louvel M, Ferrero I (1989) *RAG1* and *RAG2*: nuclear genes involved in the dependence/independence on mitochondrial respiratory function for the growth on sugars. Yeast 5: 99–106

Goffrini P, Wésolowski-Louvel M, Ferrero I, Fukuhara H (1990) RAG1 gene of the yeast *Kluyveromyces lactis* codes for a sugar transporter. Nucleic Acids Res 18: 5294

Goffrini P, Wésolowski-Louvel M, Ferrero I (1991) A phosphoglucose isomerase gene is involved in the Rag phenotype of the yeast *Kluyveromyces lactis*. Mol Gen Genet 228: 401–409

Gonçalves PM, Maurer K, Mager WH, Planta R (1992) *Kluyveromyces* contains a functional ABF-1 homologue. Nucleic Acids Res 20: 2211–2215

Gorbalenya AE, Koonin EV, Donchenko AP, Blinov VM (1989) Two related superfamilies of putative helicases involved in replication, recombination, repair and expression of DNA and RNA genomes. Nucleic Acids Res 17: 4713–4730

Griggs DW, Johnston M (1991) Regulated expression of the *GAL4* activator gene in yeast provides a sensitive genetic switch for glucose repression. Proc Natl Acad Sci USA 88: 8597–8601

Gunge N (1986) Linear DNA killer plasmids from the yeast *Kluyveromyces*. Yeast 2: 153–162

Gunge N, Kitada K (1988) Replication and maintenance of the *Kluyveromyces* linear pGKL plasmids. Eur J Epidemiol 4: 409–414

Gunge N, Sakaguchi K (1981) Intergenic transfer of deoxyribonucleic acid killer plasmids, pGKL1 and pGKL2, from *Kluyveromyces lactis* into *Saccharomyces cerevisiae* by cell fusion. J Bacteriol 147: 155–160

Gunge N, Yamane C (1984) Incompatibility of linear DNA killer plasmids pGKL1 and pGKL2 from *Kluyveromyces lactis* with mitochondrial DNA from *Saccharomyces cerevisiae*. J Bacteriol 159: 533–539

Gunge N, Tamaru A, Ozawa F, Sakaguchi K (1981) Isolation and characterization of linear deoxyribonucleic acid plasmids from *Kluyveromyces lactis* and the plasmid-associated killer character. J Bacteriol 151: 462–464

Gunge N, Murata K, Sakaguchi K (1982) Transformation of *Saccharomyces cerevisiae* with linear DNA plasmids from *Kluyveromyces lactis*. J Bacteriol 151: 462–464

Gunge N, Murakami K, Takesato T, Morijima H (1990) Mating type locus-dependent stability of the *Kluyveromyces lactis* linear pGKL plasmids in *Saccharomyces cerevisiae*. Yeast 6: 417–427

Haffter P, Fox TD (1992) Nuclear mutations in the petite-negative yeast *Schizosaccharomyces pombe* allow growth of cells lacking mitochondrial DNA. Genetics 131: 255–260

Hardy CM, Clark-Walker GD (1989) Nucleotide sequence of the structural genes for the mitochondrial cys, lys, gln, and leu tRNAs from the yeast *Kluyveromyces lactis* K8. Nucleic Acids Res 17: 1762

Hardy CM, Clark-Walker GD (1990) Nucleotide sequence of the cytochrome oxidase subunit 2 and val tRNA genes and surrounding sequences from *Kluyveromyces lactis* K8 mitochondrial DNA. Yeast 6: 403–410

Hardy CM, Clark-Walker GD (1991) Nucleotide sequence of the *COX1* gene in *Kluyveromyces lactis* mitochondrial DNA: evidence for recent horizontal transfer of a group II intron. Curr Genet 20: 99–114

Hardy CM, Galeotti C, Clark-Walker GD (1989) Deletion and rearrangements in *Kluyveromyces lactis* mitochondrial DNA. Curr Genet 16: 419–427

Heinisch J, Kirchrath L, Liesen T, Vogelsang K, Hollenberg CP (1993) Molecular genetics of phosphofructokinase in the yeast *Kluyveromyces lactis*. Mol Microbiol 8: 559–570

Hendriks L, Goris A, Van de Peer Y, Neefs Y-M, Vancanneyt M, Kersters K, Berny J-F, Hennebert GL, De Waechter R (1992) Phylogenetic relationships among ascomycetes and ascomycetes-like yeasts as deduced from small ribosomal subunit RNA sequences. Syst Appl Microbiol 15: 98–104

Herman AI (1963) Genetic control of β-glucosidase synthesis in yeast. PhD Thesis, University of Wisconsin, Madison

Herman AI, Griffin PS (1968) Respiratory-deficient mutants in *Saccharomyces lactis*, a petite-negative yeast. J Bacteriol 96: 457–461

Herman AI, Halvorson HO (1963a) Identification of the structural gene for β-glucosidase in *Saccharomyces lactis*. J Bacteriol 85: 895–900

Herman AI, Halvorson HO (1963b) Genetic control of β-glucosidase in *Saccharomyces lactis*. J Bacteriol 85: 901–910

Herman A, Roman H (1966) Allele specific determinants of homothalism in *Saccharomyces lactis*. Genetics 53: 727–740

Heslot H, Louis C, Goffeau A (1970) Segregational respiratory-deficient mutants of a "petite-negative" yeast *Schizosaccharomyces pombe* 972 h⁻. J Bacteriol 104: 482–491

Heus JJ, Zonneveld BJM, Steensma HY, Van den Berg JA (1990) Centromeric DNA of *Kluyveromyces lactis*. Curr Genet 18: 517–522

Hishinuma F, Nakamura K, Hirai K, Nishizawa R, Gunge N, Maeda T (1984) Cloning and nucleotide sequences of the linear killer plasmids from yeast. Nucleic Acids Res 12: 7581–7587

Hishinuma F, Tokunaga M, Wada N (1986) Expression of immunity gene of linear DNA killer plasmid. Yeast 2: S158

Hoekstra R, Ferreira PM, Bootsman TC, Mager WH, Planta RJ (1992) Structure and expression of the ABF1-regulated ribosomal protein S33 gene in *Kluyveromyces*. Yeast 8: 949–959

Jakobsen BK, Pelham HRB (1991) A conserved heptapeptide restrains the activity of the yeast heat shock transcription factor. EMBO J 10: 369–375

Jacoby J, Hollenberg CP, Heinisch J (1993) Transaldolase mutants in the yeast *Kluyveromyces lactis* provide evidence that glucose can be metabolized through the pentose phosphate pathway. Mol Microbiol 10: 867–876

Jacquier A, Dujon B (1983) The intron of the mitochondrial 21S rRNA gene: distribution in different yeast species and sequence comparison between *Kluyveromyces thermotolerans* and *Saccharomyces cerevisiae*. Mol Gen Genet 192:487–499

Jimenez A, Davies J (1980) Expression of a transposable antibiotic resistance element in *Saccharomyces*. Nature 287: 869–881

Johannsen E (1980) Hybridization studies with the genus *Kluyveromyces* van der Walt emend. van der Walt. Antonie Leeuwenhoek J Microbiol 46: 177–189

Johannsen E, van der Walt JP (1978) Interfertility as a basis for the delimitation of *Kluyveromyces marxianus*. Arch Microbiol 118: 45–48

Johnston JR, Contopoulou CR, Mortimer RK (1988) Karyotyping of yeast strains of several genera by field inversion gel electrophoresis. Yeast 4: 191–198

Johnston M, Hopper JE (1982) Isolation of the yeast regulatory gene *GAL4* and analysis of its dosage effect on the galactose/melibiose regulon. Proc Natl Acad Sci USA 79: 6971–6975

Jung G, Leavitt MC, Ito J (1987) Yeast killer plasmid pGKL1 encodes a DNA polymerase belonging to the family β DNA polymerases. Nucleic Acids Res 15: 9088

Kämper J, Meinhardt F, Gunge N, Esser K (1989a) New recombinant linear DNA-elements derived from *Kluyveromyces lactis* killer plasmids. Nucleic Acids Res 17: 1781

Kämper J, Meinhardt F, Gunge N, Esser K (1989b) In vivo construction of linear vectors based on killer plasmids from *Kluyveromyces lactis*: selection of a nuclear gene results in attachment of telomeres. Mol Cell Biol 9: 3931–3937

Kikuchi Y, Hirai K, Hishinuma F (1984) The yeast linear DNA killer plasmids pGKL1 and pGKL2, possess terminally attached proteins. Nucleic Acids Res 12: 5685–5692

Kikuchi Y, Hirai K, Gunge N, Hishinuma F (1985) Hairpin plasmid – a novel linear DNA of perfect hairpin structure. EMBO J 4: 1881–1886

Kitada K, Gunge N (1988) Palindrome-hairpin linear plasmids possessing only a part of the ORF1 gene of the yeast killer plasmid pGKL1. Mol Gen Genet 215: 46–52

Klebe RJ, Harriss JV, Sharp D, Douglas MG (1983) A general method for polyethyleneglycol-induced transformation of bacteria and yeast. Gene 25: 333–341

Kuger P, Gödecke A, Breunig KD (1990) A mutation in the Zn-finger of the GAL4 homolog LAC9 results in glucose repression of its target genes. Nucleic Acids Res 18: 745–751

Kurtzman CP, Phaff HJ (1989) Molecular taxonomy In: Rose AH, Harrison JS (eds) The yeasts, 2nd edn. Vol 1. Academic Press, London, pp 63–94

Kurtzman CP, Smiley MJ, Johnson CJ, Wickerham LJ, Fuson GB (1980) Two new and closely related heterothallic species, *Pichia amylophila* and *Pichia mississippiensis*: characterization by hybridization and deoxyribonucleic acid reassociation. Int J Syst Bacteriol 30: 208–216

Kuzhandaivelu N, Jones WK, Martin AK, Dickson RC (1992) The signal for glucose repression of the lactose-galactose regulon is amplified through subtle modulation of transcription of the *Kluyveromyces lactis Kl-GAL4* activator gene. Mol Cell Biol 12: 1924–1931

Lachance M-A (1989) Restriction mapping of rDNA and the taxonomy of *Kluyveromyces* van der Walt Emend. van der Walt. Yeast 5: S579–S383

Lachance M-A (1993) *Kluyveromyces* systematics since 1970. Antonie Leeuwenhoek J Microbiol 63: 95–104

Lacy LR, Dickson RC (1981) Transcriptional regulation of the *Kluyveromyces lactis* β-galactosidase gene. Mol Cell Biol 1: 629–634

Lamphier MS, Ptashne M (1992) Multiple mechanisms mediate glucose repression of the yeast *GAL1* gene. Proc Natl Acad Sci USA 89: 5922–5926

Larson GP, Rossi JJ (1991) Altered response to growth rate change in *Kluyveromyces lactis* versus *Saccharomyces cerevisiae* as demonstrated by heterologous expression of ribosomal protein 59 (CRY1). Nucleic Acids Res 19: 4701–4707

Leonardo JM, Bhairi SM, Dickson RC (1987) Identification of upstream activator sequences that regulate induction of the β-galactosidase gene in *Kluyveromyces lactis*. Mol Cell Biol 7: 4369–4376

Lewis MJ, Pelham HRB (1990) The sequence of the *Kluyveromyces lactis* BiP gene. Nucleic Acids Res 18: 6438

Lloyd AT, Sharp PM (1993) Synonymous codon usage in *Kluyveromyces lactis*. Yeast 9: 1219–1228

Lodi T, O'Connor D, Goffrini P, Ferrero I (1994) Carbon catabolite repression in *Kluyveromyces lactis*: isolation and characterization of the *KlDLD* gene encoding the mitochondrial enzyme D-lactate ferricytochrome c oxidase. Mol Gen Genet 244: 622–625

Luyten K, De Koning W, Tesseur I, Ruiz MC, Ramos J, Cobbaert P, Thevelein JM, Hohmann S (1993) Disruption of the *Kluyveromyces lactis* GGS1 gene causes inability to grow on glucose and frutose and is suppressed by mutations that reduce sugar uptake. Eur J Biochem 217: 701–713

Macreadie IG, Horaitis O, Vaughan PR, Clark-Walker GD (1991) Constitutive expression of the *Saccharomyces cerevisiae* CUP1 gene in *Kluyveromyces lactis*. Yeast 7: 127–135

Maleszka R, Clark-Walker GD (1989) A petite positive strain of *Kluyveromyces lactis* has a 300 kb deletion in the rDNA cluster. Curr Genet 16: 429–432

Maniatis T, Fritsch EF, Sambrook J (1982) Molecular cloning: a laboratory manual. Cold Spring Harbor Laboratory, Cold Spring Harbor, New York

Marmiroli N, Ferrero I, Rossi C, Donini C, Puglisi PP, Parenti F (1979) In vivo differentiation with trichodermin, axenomycin, emetine, anisomycin and erythromycin of cytoplasmic and mitochondrial protein synthesis in the yeasts *Saccharomyces cerevisiae*, *Kluyveromyces lactis* and *Endomycopsis capsularis*. Microbiologica 2: 149–165

Massardo DR, Manna F, Schäfer B, Wolf K, Del Giudice L (1994) Complete absence of mitochondrial DNA in the petite-negative yeast *Schizosaccharomyces pombe* leads to resistance towards the alkaloid lycorine. Curr Genet 25: 80–83

McEachern MJ, Blackburn E (1994) A conserved sequence motif within the exceptionally diverse telomeric sequences of budding yeasts. Proc Natl Acad Sci USA 91: 3453–3457

McNeel DG, Tamaoki F (1991) TRF1, a novel DNA-binding protein recognizing the terminal sequence of the pGKL linear DNA plasmids. Proc Natl Acad Sci USA 88: 11398–11402

Meilhoc E, Masson JM, Teissié J (1990) High efficiency transformation of intact yeast cells by electric field pulses. Bio/Technology 8: 223–227

Meyer J (1993) Galaktokinase – ein regulatorisches Protein des Laktose/Galactose-Metabolismus in Hefe. Heinrich-Heine-Universität Düsseldorf, Dissertation

Meyer J, Walker-Jonah A, Hollenberg CP (1991) Galactokinase encoded by *GAL1* is a bifunctional protein required for induction of the *GAL* genes in *Kluyveromyces lactis* and is able to suppress the *gal3* phenotype in *Saccharomyces cerevisiae*. Mol Cell Biol 11: 5454–5461

Miller J (1972) In: Experiments in molecular biology. Cold Spring Harbor Laboratory, Cold Spring Harbor, New York, pp 352–355

Morrow BE, Ju Q, Warner JR (1993) A bipartite DNA binding domain in Rebip of yeast. Mol Cell Biol 13: 1173–1182

Mulder W, Scholten IHJM, van Roon H, Grivell LA (1994a) Isolation and characterisation of the linked genes, *APA2* and *OCR7*, coding for Ap4A phosphorylase II and the 14 kDa subunit VII of the mitochondrial bc1-complex of the yeast *Kluyveromyces lactis*. Biochem Biophys Acta 1219: 719–723

Mulder W, Scholten IHJM, De Boer RW, Grivell LA (1994b) Sequence of the HAP3 transcription factor of *K. lactis* predicts the presence of a novel 4-cystein zinc-finger motif. Mol Gen Genet 245: 96–106

Mulder W, Scholten IHJM, Nagelkerken R, Grivell LA (1994c) Isolation and characterisation of the linked genes, *FPS1* and *QCR8*, coding for farnesyl-diphosphate synthase and the 11 kDa subunit VII of the mitochondrial *bcl*-complex of the yeast *Kluyveromyces lactis*. Biochem Biophys Acta 1219: 713–718

Mulder W, Winkler RA, Scholten IHJM, Zonneveld BJM, De Winde GH, Steensma HY, Grivell LA (1994d) Centromers Promotor Factors (CPF1) of the yeasts *Saccharomyces cerevisiae* and *Kluyveromyces lactis* are functionally exchangeable despite low overall homology. Curr Genet 26: 198–207

Mylin LM, Gerardot CJ, Hopper JE, Dickson RC (1991) Sequence conservation in the *Saccharomyces* and *Kluyveromyces* GAL11 transcription activators suggest funcional domains. Nucleic Acids Res 19: 5345–5350

Na JG, Hampsey M (1993) The *Kluyveromyces* gene encoding the general transcription factor IIB: structural analysis and expression in *Saccharomyces cerevisiae*. Nucleic Acids Res 21: 3413–3417

Nehlin JO, Carlberg M, Ronne H (1991) Control of yeast *GAL* genes by MIG1 repressor: a transcriptional cascade in the glucose response. EMBO J 10: 3373–3377

Nguyen C, Bolotin-Fukuhara M, Wésolowski-Louvel M, Fukuhara H (1995) The respiratory system of *Kluyveromyces lactis* escapes from HAP2 control. Gene 152: 113–115

Niwa O, Sakaguchi K, Gunge N (1981) Curing of the killer deoxyribonucleic acid plasmids of *Kluyveromyces lactis*. J Bacteriol 148: 988–990

Oberyé EHH, Maurer K, Mager WH, Planta RJ (1993) Structure of the ABF1-homologue from *Kluyveromyces marxianus*. Biochim Biophys Acta 1173: 233–236

Ogawa Y, Tatsumi H, Murakami S, Ishida Y, Murakami K, Masaki A, Kawabe H, Arimura H, Nakano E, Motai H, Toh-e A, (1990) Secretion of *Aspergillus oryzae* protease in an osmophilic yeast, *Zygosaccharomyces rouxii*. Agric Biol Chem (Tokyo) 54: 2521–2529

Osinga KA, De Haan M, Christianson T, Tabak HF (1982) A nonanucleotide sequence involved in promotion of ribosomal RNA synthesis and RNA priming of DNA replication in yeast. Nucleic Acids Res 10: 7993–8006

Pan T, Halvorsen Y-D, Dickson RC, Coleman JE (1990) The transcription factor LAC9 from *Kluyveromyces lactis* like GAL4 from *Saccharomyces cerevisiae* forms a Zn(II)2 Cys6 binuclear cluster. J Biol Chem 265: 21427–21429

Poch O, L'Hôte H, Dallery V, Debeaux F, Fleer R, Sodoyer R (1992) Sequence of the *Kluyveromyces lactis* β-galactosidase: comparison with prokaryotic enzymes and secondary structure analysis. Gene 118: 55–63

Price CW, Fuson GB, Phaff HJ (1978) Genome comparison in yeast systematics: delimitation of species within the genera *Schwanniomyces' Saccharomyces, Debaryomyces* and *Pichia*. Microbiol Rev 42: 162–193

Prior C (1993) Caractérisation génétique et moléculaire de gènes impliqués dans le métabolisme de glucose chez *Kluyveromyces lactis*. Identification d'un gène de transporteur de glucose chez *Saccharomyces cerevisiae*. PhD Thesis, University of Paris XI, Paris

Prior C, Fukuhara H, Blaisonneau J, Wésolowski-Louvel M (1993a) Low-affinity glucose carrier gene LGT1 of *Saccharomyces cerevisiae*, a homolog of the *Kluyveromyces lactis* RAG1 gene. Yeast 9: 1373–1377

Prior C, Mamessier P, Fukuhara H, Chen X-J, Wésolowski-Louvel M (1993b) The hexokinase gene is required for the transcriptional regulation of the glucose transporter gene *RAG1* in *Kluyveromyces lactis*. Mol Cell Biol 13: 3882–3889

Ragnini A, Fukuhara H (1988) Mitochondrial DNA of the yeast *Kluyveromyces*: guanine-cytosine-rich sequence clusters. Nucleic Acids Res 16: 8433–8442

Ragnini A, Fukuhara H (1989) Genetic instability of an oligomycin resistant mutation in yeast is associated with an amplification of a mitochondrial DNA segment. Nucleic Acids Res 17: 6927–6937

Riley MI, Dickson RC (1984) Genetic and biochemical characterization of the galactose gene cluster in *Kluyveromyces lactis*. J Bacteriol 158: 705–712

Riley MI, Hopper JE, Johnston SA, Dickson RC (1987a) GAL4 of *Saccharomyces cerevisiae* activates the lactose-galactose regulon of *Kluyveromyces lactis* and creates a new phenotype: glucose repression of the regulon. Mol Cell Biol 7: 780–786

Riley MI, Sreekrishna K, Bhairi S, Dickson RC (1987b) Isolation and characterization of mutants of *Kluyveromyces lactis* defective in lactose transport. Mol Gen Genet 208: 145–151

Romanos MA, Boyd A (1988) A transcriptional barrier to expression of cloned toxin genes of the linear plasmid k1 of *Kluyveromyces lactis*: evidence that native k1 has novel promoters. Nucleic Acids Res 16: 7333–7350

Romanos MA, Scorer CA, Clare JJ (1992) Foreign gene expression in yeast: a review. Yeast 8: 423–488

Ruzzi M, Breunig KD, Ficca AG, Hollenberg CP (1987) Positive regulation of the yeast β-galactosidase gene from *Kluyveromyces lactis* is mediated by a UAS element showing homology to the GAL upstream activation site of *Saccharomyces cerevisiae*. Mol Cell Biol 7: 991–997

Saliola M, Falcone C, Shuster JR (1990) The alcohol dehydrogenase system in the yeast *Kluyveromyces lactis*. Yeast 6: 193–204

Saliola M, Gonnella R, Mazzoni C, Falcone C (1991) Two genes encoding putative mitochondrial alcohol dehydrogenases are present in the yeast *Kluyveromyces lactis*. Yeast 7: 391–400

Salmeron JM, Johnston SA (1986) Analysis of the *Kluyveromyces lactis* positive regulatory gene LAC9 reveals functional homology to, but sequence divergence from the *Saccharomyces cerevisiae* GAL4 gene. Nucleic Acids Res 19: 7767–7781

Salmeron JM, Langdon SD, Johnston SA (1989) Interaction between Transcriptional Activator Protein LAC9 and Negative Regulatory Protein GAL80. Mol Cell Biol 9: 2950–2956

Sanchez M, Iglesias FJ, Santamaria C, Dominguez A (1993) Transformation of *Kluyveromyces lactis* by electroporation. Appl Environ Microbiol 59: 2087–2092

Sanders JPM, Weijers PJ, Groot GSP, Borst P (1974) Properties of mitochondrial DNA from *Kluyveromyces lactis*. Biochim Biophys Acta 374: 136–144

Schaffrath K, Stark MJR, Gunge N, Meinhardt F (1992) *Kluyveromyces lactis* killer system: ORF1 of pGKL2 has no function in immunity expression and is dispensable for killer plasmid replication and maintenance. Curr Genet 21: 357–363

Shain DH, Salvadore C, Denis CL (1992) Evolution of the alcohol dehydrogenase (*ADH*) genes in yeast: characterization of a fourth ADH in *Kluyveromyces lactis*. Mol Gen Genet 232: 479–488

Sharp PM, Li W-H (1987) Codon adaptation index, a measure of directional synonymous codon usage bias, and its potential application. Nucleic Acids Res 15: 1281–1295

Sheetz RM, Dickson RC (1980) Mutations affecting synthesis of β-galactosidase activity in the yeast *Kluyveromyces lactis*. Genetics 95: 877–890

Sheetz RM, Dickson RC (1981) LAC4 is the structural gene for β-galactosidase in *Kluyveromyces lactis*. Genetics 98: 729–745

Sherman F, Fink GR, Lawrence CW (1974) Methods in yeast genetics. Laboratory manual. Cold Spring Harbor Laboratory, Cold Spring Harbor, New York

Sherman F, Fink GR, Hicks JB (1982) Methods in yeast genetics. Laboratory manual. Cold Spring Harbour Laboratory, Cold Spring Harbor, New York

Shuster JR (1990) *Kluyveromyces lactis* glyceraldehyde-3-phosphate dehydrogenase and alcohol dehydrogenase-1 genes are linked and divergently transcribed. Nucleic Acids Res 18: 4271

Shuster JR (1991) Gene expression in yeast: protein secretion. Curr Opinion Biotechnol 2: 685–690

Shuster JR, Moyer D, Irvine B (1987) Sequence of the *Kluyveromyces lactis* URA3 gene. Nucleic Acids Res 15: 8573

Sidenberg DG, Lachance M-A (1986) Electrophoretic isoenzyme variation in *Kluyveromyces* population and revision of *Kluyveromyces marxianus* (Hansen) van der Walt. Int J Syst Bacteriol 36: 94–102

Skelly PJ, Hardy CM, Clark-Walker GD (1991) A mobile group II intron of a naturally occurring rearranged mitochondrial genome in *Kluyveromyces lactis*. Curr Genet 20: 115–120

Sor F (1988) A computer program allows the separation of a wide range of chromosome sizes by pulsed field gel electrophoresis. Nucleic Acids Res 16: 4853–4863

Sor F (1992) Electrophoretic technique for chromosome separation. Methods Mol Cell Biol 3: 65–70

Sor F, Fukuhara H (1985) Structure of a linear plasmid of the yeast *Kluyveromyces lactis*: compact organization of the killer genome, Curr Genet 9: 147–155

Sor F, Fukuhara H (1989) Analysis of chromosome patterns of the genus *Kluyveromyces*. Yeast 5: 1–10

Sor F, Wésolowski M, Fukuhara H (1983) Inverted terminal repetitions of the two linear DNA associated with the killer character of the yeast *Kluyveromyces lactis*. Nucleic Acids Res 11: 5037–5044

Sreekrishna K, Dickson RC (1985) Construction of strains of *Saccharomyces cerevisiae* that grow on lactose. Proc Natl Acad Sci USA 82: 7909–7913

Sreekrishna K, Webster TD, Dickson RC (1984) Transformation of *Kluyveromyces lactis* with the kanamycin (G418) resistance gene of Tn903. Gene 28: 73–81

Stam JC, Kwakman J, Meijer M, Stuitje AR (1986) Efficient isolation of the linear plasmid of *Kluyveromyces lactis*: evidence for location and expression in the cytoplasm and characterization of their terminally bound proteins. Nucleic Acids Res 14: 6871–6884

Stark MJR (1988) Resolution of sequence discrepancies in the ORF1 region of the *Kluyveromyces lactis* plasmid k1. Nuclec Acids Res 16: 771

Stark MJR, Boyd A (1986) The killer toxin of *Kluyveromyces lactis*: characterization of the toxin subunits and identification of the genes which encode them. EMBO J 5: 1995–2002

Stark MJR, Milner JS (1989) Cloning and analysis of the *Kluyveromyces lactis* TRP1 gene: a chromosomal locus flanked by genes encoding inorganic pyrophosphatase and histone H3. Yeast 5: 35–50

Stark MJR, Mileham AJ, Romanos MA, Boyd A (1984) Nucleotide sequence and transcription analysis of a linear DNA plasmid associated with the killer character of the yeast *Kluyveromyces lactis*. Nucleic Acids Res 12: 6011–6030

Stark MJR, Boyd A, Mileham AJ, Romanos MA (1990) The plasmid-encoded killer system of *Kluyveromyces lactis*: a review. Yeast 6: 1–29

Steensma HY, de Jongh FCM, Linnekamp M (1988) The use of electrophoretic karyotypes in the classification of yeasts *Kluyveromyces marxianus* and *K. lactis*. Curr Genet 14: 311–317

Sugisaki Y, Gunge N, Sakaguchi K, Tamasaki M, Tamaru G (1983) *Kluyveromyces lactis* killer toxin inhibits adenylate cyclase of sensitive yeast. Nature 304: 464–466

Sugisaki Y, Gunge N, Sakaguchi K, Yamasaki M, Tamura G (1984) Characterization of a novel killer toxin encoded by a double-stranded linear DNA plasmid of *Kluyveromyces lactis*. Eur J Biochem 141: 241–245

Sugisaki Y, Gunge N, Sakaguchi K, Yamasaki M, Tamaru G (1985) Transfer of DNA killer plasmids from *Kluyveromyces lactis* to *Kluyveromyces fragilis* and *Candida pseudotropicalis*. J Bacteriol 164: 1373–1375

Tanguy-Rougeau C (1991) Plasmides killer de la levure *Kluyveromyces lactis*: étude du gène nucléaire *KEX1* impliqué dans la maturation de la toxine killer; étude fonctionnelle de gènes des plasmides killer par modification in vivo. PhD Thesis, University of Paris XI, Paris

Tanguy-Rougeau C, Wésolowski-Louvel M, Fukuhara H (1988) The *Kluyveromyces lactis* *KEX1* gene encodes a subtilisin-type serine proteinase. FEBS Lett 234: 464–470

Tanguy-Rougeau C, Chen C-J, Wésolowski-Louvel M, Fukuhara H (1990) Expression of a foreign Km^R gene in linear killer DNA plasmids in yeast. Gene 91: 43–50

Thompson A, Oliver SG (1986) Physical separation and functional interaction of *Kluyveromyces lactis* and *Saccharomyces cerevisiae* ARS elements derived from killer plasmid DNA. Yeast 2: 179–191

Tingle M (1967) Biochemical and genetic studies in *Saccharomyces lactis*. PhD Thesis, University of Wisconsin Madison

Tingle M, Halvorson HO (1972) Mutants in *Saccharomyces lactis* controlling both β-glucosidase and β-galactosidase activities. Genet Res 19: 27–32

Tingle M, Herman A, Halvorson HO (1968) Characterization and mapping of histidine genes in *Saccharomyces lactis*. Genetics 58: 361–371

Tokunaga M, Wada N, Hishinuma F (1987) Expression and identification of immunity determinants on linear DNA killer plamsids pGKL1 and pGKL2 in *Kluyveromyces lactis*. Nucleic Acids Res 15: 1031–1046

Tokunaga M, Wada N, Hishinuma F (1988) A novel yeast secretion signal isolated from 28K killer precursor protein encoded on the linear DNA plasmid pGKL1. Nucleic Acids Res 16: 7499–7511

Tokunaga M, Kawamura A, Hishinuma F (1989) Expression of pGKL killer 28K subunit in *S. cerevisiae*: identification of 28K subunit as a killer protein. Nucleic Acids Res 17: 3435–3446

Tommasino M (1991) Killer system of *Kluyveromyces lactis*: the open reading frame 10 of the pGKL2 plasmid encodes a putative DNA binding protein. Yeast 7: 245–252

Tommasino M, Ricci S, Galeotti CL (1988) Genome organization of the killer plasmid pGKL2 from *Kluyveromyces lactis*. Nucleic Acids Res 16: 5863–5877

Trueman LJ, Astolfi S, Gardner DCJ, Oliver S (1990) Isolation and characterzation of autonomously replicating sequences from the lactose-fermenting yeast, *Kluyveromyces lactis*. Yeast 6: S592

Van Aelst L, Hohmann S, Bulaya B, De Koning W, Sierkstra L, Neves MJ, Luyten K, Alijo R, Ramos J, Coccetti P, Martegani E, De Magalhaes-Rocha NM, Lopes Brandao R, Van Dijck P, Vanhalewyn M, Durnez P, Jans AWH, Thevelein JM (1993) Molecular cloning of a gene involved in glucose sensing in the yeast *Saccharomyces cerevisiae*. Mol Microbiol 8: 927–943

Van den Berg JA, Van der Laken KJ, Van Ooyen AJJ, Renniers CHM, Rietveld K, Schaap A, Brake AJ, Bishop RJ, Schultz K, Moyer D, Richman M, Shuster JR (1990) *Kluyveromyces* as a host for heterologous gene expression: expression and secretion of prochymosin. Bio/Technology 8: 135–139

Van der Walt JP (1970) *Kluyveromyces lactis*. In: Lodder J (ed) The yeasts: a taxonomical study. North-Holland Publishing Co., Amsterdam, pp 316–378

Van der Walt JP, Johannsen E (1979) A comparison of interfertility and in vitro DNA-DNA ressociation as criteria for speciation in the genus *Kluyveromyces*. Antonie Leeuwenhoek J Microbiol 45: 281–291

Vasquez D (1979) Inhibitors of protein synthesis. Springer, Berlin Heidelberg New York, p 312

Vaughan-Martini A, Martini A (1987) Taxonomic revision of the yeast genus *Kluyveromyces* by nuclear deoxyribonucleic acid reassociation. Int J Syst Bacteriol 37: 380–385

Verbeet MP, van Heerikhuizen H, Klootwijk J, Fontijn RD, Planta RJ (1984) Evolution of yeast ribosomal DNA: molecular cloning of the rDNA units of *Kluyveromyces lactis* and *Hansenula wingeii* and their comparison with the rDNA units of other Saccharomycetoideae. Mol Gen Genet 195: 116–125

Viola AM, Ficarelli P, Goffrini P, Galeotti CL, Ferrero I (1992) A *Kluyveromyces lactis* gene that complements the *S. cerevisiae* op1 mutation. Yeast 8: S446

Viovy JL, Miomandre F, Miquel M-C, Caron F, Sor F (1992) Irreversible trapping of DNA during crossed-field gel electrophoresis. Electrophoresis 13: 1–6

Volkert F, Wilson C, Broach JR (1989) Deoxyribonucleic acid plasmids in yeasts. Microbiol Rev 53: 299–317

Webster TD, Dickson RC (1988) The organization and transcription of the galactose gene cluster of *Kluyveromyces lactis*. Nucleic Acids Res 16: 8011–8028

Weirich J (1992) Isolierung und Charakterisierung des Glukose Transporters RAG1 aus *Kluyveromyces lactis* und Untersuchungen zu dessen Rolle bei der Glukoserepression der β-Galaktosidase. PhD Thesis Heirich-Heine Universität, Düsseldorf

Wésolowski M, Algeri A, Fukuhara H (1981) Gene organization of the mitochondrial DNA of yeasts: *Kluyveromyces lactis* and *Saccharomycopsis lipolytica*. Curr Genet 3: 157–162

Wésolowski M, Algeri A, Goffrini P, Fukuhara H (1982a) Killer DNA plasmids of the yeast *Kluyveromyces lactis*. I. Mutation affecting the killer phenotype. Curr Genet 5: 191–197

Wésolowski M, Dumazert P, Fukuhara H (1982b) Killer DNA plasmids of the yeast *Kluyveromyces lactis*. II. Restriction endonuclease maps. Curr Genet 5: 199–203

Wésolowski M, Algeri A, Fukuhara H (1982c) Killer DNA plasmids of the yeast *Kluyveromyces lactis*. III. Plasmid recombination. Curr Genet 5: 205–208

Wésolowski-Louvel M, Fukuhara H (1990) A palindromic mutation of the linear killer plasmid k2 of yeast. Nucleic Acids Res 18: 4877–4882

Wésolowski-Louvel M, Goffrini P, Ferrero I (1988a) The *RAG2* gene of the yeast *Kluyveromyces lactis* codes for a putative phosphoglucose isomerase. Nucleic Acids Res 16: 8714

Wésolowski-Louvel M, Tanguy-Rougeau C, Fukuhara H (1988b) A nuclear gene required for the expression of the linear DNA-associated killer system in the yeast *Kluyveromyces lactis*. Yeast 4: 71–81

Wésolowski-Louvel M, Prior C, Mammesier P, Goffrini P, Ferrero I, Sor F, Fukuhara H (1990) *Kluyveromyces lactis* and its plasmids. In: Heslot H, Davies J, Florent J, Bobichon L, Durand G, Penesse L (eds) Proceedings of the 6th International Symposium on Genetics of Industrial Microorganisms (GIM90). Société Française de Microbiologie, Paris, vol 1, pp 519–532

Wésolowski-Louvel M, Goffrini P, Ferrero I, Fukuhara H (1992a) Glucose transport in the yeast *Kluyveromyces lactis*. I. Properties of an inducible low-affinity glucose transporter gene. Mol Gen Genet 33: 89–96

Wésolowski-Louvel M, Prior C, Bornecque D, Fukuhara H (1992b) Rag⁻ mutations involved in glucose metabolism in yeast: isolation and genetic characterization. Yeast 8: 711–719

White JH, Butler AR, Stark MJR (1989) *Kluyveromyces lactis* toxin does not inhibit yeast adenylate cyclase. Nature 341: 666–668

Wickerham LJ (1951) Taxonomy of yeasts. US Department of Agriculture Technical Bulletin, Washington, No 1029

Wilson C, Fukuhara H (1991) Distribution of mitochondrial r1-type introns and the associated open reading frame in the yeast genus *Kluyveromyces*. Curr Genet 19: 163–167

Wilson C, Ragnini C, Fukuhara H (1989) Analysis of the regions coding for transfer RNA genes in *Kluyveromyces lactis* mitochondrial DNA. Nucleic Acids Res 17: 4485–4491

Wilson DW, Meacock PA (1988) Extranuclear gene expression in yeast: evidence for a plasmid-encoded RNA polymerase of unique structure. Nucleic Acids Res 16: 8097–8112

Witte MM, Dickson RC (1990) The C6 Zinc finger and adjacent amino acids determine DNA-binding specificity and affinity in the yeast activator protein LAC9 and PPR1. Mol Cell Biol 10: 5128–5137

Wray LV Jr, Witte MM, Dickson RC, Riley MI (1987) Characterization of a positive regulatory gene, LAC9, that controls induction of the lactose-galactose regulon of *Kluyveromyces lactis*: structural and functional relationships to GAL4 of *Saccharomyces cerevisiae*. Mol Cell Biol 7: 1111–1121

Yeh P, Landais D, Lemaitre M, Maury I, Crenne JY, Becquart J, Murry-Brelier A, Boucher F, Montay G, Fleer R (1992) Design of yeast-secreted albumin derivatives for human therapy–biological and antiviral properties of a serumalbumin-CD4 genetic conjugate. Proc Natl Acad Sci USA 89: 1904–1908

Yuan YO, Stroke IL, Fields S (1992) STE12 and MAT α1 homologs from *Kluyveromyces lactis*: implications for the role of STE12 in the activation of α-specific genes. Yeast 8: S394

Zachariae W, Breunig KD (1993) Expression of the transcriptional activator LAC9 (K1GAL4) in *Kluyveromyces lactis* is controlled by autoregulation. Mol Cell Biol 13: 3058–3066

Zachariae W, Kuger P, Breunig KD (1993) Glucose repression of lactose/galactose metabolism in *Kluyveromyces lactis* is determined by the concentration of the transcriptional activator LAC9 (K1GAL4). Nucleic Acids Res 21: 69–77

Zaror I, Marcus F, Moyer DL, Tung J, Shuster J (1993) Fructose-1,6-bisphosphatase of the yeast *Kluyveromyces lactis*. Eur J Biochem 212: 193–199

Zenke F, Zachariae W, Lunkes A, Breunig KD (1993) Gal80 proteins of *Kluyveromyces lactis* and *Saccharomyces cervisiae* are highly conserved but contribute differently to glucose repression of the galactose regulon. Mol Cell Biol 13: 7566–7576

Zennaro E, Falcone F, Frontali L, Puglisi PP (1977) Dependence of cytoplasmic on mitochondrial protein synthesis in *K. lactis* CBS 2360. I. Biochemical analysis. Mol Gen Genet 150: 137–140

Zhang Y-P, Chen X-J, Li Y-Y, Fukuhara H (1992) *LEU2* gene homolog in *Kluyveromyces lactis*. Yeast 8: 801–804

CHAPTER 6

Pichia pastoris

Koti Sreekrishna and Keith E. Kropp

1
History of *Pichia pastoris*

Interest in the study of nonconventional yeasts (yeasts other than *Saccharomyces cerevisiae* and *Schizosaccharomyces pombe*) has increased dramatically in the past few years (Reiser et al. 1990). One such category is methylotrophic yeasts (Wegner and Harder 1986; Harder et al. 1986), e.g., *Pichia pastoris, Hansanula polymorpha, Candida boidinii*, etc. Methylotrophic yeasts have the ability to use methanol as a sole source of carbon and energy. Adaptation to growth on methanol is associated with induction of methanol oxidase, *MOX* (also referred to as alcohol oxidase, *AOX*), dihydroxy acetone synthase *DAS*, and several other enzymes involved in methanol metabolism. The most spectacular increase, however, is seen with alcohol oxidase, which is virtually absent in glucose-grown cells, but can account for over 30% of the cell protein in methanol-grown cells. Extensive proliferation of peroxisomes, accounting for over 80% of the cell volume, is also observed in methanol-grown cells (Veenhuis et al. 1983). Due to these characteristics, methylotrophic yeasts have gained the attention of biochemists, molecular biologists, cell biologists, biotechnologists, microbiologists, and chemists in academics and industry.

The present chapter focuses on one of the methylotrophic yeasts, namely *P. pastoris*. This yeast was initially developed by Phillips Petroleum Company for the production of single-cell protein for feed stock. A very efficient ultra-high cell density (>130 g dry cell weight per liter) fermentation process with high biomass productivity (>10 g/liter-hour) was developed through meticulous fermentation research (Wegner 1983). Unfortunately, the economics of this process, while impressive from a fermentation standpoint (approximately $5 per pound of protein), was clearly an order of magnitude higher in comparison to the cost of a pound of soybean. Following this setback, Phillips Petroleum Company invested its efforts in developing this yeast as an expression system for the production of recombinant proteins, and this has proved to be a worthwhile endeavor (Wegner 1990; Romanos et al. 1992; Cregg et al. 1993).

Hoechst Marion Roussel Inc., 2210 E. Galbraith Road, Cincinnati, Ohio 45215, USA

Since 1988, several pharmaceutical and biotechnology companies have licensed the *P. pastoris* expression technology. At least three products produced with this technology (Hepatitis B surface antigen, human serum albumin, and insulin-like growth factor-1) are being pursued for commercialization and many more are in the pipeline. Since 1990, Phillips Petroleum Company has made this technology available at no actual cost for research use to universities and nonprofit organizations through a materials transfer agreement. More recently, however, the system can be readily obtained for a nominal fee from Invitrogen Corporation, San Diego, California, USA. However, for commercial purposes, this technology can be licensed from Research Corporation Technologies, Tucson, Arizona, USA.

In addition to its extensive use for the expression of heterologous proteins, *P. pastoris* is also being developed as a model organism for molecular analysis of peroxisome biogenesis (Gould et al. 1992; Liu et al. 1992).

The primary intent of this chapter is to introduce investigators to practical techniques for manipulating *P. pastoris* with emphasis on its use for expression of heterologous proteins.

2
Growth and Storage

2.1
Shake Flask, Shake Tube, Plate, and Slant Cultures

P. pastoris grows well both in liquid and on solid media, on a wide variety of simple carbon sources including glucose, glycerol, fructose, sorbitol, ethanol, methanol, alanine, lysine, succinate, ethyl amine, cadaverine, glucitol, mannitol, L-rhamnose, and trehalose. The doubling time is dependent on the carbon source used and is typically 90 min on glucose and approximately 6 h on methanol. In solid media it forms white or cream-colored nonfilamentous colonies. Multilateral buds are noticed under light microscopy. The natural habitats of *P. pastoris* are the oak tree and packaged foods. The compositions of the various *P. pastoris* growth media (MD, MDH, MGy, MGyH, MM, MMH, YPD) are given in Sects. 2.2.2 and 2.2.3.

The growth temperature is 30 °C with shaking (250 rpm) for liquid cultures and with incubation for plates and slants. When minimal methanol (MM) plates are used as growth medium, 100 μl of 100% methanol (filter-sterilized) is added to the plate lid once every day to compensate for the methanol lost due to evaporation (the plate is placed inverted in the incubator). When MM liquid medium is used, methanol is added to a final concentration of 0.5% (v/v) every 2 days to compensate for methanol lost due to evaporation. Because *P. pastoris* grows well under a wide pH range of 3–6.5, buffering the growth medium is generally unnecessary, except under some special circumstances, as noted under Sect. 7.4. For large-scale, high cell-density cultivation of *P. pastoris*, refer to Sect. 5.

P. pastoris as such does not grow on galactose, arabinose, ribose, maltose, sucrose, lactose, raffinose, melibiose, cellulose, or starch. However, it will grow on autoclaved sucrose, due to breakdown of sucrose into glucose and fructose. Also, *P. pastoris* strains transformed with the sucrase (invertase) gene *SUC2* of *Saccharomyces cerevisiae* efficiently grow on sucrose with high growth yields (Sreekrishna et al. 1987).

2.2
Media

2.2.1
Stock Solutions

Note. For filter sterilization of various solutions and liquids, filter wares (disposable or reusable types) equipped with cellulose acetate or cellulose nitrate membranes (pore size 0.2 to 0.22 μm) from one of the several manufacturers (Nalgene Company, Rochester, New York, USA; Costar Corporation, Cambridge, Massachusetts, USA; Corning Glass Works, Corning, New York, USA) can be used. For filter sterilization of methanol and methanol-containing media, only cellulose acetate membranes (0.2–0.22 μm) are suitable, because methanol does not filter through cellulose nitrate membranes of pore size 0.2 μm.

10× YNB: Dissolve 13.4 g of yeast nitrogen base without amino acids (YNB, Difco labs., Detroit, Michigan, USA) in 100 ml of water (heat if necessary) and filter-sterilize. This solution can be stored for over a year at 4 °C.

500× B: Dissolve 20 mg of d-biotin (Sigma Chemicals, St. Louis, Missouri, USA) in 100 ml of water and filter sterilize.

100× H: Dissolve 400 mg L-histidine in 100 ml of water (heat if necessary) and filter sterilize.

10× D: Dissolve 20 g of D-glucose in 100 ml water. Autoclave for 15 min or filter sterilize. Stores well for years at room temperature.

10× GY: Mix 10 ml of glycerol with 90 ml of water. Filter sterilize. Stores well for years at room temperature.

10× M: Mix 5 ml of methanol (100%) with 95 ml of water. Filter sterilize and store at 4 °C.

100% methanol: Filter sterilize pure methanol (100%). Store at room temperature in a fireproof cabinet.

2.2.2
Minimal Media Compositions

MD: Mix 100 ml of 10× YNB, 2 ml of 500× B, and 100 ml of 10× D with 800 ml of autoclaved water (include 15 g Bacto agar for plates).

MM: Mix 100 ml of 10× YNB, 2 ml of 500× B, and 100 ml of 10× M with 800 ml of autoclaved water (include 15 g Bacto agar for plates).

MGY: Mix 100 ml of 10× YNB, 2 ml of 500× B, and 100 ml of 10× GY with 800 ml autoclaved water (include 15 g Bacto agar for plates).

All these liquid media and plates store well for several weeks at 4 °C.

Minimal media with other carbon sources (such as D-sorbitol, D,L-alanine) are prepared by using the desired carbon source at 10 g/l in place of glucose in MD. Minimal media containing a mixture of carbon sources can also be prepared by combining two or more desired substrates in the growth medium.

2.2.3
Supplemental Minimal Media Compositions

Minimal media are supplemented with necessary supplemental nutrients such as amino acids, depending on the specific requirement of a given strain. For example, *P. pastoris* strains GS115 and KM71, commonly used in molecular genetic manipulations, are auxotrophic for histidine. Such strains will grow in minimal media only in the presence of supplemental histidine. However, once transformed with *HIS4* (histidinol dehydrogenase gene), they readily grow in the absence of histidine.

The composition of supplemental minimal histidine media (suitable for histidine auxotrophic strains such as GS115) is as follows. Other supplemental media can be prepared depending on the need of a particular strain in use.

MDH: Mix 100 ml of 10× YNB, 2 ml of 500× B, 100 ml of 10× D, and 10 ml of 100× H with 790 ml of autoclaved water (include 15 g agar for plates).

MMH: Mix 100 ml of 10× YNB, 2 ml of 500× B, 100 ml of 10× M, and 10 ml of 100× H with 790 ml of autoclaved water (include 15 g agar for plates).

MGyH: Mix 100 ml of 10× YNB, 2 ml of 500× B, 100 ml of 10× GY, 10 ml of 100× H with 790 ml of autoclaved water (include 15 g agar for plates).

All of these liquid media and plates store well for several weeks at 4 °C.

Supplemental minimal histidine media with other carbon sources is prepared by adding a similar amount of histidine as above to the minimal media with the desired carbon source.

2.2.4
Complex Medium Composition

YPD: Dissolve 10 g of Bacto yeast extract, 20 g of peptone, and 20 g of glucose in 1000 ml of water (also include 15 g Bacto agar for slants and plates) and autoclave for 20 min.

2.3
Storage

For medium-term storage, *P. pastoris* should be kept at 4 °C in complex (YPD) liquid medium or YPD agar slants. Cells are cultured initially on a desired media (selective minimal methanol, dextrose, or glycerol media such as MM, MD, or MGY for transformants) and then transferred to YPD. Most of the commonly used strains can be stored in such media for over 1 year at 4 °C. However, the protease-deficient strains (SMD1163, SMD1165, and SMD1168) should be restreaked or regrown every 2–4 weeks, because they do not keep well.

For long-term storage, cells are suspended at an $O.D_{600\,nm}$ of 50–100 in YPD containing 50% glycerol. The cells are frozen at −80 °C or preferably kept in a liquid nitrogen freezer. Cells stored in these ways have remained viable for several years.

3
Available Strains

NRRL Y-11430-SC5 (wild type; Sreekrishna et al. 1987)

GS115 (*his4*) – this strain is also known as GTS115 (Sreekrishna et al. 1987)

KM71 (*his4*, *aox1::ARG4*; Cregg and Madden 1988)

PPF1 (*his4*, *arg4*; J.M. Cregg, pers. comm.)

Protease-deficient strains (derived by protease A (PEP4) and/or Protease B (PRB) gene disruption (M.A. Gleeson, pers. comm.):

SMD1163 (*his4*, *pep4*, *prB1*)

SMD1165 (*his4*, *prB1*)

SMD1168 (*his4*, *pep4*)

4
Genetic Techniques

4.1
Life Cycle

The members of the ascomycetous genus *Pichia* Hansen are distinguished from most other yeasts by the occurrence of hat-shaped spores. Investigations of the DNA/DNA reassociation demonstrated a narrow relatedness to members of the genus *Hansenula*. Therefore, many of these yeasts are included now also in the genus *Pichia* (Kurtzman 1984a,b; Barnett et al. 1990).

In most cases, the life cycle of the *Pinus* species is unknown. The protocols given here are for the genetic analysis of *P. pastoris* (M.E. Digan, pers. comm.; see also Digan and Lair 1986).

4.2
Mating and Sporulation

P. pastoris strains that are available are homothallic (switch mating types), thus it is essential to use selection plates against both parents used in the genetic cross prior to sporulation of mated cells. Appropriate minimal plates are used to select against parents based on amino acid or nucleotide or carbon source requirements. Cultures form four-spored asci, but the viability of the spores is low. Tetrad analysis is possible with a few asci. However, the segregation frequencies do not fit the expected $2^+:2^-$ ratio. In most cases the spores are phenotypically wild type. The low spore viability and the aberrant segregation ratios suggest that the establishment of a mating system which can be used for genetic analysis requires further research, including a backcrossing program with several strains and mutants and, perhaps, the search for heterothallic strains.

4.2.1
Mating

1. Resuspend single colonies of each of the two parental strains in $100\,\mu l$ of YPD. Mix the parental strains thoroughly together, and spread on presporulation plates (see below for composition). Incubate for 24 h at 30 °C.

2. Replica-plate onto sporulation plates (see below for composition). Incubate for 24 h at 30 °C.

3. Replica-plate onto plates which select against both parental types and allows survival of only the progeny of cross-mated cells. Incubate at 30 °C until colonies appear.

YPD (see Sect. 2.2.4)

Presporulation medium: Mix 50 g glucose, 20 g peptone, 10 g yeast extract, 5 g agar, and 23 g nutrient agar in 1000 ml water, and autoclave.

Sporulation medium: Mix 5 g NaOAc (anhydrous), 10 g KCl (anhydrous), and 20 g agar in 1000 ml water, and autoclave.

4.2.2
Sporulation

4. Disperse single colonies from step 3 in YPD. Plate on presporulation plates and after 24 h of incubation at 30 °C, replica-plate onto sporulation plates. Incubate at 30 °C for 3–5 days. At this point, asci can be dissected directly, or the mixture of cells and spores on the SPA plate can be digested extensively with cell wall-degrading enzymes to kill the vegetative cells as described below.

4.2.3
Random Spore Preparation

5. Wash sporulation plates with sterile water to harvest the spores.

6. Wash the spore suspension twice with 0.1 M sodium phosphate buffer, pH 7.4.

7. Resuspend spores in 3 ml of 0.1 M phosphate buffer, pH 7.4. Add Zymolyase 100 T, Glusulase, and β-mercaptoethanol to final concentrations of 0.5 mg/ml, 2% v/v, and 0.1% v/v respectively. Incubate for 5 h at 30 °C with occasional shaking.

8. Sonicate the spore suspension three times for 15 s and harvest by centrifugation at 3000 g for 10 min at room temperature.
   Note. Spores can be quantified under a light microscope by counting an aliquot of spore suspension placed in a counting chamber.

9. Examine spores microscopically for clumps. If clumping appears to be excessive, repeat step 4.

10. Plate the spores on YPD plates at approximately 200 spores/plate or use a micro-manipulator to separate single spores on YPD plates.

11. Let spores germinate at 30 °C on YPD and screen for the desired phenotype(s) by replica-plating onto two selective plates, each of which selects against one or the other parent in the cross.

5
Fermentation Process

The following paragraphs describe general methods for production of biomass as well as for the production of heterologous proteins using Mut$^+$ and Mut$^-$ cells in both continuous and batch modes of fermentation. The process described here can be scaled up (>1000 liters) or scaled down (0.2 l) as desired. Fermentation can be conducted over a wide pH range (3.0–5.9) at 30 °C (Wegener 1983).

Prolific growth at low pH (which reduces risk of microbial contamination) was considered as one of the advantages of using this yeast for production of single-cell protein. Interestingly, this has also been valuable in optimizing fermentation parameters for the production of recombinant proteins. For example, human serum albumin (HSA) secretion yield is improved over threefold by using pH 5.85 compared to the generally used pH of 5.0 (Sreekrishna et al. 1990). In the case of secretion of the V$_1$ domain of CD$_4$ (amino acid residues 1 to 106 of mature CD$_4$), intact product was seen only at acidic pH (2.5–3.5) (Buchholz et al. 1991). As high as a two- to fourfold increase in secreted yields of human epidermal growth factor and human insulin-like growth factor-1 are observed at pH 3.0 (Siegel et al. 1990; Brierley et al. 1992).

Other kinds of media and growth conditions are also known to improve the production of specific proteins. Addition of yeast extract and peptone increased the level of secretion of HSA and tissue plasminogen activator (TPA) (Sreekrishna et al. 1990; J.F. Tschopp, pers. comm.). Likewise, addition of 1% casamino acids improved the yield of mouse epidermal growth factor (Clare et al. 1991). In the case of intracellular production of hepatitis B surface antigen, the addition of certain trace metals (KI, $NaMoO_4 \cdot 2H_2O$, $CoCl_2 \cdot 6H_2O$ at 0.8, 0.2, and 0.5 g per liter, respectively), plus allowing the cells to sit longer in the fermentor, improved the yield of antigen particles (J.A. Cruze, pers. comm.). Thus, it is evident that some experimentation with fermentation parameters may be necessary to establish product-specific optimal conditions.

5.1
Continuous Culture of Mut+ and Mut⁻ Strains on Methanol

Fermentation is carried out in two steps. First in the batch mode on glycerol or glucose as the carbon source followed by continuous mode on methanol-containing medium. The process described here is typically with recombinant cells, where it is preferable to use glycerol rather than glucose.

5.1.1
Inoculum for the Fermentor

Grow cells to an $O.D_{600nm}$ of 2–10 in a 2-l shake flask containing 1l of one of the following growth media: MD, MGY, YPD (see Sect. 2.2.2 for composition), YMPD, YMPGy, or MGyB (see Sect. 5.1.2 for compositions). This volume of inoculum is adequate for inoculating a 20-l fermentor with a 10-l operating volume.

5.1.2
Media

YMPD: Dissolve 3 g of yeast extract, 3 g of malt extract, 5 g of peptone, and 10 g of glucose in 1 l of water, and autoclave.

YMPGy: Same as YMPD with the exception that 10 ml of 100% glycerol is used instead of 10 g of glucose.

MGyB: Dissolve 11.5 g KH_2PO_4, 2.66 g K_2HPO_4, 6.7 g YNB, pH 6.0, and 20 ml glycerol in 1 l water, and autoclave.

FM21 basal salt media
  Composition is for 1 l final volume in water

Phosphoric acid, H_3PO_4 (85%)	3.5 ml
Calcium sulfate, $CaSO_4 \cdot 2H_2O$	0.15 g
Potassium sulfate, K_2SO_4	2.4 g
Magnesium sulfate, $MgSO_4 \cdot 7H_2O$	1.95 g
Potassium hydroxide, KOH	0.65 g

Biotin stock solution

Biotin	0.2 g per liter

PTM1 trace salts
Composition is for 1 l final volume in water

Cupric sulfate ($CuSO_4 \cdot 5H_2O$)	6.0 g
Manganese sulfate ($MnSO_4 \cdot H_2O$)	3.0 g
Ferrous sulfate ($FeSO_4 \cdot 7H_2O$)	65.0 g
Zinc sulfate ($ZnSO_4 \cdot 7H_2O$)	20.0 g
Silfuric acid (H_2SO_4)	5.0 ml
Cobalt chloride ($CoCl_2 \cdot 6H_2O$)	0.5 g
Boric acid (H_3BO_3)	0.02 g
Sodium molybdate ($NaMoO_4 \cdot 2H_2O$)	0.2 g
Potassium iodide (KI)	0.1 g

BSM medium composition
Composition is for 1 l final volume in water

Phosphoric acid, H_3PO_4 (85%)	26.0 ml
Calcium sulfate, $CaSO_4 \cdot 2H_2O$	0.9 g
Potassium sulfate, K_2SO_4	18.0 g
Magnesium sulfate, $MgSO_4 \cdot 7H_2O$	14.0 g
Potassium hydroxide, KOH	4.0 g

5.1.3
Batch Phase

Sterilize a 20-l fermentor with 9 l of the basal salt medium FM21 (see Sect. 5.1.2 for composition) containing 5% v/v glycerol (higher levels may be toxic to the cells) or 5–10% glucose. Allow the system to cool to the set temperature of 30 °C. The pH of this medium will be <2. Adjust the pH to 5 with NH_3 gas or with 50% NH_4OH solution. Add 4 ml of biotin stock solution and 11 ml of trace mineral mix PTM1 (see Sect. 5.1.2 for composition). Inoculate the fermentor with 1 l of the shake flask culture (see Sect. 5.1.1). Fermentation is conducted until all of the carbon source (glucose or glycerol) is completely consumed. During the run, dissolved O_2 is maintained at >20%, with the agitator speed set between 500 to 1500 rpm and a vessel pressure of 2 to 3 psi. Foaming is controlled through the addition of a 5% Struktol J673 (Strucktol Company of America, Stow, Ohio, USA) or Mazu DF 37C (Mazer Chemicals, Inc., Gurnee, Illinois, USA). The cell yield expected for the batch phase on 5% glycerol under these conditions is 20–25 g of washed dry cell weight per liter.

5.1.4
Continuous Phase

Continuous fermentation is established by feeding FM21-methanol (15% v/v) for Mut⁺ cells or FM21-methanol (1% v/v) +15% glycerol, sorbitol, or alanine for Mut⁻ cells. The feed for continuous culture is also supplemented with PTM1 (1.1 ml of stock solution/l) and biotin (0.4 ml of the stock solution/l). Feed sterilization is carried out by filtration (Pall Ultipor disposable filter assembly DFA 4001 AR, 0.2 μm). Feed addition is achieved with a Milton-Roy positive displacement metering pump (Model 2396 Duplex). Continuous culture is performed as a chemostat under steady-state conditions where the dilution rate D is equal to the growth rate of the population. The growth rate of cells can be controlled by adjusting the flow rate of fresh medium into the fermentor. Typical D values range from 0.056/h to 0.11/h. Dissolved oxygen is held in the 45–75% air saturation range by varying air flow, vessel pressure, and/or agitator speed. The maximum cell mass achievable under these conditions is around 80 g of washed dry cell weight per liter with a cell productivity of approximately 10 g/l/h. If a higher cell mass is desired, the amount of carbon source in the feed can be raised (25% methanol for Mut⁺ and 1% methanol +25% glycerol, sorbitol, or alanine for Mut⁻). This would also necessitate a proportional increase in the amount of minerals, trace elements, and biotin. As high as 110 g/l washed dry cell weight with 11.6 g/l/h. productivity are achievable in this process.

Highest induction of the *AOX1* promoter occurs using methanol (Mut⁺ cells) or methanol+sorbitol or alanine (Mut⁻) as the carbon source in the feed. Intermediate levels of induction are seen with methanol+glycerol feed.

5.1.5
Equipment

Typically, a bench top fermentor 2–20-l capacity with a 1–10-l operation volume and equipped with monitors and controls for pH, dissolved oxygen (D.O), agitator speed, temperature, air-flow, pressure, foam, and weight is used.

Note. Fermentors can be custom-built or purchased from one of the numerous commercial sources such as: Biolaffitte, SA, France, New Brunswick Scientific, Edison, New Jersey, USA, or Porton Instruments Inc., Hayward, California, USA. More recently, a new class of fermentors called Sixfors were introduced by Infors, UK Ltd., Crewe, Cheshire, Great Britain. Sixfors bridge the gap between shake flasks and fermentor. They can be used for small-scale (typically 0.3 l) continuous, batch, or cascade processes where the growth from one vessel is passed into another.

5.1.6
Methods of Monitoring the Fermentation

Bio-Rad's portable fermentation monitoring analyzer can be used on the spot to provide fast HPLC analysis of the concentration of methanol, glycerol, ethanol, glucose, acetic acid, lactic acid, fructose, as well as maltotriose and maltose. The equipment is compatible with automatic sampling, and computerized data analysis.

Analysis takes only 10 min. The conditions used are as follows:

Instrument: Bio-Rad's fermentation monitoring analyzer (Catalog No.125-0520)

Column:	Fermentation monitoring column (150 × 7.8 mm) (Catalog No. 125-0115)
Sample:	Extra cellular broth, prefiltered through a 0.22-μm syringe filter
Sample volume:	10 to 20 μl
Injector:	Rheodyne injector valve with a 10-μl injection loop
Solvent:	0.002NH_2SO_4 in water at a flow rate of 0.8 ml/min (isocratic, so only one pump is required)
Temperature:	65 °C
Detector:	Refractive index detector

Wet weight of the culture is determined by pelleting 1 ml of the fermentor culture in a microfuge (centrifuged for 4 min), decanting the supernatant, and weighing the pellet. The washed dry cell weight of the culture is determined as follows. Harvest the cells by centrifugation (3000 g for 10 min at room temperature) from a known volume of the fermentor culture (typically 10 to 50 ml). Wash the cell pellet twice with water (10–50 ml) and dry overnight at 100 °C. Weigh the dried yeast pellet and use the value to estimate the washed dry cell weight per liter of the culture.

Mass transfer is determined by measuring the air flow to the fermentor and the composition of both the inlet and outlet air with a Perkin-Elmer gas analyzer. The O_2 transfer rate (OTR) is calculated as follows:

$$OTR = f(C_{in} - C_{out})/V,$$

where f is flow rate l/h, C_{in} and C_{out} are the concentration in mmol of oxygen in the inlet and outlet gases, and V is the ungassed broth volume.

OTR can also be established from yields using the equation:

$$OTR = \mu X/YxO_2,$$

where μ is the specific growth rate or dilution rate (h^{-1}), X is cell the density (g/l) and YxO_2 is the yield on oxygen (g cells/mmol O_2)

Heat transfer rate and heat load can be determined by measuring the temperature of the cooling water in and out of the fermentor heat exchanger.

Productivity P for the continuous culture is determined using P = DX, where D is the dilution rate (h^{-1}), and X is the cell density. An increase in D and/or X will result in higher productivity. Since D also equals the specific growth rate μ, the dilution rate is limited by the microorganism's intrinsic characteristics. Once u_{max} or D_{max}, maximum dilution rate, is achieved, P can only be increased by increasing the cell density.

Extracellular protein concentration (for secreted proteins) can be estimated by using a Lowry type of protein analysis on TCA precipitated material and analyzed for specific product by SDS-PAGE and immunological methods.

Aliquots of cells are lysed to prepare cell extracts for protein analysis.

A set of parameters determined for the high-productivity process for *P. pastoris* on methanol in a 1500-l fermentor are as follows:

Substrate concentration (g/l)	263
Dilution rate (h^{-1})	0.11
pH	3.5
Temperature (°C)	30
Cell mass (g of cells/l)	105
Productivity (g/l/h)	11.6
Yield (g of cells/g of methanol)	0.4
OTR (mmol O_2/l/h)	880
Oxygen consumption (g O_2/g cells)	2.42
Heat release	
Kcal/l/h	109
Kcal/mol O_2	123.9

5.2
Fed-Batch Fermentation of Mut $^+$ and Mut $^-$ Strains on Methanol

5.2.1
Inoculum for Fermentor

Inoculum (1l) is prepared as described above for the continuous fermentation protocol.

5.2.2
Batch Phase

Sterilize a 20-l fermentor with 5l of the basal salt medium BSM (see Sect. 5.1.2) containing 5% v/v glycerol (higher levels may be toxic to the cells) or 5–10% glucose. Allow to cool to the set temperature of 30 °C. The pH of this medium will be <2. Adjust the pH to 5 with NH_3 gas or with 50% NH_4OH solution. Add 40 ml of biotin stock solution and 40 ml of PTM1 (see Sect. 5.1.2). Inoculate the fermentor with 1l of the shake flask culture (see Sect. 5.1.1). Fermentation is conducted until all the carbon source (glucose or glycerol) is completely consumed as previously described. This phase should take 18 to 24h.

5.2.3
Fed-Batch Phase on Glycerol

Initiate by starting with a 50% w/v glycerol feed (500 ml of 100% glycerol + 480 ml water; autoclave and add a filter sterilized mixture of 10 ml each of PTM1 and biotin stock solution) at a feed rate of 18 ml/h/l initial fermentation volume with the aid of a peristalic pump. Glycerol feeding is carried out for 4h. The cell mass doubles from the value in the batch phase (Sect. 5.2.2). The cell yield at this point will be in the range of 180 to 220 g/l of wet cells (equivalent to approximately 30–55 g washed dry cell weight/l).

This phase can be manipulated by varying the concentration of glycerol and/or the duration of fermentation to achieve optimal heterologous protein yield in the fed-batch phase on methanol (Sect. 5.2.4). Some suggested cell yield ranges that should be tested for are as follows:

Mut⁻/intracellular expression: 50–100 g/l dry weight or 200–400 g/l wet weight
Mut⁺/intracellular expression: 30–80 g/l dry weight or 140–320 g/l wet weight
Mut⁻/secretion: 20–40 g/l dry weight or 80–160 g/l wet weight
Mut⁺/secretion: 12–80 g/l dry weight or 50–300 g/l wet weight

Note. It is obvious from the suggested ranges for cell density that in several instances the fed-batch phase on glycerol (Sect. 5.2.3) is unnecessary.

5.2.4
Fed-Batch Phase on Methanol

Initiate by starting a methanol feed (980 ml of 100% methanol +10 ml of PTM1 +10 ml of biotin stock solution and filter sterilized) at a rate of 6–24 ml/h (i.e., 1–4 ml/l/h of the initial culture volume) with the aid of a master flex peristaltic variable speed pump (Cole-Parmer Instrument Company, Chicago, Illinois, USA). The methanol feed flow rate is adjusted so that with Mut⁻ strains, the methanol concentration in the fermentor is maintained in the 0.2 to 0.5% v/v level, whereas with the Mut⁺ strains, methanol levels in the fermentor approach zero and the fermentor is run in a methanol limited fashion.

Dissolved oxygen in both cases is maintained at the 20–70% range. If dissolved oxygen falls below 20%, the methanol feed is stopped and nothing should be done to increase oxygen rates until an upward spike in the dissolved oxygen level is seen. At this point, adjustments (rpm, aeration, vessel pressure, oxygen feed) can be made. Maintaining the dissolved oxygen above 20% may be difficult, depending on the oxygen transfer rate (OTR) of the fermentor.

With stainless steel vessels, the system can be pressurized up to 15–30 psi to increase OTR. Also the oxygen feed (air+oxygen mixture) at 0.1 to 0.3 vvm can be used to maintain adequate levels of dissolved oxygen. The methanol fed batch phase generally lasts for 70 h. However, it may prolong to over 200 h if the feed rate is slowed down due to the fall in dissolved oxygen levels. Another factor which may also influence the fermentation time is the secretion rate of the heterologous protein. Longer times may be necessary to allow for accumulation of high levels of a slowly secreted protein in the broth.

Note. Depending on the product being produced, adjustments in the media (addition of casamino acids) and pH will have to be made to increase product yield. One technique that can be used to lower the pH is by setting the pH to the desired lower value (pH 3) and then letting the culture pH decrease to the new set point of pH 3.0 as a result of cellular metabolism (Siegel et al. 1990; Brierley et al. 1992).

6
Transformation

Introduction of recombinant DNA into *P. pastoris* can be accomplished by any of the following procedures: the spheroplast method (Cregg et al. 1985; Sreekrishna et al. 1987); the lithium chloride method (Ito et al. 1983); the PEG-1000 method (Dohmen et al. 1991); and electroporation (Becker and Guartente 1992). The introduced DNA can establish itself in two ways: integration into chromosomal DNA by homologous recombination or autonomous replication as a circular plasmid. Chromosomal integration requires homology of the introduced DNA with a chromosomal locus. Although circular DNA can integrate, the preferred template is a linear DNA with free ends homologous to a genomic locus. Autonomous replication of circular DNA plasmid requires the presence of an autonomously replicating sequence (ARS) in the circular DNA used for transformation. A detailed protocol for various transformation methods is described in this section.

6.1
Spheroplast Transformation Procedure

6.1.1
Composition of Reagents

Stock Solutions

100× HA: Dissolve 20 g of histidine assay medium (Sect. 6.1.1.4) (Difco Laboratories, Detroit, Michigan, USA) in 100 ml of water. Heat to dissolve and filter sterilize the solution. It stores at 4 °C for over a year.

Note. Lately, Difco has discontinued selling histidine assay medium. It may be available from Invitrogen Corporation (San Diego, California, USA). In any case, it can be readily prepared using the recipe given at the end of this section.

100× AA: Dissolve 500 mg each of the following amino acids glutamic acid, methionine, lysine, leucine, and isoleucine in water. Heat to dissolve if necessary. Filter sterilize. It stores at 4 °C for 1 year.

2 M Sorbitol: Dissolve 364.4 g of D-sorbitol in 1000 ml of water. Autoclave and store at room temperature.

0.5 M EDTA: Add 18.6 g of disodium ethylene diamine tetraacetate $2H_2O$ to 80 ml of water. Stir vigorously on a magnetic stirrer and titrate with 10 N NaOH to pH 8.0. Adjust the volume to 100 ml with water. Sterilize by autoclaving and store the solution at room temperature. (note: the disodium salt of EDTA will not dissolve until the pH of the solution approaches 8.0 by the addition of NaOH).

1 M Sodium citrate buffer, pH 5.8: Dissolve 29.4 g of sodium citrate $2H_2O$ in 100 ml of water to give a 1 M solution (note: the 51/2 H_2O salt is not recommended). Titrate with a 1 M citric acid solution in water (21.01 g in 100 ml) to a pH of 5.8. The titration should need approximately 10 ml of 1 M citric acid. Filter sterilize the solution and store at room temperature.

1 M Tris-HCl, pH 7.5: Dissolve 121.4 g of Tris (hydroxy methyl) amino methane in 500 ml of water to make a 2 M solution. Titrate the solution with 2 M HCl to a pH of 7.5 (this should take approximately 380 ml) and add water to 1000 ml final volume. Autoclave and store at room temperature.

1 M CaCl₂: Dissolve 14.7 g calcium chloride in 100 ml of water. Autoclave and store at room temperature.

1 M DTT: Dissolve 7.71 g in 50 ml water, filter sterilize, and store frozen at −20 °C in 0.5 ml aliquots.

10% SDS: Dissolve 10 g of electrophoresis grade SDS (sodium dodecyl sulfate, also called sodium lauryl sulfate) in 100 ml of water. Heat to dissolve if necessary. Filter sterilize (optional) and store at room temperature.

ZT-100: Suspend 3 mg Zymolyase of specific activity 100 000 units/g in 1 ml of sterile water. Transfer 20-μl aliquots into microfuge tubes and store at −20 °C. The solution is stable for several years. Storage at temperatures much lower than −20 °C is not recommended. If you are using Zymolyase of a different specific activity (e.g., 60 000 instead of 100 000 units/g), use the appropriate amounts so as to obtain a solution of 300 units/ml. Some suggested sources of Zymolyase are: Seikagaku America, Inc. (Rockville, Maryland, USA) and ICN Biomedicals.

For the other stock solutions used for making reagents for the spheroplast transformation please refer to Sect. 2.2.1.

Spheroplasting and Transformation Reagents

1 M sorbitol: Mix 50 ml of 2 M sorbitol with 50 ml of sterile water.

SED: First make SE by mixing 50 ml of 2 M sorbitol, 5 ml of 0.5 M EDTA, pH 8.0, and 45 ml of sterile water. Just prior to use, convert SE to SED by adding 0.5 ml of 1 M DTT (frozen stock) per 9.5 ml of SE.

SCE: Mix 50 ml of 2 M sorbitol, 5 ml of 1 M sodium citrate buffer, pH 5.8 with 0.2 ml of 0.5 M EDTA and dilute to 100 ml with sterile water.

CaS: Mix 50 ml of 2 M sorbitol, 1 ml of 1 M CaCl₂, 1 ml of 1 M Tris-HCl, pH 7.5, and 48 ml of sterile water.

PEG solution: Dissolve 20 g of PEG 3350 (Fisher CarbowaxR PEG 3350, Laboratory grade, P-146, molecular weight 3000–3700; Fisher Scientific, Fair Lawn, New Jersey, USA) in 98 ml of water. Add 1 ml each of 1 M CaCl₂ and 1 M Tris-HCl buffer, pH 7.5. Filter sterilize and store at room temperature. Discard the solution if pH falls below 6.0.

SOS: Mix 25 ml of 2 M sorbitol, 16.6 ml of YPD medium, 0.5 ml of 1 M CaCl₂, and 7.9 ml of water. Filter sterilize and store the solution at room temperature.

Regeneration Media

RD: Dissolve 182.2 g of sorbitol (instead of sorbitol, 44.8 g of KCl can be used) in 700 ml of water, add to it 10 g of agar (or agarose) and autoclave. Cool to 45 °C and maintain at that temperature. To this add a prewarmed (to 45 °C) mixture of 100 ml 10× D, 100 ml of 10× YNB, 2 ml of 500× B, 10 ml of 100× HA, 10 ml of 100× AA, and 78 ml of sterile water. Aliquot 8 ml each into 15-ml tubes and maintain the tubes at 45 °C.

Note. Using RD containing agarose in place of agar gives five- to tenfold more transformants; however, the transformed colonies grow slower in agarose-containing plates.

RDB: Same as RD, except use 20 g of agar (or agarose). Maintain autoclaved solution at 60 °C prior to the addition of the prewarmed (to 60 °C) mixture of stock solutions. After mixing, pour the plates immediately.

RDH and RDHB: These are prepared as described for RD and RDB except that 10 ml of 100× H is included in the stock solution mixture and the amount of water is reduced by the same volume.

Histidine Assay Medium

Amino acid mix
Dissolve the following quantities of amino acids in 100 ml of water (10×) and filter sterilize:

DL-Alanine	0.8 g
L-Arginine hydrochloride	0.92 g
Asparginine monohydrate	0.8 g
L-Aspartic acid	0.4 g
L-Cystine	0.2 g
L-Glutamic acid	1.2 g
Glycine	0.4 g
D,L-Phenylalanine	0.4 g
L-Proline	0.4 g
D,L-Serine	0.2 g
D,L-Threonine	0.8 g
D,L-Tryptophan	0.16 g
L-Tyrosine	0.4 g
D,L-Valine	0.1 g

Nucleotide mix
Dissolve the following quantities of nucleotides in 100 ml of water (10×) and filter sterilize:

Adenine sulfate	40 mg
Guanine hydrochloride	40 mg
Uracil	40 mg
Xanthine	40 mg

Vitamin mix
Dissolve the following quantities of vitamins in 100 ml of water (100×) and filter sterilize:

Thiamine hydrochloride	20 mg
Pyridoxine hydrochloride	40 mg
Pyridoxamine hydrochloride	12 mg
Pyridoxal hydrochloride	12 mg
Calcium Pantothenate	20 mg
Riboflavin	40 mg
Nicotinic acid	40 mg
Para amino benzoic acid	4 mg
D-Biotin	40 μg
Folic acid	40 μg

Mineral salt mix
Dissolve the following quantities of mineral salts in 100 ml of water (10×) and filter sterilize:

Monopotassium phosphate	240 mg
Dipotassium phosphate	240 mg
Magnesium sulfate	8 mg
Ferrous sulfate	40 mg
Manganese sulfate	80 mg
Sodium chloride	40 mg

To make 100× HA medium, dissolve 80 g of sodium acetate, 12 g of ammonium chloride and 100 g of glucose in 690 ml of water. Add to this 10 ml of the vitamin mix, 100 ml of the amino acid mix, 100 ml of the nucleotide mix, and 100 ml of the mineral salt mix. Filter-sterilize the solution. This step also removes any insoluble material that may result from mixing the ingredients; store at 4 °C. The stock stores well for several months.

6.1.2
Procedure

1. Inoculate 10 ml of YPD with a colony of GS115 or any suitable strain and grow to saturation at 30 °C. This should take less than 2 days. Store the saturated culture at 4 °C.

2. Place 200 ml of YPD in each of three 500-ml culture flasks. Inoculate flasks with 5, 10, and 20 μl of cells from step 1 and incubate overnight with shaking at 30 °C.

3. The next morning, check the $O.D_{600nm}$ of the cells. Harvest the cells from the culture which has an $O.D_{600nm}$ of 0.2 to 0.3. Cells are harvested by centrifugation at room temperature for 5 min at 1500 g. Use these cells for the transformation. Discard the other cells.

4. Wash the cells once with 20 ml of sterile water by centrifugation at 1500 g for 5 min at room temperature.

5. Wash the cells once with 20 ml of fresh sorbitol-EDTA-DTT solution (SED) as described in step 4. Do not let the cells remain in SED for any longer than is

necessary. Prolonged incubation in SED will affect the transformation frequency due to decreased cell viability.

6. Wash cells once with 20 ml of 1 M sorbitol to remove residual SED by centrifugation as in step 4.

7. Resuspend cells in 20 ml of sorbitol-citrate-EDTA (SCE) buffer and divide the suspension into two tubes (10 ml each).

8. Add 7.5 µl of Zymolyase ZT-100 stock to one tube, mix gently and incubate the cells at 30 °C. Do not shake the sample. The moment after the addition of Zymolyase, the digestion of the yeast cell wall begins. Cells without cell walls (spheroplasts or protoplasts) are extremely fragile. Handle the sample gently. This procedure is used to establish the incubation time for optimal spheroplasting as described in step 9. Reserve the second tube at room temperature for the actual transformation.

9. Monitor spheroplast formation as follows: add 200 µl cells (from step 8) to 800 µl of 5% SDS at time t = 0, 2, 4, 5, 6, 7, 8, 9, 10, 15, 20, and 30 min after adding the Zymolyase and measure $O.D._{800nm}$. Determine the percent of spheroplasting for each time point using the equation:

% spheroplasting = $100 - O.D._{800nm}$ at time t/$O.D._{800nm}$ at time 0) × 100.

Determine the time of Zymolyase treatment that results in approximately 70% of spheroplasted cells. It should take approximately 6 min of Zymolyase treatment with ZT-100. It is important to establish the minimum time required for this to happen, because prolonged incubation in Zymolyase is deleterious and hampers transformation efficiency. The optimal time of treatment determined for a given combination of P. pastoris strain, cell density, and the Zymolyase stock solution remains the same for subsequent uses over several years as long as the enzyme is stored in aliquots at −20 °C.

10. Add Zymolyase ZT-100 stock (7.5 µl) to the second 10-ml sample from step 8 and incubate at 30 °C for the time that was established in step 9 to obtain the optimal level (70%) of spheroplasts. Note that the spheroplasting will continue during centrifugation (step 11), and thus the final extent may actually exceed 70%.

11. Harvest the spheroplasts by gentle centrifugation at 750 g for 5 min at room temperature.

12. Wash the spheroplasts once with 10 ml of 1 M sorbitol and gently disperse the pellet by tapping the tube; do not vortex. Collect the spheroplasts by centrifugation as in step 11.

13. Wash the spheroplasts once with 10 ml of calcium-sorbitol (CaS) as in step 12.

14. Resuspend the spheroplasts with 0.6 ml of CaS.
    Note. Spheroplasts can be left in CaS for several days at room temperature before their use for transformation. However, their transformation efficiency

gradually decreases with time (probably due to decrease in viability). The best transformation frequencies are obtained if the spheroplasts are used the same day.

15. Dispense 100 μl of spheroplasts into sterile 6-ml snap top tubes (Falcon 2058 tubes from Becton Dickinson Labware, Lincoln Park, New Jersey, USA).

16. Add 1 to 10 μg of DNA in ≤10 μl final volume (1 μg in the case of supercoiled plasmid DNA such as pHIL-A1, or 5–10 μg in the case of linearized DNA such as Sal I or Bgl II digested pHIL-D1 (see Sect. 9.2.2), and incubate at room temperature for 10 min. Use 10 μl of buffer alone for a negative control.

17. Add 1 ml PEG-3350 solution, mix gently, and incubate for an additional 10 min at room temperature.

18. Centrifuge at 750 g for 5 min at room temperature and aspirate the PEG-3350 solution, invert the tubes, and then tap gently on a blotting paper to drain off the excess PEG. Resuspend the pellet in 150 μl of sorbitol-YPD solution (SOS) and incubate at room temperature for 20 min. Add 850 μl of 1 M sorbitol.
    Note. Transformed spheroplasts in SOS can be left for several hours to days prior to plating. However, the number of transformants that are obtained decreases with time. Spheroplasts will not regenerate their cell wall under these conditions even after several days in the complex medium, as they need to be embedded in a solid support (agar, agarose) or alginate to regenerate their cell walls.

6.1.3
Plating of Transformants

19. Add 100- to 250-μl aliquots of suspension from step 18 to 8 ml of regeneration-dextrose medium (RD) held at 45 °C. Mix by inverting the tube back and forth. Pour the contents on to RD base plate (RDB). Incubate plates at 30 °C. Transformants (His$^+$) should appear in 3–4 days for *ARS* vectors containing the *HIS4* gene (transformation frequency ≃ 10000–100000/μg DNA) and 4–5 days for integrative vectors containing *HIS4* gene (transformation frequency ≃ 100–10000/μg DNA). Negative controls will not form colonies.

6.1.4
Plating for Determination of Spheroplast Viability

20. Add a 100-μl aliquot from step 18 to 900 μl of 1 M sorbitol. Mix 100 μl of the diluted sample with 10 ml of RD supplemented with histidine (RDH) at 45 °C. Mix and pour on a RDH-base plate (RDHB). Under these conditions, both the viable spheroplasts and cells with intact cell walls (those which escaped spheroplasting) will form colonies.
    Also spread 100 μl of the diluted sample directly on to an MDH plate (see Sect. 2.2.3 for composition). Under these conditions, only cells which escaped spheroplasting will form colonies. The % of viable spheroplasts is determined

by counting the number of colonies on RDHB and MDH. Generally, the number of colonies on RDHB should be 10 to 100 times greater than the number of colonies on MDH.

% of viable spheroplasts $= 100 \ (x - y)/x$,

where x = number of colonies on RDHB and y = number of colonies on MDH.

6.1.5
Screening for AOX1 Gene Disruption

Transformation of GS115 with a linearized DNA containing the *HIS4* gene with the ends homologous to the 5′ and 3′ regions of the *AOX1* chromosomal locus (e.g., Bgl II-digested DNA of pHIL-D1 type plasmids or Not I-digested DNA of pHIL-D2 type plasmids; see Sect. 9 for plasmid diagrams) results in the site-specific eviction of the *AOX1* structural gene as illustrated in Fig. 1. Eviction of the *AOX1* gene occurs at a frequency of 1–5 per 20 His+ transformants. *AOX1*-deleted transformants show a slower growth phenotype (Mut−) on MM plates as compared to *AOX1* intact methanol-normal cells (Mut+). However, the His+ colonies from the original transformation plate (see step 19) cannot be directly toothpick patched onto MM plates to screen for the Mut− phenotype. If this is done, every colony will score as Mut+ because colonies in the transformation plate contain a heterogeneous population of cells due to differences in the mode, site, and extent of integration of the transforming DNA (tDNA). Many type of cells can be present within a given His+ colony.

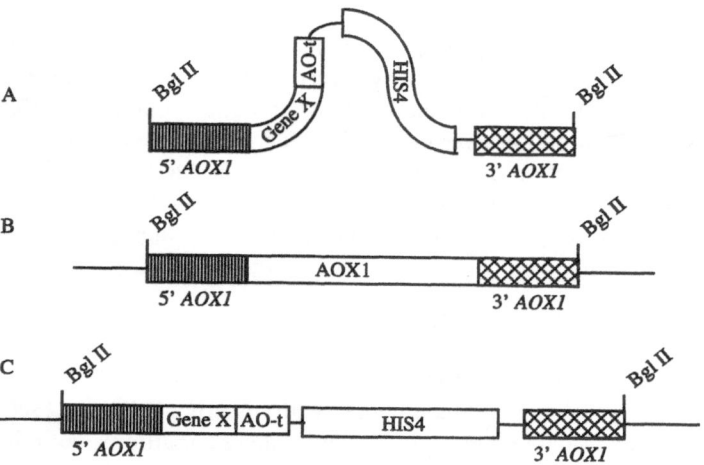

Fig. 1A–C. Site-specific eviction of *AOX1* by gene replacement. **A** Bgl II-digested DNA derived from pHIL-D1 expression plasmid used for transformation. Gene X is the gene of interest cloned at the Eco RI site of pHIL-D1 (see Sect. 9). **B** *P. pastoris* chromosomal *AOX1* locus. **C** Chromosomal structure resulting from replacement of the entire *AOX1* locus by the transforming DNA

Mut⁺, single copy of tDNA at *AOX1* locus
Mut⁺, multiple copy of tDNA at *AOX1* locus
Mut⁺, single copy of tDNA at *his4* locus
Mut⁺, multiple copy of tDNA at *his4* locus
Mut⁻, single copy of tDNA at *AOX1* locus
Mut⁻, multiple copy of tDNA at *AOX1* locus.

The following protocol involving the dispersion of cells from colonies is used to overcome the problem of colony heterogeneity in screening for the Mut⁻ phenotype.

The top-agar layer containing His⁺ colonies is transferred into a sterile 50-ml disposable plastic centrifuge tube and mixed with 20 ml of sterile water. The suspension is vortexed vigorously to disperse cells from agar and the suspension is filtered through four layers of sterilized cheesecloth (USP type VII gauze, 20 × 12 mesh from Ultimed International, Inc., Glendale Hts., Illinois, USA).

The gel chunks retained on the cheese cloth are rinsed twice with 10 ml of water to displace trapped cells. The filtrate is centrifuged at 1500 g for 5 min at room temperature. Under these conditions, the cells form a tight pellet under a fluffy layer of fine agar particles.

The agar layer is removed by gently shaking the tube to disperse the loose agar layer into water followed by the careful decanting of the milky supernatant liquid which contains most of the fine agar particles.

The cell pellet is suspended in 5 ml of water, sonicated for 10 s (optional) and serial dilutions are then plated on MD plates so as to obtain plates with <1000 colonies. To make dilutions, assume that each pooled colony will have contributed approximately 1 million cells.

Colonies that appear on the MD plate are replica-plated onto MM plate to screen for Mut⁻ transformants. Generally 5 to 35% of the colonies replica-plated will be Mut⁻.

6.2
Lithium Chloride Transformation Method

The protocol described here was communicated by J. Cregg and it is based on a procedure described for *S. cerevisiae* (Ito et al. 1983).

1. Grow a 50-ml shake flask culture of GTS115 in YPD at 30 °C to an approximate O.D.$_{600nm}$ of 1.0 (i.e., 5×10^7 cells/ml).

2. Wash cells once in 10 ml of sterile water and centrifuge (1500 g, for 10 min, at room temperature).

3. Wash the cells in 10 ml of sterile TE buffer (10 mM Tris-HCl, pH 7.5, 1 mM EDTA) and centrifuge as in step 2.

4. Resuspend cells in 20 ml of sterile LiCl + TE buffer (0.1 M LiCl, 10 mM Tris-HCl, pH 7.4, 1 mM EDTA) and incubate at 30 °C for 1 h.

5. For each transformation sample, add the following to a sterile 12×75 mm polypropylene snap top tube: 10 μg of Hae II-digested *E. coli* carrier DNA, 0.1 μg to 20 μg of transforming DNA, and 0.1 ml of GTS115 cells (step 4). The volume of DNA added should not exceed 20 μl.
   Note. Typically, vectors based on *HIS4* selection marker (e.g., pHIL-A1, pHIL-D1, pHIL-D2, see Sect. 9) are used for transformation of GTS115.

6. Incubate at 30 °C for 30 min.

7. Add 0.7 ml of 40% PEG-3350 in LiCl + TE buffer (step 4) and vortex briefly to mix.
   Note. Make sure that the pH of the PEG solution is above 6.5.

8. Incubate at 30 °C for 30 min.

9. Heat shock the cell at 37 °C for 5 min (longer duration will decrease cell viability).

10. Centrifuge the samples and resuspend in 0.1 ml of sterile water.

11. Spread on selective plates such as MD. Transformants should appear within 3 days of incubation at 30 °C. Transformation frequency is rather low with supercoiled ARS vectors (10–100/μg of DNA). Better transformation frequencies are seen with linear DNA, especially for those involving a single crossover event (e.g., Sac I- or Sal I-digested pHIL-D1) for integration (100–1000 transformants/μg of DNA).

12. Transformant colonies can be pooled from the surface of agar plates by adding 5 ml of sterile water to the plates and suspending cells with a spreader. As with the spheroplast method, it is recommended that the transformant pool be sonicated and single colonies isolated prior to screening for MUT⁻ phenotype.
   Note. Typically, LiCl method of transformation does not favor multicopy integration of linear DNA as compared to the spheroplast method of transformation (Sreekrishna et al. 1988b).

6.3
Transformation Method Using Frozen Competent Cells (PEG-1000 Method)

6.3.1
Composition of Reagents

BDES: Mix 50 ml of 2 M sorbitol, 1 ml of 1 M Bicine-NaOH buffer (pH 8.35), 3 ml of ethylene glycol (Merck and Co., Inc., Rahway, New Jersey, USA), 5 ml of DMSO

(Sigma Chemicals, St. Louis, Missouri, USA), and 41 ml of water. Filter sterilize and store at $-20\,^{\circ}$C.

Note. Use DMSO from a new unopened bottle; old DMSO contains oxidized products which inhibit transformation. For convenience, store DMSO from a new bottle in 5-ml aliquots at $-70\,^{\circ}$C.

BP: Dissolve 40 g of PEG 1000 (Roth Labs., Karlsruhe, Germany: supplier in USA is Mike Samuels, phone: 516-694-9000) in 80 ml of water and add 20 ml of 1 M bicine-NaOH, buffer (pH 8.35).

Note. The source of PEG 1000 is important for efficient yield of transformants.

BS: Dissolve 876 mg NaCl in 99 ml of water (150 mM NaCl) and add 1 ml of 1 M bicine-NaOH buffer (pH 8.35).

6.3.2
Preparation and Freezing of Competent Cells

1. Inoculate 10 ml of YPD with a single colony of GTS115 from a fresh YPD plate and incubate overnight. Measure $O.D_{600nm}$.

2. Dilute the overnight culture into 100 ml YPD to an $O.D_{600nm}$ of 0.1 and then grow to an $O.D_{600nm}$ of 0.5 to 0.8.

3. Harvest the cells ($1500\,g$, 5 min, room temperature) and wash in 50 ml of buffered sorbitol-dimethyl sulfoxide solution (BDES) by centrifugation.

4. Resuspend cell pellet in 4 ml of BDES and store in aliquots of 0.2 ml at $-70\,^{\circ}$C. Cells will remain competent for over 6 months.

6.3.3
Transformation

5. Add plasmid DNA ($1-10\,\mu g$ in $\leq 10\,\mu l$) and carrier DNA ($40\,\mu g$ of sonicated and heat-denatured salmon sperm or calf thymus DNA in $\leq 10\,\mu l$) to the top of the frozen yeast cell suspension before it melts. For a negative control use only the carrier DNA.

6. Incubate at 37 $^{\circ}$C for 5 min; mix the contents well by rapid agitation in a mixer. Efficient mixing during thawing and after thawing of the cells with DNA is essential for a high transformation frequency.

7. Add 1.4 ml of buffered polyethylene glycol-1000 solution (BP), mix and incubate for 1 h at 30 $^{\circ}$C.

8. Pellet the cells by centrifugation at $3000\,g$ for 5 min at room temperature.

9. Wash once with 1.5 ml of buffered salt solution (BS) and resuspend in 0.2 ml of the same buffer (BS).

10. Plate 50–100-μl aliquots on to selective plates (such as MD plates) and incubate at 30 °C for 3–4 days. Also plate an appropriately diluted aliquot on nonselective plates (such as MDH) to determine cell viability.

6.4
Transformation by Electroporation

1. Grow GTS115 in 500 ml of YPD placed in a 2-l flask to an O.D$_{600nm}$ of 1.3 to 1.5. **Note.** Use an overnight culture or a YPD-slant of GTS115 for the inoculation.

2. Centrifuge cells at 4 °C (1500 g for 5 min). Wash the cell pellet with 500 ml of ice-cold, sterile-deionized water.

3. Centrifuge cells (as in step 2) and resuspend the cell pellet in 250 ml of ice-cold, sterile deionized water.

4. Centrifuge cells (as in step 2) and resuspend cell pellet in 20 ml of ice-cold, sterile 1 M sorbitol.

5. Centrifuge cells (as in step 2) and resuspend in 1 ml of ice-cold, sterile 1 M sorbitol to give a final volume of 1.5 ml.

6. Mix 40 μl of cells with 1 to 5 μl of linearized DNA (0.1 to 1 μg) and transfer them to an ice-cold 0.2-cm disposable electroporation cuvette. Tap the cuvette to force the suspension to the bottom of the cuvette, and ensure that no air bubbles are trapped in the suspension (air bubbles will cause arcing when the current is applied).

7. After 5 min on ice, pulse with a Bio-Rad gene pulser (or equivalent) equipped with pulse controller set at 25 μF, 2.5 kV, and between 200–600 ohms (generally 400 ohms).

8. Immediately add 1 ml of ice-cold 1 M sorbitol to the cuvette. Transfer the contents to an ice-cold Eppendorf tube.

9. Spread aliquots of 200–500 μl on selection plates such as MD.

10. Incubate at 30 °C until colonies appear. Colonies will appear within 2 days, followed by a second wave of smaller colonies by 5 days. Transformants can be colony-purified by restreaking on MD plates for further analysis.

The transformation frequency with electroporation is comparable, and at times superior, to other whole-cell procedures (LiCl or PEG 1000). However, just as with the other whole cell procedures, this method also fails to give multicopy integrants at a high frequency (18). With the development of gene dosage-dependent selection schemes such as G418, antibiotic resistance (conferred by the kanamycin resistance *Tn903*) (Scorer et al. 1994) or with the use of the *LEU2d* gene (complements *leu2* defect in *S. cerevisiae* only when present in multiple copies; this may also be true for *leu2* strains of *P. pastoris*), it should be possible to select for multicopy transformants even if they occur at a low frequency.

7
Induction of Protein Expression

Expression vectors (e.g., pHIL-D1, pHIL-A1) used for heterologous expression in *P. pastoris* are typically derived from *AOX1* regulatory sequences (Ellis et al. 1985). One or more of the following methods can be used for activation of the promoter. These protocols also work well for activation of other methanol regulated promoters as well as for peroxisome proliferation. Refer to Sect. 5 for induction of expression in the fermentor.

7.1
Continuous Induction

The *AOX1* promoter can be maximally activated (derepression and/or induction) by growing cells on methanol as the sole source of carbon and energy (e.g., MM or MMH media) in shake flasks, shake tubes, or plates at 30 °C. The doubling time on methanol is ≃6 h as compared <2 h on glucose or glycerol medium. Even the cells in which the *AOX1* gene has been evicted (Mut⁻) can grow on methanol. However, the growth rate is slow (doubling time 18–24 h) because it is dependent solely on a secondary alcohol oxidase, *AOX2*, which is expressed only at low levels (Cregg et al. 1989). In such cases, the growth rate can be enhanced by supplementing the MM medium with 1% sorbitol or 100 mM alanine. The advantage of using sorbitol or alanine lies in the fact that these compounds do not interfere with the induction of *AOX1* promoter.

7.2
Stepwise Induction

For liquid cultures, cells are initially grown to an $O.D._{600nm}$ of 2–10 on a nonactivating carbon source such as glycerol or glucose (MGY, MD) in shake flasks or tubes at 30 °C. The cells are then harvested by centrifugation and shifted to MM and incubated with shaking at 30 °C for 1–4 days.

Induction of expression on plates is carried out by streaking or plating cells on a nitrocellulose or cellulose acetate membrane placed on a MD or MGY plate. Once colonies appear, the filter is lifted and placed on a MM plate.

7.3
Evaluation of Product Toxicity

Sometimes the product being expressed is toxic. In such cases, a stepwise induction is suggested. Whether a particular product is toxic to *P. pastoris* can be evaluated by comparing the growth of transformants in sorbitol or alanine media to that on sorbitol+methanol or alanine+methanol media. If the expressed protein is toxic, then the growth in the presence of methanol will be drastically impaired.

7.4
Efficient Secretion of Proteins

The conditions described thus far work well for the intracellular accumulation of heterologous proteins, but are rather inadequate for secreted proteins. The following protocols work well for secreted proteins for both Mut$^+$ and Mut$^-$ cells (Barr et al. 1992).

7.4.1
Secretion Media Composition

BMGY: Mix 100 ml of 1 M potassium phosphate buffer, pH 6.0, 100 ml of 10× YNB, 2 ml of 500× B (refer to Sect. 2.2.1 for composition of stock solutions), and 10 ml of glycerol. Filter sterilize and add to an autoclaved solution of 10 g yeast extract and 20 g peptone in 788 ml water (15 g Bacto agar is included for plates).

BMMY: Same as BMGY, with the exception that 5 ml of methanol is added in the place of 10 ml glycerol.
Note: Yeast extract and peptone in the above media can be replaced by 1% casamino acids. The pH 6 suggested here may not be optimal for every secreted product. Experimentation with pH values in the range 2.5–8, (by using appropriate buffers) is suggested to determine the optimal pH for a particular product. Some suggested buffers are as follows:

- Phosphate buffer for pH range 5.7 to 8.0
- Alanine-HCl buffer for pH range 2.5 to 3.6
- Aconitic acid-NaOH buffer for pH range 2.5 to 5.7
- Citrate buffer for pH range 3.0 to 6.2

Avoid buffers such as succinate buffer, because succinate can serve as a carbon source and will repress activation of the *AOX1* promoter.

7.4.2
Shake Tube Cultures

1. Grow cells to saturation in 10 ml of BMGY (see Sect. 7.4.1 for composition) placed in 50 ml tube (2–3 days), the O.D$_{600nm}$ of culture will be in the range of 10–20.

2. Harvest cells by centrifugation (1500 g, 5 min, room temperature) and resuspend the pellet with 2 ml of BMMY (see Sect. 7.4.1 for composition). Cover the tube with a sterile gauze (four layers of cheese cloth: USP type VII gauze, 20 × 12 mesh, from Ultimed International Inc. Glendale Hts. Illinois, USA) instead of the cap. If there are several tubes (generally the case), all the tubes can be covered by spreading one piece of cheesecloth over them and securing the ends with tape or a rubber band. Return the tube(s) to the 30 °C shaker.

At the end of 2–3 days, pellet the cells and analyze the supernatant media for the secreted protein. The pellets can be resuspended with 2 ml BMMY, covered with cheesecloth and returned to the shaker for renewed secretion. With *P. pastoris*/human serum albumin (HSA) strain, 10 μl of media supernatant is sufficient for analysis by SDS-PAGE followed by Coomassie brilliant blue R250 or G250 staining.

Note. Evaporation of BMMY during incubation is commonly noticed; if this happens, add BMMY as required and make sure that the cells are not allowed to go dry.

7.4.3
Shake Flask Cultures

1. Grow the cells as described above in 1 l of BMGY in a 2-l flask.

2. Harvest the cells and resuspend them with 50–75 ml of BMMY in a 300–400-ml low baffle flask and cover with four layers of cheesecloth as described above. Return the flask to 30 °C shaker and incubate for 2–4 days. Analyze the media supernatant for product (see Sect. 8). Just as with shake tubes, secretion can be renewed by rejuvenating the media.

7.4.4
Plates

Patch cells on to nitrocellulose filter (sterile) placed on a square or circular BMGY plate and incubate at 30 °C for 2–3 days. Once colonies have grown to ≃2 mm in size, remove the filter, and place it over a BMMY plate. Cover the filter containing colonies with a sterile nitrocellulose filter. If desired, a cellulose acetate filter can be interfaced between the two nitrocellulose filters. Replace the lid and incubate at 30 °C for 2–3 days. Make sure to add 100 μl of 100% methanol to the lid once every day. Remove the nitrocellulose filter that was placed on top, wash, and analyze for secreted protein.

8
Analysis of Protein Expression

This section deals with physical, chemical, and enzymatic disruption of cells for the recovery and analysis of intracellular proteins. For a more detailed analysis of the cell disruption methods, refer to Hopkins (1991).

8.1
Mechanical Lysis of Cells

1. Cells are washed once in breaking buffer composed of 50 mM sodium phosphate, pH 7.4, 1 mM EDTA, 5% glycerol, and 1 mM PMSF; a 100 mM stock solution of PMSF in isopropanol stored at −20 °C may be used to adjust the final

PMSF concentration. The cells are resuspended at an $O.D_{600nm}$ of 50–100. An equal volume of acid-washed glass beads are added (size 0.45 mm, Sigma Chemicals, St. Louis, Missouri, USA). The mixture is vortexed vigorously for a total of 4–8 min in increments of 30 s, followed by 30 s on ice (total time is 8–16 min) using a Vortex Genie 2 mixer (Fisher Scientific, Bohemia, New York, USA).

A Mini-Bead Beater can also be used instead of manually vortexing the cells. Both single and multiple sample handling devices are available (BioSpec Products, Bartlesville, Oklahoma, USA). Another source for small-volume, shaking bead mill disrupters is Taiyo Scientific Industrial Co. (Tokyo, Japan).

For samples of larger volumes (up to 250 ml) it is desirable to use the Bead Beater (BioSpec Products, Bartlesville, Oklahoma, USA). This equipment agitates the beads with a rotor rather than by shaking and it also comes with an outer cooling jacket. Very large volumes will require a Dyno-mill (Glen Mills Inc., Clifton, New Jersey, USA: Willy A. Bachofen Maschinen-Fabrik, Basel, Switzerland: Netzsch Incorporated, Exton, Pennsylvania, USA).

2. The sample is centrifuged at 12 000 g for 10 min at 4 °C.

3. The clear supernatant solution is transferred to a fresh tube and analyzed for the desired product. The total Lowry protein concentration in this lysate will be 5–10 mg/ml.

4. The pellet fraction contains cell debris, unbroken cells, and proteins not solubilized by the breaking buffer. This pellet can be extracted in the presence of nonionic and ionic detergents, chaotropic salts, reducing agents, and general denaturants available for protein extraction.

   For a quick look at whether or not the protein in question is present in the insoluble fraction, wash the pellet two times with breaking buffer and boil the washed pellet for 5 min at 100 °C with an appropriate aliquot of electrophoresis sample buffer (0.25 M Tris-HCl, pH 6.8, 5% SDS, 0.025% bromophenol blue, 10% glycerol, and 2.5% β-mercaptoethanol). Use 200 μl of sample buffer for an insoluble pellet resulting from 500 μl of original cell suspension in the breaking buffer. Analyze the boiled material by SDS-PAGE. The proteins separated on the gel can be subjected to chemical and immunological analyses.

8.2
Alkaline Lysis of Cells

This procedure is useful for rapid preparation of cell extracts for analysis of protein expression. The protein extracted by this procedure is not expected to retain biological activity.

1. Use log phase (exponential) cells captured at an $O.D_{600nm}$ of 0.4 to 4.0.

2. Suspend the cells at an O.D$_{600nm}$ of 2–10 in 200 μl of alkaline lysis solution (50 mM NaOH, pH 10.5, containing 2 mM EDTA, 1 mM PMSF, 5% SDS, 10% glycerol, and 5% β-mercaptoethanol).

3. Boil for 5 min at 100°C. Cool and adjust the pH to 7.5 with 1 M HCl or 0.5 M NaH$_2$PO$_4$.

4. Centrifuge the sample for 4 min in a microfuge.

5. Analyze 10–20 μl of clear supernatant by SDS-PAGE. The separated proteins can also be electroblotted and subjected to NH$_2$-terminal amino acid sequence and immunological analyses.

8.3
Acid Lysis of Cells

This method of lysis is generally independent of the growth phase of cells (exponential or stationary cultures) as well as the cell density used. This is the method of choice for determining incorporation of radioactive precursors into cellular macromolecules.

1. A 500-μl aliquot of culture (O.D$_{600nm}$ of 0.5–20) was mixed with an equal volume of ice-cold 20% trichloroacetic acid (TCA) in a 1.5-ml microfuge tube and incubated on ice for 60 min.

2. Harvest the pellet by centrifugation in a microfuge (12 000 g, 10 min at 4 °C) and discard the supernatant.

3. Resuspend the pellet in 1 ml of cold (−20 °C) ethanol and incubate on ice for 30 min.

4. Harvest the pellet as in step 2 and repeat step 3.

5. Harvest the pellet as above and dry it briefly under vacuum to remove most of the residual ethanol.

6. Resuspend the dry pellet in SDS-PAGE sample buffer (0.25 M Tris-HCl, pH 6.8, 5% SDS, 0.025% bromophenol blue, 10% glycerol, and 2.5% β-mercaptoethanol). Again, use an appropriate volume of sample buffer depending on the cell density of the culture used for lysis. Generally, 20 μl of sample buffer per 1 O.D$_{600nm}$ unit of the culture is appropriate. Incubate the sample in a boiling water bath for 5 min, centrifuge for 2 min in a microfuge to pellet any residual cell debris and analyze the clarified fraction by SDS-PAGE.
Note. Samples prepared by TCA generally take a longer time for electroblotting.

8.4
Enzymatic Lysis of Cells

1. Harvest cells from a 10-ml culture by centrifugation (1500 g, 5 min at room temperature) and wash once with 10 ml of sterile water by resuspending and centrifuging.

2. Resuspend the washed cell pellet with 0.5 ml of 50 mM sodium phosphate buffer, pH 7.4, and transfer it to a microfuge tube.

3. Add DTT to a final concentration of 10 mM (added from a 1 M frozen stock). **Note.** DTT can be replaced by L-cysteine. If desired, protease inhibitors such as 1 mM PMSF can also be included.

4. Add 30 μl of Zymolyase ZT-100 suspension (3 mg/ml suspension in water). This is equivalent to adding 9 units of the enzyme.

5. Incubate at 30 °C for 30 min. At this stage, an aliquot of sample is boiled with SDS sample buffer, clarified by centrifugation and analyzed by SDS-PAGE. This should give the profile of the total cellular protein, with the exception of proteins covalently held by the insoluble debris.

6. Mix the remaining sample vigorously in a vortex mixer. Centrifuge in a microfuge at 12 000 g for 5 min at 4 °C. Most spheroplasts will lyse under these conditions.

7. Analyze the supernatant and the solubilized pellet for protein expression by boiling with SDS-PAGE sample buffer for 5 min.

9
Vectors

9.1
Compilation of Vectors and Their Origins

pHIL-A1: Autonomously replicating vector (K. Sreekrishna and K.A. Parker, unpubl. observ.; Fig. 2).

pHIL-D1: Integration vector with or without deletion of AOX1 structural gene (Cregg et al. 1987; Fig. 3).

pPIC3: pHIL-D1 type vector with multiple cloning sites (Clare et al. 1991b; Fig. 4).

pAO815: pHIL-D1 type vectors for making multi-copy expression units in vitro (Thill et al. 1990; Fig. 5).

pAO856: Similar to pAO815 for making multi-copy expression units in vitro (Thill et al. 1990; Fig. 6).

pHIL-D2: Modified pHIL-D1 with Not I site and f1 ori (Sreekrishna et al. 1990; Fig. 7).

PHIL-D3: Derived from pHIL-D2 for making constructs with exact 5′UTR (Sreekrishna et al. 1990; Fig. 8).

pHIL-D4: pHIL-D1 with kanamycin resistance marker (K. Sreekrishna and S. Hopkins, unpubl. observ.; Fig. 9).

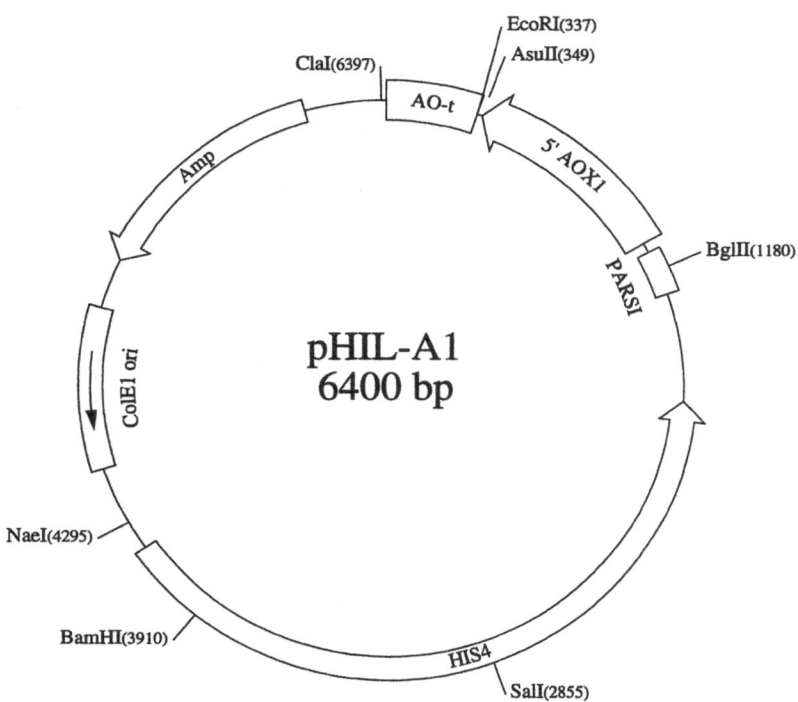

Fig. 2. pHIL-A1. Plasmid pHIL-A1, is an *E. coli – P.pastoris* shuttle vector, with sequences required for selection and autonomous replication in each host. The left half of the plasmid is a modified portion of plasmid pBR322 containing the ampicillin resistance gene and the origin of replication (*ColE1 ori*). The regions between nucleotides 1100 and 2485 of pBR322 and between *Nae* I sites 404 and 932 were deleted to eliminate poison sequences and the *Sal* I site, respectively.

The DNA elements comprising the rest of the plasmid are derived from the genome of *P. pastoris*, except for short regions of pBR322 used to link the yeast elements. The yeast elements are as follows, proceeding clockwise: *AO-t*, approximately 300-bp segment of AO terminating sequence. 5′ *AOX1*, approximately 750-bp segment of the alcohol oxidase promoter. The alcohol oxidase coding sequences following the A of the ATG initiating methionin-codon have been removed, and a synthetic linker used to generate a unique *Eco RI* site, as described for pHIL-D1. *PARS1*, approximately 190-bp segment of a *P. pastoris* autonomous replication sequence. *HIS4*, approximately 2.8-kb segment of *P. pastoris* histidinol dchydrogenase gene to complement the defective *his4* gene in GTS115.

pHIL-A1 remains autonomous for several generations, and then integrates spontancously. Integration can be directed to the *HIS4* locus by using pHIL-A1 linearized with either *Sal* I or *Stu* I, which cut within the *HIS4* gene.

Key restriction endonuclease sites are indicated on the plasmid map

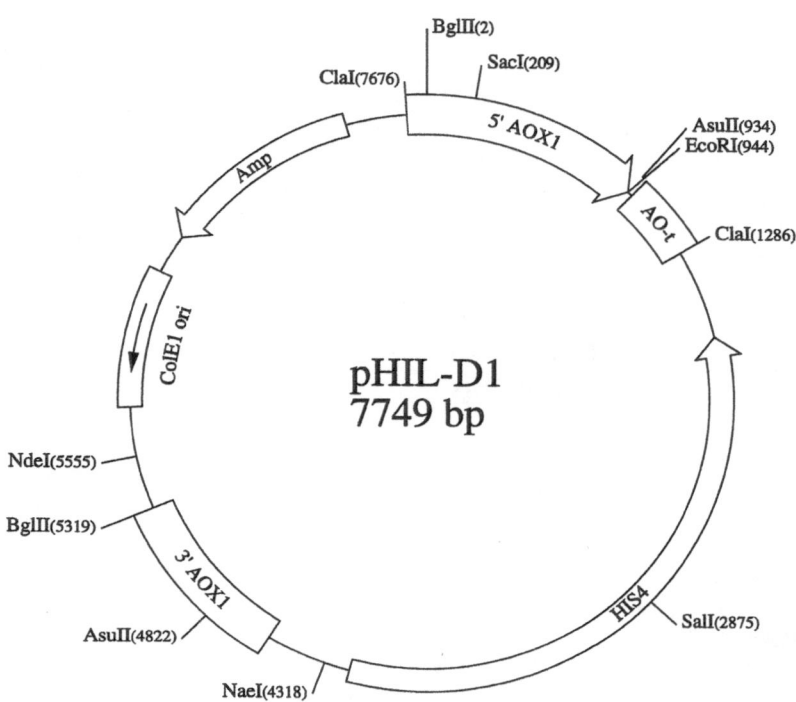

Fig. 3. pHIL-D1. Plasmid pHIL-D1, is an *E. coli* – *P. pastoris* shuttle vector, with sequences required for selection in each host. The left half of the plasmid is a portion of pBR322, from *Cla I* site through the *Pvu II* site (modified to a *Bgl II* site). This segment of pBR322 contains the ampicillin resistance gene (*Amp*) and the origin of replication (*ColE1 ori*). The *Eco RI* site in this segment has been eliminated.

The DNA elements comprising the rest of the plasmid are derived from the genome of *P. pastoris*, except for short regions of pBR322 used to link the yeast elements. These elements from pBR322 are *Cla I* to *Bam HI* (352-bp fragment), and *Bam HI* to *Sal I* (276-bp fragment). Note that both *Bam HI* and *Sal I* sites are lost in linking with *Bgl II* and *Xho I* – ended *Pichia* fragments. The portion of pBR322 from *Sal I* to *Pvu II* is absnet in pHIL-D1.

The *P. pastoris* elements in the plasmid are as follows:

5′ *AOX1*, approximately 1000-bp segment of the alcohol oxidase promoter. The alcohol oxidase coding sequences following the A of the ATG initiating methionine codon have been removed by *Bal 31* digestion, and the synthetic linker 5′-GGAATTC added to generate a unique *Eco RI* cloning site. AO-t, approximately 300-bp segment of the alcohol oxidase terminating sequence. *P. pastoris* histidinol dehydrogenase gene. *HIS4*, containd on a 2.8-kb fragment to complement the defective *his4* gene in host GTS115. Region of 3′ *AOX1* DNA approximately 650 bp in size, which, together with the 5′ *AOX1* region, is necessary for site-directed integration

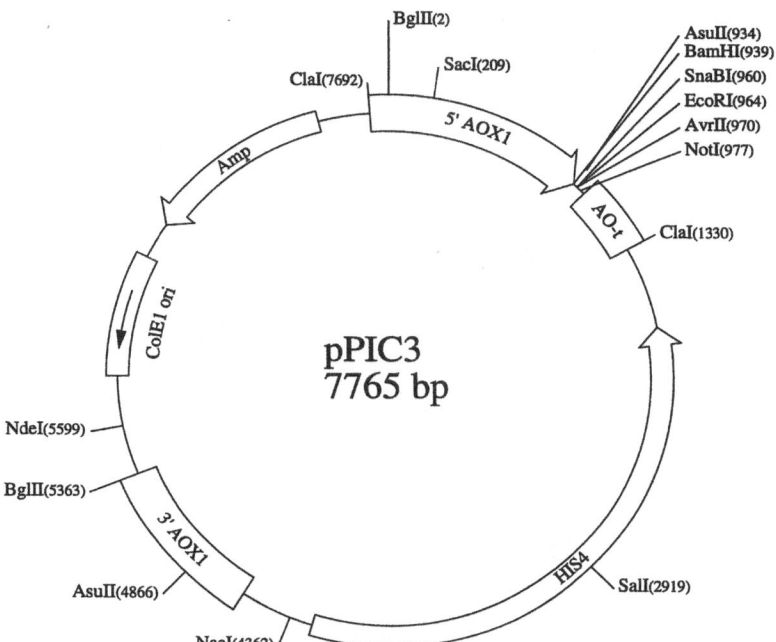

Fig. 4. pPIC3

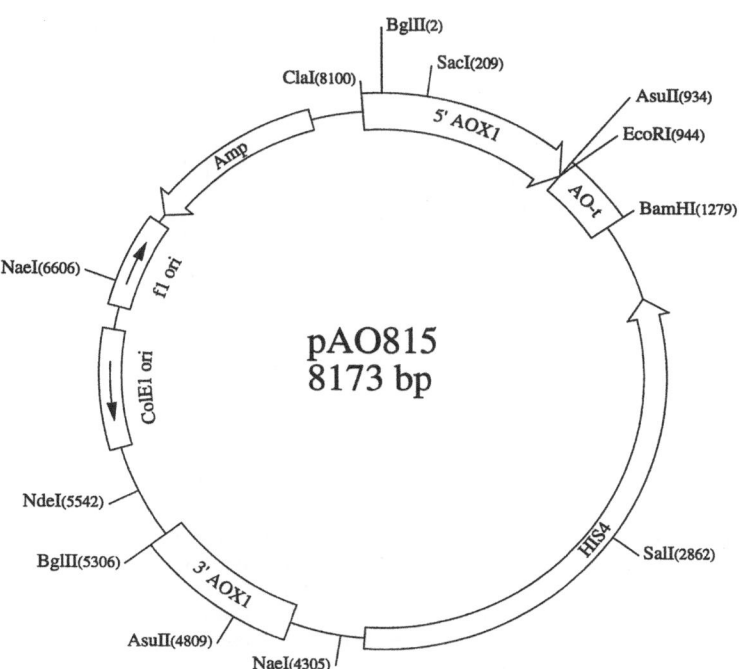

Fig. 5. pAO815

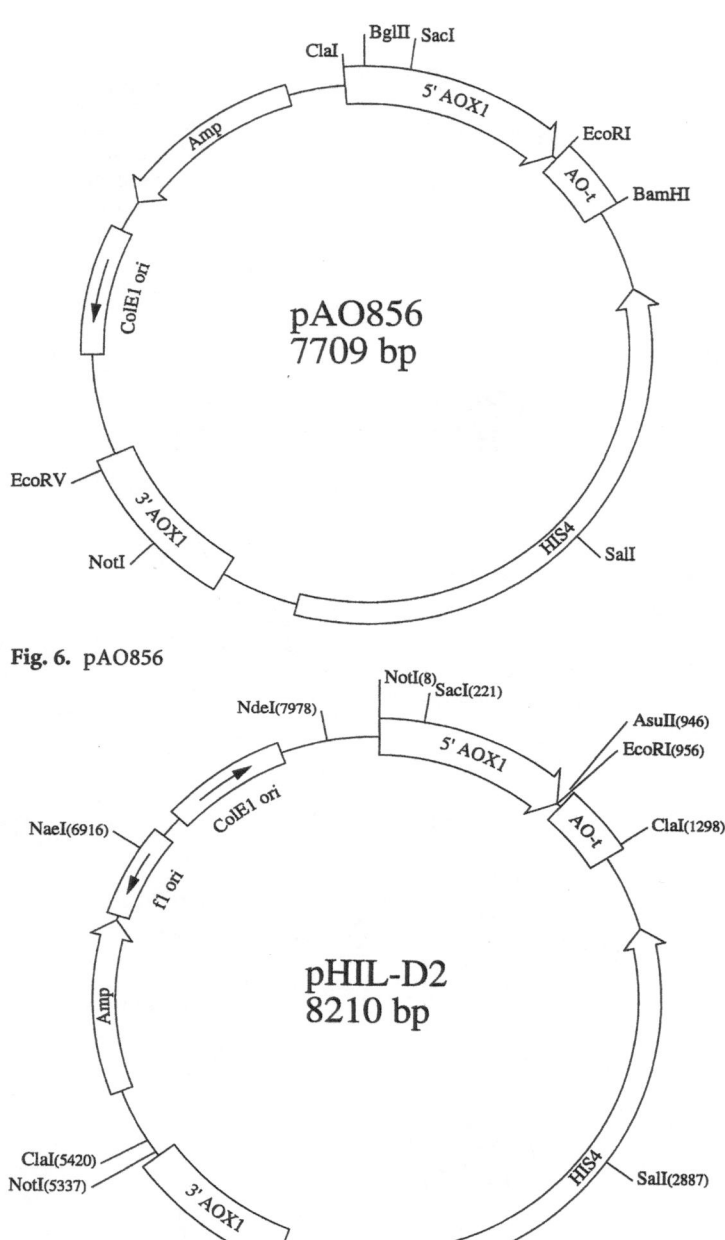

Fig. 6. pAO856

Fig. 7. pHIL-D2. Plasmid pHIL-D2 is an *E. coli–P. pastoris* shuttle vector, with sequences required for selection in each host. The left half of the plasmid is a portion of pBR322, from *Cla I* site through the *PvuII* site (modified to *Bgl II* site). This segment of pBR322 contains the ampicillin resistance gene (*Amp*) and the origin of replication (*ColE1 ori*). The *Eco RI* site in this segment has been eliminated. The pBR322 portion also contains a 458 kb DNA containing the fl-bacteriophage origin of replication at the Dra I sites at positions 3232 and 3251 of pBR322.

The DNA elements comprising the rest of the plasmid are derived from the genome of *P. pastoris*, except for short regions of pBR322 used to link the yeast elements as follows: *Cla*

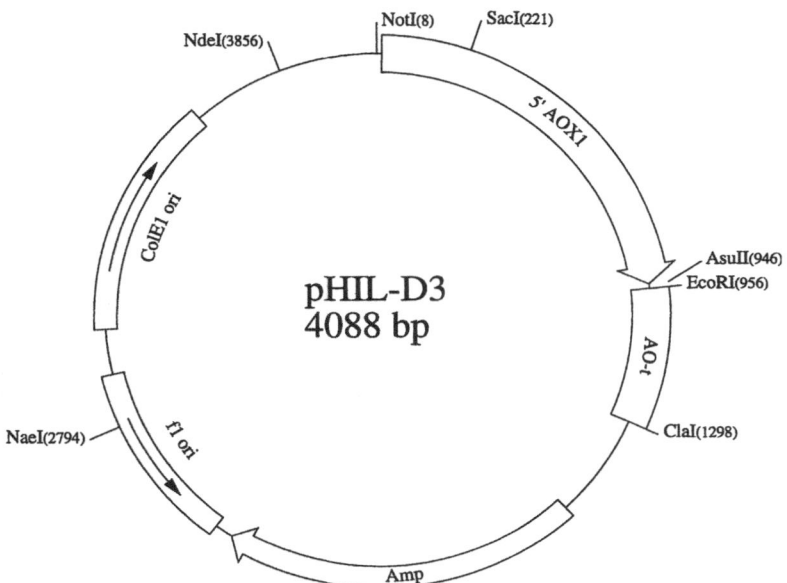

Fig. 8. pHIL-D3. pHIL-D3 is an intermediate vector helpful in making constructs in which the 5′ untranslated region (UTR) is completely devoid of extraneous sequences. In pHIL-D2 (as well as in the other *Pichia* expression vectors) the cloning site for expression is *Eco* RI. The *AOX1* promoter was modified to generate the *Eco* RI cloning site in the consruction of expression vectors (see below). This modification can have significant effect on the expression level in some instances (e.g., HSA).

AOX1 promoter:.................*TTCGAA* ACG
AOX1 promoter with the *Eco* RI site (e.g., as in pHIL-D2):
.................*TTCGAA* ACGAGGAATTC
A construct 100% devoid of extraneous sequences in the 5′UTR should be:
.............*TTCGAA* ACG atg......

The *Asu* II site (TTCGAA) in the 5′ *AOX1* can be usd to make 100% exact construct. pHIL-D2 has two *Asu* II sites. One in the 5′ *AOX1* and another in the 3′ *AOX1* (0.65-kb segment used for recombination). Thus we have made pHIL-D2 Δ Cla vector by deleting the *Cla* I fragment containing *HIS4* and 3′ *AOX1* (see Fig. 7). This vector is named pHIL-D3. pHIL-D3 has a unique *Asu* II site and serves as a useful intermediate vector in making 100% exact constructs. Once the exact construct has been made in pHIL-D3, the *Cla* I fragment that was excised previously can be returned to the *Cla* I site to complete the vector. We have noticed that pHIL-D3 does not dimerize, thus giving virually no background during the re-insertion of the *Cla* I fragment carrying *HIS4* and 3′ *AOX1*

I to *Bam* HI (352-bp fragment), and *Bam* HI to *Sal* I (276-bp fragment). Note that both *Bam* Hl and *Sal* I sites are lost in linking with *Bgl* II and *Xho* I – ended *Pichia* fragments. The portion of pBR322 from *Sal* I to *Pvu* II is absent in pHIL-D2.

The *P. pastoris* elements in the plasmid are as follows:

5′ *AOX1* approximately 100-bp segment of the alcohol oxidase promoter in which the Bgl II site has been changed to Not I. 3′ *AO-t*, approximately 300-bp segment of the alcohol oxidase terminating sequence. *P. pastoris* histidinol dehydrogenase gene. *HIS4*, contained on a 2.4-kb fragment to complement the defective *his4* gene in *Pichia* host strains. Region of 3′ *AOX1* DNA approximately 650 bp in size, in which the Bgl II site has been changed to Not I site. This fragment together with the 5′ *AOX1* region is necessary for site-directed integration

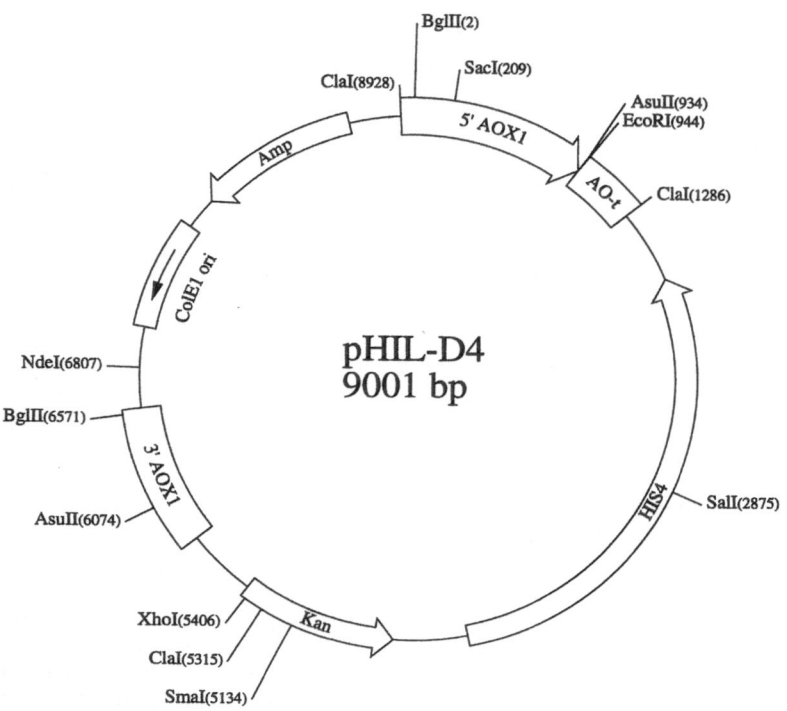

Fig. 9. pHIL-D4. pHIL-D4 allows easy screening for *Pichia* transformants with multiple copies of the transforming DNA. pHIL-D4 is a derivative of pHIL-D1. Bacterial kanamycin resistance gene from pUC-4K (PL Biochemicals) released as a *Hinc II* fragment is inserted at the unique *Nae I* site in the pBR322 derive region between *HIS4* and 3′ *AOX1* of pHIL-D1 (refer to Fig. 3).

pHIL-D4 can be used to screen for *Pichia* transformants with multiple copies of the expression cassette by screening for increased level of resistance to antibiotic G418 (sold as genticin, Sigma). One of the ways of using pHIL-D4 vector is shown below.

Pichia was transformed with *Bgl II*-digested pHIL-D4 TNF (human TNF cNDA placed at the *Eco RI* site of pHIL-D4). His⁺transformants were pooled and screened for methanol-slow phenotype. Two hundred methanol-slow transformants were secreened for antibiotic G418 resistance on YPD plates containing G418 at levels ranging from 100 to 200 μg/ml. One transformant was resistant to G418 even at 200 μg/ml. This highly resistant transformant was shown to produce TNF at levels >30% of the total soluble protein. We used TNF as an example, because its expression level in *Pichia* was known to be copy number-dependent. This approach for screening is useful in identifying transformants with presumably multicopy integrants. In several cases, it has been noted by us as well as others, that the gene dosage has a profound effect on the expression level in *Pichia* (Sreekrishna 1993) (e.g., TNF, EGF, salmon growth hormone, *Clostridium tetani* toxin fragment C, *Bordetella pertussis* p69 antigen. etc.)

pPIC3K: pPIC3 with kanamycin resistance gene (M.A. Romanos, pers. comm.; Fig. 10).

pHIL-D5: pHIL-D2 with kanamycin resistance gene (K. Sreekrishna and S. Hopkins, unpubl. observ.; Fig. 11).

pHIL-D6: pHIL-D5 with unique ASUII site and multiple cloning sites (R. Belagaje, pers. comm.; Fig. 12).

pHIL-D7: pHIL-D5 with unique ASUII site (K. Kropp, unpubl. result; Fig. 13).

pHIL-S1: Secretion vector with *P. pastoris* acid phosphatase secretion signal; Fig. 14.

Signal sequence including cloning junction of pPIC9 and pPIC9K (Fig. 15).

pPIC9: Secretion vector with *S. cerevisiae* alpha mating factor pre-pro signal (Clare et al. 1991b; Fig. 16).

pPIC9K: pPIC9 with kanamycin resistance gene (Scorer et al. 1994; Fig. 17).

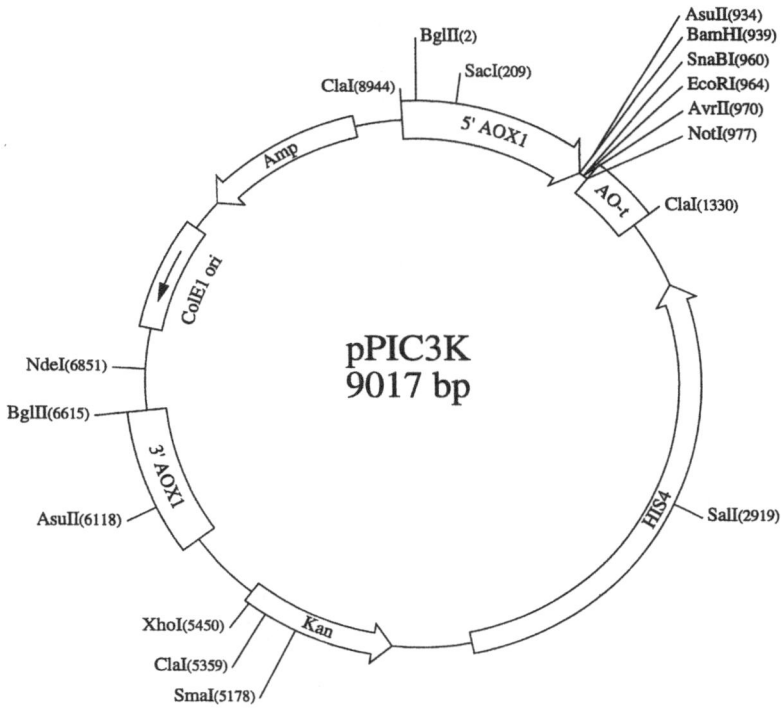

Fig. 10. pPIC3K

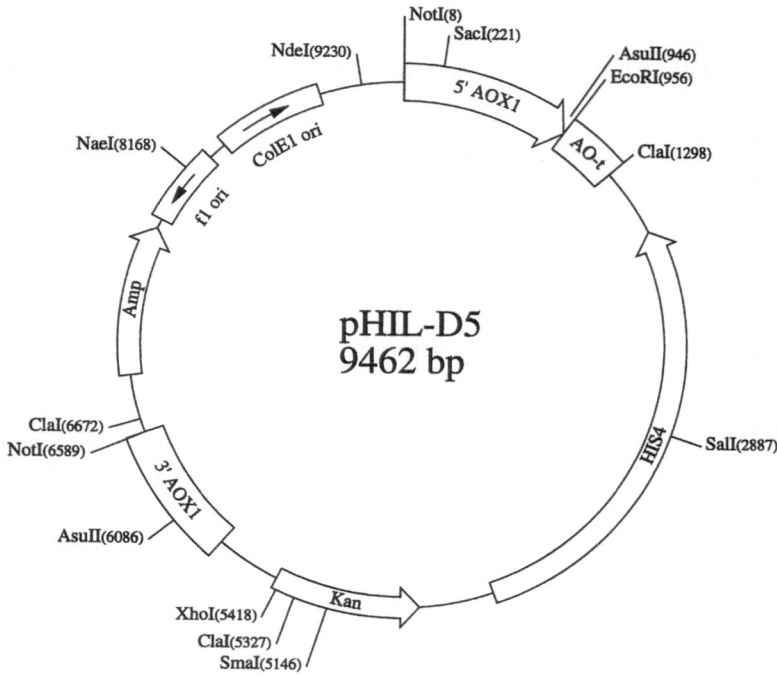

Fig. 11. pHIL-D5

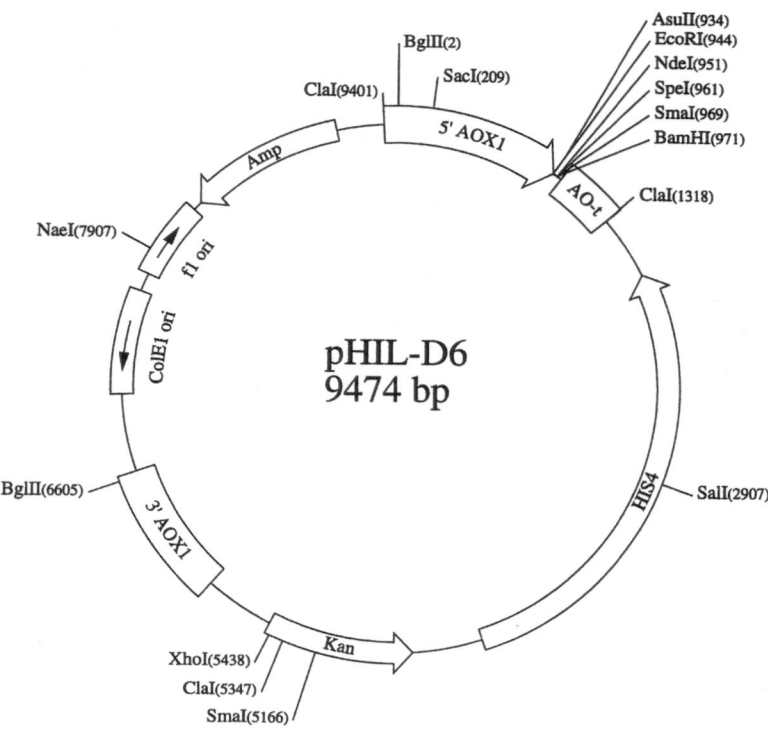

Fig. 12. pHIL-D6

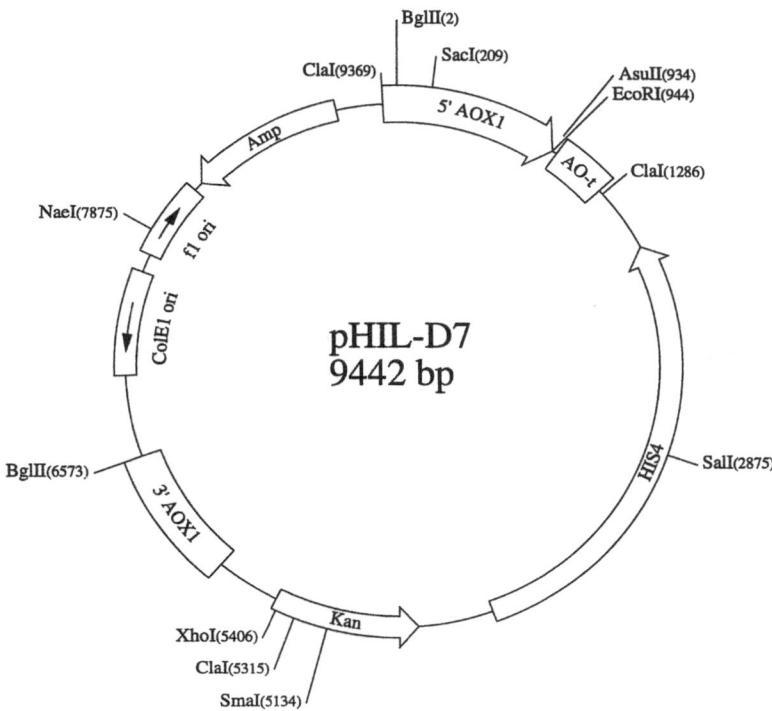

Fig. 13. pHIL-D7

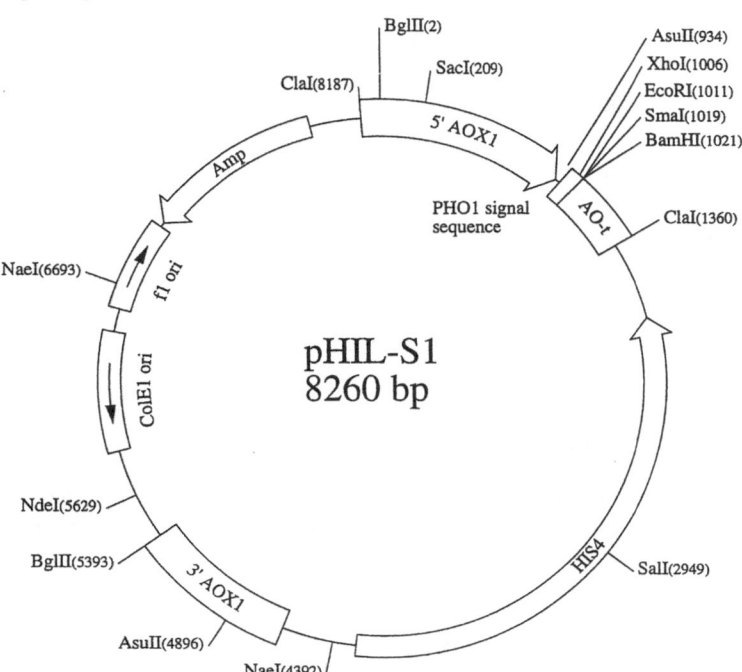

A

Fig. 14. pHIL-S1. Plasmid pHIL-S1 is a secretion vector and other features are similar to pHIL-D1. This vector also contains f1 origin of replication, similar to that in pHIL-D2. The junction sequence for making fusions to the acid phosphatase secretion signal sequence is shown

--//--5'*AOX1*----TTATTCGAAACG ATG TTC TCT CCA ATT TTG TCC TTG
                                M   F   S   P   I   L   S   L
                         <---*PHO1* Secretion Signal Sequence------

xho I

GAA ATT ATT TTA GCT TTG GCT ACT TTG CAA TCT GTC TTC <u>GCT</u> **V**
 E   I   I   L   A   L   A   T   L   Q   S   V   F   A
-------------------------- Signal Sequence continued-------------------------><

<u>Signal peptide cleavage site</u>

EcoR I   *Sma I*   *Bam HI*
<u>CGAGAATTCCCCGGG ATC C</u>TTAGA CAT........//....
 R   E   F   P   G   I   L
-------- Multi Cloning Site-----------><----3' *AO-t*-----//----

Fig. 14B

......//........................5'*AOX1*..........................TTCGAAGGATCCAAACG

ATG AGA TTT CCT TCA ATT TTT ACT GCA GTT TTA TTC GCA GCA TCC TCC
 M   R   F   P   S   I   F   T   A   V   L   F   A   A   S   S
<------------------------------Mating Factor-α Pre Sequence--------------------------

<u>Pre Sequence cleavage site</u>

GCA TTA GCT **V**GCT CCA GTC AAC ACT ACA ACA GAA GAT GAA ACG GCA
 A   L   A   A   P   V   R   I   T   T   E   D   E   T   A
-------------------->   <-----------Mating factor-α Pro sequence --------------------
CAA ATT CCG GCT GAA GCT GTC ATC GGT TAC TCA GAT TTA GAA GGG GAT
 Q   I   P   A   E   A   V   I   G   Y   S   D   L   E   G   D
-------------------------------Pro sequence continued--------------------------------------
TTC GAT GTT GCT GTT TTG CCA TTT TCC AAC AGC ACA AAT AAC GGG TTA
 F   D   V   A   V   L   P   F   S   N   S   T   N   N   G   L
-------------------------------Pro sequence continued----------------------------------
TTG TTT ATA AAT ACT ACT ATT GCC ACC ATT GCT GCT AAA GAA GAA GGG
 L   F   I   N   T   T   I   A   S   I   A   A   K   E   E   G
----------------------------- Pro sequence continued----------------------------------
         Xho I **KEX2-<u>Protease cleavage site</u>**   *Sna BI*   *Eco RI*

GTA TCT <u>CTC GAG</u> AAA AGA **V** GAG GCT GAA GCT <u>TAC GTA</u> <u>GAA TTC</u>
 V   S   L   E   K   R   E   A   E   A   Y   V   E   F
--- Pro sequence continued ---><---Multiple Cloning Sites-(MCS)---------

Avr II   *Not I*
<u>CCT AGG</u> <u>GCG GCC GCG</u> ....//..
 P   R   A   A   A
------MCS continued------><---3'*AO-t*---//--

Fig. 15. The signal sequence including cloning junction, same for vectors pPIC9 (Fig. 16)
and pPIC9K (Fig. 17). **Note:** Xho I is an unique site only in pPIC9 but not in pPIC9K

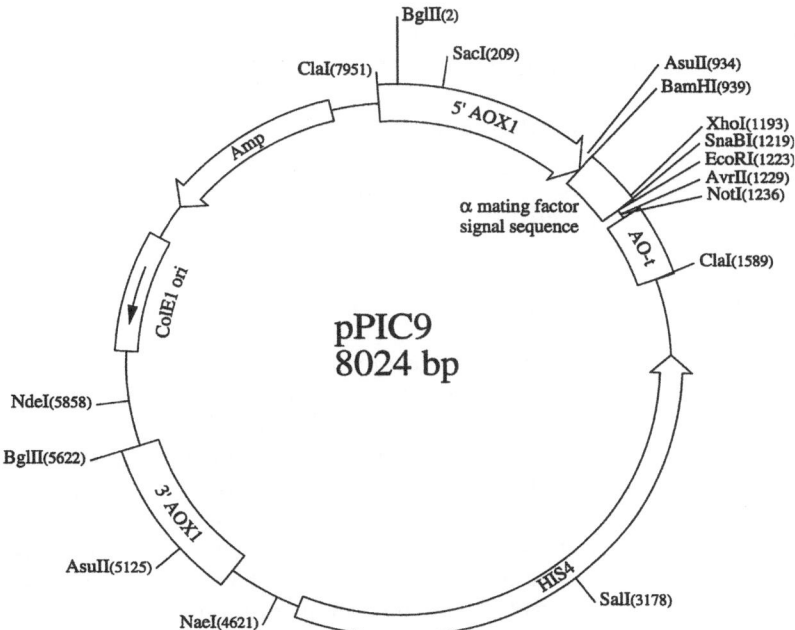

Fig. 16. pPIC9

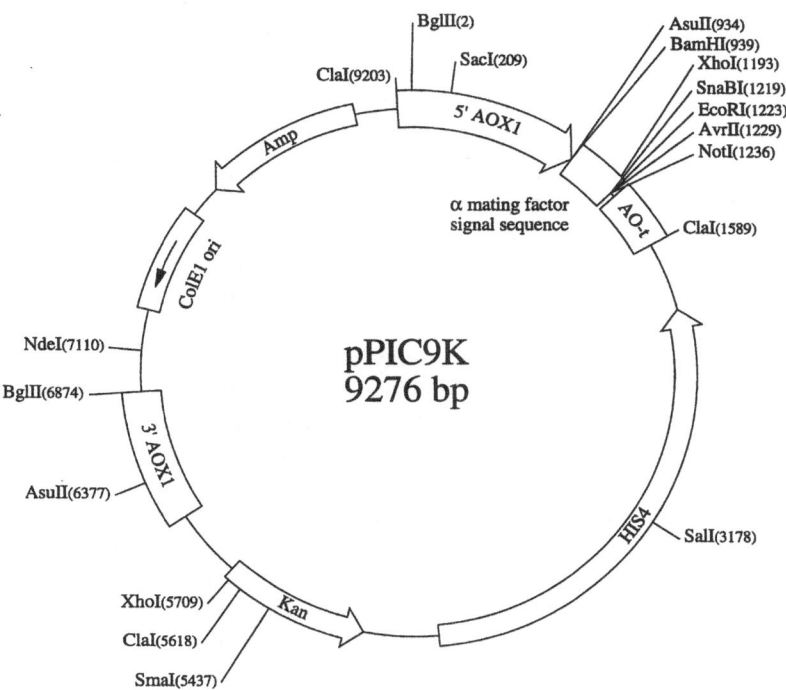

Fig. 17. pPIC9K

10
Optimization of Protein Expression

Numerous proteins have been expressed in *P. pastoris* with varying degrees of success (see Table 1 for a partial list). Nearly every protein expression project has given a new insight, and as a result the following list of factors should be considered for expression optimization.

10.1
Autonomous Replication or Integration?

Undoubtedly, the preferred mode for expression in *P. pastoris* is by integration. Such constructs are stable, and methods are available for obtaining stable multi-copy integrants (see next section below). Autonomous vectors such as pHIL-A1 (see Sect. 9) are low copy, unstable, and eventually integrate (at one or more of the chromosomal loci, namely *AOX1*, *HIS4*, or *ARS1*) especially once the promoter is activated. The constructs for LACZ, SUC2, and streptokinase expressions were initially based on the pHIL-A1 type autonomous vectors. Most other reported expression work has used integration vectors, and this trend will continue until stable, high-copy number autonomous vectors (of the sort available for *S. cerevisiae* and *Kluyveromyces lactis*) are established for *P. pastoris*.

Nevertheless, autonomous vectors, owing to their high frequency of transformation capability, are useful in mastering the spheroplast transformation procedure and in the isolation of genes by complementation.

10.2
Gene Dosage

The effect of gene copy number on expression is unpredictable. In some cases one copy, is sufficient, whereas in other cases 2 or more copies (>10) are necessary (see Table 1). There are also instances where an increase in copy number has deleterious effect on expression (Thill et al. 1990). Thus, the expression level should be examined over a wide range of copy numbers to arrive at the best production strain.

In practical terms, this can be readily accomplished by analysis of several individual transformants (24–100) obtained by using the spheroplast method of transformation. Desired colonies can be preselected by colony or dot hybridization with DNA probes prior to screening for expression. Only the spheroplast method of transformation yields a wide range of multicopy integrants. Other methods of transformation do not yield multicopy integrants at a high frequency, and thus require high throughput assays for expression or efficient selection/ screening schemes for multicopy integrants. Recently, a scheme based on increased level of resistance to antibiotic G418 as a function of gene dosage has been developed by using electroporated transformants with pPIC3 or pPIC9 type vectors (Scorer et al. 1994).

Table 1. Heterologous proteins expressed in *P. pastoris*

Protein	Location	Copy number	Amount	Reference
E. coli β-galactosidase	I-s	?	20% of soluble protein	1
HBsAg	I-p	1	0.3 g/l	2
Human TNF	I-s	>20	10.0 g/l	3, 4
		10	4.0 g/l	4
		1	0.05 g/l	4
Salmon growth hormone	I-i	18	0.3 g/l	4
Streptokinase	I-s	?	0.077 g/l	5
Tetanus toxin fragment C	I-i	14	12.0 g/l	6
Bordetella pertussus				
Pertactin P69	I-i	>10	3.0 g/l	7
Human Interleukin-2	I-i	1	0.5 g/l	8
Human Interleukin-2	I-i	?	4.0 g/l	9
Human Interleukin-2	S	?	1.0 g/l	9
B. sphaericus mosquitocidal components:				
BSP1 (42 kDa)	I-s	1	5% of soluble protein	10
BSP1 (42 kDa)	I-s	10	5% of soluble protein	10
BSP2 (51.4 kDa)	I-s	1	<0.1% of soluble protein	10
BSP2 (51.4 kDa)	I-s	>20	20% of soluble protein	10
(BSP1 + BSP2)	I-i	Multicopy	14% of total protein	10
Superoxide dismutase	I-s	1	0.33 g/l	11
	I-s	2	0.74 g/l	11
Aprotinin	S	1	0.14 g/l	11
	S	2	0.32 g/l	11
	S	5	0.93 g/l	11
Bovine lysozyme	S	1	0.46 g/l	11
	S	3	0.25 g/l	11
Human EGF	S	Multicopy	0.50 g/l	12
Mouse EGF	S	19	0.45 g/l	13
Kunitz protease inhibit	S	?	1.0 g/l	14
Invertasé	S	1	2.5 g/l	15
Human TNF-analogues:				
TNF 9 (aa 135–153 deleted)	I-i	>20	35% of total protein	16
TNF 16 (aa 7–47 deleted)	I-i	>20	32% of total protein	16
TNF-Arg-(His-Ala)₃-His tail	I-i	>20	33% of total protein	16
Human serum albumin	S	1	4.0 g/l	17
	I-i	1	0.02 g/l	17
Insulin-like growth factor-1	S	?	0.5 g/l	18
CD4-V1 domain	S	2	0.1 g/l	19
HIV gp120	I-i	12	1.25 g/l	20
A single chain antibody	S	?	>1 g/l	21
IgE Receptor (α-subunit)	S	?	0.02 g/l	22
Tick anticoagulant peptide	S	1	1.7 g/l	23

Table 1 (*contd.*)

Protein	Location	Copy number	Amount	Reference
Cathepsin E	S	?	?	24
Bm 86 antigen	S	?	1.5 g/l	25
D-alanine carboxy peptidase	S	1 or multicopy	100 mg/l	26

I-s = Intracellular soluble; I-i = Intracellular insoluble; I-p = Intracellular particle; Ss = Secreted soluble; HBsAg = Hepatitis B viral surface antigen; TNF = Tumor Necrosis Factor-α; EGF = Epidermal Growth Factor.
References: (1) Cregg and Madden (1988); (2) Cregg et al. (1987); (3) Sreekrishna et al. (1989); (4) Hagenson et al. (1989); (5) K. Sreekrishna and K.A. Parker, unpubl. observ.; (6) Clare et al. (1991a); (7) Romanos et al. (1991); (8) W.R. McCombie pers. comm.; (9) Cregg et al. (1993); (10) Sreekrishna et al. (1993); (11) Wagner et al. (1992); (12) Siegel et al. (1990); (13) Clare et al. (1991); (14) Wagner et al. (1992); (15) Tschopp et al. (1987); (16) Sreekrishna et al. (1988a); (17) Sreekrishna et al. (1990); (18) Brierley et al. (1992); (19) Buckholz et al. (1991); (20) M.A. Romanos, pers. comm.; (21) P. Mezes, pers. comm.; (22) Basu et al. (1992); (23) Laroche et al. (1994); (24) Yamada et al. (1994); (25) Rodriguoz et al. (1994); (26) Despreaux and Manning (1993).

Alternatively, multicopy construction vectors such as pAO856 and pAO815 (Thill et al. 1990) can be used to make 1,2,4,6, or 8-copy constructs in vitro prior to transformation. These constructs can be introduced by using any one of the transformation procedures to obtain strains having up to eight copies.

10.3
Mut⁺ or Mut⁻ Host?

For intracellular expression, it makes sense to use Mut⁻ cells because of increased specific yield of heterologous protein (lower levels of alcohol oxidase). For secretion, it probably does not make much difference whether Mut⁺ or Mut⁻ cells are used, although some investigators have preferred to use Mut⁺ cells for secretion. Ideally, for any given product it is better to test expression in both Mut⁺ and Mut⁻ backgrounds.

10.4
Site of Integration

Both *AOX1* and *HIS4* sites have been used successfully for expression of several proteins. While it appears that *AOX1* site is the inevitable site for creating Mut⁻ constructs, it does not have to be so because of the availability of inherently Mut⁻ hosts such as strain KM71 (see Sect. 3). Thus an expression cassette can be integrated at the *his4* site by transforming the strain KM71 with the Sal I digested pHIL-D1 type plasmid (Sal I cleaves within the *HIS4* of pHIL-D1). Instability of the integrated expression cassette due to gene conversion between chromosomal mutant copy of the *his4* and the good *HIS4* gene of the expression cassette has been noticed in only one instance, where the *LACZ* construct was integrated at the *his4* locus.

10.5
mRNA 5' and 3' Untranslated Sequences

The nucleotide sequence and the length of the 5' untranslated (leader) sequence (5'-UTR) can be detrimental to high-level gene expression. The leader length of the highly expressed AOX1 mRNA is 114 nt and the sequence is A+U rich. For optimal expression of heterologous products, it is essential to retain the 5'UTR of the resulting constructs as closely as possible to that of the AOX1 mRNA. Ideally, it is preferable to make it identical to that of the AOX1 mRNA (i.e., 100%-exact construct). In fact, the expression level of human serum albumin (HSA) is increased over 50-fold by adjusting the 5'-UTR to be identical to that of the AOX1-mRNA (Sreekrishna et al. 1990; Sreekrishna 1993). Such a dramatic effect on expression has not been observed by deleting the extraneous 3'-untranslated region. However, it is desirable to trim the 3'-portion of the gene construct so as to have no or limited extraneous 3'-UTR.

10.6
Translation Initiation Codon (AUG) Context

AUG sequence should be avoided in the 5'-UTR to ensure efficient translation of mRNA from the actual translation initiation AUG. Also, the mRNA secondary structure around the initiation AUG may be adjusted so that AUG is relatively free of secondary structure as predicted by the RNA-fold analysis (PC/gene Software, Intelligenetics, Mountain View, California, USA). This can be accomplished by redesigning the initial portion of the coding region with alternate codons (Sreekrishna 1993).

10.7
A+T Composition

Genes with a high A+T nucleotide content are not transcribed efficiently due to premature terminations (Romanos et al. 1992). One such sequence that has been identified for *P. pastoris* is the ATTATTTTATAAA stretch present in HIV-gp120 (M.A. Romanos, pers. comm.). When this stretch is altered to TTTCTTCTACAAG, the premature termination at this site is abolished. However, there are many yet unidentified AT-rich stretches that act as transcription terminators. A general strategy to overcome the transcriptional terminators in the coding region is by redesigning the gene to have an A+T content in the range of 30–55%. The expression problem of several A+T rich genes has successfuly been overcome by using this approach (Romanos et al. 1992; Clare et al. 1991a; Sreekrishna Sreekrishna et al. 1993). *P. pastoris* preferred codons are given which can be used to redesign genes. Genes with a high G+C content are efficiently transcribed (Romanos et al. 1991; Table 2)

10.8
Secretion Signal

Secretion is the preferred mode of expression for naturally secreted proteins. Most native secretion signals efficiently direct secretion in *P. pastoris*. If a native secre-

Table 2. Frequency of codon usage in highly expressed *P. pastoris* genes[a] (G.P. Thill, pers. comm.)

Amino acid	Codon	Number	Fraction	Amino acid	Codon	Number	Fraction
Gly	GGG	0.00	0.00	Trp	TGG	39.00	1.00
Gly	GGA	59.00	0.22	End	TGA	0.00	0.00
Gly	GGT	197.00	0.74	Cys	TGT	35.00	0.83
Gly	GGC	9.00	0.03	Cys	TGC	7.00	0.17
Glu	GAG	112.00	0.58	End	TAG	1.00	0.20
Glu	GAA	80.00	0.42	End	TAA	4.00	0.80
Asp	GAT	56.00	0.32	Tyr	TAT	18.00	0.12
Asp	GAC	118.00	0.68	Tyr	TAC	128.00	0.88
Val	GTG	10.00	0.05	Leu	TTG	120.00	0.52
Val	GTA	8.00	0.04	Leu	TTA	21.00	0.09
Val	GTT	107.00	0.50	Phe	TTT	24.00	0.18
Val	GTC	87.00	0.41	Phe	TTC	104.00	0.81
Ala	GCG	1.00	0.00	Ser	TCG	6.00	0.03
Ala	GCA	25.00	0.10	Ser	TCA	14.00	0.07
Ala	GCT	147.00	0.60	Ser	TCT	89.00	0.47
Ala	GCC	71.00	0.29	Ser	TCC	71.00	0.37
Arg	AGG	2.00	0.01	Arg	CGG	2.00	0.01
Arg	AGA	111.00	0.79	Arg	CGA	0.00	0.00
Ser	AGT	8.00	0.04	Arg	CGT	26.00	0.18
Ser	AGC	3.00	0.02	Arg	CGC	0.00	0.00
Lys	AAG	145.00	0.79	Gln	CAG	31.00	0.34
Lys	AAA	38.00	0.21	Gln	CAA	59.00	0.66
Asn	AAT	18.00	0.13	His	CAT	11.00	0.13
Asn	AAC	119.00	0.87	His	CAC	77.00	0.88
Met	ATG	60.00	1.00	Leu	CTG	35.00	0.15
Ile	ATA	0.00	0.00	Leu	CTA	7.00	0.03
Ile	ATT	93.00	0.56	Leu	CTT	43.00	0.18
Ile	ATC	72.00	0.44	Leu	CTC	7.00	0.03
Thr	ACG	5.00	0.03	Pro	CCG	0.00	0.00
Thr	ACA	8.00	0.05	Pro	CCA	97.00	0.57
Thr	ACT	86.00	0.50	Pro	CCT	66.00	0.39
Thr	ACC	74.00	0.43	Pro	CCC	7.00	0.04

[a] The data were compiled based on the deduced sequence of the following *P. pastoris* genes: alcohol oxidase genes *AOX1* and *AOX2*, dihydroxy acetone synthase genes *DAS1* and *DAS2*, and glyceraldehyde phosphate dehydrogenase gene *GAP*.

tion signal is not available, then signal sequences based on *S. cerevisiae* invertase secretion signal or pre-pro mating factor α (AMF) (see pPIC9 in Sect. 9.1) can be used. The AMF signal works very efficiently and is particularly valuable in secreting smaller-sized products (<10 kDa). In making protein fusions with the AMF-secretion signal (see Sect. 9.1), it is preferable to retain the Glu-Ala spacers adjacent to the Lys-Arg processing site (protease KEX2 type cleavage site) to prevent potential processing problems due to steric factors contributed by the fused protein (Thill et al. 1990).

10.9
Glycosylation

P. pastoris is capable of making both N-(aspargine) and 0-(serine or threonine)-linked carbohydrate additions to secreted proteins. As in the case of *S. cerevisiae*, the N-linked oligosaccharide chain is of the high mannose (Kukuruzinska et al. 1987; Grinna and Tschopp 1989). However, two structural differences in the oligosaccharide chain length and linkage have been noticed between *P. pastoris* and *S. cerevisiae*-secreted invertases. In the invertase from *P. pastoris*, carbohydrate moiety consists of predominantly 8–12 mannose residues with a small proportion of chains containing up to 30 mannose residues. Compared to this bulk of the carbohydrate in *S. cerevisiae*-secreted invertase, is composed of 50–150 mannose units (Grinna and Tschopp 1989). The differences in the oligosaccharide chain length between the two yeasts do not appear to be a general phenomena, because HIV-gp 120 secreted from *P. pastoris* is hyperglycosylated (Scorer et al. 1992). In fact, even the invertase secreted from *P. pastoris* using the AMF-secretion signal is hyperglycosylated (R.G. Buckholz, pers. comm.). Thus, the oligosaccharide chain length of the *P. pastoris*-secreted glycoproteins tends to be unpredictable.

Perhaps a more significant difference in the oligosaccharide structure of *Pichia*-derived invertase is the absence of terminal α 1,3-linked mannose residues (Trimble et al. 1991). This appears to be a general phenomenon for *P. pastoris*, as it does not appear to have any α 1,3 mannosyl transferase activity (Cregg et al. 1993). In contrast to this, α 1,3-mannose linkage is present in *S. cerevisiae* invertase and appears to be a characteristic feature of N-linked oligosaccharides of this yeast (Trimble et al. 1991). The α 1,3-mannose side chains are highly immunogenic (Romanos et al. 1992) and thus may be unsuitable as therapeutics. Whether the absence of the α 1,3-mannosyl linkage in *P. pastoris*-secreted glycoproteins is less immunogenic remains to be tested.

10.10
Product Stability

Proteolytic stability can be improved by using protease-deficient host strains such as SMD1168 (Sect. 3). Stability can also be improved by making fusions to other stably expressed proteins such as human serum albumin. Changing the media pH and growth conditions also can improve product yield, as already discussed in Sect. 5.

10.11
Future Perspectives

10.11.1
Expression Without Methanol

There has been some hesitation in using methanol, which is absolutely essential for activating the *AOX1*, *DAS1*, and related promoters in *P. pastoris*. To overcome this problem, mutant strains of *P. pastoris* that express the *AOX1* promoter activity in glucose medium have recently been generated and are being improved (A. Sibirny, pers. comm.). As ethanol can still repress the *AOX1* promoter in these mutants, the expression can be turned off at will by using ethanol.

Alternatively, one can use the glyceraldehyde-3-phosphate dehydrogenase (GAPDH) regulatory sequences for heterologous expression on glucose or glycerol medium (Digan 1990).

10.11.2
Improved Posttranslational Modifications in Yeast

A major limitation of *P. pastoris* and other yeast expression systems in the production of therapeutic proteins is that they are unable to carry out several posttranslational protein modifications found in mammalian proteins. These include mammalian-type protein glycosylation, vitamin K-dependent γ-carboxylation of glutamic acid, COOH-terminal amidation, etc. One would anticipate that there will be some effort in the future to engineer yeasts with mammalian genes to enable them to perform several of these post-translational modifications.

11
Miscellaneous Procedures

Readers are referred to the protocols available for *S. cerevisiae* (Rose et al. 1990; Guthrie and Fink 1991), as they can be directly adapted to *P. pastoris* for a variety of applications including, DNA isolation, RNA isolation, mutagenesis, colony hybridization, etc. Readers are also encouraged to refer to published mutagenesis protocols for *P. pastoris* (Gould et al. 1992; Liu et al. 1992).

Acknowledgments. We wish to thank all the investigators listed in the references who shared unpublishd results, Alan D. Cardin for many valuable suggestions and Robert Brankamp for a thorough perusal of the manuscript.

References

Barnett JA, Payne RW, Yarrow D (1990) Yeasts. Characteristics and identification. Cambridge University Press, Cambridge

Barr KA, Hopkins SA, Sreekrishna K (1992) Protocol for efficient secretion of HSA developed from *Pichia pastoris*. Pharm Eng 12: 48–51

Basu M, Nettleton MY, Dharm E, Manning RF, Despreaux CW, Hakimi J, Kochan JP (1992) Purification and characterization of a soluble form of the human IgE-receptor (α-subunit) expressed in *Pichia pastoris*. American Society for Biochemistry and Molecular Biology/Biophysical Society Joint meeting, Abstr #1507, February 9–13, Houston

Becker DM, Guarente L (1992) Protocol for high efficiency yeast transformation In: Chang DC, Chassy BM, Saunders JA, Sowers AE (eds) Guide to electroporation and electrofusion Academic Press, New York, pp 501–505

Brierley RA, Davis GR, Holtz GC, Gleeson MA, Howard BD (1992) Production of insulin-like growth factor-1 in methylotrophic yeast cells. International Patent Application, No WO 92/04363

Buckholz RG, Brierley RE, Odiorne MS, Siegel RS, Wondrack LM (1991) CD_4 production in *Pichia pastoris*. International Patent Application, No WO 91/05057

Clare JJ, Rayment FB, Ballantine SP, Sreekrishna K, Romanos MA (1991a) High-level expression of tetanus toxin fragment C in *Pichia pastoris* strains containing multiple tandem integration of the gene. Bio/Technology 9: 455–460

Clare JJ, Romanos MA, Rayment FB, Rowedder JE, Smith MA, Payne MM, Sreekrishna K, Henwood CA (1991b) Production of mouse epidermal growth factor in yeast: high-level secretion using *Pichia pastoris* strains containing multiple gene copies. Gene 105: 205–212

Cregg JM, Madden KR (1988) Development of the methylotrophic yeast, *Pichia pastoris* as a host for the production of foreign proteins. Dev Ind Microbiol 29: 33–41

Cregg JM, Barringer KJ, Hessler AY, Madden KR (1985) *Pichia pastoris* as a host system for transformations. Mol Cell Biol 5: 3376–3385

Cregg JM, Tschopp JF, Stilman C, Siegel R, Akong M, Craig WS, Buckholz RG, Madden KR, Kellaris PA, Davis GR, Smiley BL, Cruze J, Toregrossa R, Velicelebi G, Thill GP (1987) High-level expression and efficient assembly of hepatitis B surface antigen in methylotrophic yeast, *Pichia pastoris*. Bio/Technology 5: 479–485

Cregg JM, Madden KR, Barringer KJ, Thill GP, Stilton CA (1989) Functional characterization of the two alcohol oxidase genes from the yeast *Pichia pastoris*. Mol Cell Biol 9: 1316–1323

Cregg JM, Vedvick TS, Raschke WC (1993) Recent advances in the expression of foreign genes in *Pichia pastoris*. Bio/Technology 11: 905–910

Despreaux CW, Manning RF (1993) The dacA gene of *Bacillus stearothermophilus* coding for D-alanine carboxy peptidase: cloning, structure and expression in *E. Coli* and *Pichia pastoris*. Gene 131: 35–41

Digan ME (1990) Pichia pastoris glyceraldehyde-3-phosphate dehydrogenase gene. European Patent Application, No 0374 913 A1

Digan ME, Lair SV (1986) Genetic methods for the methylotrophic yeast, *Pichia pastoris*. In: von Borstel RC, Cooper TG (eds) Abstr 13th Int Conf on Yeast Genetics and Molecular Biology Yeast 2 (1986): S89 (Spec Issue)

Dohmen JR, Strasser AWM, Honer CB, Hollenberg CP (1991) An efficient transformation procedure enabling long term storage of competent cells of various yeast genera. Yeast 7: 691–692

Ellis SB, Brust PF, Koutz PJ, Waters AF, Harpold MM, Gingeras TR (1985) Isolation of alcohol oxidase and two other methanol-regulatable genes from the yeast *Pichia pastoris*. Mol Cell Biol 5: 1111–1121

Gould SJ, McCollum D, Spong AP, Heyman JA, Subramani S (1992) Development of the yease *Pichia pastoris* as a model organism for genetic and molecular analysis of peroxisomal assembly. Yeast 8: 613–628

Grinna LS, Tschopp JF (1989) Size distribution and general structural features of N-linked oligosaccharides from the methylotrophic yeast *Pichia pastoris*. Yeast 5: 107–115

Guthrie C, Fink GR (eds) (1991) Guide to yeast genetics and molecular biology. Methods in Enzymology, vol 194. Academic Press, New York

Hagenson MJ, Holden KA, Parker KA, Wood PJ, Cruze JA, Fuke M, Hopkins TR, Stroman DW (1989) Expression of streptokinase in *Pichia pastoris* yeast. Enzyme Microb Technol 11: 650–656

Harder W, Trotsenko YA, Bystrykh LV, Egli T (1986) Microbial growth on C_1 compounds. In: Van Verseveld HW, Duine JA (eds) Proceedings of the 5th International Symposium Martinus Nijhoff, Dordrecht, pp 138–157

Hopkins TR (1991) Physical and chemical cell disruption for the recovery of intracellular proteins. In: Seetharam R, Sharma SK (eds) Purification and analysis of recombinant proteins Mercel Dekker, New York, pp 57–83

Ito H, Fukuda Y, Murata K, Kimura A (1983) Transformation of intact yeast cells treated with alkali cations. J Bacteriol 153: 161–168

Kukuruzinska MA, Bergh MLE, Jackson BJ (1987) Protein glycosylation in yeast. Annu Rev Biochem 56: 915–944

Kurtzman CP (1984a) *Pichia* Hansen. In: Kreger-van Rij NJW (ed) The yeasts: a taxonomic study. Elsevier Amsterdam, pp 295–378

Kurtzman CP (1984b) Synonomy of the yeast genera *Hansenula* and *Pichia* demonstrated through comparisons of the deoxyribonucleic acid relatedness. Antonie Leeuwenhoek 50: 209–217

Laroche Y, Storme V, Meutter JD, Messens J, Lauwerey SM (1994) High-level secretion and very efficient isotopic labelling of tick anti-coagulant peptide (TAP) expressed in the methylotrophic yeast, *Pichia pastoris*. Bio/Technology 12: 1119–1124

Liu H, Tan X, Veenhuis M, McCollum D, Cregg JM (1992) An efficient screen for peroxisome-deficient mutants of *Pichia pastoris*. J Bacteriol 174: 4943–4951

Reiser J, Glumoff V, Kalin M, Urs O (1990) Transfer and expression of heterologous genes in yeasts other than *Saccharomyces cerevisiae*. In: Fiechter A (ed) Advances in biochemical engineering/biotechnology, vol 43. Springer, Berlin Heidelberg New York, pp 76–97

Rodrinquez M, Rubiera R, Penichet M, Montesinos R, Cremata J, Falcon V, Sanchez G, Bringas R, Cordoves C, Valdes M, Lleonart R, Herrera L, dela Fuente J (1994) High level expression of the β microplus Bm86 antigen in the yeast *Pichia-pastoris* forming highly immunogenic particles for cattle. J Biotechnol 33: 135–146

Romanos MA, Clare JJ, Beesley KM, Rayment FB, Ballantine SP, Makoff AJ, Dougan G, Fairweather NF, Charles IG (1991) Recombinant *Bordetella pertussis* (P69) from the yeast *Pichia pastoris*: high-level production and immunological properties. Vaccine 9: 901–906

Romanos MA, Scorer CA, Clare JJ (1992) Foreign gene expression in yeast: a review. Yeast 8: 423–488

Rose MD, Winston F, Hieter P (1990) Methods in yeast genetics: a laboratory course manual. Cold Spring Harbor Laboratory Press, Cold Spring Harbor, New York

Scorer CA, Buckholz RG, Sreekrishna K, Clare JJ, Romanos MA (1992) Expression of HIV gp120 in *Pichia pastoris*: selection for multi-copy transformants using drug resistance. 16th Int Conf on Yeast Genetics and Molecular Biology, Vienna, Austria, Topic No 14-49B

Scorer CA, Clare JJ, McCombie WR, Romanos MA, Sreekrishna K (1994) Rapid selection using G418 of high copy number transformants of *Pichia pastoris* for high-level foreign gene expression. Bio/Technology 12: 181–184

Siegel RS, Buckholz RG, Thill GP, Wondrack LM (1990) Production of epidermal growth factor in methylortrophic yeasts. International Patent Application, No WO 90/10697

Sreekrishna K (1993) Strategies for optimizing protein expression and secretion in the methylotrophic yeast *Pichia pastoris*. In: Baltz RH, Hegeman GD, Skatrud PL (eds) "Industrial microorganisms: basic and applied molecular genetics" American Society of Microbiology, Washington DC, pp 119–126

Sreekrishna K, Prevatt WD, Thill GP, Davis GR, Koutz P, Barr KA, Hopkins SA (1993) Production of *Bacillus* entomotoxins in methylotrophic yeast. European Patent Application, Publication no. EP 0586 892 A1

Sreekrishna K, Tschopp JF, Fuke M (1987) Invertase gene (*SUC2*) of *Saccharomyces cerevisiae* as a dominant marker for transformation of *Pichia pastoris*. Gene 59: 115–125

Sreekrishna K, McCombie WR, Potenz R, Parker KA, Mazzaferro PK, Maine GM, Lopez JL, Divelbiss DK, Holden KA, Barr RD, Fuke M (1988a) Clonal variation in the expression of human tumor necrosis factor (TNF) in the methylotrophic yeast *Pichia pastoris*, Presented at the 5th Annual Biotech USA Industry Conference and Exhibition, San Francisco

Sreekrishna K, Potenz R, Cruze JA, McCombie WR, Rarker KA, Nelles L, Mazzaferro PK, Holden KA, Harrison RG, Wood PJ, Phelps DA, Hubbard CE, Fuke M (1988b) High level expression of heterologous proteins in methylotrophic yeast *Pichia pastoris*. J Basic Microbil 28: 265–278

Sreekrishna K, Nelles L, Potenz R, Cruze J, Mazzaferro P, Fish W, Fuke M, Holden K, Phelps D, Wood P, Parker K (1989) High-level expression, purification, and characterization of recombinant human tumor necrosis factor synthesized in the methylotrophic yeast *Pichia pastoris*. Biochemistry 28: 4117–4125

Sreekrishna K, Barr KA, Hoard SA, Prevatt WD, Torregrosa RE, Levingston RE, Cruze JA, Wegner GH (1990) Expression of human serum albumin in *Pichia pastoris*. 15th Int Congr on Yeast Genetics and Molecular Biology, Hague, The Netherlands, Topic No 09-37B

Thill GP, Davis GR, Stillman GR, Holtz C, Brierly R, Engel M, Buckholz R, Kinney J, Provo S, Vedvick T, Siegel RS (1990) Positive and negative effects of multi-copy integrated expression vectors on protein expression in *Pichia pastoris*. In: Proc 6th Int Symp on Genetic of Industrial Microorganisms, Strasbourg, pp 477–490

Trimble RB, Atkinson PH, Tschopp JF, Townsend RR, Maley F (1991) Structure of oligosaccharides on *Saccharomyces SUC2* invertase secreted by the methylotrophic yeast *Pichia pastoris*. J Biol Chem 266: 22807–22817

Tschopp JF, Svelow G, Kosson R, Craig W, Grinna L (1987) High-level secretion of glycosylated invertase in the methylotrophic yeast, *Pichia pastoris*. Bio/Technology 5: 1305–1308

Veenhuis M, van Dijken JP, Harder W (1983) The significance of peroxisomes in the metabolism of one-carbon compounds in yeasts. Adv Microb Physiol 24: 1–82

Wagner SL, Siegel RS, Vedvick TS, Raschke WC, van Nostrand WE (1992) High-level expression, purification and characterization of the Kunitz-type protease inhibitor domain of protease nexin-2/amyloid β-protein precursor. Biochem Biophys Res Commun 186: 1138–1145

Wegner GH (1983) Biochemical conversions by yeast fermentation at high cell densities. US Patent 4: 329,414

Wegner GH (1990) Emerging applications of the methylotrophic yeasts. FEMS Microb Rev 87: 279–284

Wegner GH, Harder W (1986) Methylotrophic yeasts-1986. Microbial growth on C_1 compounds. In: Van Verseveld HW, Duine JA (eds) Proc 5th Int Symp Martinus Nijhoff Publishers, Dordrecht, pp 139–149

Yamada M, Azuma T, Matsuba T, Iida H, Suzuki H, Yamamoto K, Kohli Y, Hori H (1994) Secretion of human intracellular aspartic proteinase Cathepsine expressed in the methylotrophic yeast, *Pichia pastoris* and characterization of produced recombinant cathepsine. Bichom Biophy Acta 1206: 279–285

CHAPTER 7

Pichia guilliermondii

Andrei A. Sibirny

1
History of *Pichia guilliermondii* Research

Pichia guilliermondii Wickerham represents a collection of sporogenous strains which formerly belonged to the asporogenous species *Candida guilliermondii* (Cast.) Langeron a. Guerra (Wickerham and Burton 1954; Wickerham 1966; Kreger van Rij 1970). This means that each strain of *C. guilliermondii* which is able to hybridize with any strain of *P. guilliermondii* must be transferred to the latter species. For example, even the strain type *C. guilliermondii* ATCC 9058 must now be considered as *P. guilliermondii* (Sibirny et al. 1977b).

The genetics of *P. guilliermondii* was studied almost exclusively in the former "socialist countries", i.e., the former USSR (mostly in Lviv and Kiev, Ukraine), and the former GDR (in Greifswald). The interest in genetic studies of this species lies firstly in its ability to utilize hydrocarbons as sole carbon and energy source (Shchelokova et al. 1974) and especially as an industrial producer of single-cell protein from hydrocarbons. Formerly erronously identified in the USSR as *Candida guilliermondii*, it has now been reidentified as *Candida maltosa* (see Bykov et al. 1987). The second reason is the remarkable ability of *P. guilliermondii* strains to oversynthesize riboflavin (vitamin B_2) during growth in iron-deficient media (Shavlovsky and Logvinenko 1988b). Such features are known to be characteristics of a limited number of yeast species, and *P. guilliermondii* can be considered as the model organism for this group, due to its genetic qualities.

2
Physiology

Cells of *P. guilliermondii* are heterogenous, mostly elongate in shape (approx. $2 \times 10\mu$), sometimes forming a pseudomycelium (Kreger van-Rij 1970). The natural habitat of the species is diverse. A study of 140 strains of *P. guilliermondii* isolated from natural habitats showed that the most frequent source of their isolation is oil-

Division of Regulatory Cell System, Institute of Biochemistry, Ukrainian Academy of Sciences, Drahomanov Street, 14/16, Lviv 290005, Ukraine containing soil (123 strains); others were isolated from plant leaves, lake water, and cow paunch (Zharova et al. 1977; Zharova 1980).

All known strains do not utilize lactose, starch, and inosite, whereas they differ in their ability to utilize D-ribose, D-arabinose, D-cellobiose, D-melibiose, salicin, L-rhamnose, L-sorbose, and dulcite. All known strains utilize hydrocarbons (natural mixtures or n-hexadecane) as sole source of carbon and energy (Kreger van-Rij 1970; Zharova 1980). Growth on other respiratory substrates, such as ethanol, glycerol, succinate, or citrate, is satisfactory. *P. guilliermondii* is a typical representative of aerobic yeasts, and cannot grow under strictly anaerobic conditions.

The standard growth temperature for *P. guilliermondii* is 30 °C. The upper limit is near 42 °C.

Standard media for yeast cultivation can be used for laboratory cultivation of *P. guilliermondii* (YEPD or YEPS in which glucose is substituted by sucrose, are used as "complete" media). As minimal medium, modified Burkholder medium is used (Burkholder 1943; Shavlovsky et al. 1978). Other standard media can also be used in work with *P. guilliermondii* (Sibirny et al. 1977b).

The remarkable feature of almost all known strains of *P. guilliermondii* is their ability to overproduce riboflavin during cultivation in iron-deficient media (Demain 1972; Dikanskaya 1972; Shavlovsky et al. 1978; Shavlovsky and Logvinenko 1988a). Thus, 146 strains out of 147 analyzed were able to excrete a yellow pigment, identified as riboflavin, during cultivation in iron-deficient medium (Shavlovsky et al. 1978). One strain, which did not synthesize elevated quantities of riboflavin during iron starvation, apparently represents a species very similar to but still distinct from *P. guilliermondii*. This strain is able to mate with the type strain of this species but the resulting hybrids, in contrast to all other hybrids in the *P. guilliermondii* species, were not able to sporulate (Sibirny 1986). Thus, the ability to overproduce riboflavin and hence to accumulate yellow pigment in the cultural liquid during cultivation in iron-deficient media, is a species characteristic of *P. guilliermondii*.

3
Available Strains

The following strains have been used in genetic studies:

NRRL Y-2075, mating type *mat+*;
NRRL Y-2075, mating type *mat−*;
ATCC 9058 (indicated as *C. guilliermondii*) mating type *mat+*.

The last two strains were used for the isolation of a fertile genetic line of *P. guilliermondii* (Sibirny et al. 1977c).

German authors also used the strains *P. guilliermondii* 799 and 809 (both *mat−*) obtained from Dr. Zsolt (Hungary), a local strain H17 (*mat−*), and some others (Prahl et al. 1980).

The strains can be stored on wort agar slants at 4 °C for months.

4
Genetic Techniques

4.1
Life Cycle

All known natural or collection strains of *P. guilliermondii* are heterothallic (Wickerham and Burton 1954; Kreger van-Rij 1970; Shchelokova et al. 1974; Zharova et al. 1977). The mating types were designated *mat+* and *mat-* (Sibirny et al. 1977b). Homothallic natural or collection strains are not described, though some of meiotic segregants of the hybrids obtained by protoplast fusion manifest a homothallic phenotype (Sibirny 1986). Such a phenotype was designated as pseudohomothallic (*mat+,-*) as these strains segregate *mat+* and *mat-* clones during cultivation in complete synthetic medium. The pseudohomothallic phenotype was completely eliminated in the meiotic pedigree of *mat+,-*, × *mat+* or *mat+,-* × *mat-* hybrids (Sibirny 1986). It was hypothesized that such pseudohomothallic *mat+,-* strains are aneuploids.

4.2
Sexual Crosses

Cell conjugation between prototrophic strains of opposite mating types was observed on wort agar (Zharova et al. 1977). Study of sexual hybridization between auxotrophic mutants showed that most efficient matings occurred in solid media with sodium acetate or tomato juice (Sibirny et al. 1977b). Hybridization was not observed in complete media. Thus, optimal conditions for crossing of *P. guilliermondii* appeared to be poor starvation media. Later, such a conclusion found support during the investigation of hybridization conditions for other yeast species. It was shown that yeast species found in natural habitats as haploids (haplonts) efficiently cross in poor media, whereas diplontic species (found in nature as diploids) hybridize predominantly in rich media (Naumov et al. 1980, 1981; Vustin 1981).

The following procedure of sexual hybridization is generally used for *P. guilliermondii* (Sibirny et al. 1977b).

Auxotrophic strains of opposite mating types with complementary nutritional requirements are grown as a streak on complete YEPD or YEPS media. The strains are crossed on plates with acetate medium (sodium acetate, 1%; KCl, 0.5%), incubated for 2–3 days, and then are replica-plated onto minimal medium. The prototrophic hybrids are formed at the contact sites of the streak cross.

4.3
Protoplast Fusion

Hybridization can also be obtained by protoplast fusion (Klinner et al. 1980; Sibirny et al. 1982; Klinner and Böttcher 1984). In this case, hybrids were obtained between auxotrophs belonging to opposite or to the same mating type. The maximal frequency of hybridization by protoplast fusion was near 2×10^{-2} (Sibirny et al. 1982).

The protocol for protoplast isolation and fusion in *P. guilliermondii* is given in Sect. 4.4.

In addition to intraspecific *P. guilliermondii* hybrids, protoplast fusion was used for isolation interspecific hybrids with *Pichia kudriavzevii* and *Hansenula polymorpha* (Sibirny et al. 1982; Kashchenko et al. 1987).

Sexual or protoplast fusion hybrids of *P. guilliermondii* appeared to be very stable mononuclear diploids during growth on synthetic or complete media, and did not sporulate under such conditions (Zharova 1980; Klinner and Böttcher 1984; Sibirny 1986). It is interesting to note that ploidy of the protoplast fusion hybrids never exceeded the diploid level (Klinner and Böttcher 1984; Büttner et al. 1985). It was suggested that protoplast fusion hybrids appeared to be aneuploids in many cases.

Hybrids of *P. guilliermondii* were able to sporulate on acetate media. Most hybrids produced only one to two spores after 5–6 days of incubation at room temperature, but some pairs of strains gave diploids, which produced up to 40% asci with spores (Sibirny et al. 1977b; Sibirny 1986). Incubation at 30 °C depressed sporulation. Five other media tested did not induce sporulation.

4.4
Protocol for Isolation and Fusion of Protoplasts
(see also Chap. 2, this Vol.)

Cells are cultivated in sugar-mineral medium containing yeast extract (0.5%), 0.05 potassium phosphate buffer, pH 6.0, and growth factors for auxotrophs (40 µg/ml). Cultivation is run for 16–24 h to middle exponential growth phase (cell mass 0.7–1.0 mg dry weight/ml). Cells are sedimented by centrifugation and washed twice with water.

Cells (50–100 mg dry weight/ml) are incubated in the following mixture: 0.05 M Tris-HCl, pH 7.0; 0.4 M $CaCl_2$; 0.01 M dithiothreitol; Zymolyase 20 000 (2 mg/ml; Seikagaku Corp.) or β-Glucuronidase (10–15 mg/ml; Sigma) for 30–40 min at 37 °C with periodic mixing. Control of protoplasting is monitored using phase contrast microscopy. Yield of protoplasts usually reached 100%.

Protoplasts are separated from incubation medium by centrifugation at 2000 rpm for 5 min at 4 °C, twice washed with cold (4 °C) 0.4 M $CaCl_2$, and then are resuspended in the same solution at 7.5 mg dry weight/ml.

Suspensions of a pair of auxotrophic mutants containing 50×10^6 of protoplasts/ml are mixed in a 1:1 ratio, sedimented by centrifugation, and resuspended in the medium inducing protoplast fusion (25% solution of polyethylene glycol with molecular weight 6000, containing 0.1 M $CaCl_2$ in 0.05 M Tris-HCl buffer, pH 8.6). Final concentration of the protoplasts of both strains is 50×10^6/ml mixture. The obtained suspension is incubated for 20 min at 30 °C with periodic mixing.

The protoplast mixture is spread onto a surface of an agar medium containing an osmotic stabilizer (1 M sucrose or 1 M sorbitol). Any additional pouring of top-overlaid mild agar is not necessary for *P. guilliermondii*.

4.5
Analysis of Meiotic Segregants

Asci of hybrids contain as a rule two spores, sometimes one-spored asci appeared, while three- or four-spored asci were seldom found. Asci have oval or elongate forms, while spores are characterized by round or hat-shaped forms (Sibirny et al. 1977b). Preferential production of two-spored asci is the result of degeneration of several nuclei formed during meiosis (Zharova 1980).

Meiotic segregants of sexual hybrids which produced abundant amounts of spores frequently gave diploids characterized by low spore frequency. A genetic line, i.e. haploid strains hybrids of which produced a large amount of spores, was selected by inbreeding, using sister crosses between several consecutive pedigrees of meiotic segregants (Sibirny et al. 1977c). The strains denied from this genetic line easily crossed and sporulated, producing up to 60% of two-spored asci. Unfortunately, strains which appeared to be capable of producing four-spored asci were not isolated. Thus, tetrad analysis is impossible for *P. guilliermondii*. Electron microscopic studies showed that, during meiosis, diploid cells formed three or four nuclei but only part of them were surrounded by a spore envelope and formed spores (Zharova 1980). Apparently the other nuclei degenerated.

Several methods of elimination of vegetative diploid cells for random spore analysis were developed. The method, based on selective killing of diploid vegetative cells by elevated temperature (55 and 60 °C), was unsuccessful. More appropriate appeared to be the method based on eliminating vegetative cells by vaseline oil or killing diploid cells by ethanol or diethyl ester (Sibirny et al. 1977b,c; Sibirny 1986). The most suitable method, which used 20% ethanol to eliminate non-sporulated hybrid cells for random spore analysis, is given in Sect. 4.6.

Survival of spores of *P. guilliermondii* genetic line is equal to 86% (Sibirny et al. 1977c). Most frequently, UV-irradiation at a 1–10% survival rate was used for mutation induction. Mutants were also selected using several chemical mutagens, e.g., nitrosoguanidine (Sibirny 1986).

4.6
Protocol for Random Spore Analysis

A sporulating diploid cell suspension (2–2.5 mg dry weight/ml) is incubated at 30 °C for 2 h in Helicase (β-Glucuronidase) solution (10–15 mg/ml) for digestion of asci envelopes. Then the suspension is gently homogenized in a glass homogenizer to separate the spores, diluted to a cell concentration of approx. 0.6 mg dry weight/ml, and 1/5 part (by volume) of ethanol is added.

The suspension is incubated with permanent shaking for 10–12 min and, after dilution with 20% ethanol, is spread onto YEPD medium.

5
Chromosomes, Genes, and Genetic Markers

5.1
Pulsed Field Electrophoresis

Pulsed field electrophoresis of *P. guilliermondii* chromosomes revealed six to seven chromosomal bands, depending on the electrophoretic parameters (Fig. 1; A.A. Sibirny and P. Philippsen, unpubl. observ.).

The sizes of chromosomes were estimated by calibrating the pulsed field gels with accepted chromosome sizes of *S. cerevisiae* (Link and Olson 1991) and *Ashbya gossypii* (K. Gaudenz, Diploma Thesis, Biocenter of Basel University, 1994). The individual sizes of *P. guilliermondii* chromosomes are 590, 1140, 1220, 1450, 1800, 2050, and 2150 kb.

Thus, the minimal calculated genome size of *P. guilliermondii* is 10 400 kb.

5.2
Genetic Mapping

Genetic mapping was conducted using random spore analysis and benomyl-induced mitotic segregation. Auxotrophic mutants defective in synthesis of amino acids, purines, and vitamins were used. Six linkage groups in *P. guilliermondii* were identified (N. Prahl, pers. comm.). Centromeres or genes tightly linked to the centromeres have not been reported. Together, 21 genetic markers were localized on the genetic map of *P. guilliermondii* (N. Prahl, pers. comm.). Unfortunately, the impossibility of conducting tetrad analysis has hampered genetic mapping in this species.

6
DNA Isolation and Transformation

6.1
Isolation of Chromosomal DNA and Construction of a Gene Bank

An overnight culture of *P. guilliermondii* grown in YEPD medium is sedimented by centrifugation, resuspended in the mixture of 20 ml 0.1 M Tris-sulfate buffer, pH 9.3; 2 ml 0.5 M EDTA, pH 9.0; 0.5 ml 2-mercaptoethanol, and incubated for 15 min at room temperature.

Cells are centrifuged, resuspended in 20 ml 0.05 M Tris-HCl buffer, pH 7.8, containing 0.15 M NaCl and 0.05 M EDTA; 20 ml 2% solution of sodium dodecylsulfate and Proteinase K (50 μg/ml) are added to this suspension and incubated first for 10 min at 65 °C and then for 30 min at 37 °C.

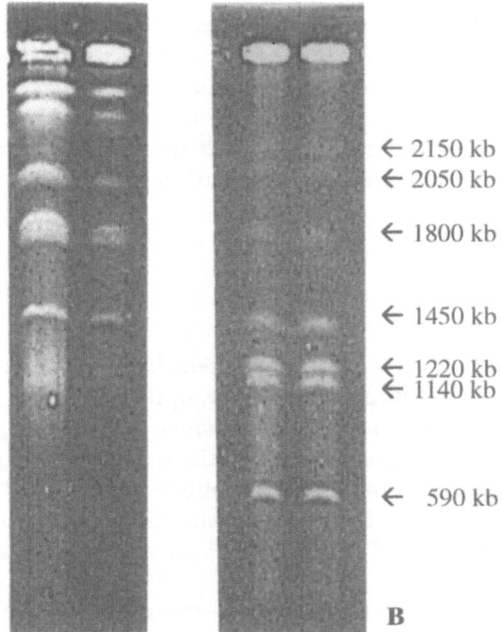

Fig. 1A,B. Pulsed-field gel electrophoretic separation of chromosomal DNA of *P. guilliermondii*. The new procedure for accelerated DNA isolation (A.A. Sibirny and P. Philippsen, unpubl.) was used. A stationary yeast culture after 2 days cultivation in YPD medium (strain of genetic line L2) was washed with 50 mM EDTA, pH 8.0, and resuspended for protoplasting in the following mixture (approx. 300 mg dry weight per 3 ml reaction mixture): 0.05 M Tris-HCl, pH 7.0; 0.4 M CaCl$_2$; 0.01 M dithiothreitol; Zymolyase Seikagaku Corp. 20 000 (2 mg/ml). After 30–40 min of incubation at 37 °C with periodic mixing, most cells were converted to protoplasts. Protoplasts were mixed with 5 ml of 1% low-melting agarose (42 °C) in 0.125 M EDTA, pH 8.0 and poured in to a 4-cm petri dish. After agarose hardening, the following mixture was added onto the surface of agarose layer: 0.5 M EDTA, pH 8.0; 0.01 M Tris; lauroylsarcosinate (10 mg/ml); proteinase K (2 mg/ml). The plate was incubated at 52 °C for 4 h. The conditions used for pulse-field gel electrophoresis in a Rotaphor apparatus (Biometra) were as following. A Voltage: 185 V (constant); angle: 110° to 95° (logarithmic change); pulse time: 120 to 42 s (linear change); countdown: 24 h; temperature: 12 °C; gel: 1% agarose; buffer: 0.025 M Tris-borate-EDTA, pH 8.0. B Voltage: 150 to 120 V (linear change); angle: 120°C (constant); pulse time: 210 to 190 s (linear change); countdown: 36 h; temperature: 12°C; gel: 0.9% agarose; buffer: 0.025 M Tris-borate-EDTA, pH 8.0

Proteins are extracted by adding an equal volume of phenol/chlorophorm/isoamyl alcohol mixture, and DNA is precipitated from the water phase by addition of an equal volume of isopropanol. The DNA precipitate is dissolved in 10 ml of 0.01 M Tris-HCl buffer, pH 8.0, containing 0.001 M EDTA. The DNA solution is dialyzed against the same buffer for 16 h. DNA concentration in such preparation is near 0.1 mg/ml.

Isolated DNA is partially digested with Sau3A endonuclease. The obtained fragments are fractionated using preparative electrophoresis in 1% agarose. The fragments of 6 to 15 kb in size are eluted from the gel by electroelution and ligated with a BamHI digested pFL38 shuttle vector (see Fig. 2) treated with calf intestinal alkaline phosphatase. The ligated DNA mixture is used for transformation of *E. coli* C600. Approximately 10^4 independent recombinant pFL38 plasmids were isolated as *P. guilliermondii* gene bank.

6.2
Transformation

P. guilliermondii does not contain its own plasmids. A gene bank of this species was constructed using Sau3A-digested chromosomal DNA ligated into the BamHI site of the shuttle vector pFL38 (Zakalsky et al. 1990; Logvinenko et al. 1993). The protocol for chromosomal DNA isolation from *P. guilliermondii* cells is given below. Together, 10 000 independent recombinant plasmids were obtained after transformation of *Escherichia coli* C600. Schemes for construction of recombinant plasmids containing structural genes *RIB1* and *RIB7* of *P. guilliermondii* are presented in Figs. 2 and 3.

Isolated recombinant plasmids were used for transformation of *P. guilliermondii* by the lithium method (Ito et al. 1983). The genes RIB1 and RIB7 coding GTP cyclohydrolase and riboflavin synthase, respectively, were used as selective markers; recipient strains contained *rib1* or *rib7* mutations (Logvinenko et al. 1993; A. Voronovsky and G. Shavlovsky, unpubl. observ.). Other standard selective markers for yeast transformation (amino acids, pyrimidine biosynthesis, G418 resistance) were have not yet been used. The *P. guilliermondii* transformation yields 80–200 transformants/µg DNA. The fate of transforming DNA in the cells is not known. Most *RIB1* transformants (70–80%) were stable, whereas a minority was unstable and segregated after spreading on complete medium 44–88% of auxotrophic *rib1* clones (A. Voronovsky and G. Shavlovsky, pers. comm.). Stable transformants appeared to be integrants, while unstable cells possibly contain autonomously replicating vectors. One may assume that some of the *Sacharomyces cerevisiae* portion of the shuttle vector or the cloned fragment of *P. guilliermondii* contain sites acting as replicator sequence in *P. guilliermondii*. The plasmid copy number in unstable transformants is not known.

Transformation of *P. guilliermondii* adenine auxotrophs (*ade2*) by the circular bireplicon vector pYE(ADE2)2 containing the *ADE2* gene of *S. cerevisiae* was also performed in another laboratory (Neistat et al. 1986), though the frequency of transformation using the same method was very low (one to three transformants per µg DNA) for unknown reasons. All transformants in this case were stable.

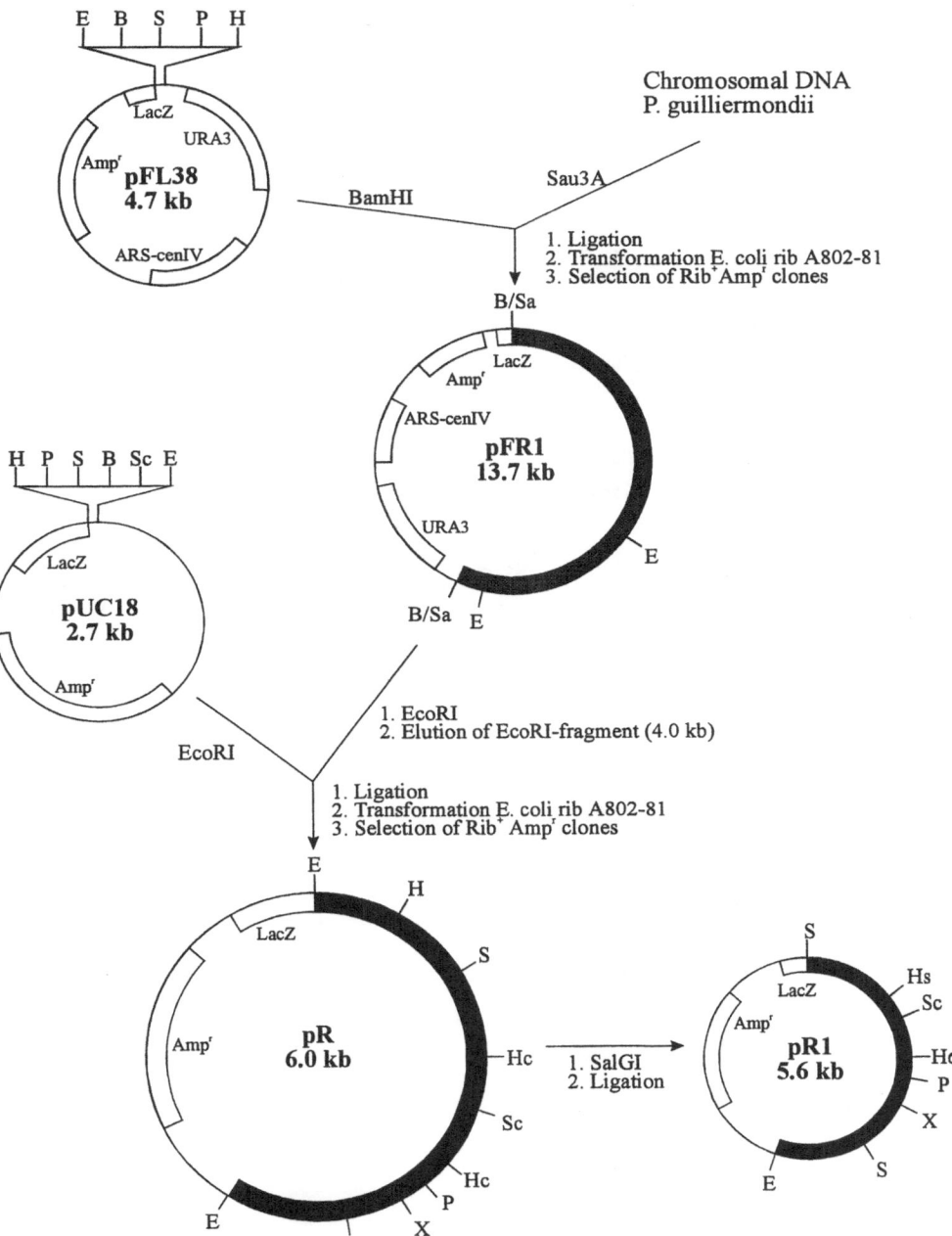

Fig. 2. Scheme showing the construction of recombinant plasmids containing the *RIB1* gene of *P. guilliermondii*. Sites of restriction endonucleases: *B* BamHI; *E* EcoRI; *H* HindIII; *Hc* HincII; *P* PstI; *S* SalGI; *Sa* Sau3A; *Sc* SacI *X* XhoI

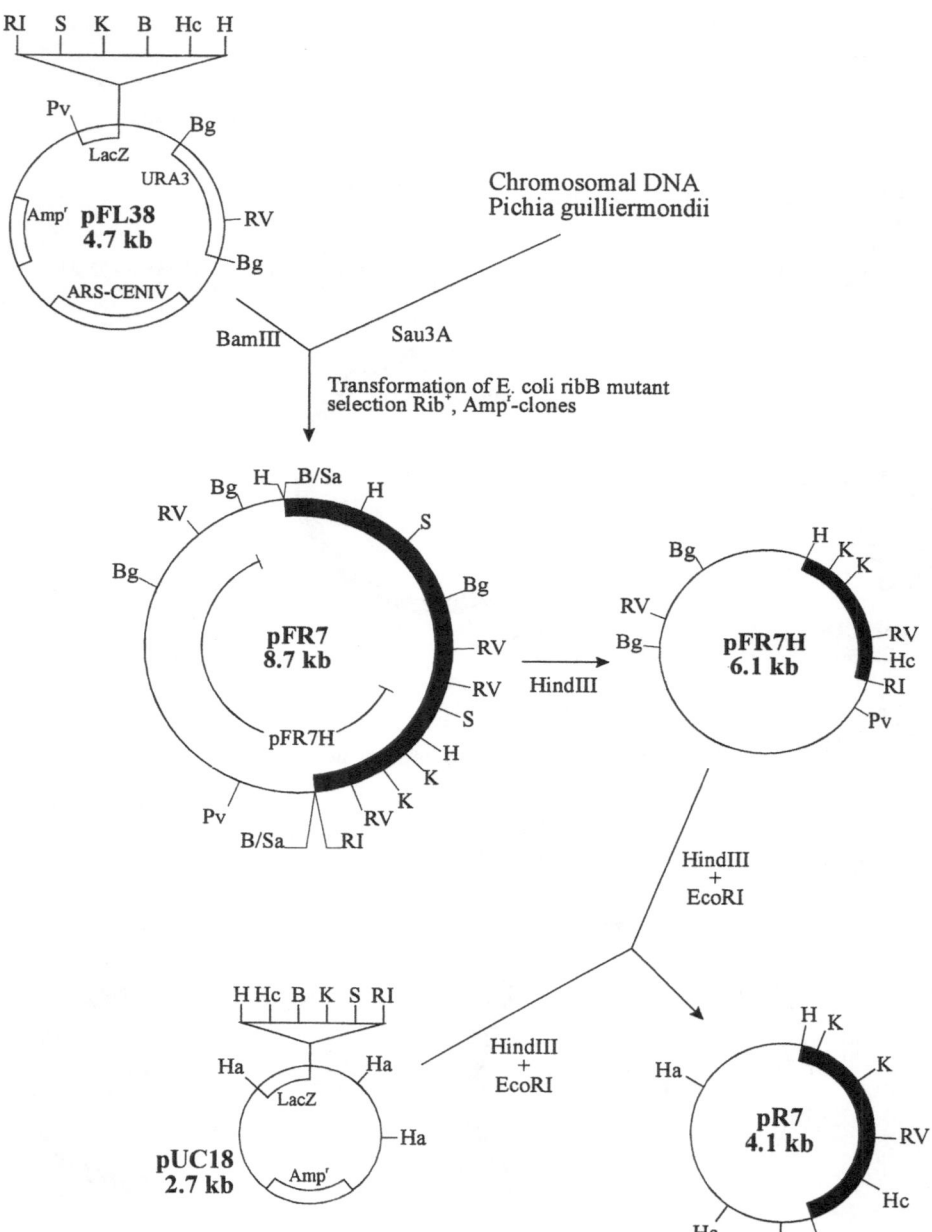

Fig. 3. Scheme showing the construction of recombinant plasmids containing the *RIB7* gene of *P. guilliermondii*. Sites of restriction endonucleases: *B* BamHI; *Bg* BglII; *RI* EcoRI; *RV* EcoRV; *H* HindIII; *Ha* HaeII; *Hc* HincII; *K* KpnI; *Pv* PvuII; *S* SacI; *Sa* Sau3A

7
Biochemical Genetics

7.1
Hydrocarbon Utilization

Mutants unable to utilize hydrocarbons as sole source of carbon and energy were isolated from the strains of a genetic line of *P. guilliermondii*. They were designated *alc* (*alc*ane nonutilizable) mutants. Together, several tens of *alc* mutants were selected, using nitrosoguanidine as mutagen. Approximately half of them were unable to grow on glycerol, others lost the ability to utilize only hydrocarbons (Zharova 1980; Zharova et al. 1980). Genetic analysis of 22 specific *alc* mutations showed that they are all recessive and can be divided into at least three complementation groups (*alc1*, *alc2*, *alc3*). The biochemical lesions of the mutants were not studied.

7.2
Riboflavin Biosynthesis

As mentioned before, *P. guilliermondii* is the only genetically studied eukaryote able to regulate riboflavin biosynthesis. This organism was used for identification of strutural and regulatory genes involved in flavinogenesis.

Using strains of opposite mating types of a *P. guilliermondii* genetic line, 114 riboflavin-deficient mutants were isolated after UV irradiation. Biochemical study based on analysis of accumulated riboflavin intermediates divided them into five biochemical groups. Complementation analysis of 106 mutants revealed 7 complementation classes or genes (*RIB1* to *RIB7*). The strains of the biochemical group, accumulating no specific products, corresponded to complementation group *rib1*; group II, accumulating 2,4,5-triaminopirimidine to complementation group *rib2*; group III, accumulating 2,6-dihydroxy-4-ribitylaminopyrimidine, to complementation group *rib3*; the mutants of group IV, accumulating 2,6-dihydroxy-5-amino-4-ribitylaminopyrimidine, were divied into three complementation groups, *rib4*, *rib5*, and *rib6*; the mutants of group V, accumulating 6,7-dimethyl-8-ribityl lumazine, corresponded to complementation group *rib7* (Shavlovsky et al. 1979). Biochemical and genetic studis of *P. guilliermondii* riboflavin auxotrophs showed similarity in riboflavin biogenesis between this species and *S. cerevisiae* (Oltamnns et al. 1969). *P. guilliermondii* appeared to be a much more convenient species for study of riboflavin biosynthesis when compared with *S. cerevisiae* due to the ability to overproduce riboflavin or its precursors in iron-deficient media (Sibirny 1986; Shavlovsky and Logvinenko 1988a).

Later, detailed enzymological analysis of *P. guilliermondii* riboflavin auxotrophs was carried out. Such analysis, by the way, helped to identify new enzymes, unknown at that time, that participate in riboflavin biosynthesis. It was found that the gene *RIB1* encodes the enzyme of the first step of flavinogenesis, GTP cyclohydrolase (Shavlovsky et al. 1976, 1980a); the gene *RIB2* encodes the enzyme of the second step, 2,5-diamino-6-hydroxy-4-ribosylaminopyrimidine-5′-phosphate reductase (Shavlovsky et al. 1981); the gene *RIB3* encodes the enzyme of the third reaction, 2,5-diamino-6-hydroxy-4-ribytilaminopyrimidine-5′-phosphate deaminase (Nielsen and Bacher 1981). The gene *RIB5* encodes 6,7-dimethyl-8-ribityllumazine synthase (Logvinenko et al. 1985, 1987) and the gene *RIB6* encodes the synthesis of the aliphatic precursor of the pteridine structure of 6,7-dimethyl-8-ribityllumazine, 3,4-dihydroxy-2-butanone-4-phosphate, which is produced from ribulose-5-phosphate (Logvinenko et al. 1982, 1985, 1987; Volk and Bacher 1988, 1990; Bacher 1991). The gene *RIB7* encodes riboflavin synthase (Logvinenko et al. 1987). The exact role of the gene *RIB4* was not elucidated. A gene encoding the enzyme dephosphorylating 2,4-dihydroxy-5-aminopyrimidine-5'-phosphate (specific phosphatase?) has not yet been identified. Most structural genes involved in riboflavin biosynthesis in *P. guilliermondii* are not linked (Shavlovsky et al. 1979). Thus, the study of riboflavin auxotrophs of *P. guilliermondii* allowed elucidation of the biochemistry of a new metabolic pathway, i.e., the riboflavin biosynthesis in microorganisms. A scheme of riboflavin biosynthesis is given in Fig. 4.

As mentioned above, the genes *RIB1* and *RIB7* of *P. guilliermondii* were cloned by complementation of the corresponding mutations of *E. coli* and used for transformation of *P. guilliermondii rib1* and *rib7* mutants (Zakalsky et al. 1990; Logvinenko et al. 1993; A. Voronovsky, pers. comm.). Cloned *RIB1* and *RIB7* genes were also used for heterologous transformation of riboflavin-deficient mutants of *Candida famata* (this asporogenous yeast species, similarly to *P. guilliermondii*, is able to overproduce riboflavin in iron-deficient media).

The frequency of *H. polymorpha rib7* mutant transformation was very low. Maximal frequency was observed in the experiments with a linearized *RIB7* gene contained in the pR7 plasmid (up to 40 transformants/μg DNA). Riboflavin synthase activity of transformants was very low. Mitotic stability of *H. polymorpha* transformants was also rather low, and was eliminated in nonselective media with riboflavin. Apparently, the cloned *RIB7* gene is located in an autonomously replicating plasmid dut to the fact that some fragments of the vector can be used as replicator in *H. polymorpha* cells (A. Voronovsky, G. Shavlovsky, pers. comm.).

Frequency of transformation of the *rib1* mutant of *C. famata* with the cloned *RIB1* gene of *P. guilliermondii* was from 50 (circular plasmid) to 600 (HindIII-EcoRI linearized fragments) transformants/μg DNA. All transformants of *C. famata* appeared to be stable, and produced elevated amounts of riboflavin in iron-deficient media (A. Voronovsky and G. Shavlovsky, pers. comm.).

Starvation by iron causes riboflavin oversynthesis and derepression of enzymes coding by *RIB1*, *RIB3*, *RIB5*, *RIB6*, and *RIB7* genes in *P. guilliermondii*, while the *RIB2* gene product, the enzyme reductase, is produced constitutively (Shavlovsky and Logvinenko 1988b). The enzymes involved in biosynthesis of flavin coenzymes, riboflavin kinase and FAD synthetase, are also insensitive to regulation by iron content in *P. guilliermondii* (Shavlovsky et al. 1975; Shavlovsky and Fedorovich 1977).

Mutants affected in the regulation of flavinogenesis and of corresponding enzyme syntheses by iron ions were isolated in *P. guilliermondii*. Mutatios *rib80* and

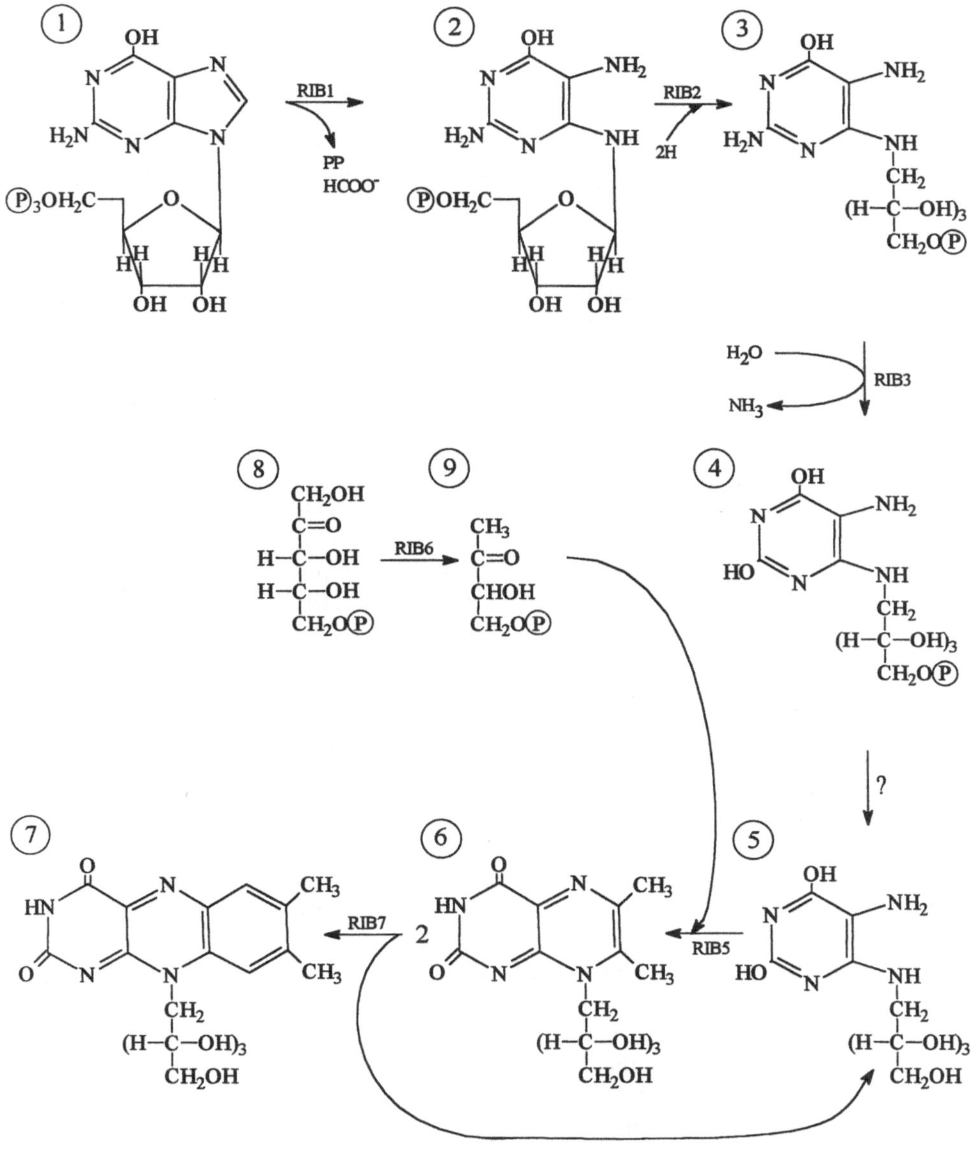

Fig. 4. Scheme of riboflavin biosynthesis in the yeast *P. guilliermondii*. *I* GTP; *II* 2,5-diamino-4-hydroxy-6-ribosylaminopyrimidine-5'-phosphate; *III* 2,5-diamino-4-hydroxy-6-ribitylaminopyrimidine-5'-phosphate; *IV* 2,4-dihydroxy-5-amino-6-ribitylaminopyrimidine-5'-phosphate; *V* 2,4-dihydroxy-5-amino-6-ribitylaminopyrimidine; *VI* 6,7-dimethyl-8-ribityl-lumazine; *VII* riboflavin; *VIII* ribulose-5-phosphate; *IX* 3,4-dihydroxy-2-butanone-4-phosphate. *RIB1* to *RIB7* strutural genes of riboflavin biosynthesis

rib81 appeared to be recessive, and caused oversynthesis of riboflavin and derepression of the corresponding enzymes in iron-rich media (Shavlovsky et al. 1982, 1985a,b).

Several approaches were used for isolation of derepressed mutants. One of them is based on selection of the mutants resistant to the riboflavin analogue, 7-methyl-8-trifluoromethyl-10-(1'-D-ribityl) isoalloxazine. Since wild-type strains of *P. guilliermondii* are resistant to riboflavin analogues due to slow penetration into the cell, mutants with multiple sensitivity to antibiotics and antimetabolites (Sibirny et al. 1977d, 1978b) were used as initial strains for selection (Shavlovsky et al. 1980b). Besides, it was found that high concentrations of sulfate or phosphate anions induce increasing sensitivity to antibiotics and antimetabolites (including riboflavin analogues) of wild-type strains of *P. guilliermondii* and of other yeast species (Sibirny and Shavlovsky 1981); it was possible to isolate analogue-resistant mutants directly from wild-type strains on high sulfate concentration (Shavlovsky et al. 1985a). Additionally, derepressed mutants were isolated from previously selected ts *rib1* mutants unable to grow at the restrictive temperature in iron-rich media without derepression of the thermolabile GTP cyclohydrolase, as in pseudorevertants able to grow in such conditions (Shavlovsky et al. 1985a; Sibirny 1986). A third method for isolation of riboflavin-derepressed mutants uses specially constructed leaky *rib2* auxotrophs as initial strains. Activity of the second enzyme of flavinogenesis in the leaky *rib2* strain was so low that they grew without exogenous riboflavin only in iron-deficient media, where other enzymes were derepressed. Selection of leaky *rib2* mutants capable of growing in iron-rich medium without riboflavin allowed the isolation of regulatory mutants (Shavlovsky et al. 1982). As a result, the regulatory genes *RIB80* and *RIB81* were identified, which appeared to be unlinked. Interallelic complementation was found in the *RIB80* gene (Shavlovsky and Logvinenko 1988a). It was suggested that the *RIB80* and *RIB81* genes together encode a heterooligomeric regulatory protein complex of negative action (apoprotein) which is activated after binding of iron ions and represses transcription of the corresponding structural genes (Shavlovsky et al. 1982; Sibirny 1986).

The regulatory mutants *rib83* and *rib84*, unable to overproduce riboflavin in iron-deficient medium, were also isolated (Shavlovsky et al. 1989). They were selected as mutants which have lost the ability to excrete riboflavin in iron-deficient medium. Apparently, the corresponding genes encode regulatory proteins of positive control. Double *rib80 rib83* and *rib81 rib83* mutants were unable to overproduce riboflavin, which suggests an epistasis of the *RIB83* gene over the genes *RIB80* and *RIB81*. Possibly, genes of positive and negative types of regulation form a cascade system for transmission of the regulatory signal. The molecular mechanisms of the action of identified regulatory genes remain unknown. It is interesting to note that *rib80* mutants manifest simultaneously a derepression of the iron transport system. One may assume that the regulatory gene *RIB80* regulates the expression of riboflavin biosynthesis structural genes and of gene(s) responsible for uptake of the corepressor of riboflavin production, the iron ions (Shavlovsky et al. 1985b; Shavlovsky and Logvinenko 1988b).

The cloned *RIB1* and *RIB7* structural genes of *P. guilliermondii* were efficiently expressed in the heterologous *E. coli* system, though iron did not regulate synthesis of yeast enzymes in bacterial cells (Zakalsky et al. 1990; Logvinenko et al. 1993). This means that *P. guilliermondii* promoters are functioning efficiently in *E. coli* and the corresponding enzymes are properly folded in the bacteria. However, bacterial regulatory iron-dependent proteins (e.g., involved in siderophore biosynthesis) do not recognize specific sites in yeast *RIB1* and *RIB7* promoters. In contrast to *E. coli*, heterologous expression of the *RIB7* gene in *H. polymorpha* was inefficient, and was not regulated by the iron ion content of the medium (A. Voronovsky and G. Shavlovsky, pers. comm.). Heterologous expression of the *RIB1* gene of *P. guilliermondii* in *C. famata* cells was efficient, and some transformants possessed an up to threefold elevated riboflavin production compared with the *C. famata* wild-type strain. Biosynthesis of riboflavin was regulated by iron deficiency in *C. famata* transformants (A. Voronovsky and G. Shavlovsky, pers. comm.). One may suggest that *C. famata* possesses regulatory sequences homologous to *P. guilliermondii* which regulate riboflavin structural gene expression by iron ions.

7.3
Riboflavin Transport

Most of the wild-type strains and the riboflavin-deficient mutants of *P. guilliermondii* are incapable of riboflavin-mediated transport from the medium (Sibirny et al. 1977d; Shavlovsky et al. 1977). Such ability is inherent, however, to some riboflavin-deficient mutants of this species able to grow in the media containing very low (0.1 to 0.3 μg/ml) concentrations of riboflavin and to prototrophs isolated from them. Various riboflavin-deficient mutants possessed riboflavin uptake systems with different properties (riboflavin permeases I and II; Shavlovsky et al. 1985b). The mechanism of evolution of these transport systems, which are cryptic in the wild-type strains, is not known. It is worth mentioning that the mutants capable of riboflavin permease activity have evolved during adaptation of riboflavin-deficient mutants of *P. guilliermondii* to low concentration of riboflavin in the medium; such mutants were never discovered after mutagenesis in nonselective media. In addition to riboflavin permeases, *P. guilliermondii* possesses a riboflavin-excreting system which is present also in the wild-type strains (Sibirny et al. 1977d, 1978a).

Riboflavin permease II and riboflavin excretase are apparently synthesized constitutively. Riboflavin permease I is characterized by peculiar properties. It catalyzes riboflavin uptake against a concentration gradient and accumulation of huge concentrations of vitamin (up to 20 mg/g dry cells) which is crystallized inside the vacuoles. Synthesis of this transport system occurs only in the presence of α-glucosides; induction is depressed by translation and transcription inhibitors (Sibirny et al. 1977d, 1978a, 1979; Shavlovsky and Sibirny 1985).

Coordinate regulation of riboflavin permease I and α-glucosidase was found (Sibirny et al. 1979). Mutants constitutively synthesizing riboflavin permease and α-glucosidase were selected as riboflavin auxotrophs capable of growing at low riboflavin concentrations in glucose medium (without α-glucosides). Genetic analysis revealed three unlinked regulatory genes: *RFP80*, *RFP81* of negative action (recessive mutations *rfp80*, *rfp81*), and the gene *RFP82* of positive action (dominant mutations *RFP82*). The recessive mutations *rfp82* were isolated as meiotic segregants from an intragenic recombination between two *RFP82* alleles or as mutants resistant to riboflavin analogues in a medium with the riboflavin permease inducer sucrose. Interallelic complementation was found within the *RFP80* and *RFP82* loci, and the corresponding maps were constructed (Sibirny and Shavlovsky 1984a; Sibirny 1986). The epistasis-hypostasis test showed that the gene *RFP82* acts after the gene *RFP80*. A model for the action of regulatory products of the identified genes, which form a cascade system in expression of riboflavin permease and α-glucosidase, was presented (Sibirny and Shavlovsky 1984a).

8
Biotechnological Applications

Even the most active mutants of *P. guilliermondii* produce much less riboflavin than the fungi *Eremothecium ashbyii* and *Ashbya gossypii* or mutant strains of the yeast *C. famata* or the bacterium *Bacillus subtilis* (Demain 1972; Shavlovsky and Logvinenko 1988a; Heefner et al. 1988, 1993).

By combining the mutations which lead to riboflavin oversynthesis in iron-rich media and the mutation which specifically impairs the riboflavin-excreting system, mutant strains of *P. guilliermondii* were isolated that accumulated during growth in synthetic or industrial media high amounts of riboflavin (up to 12 mg/g dry cells, i.e., 500 times more than the normal content of riboflavin in the yeast cells; Sibirny and Shavlovsky 1984b; Sibirny 1986). Excretion of vitamin B_2 into the medium was negligible. Such mutants may be of interest for production of fodder riboflavin preparations as dried yeast biomass, because drying of yeast pellets may be cheaper than evaporation and drying of microbial biomass together with cultural liquid (all known industrial producers of riboflavin excrete most of riboflavin into the cultural medium).

The mutants of *P. guilliermondii* capable of active riboflavin transport efficiently accumulate this vitamin from diluted solutions inside the cells. It is possible to use such mutants for concentration of riboflavin from cultural liquids of weak or moderate producers which accumulate up to 100 μg riboflavin/ml in the medium (Sibirny et al. 1984). It is suggested that obtaining of riboflavin-enriched yeast biomass as a source of riboflavin would be a much cheaper procedure than other possible methods for concentration of this vitamin from diluted solutions.

The availability of mutant strains capable of riboflavin uptake from the medium and accumulation inside the cells allowed a new method of riboflavin kinase assay to be developed (Kashchenko et al. 1991). It was found that riboflavin permease is absolutely unable to catalyze flavin mononucleotide uptake (Sibirny et al. 1977d). This observation was the basis for removing riboflavin which is not converted into flavin mononucleotide by the riboflavin kinase reaction. After finishing the reaction, the mixture is incubated for 1 h with washed cells possessing riboflavin permease activity, and all the riboflavin is taken up by the cells. The concentration of flavin mononucleotide is determined by direct fluorescent analysis of the reaction mixture after cell removal by centrifugation or filtration. The method is much simpler, cheaper, and more convenient than other methods used before.

The riboflavin-deficient mutant *P. guilliermondii*, 9i, which possesses riboflavin permease activity and is able to grow at very low concentration of exogenous riboflavin (growth at 0.005 μg of riboflavin/ml), allows the development of a new auxonographic method of a microbiological assay of vitamin B_2. The sensitivity of the method is similar to that known for *Lactobacillus casei*, but the yeast method is more convenient due to the much simpler nutritional requirements of *P. guilliermondii* (Sibirny 1986; A. Sibirny, unpubl. observ.). The immediate precursor of riboflavin, 6,7-dimethyl-8-ribityllumazine, which is used for reasearch, has not yet been produced commercially. The strain of *P. guilliermondii* which contains mutations leading to riboflavin oversynthesis in iron-rich medium and leaky *rib7* (riboflavin synthase) deficiency, has been patented. This mutant is able to grow in minimal medium without riboflavin and to accumulate nearly 100 μg of 6,7-dimethyl-8-ribityllumazine per ml in the cultural liquid. The produced lumazine can be used as substrate of the riboflavin synthase reaction and for other purposes (Sibirny and Shavlovsky 1984c).

9
Concluding Remarks

Today, *P. guilliermondii* seems to be an attractive model for studying the genetic control of riboflavin biosynthesis and transport in yeasts. The main obstacles for further investigations are (1) the impossibility of conducting tetrad analysis due to production of two-spored asci and (2) the insufficiently known molecular genetics of *P. guilliermondii*, including the absence of convenient vectors and even of standard selective markers for transformation. Most difficult appears the isolation of strains producing four-spored asci. Hybrids of some pairs of *P. guilliermondii* natural strains rarely produce tetrads (A. Sibirny, unpubl. obser.). They may be used for isolation of corresponding genetic lines.

Selection of more active *P. guilliermondii* riboflavin producers is possible, but will demand a strong increase in specific activity of GTP cyclohydrolase, which apparently limits flavinogenesis and the enhancement of this process by the purine precursor, GTP.

Acknowledgments. The work on chromosome separation of *P. guilliermondii* was supported by a grant from Swiss National Science Foundation (No. 5002-41029) awarded to P. Philippsen.

References

Bacher A (1991) Biosynthesis of flavins. In: Müller F (ed) Chemistry and biochemistry of flavoenzymes. CRC Press, Boca Raton, vol 1, pp 215–259

Burkholder PR (1943) Influence of some environmental factors upon the production of riboflavin by yeast. Arch Biochem 3: 121–130

Büttner M, Klinner U, Birnbaum D, Böttcher F (1985) Auswirkungen von Mutanten-induktion und Protoplastenfusion auf den DNA-und Proteingehalt von Pichia guilliermondii. J Basic Microbiol 25: 13–19

Bykov VA, Manakov MN, Panfilov BI, Svitsov AA, Tarasova NV (1987) Production of protein compounds. In: Yegorov NS, Samuilov VD (eds) Biotekhnologiya, vol 5. Vysshaya Shkola, Moscow (in Russian)

Demain AL (1972) Riboflavin oversynthesis. Annu Rev Microbiol 26: 369–388

Dikanskaya EM (1972) Flavin biosynthesis in microorganisms. In: Advances in science and technology. Series Virology and microbiology. USSR Acad Sci 3: 5–55 (in Russian)

Heefner DL, Weaver CA, Yarus MJ, Burdzinsky LA, Gyure DC, Foster EW (1988) Riboflavin producing strains of microorganisms, methods for selecting, and method for fermentation. International Patent, No W088/09822

Heefner DL, Boyts A, Burdzinsky L, Yarus M (1993) Efficient riboflavin production with yeast. United States Patent No 5231007

Ito H, Fukuda Y, Murata K, Kimura A (1983) Transformation of intact yeast cells treated with alkali cations. J Bacteriol 153: 161–168

Kashchenko VE, Shavlovsky GM, Kutsiaba VI (1987) Interspecific and intergeneric hybridization of methylotrophic yeast Hansenula polymorpha by protoplast fusion. In: Abstr XII Int Spec Symp on Yeasts. Genetics of Non-conventional Yeasts. Weimar, p 13

Kashchenko VE, Preobrazenskaya EN, Sibirny AA (1991) The method for determination of riboflavin kinase. Soviet Patent (Author's Certificate) No 1631089 (in Russian)

Klinner U, Böttcher F (1984) Hybridization of yeasts by protoplast fusion: ploidy level of hybrids resulting from fusions in haploid strains fo Pichia guilliermondii. Z Allg Mikrobiol 24: 533–537

Klinner U, Böttcher F, Samsonova IA (1980) Hybridization of Pichia guilliermondii by protoplast fusion. In: Ferenczi L, Farkas GL (eds) Advances in protoplast research. Academiai Kiado, Budapest, pp 113–118

Kreger-van Rij NJW (1970) Genus 15. Pichia Hansen. In: Lodder J (ed) The yeasts. A taxonomic study, 2nd edn. North-Holland Publ, Amsterdam, pp 455–458

Kvasnikov EI, Zharova VP, Shchelokova IF (1984) Taxonomic characteristics of the yeast Pichia guilliermondii. Mikrobiologiya 53: 305–307 (in Russian)

Link AJ, Olson MV (1991) Physical map of the Saccharomyces cerevisiae genome at 110-kilobase resolution. Genetics 127: 681–698

Logvinenko EM, Shavlovsky GM, Zakalsky AE, Zakhlodylo IV (1982) Biosynthesis of 6,7-dimethyl-8-ribityllumazine in yeast extracts of Pichia guilliermondii. Biokhimiya 47: 931–936 (in Russian)

Logvinenko EM, Shavlosky GM, Tsarenko NY (1985) The proteins of 6,7-dimethyl-8-ribityllumazine system of Pichia guilliermondii yeast and regulation of their biosynthesis. Biokhimiya 50: 744–748 (in Russian)

Logvinenko EM, Shavlovsky GM, Kontorovskaya NY (1987) On biochemical functions of the products of genes RIB5 and RIB6 involved in riboflavin biosynthesis in the yeast Pichia guilliermondii. Genetika 23: 1699-1701 (in Russian)

Logvinenko EM, Stasiv YZ, Zlochevsky ML, Voronovsky AY, Beburov MY, Shavlovsky GM (1993) Cloning of the RIB7 gene encoding the riboflavin synthase of the yeast Pichia guilliermondii. Genetika 29: 922–927 (in Russian)

Naumov GI, Vustin MM, Babieva IP (1980) Sexual divergence of yeast genera Williopsis Zender, Zygowilliopsis Kudriavzev and Hansenula H. et P. Sydow. Dokl Akad Nauk SSSR 255: 468–471 (in Russian)

Naumov GI, Vustin MM, Naumova TI (1981) Hybridologic study of the yeast species *Zygowilliopsis californica, Williopsis saturnus, W. beierinckii* comb. nov., *W. mrakii* comb. nov. Dokl Akad Nauk SSSR 259: 718–722 (in Russian)

Neistat MA, Alenin VV, Tolstorukov II (1986) Transformation of *Hansenula polymorpha, Pichia guilliermondii, Williopsis saturnus* yeasts with the plasmid-containing *ADE2* gene of *Saccharomyces cerevisiae.* Mol Genet Mikrobiol Virusolog 12: 19–23 (in Russian)

Nielsen P, Bacher A (1981) Biosynthesis of riboflavin. Characterization of the product of the deaminase. Biochim Biophys Acta 662: 312–317

Oltmanns O, Bacher A, Lingens F, Zimmermann FK (1969) Biochemical and genetic classification of riboflavine-deficient mutants of *Saccharomyces cerevisiae.* Mol Gen Genet 105: 306–313

Prahl N, Samsonova IA, Bottcher F (1980) Sexuelle Hybridisierung von *Pichia guilliermondii.* Wiss Z E-M-Arndt-Universitat Greifswald. Math-Naturwiss Reihe 29: 63–64

Shavlovsky GM, Fedorovich DV (1977) The activity of enzymes involved in synthesis and hydrolysis of flavin adenine dinucleotide in *Pichia guilliermondii* studied at different level of flavinogenesis. Mikrobiologiya 46: 904–911 (in Russian)

Shavlovsky GM, Logvinenko EM (1988a) Flavin overproduction and its molecular mechanisms is microorganisms. Prikl Biokhim Mikrobiol 24: 435–447 (in Russian)

Shavlovsky GM, Logvinenko EM (1988b) The role of iron in the regulation of protein synthesis by microorganisms. Usp Sovr Biol 29: 108–133 (in Russian)

Shavlovsky GM, Sibirny AA (1985) Riboflavin transport in yeasts and its regualtion. In: Kulaev IS, Dawes EA, Tempest DW (eds) Environmental regulation of microbial metabolism. Academic Press, London, pp 385–392

Shavlovsky GM, Logvinenko EM, Strugovshchikova LP, Kashchenko VE (1975) Biosynthesis of flavins and its regulation in the yeast *Pichia guilliermondii.* Ukr Biokhim Zh 47: 649–660 (in Russian)

Shavlovsky GM, Logvinenko EM, Kashchenko VE, Koltun LV, Zakalsky AE (1976) Detection of the enzyme of the first step of flavinogenesis, GTP cyclohydrolase, in the yeast *Pichia guilliermondii.* Dokl Akad Nauk SSSR 230: 1485–1487 (in Russian)

Shavlovsky GM, Sibirny AA, Ksheminskaya GP (1977) Permease and "excretase" for riboflavin in the mutants of *Pichia guilliermondii* yeast. Biochem Physiol Pflanz 171: 139–145

Shavlovsky GM, Zharova VP, Shchelokova IF, Trach VM, Sibirny AA, Ksheminskaya GP (1978) Flavinogenic activity of natural strains of the yeast *Pichia guilliermondii.* Prikl Biokhim Mikrobiol 14: 184–189 (in Russian)

Shavlovsky GM, Sibirny AA, Kshanovskaya BV, Koltun LV, Logvinenko EM (1979) Genetic classification of the riboflavin – deficient mutants of the yeast *Pichia guilliermondii.* Genetika 15: 1561–1568 (in Russian)

Shavlovsky GM, Logvinenko EM, Benndorf R, Koltun LV, Kashchenko VE, Zakalsky AE, Schlee D, Reinbothe H (1980a) First reaction of riboflavin biosynthesis – catalysis by a guanosine triphosphate cyclohydrolase from yeast. Arch Microbiol 124: 255–259

Shavlovsky GM, Sibirny AA, Ksheminskaya GP, Pinchuk GE (1980b) Oversynthesis of riboflavin in the mutants of the yeast *Pichia guilliermondii* resistant to 7-methyl-8-trifluoromethyl-10-(1'-D-ribityl) isoalloxazine. Mikrobiologiya 49: 702–707 (in Russian)

Shavlovsky GM, Logvinenko EM, Sibirny AA, Fedorovich DV, Zakalsky AE (1981) The activity of 2,5-diamino-6-hydroxy-4-ribosylaminopyrimidine-5'-phosphate reductase, an enzyme of the second step of flavinogenesis, studied in the yeast *Pichia guilliermondii.* Mikrobiologiya 50: 1008–1011 (in Russian)

Shavlovsky GM, Sibirny AA, Fedorovich DV, Senyuta EZ (1982) Selection and properties of the mutants of *Pichia guilliermondii* yeast with derepressed enzyme of the first step of flavinogenesis, GTP cyclohydrolase. Mikrobiologiya 51: 96–101 (in Russian)

Shavlovsky GM, Babyak LY, Sibirny AA, Logvinenko EM (1985a) Genetic control of riboflavin biosynthesis in yeast *Pichia guilliermondii.* Detection of the new regulatory gene *RIB81.* Genetika 21: 368–374 (in Russian)

Shavlovsky GM, Fedorovich DV, Logvinenko EM, Koltun LV (1985b) Isolation and characterization of flavinogenic strains *Pichia guilliermondii* manifesting regulatory mutation *rib80/ribR*. Mikrobiologiya 54: 919–926 (in Russian)

Shavlovsky GM, Koltun LV, Kshanovskaya BV, Logvinenko EM, Stenchuk MM (1989) Regulation of biosynthesis of riboflavin by elements of the positive control in *Pichia guilliermondii* yeast. Genetika 25: 250–258 (in Russian)

Shchelokova IF, Zharova VP, Kvasnikov EI (1974) Obtaining of the hybrids in haploid strains of *Pichia guilliermondii* Wickerham assimilating oil hydrocarbons. Mikrobiol Zh 34: 275–278 (in Ukrainian)

Sibirny AA (1986) Genetic control of biosynthesis and transport of riboflavin in the yeast *Pichia guilliermondii*. Doctor of Biological Sciences Thesis. Leningrad State University (in Russian)

Sibirny AA, Shavlovsky GM (1981) Potentiation of sensitivity of yeasts and bacteria to some inhibitors and riboflavin due to the action of a high sulfate and phosphate concentrations. Mikrobiologiya 50: 242–248 (in Russian)

Sibirny AA, Shavlovsky GM (1984a) Identification of regulatory genes of riboflavin permease and α-glucosidase in the yeast *Pichia guilliermondii*. Curr Genet 8: 107–144

Sibirny AA, Shavlovsky GM (1984b) Strain *Pichia guilliermondii* ss16–8 accumulating a large amounts of riboflavin inside the cells. Soviet Patent (Authors Certificate) No 207914 (in Russian)

Sibirny AA, Shavlovsky GM (1984c) The method for preparing of 6,7-dimethyl-8-ribithyllumazine. Soviet Patent (Authors Certificate) No 1092952 (in Russian)

Sibirny AA, Shavlovsky GM, Goloshchapova GV (1977a) The mutants of the yeast *Pichia guilliermondii* with multiple sensivity to antibiotics and antimetabolites. Selection and some properties of the mutants. Genetika 13: 872–879 (in Russian)

Sibirny AA, Shavlovsky GM, Kshanovskaya BV, Naumov GI (1977a) Hybridization and meiotic segregation in the paraffin-utilizing yeast *Pichia guilliermondii* Wickerham. Genetika 13: 314–321 (in Russian)

Sibirny AA, Shavlovsky GM, Ksheminskaya GP, Orlovskaya AG (1977d) Active riboflavin transport in the yeast *Pichia guilliermondii*. Detection and properties of the cryptic riboflavin permease. Biokhimiya 42: 1841–1851 (in Russian)

Sibirny AA, Zharova VP, Kshanovskaya BV, Shavlovsky GM (1977c) Selection of genetic line of the yeast *Pichia guilliermondii* capable of producing large amounts of spores. Tsitologiya i Genetika 11: 330–333 (in Russian)

Sibirny AA, Ksheminskaya GP, Shavlovsky GM (1978a) The mutants of the yeast *Pichia guilliermondii* impaired in permeability barrier. In: Molekularnaya Biologiya. Naukova Dumka, Kiev 19: 42–46 (in Russian)

Sibirny AA, Shavlovsky GM, Ksheminskaya GP, Orlovskaya AG (1978b) The influence of glucose and some its derivates on the systems for uptake and excretion of riboflavin in yeast *Pichia guilliermondii*. Biokhimiya 43: 1414–1422 (in Russian)

Sibirny AA, Shavlovsky GM, Ksheminskaya GP, Orlovskaya AG (1979) Coordinate regulation of riboflavin permease and α-glucosidase synthesis in the yeast *Pichia guilliermondii*. Biokhimiya 44: 1558–1568 (in Russian)

Sibirny AA, Shavlovsky GM, Kshanovskaya BV, Kutsiaba VI (1982) Intraspecific and interspecific hybridization of yeasts by protoplast fusion. In: Molekularnaya Biologiya. Naukova Dumka, Kiev 32: 16–24 (in Russian)

Sibirny AA, Trach VM, Shavlovsky GM (1984) The strain of the yeast *Pichia guilliermondii* 72–6 capable of efficient riboflavin accumulation from solutions. Soviet Patent (Authors Certificate) No 209240 (in Russian)

Volk R, Bacher A (1988) Biosynthesis of riboflavin. The structure of the four-carbon precursor. J Am Chem Soc 110: 3651–3653

Volk R, Bacher A (1990) Studies on the 4-carbon precursor in the biosynthesis of riboflavin. Purification and properties of L-3,4-dihydroxy-2-butanone-4-phosphate synthase. J Biol Chem 265: 19479–19485

Vustin MM (1981) Taxonomic study of the soil yeasts with Saturn-like spores. PhD Thesis, Moscow State University (in Russian)

Wickerham LJ (1966) Validation of the species *Pichia guilliermondii*. J Bacterol 92: 1269–1273

Wickerham LJ, Burton KA (1954) A clarification of relationship of *Candida guilliermondii* to other yeasts by a study of their mating types. J Bacteriol 68: 594–597

Zakalsky AE, Zlochevsky ML, Stasiv YZ, Logvinenko EM, Beburov MY, Shavlovsky GM (1990) Cloning of gene *RIB1* encoding the enzyme of the first stage of the yeast *Pichia guilliermondii* flavinogenesis, GTP cyclohydrolase, in *Escherichia coli*. Genetika 26: 614–620 (in Russian)

Zharova VP (1980) Taxonomi and genetic study of hydrocarbon-utilizing yeast *Pichia guilliermondii* Wickerham. PhD Thesis Institute of Genetics and Selection of Industrial Microorganisms, Moscow (in Russian)

Zharova VP, Shchelokova IF, Kvasnikov EI (1977) Genetic study of the hydrocarbon-utilizing yeast *Pichia guilliermondii* Wickerham. Identification of haploid cultures by mating types and hybrid selection. Genetika 13: 309–313 (in Russian)

Zharova VP, Kvasnikov EI, Naumov GI (1980) Obtaining and genetic analysis of *Pichia guilliermondii* Wickerham mutants non-assimilating hexadecane. Mikrobiol Zh 42: 167–171 (in Russian)

CHAPTER 8

Pichia methanolica (Pichia pinus MH4)

Andrei A. Sibirny

1
History of *Pichia methanolica* Research

Pichia methanolica was first described by Kato and coworkers (Kato et al. 1974). The yeast strain used in the laboratory of Dr. I.I. Tolstorukov (Moscow) for thorough genetic study was isolated in the Institute of Biotechnology, Leipzig, Germany, and was formerly identified by German scientists as *Pichia pinus*, strain MH4 (Tolstorukov 1994; see below).

Three species of methylotrophic yeasts were used for genetic studies: *Pichia pastoris*, *Hansenula polymorpha*, and *P. methanolica* (*P. pinus* MH4). The study of the last species began in 1977 (Tolstorukov et al. 1977) and is being conducted mainly in the laboratory of Dr. I.I. Tolstorukov in the Scientific Center for Microbial Genetics and Bioengineering in Moscow. Biochemical genetics of *P. methanolica* is being studied mainly in the Lviv Division of the Institute of Biochemistry and the St. Petersburg State University. Thus, the genetic study of *P. methanolica* (*P. pinus* MH4) is almost exclusively carried out in the former Soviet Union.

What is the need for genetic study of *P. methanolica*? This question is quite reasonable, as the two other species above are used in molecular genetic investigations. The answer is as follows: only the *P. methanolica* system permits a convenient and useful genetic analysis using the methods of classical genetics, including gene mapping. As a result, *P. methanolica* today is the genetically best studied species of methylotrophic yeasts, and its genetic map includes more than 20 markers localized at 4 chromosomes and 1 fragment.

2
Physiology

P. methanolica MH4 cells are spherical to oval. This organism, similar to other methylotrophic yeast species, seems to be essentially aerobic. It assimilates methanol as well as many polycarbon substrates. Optimal temperature for growth is 30 °C. For routine laboratory studies, the basic culture conditions and media are

Division of Regulatory Cell Systems of the Institut of Biochemistry, Ukrainian Academy of Sciences, Drahomanov Street, 14/16, Lviv 290005, Ukraine the same as those generally used for *Saccharomyces cerevisiae* (see also Tolstorukov et al. 1977).

3
Available Strains

Most available strains were isolated from the wild-type strain MH4 in the laboratory of Dr. I.I. Tolstorukov. Most mutants defective in regulation of C1-metabolism were isolated in the laboratory of the author of this chapter.

A total of several hundreds of auxotrophic mutants (mono-, di- and triauxotrophs) were isolated. Mutants defective in utilization of C_2 compounds, as well as those impaired in mating type switching, were selected (Tolstorukov 1988). Nitrosoguanidine appeared to be the most efficient mutagen. Some mutants were isolated using UV irradiation and 8-hydroxylaminopurine as mutagens.

4
Genetic Techniques

P. methanolica is homothallic: haploid cells are able to perform self-diploidization in minimal medium. Diploid cells are able to propagate for a long period in minimal medium. After transfer into Rg medium, abundant sporulation (up to 90% asci, most of which contained four spores) occurs.

Diploid cells are seldom able to cross with haploid or diploid cells with production of tri- and tetraploids. Vegetatively growing diploid cells can segregate chromosomes due to spontaneous haploidization. This process is sufficiently induced by γ-irradiation and, under optimal conditions, up to 10% of cells appear to be aneuploids (Tolstorukov and Benevolensky 1980, 1981, 1982). Segregating aneuploids form slow-growing colonies. This feature facilitates identification of aneuploids, and was used for localization of the markers on the chromosomes (see below).

5
Chromosomes

Pulsed-field gel electrophoresis performed using a CHEF system (Countour Clamped Homogenous Electrophoresis Field) clearly distinguishes four DNA bands for strains *P. methanolica* (*P. pinus*) MH4 and for the type strain *P. methanolica* NRRL Y-7685, whereas the type strain *P. pinus* NRRL Y-11582 shows up to nine chromosomal bands (Tolstorukov 1994). The chromosomal sizes of *P. methanolica* MH4 and NRRL Y-7685 are identical and are equal to 6, 4.2, 3.6 and 3.1 Mb (see also Chap. 3, this Vol.).

It was shown that the cloned *ADE1*, *ADE5*, and *SUP2* genes of *P. methanolica* MH4 used as probes yield positive Southern blot hybridization patterns with separate DNA bands, identical for both *P. methanolica* MH4 and NRRL Y-7685 strains (Tolstorukov 1994).

6
Genes and Genetic Markers

6.1
Meiosis

Auxotrophic mutants isolated from MH4 and NRRL Y-7685 strains freely crossed and resulted in diploids sporulating on Rg medium, producing a high percentage of four-spored asci. Tetrad analysis of several interstrain hybrids showed a regular $2^+ : 2^-$ segregation. Contrary to this, no hybrids between *P. methanolica* (*P. pinus*) MH4 and the type strain *P. pinus* NRRL Y-11582 were ever obtained. It was concluded that strain MH4 represents the species *P. methanolica* (Tolstorukov 1994).

6.2
Nomenclature

The basic rules used for locus designation in *S. cerevisiae* have been adopted for *P. methanolica* MH4: a locus is written by three letters followed by a locus number. Locus numbers are provisory and are often different from corresponding equivalents of *S. cerevisiae*.

6.3
Induced Haploidization

Linkage groups in *P. methanolica* MH4 were identified using two methods: induced haploidization and tetrad analysis.

Mitotic loss of chromosomes was induced by γ-rays. Aneuploids were picked as slow-growing small colonies. It was found that the markers localized in different linkage groups segregate independently, and those localized in the same linkage group, even far one from another, show linkage (Tolstorukov et al. 1983).

6.4
Tetrad Analysis

Mapping by tetrad analysis was done for about 30 markers (auxotrophic mutations, mating type locus mutations, mutations affecting C_1- and C_2 compound utilization). As a result, four centromeres (chromosomes) and one chromosomal fragment (apparently the distal part of one of these linkage groups) were identified (Tolstorukov and Efremov 1984; Tolstorukov 1988). The data of genetic mapping are in good agreement with the results of pulsed-field gel electrophoresis, which identified four DNA bands (see above). The genetic map of *P. methanolica* MH4 is presented in Fig. 1.

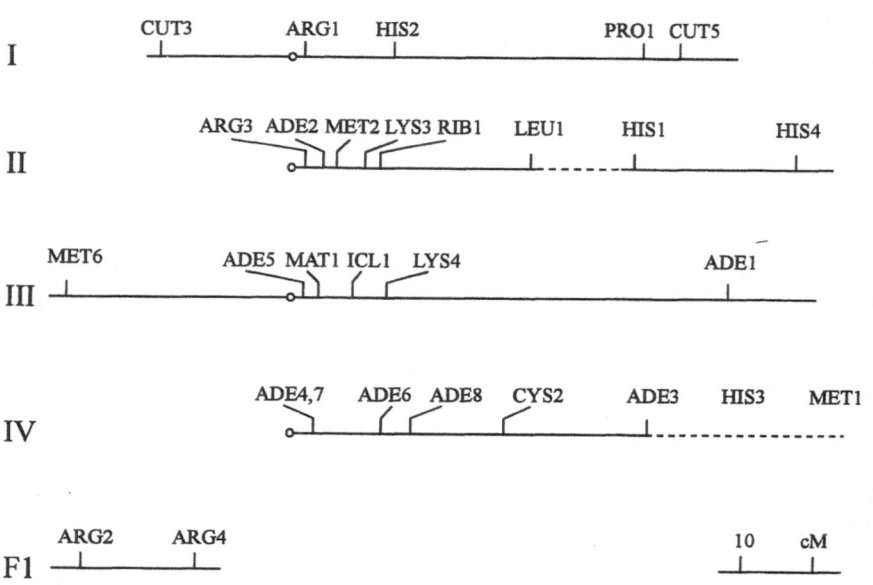

Fig. 1. Genetic map of the yeast *P. methanolica* MH4. *Continuous lines* indicate linkage determined by tetrad analysis; *dotted lines* indicate linkage determined by induced haploidization. *Circles* designate centromeres

7
Transformation

7.1
Transformation Procedure

As transformation markers, *ade1*, *ade2*, and *leu1 P. methanolica* MH4 mutations were used which correspond to *ade2*, *ade1*, and *leu2* mutations of *S. cerevisiae*. *S. cerevisiae* episomal vectors pYE (ADE2)2, pYE(ADE1) (Sasnauskas et al. 1987) and the standard YEp13 vector, containing the *LEU2* gene, were used as source of selectable genes. Transformation was performed using the lithium method (Ito et al. 1983). No transformants were obtained using genes of purine biosynthesis (Neistat et al. 1987). When YEp13 was used as transforming vector, *leu1* mutants of *P. methanolica* MH4 were trnasformed, and leucine prototrophs appeared. It was shown that the transformation frequency is increased when cells from early exponential growth phase were used. Deproteinization and adding of carrier DNA also stimulated transformation. Maximal frequency of transformation using the intact YEp13 vector was near to 20 transformants per μg of DNA (Tarutina and Tolstorukov 1994). Linearization of the YEp13 plasmid at the unique BamHI or XhoI sites increased the transformation frequency by more than one order. An increase of two further orders of transformation frequency was achieved using small linear fragments of the YEp13 plasmid after digestion with BglII or a mixture of XhoI and SalGI (sizes of the fragments were 3.0 and 2.2 kb, respectively). Circularization of such fragments diminished the frequency of transformation by more than one order (Tarutina and Tolstorukov 1994).

Two types of transformants were obtained: large colonies with stable Leu⁺ phenotype, and small or moderate colonies which were unstable and segregated Leu⁻ cells on complete medium. Stable transformants appeared to be integrants, whereas unstable transformants contained sequences of smaller size than the vector YEP13, which could replicate autonomously in *P. methanolica*. It was suggested that the *LEU2* gene of *S. cerevisiae* can fulfil replicator function in *P. methanolica* cells similar to the situation postulated earlier for *Hansenula polymorpha Leu⁺* transformants (Berardi and Thomas 1991).

Very similar results were obtained using the previously cloned *ADE1* gene from *P. methanolica* MH4 (homologous to the *ADE2* gene of *S. cerevisiae*). In this case the transformation frequency of *ade1* mutants of *P. methanolica* MH4 was also sufficiently increased after linearization of the vector (Hiep et al. 1992).

7.2
Molecular Cloning of SUP2 *and* ADE1

Two *P. methanolica* MH4 structural genes were cloned by functional complementation of the corresponding *S. cerevisiae* mutations, *SUP2* and *ADE1*.

SUP2 is an omnipotent suppressor, coding for an EF-1a-like protein factor involved in translational accuracy (Surguchov et al. 1984). Cloning of the *P. methanolica SUP2* gene was based on the construction of a genomic library of *P. methanolica* chromosomal DNA using partial digestion, with Sau3A insertion into the YEp13 vector and selection of *S. cerevisiae leu2 sup2* (ts) double revertants growing in leucine-free medium at the restrictive (36 °C) temperature (Kushnirov et al. 1990). A similar approach was used for *ADE1* gene cloning of *P. methanolica* MH4 (Hiep et al. 1991).

The complete nucleotide sequence of the *P. methanolica* MH4 *SUP2* gene was determined and compared with that of *S. cerevisiae*. The comparison of deduced amino acid sequences of both proteins revealed high conservation (76%) of the C-terminal part region homologous to EF-1a, and low conservation (36%) of the N-terminal part (Kushnirov et al. 1990).

Codon usage of both *SUP2* genes appeared to be similar. A single exception may be leucine, for which codon TTA is more frequent in *SUP2* of *P. methanolica* MH4, whereas in *SUP2* of *S. cerevisiae* TTG is preferred (Kushnirov et al. 1990).

8
Genetic Control of Mating

Although all cells of *P. methanolica* MH4 can be crossed in each combination, the efficiency of copulation was higher in some combinations than in others. By this criterion, all strains were divided into two groups or mating types designated a and

α or *MAT1a* and *MAT1α*. Tetrad analysis showed in all cases monogenic segregation 2a:2α (Tolstorukov and Benevolensky 1978). It was shown that unbudded cells keep their initial mating type, while budding cells undergo mating type switching. In contrast to *S. cerevisiae*, such switching has accidental character: this means that daughter cells can possess either the parental or the opposite mating type. The gene *HTH1* was identified by *hth1* mutation, which, similarly to the *ho* allele of *S. cerevisiae*, decreases switching of mating type in *P. methanolica* MH4 (Benevolensky and Tolstorukov 1980). A similar phenotype is manifested in the *mat1α-46* mutation (Tolstorukov et al. 1981, 1982). It was suggested that *MAT1* locus consists of two regions interacting between each other, and with hypostatic genes controlling mating, meiosis, and sporulation.

9
Biochemical Genetics of Purine Biosynthesis

A large collection of Ade$^-$ mutants of *P. methanolica* was isolated. Complementation analysis divided them into seven groups: *ade1*, *ade2* (pink), *ade3* to *ade7* (white). Tetrad analysis showed that most mutations are not linked, while the *ade4* and *ade7* mutations are mapped at the same locus designated *ade4.7* and do not recombine (Tolstorukov et al. 1987).

Among the pink Ade-mutants, most belonged to *ade1*. Interallelic complementation was observed in the *ADE1*, but not in the *ADE2* locus. Mutants in *ADE1* were devoid of AIR carboxylase and mutants in *ADE2* were deficient in SAICAR synthetase (Tolstorukov et al. 1987). Thus, the gene *ADE1* of *P. methanolica* MH4 is homologous to the *ADE2* gene of *S. cerevisiae* and vice versa.

10
Genetic Control of Methanol Metabolism

It appeares that *P. methanolica* (*P. pinus*) MH4 uses the same enzymes of methanol oxidation and assimilation as the other species of methylotrophic yeasts (Veenhuis et al. 1983; Sibirny et al. 1986a, 1988; Motruk et al. 1989; Titorenko et.al. 1990a).

Methanol is oxidized by the air oxygen to formaldehyde; the reaction is catalyzed by alcohol oxidase and leads to the formation of H_2O_2, which is decomposed by catalase. Formaldehyde is oxidized by the NAD- and glutathione-dependent formaldehyde dehydrogenase to S-formylglutathione. The latter compound is either oxidized by the NAD-dependent formate dehydrogenase, leading to the formation of CO_2 and regeneration of free glutathione, or formaldehyde is hydrolyzed, first by the formylglutathione hydrolase to glutathione and formate, which is then oxidized to CO_2 by the formate dehydrogenase. Formaldehyde is the substrate of a specific transketolase, the dihydroxyacetone synthase, which catalyzes the transfer of a glycolaldehyde moiety from xylulose-5-phosphate to formaldehyde with the formation of glyceraldehyde-3-phosphate and dihydroxyacetone. The latter compound is phosphorylated by the corresponding kinase. After aldolic condensation of the two triose phosphates molecules, fructose-1, 6-bisphosphate is formed, which is converted into fructose-6-phosphate by a specific bis-phosphatase. Three molecules of xylulose-5-phosphate are regenerated from two molecules of fructose-6-phosphate and one molecule of triose phosphate, in the reactions of the nonoxidative part of the pentose phosphate cycle, catalyzed by transaldolase, transketolase, ribose-5-phosphate isomerase, and ribulose-5-phosphate epimerase (rearrangement reactions). The formaldehyde-assimilating reactions form the so-called xylulose monophosphate (dihydroxyacetone) cycle. As a result, one molecule of triose phosphate is synthesized from three molecules of formaldehyde and ATP. Formaldehyde is reduced to methanol in an NAD-dependent formaldehyde reductase reaction (apparently the inverse reaction of alcohol dehydrogenase). Alcohol oxidase, catalase, and dihydroxyacetone synthase are localized in special microbodies, peroxisomes, which can occupy up to 80% of the cellular volume during methylotrophic growth. Other enzymes of methanol metabolism appear to be cytosolic (Veenhuis et al. 1983; Harder and Veenhuis 1989). The scheme of methanol metabolism in yeasts is given in Fig. 2.

It is generally accepted that the energy supply of yeast methylotrophic growth occurs exclusively in a direct pathway of formaldehyde oxidation to CO_2 (Veenhuis et al. 1983). This view is not in agreement with some more recent observations, e.g., that 2-fluoroacetate inhibits growth of the wild-type strain *P. methanolica* MH4 and that malonate blocks growth of the *icl1* mutant deficient in isocitrate lyase on solid media with methanol (and also with glycerol or succinate), but not with glucose (Sibirny et al. 1986a). It was found that the ability of malonate to inhibit growth on methanol-containing medium correlates with the *icl1* marker. This was observed, when segregants of an *ICL1* × *icl1* hybrid were analyzed (Sibirny et al. 1990). These, as well as other observations (including isolation of formaldehyde- and formate dehydrogenase-deficient mutants), permitted an alternative hypothesis, according to which the main role in energy supplying of yeast methylotrophic growth belongs to the xylulose monophosphate pathway of formaldehyde assimilation plus the tricarboxylic acid cycle (Sibirny 1987; Sibirny and Ubiyvovk 1988; Sibirny 1990; Sibirny et al. 1990; Sibirny and Gonchar 1990).

Methanol metabolism is tightly regulated at the level of enzyme activities, enzyme synthesis, and degradation (Sahm 1977; Veenhuis et al. 1983). In the experiments with wild-type strains it became obvious that methanol, formaldehyde, and formate induce the synthesis of the enzymes involved in oxidation and assimilation of methanol. Multicarbon substrates, especially glucose and ethanol, strongly repress such induction. Similarly, one-carbon compounds induce biogenesis of peroxisomes, whereas glucose and ethanol repress this process (Veenhuis et al. 1983; Harder and Veenhuis 1989).

Glucose and ethanol added to methanol-grown cells, in addition to repression, cause degradative catabolite inactivation of peroxisomal enzymes (alcohol oxidase, catalase, dihydroacetone synthase) and degradation of peroxisomes, apparently due to vacuole fusion with peroxisomes and ensuing proteolysis of peroxisomal constituents (Bormann and Sahm 1978; Veenhuis et al. 1983; Hill et al. 1985; Tuttle et al. 1993). The genes involved in catabolite repression and catabolite inactivation of methanol-metabolizing enzymes were not identified.

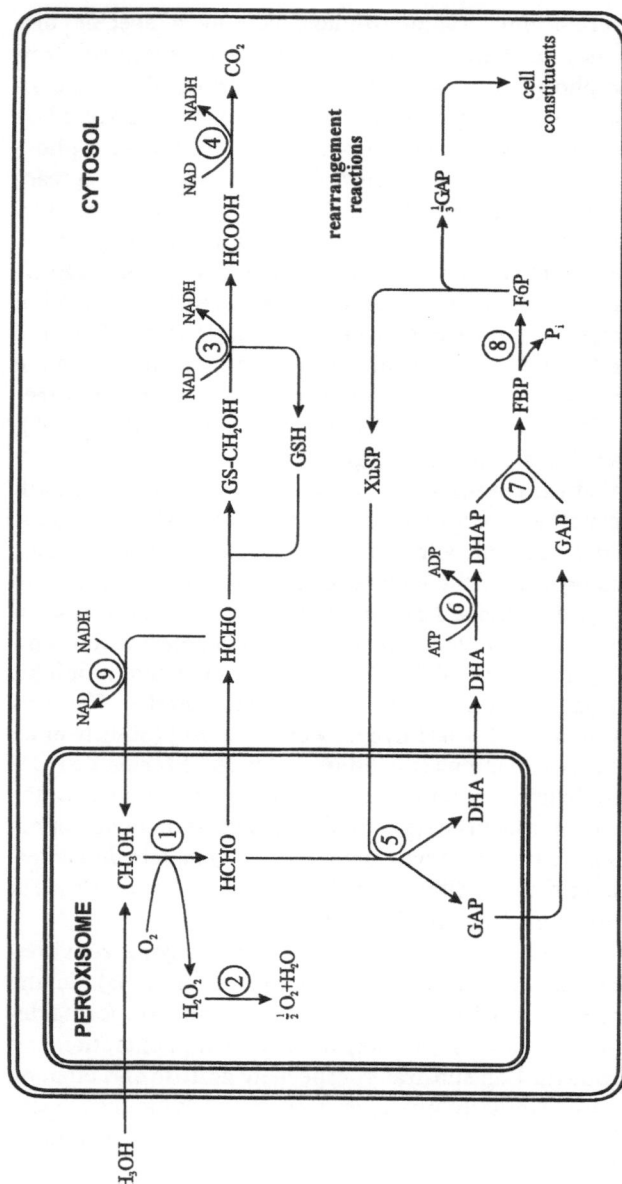

Fig. 2. Methanol metabolism pathways and their compartmentation in methylotrophic yeasts. Designations: *1* alcohol oxidase; *2* catalase; *3* formaldehyde dehydrogenase; *4* formate dehydrogenase; *5* dihydroxyacetone synthase; *6* dihydroxyacetone kinase; *7* fructose-1,6-bisphosphate aldolase; *8* fructose-1,6-bisphosphatase; *9* formaldehyde reductase

The model of *P. methanolica* MH4 was used for studying several aspects of methylotrophic metabolism regulation. The following problems were studied:

- The nature of ethanol intermediate(s) causing catabolite repression and catabolite inactivation of enzymes involved in methanol metabolism.

- Identification of the genes controlling catabolite repression.

10.1
Catabolite Repression and Catabolite Inactivation of Enzymes Involved in Methanol Metabolism

Glucose is the "classic" compound which initiates catabolite repression and inactivation in yeasts (Holzer 1976; Entian 1986; Sibirny and Titorenko 1990), whereas an analogous action of ethanol occurs only in relation to enzymes of C_1-metabolism in methylotrophic yeasts (Sibirny and Titorenko 1990; Bormann and Sahm 1978). The elucidation of the mechanisms of such a unique regulatory mechanism is a very interesting problem. The possibility of isolating mutants of *P. methanolica* blocked in subsequent steps of ethanol utilization can help to reveal the nature of the effectors causing catabolite repression and inactivation in the medium with ethanol.

P. methanolica mutants defective in particular steps of ethanol metabolism were isolated. Some of them were isolated by selection of ethanol nonutilizing mutants (Tolstorukov et al. 1989). All 24 such mutants were unable to grow on ethanol or acetate, but retained the ability to grow on methanol and multicarbon substrates. The mutations were recessive and were divided into four complementation groups. The specific activities of isocitrate lyase, malate synthase, phosphoenolpyruvate carboxykinase, and "malic" enzyme were impaired in the mutants of the corresponding groups. Mutant genes were designated *icl1*, *mls1*, *pck1*, and *mdd1*, respectively. In another series of experiments, 106 2-fluoroacetate-resistant mutants unable to utilize ethanol or acetate were isolated. Among them, three complementation groups were revealed. All mutants were devoid of acetyl-CoA synthetase activity (designated *acs1*, *acs2*, and *acs3*). No linkage was found among the identified mutations.

Among the mutants of *P. methanolica* MH4 resistant to allyl alcohol, a mutant with a 30–40 times decreased alcohol dehydrogenase activity was isolated. The mutation appeared to be recessive, monogenic, and was designated *adh1* (Titorenko et al. 1990a). From this strain, the double mutant *adh1 adh2* was isolated, which was characterized by the total absence of alcohol dehydrogenase activity (Titorenko et al. 1990b). Additionally, a mutant was isolated, which grew on a medium with acetate but not with ethanol as a sole carbon source. It is totally devoid of acetaldehyde dehydrogenase activity and was designated *aldX* (V. Titorenko and A. Sibirny, unpubl. observ.).

Electron microscopic study of *P. methanolica* MH4 mutants defective in different steps of C_2-metabolism showed that *acs1*, *acs2*, *acs3*, *icl1*, *mls1*, *pck1*, and *mdd1* mutations severely (three- to fourfold) inhibited propagation of the glyoxysomal type of microbodies, although they did not influence the specific activity of glyoxysomal enzymes, isocitrate lyase, or malate synthase (of course, with the exception of *icl1* or *mls1* mutations, which block the corresponding enzymes (Kulachkovsky et al. 1990)). Possibly, mutations affecting cell growth on ethanol limit to some extent the propagation of microbodies in a medium with C_2 compounds.

The effect of ethanol and acetate on repression and inactivation of alcohol oxidase and catalase was studied. It was found that ethanol normally repressed alcohol oxidase, and catalase in *acs1*, *acs2*, *icl1*, *mls1*, *pck1*, and *mdd1* strains, and does not repress their synthesis in *adh1 adh2* and *aldX* mutants. On the other hand, acetate normally repressed alcohol oxidase and catalase in all mutants studied. It was suggested that acetate is the true effector (corepressor) which evokes catabolite repression of C_1-metabolizing enzymes in the medium with ethanol (Sibirny 1990).

Studies of catabolite inactivation of alcohol oxidase and catalase in the isolated mutants showed that such inactivation is impaired in *acs1*, *acs2*, *icl1*, *adh1 adh2*, and *aldX* mutants in a medium with ethanol, and only in *acs1*, *acs2*, and *icl1* mutants in a medium with acetate (Sibirny et al. 1986a, 1988; Tolstorukov et al. 1989; Sibirny 1990). It was suggested that the immediate effector of ethanol-induced catabolite inactivation of peroxisomal enzymes is glyoxylate, since production of this compound is defective in mutants with impaired acetyl-CoA synthetase and isocitrate lyase.

10.2
Identification of Genes Controlling Catabolite Repression

In another part of the work, mutants of *P. methanolica* MH4 defective in carbon catabolite repression were isolated and studied. Mutants resistant to the nonmetabolized glucose analogue 2-deoxyglucose during growth in a medium with methanol were isolated (Alamyae et al. 1985; Sibirny et al. 1986b, 1987; Titorenko et al. 1991b). Genetic analysis of the mutants divided them into four classes: *gcr1*, *gcr2* (both recessive) *GCR3^c*, and *GCR4^c* (both dominant). The mutations appeared to be monogenic and nonlinked. It was hypothesized that the genes *GCR1* and *GCR2* act as negative and the genes *GCR3* and *GCR4* as positive regulators of the methanol-metabolizing enzymes. A part of these mutations affected only catabolite repression and methanol was needed for enzyme synthesis, while others, even some represented by the same gene (*GCR1*), manifested constitutive synthesis of enzymes in glucose medium without methanol. Ethanol normally repressed alcohol oxidase and catalase synthesis in all mutants studied. Catabolite inactivation of alcohol oxidase and catalase in glucose medium was also quite normal in the mutants.

For isolation of *P. methanolica* mutants defective in ethanol-induced catabolite repression, the following approach was used. Ethanol normally represses alcohol oxidase synthesis in *icl1* mutants deficient in isocitrate lyase; hence such mutants cannot grow on methanol/ethanol mixture. *icl1* mutants able to grow on a mixture of both alcohols contained a regulatory recessive monogenic mutation designated *ecr1* (Sibirny et al. 1986b, 1987). Glucose-induced catabolite repression was quite normal in the *ecr1* mutant. Catabolite inactivation of alcohol oxidase and catalase in the medium with ethanol was also unaffected.

Ethanol-induced catabolite repression was also impaired in *adh1* mutants, which are characterized by a 30–40-fold decrease of alcohol dehydrogenase activity (Titorenko et al. 1990a). Glucose normally repressed alcohol oxidase synthesis in the mutants.

Carbon compounds repressing alcohol oxidase synthesis in the wild-type strain were divided into four groups, depending on their effect on *gcr1* and *ecr1* mutants. The hexoses and xylose form the first group: the repressive action of these compounds is impaired only in the *gcr1* mutant (Sibirny et al. 1987). The second group includes ethanol, acetate, and 2-oxoglutarate, whose action is impaired only in the *ecr1* mutant. Malate and dihydroxyacetone form the third group: their repressive action is not eliminated in either *ecr1* and *gcr1* mutants. The fourth group consists of compounds whose repressive effect is partly abolished in both mutant classes. Thus, several independent mechanisms of catabolite repression, depending on the nature of corepressor, exist in *P. methanolica* MH4.

Mutants *gcr1*, *gcr2*, *GCR3*^{c}, and *GCR4*^{c} are characterized by a significant decrease in phosphofructokinase activity, an increase in the intracellular pools of hexose phosphates and a drop in fructose-1, 6-bisphosphate activity (Sibirny et al. 1987; Titorenko et al. 1991a). Mutant *ecr1* manifested a decrease in 2-oxoglutarate dehydrogenase activity and an increase in the intracellular pool of 2-oxoglutarate. It was suggested that phosphofructokinase and 2-oxoglutarate dehydrogenase participate in glucose and ethanol catabolite repression, respectively (Sibirny et al. 1987). Data also indicate different genetic mechanisms of catabolite repression and catabolite inactivation of methanol-metabolizing enzymes.

A detailed study of the properties of the *P. methanolica ecr1* mutant quite unexpectedly showed that in this mutant, contrary to the *ECR1* wild-type strain, methanol causes an almost complete block of induction by ethanol of glyoxysomal enzymes involved in C_2 metabolism, isocitrate lyase, and malate synthase (A. Sibirny and V. Titorenko 1988). During cultivation of the *ECR1* wild-type strain and the *ecr1* mutant on a methanol/ethanol-mixture, diauxic growth was observed. The *ECR1* strain first utilized ethanol and then synthesized glyoxysomal enzymes, whereas the *ecr1* mutant first utilized methanol and then synthesized peroxisomal enzymes (Sibirny et al. 1991). It was suggested that the succession of ethanol and methanol utilization from a mixture is determined by the sequence development of two different microbody types: in the wild-type strain, glyoxysomes are formed first whereas in the *ecr1* mutant the development of peroxisomes is observed first. It is possible that the *ECR1* gene plays a central regulatory role and determines the type of microbody to be formed first. In the normal state, the *ECR1* gene directs microbody biogenesis to the "glyoxysomal" type of development while in mutants defective in this gene, biogenesis is guided to the "peroxisomal" type. Thus, the *ECR1* gene may be formally compared with homeoboxes, which regulate the succession of development in higher eukaryotes (Wright et al. 1989).

In contrast to *ecr1*, the *adh1* mutation allowed the simultaneous utilization of ethanol and methanol by permitting joint synthesis of both glyoxysomal and peroxisomal enzymes (Sibirny et al. 1991). Growth of the *adh1* mutant on an ethanol/methanol mixture occurred without diauxy. Ethanol-growing cells of the *adh1* mutant contained hybrid microbodies, glyoxyperoxisomes, possessing simultaneously alcohol oxidase and malate synthase activites (Sibirny et al. 1996).

Thus, the methylotrophic yeast *P. methanolica* MH4 possesses a very complex system which regulates catabolite repression of the methanol-metabolizing enzymes in media with various corepressors. Phosphofructokinase and 2-oxoglutarate dehydrogenase could be involved in catabolite repression.

The new phenomenon of methanol-induced catabolite inactivation of glyoxysomal enzymes (isocitrate lyase, malate synthase), alcohol dehydrogenase, and acetaldehyde dehydrogenase was found using *P. methanolica* MH4 as model system (Sibirny 1990; V. Titorenko and A. Sibirny, unpubl. observ.). Addition of methanol to ethanol-grown cells of a *P. methanolica* wild-type strain causes a rapid decrease in specific activities of the above-mentioned enzymes of C_2 metabolism. Total inactivation of the enzymes was completed 5 to 7h after incubation with methanol. Formaldehyde and formate also caused such inactivation. A methanol/ethanol mixture did not inactivate isocitrate lyase and alcohol dehydrogenase in the wild-type strain, but actively caused such inactivation in the *ecr1* mutant defective in ethanol catabolite repression of C_1-metabolizing enzymes. Apparently, metabolism of methanol is necessary to initiate inactivation of C_2-metabolizing enzymes. The mechanism of this new phenomenon remains unknown.

11
Biotechnological Applications

A mutant AOS-3 was isolated which secretes alcohol oxidase into the culture medium (Titorenko et al. 1992). Specific activity of alcohol oxidase was 0.09U/mg in cells and 0.08U/ml in the culture medium. This represents 2.9U/mg protein of the culture medium, which comprises 45% of total alcohol oxidase activity of the cells plus culture medium. Catalase activity in the culture medium was totally absent. This strain may be of interest for alcohol oxidase production, since isolation of the enzyme from the medium is a much easier procedure compared with the procedure of intracellular enzyme isolation.

It was found that *P. methanolica* mutants defective in acetyl-CoA synthetase, in contrast to the wild-type strain, efficiently acidify the medium after addition of ethanol due to extrusion of acetic acid into the medium (Gonchar et al. 1990). It is possible to immobilize intact washed cells of *acs* mutants of *P. pinus* on the gate of pH-sensitive field effect transistors (pH SFET) which react on the local acidification of the medium due to specific consumption of ethanol and extrusion of acetic acid (Korpan et al. 1992, 1993). The assay time is a few minutes. Such cell biosensors appeared to be more specific when compared with a similar cell sensor which used *Acetobacter aceti* (Kitagawa et al. 1987) or an enzyme alcohol oxidase-based sensor (Kitagawa et al. 1987).

References

Alamyae T, Teugyas K, Soom J, Simisker J (1985) A partly resistant to catabolite repression mutant of the yeast *Pichia pinus* assimilating methanol. Mikrobiologiya 54: 634–640 (in Russian)

Benevolensky SV, Tolstorukov II (1980) Study of the mechanisms of mating and self-diploidization in the haploid yeast *Pichia pinus*. III. Study of heterothallic mutants. Genetika 16: 1342–1349 (in Russian)

Berardi E, Thomas DY (1991) An effective transformation method for *Hansenula polymorpha*. Curr Genet 18: 169–170

Bormann C, Sahm H (1978) Degradation of microbodies in relation to activities of alcohol oxidase and catalase in *Candida boidinii*. Arch Microbiol 117: 67–72

Entian K-D (1986) Glucose repression: a complex regulatory system in yeast. Microbiol Sci 3: 366–371

Gonchar MV, Titorenko VI, Hladarevska NN, Sibirny AA (1990) The phenomenon of medium acidification by the cells of methylotrophic yeast and its biochemical nature. Biokhimia 55: 2148–2158 (in Russian)

Harder W, Veenhuis M (1989) Metabolism of one-carbon compounds. In: Rose AH, Harrison JS (eds) The yeasts. Academic Press, London, 2nd edn, vol 3, pp 289–316

Hiep TT, Kulikov VN, Noskov VN, Pavlov YI (1991) The structure of the *ADE1* gene from methylotrophic yeast *Pichia pinus*: comparison with the AIR-carboxylase genes from other species. Abstr 15th Int Spec Symp on Yeasts, Riga, p 55

Hiep TT, Noskov VN, Pavlov YI (1992) Transformation system in the methylotrophic yeast *Pichia methanolica*. In: Abst Int Workshop on Genetics and Mol Biol of Non-convertional Yeasts, Basel

Hill DJ, Hann AC, Lloyd D (1985) Degradative inactivation of the peroxisomal enzyme, alcohol oxidase, during adaptation of methanol-grown *Candida boidinii* to ethanol. Biochem J 232: 743–750

Holzer H (1976) Catabolite inactivation in yeast. Trends Biochem Sci 1: 178–181

Ito H, Fukuda Y, Murata K, Kimura A (1983) Transformation of intact yeast cells treated with alkali cations. J Bacteriol 153: 161–168

Kato K, Kurimura Y, Makiguchi N, Asai Y (1974) Determination of strongly methanol assimilating yeasts. J Gen Appl Microbiol 20: 123–127

Kitagawa Y, Tamiya E, Karube I (1987) Microbial-FET alcohol sensor. Anal Lett 20: 81–96

Korpan YI, Gonchar MV, Soldatkin AP, Starodub NF, Sandrovsky AK, Sibirny AA, Elskaya AV (1992) Cellular microbiosensors on the base of pH-sensitive field effect transistors for methanol and ethanol assay. Ukr Biokhim Zh 64(3): 96–100 (in Russian)

Korpan YI, Soldatkin AP, Starodub NF, Elskaya AV, Gonchar MV, Sibirny AA, Shulga AA (1993) Methylotrophic yeast microbiosensor based on ion-selective field effect transistors for methanol and ethanol determination. Anal Chim Acta 271: 203–208

Kulachkovsky AR, Titorenko VI, Sibirny AA (1990) Electron microscopic study of the mutants of *Pichia pinus* yeast defective in various steps of ethanol metabolism. Tsitologiya i Genetika 24(6): 17–20 (in Russian)

Kushnirov VV, Ter-Avanesyan MD, Didichenko SA, Smirnov VN, Chernoff YO, Derkach IL, Novikova ON, Inge-Vechtomov SG, Neistat MA, Tolstorukov II (1990) Divergence and conservation of *SUP2* (*SUP35*) gene of yeasts *Pinchia pinus* and *Saccharomyces cerevisiae*. Yeast 6: 461–472

Motruk OM, Tolstorukov II, Sibirny AA (1989) Selection of alcohol oxidase deficient mutants of methylotrophic yeast *Pichia pinus*. Biotekhnologiya 5: 692–698 (in Russian)

Neistat MA, Benevolensky SV, Dutova TA, Tolstorukov II (1987) "Illegal" transformation of red adenine-deficient mutants of the yeast *Pichia pinus* with pYE (ADE2)2 plasmid. Genetika 23: 1525–1534 (in Russian)

Sahm H (1977) Metabolism of methanol by yeast. Adv Biochem Eng 6: 77–103

Sasnauskas KV, Giadvilaite AA, Janulaitis AA (1987) Cloning of *Saccharomyces cerevisiae* ADE2 gene and localization of ARS sequence. Genetika 23: 1141–1148 (in Russian)

Sibirny AA (1987) Some aspects of methanol metabolism regulation in yeast. In: Biokhimiya Zhivotnykh i Cheloveka. Biotekhnologiya. Naukova Dumka, Kiev, vol 11, pp 63–75 (in Russian)

Sibirny AA (1990) Genetic control of methanol and ethanol metabolism in the yeast *Pichia pinus*. In: Heslot H (ed) Proc 6th Int Symp on Genetics of Ind Microorg Strasbourg. Soc Franc Microbiol, vol 1, pp 545–554

Sibirny AA, Gonchar MV (1990) The influence of citrate and dihydroxyacetone on methanol oxidation in yeast. Ukr Biokhim Zh 62: 112–115 (in Russian)

Sibirny AA, Titorenko VI (1988) The inhibition by methanol of isocitrate lyase and malate synthase induction in the yeast *Pichia pinus*. Biotekhnologiya 4: 194–196 (in Russian)

Sibirny AA, Titorenko VI (1990) Molecular mechanisms of catabolite regulation in yeast. In: Itogi Nauki i Tekhniki. Molekularnaya Biologiya, vol 33, pp 3–214 (in Russian)

Sibirny AA, Ubiyvovk VM (1988) Isolation of methylotrophic yeast *Hansenula polymorpha* mutants deficient in formaldehyde dehydrogenase. Biotekhnologiya 4: 723–725 (in Russian)

Sibirny AA, Titorenko VI, Benevolensky SV, Tolstorukov II (1986a) On the regulation of methanol metabolism in the mutant of the yeast *Pichia pinus* defective in isocytrate lyase. Biokhimiya 51: 16–22 (in Russian)

Sibirny AA, Titorenko VI, Benevolensky SV, Tolstorukov II (1986b) On the differencies in the mechanisms of ethanol and glucose catabolite repression involved in enzymes of methanol metabolism in the yeast *Pichia pinus*. Genetika 22: 584–592 (in Russian)

Sibirny AA, Titorenko VI, Efremov BD, Tolstorukov II (1987) Multiplicity of mechanisms of carbon catabolite repression involved in the synthesis of alcohol oxidase in the methylotrophic yeast *Pichia pinus*. Yeast 3: 233–241

Sibirny AA, Titorenko VI, Gonchar MV, Ubiyvovk VM, Ksheminskaya GP, Vitvitskaya OP (1988) Genetic control of methanol utilization in yeasts. J Basic Microbiol 28: 293–319

Sibirny AA, Ubiyvovk VM, Gonchar MV, Titorenko VI, Voronovsky AY, Kapultsevich YG, Bliznik KM (1990) Reactions of direct formaldehyde oxidation to CO_2 are non-essential for energy supply of yeast methylotrophic growth. Arch Microbiol 154: 566–575

Sibirny AA, Titorenko VI, Teslyar GE, Petrushko VI, Kucher MM (1991) Methanol and ethanol utilization in methylotrophic yeast *Pichia pinus* wild-type and mutant strains. Arch Microbiol 156: 455–462

Sibirny AA, Kulachkovsky AR, Moroz OM (1996) Appearance of hybrid microbodies, glyoxyperoxisomes, containing alcohol oxidase and malate synthase in the regulatory mutant of the methylotrophic yeast *Pichia methanolica*. Mikrobiologiya 65 (in press)

Surguchov AP, Smirnov VN, Ter-Avanesyan MD, Inge-Vechtomov SG (1984) Ribosomal suppression in eukaryotes. Physicochem Biol Rev 4: 147–205

Tarutina M, Tolstorukov II (1994) Elaboration of the method for vector transformation of methylotrophic yeast *Pichia methanolica*. Genetika 30: 783–790

Titorenko VI, Motruk OM, Efremov BD, Petrushko VI, Teslyar GE, Kulachkovsky AR, Sibirny AA, Tolstorukov II (1990a) Selection and properties of the mutants of the methylotrophic yeast *Pichia pinus* deficient in alcohol dehydrogenase and formaldehyde reductase. Genetika 26: 1749–1759 (in Russian)

Titorenko VI, Petrushko VI, Kulachkovsky AR, Sibirny AA (1990b) Regulatory properties of the mutants of *Pichia pinus* impaired in alcohol dehydrogenase. In: Abstr 6th Int Symp on Genetics of Ind Microorg Strasbourg, p 151

Titorenko VI, Khodursky AB, Sibirny AA (1991a) The properties of new mutants of the yeast *Pichia pinus* with defective catabolite repression. Genetika 27: 791–800 (in Russian)

Titorenko VI, Khodursky AB, Teslyar GE, Sibirny AA (1991b) Identification of new genes controlling catabolite repression of alcohol oxidase and catalase synthesis in the methylotrophic yeast *Pichia pinus*. Genetika 27: 625–635 (in Russian)

Titorenko VI, Sapozhenkova EY, Sibirny AA (1992) The strain of the yeast *Pichia pinus*, producer of alcohol oxidase. Soviet Patent (Author's Certificate) No 1770357 (in Russian)

Tolstorukov II (1988) Genetics of the methylotrophic yeast *Pichia pinus*. Doct Biol Sci Thesis Moscow (in Russian)

Tolstorukov II (1994) Genome structure and reidentification of the taxonomic status of *Pichia pinus* MH4 genetic lines. Genetika 30: 635–640 (in Russian)

Tolstorukov II, Benevolensky SV (1978) Study of the mechanisms of mating and self-diploidization in the haploid yeast *Pichia pinus*. I. Bipolarity of mating. Genetika 14: 519–526 (in Russian)

Tolstorukov II, Benevolensky SV (1980) Study of the mechanisms of mating and self-diploidization in the haploid yeast *Pichia pinus*. II. Study of heterothallic mutants. Genetika 16: 1342–1349 (in Russian)

Tolstorukov II, Efremov BD (1984) Construction of the genetic map of the yeast *Pichia pinus*. II. Tetrad analysis mapping. Genetika 20: 1099–1107 (in Russian)

Tolstorukov II, Dutova TA, Benevolensky SV, Soom YO (1977) Hybridization and genetic analysis of methanol-oxidizing yeast *Pichia pinus* MH4. Genetika 13: 322–329 (in Russian)

Tolstorukov II, Benevolensky SV, Efremov BD (1981) Study of the mechanisms of mating and self-diploidization in the haploid yeast *Pichia pinus*, YI. Functional structure of the mating type locus. Genetika 17: 1019–1028 (in Russian)

Tolstorukov II, Benevolensky SV, Efremov BD (1982) Genetic control of cell type and complex organization of mating type locus in the yeast *Pichia pinus*. Curr Genet 5: 137–142

Tolstorukov II, Bliznik KM, Efremov BD (1983) Construction of the genetic map of the yeast *Pichia pinus*. I. Identification of the linkage groups by using induced mitotic haploidization. Genetika 19: 897–902 (in Russian)

Tolstorukov II, Dutova TA, Benevolensky SV, Efremov BD, Milgrom EM, Alenin VV, Domkin VD (1987) Genetic control of purine biosynthesis in the yeast *Pichia pinus*. I. Genes controlling YI and YII steps. Genetika 23: 1399–1406 (in Russian)

Tolstorukov II, Efremov BD, Benevolensky SV, Titorenko VI, Sibirny AA (1989) Mutants of the methylotrophic yeast *Pichia pinus* defective in C$_2$ metabolism. Yeast 5: 179–186

Tuttle DL, Lewin AS, Dunn WA Jr (1993) Selective autophagy of peroxisomes in methylotrophic yeasts. Eur J Cell Biol 60: 283–290

Veenhuis M, van Dijken JP, Harder W (1983) The significance of peroxisomes in the metabolism of one-carbon compounds in yeast. Adv Microb Physiol 24: 1–82

Wright CVE, Cho KWY, Oliver G, de Robertis EM (1989) Vertebrate homeodomain proteins: families of region-specific transcription factors. Trends Biochem Sci 14: 52–56

CHAPTER 9

Hansenula polymorpha (Pichia angusta)

Hans Hansen and Cornelis P. Hollenberg

1
History of *Hansenula polymorpha* Research

A small and closely related group of yeasts is capable of using methanol as sole source of carbon and energy. Among these methylotrophic organisms, *Hansenula polymorpha* in particular has gained increasing attention in recent years, in both basic and applied sciences. This is mainly due to two physiological features that are an integral part of the methanol-utilizing machinery of this yeast. One is the expression of huge amounts of the key methanol-metabolizing enzymes, which can, in the case of the methanol oxidase, amount to up to one third of the total cellular protein. The expression is driven by very strong promoters that form the basis for a highly competitive system to produce foreign proteins at industrial scale (Gellissen et al. 1994). The other prominent characteristic of *H. polymorpha* is that the expression is accompanied by a dramatic growth and proliferation of microbodies, i.e., peroxisomes (Veenhuis and Harder 1987). *H. polymorpha* thus served and, in addition to the recent introduction of other systems such as *Saccharomyces cerevisiae*, still continues to serve as a valuable model organism to study the biogenesis of these organelles. Although we will briefly discuss the contributions of *H. polymorpha* to the field of peroxisome research, our main focus within this chapter belongs to the practical aspects that are of importance in working with this yeast.

As an ascosporogenous yeast that does not form a mycelium, *H. polymorpha* is placed into the order of Endo- or Saccharomycetales (Lodder 1970). It is a member of the family of Saccharomycetaceae because it propagates asexually by multilateral budding on a narrow base, whereafter the vegetative cells become separated. An important taxonomic feature, other than methylotrophy, which separates *Hansenula* from other yeasts, with the exception of *Pichia* and *Pachysolen*, is the production of extracellular phosphomannans. Phosphomannans are produced in large amounts by the more primitive members of this genus, which gives them a mucoid appearance (Wickerham 1970). The genera *Hansenula* and *Pichia* are closely related, especially their methylotrophic members (Lee and Komagata 1980). The ability of *H. polymorpha* to assimilate nitrate places this species in the

Institut für Mikrobiologie, Heinrich-Heine-Universität Düsseldorf, Universitätsstr. 1, Gebäude 26.12, 40225 Düsseldorf, Germany genus *Hansenula* H. et P. Sydow (Wickerman 1970). Some taxonomists (see Chap. 1, this Vol.) dispute the significance of this classification based on hybridization studies, and propose to combine *Pichia* and *Hansenula*. In this context, *H. polymorpha* has been renamed *Pichia angusta* according to the earlier description of the genus *Pichia* Hansen (Kurtzman 1984).

2
Physiology

The capability to grow on media which contain methanol as sole source of carbon and energy distinguishes methylotrophic from other yeasts. This feature is much rarer among yeasts than among bacteria. Methylotrophic yeasts can be grouped into four genera, *Hansenula*, *Pichia*, *Candida*, and *Torulopsis*, which are closely related (Harder and Brooke 1990). All of them are facultative methylotrophs and unable to grow on methane, methylamine, formaldehyde, or formate as a sole carbon and energy source. However, methylated amines can be utilized as the nitrogen source (Van Dijken and Bos 1981).

The first step in methanol utilization is shared by the assimilatory and the dissimilatory pathways. A methanol oxidase that is located in the peroxisomal matrix converts methanol to formaldehyde (Sahm and Wagner 1973a,b). The enzyme is present as a homo-dimer with flavin adenine dinucleotide (FAD) as the prosthetic group, bound to each subunit (Kato et al. 1976). When large amounts of methanol oxidase are synthesized, it forms a crystalloid inside the peroxisome (Vonck and van Bruggen 1990). No energy is provided for the cell during the first step. Instead, oxygen is used as an electron acceptor, which results in the formation of hydrogen peroxide. H_2O_2 is then removed by a peroxisomal catalase (Sahm and Wagner 1973a,b; Roggenkamp et al. 1974, 1975; Fujii and Tonomura 1975).

Formaledehyde is either assimilated enzymatically via a peroxisomal dihydroxyacetone synthase, or it leaks into the cytosol, where it reacts non-enzymatically with glutathione (Uotila and Koivusalo 1974; Waites and Quayle 1981). The resulting S-hydroxymethylglutathione is dissimilated to carbon dioxide in two energy-providing reactions (Schütte et al. 1976). A formaldehyde dehydrogenase converts S-hydroxymethylglutathione to S-formylglutathione (Uotila and Koivusalo 1974; Schütte et al. 1976), which in *H. polymorpha* is the substrate of a formate dehydrogenase (Schütte et al. 1976; van Dijken et al. 1976a). In both steps, nicotinamide adenine dinucleotide (NAD^+) serves as the electron acceptor. The role of the dissimilatory pathway as a major energy supply is controversial. Sibirny et al. (1990) have demonstrated that both formaldehyde dehydrogenase and formate dehydrogenase are not indispensable for growth on methanol, but the cells become more sensitive to higher methanol concentrations. Metabolic labeling studies have shown that only if the capacity of the assimilatory cycle to remove formaldehyde is decreased, can significant amounts of formaldehyde enter the oxidative pathway (Jones and Bellion 1991), suggesting a role in detoxification.

Formaldehyde is assimilated by the xylulose monophosphate pathway (van Dijken et al. 1978). The initial step is catalyzed by a special transketolase that is located in the peroxisomal matrix (Douma et al. 1985; Goodman 1985). This enzyme, termed dihydroxyacetone synthase, transfers a glycolaldehyde group from xylulose-5-phosphate onto formaldehyde, thereby generating dihydroxyacetone and glyceraldehyde-3-phosphate (Waites and Quayle 1981; Bystrykh et al. 1981; Kato et al. 1982). The former metabolite is phosphorylated by a cytoplasmic dihydroxyacetone kinase (Douma et al. 1985), before it is condensed by a fructose-1,6-bisphosphate aldolase with glyceraldehyde-3-phosphate to fructose-1,6-bisphosphate (FBP). During growth on methanol, increased amounts of FBPase, an enzyme essential for the regeneration of xylulose-5-phosphate, are formed (Harder and Brooke 1990). Methanol obviously represents a suboptimal carbon source for the growth of methylotrophic yeasts. This is reflected in the fact that in methanol-grown cells, trehalose, generally associated with poor growth rates, starvation, and differentiation, is the main assimilatory product (Jones and Bellion 1991).

3
Media

As for *Saccharomyces cerevisiae*, YEP (1% yeast extract, 2% Bacto peptone) is used as rich medium for *H. polymorpha*. In general, for selective purposes the yeast is grown in the presence of 0.67% Yeast Nitrogen Base supplemented with amino acids or bases at $20 \mu g/ml$ if required; 2% glucose is commonly used as carbon source. Since glucose and also ethanol are most effective in catabolite repression (Egli et al. 1980; Harder and Brooke 1990), they have to be replaced if expression from peroxisomal promoters is required. Whereas glycerol (usually 1–2%) is used for derepression of peroxisomal enzymes, methanol has a strong inducing effect (Eggeling and Sahm 1981). Methanol should be added at a concentration of 0.5–1%. For most purposes the loss of methanol by evaporation poses no problem. On the other hand, too high concentrations (over 2%) can lead to significant reductions in growth rate (Berry et al. 1987). In continuous culture, expression from the methanol oxidase promoter can also be effectively induced by formaldehyde and formic acid (Giuseppin et al. 1988). Mixed substrates have been used to induce peroxisomal biogenesis under more favorable growth conditions, such as glucose and methanol, or glucose and formaldehyde for carbon-limited cultivation of cells, and glycerol and methanol, or sorbitol and methanol in batch culture (Egli et al. 1982; Veale et al. 1992).

Optimal peroxisomal proliferation and highest expression of alcohol oxidase can be achieved in methanol-limited chemostat culture, where *H. polymorpha* is grown in a mineral medium that consists of 0.25% ammonium sulfate, 0.1% KH_2PO_4, 0.02% $MgSO_4 \cdot 7H_2O$, 0.05% yeast extract, trace elements, 0.5 mg/l thiamin, and $50 \mu g/l$ biotin (van Dijken 1976). Under these conditions, a single cell may harbor up to 20 peroxisomes which occupy up to 80% of the cell lumen (Veenhuis et al. 1978, 1983), and the alcohol oxidase can constitute up to 35% of soluble cell protein (Roggenkamp et al. 1984).

For mating and sporulation, *H. polymorpha* is plated onto 3% malt extract (Gleeson and Sudbery 1988; see Sect. 5). Under these conditions, diploid cells appear as colonies which become dark pink upon sporulation within 3 to 4 days. The sporulation of haploids is visible after 8 days.

4
Available Strains

Although a variety of wild-type strains of H. polymorpha can be obtained via the major culture collectors, only a few are routinely used for scientific or industrial purposes. Strain CBS 4732 (Centraalbureau voor Schimmelcultures, Delft, The Netherlands) or ATCC 34438 (American Type Culture Collection, Rockville, MD USA) is historically the most widely employed strain. It is used in most studies of peroxisomal biogenesis, and it served as the basis for the majority of available mutant strains that are utilized for expression of recombinant proteins. Unfortunately, the fertility of H. polymorpha laboratory strains varies considerably (Gleeson and Sudbery 1988) and CBS 4732 has been shown to have an extremely low mating efficiency. Therefore, it is not easily amenable to classical genetic studies (see below). In contrast, strain NCYC 495 from the National Yeast Culture Collection (Norwich, UK) has much better mating properties and has been used extensively in genetic analyses, in particular by Gleeson and Sudbery (1988).

An increasing number of mutant strains with nutritional requirements comparable to those of S. cerevisiae have been published. Essentially two types of auxotrophic strains are used presently for transformation purposes. Strains with defects in the orotidine-5'-phosphate decarboxylase gene (odc1), such as LR9 derived from ATCC 34438 (Roggenkamp et al. 1986), can be complemented either with the URA3 gene from S. cerevisiae or from H. polymorpha (Merckelbach et al. 1993). Strains which lack β-isopropyl malate dehydrogenase activity, termed A16 (from CBS 4732; Fellinger et al. 1991), leu1-1 (from NCYC 495; Gleeson 1986) or DL-1 (Tikhomirova et al. 1986), are available which are complementable with both the S. cerevisiae LEU2 gene or its H. polymorpha homologue HLEU2 (Agaphonov et al. 1994). For production purposes, special mutants such as RB10 (Janowicz et al. 1991) have been selected.

Since a major focus in working with H. polymorpha is the peroxisome, the development of genetic techniques has led to a steady expansion in the number of strains with defects in methanol utilization. These so-called Mut⁻ mutants are frequently found and can be isolated based on their inability to grow on methanol. They fall into three categories: (1) mutants defective in methanol utilization, (2) regulatory mutants, (3) mutants with defects in peroxisome biogenesis. Methanol oxidase mutants have been isolated by Sibirny et al. (1988) by selecting for growth on allyl alcohol. Mutant strains lacking, for example, catalase, formaldehyde reductase, or dihydroxyacetone kinase activity, or have defects in the tricarboxylic acid cycle, have been generated by different groups (Eggeling and Sahm 1980; de Koning et al. 1987; Sibirny et al. 1988; Hansen and Roggenkamp 1989). Gleeson (1986) has analyzed 33 mutants, which he generated by mutagenizing strain CBS 4732. About half of them belong to five complementation groups, only one (MUT3) lacks activity of one of the methanol-metabolizing enzymes, the dihydroxyacetone synthase. The others do not show a specific enzymatic defect and are presumably mutants of class (3). The second half of isolated mutants (*acu* mutants) are unable to grow on methanol and a variety of other carbon sources, such as ethanol, glycerol, maltose, or dihydroxyacetone, indicating defects in carbon catabolite repression. Revertants, isolated on maltose medium, belong to two groups. One is relieved of glucose repression independent of the carbon source, suggesting a common regulatory element, the other is resistant to 2-deoxyglucose, albeit to various degrees, which indicates the existence of regulatory mechanisms particular to a specific carbon source (Sudbery and Gleeson 1989).

To date, relatively little is known about the biogenesis of peroxisomes. Information about the participating genes can be obtained with a group of strains isolated form Mut⁻ mutants by electron microscopic examination of peroxisomal structures (Cregg et al. 1990; Titorenko et al. 1993a,b). Briefly, they have three distinct phenotypes. No peroxisomes could be demonstrated in Per⁻ mutants, but peroxisomal matrix and aggregated membrane proteins are present in the cytosol (Sulter et al. 1993a,b); Pim⁻ mutants have small peroxisomes, but are unable to import peroxisomal proteins; Pss⁻ mutants show an aberrant substructure of the crystalline peroxisomal matrix (Titorenko et al. 1993a). All mutants can be divided into 12 complementation groups, designated *Per1* to *Per12*. In many cases, primarily combined with a cold-sensitive phenotype, diploids with wild-type and mutant alleles are unable to grow on methanol, and still show defects in peroxisome biogenesis (Titorenko et al. 1993). This so-called unlinked noncomplementation suggests that at least five of the *PER* genes participate as essential components in a protein complex controlling major steps in peroxisome formation (Titorenko et al. 1993). The first two members of this set of genes have been isolated by functional complementation of *per1*, a mutant with Pim⁻ phenotype, and the Per⁻ strain *per8* (Waterham et al. 1994; Tan et al. 1995; see Sect. 7).

5
Genetic Techniques

5.1
Life Cycle

Methylotrophic yeasts are commonly isolated from decaying plant material and soil samples that are rich in organic matter. The species of *Hansenula* can be haploid and/or diploid (Teunisson et al. 1960). *H. polymorpha* exists in nature predominantly as haploid cells, but upon isolation frequently produces diploids (Wickerham 1970). Haploid cells have an ellipsoidal shape with diameters of (1.0– 4.3) × (2.1–4.3) μm, whereas diploids are more spherical, measuring from 3.4 to 5.2 μm (Wickerham 1970). The latter may be agglutinated. Diploid cells sporulate faster than the haploids, which makes them easily distinguishable on sporulation plates by the earlier formation of pink-colored colonies (Wickerham 1970). The ascospores are small, hemispheroidal or hat-shaped with narrow brims, and connected to each other by a thread-like structure. *H. polymorpha* can ferment glucose, but not galactose, sucrose, maltose, lactose, or raffinose (Wickerham 1970). The yeast does not ferment under aerobic conditions, that means it is Crabtree-negative, and does not form petites (Verduyn et al. 1992). It is thermotolerant with a temperature optimum of 37 °C, which is unusual for methylotrophic yeasts (Levine and Cooney 1973).

The life cycle of the homothallic yeast *H. polymorpha* is very similar to that of *S. cerevisiae* and especially to the closely related *Pichia pinus* (Gleeson 1986), which has been studied in detail by Tolstorukov and Benevolenskii (1978, 1980). Although stable in rich medium, diploid cells start to sporulate when nutrients become limited. They form four ascospores. The spores belong to two different mating types, which segregate equally. Like the diploids, the resulting haploid cells are maintained stably in rich medium (Sudbery and Gleeson 1989). Starvation, such as nitrogen deprivation, triggers some cells to switch their mating type. In contrast to *S. cerevisiae*, the switch is independent of the mating type of the mother cell or the position of the cell in the last generation cycle. The switch is random and can generate the same or a different mating type. In the latter case only, a conjugation tube is formed which initiates the mating between cells of opposite mating types. During prolonged cultivation, some commonly used laboratory strains of *H. polymorpha* (such as CBS 4732, syn. ATCC 34438), however, have become semisterile (Gleeson 1986), presumably because they have lost the ability to switch their mating type.

5.2
Induction of Mutants

H. polymorpha is in principle amenable to all classical techniques that were originally developed for the genetic manipulation of *S. cerevisiae*. Since haploid cells do not mate on rich medium, conventional mutagenesis can be used to introduce mutations. In most reported cases, either nitrosoguanidine (NTG) or ethylmethanesulfonate (EMS) have been employed successfully as mutagens, sometimes followed by an enrichment step with nystatin (Sanchez and Demain 1977; Gleeson 1986; Roggenkamp et al. 1986; Cregg et al. 1990). *ODC1* mutants, suitable for complementation with the *URA3* gene, have been isolated using 5-fluoro-orotic acid (Roggenkamp et al. 1986). Following mutagenesis, the auxotrophic phenotype, that is failure to grow on nutrient-deficient medium, is not always immediately visible (Sudbery and Gleeson 1989; H. Hansen and R. Roggenkamp, unpubl. results), which often makes it necessary to replate the colonies. If a fertile strain of *H. polymorpha* such as NCYC 495 has been used, the generated mutants can easily be analyzed for complementation groups. Mutants grown on the poor nutrient malt extract undergo a switch of their mating type which facilitates crossing of different mutants. On the other hand, diploids have to be removed and cultivated on rich medium immediately after 2 days to prevent initiation of meiosis and sporulation (Gleeson 1986). Tetrad analysis is possible, but the dissection is hampered by the small size of the spores and thread-like structures which link them together. Unfortunately, some laboratory strains, such as the popular ATCC 34438, are semisterile and have to be backcrossed with a fertile strain (Gleeson 1986).

6
Chromosomes

Little information is available about the genome of *H. polymorpha*. A recent study used contour-clamped homogenous field electrophoresis to separate the chromosomes of different strains (Marri et al. 1993). Up to six *H. polymorpha* chromosomes could be separated by this technique. They range from around 650 to 2200 kb with the majority of bands running above 1000 kb. The strains employed exhibited an extremely high degree of polymorphism of the electrophoretic pattern. This holds true even in a comparison between strain LR9, which is an uracil auxotrophic mutant, and its parental strain ATCC 34438 (Roggenkamp et al. 1986). The cause of these changes is presently undetermined. It has been shown, however, that in *Candida albicans* similar rearrangements involved recombinations between the ribosomal DNA (rDNA) genes (Marri et al. 1993).

7
Genes and Genetic Markers

The list of reported genes from *H. polymorpha* is still short (Table 1). The majority of sequences encode enzymes that reside in peroxisomes. In 1985 the genes for the two key enzymes in methanol metabolism were cloned, the methanol oxidase (*MOX*) and the dihydroxyacetone synthase (*DAS*) gene (Ledeboer et al. 1985; Janowicz et al. 1985). Both genes are highly expressed and tightly regulated. The *DAS* mRNA for example amounts to 7% of the polyA+ RNA in methanol-induced mid-log phase cells, but is undetectable under repressed conditions (Janowicz et al. 1985). Not surprisingly, *MOX* and *DAS* share common features in the 5′ and 3′ regions, such as TATA box-like sequences around −55, some homologies to highly expressed eukaryotic genes around the translation start, and regions of dyad symmetry in the promoter and terminator regions (Janowicz et al. 1985; Ledeboer et al. 1985). Interestingly, two putative stem-loop structures flank a UAS sequence inside the *MOX* promoter that is also present in the coregulated catalase promoter (Didion and Roggenkamp 1992; Gödecke et al. 1994; see below). The similarities to genes from *S. cerevisiae* are limited. Many features that are common for promoter regions in *S. cerevisiae* (Dobson et al. 1982) are missing. The codon usage resembles more that of *E. coli* genes and is less biased than in *S. cerevisiae* (Janowicz et al. 1985; Ledeboer et al. 1985). Another major enzyme in cells grown in the presence of methanol, the gene for formate dehydrogenase, has also been cloned (Hollenberg and Janowicz 1989). Together with the *MOX* promoter, it is used for the high-yield expression of heterologous proteins.

The first step in the oxidation of primary amines is catalyzed by a peroxisomal enzyme, the amine oxidase (*AMO*). The *AMO* gene of *H. polymorpha* has been cloned by Bruinenberg et al. (1989). The enzyme contains an unusual enzymatic

Table 1. Cloned genes of *H. polymorpha*

Gene	Function	ORF (amino acids)	Compartment	Reference
MOX	Methanol oxidase	644	Peroxisome	Ledeboer et al. (1985)
DAS	Dihydroxyacetone synthase	710[a]	Peroxisome	Janowicz et al. (1985)
FMD	Formate dehydrogenase	360	Cytoplasm	Hollenberg and Janowicz (1989)
CAT1	Catalase	507	Peroxisome	Didion and Roggenkamp (1992)
AMO	Amine oxidase	692	Peroxisome	Bruinenberg et al. (1989)
MAS	Malate synthase	555	Peroxisome	Bruinenberg et al. (1990)
HSA1	Heat-shock protein (HSP70)	645	Cytoplasm	Diesel and Roggenkamp (1994)[b]
PER1	Peroxisomal matrix protein	650	Peroxisome	Waterham et al. (1994)
PER8	Peroxisomal membrane protein	295	Peroxisome	Tan et al. (1995)
URA3	Orotidine-5′-phosphate decarboxylase	263	Cytoplasm	Merckelbach et al. (1993)
HLEU2	β-Isopropylmalate dehydrogenase	363	Cytoplasm	Agaphonov et al. (1994)

[a] Updated in Hansen et al. (1992).
[b] Unpublished, submitted to the EMBL/GenBank/DDBJ databases.

redox cofactor, a topa quinone, that is derived from a tyrosine codon at the active site (Mu et al. 1992). Recently, the *H. polymorpha* catalase gene (*CAT1*) has been isolated (Didion and Roggenkamp 1992). It is essential for growth on methanol and cannot be replaced with a cytoplasmic catalase activity (Hansen and Roggenkamp 1989; Didion and Roggenkamp 1992).

Two of the *PER* genes mentioned above have been cloned. *PER1* encodes a protein that is located in the peroxisomal matrix and seems to have a role in protein import into peroxisomes (Waterham et al. 1994). The product of the *PER8* gene, called Per8p, is an integral protein of the peroxisomal membrane (Tan et al. 1995). Upon overexpression, the transformed cells show an enhanced increase in peroxisome numbers, suggesting that Per8p is involved in peroxisome proliferation (Tan et al. 1995).

The sequences of two genes which allow plasmid selection in *H. polymorpha* have been published. They encode the homologues for the *URA3* and *LEU2* markers from *S. cerevisiae* that are currently mainly used. Whereas the *H. polymorpha* *URA3* functions in *S. cerevisiae* (Merckelbach et al. 1993), the cloned sequence of the *LEU2* homologue, termed *HLEU2*, does not complement the mutation in *S. cerevisiae* (Agaphonov et al. 1994).

8
Vector Systems

Successful transformation of *H. polymorpha* was independently reported in 1986 by three groups (Roggenkamp et al. 1986; Tikhomirova et al. 1986; Gleeson et al. 1986). In all protocols, strategies similar to those previously developed for the transformation of *S. cerevisiae* have been employed. To allow for the selection of transformed cells, mutations have been created in genes corresponding to those which serve as selectable markers of *S. cerevisiae*. Roggenkamp et al. (1986) have isolated mutants lacking orotidine-5'-phosphate, named *odc1*, that can be complemented by the *URA3* gene of *S. cerevisiae*. In the other studies, mutants deficient in β-isopropyl malate dehydrogenase and recombinant plasmids containing the *LEU2* gene of *S. cerevisiae* have been used (Tikhomirova et al. 1986; Gleeson et al. 1986). Different procedures have been developed to introduce the DNA into the cells, all of which represent modifications of methods described for the transformation of *S. cerevisae*. The transformation efficiencies are strain-dependent. Whereas ATCC 34438 shows only low transformation rates with spheroplasted cells (Roggenkamp et al. 1986). DL-1 is rather amenable to this method (Tikhomirova et al. 1988), A relatively simple protocol using whole cells (Klebe et al. 1983; Dohmen et al. 1991) has been employed with modifications by Roggenkamp et al. (1986). In any case, the transformation rates range form 200 transformants per microgram DNA, with sequences for autonomous replication (ARS) from *S. cerevisiae*, up to 3000 with ARS from *H. polymorpha* or *Candida utilis* (Roggenkamp et al. 1986; Tikhomirova et al. 1986). Significantly higher efficiencies ($2 \times 10^4 - 1.7 \times 10^6$) have recently been achieved with electroporation, reaching maximum rates with linearized plasmids (Faber et al. 1994).

Sequences that confer autonomous vector replication in *H. polymorpha* can be of homologous or heterologous origin. The ARS1 element of *S. cerevisiae* allows unstable plasmid replication with a loss of 99% of plasmids after growth for ten generations without marker selection (Roggenkamp et al. 1986), as does the replicon-like sequence of the *S. cerevisiae LEU2* marker (Berardi and Thomas 1990). The origin of the 2-μm DNA of *S. cerevisiae* has no replicative function in *H. polymorpha* (Roggenkamp et al. 1986). Roggenkamp et al. (1986) have cloned two ARS (HARS1 and HARS2) elements from *H. polymorpha* fragments which allow autonomous replication with high copy numbers of 30 to 40 per cell, about 10 times higher than with ARS1. In a similar approach, Tikhomirova et al. (1986) have isolated *H. polymorpha* ARS elements from mitochondrial (mt) DNA after they had shown autonomous replication with mt DNA fragments from *Candida utilis*. HARS1 has been sequenced (Roggenkamp et al. 1986). It is an AT-rich fragment of 0.5 kb, which shows some general similarity to ARS elements of *S. cerevisiae* (Broach et al. 1982; Kearsey 1984), but no ARS consensus. Consequently, it is not surprising that, like HARS2, it does not function in *S. cerevisiae*. Using integrative plasmids, Bogdanova et al. (1995) have captured recently several new HARS elements from the *H. polymorpha* genome. After transformation, the plasmids recombined spontaneously with different chromosomal regions and aquired an increased mitotic stability. One of the new elements has been sequenced and shows a HARS1-like composition including several blocks with similarity to the 5'-ATAATAATA-3' core sequence described in Roggenkamp et al. (1986).

As is the case with ARS-containing plasmids of *S. cerevisiae*, all autonomously replicating vectors segregate poorly in *H. polymorpha*. Therefore the transformation is very unstable under nonselective conditions (Roggenkamp et al. 1986; Thikomirova et al. 1986). However, upon growth over several generations, these plasmids tend to spontaneously form tandem polymers with up to 100 copies and extremely high mitotic stability (Roggenkamp et al. 1986), a feature that is particularly useful for high-level expression of foreign proteins (Gellissen et al. 1992). Most probably, these plasmid multimers are integrated into the genome (Gatzke et al. 1994). The underlying mechanism is poorly understood, although preliminary data suggest that plasmid multimerization might occur via extrachromosomal intermediates of oligomerized plasmids (Tikhomirova et al. 1986; Gatzke et al. 1994).

Targeted vector integration into the genome of *H. polymorpha* is sequence-dependent and difficult to achieve, due to a high percentage of nonhomologous recombination (Hansen 1990; Faber et al. 1992). However, frequencies of 1–22% could be reached with the methanol oxidase gene linearized within the homologous region (Faber et al. 1992).

9
Heterologous Gene Expression

The methanol oxidase (*MOX*) promoter, which is known as one of the most powerful promoters in yeast (van Dijken et al. 1976b; van der Klei et al. 1991), is currently almost exclusively employed to express heterologous genes in *H. polymorpha* (Gellissen et al. 1992). Expression from this promoter can be tightly regulated by changing the carbon source. Whereas glucose and ethanol shut the promoter off completely (Egli et al. 1980), glycerol is routinely used to obtain derepression. In contrast, methanol has a strong inducing effect (van Dijken et al. 1976b; Veenhuis et al. 1983; Guiseppin et al. 1988). The expression is regulated at the level of transcription (Roggenkamp et al. 1984). Only limited data are available about the mechanisms of regulation. Wild-type cells do not allow overexpression of methanol oxidase. Roggenkamp et al. (1989) have generated mutants which are able to synthesize *MOX* protein at quantities of about 70% of the total cellular protein. The mutation is undefined, but is most likely at the level of promoter regulation (Roggenkamp et al. 1989). Only recently, a detailed functional analysis of the *MOX* promoter region has been performed by measuring activity of a reporter protein expressed from different promoter deletions and by DNA-binding studies (Gödecke et al. 1994). The data suggest a complex promoter structure with one upstream repression sequence (URS) and two upstream activation sites (UAS), including one that is also present in the catalase promoter (Didion and Roggenkamp 1992). A promoter region encompassing 1.5 kb upstream of the *MOX* ATG is required for optimal expression of foreign proteins.

Besides the methanol oxidase, two other genes from *H. polymorpha* are known to be induced by methanol to comparably high levels, the dihydroxyacetone synthase (*DAS*) and the formate dehydrogenase (*FMD*). Both genes have been cloned (Janowicz et al. 1985; Hollenberg and Janowicz 1989), but so far only the *FMD* promoter has been used in a systematic way for the expression of heterologous proteins. Recent results indicate that, using the *FMD* promoter, expression levels can be reached which are even higher than those obtained with the *MOX* promoter (Gellissen et al. 1992).

10
Peroxisomal Biogenesis

Peroxisomes are almost ubiquitous organelles that exhibit an astounding versatility with regard to metabolic function and morphology (Tolbert 1981). Yeasts are ideal model organisms to study peroxisomes, because proliferation and enzymatic composition of these organelles are dependent on the growth medium provided and can thus easily be manipulatec. In this way, yeasts are able to metabolize substrates that can range from reduced C_1-compounds such as methanol, to long-chained fatty acids, n-alkanes, and alkylated amines (Veenhuis and Harder 1987). Before the discovery that peroxisome proliferation can be induced in *S. cerevisiae* by growth on oleate (Veenhuis et al. 1987; Kunau et al. 1987), methylotrophic yeasts were the only organisms of choice for peroxisomal studies. Especially in *H. polymorpha*, peroxisomes, just like the key methanol-metabolizing enzymes, can be induced to impressive levels. For example, in methanol-limited chemostat cultures they can occupy up to 80% of the cytoplasmic volume (Veenhuis et al. 1978; Roggenkamp et al. 1984).

Peroxisomes are surrounded by a single membrane, generally 0.5 to 1.5 μm in diameter. For many years, it was suggested that they form by budding from the endoplasmic reticulum (Tolbert 1981). More recently, however, biochemical and molecular studies revealed that peroxisomes develop from preexisting organelles (Lazarow and Fujiki 1985). In *H. polymorpha* under repressing conditions, that is if the cells are grown in the presence of glucose or ethanol, only one very small peroxisome is present (Veenhuis et al. 1979). If the cells are shifted to conditions which require peroxisomal enzymes, for example methanol, the organelle increases in size and subsequently new, small peroxisomes are formed by division (Veenhuis et al. 1978). Interestingly, Waterham et al. (1993) could isolate temperature-sensitive mutants of *H. polymorpha*, which completely lacked peroxisomes at the restrictive temperature. After a shift to the permissive temperature, peroxisomes were formed. Apparently they did not incorporate matrix or membrane proteins that were present in the cytosol. This raises the question whether *H. polymorpha* might be able to form peroxisomes by a process other than fission.

Peroxisomes have an important function in segregating the highly reactive hydrogen peroxide from the cytosol. It has been demonstrated that a peroxisomal, but not a cytoplasmic catalase can complement the inability of a catalase-negative

H. polymorpha mutant to grow on methanol (Hansen and Roggenkamp 1989). This is in line with the finding that most of the catalase is located in a narrow zone on the inner side of the peroxisomal membrane, likely to prevent leaking of H_2O_2 into the cytosol (Keizer et al. 1992).

Peroxisomes do not contain nucleic acids and ribosomes. All peroxisomal proteins are encoded in nuclear DNA. The proteins are synthesized on free ribosomes inside the cytoplasm and imported into peroxisomes posttranslationally (Borst 1989). Veenhuis et al. (1989) have shown that not all peroxisomes incorporate newly synthesized matrix proteins, but only those which are small and capable of fission.

The import of peroxisomal proteins is a process which, in comparison to other organelles, has only recently begun to be elucidated. The main reason for this is that almost without exception these proteins do not undergo proteolytic processing, which made the identification of a targeting signal a difficult task (Borst 1989). In vitro studies have indicated that the import process requires ATP but no membrane potential (Imanaka et al. 1987). The first peroxisomal targeting signal (PTS) has been identified on the firefly luciferase by expression in mammalian cells (Gould et al. 1989) and by import into peroxisomes in vitro (Myazawa et al. 1989). It consists of a carboxyl terminal tripeptide with the sequence Ser-Lys-Leu-COOH that is both necessary and sufficient to route cytosolic marker proteins into peroxisomes. The targeting function of this motif has been demonstrated to be highly conserved in evolution (Gödecke et al. 1989; Hansen and Roggenkamp 1989; Hansen et al. 1992; Gould et al. 1990). Recently, the topogenic signals of the key enzymes in methanol metabolism of *H. polymorpha*, dihydroxyacetone synthase, methanol oxidase, and catalase have been characterized (Hansen 1990; Hansen et al. 1992; Didion and Roggenkamp 1992). They all reside at the extreme C-terminus and exhibit certain similarities to the luciferase signal (DAS: Asn-Lys-Leu, MOX: Ala-Arg-Phe, catalase: Ser-Lys-Ile). There results and mutagenesis studies on the firefly PTS (Gould et al. 1989) have shown that a functional signal can be degenerated to a certain extent, but that there should always be a basic amino acid at the second position and a hydrophobic residue at the third (Roggenkamp 1992). Not all proteins, however, have a carboxyl terminal signal. For example, the rat 3-ketoacyl-CoA thiolase is synthesized as a larger precursor with an amino-terminal pre-piece, that is cleaved upon translocation into peroxisomes (Swinkels et al. 1991). The pre-piece contains a targeting signal different from the luciferase consensus, but similar to sequences at the N-termini of the watermelon glyoxysomal malate dehydrogenase and the amine oxidase of *H. polymorpha* (Faber et al. 1993; van der Klei et al. 1993). The signal of the malate dehydrogenase is functional in *H. polymorpha*, but processing of the precursor does not occur (van der Klei et al. 1993). So far, proteolytic processing has been observed only in higher eukaryotes. The amine oxidase of *H. plymorpha* does not contain a cleavable pre-piece (Faber et al. 1993). Recently, a consensus sequence (Ser-Leu/Ile-X_5-His-Leu) for the amino-terminal topogenic signal has been determined by import studies in *H. polymorpha* (Gietl et al. 1994). The import pathway for proteins containing an N-terminal PTS is independent from the targeting machinery for proteins with C-terminal signals, and both appear to be inducible by specific growth substrates

(McCollum et al. 1993; Faber et al. 1994). Interestingly, AMO is not routed into peroxisomes when expressed in the closely related *S. cerevisiae*, which shows that the amino-terminal targeting sequence is not universal (de Hoop et al. 1992).

11
Applied Aspects

Since Ogata et al. reported the first methylotrophic yeast in 1969, these yeasts have been the focus of constant attention from the biotechnological industry, albeit over the years for different reasons. The initial interest in using methylotrophic yeasts for the production of "single-cell protein" has decreased considerably, in part caused by low protein prizes and higher costs for methanol (Berry et al. 1987). Several processes have been described using *H. polymorpha* to produce metabolites, involving amino acids, nucleotide coenzymes, methylketones, dihydroxyacetone, and glycerol (for review: Harder and Brooke 1990). In recent years, however, with the cloning of the strong promoters of the methanol-metabolizing enzymes and the development of a transformation system, *H. polymorpha* has become, together with *Pichia pastoris*, a major focus for the high-yield expression of foreign proteins. *H. polymorpha* offers several advantages over other expression systems. *H. polymorpha* requires short fermentation times, and high levels of expression can already be obtained in glycerol when using the appropriate promoters. It is a Crabtree-negative yeast, yielding thus very high cell densities (>100 g dry weight per liter), the media used are comparably cheap, and since the amount of secreted proteins is low, it is possible to achieve purities of more than 90% heterologous protein in the medium with little proteolysis (Gellissen et al. 1992). Another advantage is the formation of integrants in extremely high copy number that are mitotically stable (Roggenkamp et al. 1986; Tikhomirova et al. 1986). Unlike *S. cerevisiae*, *H. polymorpha* rarely hyperglycosylates (Gellissen et al. 1992).

Highest productivities have been obtained in experiments which used fed-batch fermentation or continuous cultures, expression from the methanol oxidase or the formate dehydrogenase promoter, and multiple plasmid integration (Janowicz et al. 1991; Gellisen et al. 1992; Giuseppin et al. 1988). Janowicz et al. (1991) have produced S and L surface antigens of the hepatitis B virus intracellular to up to 2–8% of the total soluble cell protein. Expressed simultaneously in the same transformants, L and S antigens still constituted about 0.6 and 3.0% of cellular protein, respectively. The antigens sponstaneously assembled into particles that are suitable for vaccine development. Expression levels of around 1.5 g/l medium have been reported for hirudin, when the *S. cerevisiae* alpha factor (MFα1) leader sequence was used for protein secretion, and the *Schwanniomyces occidentalis* glucoamylase (Gellissen et al. 1991). In continuous culture, productivities of 5.5 mg/g biomass × 1 × h have been reached, expressing the *Cyamopsis tetragonoloba* (guar) α-galactosidase fused to the MFα1 leader (Giuseppin et al. 1988). Altogether, the protein yields of *H. polymorpha* are among the highest reported for yeast systems, and the productivity seems to be even better than what has been shown for *Pichia* sp. (Giuseppin et al. 1993).

12
Concluding Remarks

Within this chapter we have described briefly the system *H. polymorpha* with emphasis on the genetic and the molecular-biological aspects. We have tried to put it into a context with some of the major contributions to research and biotechnology made by this yeast in recent years, to point out the areas where *H. polymorpha* has particular advantages over other yeast systems. An increasing amount of available mutants, together with constantly improving tools for genetic manipulation, is expected to facilitate the difficult studies of peroxisomal biogenesis in the future. Powerful vector systems already make it possible to produce heterologous proteins at highest levels. During its relatively short scientific career, *H. polymorpha* has become an extremely valuable supplement to *S. cerevisiae*.

References

Agaphonov MO, Poznyakovski AI, Bogdanova AI, Ter-Avanesyan MD (1994) Isolation and characterization of the *LEU2* gene of *Hansenula polymorpha*. Yeast 10: 509–513

Berardi E, Thomas DY (1990) An efficient transformation method for *Hansenula polymorpha*. Curr Genet 18: 169–170

Berry DR, Russel I, Stewart GG (1987) Yeast biotechnology. Allen and Unwin, London

Bogdanova AI, Agaphonov MO, Ter-Avanesyan MD (1995) Plasmid reorganization during integrative transformation in *Hansenula polymorpha*. Yeast 11: 343–353

Borst P (1989) Peroxisome biogenesis revisited. Biochim Biophys Acta 1008: 1–13

Broach JR, Li YY, Feldmann J, Jayaram M, Abraham J, Nasmyth KA, Hicks JB (1982) Localizaton and sequence analysis of yeast origins of replication. Cold Spring Harbor Symp Quant Biol 47: 1165–1173

Bruinenberg PG, Evers M, Waterham HR, Kuipers J, Arnberg AC, AB G (1989) Cloning and sequencing of the peroxisomal amine oxidase gene from *Hansenula polymorpha*. Biochim Biophys Acta 1008: 157–167

Bruinenberg PG, Blaauw M, Kazemier B, AB G (1990) Cloning and sequencing of the malate synthase gene from *Hansenula polymorpha*. Yeast 6: 245–254

Bystrykh LV, Sokolov AP, Trotsenko YA (1981) Purification and properties of dihydroxyacetone synthase from the methylotrophic yeast *Candida boidinii*. FEBS Lett 132: 324–328

Cregg JM, van der Klei IJ, Sulter GJ, Veenhuis M, Harder W (1990) Peroxisome-deficient mutants of *Hansenula polymorpha*. Yeast 6: 87–97

de Hoop MJ, Valkema R, Kienhuis CBM, Ab G (1992) The peroxisomal import signal of amine oxidase from the yeast *Hansenula polymorpha* is not universal. Yeast 8: 243–252

de Koning W, Gleeson MA, Harder W, Dijkhuizen L (1987) Regulation of methanol metabolism in the yeast *Hansenula polymorpha*. Isolation and characterization of mutants blocked in methanol assimilatory enzymes. Arch Microbiol 17: 375–382

Didion T, Roggenkamp R (1992) Targeting signal of the peroxisomal catalase in the methylotrophic yeast *Hansenula polymorpha*. FEBS Lett 303: 113–116

Dobson MJ, Tuite MF, Roberts NA, Kingsman AJ, Kingsman SM, Perkins RE, Conroy SC, Fothergill LA (1982) Conservation of high efficiency promoter sequences in *Saccharomyces cerevisiae*. Nucleic Acids Res 10: 2625–2637

Dohmen RJ, Strasser AWM, Höner CB, Hollenberg CP (1991) An efficient transformation procedure enabling long-term storage of competent cells of various yeast genera. Yeast 7: 691–692

Douma AC, Veenhuis M, de Koning W, Evers M, Harder W (1985) Dihydroxyacetone synthase is localized in the peroxisomal matrix of methanol-grown *Hansenula polymorpha*. Arch Microbiol 143: 237–243

Eggeling L, Sahm H (1980) Direct enzymatic assay for alcohol oxidase, alcohol dehydrogenase and formaldehyde dehydrogenase in colonies of *Hansenula polymorpha*. Appl Environ Microbiol 42: 268–275

Eggeling L, Sahm H (1981) Regulation of alcohol oxidase synthesis in *Hansenula polymorpha*: oversynthesis during growth on mixed substrates and induction by methanol. Arch Microbiol 127: 119–124

Egli H, van Dijken JP, Veenhuis M, Harder W, Fiechter A (1980) Methanol metabolism in yeasts: regulation of the synthesis of catabolic enzymes. Arch Microbiol 124: 115–121

Egli H, Kaeppeli O, Fiechter A (1982) Regulatory flexibility of methylotrophic yeasts in chemostat cultures; simultaneous assimilation of glucose and methanol at fixed dilution rates. Arch Microbiol 131: 1–7

Faber KN, Swaving GJ, Faber F, AB G, Harder W, Veenhuis M, Haima P (1992) Chromosomal targeting of replicating plasmids in the yeast *Hansenula polymorpha*. J Gen Microbiol 138: 2405–2416

Faber KN, Haima P, de Hoop MJ, Harder W, Veenhuis M, AB G (1993) Peroxisomal amine oxidase of *Hansenula polymorpha* does not require its SRL-containing C-terminal sequence for the targeting. Yeast 9: 331–338

Faber KN, Haima P, Harder W, Veenhuis M, AB G (1994a) Highly-efficient electrotransformation of the yeast *Hansenula polymorpha*. Curr Genet 25: 305–310

Faber KN, Haima P, Gietl C, Harder W, AB G, Veenhuis M (1994b) The methylotrophic yeast *Hansenula polymorpha* contains an inducible import pathway for peroxisomal matrix proteins with an N-terminal targeting signal (PTS2 proteins). Proc Natl Acad Sci USA 91: 12985–12989

Fellinger AJ, Verbakel JMA, Veale RA, Sudberry PE, Bom IJ, Overbeke N, Verrips CT (1991) Expression of α-galactosidase from *Cyamposis tetragonobia* (guar) by *Hansenula polymorpha*. Yeast 7: 463–474

Fujii T, Tonomura K (1975) Purification and properties of catalase from a methanol-utilizing yeast, *Candida* sp. N-16. Agric Biol Chem 39: 2325–2330

Gatzke R, Weydemann U, Janowicz ZA, Hollenberg CP (1995) Stable multiple integration of vector sequences in *Hansenula polymorpha* (in press)

Gellissen G, Janowicz ZA, Merckelbach A, Piontek M, Keup P, Weydemann U, Hollenberg CP, Strasser AWM (1991) Heterologous gene expression in *Hansenula polymorpha*: efficient secretion of glucoamylase. Bio/Technology 9: 291–295

Gellissen G, Weydemann U, Strasser AWM, Piontek M, Janowicz ZA, Hollenberg CP (1992) Progress in developing methylotrophic yeasts as expression systems. Trends Biotechnol 10: 413–417

Gellissen G, Hollenberg CP, Janowicz ZA (1994) Gene expression in methylotrophic yeasts. In: Smith A (ed) Gene expression in recombinant micoroorganisms. Marcel Dekker, New York, pp 195–239

Gietl C, Faber KN, van der Klei IJ, Veenhuis M (1994) Mutational analysis of the N-terminal topogenic signal of watermelon glyoxysomal malate dehydrogenase using the heterologous host *Hansenula polymorpha*. Proc Natl Acad Sci USA 91: 3151–3155

Giuseppin MLF, van Eijk HML, Bes BCM (1988) Molecular regulation of alcohol oxidase in *Hansenula polymorpha* in continuous cultures. Biotechnol Bioeng 32: 577–583

Gleeson MA (1986) The genetic analysis of the methylotrophic yeast *Hansenula polymorpha*. PhD Thesis, University of Sheffield, Sheffield

Gleeson MA, Sudbery PE (1988) Genetic analysis in the methylotrophic yeast *Hansenula polymorpha*. Yeast 4: 293–303

Gleeson MA, Ortori GS, Sudbery PE (1986) Transformation of the methylotrophic yeast *Hansenula polymorpha*. J Gen Microbiol 132: 3459–3465

Gödecke A, Veenhuis M, Roggenkamp R, Janowicz ZA, Hollenberg CP (1989) Biosynthesis of the peroxisomal dihydroxyacetone synthase form *Hansenula polymorpha* in *Saccharomyces cerevisiae* induces growth but not proliferation of peroxisomes. Curr Genet 16: 13–20

Gödecke S, Eckart M, Janowicz ZA, Hollenberg CP (1994) Identification of sequences responsible for transcriptional regulation of the strongly expressed methanol oxidase-encoding gene in *Hansenula polymorpha*. Gene 139: 35–42

Goodman J (1985) Dihydroxyacetone synthase is an abundant constituent of the methanol-induced peroxisome of *Candida boidinii*. J Biol Chem 260: 7108–7114

Gould SJ, Keller G-A, Hosken N, Wilkinson J, Subramani S (1989) A conserved tripeptide sorts proteins to peroxisomes. J Cell Biol 108: 1657–1664

Gould SJ, Keller G-A, Schneider M, Howell SH, Garrard LJ, Goodman JM, Distel B, Tabak H, Subramani S (1990) Peroxisomal import is conserved between yeast, plants, insects and mammals. EMBO J 9: 85–90

Hansen H (1990) Struktur und Funktion von peroxisomalen Importsignalen in der methylotrophen Hefe *Hansenula polymorpha*. PhD Thesis, Heinrich-Heine-Universität Düsseldorf, Düsseldorf

Hansen H, Roggenkamp R (1989) Functional complementation of catalase-defective peroxisomes in a methylotrophic yeast by import of the catalase A from *Saccharomyces cerevisiae*. Eur J Biochem 184: 173–179

Hansen H, Didion T, Thiemann A, Veenhuis M, Roggenkamp R (1992) Targeting sequences of the two major peroxisomal proteins in the methylotrophic yeast *Hansenula polymorpha*. Mol Gen Genet 235: 269–278

Harder W, Brooke AG (1990) Methylotrophic yeasts. In: Verachtert H, de Mot R (eds) Yeast biotechnology and biocatalysis. Marcel Dekker, New York, pp 395–428

Hollenberg CP, Janowicz ZA (1989) DNA molecules coding for *FMDH* control region and structured gene for a protein having *FMDH*-activity and their uses. European Patent, EP 0299108-A1

Imanaka T, Small GM, Lazarow PB (1987) Translocation of acyl-CoA-oxidase into peroxisomes requires ATP hydrolysis but not a membrane potential. J Cell Biol 105: 2915–2922

Janowicz Z, Eckart M, Drewke C, Roggenkamp R, Hollenberg CP, Maat J, Ledeboer AM, Visser C, Verrips CT (1985) Cloning and characterization of the DAS gene encoding the major methanol assimilatory enzyme from the methylotrophic yeast *Hansenula polymorpha*. Nucleic Acids Res 13: 3043–3062

Janowicz ZA, Melber K, Merckelbach A, Jacobs E, Harford N, Comberbach M, Hollenberg CP (1991) Simultaneous expression of the S and L surface antigen of hepatitis B, and formation of mixed particles in the methylotrophic yeast *Hansenula polymorpha*. Yeast 7: 431–443

Jones JG, Bellion E (1991) Methanol oxidation and assimilation in *Hansenula polymorpha*. Biochem J 280: 475–481

Kato N, Omori Y, Tani Y, Ogata K (1976) Alcohol oxidase of *Kloeckera* sp. and *Hansenula polymorpha*. Catalytic properties and subunit structures. Eur J Biochem 64: 341–350

Kato N, Higuchi T, Sakazawa C, Nishizawa T, Tani Y, Yamada H (1982) Purification and properties of a transketolase responsible for formaldehyde fixation in a methanol-utilizing yeast, *Candida boidinii* (*Kloeckera* sp.) No 2201. Biochim Biophys Acta 715: 143–150

Kearsey S (1994) Structural requirements for the function of a yeast chromosomal replicator. Cell 37: 299–307

Keizer I, Roggenkamp R, Harder W, Veenhuis M (1992) Location of catalase in crystalline peroxisomes of methanol-grown *Hansenula polymorpha*. FEMS Mirobiol Lett 71: 7–11

Klebe RJ, Harriss JV, Sharp ZD, Douglas MG (1983) A general method for polyethylene-glycol-induced genetic transformation of bacteria and yeast. Gene 25: 333–341

Kunau WH, Kionka C, Ledebur A, Mateblowski M, Morena de la Garza M, Schultz-Borchard U, Thieringer R, Veenhuis M (1987) β-Oxidation systems in eukaryotic microorganisms.

In: Fahimi HD, Sies H (eds) Peroxisomes in biology and medicine. Springer, Berlin Heidelberg New York, pp 128–140

Kurtzman CP (1984) Synonomy of the yeast genera *Hansenula* and *Pichia* demonstrated through comparisons of deoxyribonucleic acid relatedness. Antonie Leeuwenhoek J Microbiol 50: 209–217

Lazarow PB, Fujiki Y (1985) Biogenesis of peroxisomes. Annu Rev Cell Biol 1: 489–530

Ledeboer AM, Edens L, Maat J, Visser C, Bos JW, Verrips CT, Janowicz Z, Eckart M, Roggenkamp R, Hollenberg CP (1985) Molecular cloning and characterization of a gene coding for methanol oxidase in *Hansenula polymorpha*. Nucleic Acids Res 13: 3063–3082

Lee J-D, Komagata K (1980) Taxonomic study of methanol-assimilating yeasts. J Gen Appl Microbiol 26: 133–158

Levine DW, Cooney CL (1973) Isolation and characterization of thermotolerant methanol-utilizing yeasts. Appl Microbiol 26: 982–990

Lodder J (1970) The yeasts, a taxonomic study. North-Holland Publishing, Amsterdam

Marri L, Rossolini GM, Satta G (1993) Chromosome polymorphisms among strains of *Hansenula polymorpha* (syn. *Pichia angusta*). Appl Environ Microbiol 59: 939–941

McCollum D, Monosov E, Subramani S (1993) The *pas8* mutant of *Pichia pastoris* exhibits the peroxisomal protein import deficiencies of Zellweger syndrome cells – the PAS8 protein binds to the COOH-terminal tripeptide peroxisomal targeting signal, and is a member of the TPR protein family. J Cell Biol 121: 761–774

Merckelbach A, Gödecke S, Janowicz ZA, Hollenberg CP (1993) Cloning and sequencing of the *URA3* locus of the methylotrophic yeast *Hansenula polymorpha* and its use for the generation of a deletion by gene replacement. Appl Microbiol Biotechnol 40: 361–364

Mu D, Janes SM, Smith AJ, Brown DE, Dooley DM, Klinman JP (1992) Tyrosine codon corresponds to topa quinone at the active site of copper amine oxidase. J Biol Chem 267: 7979–7982

Myazawa S, Osumi T, Hashimoto T, Ohno K, Miura S, Fujiki Y (1989) Peroxisome targeting signal of rat liver acyl-coenzyme A oxidase resides at the carboxy terminus. Mol Cell Biol 9: 83–91

Ogata K, Nishikawa H, Ohsugi M (1969) A yeast capable of utilizing methanol. Agric Biol Chem 33: 1519–1520

Roggenkamp R (1992) Targeting signals for protein import into peroxisomes. Cell Biochem Funct 10: 193–199

Roggenkamp R, Sahm H, Wagner F (1974) Microbial assimilation of methanol. Induction and function of catalase in *Candida boidinii*. FEBS Lett 41: 283–286

Roggenkamp R, Sahm H, Hinkelmann W, Wagner F (1975) Alcohol oxidase and catalase in peroxisomes of methanol-grown *Candida boidinii*. Eur J Biochem 59: 231–236

Roggenkamp R, Janowicz Z, Stanikowski B, Hollenberg CP (1984) Biosynthesis and regulation of the peroxisomal methanol oxidase from the methylotrophic yeast *Hansenula polymorpha*. Mol Gen Genet 194: 489–493

Roggenkamp R, Hansen H, Eckart M, Janowicz Z, Hollenberg CP (1986) Transformation of the methylotrophic yeast *Hansenula polymorpha* by autonomous replication and integration vectors. Mol Gen Genet 202: 302–308

Roggenkamp R, Didion T, Kowallik KV (1989) Formation of irregular giant peroxisomes by overproduction of the crystalloid core protein methanol oxidase in the methylotrophic yeast *Hansenula polymorpha*. Mol Cell Biol 9: 988–994

Sahm H, Wagner F (1973a) Microbial assimilation of methanol. The ethanol- and methanol-oxidizing enzymes of *Candida boidinii*. Eur J Biochem 36: 250–256

Sahm H, Wagner F (1973b) Microbial assimilation of methanol. Properties of formaldehyde dehydrogenase and formate dehydrogenase from *Candida boidinii*. Arch Microbiol 90: 263–268

Sanchez S, Demain AL (1977) Enrichment of auxotrophic mutants of *Hansenula polymorpha*. Eur J Appl Microbiol 4: 45–49

Schütte H, Flossdorf J, Sahm H, Kula MR (1976) Purification and properties of formaldehyde dehydrogenase and formate dehydrogenase from *Candida boidinii*. Eur J Biochem
62: 151-160

Sibirny AA, Titorenko VI, Gonchar MV, Ubiyvovk VM, Ksheminskaya GP, Vitvitskaya
OP (1988) Genetic control of methanol utilization in yeasts. J Basic Microbiol 28: 293-
319

Sibirny AA, Ubiyvovk VM, Gonchar MV, Titorenko VI, Voronovsky AY, Kapultsevich JG,
Bliznik KM (1990) Reactions of direct formaldehyde oxidation to CO_2 are non-essential
for energy supply of yeast methylotrophic growth. Arch Microbiol 154: 566-575

Sudbery PE, Gleeson MAG (1989) Genetic manipulation of methylotrophic yeasts. In:
Walton EF, Yarranton GT (eds) Molecular and cell biology of yeasts. Blackie, London, pp
304-329

Sulter GJ, Vrieling EG, Harder W, Veenhuis M (1993a) Synthesis and subcellular location of
peroxisomal membrane proteins in a peroxisome-deficient mutant of the yeast
Hansenula polymorpha. EMBO J 12: 2205-2210

Sulter GJ, Waterham HR, Vrieling EG, Goodman JM, Harder W, Veenhuis M (1993b)
Expression and targeting of a 47 kDa integral peroxisomal membrane protein of *Can-
dida boidinii* in wild-type and a peroxisome-deficient mutant of *Hansenula polymorpha*.
FEBS Lett 315: 211-216

Swinkels BW, Gould SJ, Bodnar AG, Rachubinski RA, Subramani S (1991) A novel, cleavable
peroxisomal targeting signal at the amino-terminus of the rat 3-ketoacyl-CoA thiolase.
EMBO J 10: 3255-3262

Tan X, Waterham HR, Veenhuis M, Cregg JM (1995) The *Hansenula polymorpha* PER8 gene
encodes a novel peroxisomal integral membrane protein involved in proliferation. J Cell
Biol 128: 307-319

Teunisson DJ, Hall HH, Wickerham LJ (1960) *Hansenula angusta*, an excellent species for
the demonstration of the coexistence of haploid and diploid cells in homothallic yeast.
Mycologia 52: 184-188

Tikhomirova LP, Ikonomova RN, Kuznetsova EN (1986) Evidence for autonomous replication and stabilization of recombinant plasmids in the transformants of yeast *Hansenula
polymorpha*. Curr Genet 10: 741-747

Tikhomirova LP, Ikonomova RN, Kuznetsova EN, Fodor II, Bystrykh LV, Aminova LR,
Trotsenko YA (1988) Transformation of methylotrophic yeast *Hansenula polymorpha*:
Cloning and expression of genes. J Basic Microbiol 28: 353-351

Titorenko VI, Waterham HR, Haima P, Harder W, Veenhuis M (1993a) Peroxisome biogenesis in *Hansenula polymorpha*: different mutations in genes, essential for peroxisome
biogenesis, cause different peroxisomal mutant phenotypes. FEMS Microbiol Lett 74:
143-148

Titorenko VI, Waterham HR, Cregg JM, Harder W, Veenhuis M (1993b) Peroxisome biogenesis in the yeast *Hansenula polymorpha* is controlled by a complex set of interacting
gene products. Proc Natl Acad Sci USA 90: 7470-7474

Tolbert NE (1981) Metabolic pathways in peroxisomes and glyoxysomes. Annu Rev
Biochem 50: 133-157

Tolstorukov II, Benevolenskii SV (1978) Study of the mechnism of mating and self-
diploidization in haploid yeasts of *Pichia pinus*. Communication I. Bipolarity of mating.
Genetika 14: 519-526

Tolstorukov II, Benevolenskii SV (1980) Study of the mechanism of mating and self-
diploidization in haploid yeasts of *Pichia pinus*. Communication II. Mutations in the
mating type-locus. Genetika 16: 1335-1342

Uotila L, Koivusalo M (1974) Formaldehyde dehydrogenase from human liver. Purification,
properties and evidence for the formation of glutathione thioesters by the enzyme. J Biol
Chem 249: 7653-7663

van der Klei IJ, Harder W, Veenhuis M (1991) Biosynthesis and assembly of alcohol oxidase,
a peroxisomal matrix protein in methylotrophic yeasts: a review. Yeast 7: 195-209

van der Klei IJ, Faber KN, Keizer-Gunnink I, Gietl C, Harder W, Veenhuis M (1993) Water-melon glyoxysomal malate dehydrogenase is sorted to peroxisomes of the methylotrophic yeast *Hansenula polymorpha*. FEBS Lett 334: 128–132

van Dijken JP (1976) Oxidation of methanol by yeasts. PhD Thesis, University of Groningen, Groningen

van Dijken JP, Bos P (1981) Utilization of amines by yeasts. Arch Microbiol 128: 320–324

van Dijken JP, Oostra-Demkes GJ, Otto R, Harder W (1976a) S-Formylglutathione: the substrate for formate dehydrogenase in methanol utilizing yeasts. Arch Microbiol 111: 77–83

van Dijken JP, Otto R, Harder W (1976b) Growth of *Hansenula polymorpha* in a methanol-limited chemostat. Physiological responses due to the involvement of methanol oxidase as a key enzyme in methanol metabolism. Arch Microbiol 111: 137–144

van Dijken JP, Harder W, Beardsmore AJ, Quale RR (1978) Dihydroxyacetone: an interme-diate in the assimilation of methanol by yeasts? FEMS Microbiol Lett 4: 97–102

Veale RA, Giuseppin LF, van Eijk HMJ, Sudbery PE, Verrips CT (1992) Development of a strain of *Hansenula polymorpha* for the efficient expression of guar α-galactosidase. Yeast 8: 361–372

Veenhuis M, Harder W. (1987) Metabolic significance and biogenesis of microbodies in yeasts. In: Fahimi HD, Sies H (eds) Peroxisomes in biology and medicine. Springer, Berlin Heidelberg New York, pp 436–457

Veenhuis M, van Dijken JP, Pilon SAF, Harder W (1978) Development of crystalline peroxi-somes in methanol-grown cells of the yeast *Hansenula polymorpha* and its relation to environmental conditions. Arch Microbiol 117: 153–163

Veenhuis M, Keizer I, Harder W (1979) Characterization of peroxisomes in glucose-grown *Hansenula polymorpha* and their development after transfer of cells into methanol-containing media. Arch Microbiol 120: 167–175

Veenhuis M, van Dijken JP, Harder W (1983) The significance of peroxisomes in the metabolism of one-carbon compounds in yeast. Adv Microbiol Physiol 24: 1–82

Veenhuis M, Mateblowski M, Kunau WH, Harder W (1987) Proliferation of microbodies in *Saccharomyces cerevisiae*. Yeast 3: 77–85

Veenhuis M, Sulter G, van der Klei I, Harder W (1989) Evidence for functional heterogeneity among microbodies in yeasts. Arch Microbiol 151: 105–110

Verduyn C, Postman E, Scheffers WA, van Dijken JP (1992) Effect of benzoic acid on metabolic fluxes in yeasts: a continuous-culture study on the regulation of respiration and alcoholic fermentation. Yeast 8: 501–517

Vonck J, van Bruggen EFJ (1990) Electron microscopy and image analysis of two-dimen-sional crystals and single molecules of alcohol oxidase from *Hansenula polymorpha*. Biochim Biophys Acta 1038: 74–79

Waites MJ, Quayle JR (1981) The interrelationship between transketolase and dihydroxyac-etone synthase activities in the methylotrophic yeast *Hansenula polymorpha*. J Gen Microbiol 124: 309–316

Waterham HR, Titorenko VI, Swaving GJ, Harder W, Veenhuis M (1993) Peroxisomes in the methylotrophic yeast *Hansenula polymorpha* do not necessarily derive from pre-existing organelles. EMBO J 12: 4785–4794

Waterham HR, Titorenko VI, Haima P, Cregg JM, Harder W, Veenhuis M (1994) The *Hansenula polymorpha* PER1 gene is essential for peroxisome biogenesis and encodes a peroxisomal matrix protein with both carboxy- and amino-terminal targeting signals. J Cell Biol 127: 737–749

Wickerham LJ (1970) *Hansenula* H. et P. Sydow. In: Lodder J (ed) The yeasts. North-Holland Publishing, Amsterdam, pp 226–315

CHAPTER 10

Yarrowia lipolytica

Gerold Barth[1] and Claude Gaillardin[2]

1
History of *Yarrowia lipolytica* Research

Interest in *Candida lipolytica* (Harrison) Diddens et Lodder 1942 initially arose from its rather uncommon physiological characteristics. Strains of this species were more often isolated from lipid- or protein-containing substrates like cheese or sausage than from sugar-containing substrates. Indeed, strains of *Candida lipolytica* used few sugars (mainly glucose) as carbon source, but did readily assimilate various polyalcohols, organic acids, or normal paraffins. They were noted in the late 1940s by dairy technologists (Peters and Nelson 1948a,b) for their high extracellular protease and lipase activities, although these purified enzymes were never put to work industrially.

With the emergence of single-cell protein projects in the mid 1960s, a strong industrial interest stemmed from the fact that strains of this species were able to use n-paraffins which were cheap and abundant at that time, as sole carbon source. It was also observed that *C. lipolytica* was able to produce high amounts of organic acids (2-ketoglutaric acid and citric acid) when grown on these substrates (Tsugawa et al. 1969). Large-scale industrial production of citric acid or SCP using *Y. lipolytica* thus permitted the accumulation of extensive data on its behavior in very large fermentors.

The species was classified as a *Candida* at that time, since no sexual state had been described. The perfect form of *C. lipolytica* was identified in the late 1960s by Wickerham at the Northern Regional Research Laboratory of the USDA at Peoria. A culture isolated in 1945 from a jar of fiber tailings in a corn processing plant was found to form asci attached to hyphal elements when put on suitable media. One to four spores of various size and shape could be isolated from these asci, but spore viability was found to be very low. Two mating types, called A and B, were identified among the progeny. Nearly all other wild-type isolates from the species would mate to one of these two types, albeit at very low frequency, suggesting that most

[1] Institut für Mikrobiologie, Technische Universität Dresden, Mommsenstr. 13, 01062 Dresden, Germany
[2] Laboratoire de Génétique Moléculaire et Cellulaire, Institut National de la Recherche Agronomiques – Centre National de la Recherche Scientifique, Institut National Agronomique Paris-Grignon, 78850 Thiverval-Grignon, France

natural isolates are haploid (or near haploid). The perfect form was reclassified first as *Endomycopsis lipolytica* (Wickerham et al. 1970), then as *Saccharomycopsis lipolytica* (Yarrow 1972), and finally as *Yarrowia lipolytica* (van der Walt and von Arx 1980). Further details on its taxonomic position will be discussed below.

Wickerham's strains were sent both to Mortimer's laboratory in Berkeley and to Heslot's group in Paris, which independently started genetic studies on this species. Initial studies were plagued by low mating frequencies, poor sporulation ability, and low ascospore viability. These defects could be partially alleviated by inbreeding programs, but no perfect set of strains could be obtained: mating frequencies remained low, and spore viability plateaued at around 80% (Gaillardin et al. 1973; Ogrydziak et al. 1978). Protoplast fusion was attempted to overcome some of these difficulties (Esser and Stahl 1976; Weber et al. 1980). Later groups joining the *Y. lipolytica* club initiated new and independent inbreeding programs, starting with different strains, in Poland (Bojnanska 1977), East Germany (Kurischko et al. 1983; Barth and Weber 1985), and in private companies (British Petroleum or Pfizer USA). As a result, several inbred lines of *Y. lipolytica* exist nowadays, and linkage groups were defined by tetrad dissection or chromosome loss in several of these lines (Ogrydziak et al. 1978, 1982; Kurischko 1984, 1986; De Zeeuw, pers. comm.). It is still unclear today how conserved these linkage groups are across the different lines, which probably differ grossly at the level of chromosome structure (see below); clearly, the different lines are poorly interfertile; genetic exchange between them is difficult and commonly requires several rounds of backcrosses before acceptable behavior is restored.

Happily, however, these various groups entered *Y. lipolytica* genetics for different reasons and focused on specific aspects of its biology: Krebs cycle (Akiyama et al. 1973a,b; Finogenova et al. 1982; Barth 1985), lysine metabolism (Gaillardin et al. 1975), secretion of extracellular enzymes (Ogrydziak and Mortimer 1977), alkane degradation (Bassel and Mortimer 1982), mating and parasexual processes (Weber 1979; Barth and Weber 1984; Kurischko 1986), mitochondrial genetics (Matsuoka et al. 1982), virus-like particle (Groves et al. 1983; Tréton et al. 1985), ribosomal RNA genes (van Heerikhuizen et al. 1985). An exhaustive review on this early work has been published (Heslot 1990).

A fair amount of data both on genetic and physiological aspects of this species was thus accumulated in the mid 1980s, when finally a transformation system became available (Davidow et al. 1985; Gaillardin et al. 1985). Gene cloning and successful expression of heterologous proteins, which appeared to be efficiently secreted to the growth medium, initiated a second wave of interest in this yeast (for an early review, see Gaillardin and Heslot 1988). The availability of an amazingly efficient and precise system for integrative transformation, the fine dissection of the processing pathway of an abundantly secreted extracellular protease (Matoba et al. 1988), the discovery of a signal recognition particle closely resembling that of higher eukaryotes (Poritz et al. 1988), all seemed to pave the way for the rapid development of *Y. lipolytica* as a model organism for studies on protein secretion. Some tools were, however, still missing, and forcefully resisted availability: for instance, and for reasons very unclear at that time, no replicative vector could be built for *Y. lipolytica*, although their construction from chromosomal origins of replication seemed straightforward in all other yeast species (Wing and Ogrydziak 1985). This problem was solved in a typical *Y. lipolytica* way, by looking for one thing and finding another (Fournier et al. 1991); centromeric *ARS* were isolated and filled one of the last gaps in the genetic tools required. With the recent development of powerful gene amplification systems (Le Dall et al. 1994), the identification of a retrotransposon (Schmid-Berger et al. 1994), and the character-ization of strong regulated or constitutive promoters (Strick et al. 1992; Barth and Scheuber 1993; Blanchin-Roland et al. 1994), the stage is now set for concentrating on specific biological problems in this yeast.

Areas of current interest include mechanisms of gene expression and protein secretion (contrary to most other yeast species, *Y. lipolytica* appears dedicated to that function under certain conditions) which are obviously application-linked. More fundamental topics concern the structure and functioning of the genome, which in some aspects radically differ from what is seen in other yeasts: structure and maintenance of polymorphic and dispersed rRNA genes, absolute require-ment for a centromeric function for the maintenance of extrachromosomal plas-mids, and genetic structure of natural populations which consist of widely divergent haploid lines. Other aspects of *Y. lipolytica* biology have begun to be explored, such as determinants of the mating type, control of the dimorphic transition from yeast to hyphae, alkane and fatty acid metabolism, glyoxylic path-way, peroxisome biogenesis, etc.

In several of these fields, *Y. lipolytica* proves to be a superb model organism, both exquisitely original and generally meaningful, thus plainly rewarding the often painful days needed to develop its genetic system. It was a common joke among the few *Y. lipolytica* aficionados who struggled over 20 years or so to tame the beast, that the yeast would never do anything as expected. The final message, however, was never dull and becomes increasingly exciting as we now rapidly proceed in deciphering it.

2
Physiology/Biochemistry/Cell Structure

2.1
Occurrence in Nature

Yarrowia lipolytica strains are readily isolated from dairy products such as cheese, yoghurts, but also from kefir, shoyu, or from salads containing meat or shrimps. This inability to survive under anaerobic conditions permits their easy elimination from dairy products, in contrast to *Kluyveromyces marxianus* strains, for example (Mc Kay 1992). *Y. lipolytica* is not considered as a pathogenic species, probably (among other reasons) because its maximum growth temperature seldom exceeds 32–34 °C. Intravenous injection of *Y. lipolytica* cells (Holzschu et al. 1979) into normal or cortisone-stressed mice was not associated with any mortality (in con-trast to *Candida albicans* or *Candida tropicalis*), and no viable cell could be rescued 6 days after the injection (whereas occasional persistence was observed up to 30 days with, e.g., *C. utilis*, *C. maltosa*, or *K. fragilis*).

2.2
Main Substrates and Biochemical Tests

The most common assimilated substrates (data from Kreger-van Rij 1984; Barnett et al. 1990) are listed in Table 1. Very few compounds may actually be used as carbon or nitrogen sources. Numerical taxonomy was performed on the basis of classical biochemical tests, and included *C. lipolytica* among a heterogeneous subgroup of *Candida* nonfermenting species (Campbell 1975). Poncet and Arpin (1965) devised a set of six substrates for rapid identification of *C. lipolytica* among non-fermenting *Candida* which do not assimilate nitrates, whereas Barnett et al. (1990) suggest a minimal set of 13 substrates, from which five rapid tests can be derived to distinguish *Yarrowia* from any other yeast species. Some controversy exists concerning urease production by *Y. lipolytica* strains: whereas Sen and Komagata (1979) and Barnett et al. (1990) describe the species as urease-positive, Booth and Vishniac (1987) have found no evidence for urease activity in *Y. lipolytica* strains.

2.3
Phylogenetic Relationships

Several observations early suggested that *Y. lipolytica* may have diverged considerably from other ascomycetous yeasts: high GC content (see Sect. 4.2.1), unusual structure of rDNA genes (see Sect. 4.2.4), coupled with a lack of RNA polymerase I consensus sequences found in other yeasts (van Heerikhuizen et al. 1985), higher eukaryotic-like size of snRNA (Roiha et al. 1989) and of 7S RNA (Poritz et al. 1988). Homologous genes tend to display a low level of similarity (typically in the 50–60% range at amino acid level) with their counterparts in *S. cerevisiae*, *K. lactis*, or *C. albicans*, thus hindering in most cases attempts to clone homologues by hybridization. Naumova et al. (1993) showed that *Y. lipolytica* genes did not hybridize detectably with DNA from species of the genera *Saccharomycopsis*, *Endomyces*, or *Endomycopsella*, some of which were formerly classified in the same genus as *Y. lipolytica*. Similarly, few genes of *Y. lipolytica* seem to be directly expressed in *S. cerevisiae*, suggesting that RNA polymerase II promoters and/or associated transcriptional factors have diverged considerably. Recent data on 7S RNA genes (see Sect. 4.2.4), on intron structure, or on *ARS* functioning (see Sect. 4.2.3) all confirmed that this species was quite peculiar when compared to most other yeasts. Evolutionary trees based on sequence comparison of genes encoding well-conserved functions (glycolytic genes, ribosomal RNA genes) locate *Y. lipolytica* on an isolated branch, clearly separated from *S. pombe* on one hand, and from the bulk of other ascomycetous yeasts on the other (Barns et al. 1991; Okuma et al.1993). A truly positive aspect of this divergence is that the observation of structural conser-

Table 1

Author	Place	Strain	Origin	Comment
Barth	Basel/Dresden	B204-12C B204-12D	YB423-3 YB423-12	Inbreeding program see Table 2
Chattoo	Baroda	B204 series	From Barth	
Dominguez	Salamanca	E129 series CX series	From Gaillardin From Ogrydziak	
Gaillardin	Grignon	E129 series E150 series	W29 (Paris, waste water) YB423-12	Inbreeding program see Table 2
Kohlwein	Graz	B204 series	From Barth	
Kurischko	Jena	K35-6 KF series	H222 (Leipzig, soil) CBS6124-2, CX series	Inbreeding program see Table 2
Matsuoka	Osaka	ATCC44601		
Nga	Singapore	ML15-29 21501-4	From Gaillardin and Ogrydziak series Yeast collection (Bratislava)	
Novotny	Prag	CCY29-26-3 CCY29-26-8		
Ogrydziak	Davis	DX series CX series	YB423-3, YB423-12, YB421	Inbreeding program see Table 2
Paszewski	Warsaw	CL1, CL2 A101, 115	Isolates from Polish areas	
Rachubinski	Edmonton	E129 series	From Gaillardin	
Schweizer	Erlangen	B204 series	From Barth	
Strick	Groton			
Young	Birmingham	E129 series Y148	From Gaillardin	Inbreeding program (unpubl.)

vation between *Y. lipolytica* and other yeast genes is much more likely to reflect functional constraints than when it happens between closely related yeasts.

2.4
Dimorphism

Wild-type strains of *Y. lipolytica* exhibit various colony shapes, ranging from smooth and glistening to heavily convoluted and mat. The colonial morphology is determined both by the growth conditions (aeration, carbon and nitrogen sources, pH, etc.) and by the genetic background of the strain. Wild-type isolates may give rise to various colonial types, each giving rise to the other types: the basis for this instability, which is not seen with laboratory strains, is unknown.

Y. lipolytica is a natural dimorphic fungus, which forms yeast cells, pseudohyphae, and septate hyphae (van der Walt and von Arx 1980; Barnett et al. 1990). True mycelium consists of septate hyphae 3 to 5 μm in width and up to several mm in length. Apical cells often exceed 100 μm, whereas articles are 50 to 70 μm long. There is a single nucleus per article, and septa show a minute, ascomycete-type central pore, unusual for other filamentous yeasts, with endoplasmic reticulum extending through it from one article to the next (Kreger van-Rij and Veenhuis 1973).

The proportion of the different cell forms depends on the strain used, probably accounting for the morphological differences observed at the colony level. Certain conditions are known, however, which cause preferential formation of yeast cells, or induce mycelial development (Ota et al. 1984; Rodriguez and Dominguez 1984; Guevara-Olivera et al. 1993). Ota et al. (1984) observed that carbon sources such as olive oil, oleic acid, oleyl alcohol, linoleic acid, or triolein, together with the nitrogen sources bovine milk casein, soybean fraction, or meat extract, strongly induce formation of hyphae in several strains of *Y. lipolytica*. Mycelial development was inhibited by a deficiency of magnesium sulfate and ferric chloride or by the addition of cysteine or reduced glutathione. Rodriguez and Dominguez (1984) could induce a complete and reproducible yeast-mycelium transition in minimal medium containing N-acetylglucosamine as sole carbon source, but this seems to be strain-dependent. Guevara-Olvera et al. (1993) enhanced the yeast-mycelial transition by heat shocking the cells during the inoculation in medium containing N-acetylglucosamine (see Sect. 2.4).

Some studies were done to detect physiological changes occurring during yeast-hyphae transition (Vega and Dominguez 1986; Rodriguez et al. 1990; Guevara-Olvera et al. 1993). Comparison of the composition of yeast and hyphal cells has shown that hyphal walls exhibit a higher content of amino sugars and a reduced content of protein (Vega and Dominguez 1986). Furthermore, ornithine decarboxylase activity and polyamine cell pools increased in hyphal cells grown on N-acetylglucosamine-containing medium (Guevara-Olivera et al. 1993). Mutations in the genes *SEC14* (Lopez et al. 1994), *GPR1* (G. Barth, unpubl.), and deletion of XPR6 (Enderlin and Ogrydziak 1994) have strong effects on the yeast to hyphae transition. However, it is not known whether the proteins encoded by these genes are directly involved in the regulation of this transition. Morphological mutants completely unable to form hyphae can be easily selected after visual inspection of the colonies (Fournier et al. 1991) and an analysis of the genes complementing these mutations has been undertaken (A. Dominguez, pers. comm.).

2.5
Studied Metabolic Pathways

2.5.1
Utilization of Hydrocarbons as Carbon Source

It has long been known that *Y. lipolytica* can use n-alkanes and l-alkenes as carbon sources (Bruyn 1954; Ishikura and Forster 1961; Klug and Markovetz 1967; Nyns et al. 1967). Polymethylated and chlorinated alkanes are also assimilated by this fungus (Hagihara et al. 1977; Murphy and Perry 1984). However, only few enzymes are characterized, and no gene involved in the early steps of degradation of these compounds has so far been cloned.

The first step in the assimilation is likely to be an emulsification at the cell surface to form small droplets which can be internalized. A 27-kDa extracellular emulsifier, called liposan, is induced in cells growing on n-alkanes (Cirigliano and Carman 1984, 1985). Liposan contains 88% carbohydrate and 12% protein. After entry into the cell, n-alkanes are hydroxylated by a cytochrome P-450 monooxygenase system. Cytochrome P-450 was detected in several strains of *Y. lipolytica* (Ilchenko et al. 1980; Mauersberger and Matyashova 1980; Delaisse et al. 1981). The synthesis of this enzyme is induced during growth on n-alkane, but not on glucose, ethanol, or acetate. Differences in sensitivity to glucose repression during growth on hexadecane and decane indicate the presence of different regulated cytochrome P-450 genes in *Y. lipolytica* (Mauersberger et al. 1991). Cytochrome P-450 is localized in the endoplasmic reticulum, as shown by subcellular fractionation of alkane-grown cells of *Y. lipolytica* (Mauersberger et al. 1991). The l-alkanol formed after the first step is further oxidized by a membrane-bound fatty alcohol oxidase (Kemp et al. 1990; Mauersberger et al. 1992) and not by NAD(P)-dependent alcohol dehydrogenases, which are also present in *Y. lipolytica* (Barth and Kunkel 1979; see Sect. 2.5.3).

Several alkane-nonutilizing mutants have been isolated and characterized (Bassel and Mortimer 1982, 1985; Mauersberger 1991; Kujau et al. 1992). Bassel et al. (1985) defined five phenotypic classes of alkane-nonutilizing mutants (alkA to alkE), depending on which intermediates of the alkane degradation pathway they could use. They have also shown that the uptake of n-alkanes is inducible and due to active transport. Mutations in 16 loci caused a significant reduction in n-alkane uptake (Bassel and Mortimer 1985). Mauersberger (1991) characterized three subtypes of alkA mutants, which were unable to utilize either all types of n-alkanes (alkAa), or only short ones in the C8 to C12 range (alkAb), or only long alkanes in the C16 range (alkAc). These data indicate that in *Y. lipolytica* there exist several length-specific alkane-uptake systems, or specific cytochrome P-450 monooxygenases.

2.5.2
Fatty Acid Biosynthesis and Degradation

Fatty acid biosynthesis and degradation were initially investigated by Kamiryo et al. (1979), and a few reports were issued on using *Y. lipolytica* for the stereospecific conversion of hydroxylated long chain fatty acids into γ-lactones (Ercoli et al. 1992).

Kohlwein and Paltauf (1983) detected at least two fatty acid carrier systems in *Y. lipolytica*, one being specific for fatty acids with 12 and 14C atoms, the other for C_{16} and C_{18} saturated or unsaturated fatty acids. Octanoic acid and decanoic acid are not taken up by this yeast.

An anabolic acyl CoA synthetase I activity was found in peroxisomes, mitochondria, and cytoplasm, and was shown to be required for the incorporation of exogenous fatty acids into cellular lipids. Mutants were reported as being deficient in this activity (Kamiryo et al. 1979). The gene *FAS1*, encoding the pentafunctional β-subunit of the fatty acid synthetase, was cloned and sequenced (Schweizer et al. 1988; Köttig et al. 1991). Its overall structure was very similar to that of *S. cerevisiae*. A catabolic acyl CoA synthetase II was found exclusively in peroxisomes, and was required for the β-oxidation of fatty acids. A mutant initially reported as affecting this activity was latter shown to affect another step (Kamiryo, pers. comm.). Contrary to what happens in *S. cerevisiae*, an intracellular 100-kDa fatty acid-binding protein is readily induced in *Y. lipolytica* when grown on palmitate (Dell'Angelica et al. 1992).

Use of *Y. lipolytica* cells for the production of gamma-decalactone from alkyl ricinoleate (a derivative of castor oil) has been patented by the BASF (DE4126997, 1993; see also Cardillo et al. 1991; Ercoli et al. 1992). Other possible uses include production of wax esters from fatty alcohols (Sekula 1991).

The *POT1* gene encoding a catabolic 3-oxoacyl-CoA-thiolase, which catalyzes the thiolytic cleavage of 3-oxoacyl-CoA thioesters during ß oxidation of fatty acids, was cloned and characterized by Berninger et al. (1993). Disruption of *POT1* inhibited the utilization of oleate, but not the elongation of externally added tridecanoic acid to higher-chain-length homologues. Their data suggest that beside the catabolic 3-oxoacyl-CoA thiolase an additional biosynthetic 3-oxoacyl-CoA thiolase must be present in *Y. lipolytica*.

2.5.3
Assimilation of Alcohols

Y. lipolytica does not produce ethanol, but uses ethanol as carbon source at concentrations up to 3%. Higher concentrations of ethanol are toxic. Several NAD$^+$- and NADP$^+$-dependent alcohol dehydrogenases were observed in *Y. lipolytica* (Barth and Kunkel 1979). There probably exist two NAD$^+$-dependent alcohol dehydrogenases of a molecular weight of 240 000 Da which differ in their substrate specificity. Synthesis of both enzymes seems not to be repressible by glucose or inducible by ethanol (Barth and Kunkel 1979). Three NADP$^+$-dependent alcohol dehydrogenases (ADHII, III, and IV) were detected which exhibited different substrate specificities. Furthermore, the occurrence of these enzymes varies, depending on growth phase and carbon source of the media (Barth and Kunkel 1979).

2.5.4
Assimilation of Acetate

Most strains of *Y. lipolytica* grow very efficiently on acetate as sole carbon source. Concentrations up to 0.4% sodium acetate are well tolerated, higher concentrations reduce the growth rate and concentrations above 1.0% inhibit the growth.

Several mutants have been isolated and characterized, which are blocked in the utilization of acetate (Matsuoka et al. 1980, 1984; Bassel and Mortimer 1982; Barth 1985; Barth and Weber 1987; Kujau et al. 1992). Nothing is known about the uptake of acetate by this yeast. Mutants blocked in the activity of acetyl-Coenzyme A synthetase are characterized by Kujau et al. (1992) in more detail. Acetyl-Coenzyme A is needed for the induction of the glyoxylate cycle, which is not induced in acetyl Coenzyme A-deficient mutants (Matsuoka et al. 1980; Kujau et al. 1992).

The induction of the glyoxylate pathway is necessary for the utilization of alkanes, fatty acids, alcohols, and acetate. Isocitrate lyase, the key enzyme of this cycle, and its encoding gene are well studied (Matsuoka et al. 1980, 1984; Barth 1985; Barth and Weber 1987; Hoenes et al. 1991; Kujau et al. 1992). The structural gene has been cloned and sequenced (Barth and Scheuber 1993).

Dominant mutations in the gene *GPR1* (glyoxylate pathway regulator) (Kujau et al. 1992) make the cells sensitive to low concentrations of acetate or ethanol also in the presence of glucose (Barth et al., to be published).

2.6
Secretion of Metabolites

2.6.1
Citrate and Isocitrate

Wild-type strains of *Y. lipolytica* secrete a mixture of citric and isocitric acid when grown on n-paraffins as carbon source: a total yield of 130% and a ratio of 60:40 (citric to isocitric) was reported for strain ATCC 20114 by Akyiama et al. (1973a,b). When the cells were supplied with monofluoroacetate, this compound was transformed into monofluorocitrate, which competitively inhibited aconitase: this, in turn, improved the ratio of citric to isocitric acid to 85:15. In an effort to obtain a strain with low aconitase activity, a mutant unable to use citrate as carbon source was first isolated, and further mutagenized in order to obtain a fluoroacetate-sensitive strain: this last strain yielded citric and isocitric acids in a ratio of 97:3, with a yield of 145% (w/w).

Tréton and Heslot (1978) observed that on glucose or glycerol medium, the ratio of citrate to isocitrate was 92:8 vs. 67:33 on n-paraffin. The intracellular ratio of the two acids was however roughly the same (90 : 10) on all media. These authors further showed that both acids were equally well secreted in the growth medium, but that citric acid alone was reconsumed on paraffins, whereas neither was on glucose.

Finogenova and Glasunova (1976), Finogenova et al. (1982) made a detailed analysis of enzymes from the citrate and glyoxylate cycles in strains grown on hexadecane. They studied both a wild-type strain and two mutants with respectively increased or decreased ratios of secreted citrate to isocitrate. They concluded that the ratio of two key enzymes, isocitrate lyase and aconitate hydratase, was controlling the ratio of the two acids in the growth medium. No data are available, unfortunately, on the genes which had been mutated in the strains studied. Further details on isocitrate lyase control and the glyoxylic cycle were discussed in Sect. 2.6.1. Use of mutants unable to use acetate or citrate as carbon sources has been evaluated by Wojatatowicz et al. (1991); direct selection of citrate overproducing strains has been described by McKay et al. (1990).

Processes have been described for producing citric acid from glucose or n-paraffins in liquid cultures (for a review see Mattey 1992) or with immobilized cells (Kautola et al. 1991), but also from date coats (Abou-Zeid and Khoja 1993), tapioca starch hydrolyzates (Shah and Chattoo 1993), and glucose hydrol (Wojatatowicz et al. 1991).

2.6.2
α-Ketoglutarate and Other Organic Acids

Y. lipolytica strains grown on n-alkanes in the presence of a limited thiamine supply (a vitamin not synthesized by this yeast and required for α-ketoglutarate dehydrogenase activity) secrete large amounts of α-ketoglutarate. Use of a diploid strain for α-ketoglutarate production has been patented (Maldonado and Gaillardin 1972). Furthermore, a haploid strain selected as a suppressor mutant of a mutation in the methionine biosynthesis secretes very large amounts of this organic acid during growth on glucose (patented by Weißbrodt et al. 1988).

Production of 2-hydroxyglutaric and 2-ketoglutaric acids from glucose has been described by Oogaki et al. (1983). A process for very high yield production of isopropylmalic acid using a leucine auxotroph has been patented (De Zeeuw and Stasko 1983).

2.6.3
Lysine

Mutations affecting all eleven biosynthetic steps but one have been identified. A single locus LYS1 controls the first step, homocitrate synthase (Gaillardin et al. 1976a; Gaillardin and Heslot 1979), whereas the existence of isoenzymes has up to now prevented characterization of such mutants in S. cerevisiae. Data on intragenic complementation among lys1 mutants suggested that homocitrate synthase might be an oligomer. Semidominant mutations of *LYS1* lead to desensitization to lysine feedback inhibition (*LYS1.5* mutation), and to lysine accumulation in the cell, identifying homocitrate synthase as a major check point of the pathway. Whereas homocitrate synthase activity is strongly inhibited by lysine, its rate of synthesis is not controlled by the lysine content of the cell, but by intermediates of the pathway, probably aminoadipate semialdehyde. Mutations at the *LYS11* locus, unlinked to *LYS1*, were initially isolated as suppressors of *LYC1* mutations (Gaillardin et al. 1979). They completely abolish homocitrate synthase activity. It is unknown at present if *LYS11* defines a regulator of *LYS1*, or a second subunit of homocitrate synthase. The *LYS1* gene has recently been cloned and sequenced; it corresponds to only one of the two domains described for bacterial homocitrate synthases (A. Dominguez, to be published).

No mutant affecting homoaconitate hydratase has been identified up to now. This activity could not be purified away from aconitate hydratase (Gaillardin, unpubl.). *LYS6* and *LYS7* mutations are defective in the third step of the pathway (homoaconitase), thermosensitive mutants at the *LYS6* locus identified it as the homoaconitase structural gene (unpubl.). Homoisocitrate dehydrogenase was purified to homogeneity; two tightly linked loci, *LYS9* and *LYS10*, actually define the same structural gene encoding a bifunctional protein involved in successive dehydrogenation and decarboxylation of homoisocitrate (Gaillardin et al. 1982). *LYS8* mutants are unable to convert α-ketoadipic acid into α-aminoadipic acid (unpubl.), suggesting that a single enzyme catalyzes this step in *Y. lipolytica* vs. probably two isoenzymes in *S. cerevisiae*. *LYS2* and *LYS3* are two closely linked loci (less than 2% recombination), which control adenylylation of α-aminoadipic acid, a rate-limiting step in vivo, and the only one in the pathway which is insensitive to the general control of amino acid biosynthesis (Gaillardin et al. 1979). The last two steps of the pathway are controlled by *LYS4* and *LYS5*, respectively. The latter gene, encoding saccharopine dehydrogenase, has been cloned and sequenced (Xuan et al. 1988, 1990). Saccharopine dehydrogenase levels are depressed in cells grown on lysine-containing medium, except in strains deficient in the lysine catabolic pathway (*lyc⁻* mutants). Experiments with *LYS5:lacZ* fusions evidenced a slight repressive effect of lysine, but failed to demonstrate an effect of the *LYC* context (Xuan, unpubl.). Whether lysine repression of saccharopine dehydrogenase involves "catabolic inactivation" by a complex between lysine-forming and -degrading activities (as in the arginine pathway of *S. cerevisiae*) or another posttranscriptional mechanism, remains thus an open question.

Y. lipolytica uses lysine as both a nitrogen and a carbon source, via the N-6-acetyllysine-5 aminovalerate pathway, like *Hansenula saturnus* (Rothstein et al. 1962; Gaillardin et al. 1976b). The *LYC1* gene encoding the first step of the pathway, N-6-lysine acetyl transferase (LAT), was cloned and sequenced (Beckerich et al. 1994), and shown to be induced by lysine.

Combining mutations desensitizing homocitrate synthase (*LYS1.5*), as well as mutations simultaneously preventing lysine catabolism and relieving the lysine effect on saccharopine dehydrogenase (*lyc1.5*), yielded strains accumulating 40 times more lysine than the wild-type strain (Heslot et al. 1979). No lysine secretion in the growth medium was observed, indicating that very efficient systems existed for uptake and/or retention within the cell.

Both a high and a low affinity transport system have been described for lysine uptake (Beckerich and Heslot 1978), and mutants affecting either or both systems were identified. Intracellular lysine is stored in the vacuole (Sawnor-Korszynska et al. 1977). Mutants affecting vacuolar storage of lysine were identified among strains selected for reduced content of polyphosphates in the vacuole (Beckerich et al. 1981, 1986), although no direct relationship could be established between lysine and polyphosphate pools. A direct selection of mutations preventing lysine retention within the cell was carried out, starting with a *LYS1.5, lyc1.5* strain: mutants displayed a complex phenotype including reversion of the *lyc1.5* mutation, mating-type switch, and mating-type context-dependent phenotype (Beckerich et al. 1984).

2.7
Secretion of Proteins

2.7.1
Extracellular Alkaline Protease (AEP)

Y. lipolytica strains grown on rich (YPD-type) medium at pH 6.8 secrete large amounts (1–2 g/l) of an alkaline extracellular protease (AEP). As excellent reviews on this protease have been recently published (Ogrydziak 1988b, 1993), only the main results will be recorded here. AEP is encoded by the *XPR2* gene, which was identified among at least 11 genes controlling AEP synthesis, secretion and/or activity (Ogrydziak and Mortimer 1977; Simms and Ogrydziak 1981). The gene has been cloned and sequenced from three different strains (Davidow et al. 1987; Matoba et al. 1988; Nicaud et al. 1989a). AEP is a 32-kDa protease of the subtilisin family, which is intracellularly processed from a 55-kDa glycosylated precursor (Matoba and Ogrydziak 1989; Matoba et al. 1988). Processing includes cleavage of a 15 amino acid presequence and glycosylation of a unique site on the propeptide (both occurring cotranslationally during translocation into the endoplasmic reticulum), processive hydrolysis of a row of nine N-terminal dipeptides (X-Ala or X-Pro) by a Golgi dipeptidyl aminopeptidase, followed by the cleavage of the propeptide at a Lys-Arg junction in front of the mature part. The endopeptidase is encoded by the *XPR6* gene, a *KEX2* homologue which has recently been cloned and sequenced (Enderlin and Ogrydziak 1994). The propeptide has been shown to be required for folding, transit, and activation of the mature part (Fabre et al. 1991, 1992), a finding which has since been extended to various proezymes in *S. cerevisiae*.

Regulation of AEP synthesis is complex and reflects the carbon, nitrogen, sulfur, and pH status of the cell, among others. A study of the promoter has been initated; internal deletions and in vivo footprints demonstrate there are two main UASs located 700 bp (UAS1) and 40 bp (UAS2) upstream from the TATA box (Blanchin-Roland et al. 1994). Each UAS seems to contain targets for general transcription factors (RAP1- and possibly ABF1-like). These UAS are able to activate a minimal promoter consisting of the *LEU2* TATA box and initiation site (Madzak et al., to be published). Mutations affecting transcriptional control of *XPR2* have been selected and identify at least four unlinked loci (Lambert et al., to be published).

2.7.2
Extracellular RNase

A secreted RNase activity is detected in *Y. lipolytica* cultures grown under conditions leading to alkaline protease secretion, the major 45-kDa species being partially degraded by AEP into 43-kDa and 34-kDa species (Cheng and Ogrydziak 1986). RNase is probably synthesized as a 73-kDa glycosylated precursor (Cheng and Ogrydziak 1987), but definite confirmation of the maturation process awaits determination of the sequence of the gene.

2.7.3
Acid Extracellular Protease

On rich YPD medium at pH 4.0, an acid protease activity is detected in the growth medium. Three protein species were initially described, of 28, 32, and 36kDa (Yamada and Ogrydziak 1983). More recent data suggest that there is only one acid protease, which may undergo partial proteolysis in some strains and/or under certain growth conditions (T. Young, to be published). Interestingly, induction of the acid protease occurs under conditions very similar to those used for inducing AEP, except for the pH of the medium. The gene-encoding acid protease has been cloned and sequenced (T. Young, to be published), and its disruption abolishes acid protease activity. It encodes a preproenzyme with an unusually large 5' upstream region. Interestingly, some conservation of the promoter elements of both acid and alkaline protease seems to exist. Comparing the determinants of the regulation of these two genes might thus be quite revealing.

2.7.4
Extracellular Phosphatases

A cell wall-bound acid phosphatase activity is induced when *Y. lipolytica* is grown on media depleted for inorganic phosphate sources. A glycosylated protein of 90 000–200 000 Da has been purified, which is converted into a 60 000-Da species upon endoglycosidase H treatment. Antibodies directed against it cross-react with *S. cerevisiae* major acid phosphatase (Lopez and Dominguez 1988). A kinetic study of this enzyme has been published (Moran et al. 1989).

Tréton et al. (1992) tried to clone its structural gene by functional complementation of a *pho5*, *pho3* mutant of *S. cerevisiae*. This approach failed to identify the gene searched for, but yielded *PHO2*, the gene of a secreted minor phosphatase. The *PHO2* gene product has no homology to other acid phosphatases; it corresponds to an acid phosphatase activity with a narrow substrate spectrum, the synthesis of which is induced in cells starved for inorganic phosphate.

2.7.5
Extracellular Lipase and Esterase

Y. lipolytica strains display a lipase activity which acts preferentially on oleyl residues at positions 1 and 3 of the glyceride. Lipase activities were examined by different investigators on different strains, and it is unclear whether the reported discrepancies reflect differences in the methodologies used, or the existence of different types of strains. Peters and Nelson (1948a,b) described a single glucose-repressible activity with a pH optimum around pH 6.2–6.5. Zviagintzeva et al. (1980) also reported on a single cell wall anchored lipase, wherease Kalle et al. (1972) described two cell bound activities: a constitutive, glucose-insensitive lipase, and a second activity induced by sorbitan monooleate. Ota et al. (1978) described both an extracellular activity in cultures supplemented with a protein-like fraction derived from soybean, and cell bound lipases. The extracellular lipase required oleic acid as a stabilizer/activator, whereas the cell bound lipases did not, and differed by several properties from the extracellular enzyme. Sugiura et al. (1976) and Ota et al. (1982) solubilized two cell bound, monomeric lipases of 39 and 44 kDa, which differed in their pH optima. Gomi et al. (1986) showed that the lipase(s) exist in the cell wall as an activator bound complex, which is rapidly dissociated upon enzyme purification. The extracellular, activator-dependent activity may thus reflect release of (up to 50% of) the cell bound acivity, perhaps under conditions leading to extracellular protease(s) induction. Further details on *Y. lipolytica* lipase(s) can be found in a recent review (Hadeball 1991). Mutants unable to use tributyrin as a carbon source were isolated after nystatin enrichment (Nga et al. 1988), and were shown to define three complementation groups (Nga et al. 1989). A gene encoding a secretory protein highly homologous to fungal lipases has been recently cloned in an independent approach (A. Dominguez, pers. comm.): this should rapidly permit assessing if there is more than one lipase activity in *Y. lipolytica*.

Y. lipolytica lipase may be of interest for synthesis of 2,4-dimethylglutaric acid monoesters (Ozegowski et al. 1993), transesterification of meso-cyclopertane diols (Theil et al. 1991), or use in the leather inductry or in cheese manufacturing (German patent DD-272867).

An extracellular thermostable esterase of low molecular weight (10000 Da), specific for the 1-position of triglycerides has recently been described (Mattey and Adoga 1991).

2.7.6
α-Mannosidase

A cell wall bound α-mannosidase, sensitive to catabolite repression, can be solubilized by digitonine treatment from the cell wall of *Y. lipolytica*. A kinetic study of the enzyme has been published (Vega and Dominguez 1988).

2.7.7
Genes and Mutations of the Secretory Pathway

Screening of thermosensitive mutants identified a few strains possibly affected in the secretion pathway (D. Ogrydziak, pers. comm.), but these were not characterized further. Enrichment for hyperdense cells on velocity gradients (as done in *S. cerevisiae* for *sec* mutants) is not practicable in *Y. lipolytica* due to large variations of cell size and shape in a wild-type culture (J.M. Beckerich, pers. comm.). Thus, genes affecting protein secretion have so far been isolated on the basis of functional/structural similarity with known *SEC* genes. Mutated versions of these genes were then created, which in turn served to isolate secondary mutations (suppressors, colethal mutations).

An essential gene called *RYL1* has been isolated by hybridization (Pertuiset et al. 1994) and may represent a homologue of *SEC4*, which encodes a small G protein involved in the control of secretory vesicle fusion with the plasma membrane. A homologue of *S. cerevisiae SEC14* was identified both by PCR and by functional complementation of a *S. cerevisiae sec14* mutant using a cDNA expression library of *Y. lipolytica* mRNA made in a *S. cerevisiae* expression vector (Lopez et al. 1994): unexpectedly, the deletion of this gene in *Y. lipolytica* proved to be viable and did not affect protein secretion, although recent electron microscopy data suggest that (derivatives of) the Golgi apparatus look abnormal (Rambourg et al., unpubl.). This deletion, however, prevented the yeast-mycelium transition characteristic of this species, suggesting the existence of some link between Golgi function(s) and cellular differentiation.

Several genes controlling early steps of the pathway have been characterized. A homologue of *SEC61* (one of the subunits of the translocation pore into the endoplasmic reticulum) was cloned by PCR and shown to be highly conserved between *S. cerevisiae* and *Y. lipolytica* (68% identity at amino acid level). A homologue of *SEC62* was isolated using the cDNA expression library to complement a *sec62* thermosensitive mutant of *S. cerevisiae*. Its predicted product shows only 37% identity with its *S. cerevisiae* counterpart (D. Swennen, to be published).

Rather close homologues of a signal recognition particle (SRP), involved in the targeting of secretory polypeptides to the endoplasmic reticulum, have been observed in *Y. lipolytica* and in *S. pombe* (Poritz et al. 1988). On the contrary, *S. cerevisiae* SRP seems to have diverged considerably and was identified later (Hann and Walter 1991). *Y. lipolytica* SRP contains a typical 7S RNA, which can be immunoprecipitated from whole cell extracts by human anti SRP antibodies (He et al. 1992). This RNA is encoded by two unlinked genes, *SCR1* and *SCR2*, whose products share 94% identity (He et al. 1989). Either gene can be deleted, but the double deletion is lethal (He et al. 1990), indicating that essential secretory proteins strictly require the SRP-dependent pathway for secretion. This also seems to be the case for AEP, synthesis and secretion of which are severely depressed under nonpermissive conditions in conditional mutants of *SCR1* or *SCR2* (Yaver et al. 1992, He et al. 1992). Such mutants were created by site-directed mutagenesis of conserved nucleotides in a loop binding Srp19p, one of the six polypeptides associated to mammalian 7S RNA. The genes encoding two of these polypeptides,

Srp19 (*SEC65*) and Srp54, have been cloned and sequenced recently (M. Sanchez et al., unpubl.; D. Ogrydziak, unpubl.).

In an attempt to identify new genes whose product may (directly or not) interact with the SRP, mutations displaying colethality with conditional mutations within *SCR2* were selected (Boisramé et al., to be published). One of these mutations identifies a gene (tentatively named *226*) whose product is predicted to be a resident protein in the lumen of the endoplasmic reticulum (ER). Interestingly, deletion of this gene gives a thermosensitive phenotype, which is further exaggerated in the presence of a 7S RNA thermosensitive mutation, and results in a dramatic decrease of the amount of AEP precursor synthesis: this suggests that 226p may facilitate SRP-dependent translocation of secretory precursors from the lumenal side of the ER membrane.

2.8
Peroxisome Biosynthesis

Y. lipolytica grows well on oleic acid, and this is accompanied by an extensive proliferation of peroxisomes. Antibodies directed against the targeting signal Ser-Lys-Leu COOH (anti-SKL) of peroxisomal proteins of several yeasts, react strongly with three peroxisomal proteins of *Y. lipolytica* (Atchinson et al. 1992). Using these two features, studies on peroxisome biosynthesis have recently begun with the identification of 17 oleic acid nonutilizer mutants. Three of them were shown to affect peroxisomal assembly (*pay* mutants) by a rapid immunofluorescence assay, using anti-SKL antibodies (Nuttley et al. 1993). The punctate labeling pattern observed with anti-SKL in wild-type cells was lost in the mutants, no peroxisomal structures could be observed, and two peroxisomal marker enzymes (β-hydroxyacyl-CoA dehydrogenase and catalase) failed to localize to the particulate fraction in mutant cells. Genes complementing two *pay* mutations (*pay2* and *pay4*) were isolated by complementation, using a genomic library made in a replicating vector. The *PAY4* gene has been sequenced (Nuttley et al. 1994). It encodes a 112-kDa hydrophilic protein, presumably localized in the cytoplasm, which is expressed at low levels on glucose-containing medium and strongly induced upon shifting to oleic acid. Pay4p exhibits an ATP-binding site and appears closely related to Pas5p of *Pichia pastoris* (59% identity) and less to Pas1p of *S. cerevisiae*, both involved in peroxisome assembly in their respective hosts.

3
Life Cycle

3.1
Heterothallism and Mating Type Alleles

All natural isolates of *Y. lipolytica* so far tested are heterothallic. However, it was reported that mating type switching occurred in conjunction with reversion of the *lyc1.5* mutation in haploid and diploid strains (Beckerich et al. 1984). It must be investigated whether such mating type switching takes place in nature to create homothallic strains.

The mating type is determined by the two alleles *MATA* and *MATB* (Bassel et al. 1971). The *MATA* allele has been already cloned (Kurischko et al. 1992). Similarly to heterothallic strains of *S. cerevisiae*, haploid and diploid cells are vegetatively stable with some exceptions (see Sect. 3.5). Extensive overviews on the life cycle, including mating processes and sporulation, are given by Weber and Barth (1988), Weber et al. (1988), and Heslot (1990).

3.2
Mating Frequency

Mating frequencies of natural isolates are always very low (1% viable zygotes/cell or less). Attempts to increase the mating response by inbreeding (Gaillardin et al. 1973; Ogrydziak et al. 1978; Kurischko et al. 1983; Barth and Weber 1985) failed to raise mating frequency above 15% viable zygotes/cell (see also Sect. 4.1). Mating frequency is dependent on several parameters besides the genetic background of strains: medium, cell density, growth phase, and temperature (Gaillardin et al. 1973; Barth and Weber 1984). Several media were tested for induction of conjugation. High values of mating were obtained using McClary medium (Gaillardin et al. 1973), restricted growth medium (RG; Ogrydziak et al. 1978), or a yeast extract-malt medium (YM; Barth and Weber 1984). A slight increase in mating was observed by adding 0.05% sodium citrate to YM (Weber et al. 1988). Cell densities of about 10^7 cells/ml resulted in highest mating frequencies for most strains (Gaillardin et al. 1973; Weber et al. 1988) but some strains need higher cell densities (about 10^8 cells/ml; Weber et al. 1988).

Temperatures below 20 °C and above 28 °C reduce mating dramatically (Gaillardin et al. 1973). The age of the preculture before mixing in or on conjugation medium is important for the mating frequency. Highest values were observed with cells taken from the late logarithmic growth phase (Barth and Weber 1984).

3.3
Sporulation

Y. lipolytica does not require nitrogen limitation for induction of sporulation in contrast to *S.cerevisiae* and some other yeasts. Diploid strains sporulate on solid or in liquid complete medium when glucose is exhausted. Highest sporulation frequencies were obtained with many strains when 1.5% sodium citrate was used as a carbon source (Barth and Weber 1985) at temperatures between 20 and 30 °C. High sporulation frequencies can also be obtained in liquid or on solid YM and V8 media (Gaillardin et al. 1973).

Some natural isolates of *Y. lipolytica* sporulate after mating quite well, but only inbred strains form a high proportion of complete tetrads and have a satisfactory level of germinating spores (see Sect. 3.4).

Table 2

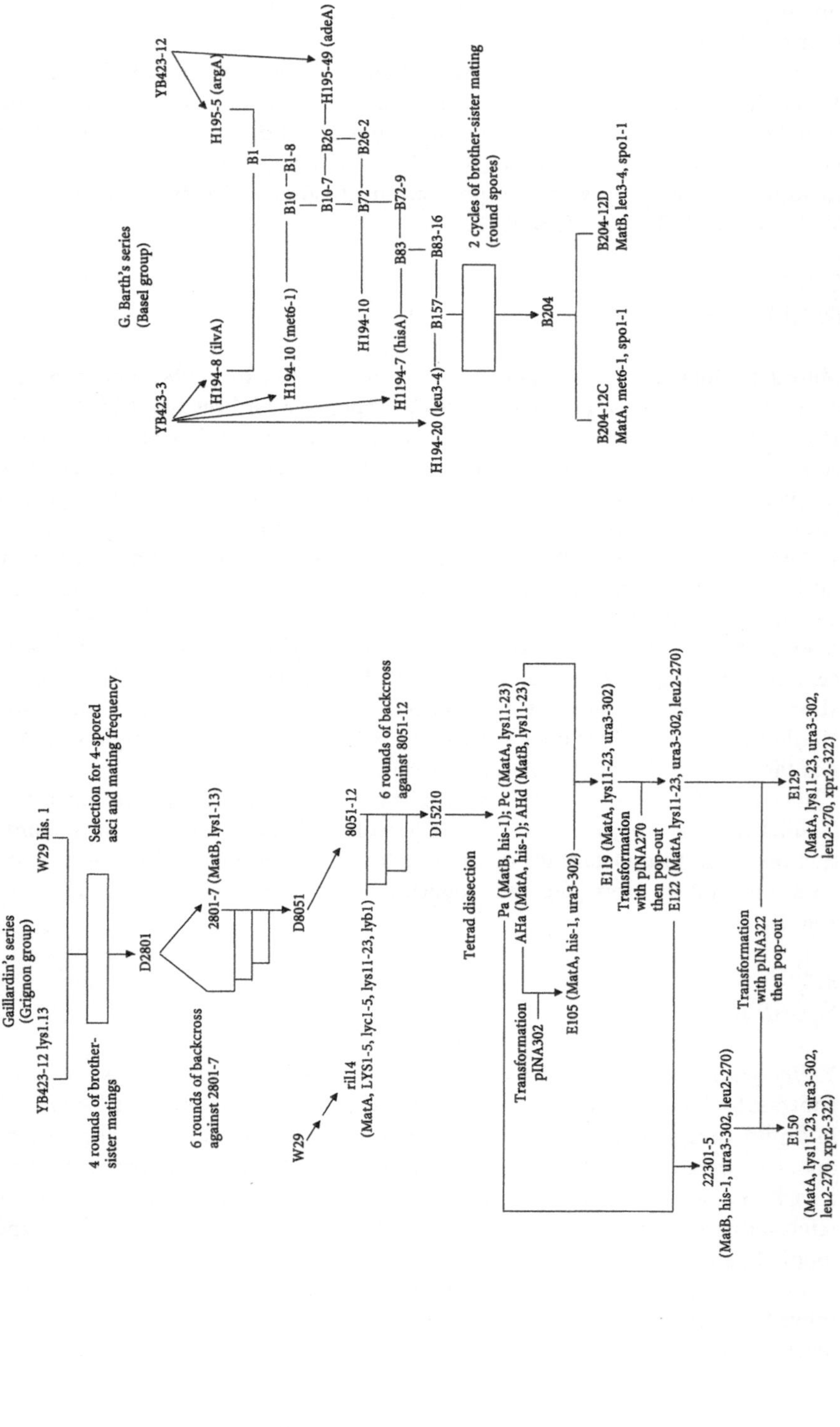

Gaillardin's series
(Grignon group)

YB423-12 lys1.13   W29 his. 1

4 rounds of brother-   Selection for 4-spored
sister matings   asci and mating frequency

D2801

6 rounds of backcross   2801-7 (MatB, lys1-13)
against 2801-7

D8051

W29

ril14
(MatA, LYS1-5, lyc1-5, lys11-23, lyb1)   8051-12

6 rounds of backcross
against 8051-12

D15210

Tetrad dissection

Pa (MatB, his-1); Pc (MatA, lys11-23)
AHa (MatA, his-1); AHd (MatB, lys11-23)
Transformation
pINA302

E105 (MatA, his-1, ura3-302)

E119 (MatA, lys11-23, ura3-302)
Transformation
with pINA270
then pop-out
E122 (MatA, lys11-23, ura3-302, leu2-270)

22301-5
(MatB, his-1, ura3-302, leu2-270)

Transformation
with pINA322
then pop-out

E150
(MatA, lys11-23, ura3-302,
leu2-270, xpr2-322)

E129
(MatA, lys11-23, ura3-302,
leu2-270, xpr2-322)

G. Barth's series
(Basel group)

YB423-3   YB423-12

H194-8 (ilvA)   H195-5 (argA)

B1

H194-10 (met6-1) — B10 — B1-8

B10-7 — B26 — H195-49 (adeA)

H194-10 — B72 — B26-2

H1194-7 (hisA) — B83 — B72-9

H194-20 (leu3-4) — B157 — B83-16

2 cycles of brother-sister mating
(round spores)

B204

B204-12C   B204-12D
MatA, met6-1, spo1-1   MatB, leu3-4, spo1-1

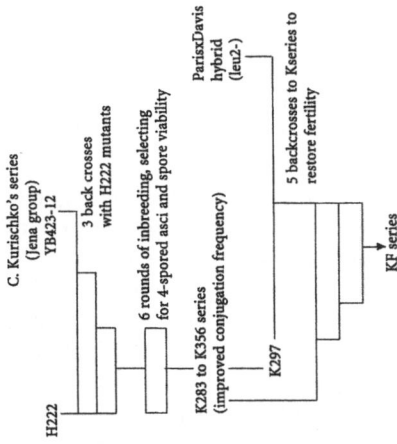

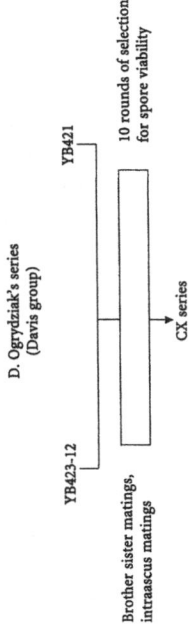

3.4
Inbred Strains

The origin of the strains currently used by different groups working on *Y. lipolytica* is given in Table 2, and a tentative genealogy of the major inbred lines is summarized in Table 3.

Mating frequencies within these lines varies between 3 and 15% (viable zygotes/ number of cells), which allows complementation analysis. The proportion of complete tetrads is high, and the spore germination of these strains reaches 80– 90%, which makes tetrad analysis possible. There exists one inbred line which forms spherical spores in contrast to the normally formed helmet-shaped spores (Barth and Weber 1985). These strains turn out to be more suitable for asci dissection.

As stated before, the main difficulty is the poor interfertility between these inbred lines. Unfortunately, there is currently no consensus on the reference strains to be used. Since genetic data are mainly gathered currently on the E129- E150 series, it might be advisable to localize new genes on this map (see Sect. 4.2.2). In any case, the existence of excellent methodologies for gene transfer/replace- ment, and for retrieval of chromosomal mutations once the gene has been cloned, make it likely that exchange of markers between different inbred lines will be much easier than in the past.

3.5
Spontaneous Haploidization and Stability of Diploid Strains

Kurischko and Weber (1986) reported spontaneous haploidization of some diploid cells after formation of zygotes. Furthermore, it was observed in several laborato- ries that diploid strains spontaneously sporulate during prolonged storage on complete media. Therefore, it is recommended to prepare fresh diploid strains for genetic analysis.

4
Genetic and Molecular Data

4.1
Mutagenesis and Mutants

Several commonly used chemical and physical mutagens have been tested to induce mutations in wild-type and inbred strains of *Y. lipolytica*, and many differ- ent mutants have been isolated and characterized. This is not the place to describe all these mutants, but some main classes and general aspects will be discussed. Results of extensive studies on cell inactivation, frequency of mutants or of certain classes of mutants, spontaneous mutability, and frequency of reversions in differ- ent strains of *Y. lipolytica* have been published by Bassel et al. (1971), Gaillardin and Heslot (1971), Bassel and Mortimer (1973), Gaillardin et al. (1973), Ogrydziak et al. (1978), and Barth and Weber (1983). Mutants defective in the utilization of certain carbon sources or modified for the secretion of metabolites and proteins are described in Sects. 2.4 to 2.6 of this chapter.

Many mutants have been described which exhihit nutritional requirements for most of the amino aicds, adenine, and uracil, and the vitamins biotin, nicotinic acid, pantothenic acid, riboflavin, and thiamine (Bassel et al. 1971; Gaillardin et al. 1973; Mortimer and Bassel 1973; Ogrydziak et al. 1978; Barth and Weber 1983; Heslot 1990). The data on thiamine requirement are conflicting. Most, if not all, natural isolates of *Y. lipolytica* need thiamine for normal growth but, nevertheless, they can continue to grow for some generations after exhaustion of thiamine. The isolation of complete thiamine-dependent and vice versa of thiamine prototrophic mutants (Barth and Weber 1983) indicates that thiamine can be synthesized by this yeast, but its synthesis seems to be restricted at a low level.

There are several color mutants described which turn the colonies green, brown, or red (Mortimer and Bassel 1973; Bassel et al. 1975; Barth and Weber 1983) but no red adenine-requiring mutants have been published.

Several mutants have been described which are resistant to antifungal agents or different analogues of amino acids and carbon sources (Barth and Weber 1983; Morzycka et al. 1976; Gaillardin et al. 1976a, 1976b). Analysis of mutants resistant to the antifungal agent nystatin resulted in the detection of recessive, semidominant, and dominant mutations in different genes which cause this phenotype (Barth and Weber 1983). Dominant and recessive mutations have also been detected in mutants resistant to D(+)-glucosamine and 2-deoxyglucose (Barth and Weber 1983). Some ethionine-resistant mutants exhibited a 1.5 to 18 times enhanced pool of free methionine in the cells (Morzycka et al. 1976). Canavanine-resistant mutants have been used for random spore analysis (Bassel et al. 1971) and determination of spontaneous reversion frequencies (Barth and Weber 1983). Lysine analogues such as transdehydrolysine or amino-ethylcysteine have been used to isolate mutants deregulated for lysine biosynthesis and uptake (Gaillardin et al. 1976a; Beckerich and Heslot 1978).

Copper resistance varies widely among natural isolates of *Y. lipolytica*. Two tightly linked metallothionein genes, transcribed by a bidirectional promoter, have been cloned and sequenced (Dominguez et al., to be published).

Most of the isolated mutants reverted at the expected frequency of 10^{-7} to 10^{-8} or less. However, some of the mutants reverted at much higher frequencies (Barth and Weber 1983; Beckerich et al. 1984; Kujau et al. 1992). The reason for this high revertibility has not been clarified in most cases. The cloning and sequencing of one of such a highly revertible allele (*GPR1-112*) resulted in the detection of a retrotransposon (see Sect. 4.3) which is involved in the reversion process (Barth, to be published).

The nystatin method for enrichment of mutants (Moat et al. 1959; Snow 1966) has been successfully adapted to concentrate auxotrophic and citrate-nonutilizing mutants of *Y. lipolytica* (Gaillardin et al. 1973; Barth and Weber 1983).

4.2
Genome and Gene Structure

4.2.1
Genome Structure

A G+C content of 49.6–51.7% has been reported for *Y. lipolytica* (Nakase and Komagata 1971; Kurtzman and Phaff 1987; Kück et al. 1980). Determination of deoxyadenosine content of exponentially growing haploid cells yielded and estimated genome size of 4×10^9 Da or about 11 Mb (Gaillardin et al. 1973), which would be smaller than that of *S. cerevisiae*. It was later reported that different haploid strains of *Y. lipolytica* exhibit wide variations in their DNA content estimated by this method (Shah et al. 1982), but no absolute values were given.

Chromosome separation by pulsed field electrophoresis yielded much larger genome estimates in the range of 12.7 to 22.1 Mb for wild-type or laboratory strains (Naumova et al. 1993). A pronounced chromosome length polymorphism was observed between different isolates, as well as a variation of the number of chromosomal bands detected (four to six), which may reflect both gel artifacts (chromosome comigration) and aneuploidy (evidence for a duplication of the *URA3* marker on separate chromosomes was observed in a strain displaying six bands). Natural isolates of *Y. lipolytica* appeared to have widely divergent genetic structure, which may account for the low spore viability observed in the initial inbreeding programs: interestingly, Wickerham's first monosporic isolates from the diploid-type strain displayed chromosomes of completely different sizes; one looked like an aneuploid, and genetic markers were distributed differently in the two clones, as shown by hybridization.

4.2.2
Genetic Linkage Groups and Chromosome Maps

Based on tetrad dissection and random spore analysis, a genetic map encompassing more than 60 markers was established for the American lines (Ogrydziak et al. 1978, 1982). Linkage was observed eight to ten times more frequently between random markers than in *S. cerevisiae*, suggesting that the average recombination per kb was lower than in *S. cerevisiae*. Five linkage groups were defined at that time, but linkage studies have proceeded at a slow pace since then (Ogrydziak 1988a). Linkage groups were also defined for former East German lines (Kurischko 1984) and for industrial strains (De Zeeuw, pers. comm.), but since marker exchange between the different lines is difficult, no pooled set of data is available so far. No centromere-linked marker has been conclusively identified, but strains carrying centromeres tagged with a marker have been recently built (Fournier et al. 1993) and should help in this regard.

Inbred strains of *Y. lipolytica* tend to display five chromosomes in the range of 2 to 5 Mb (P. Fournier, S. Casaregola, unpubl.). It should be cautioned that at this stage, five is still an estimate for the chromosome number of the species:

comigration may occur in this size range; that some bands are fuzzy may be due to the presence of rDNA repeats, so that specific chromosome breakage will be required in order to obtain definite results. As expected from the study of the initial isolates from which the inbred lines were derived (see Sect. 4.2.1), each line shows a completely different chromosome pattern. An average size of 15 to 18 Mb has been estimated for inbred lines (Fournier and Nguyen, unpubl.). A highly inbred strain (E150) consistently giving five well-resolved chromosomal bands has been selected to assign 36 cloned genes by hybridization. Ribosomal RNA genes were found on four of the five chromosomes. The linkage groups appear not to be conserved among the different lines. Whether this reflects exchanges between rDNA repeats present on nonhomologous chromosomes has to be assessed, but it clearly questions the very existence of a consensus genetic map for this species (S. Casaregola et al., to be published).

4.2.3
ARS and Centromeres

Initial attempts at isolating chromosomal origins of replication were based on a strategy well established for several other yeast species: a genomic library was made in an integrating vector. Since origins of replication are supposed to be abundant (one for every 5 to 10 plasmids in the library) and to confer high frequency of transformation (100 to 1000 times above the frequency obtained with the uncut integrative vector), most of transformants obtained with such a library should reflect cloning of an origin. Beside abundance and aptitude to confer high transformation frequency, *ARS* plasmids were supposed to be highly unstable, being virtually lost from the transformed clones after a few generations on nonselective medium. Among these three properties, only the second was true for *Y. lipolytica*, which rendered identification of *ARS* quite difficult. We know now that an origin of replication alone is unable to maintain a plasmid extrachromosomally in this species, and that in addition to it, a centromeric sequence is absolutely required (Fournier et al. 1991, 1993).

Three different *ARS* have been isolated up to now (Fournier et al. 1991; Matsuoka et al. 1993), each carrying a centromere (*CEN*) and a nearby chromosomal origin of replication (*ORI*): *ORI3018/CEN3*, *ORI1068/CEN1*, *ORI4002/CEN4*.[3] *ORI1068* and *ORI3018* have been shown by two-dimensional gel electrophoresis to be included in regions of initiation of replication in the chromosome (Fournier and Vernis, to be published), whereas all three *CEN* sequences are able to induce chromosome breakage when integrated at the *LEU2* locus. *CEN1* and *CEN3* were shown additionally to be closely linked genetically to a chromosomal centromere,

---

[3] Nomenclature of *ORI* and *CEN*. The *CEN* number refers to the chromosome number in strain E150 map. Four digits are used to define *ORIs*: the first refers to the chromosome number in strain E150 map, the three following are specific of each *ORI*. For *ORIs* which are repeated in the genome, the first digit is replaced by an X.

and to correspond to a strong pausing site of the polymerase. *ORI* and *CEN* functions are carried by two independent regions of the original *ARS* inserts, and can be exchanged between different *ARS*: a given *CEN* sequence can thus apparently confer *ARS* function to any chromosomal *ORI*, thus opening the way to the cloning of many *ORI*. Several new *ORIs* were recently isolated from a DNA library made in a *CEN* nontransforming vector (Chasles and Fournier, to be published). One of them (*ORIX009*) was shown to be repeated at several places in the genome, another originated from rDNA spacer.

ORI and CEN sequences of *Y. lipolytica* show no homology to corresponding *S. cerevisiae* or *K. lactis* sequences (or to *S. pombe*), and the existence of a clear functional consensus for these sequences in *Y. lipolytica* has still to be assessed.

4.2.4
Ribosomal RNA and Other RNA Genes

Genetic data and hybridization of separated chromosomes showed that most, if not all, wild-type isolates of *Y. lipolytica* contained several clusters of rDNA units located on one to four chromosomes (van Heerikhuizen et al. 1985; Fournier et al. 1986). Up to five types of repeated units were identified in a single strain, differing by the length and the structure of the nontranscribed spacer DNA. All isolates so far tested showed this polymorphic rDNA structure, but one strain (ATCC 18944) was reported to contain a single type of unit (Clare et al. 1986). Genetic recombination inside the clusters occurs during meiosis, leading to expansion or reduction of individual clusters. Thus, whereas independent wild-type isolates differ by the number and types of rDNA units they carry, crossing individual strains within a given *Y. lipolytica* inbred line results in the progeny in a strain-specific reshuffling of rDNA genes supplied by each parent (Fournier et al. 1986). Strain typing can thus be done by simple inspection of hyperdense bands on digests of total chromosomal DNA.

The total number of repeated units varies around a mean value of 50–60 (M.T. Le Dall, unpubl.; Clare et al. 1986). The possible contribution of rDNA repeats to interchromosome exchange has been discussed in Sect. 4.2.2. The 5S RNA genes are not part of the rDNA clusters as in most other Saccharomycetoideae, but are dispersed throughout the genome as in *S. pombe* (van Heerikhuizen et al. 1985; Clare et al. 1986).

Small nuclear RNAs U1, U2, U4, U5, and U6 were analyzed by Roiha et al. (1989): their sizes appeared much closer to that of human snRNA than to those of *S. cerevisiae*. No gene has been cloned so far.

Two genes encoding an abundant 7S RNA molecule, which forms part of signal recognition particle, have been identified (He et al. 1989). Both genes share 94% identity, both are expressed, and lie on different chromosomes. Their product is structurally much closer to higher eukaryotic 7S RNA than is the homologous 11S RNA in *S. cerevisiae*. The deletion of either gene is viable, but the double deletion is not, suggesting that the SRP-dependent pathway of protein secretion is essential in *Y. lipolytica*.

No other genes encoding small RNA have so far been identified; in particular, no suppressor tRNA is known in this yeast.

4.2.5
Features of Structural Genes

A tentative list of cloned genes is given on the following table (Table 3). Several motifs similar to consensus sequences for transcriptional factor binding sites in *S. cerevisiae* can be observed in *Y. lipolytica* promoters, including TATA boxes, CT-rich blocks, TUF/RPG- or ABF1-binding sites, or UAS_{GCN} and UAS_{LEU}. The function of these sequences has not been assessed in most cases (see, however, Blanchin-Roland et al. 1994), and the corresponding transcription factors have not been isolated. Three genes encoding a transcription factor have recently become available: *CRF1*, HOY1, and RIM1 (see Table 3).

Transcription starts have been mapped in a few cases, and often occur within a CCAAA type of structure, 20–30 bp downstream from the TATA box.

Introns are relatively rare in *Y. lipolytica* genes (six intron-containing genes among 49 known sequences, see Table 3). Two genes show two introns (*CDC42* and *SEC14*) separated by a short exon. The 5' end of the intron (donor site) is GTGAGTPu in all cases. The 3' internal consensus (branch site) is TACTAAC in all cases but one (cgCTAAC in the first intron of *SEC14*), and is separated by one or two nucleotides only from the 3' end of intron: CAG.

Most *Y. lipolytica* genes show a typical signal for transcription termination TAG . . . TA(T)GT . . . TTT, which is located upstream from the site of poly A addition (Lopez et al. 1994).

Positions-1 and -3 are strongly conserved upstream from the initiator ATG: an A is observed in 12/15 cases at position-1 and in 11/15 cases at position -3. An A is found at both positions in strongly expressed genes such as *XPR2*, *PGK1*, *TPI1*, *PYK1*, *ICL1*, and *LEU2*.

Codon usage appears to be different from that of *S. cerevisiae* (Gaillardin and Heslot 1988), and similar to that of *Aspergillus* (A. Dominguez, to be published).

4.3
Transposon

A retrotransposon, called Ylt1, was detected in the genome of *Y. lipolytica* (Schmid-Berger et al. 1994). It is 9.4 kb long and can transpose in the genome. This retrotransposon is bounded by a long terminal repeat (LTR), the zeta element, which is 714 bp long, highly conserved, and can exist also as a solo element. Ylt1 and solo zeta elements are flanked by a 4-bp directly repeated genomic sequence. The copy numbers of Ylt1 and solo zeta are dependent on the strain examined, but at least 35 copies of Ylt1 and more than 30 copies of the solo zeta element per haploid genome have been observed (Schmid-Berger et al. 1994).

Ylt1 belongs to the Ty3/gypsy group of retrotransposons and contains two overlapping large open reading frames (*YltA* and *YltB*). The predicted *YltA* gene

Table 3. List of cloned genes (July 95)

Gene nuclear	Function genes	Homologues	Sequenced	EMBL accession number	Reference
ADE1	Glycinamide ribotide synthetase; aminoimidazole ribotide synthetase	ADE1 *S. pombe* ADE5,7 *S. cerevisiae*	Yes Yes		C. Strick, pers. comm. G. Barth, unpubl.
ASR1	Acidic serin rich protein	–	Yes		T. Young, pers. comm.
AXP1	Acid extracellular protease	–	Yes		
BIO-6	Biotin synthesis		No		
CRF1	Transcriptional activator of MTPs	CUP2 *S. cerevisiae*	Yes	Z23265	A. Dominguez, pers. comm.
FAS1	Fatty acid synthetase	FAS1 *S. cerevisiae*	Yes	X59690	Köttig et al. (1991)
FBA1	Fructose biphosphate aldolase	FBA1 *S. cerevisiae* FUN34 *S. cerevisiae*	Partial		D. Swennen, pers. comm.
GPR1	Glyoxylate pathway regulator	YCR10C *S. cerevisiae*	Yes	X74146	G. Barth, unpubl.
HIS1	ATP PR-transferase	HIS1 *S. cerevisiae*	Yes		C. Strick, pers. comm.
HIS3	Imidazoleglycerol-P dehydratase	HIS3 *S. cerevisiae*	No		Prodromou et al. (1991)
ICL1 (1)	Isocitrate lyase	ICL1 *S. cerevisiae*	Yes	X72848	Barth and Scheuber (1993)
LEU2	Isopropyl malate dehydrogenase	LEU2 *S. cerevisiae*	Yes	M35579	Davidow et al. (1987)
LYC1	Lysine N6 acetyltransferase	–	Yes	X63548	Beckerich et al. (1994)
LYS1	Homocitrate synthetase	–	Yes		Dominguez, pers. comm.
LYS5	Saccharopine dehydrogenase	LYS1 *S. cerevisiae*	Yes	M43929	Xuan et al. (1990)
MTP1	Metallothionein	–	Yes	Z23264	A. Dominguez, pers. comm.
MTP2	Metallothionein	–	Yes	Z23264	A. Dominguez, pers. comm.
PAY2	Peroxisome assembly	PAS2 *S. cerevisiae* PAS1 *S. cerevisiae*	Yes		Rachubinski, pers. comm.
PAY4	Peroxisome assembly	PAS5 *P. pastoris*	Yes	L23858	Nuttley et al. (1993)
PGK1	Phosphoglycerate kinase	PGK1 *S. cerevisiae*	Yes	M91598	B. Treton, pers. comm.
PHO2	Secreted phosphatase	–	Yes	X65225	Treton et al. (1992)
POT1	Peroxisomal oxoacyl-CoA thiolase	POT1 *S. cerevisiae*	Yes	X69988	Berninger et al. (1993)
PYK1 (1)	Pyruvate kinase	PYK1-cdc19 *S. cerevisiae*	Yes	M86863	Strick et al. (1992)

Gene	Protein/function	Homolog		Accession	Reference
RYL1	Small G protein	SEC4 _S. cerevisiae_	Yes	L06969	Beckerich et al., submitted
RYL2	Small G protein	Rab2 & Rab5 from rat	Yes	L06970	Beckerich et al., submitted
RYL3 (1)	Small G protein	CDC42 _S. cerevisiae_	Yes		Beckerich et al., submitted
RYL4	Small G protein	Rho family	Yes		Beckerich et al., submitted
SEC14 (1)[a]	Phospholipid transporter	SEC14 _S. cerevisiae_	Yes	L20972	Lopez et al. (1994)
SEC61	ER translocation pore	SEC61 _S. cerevisiae_	Yes		D. Swennen, pers. comm.
SEC62	ER translocation apparatus	SEC62 _S. cerevisiae_	Yes		D. Swennen, pers. comm.
SEC65	SRP subunit	SEC65 _S. cerevisiae_	Yes	Z22570	A. Dominguez, pers. comm.
SCR1	7S RNA gene	scr1 _S. pombe_	Yes	M20837	Poritz et al. (1988)
SCR2	7S RNA gene	scr1 _S. pombe_	Yes	X51658	He et al. (1989)
SRP54	SRP subunit	SRP54 _S. cerevisiae_	Yes		D. Ogrydziak, pers. comm.
TPI1 (1)	Triose phosphate isomerase	TPI1 _S. cerevisiae_	Yes		C. Strick, pers. comm.
URA3	OMP decarboxylase	URA3 _S. cerevisiae_	Yes		L. Davidow, pers. comm.
URA5	orotate PR transferase	URA5 _S. cerevisiae_	Yes	Z22571	A. Dominguez, pers. comm.
XPR2	Alkaline extracellular protease	Subtilisine	Yes	M17741	Davidow et al. (1987)
XPR6	Dibasic endoprotease	KEX2 _S. cerevisiae_	Yes	L15238	Enderlin and Ogrydziak (1994)
XRN1	Extracellular RNase	–	Partial		D. Ogrydziak, pers. comm.
YltA	Gag-protein (retrotransposon)	GAG of retrotransposons	Yes		G. Barth, to be published
YltB	Polyprotein (retrotransposon)	POL of retrotransposons	Yes		G. Barth, to be published

Mitochondrial genes

Gene	Protein/function	Homolog		Accession	Reference
tRNA-Trp, tRNA-Thr, tRNA-Pro			Yes	L15359	Matsuoka et al. unpubl.
atp9	ATP synthase subunit 6, 8, 9	oli1, oli2 _S. cerevisiae_	Yes	L15359	Matsuoka et al., unpubl.
cox3	Cytochrome oxydase subunit 3	oxi2	Partial	L15359	Matsuoka et al., unpubl.
	NADH oxidoreductase subunit 4		Yes	L15359	Matsuoka et al., unpubl.

[a](1) = intron containing gene.

product showes high similarity to retroviral Gag-encoded proteins. *YltB* encodes protein domains with some similarity to retroviral gene products encoded by the *pol* gene. In the *YltB* gene, these predicted products occur in the order protease, reverse transcriptase, RNase H, and integrase (Barth et al., in prep.).

4.4
Plasmids and VLPs

No DNA plasmid was detected in a systematic survey of 24 wild-type isolates (B. Tréton, unpubl.), but a linear dsRNA of 4.9 kb was observed in several strains (Groves et al. 1983; Tréton et al. 1985). This RNA is encapsidated within a 50-nm virus-like particles (VLP). The capsid seems to be composed of two major polypeptides of 83 and 77 kDa (Tréton et al. 1985; El-Sherbeini et al. 1987). Based on hybridization data, there seem to exist at least two types of dsRNAs in different strains, which show little homology with one another (Tréton et al. 1985). An additional linear dsRNA molecule of about 6 kb was detected in some strains besides the smaller 4.9-kb-long dsRNA (Barth, unpubl.) similar to the situation in killer strains of *S. cerevisiae*. No homology was found with genomic DNA of *Y. lipolytica*, nor with ds-RNAs from *S. cerevisiae*. No killer phenotype associated with these VLPs could be evidenced, and curing after UV treatment (Tréton et al. 1987) did not lead to any phenotypic change.

4.5
Mitochondrial Genome

The mitochondrial genome of *Y. lipolytica* has a buoyant density of 1.687g/cm^3 and a GC content of 24.9% (Kück et al. 1980). It consists of circular molecule of $14.5 \mu\text{m}$ and its restriction map was established by Wesolowski et al. (1981) on strain W29. The arrangement of genes for ATPase subunits, rRNA, and 4S RNA is conserved with respect to *S. cerevisiae* and *K. lactis*. Several mitochondrial genes, including ATPase subunit 6, 8, and 9, cytochrome oxidase subunit 3, NADH oxidoreductase (ubiquinone) subunit 4 and tRNA genes, have been sequenced (Matsuoka et al., unpubl.; see Table 1). The restriction map of mitDNA seems to be well conserved among different isolates, as judged from the mobility of the corresponding hyperdense bands observed on digests of total genomic DNA stained with ethidium bromide (unpubl.).

A single mitochondrial mutation leading to oligomycin resistance was described by Matsuoka et al. (1982). It has been used in protoplast fusion experiments (Tréton et al. 1987).

5
Transformation, Vectors, and Expression Systems

The development of integrative and replicative vectors has been previously reviewed (Gaillardin and Heslot 1988; Heslot 1990; see also Fig. 1A and B). Marker

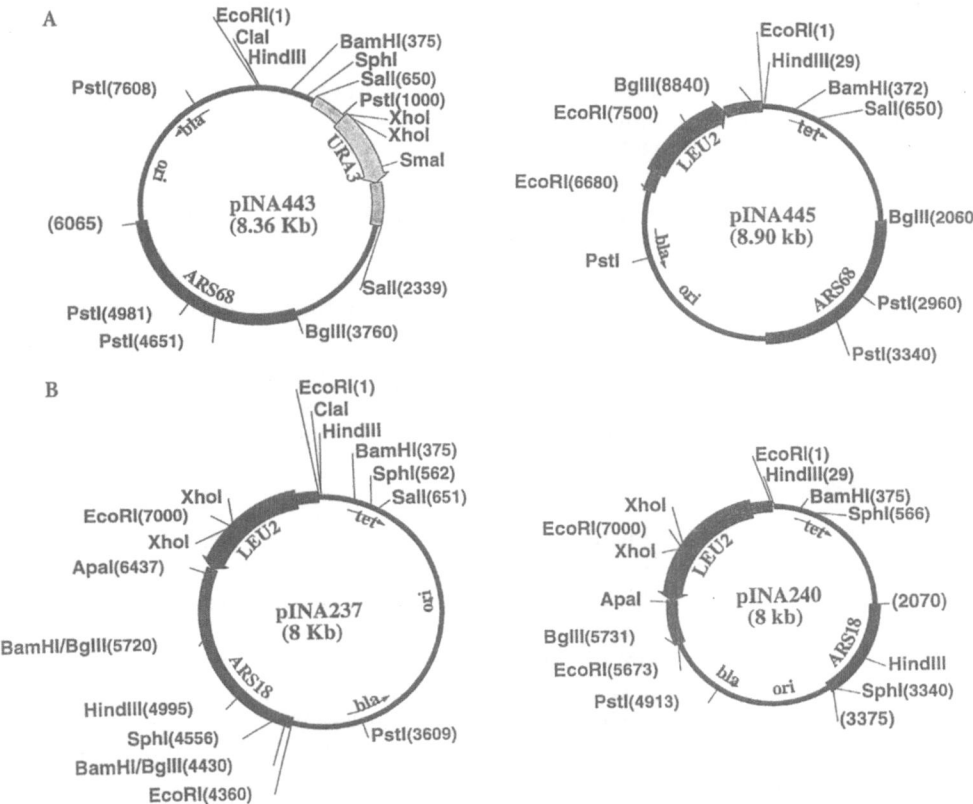

Fig. 1A,B. Integrative (A) and replicative (B) vectors. All vectors are based on a pBR322 backbone (*black line*). Known coding sequences of *Y. lipolytica* genes are depicted as *thick arrows* indicating the direction of transcription. Noncoding sequences of yeast DNA are indicated by *boxes*

genes are usually *Y. lipolytica* wild-type biosynthetic genes, which complement the corresponding chromosomal mutation (for these and other available markers, see Sect. 5.4).

5.1
Integrative Transformation

5.1.1
Single-Copy Integration

Integrative vectors require a homologous region with chromosomal DNA. Integrative transformation is best achieved by the lithium acetate procedure (see below) and at a lower level by electroporation (Barth, unpubl.).

Integration occurs almost exclusively by homologous recombination. Integration is stimulated by linearizing the vector within the homology region (100 to more than 1000fold enhancement over the frequency obtained with uncut vectors), and results in very high transformation frequencies (up to 10^6 transformants per μg of DNA, Xuan et al. 1988). Homology should extend a few hundred bp on each side of the cut (preferably more than 300bp). Gaps of up to 500bp within the homology region still direct efficient integration and are efficiently repaired on the chromosome.

With such a linearized vector, transformation results in more than 80% of the cases from integration of a single copy of the vector at the chosen site. A low level of double integrants at the site is observed (about 10%). Remaining events include conversion of the chromosomal marker or out of site integrations: these last events may, however, represent the major outcome when low frequency transformation is observed (for instance, if homology is imperfect or extends over less than 300bp).

Gene Disruption by Double Crossover

Gene disruption can be achieved easily by inserting a selective cassette within the target gene cloned on a plasmid: integration then occurs by a double crossover. Trimming of the ends is highly efficient, and the presence of vector sequences (without homology to the target) does not affect the efficiency of targeting. A possible drawback to this mechanism is that short regions of homology to the target might become lost: transformants in that case will result mainly from conversion of the chromosomal allele by the cassette marker. We recommend flanking regions of 500bp or more to reduce the frequency of these events. We observed, however, that very efficient gene disruption could be obtained as long as one of the flanking region was large (500 to 1000bp), the second being as short as 150bp. Checking the structure of the transformants by Southern analysis or PCR is strongly advised. In order to minimize conversion events, strains carrying chromosomal deletions of commonly used marker genes have been constructed (see Sect. 5.3). Gene disruption of possibly essential genes is best achieved in a diploid context, which is subsequently sporulated and dissected, or in a strain carrying a second copy of the target gene on a replicative vector (He et al. 1990; Lopez et al. 1994). In this last case, either the plasmidic or the chromosomal copy can act as a target for the disruption cassette; we observed between 10 and 80% disruption of the plasmid copy, depending on the gene considered.

Gene Conversion and Retrieval of Chromosomal Mutations

Since trimming of the free ends is likely to occur even when a linearized plasmid is targeted by a simple cut, eventual incorporation of chromosomal mutations(s) in the plasmid-borne copy is likely to occur during repair. This was checked directly with a plasmid carrying both a *LEU2* and a *LYS5* marker, which could be targeted at either locus depending on the cut made. The plasmid was cut at different positions within the *LYS5* gene, and the transformants were selected (e.g.) for the Leu+ phenotype (Fig. 2). The frequency of Leu+Lys− transformants varied between

Plasmid/Restriction	Integration site	Selection for Leu⁺		Selection for Lys⁺	
		Transfo Freq.	% of Lys⁺	Transfo Freq.	% of Lys⁺
pINA128/*NotI*	*leu2*	2 10⁵	98%	2.5 10⁵	99%
pINA128/*KpnI*	*lys5*	6 10⁴	48%	2 10⁵	99%
pINA128/*BamHI*	*lys5*	3 10⁴	7%	4.1 10⁵	54%
pINA128/*ApaI*	*leu2*	9.4 10⁵	98%	2.3 10⁵	79%

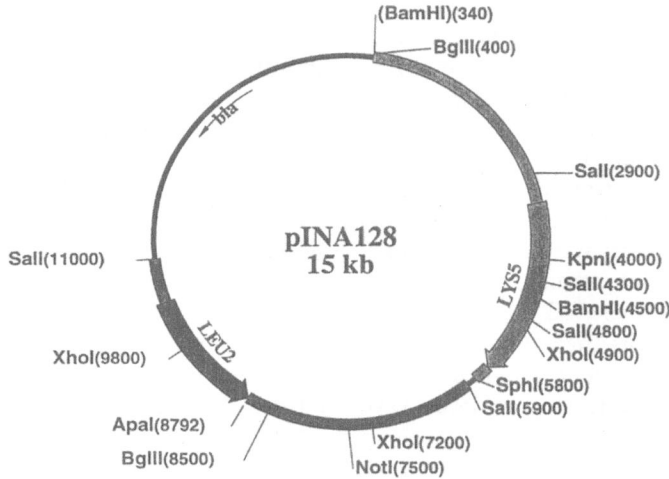

Fig. 2. Transformation efficiency of pINA128 (*below*) carrying the two selective markers *LEU2* and *LYS5*. The plasmid was targeted to the genome of a *lys5, leu2* recipient strain, either at the *LEU2* or at the *LYS5* locus, selecting for one marker and then checking complementation for the second one. See text for details

93 and 6%, depending on the position of the cut relative to the *lys5-12* chromosomal mutation (which is near the *Bam*HI site), indicating that transfer of the chromosomal *lys5.12* allele to the plasmid copy was still detectable when the cut was made some 3 kb away from the mutated site. Similar results were observed at the *LEU2* site (Ribet and Gaillardin, unpubl.). This observation obviously permits recovery of chromosomal mutations, by digesting the DNA of Lys⁻ transformants, religating it on itself and transforming an *E. coli* recipient. The *xpr2-7* mutation which directs the synthesis of a thermosensitive alkaline protease has been rescued in this way (unpubl.). One should also be aware of this fact when a vector is targeted to a mutated gene, and when integration is selected by another marker of the plasmid: integrants at the site may contain two identical copies of the chromosomal allele, particularly if the cut is made close to the mutated site.

Allele Substitution

Several in vitro-generated mutations can be substituted for a wild-type allele in a given cell line by the following procedure (Beckerich et al. 1994). The wild-type copy is first disrupted by a *URA3* cassette. Mutated copies of the gene of interest are then supplied as linear fragments which will recombine with the disrupted copy, thus excising the *URA3* cassette, an event which is easily selected on 5FOA containing plates. Allele exchange can also be obtained by the two-step "pop out method" (see Sect. 5.3).

5.1.2
Multiple-Copy Integration

Defective versions of the *URA3* gene have been constructed by deleting most of its promoter sequence and/or by creating mutations within the coding sequence (Le Dall et al. 1994). Single copies of the most defective constructs are apparently unable to confer a Ura⁺ phenotype to a strain carrying the *ura3.302* allele (this is a nonreverting *ura3* mutation).

Three targets of integration have been tested: tandemly repeated sequences of rDNA, single-copy genes like *XPR2*, or dispersed repeated sequences like the Ylt1 transposon. Transformants were obtained in all cases, albeit at a much reduced frequency and with a long delay in appearance as compared to transformants obtained with nondefective markers: 1 to 2 weeks vs. 4 days, 1 to 10 transformants per μg of plasmid vs. more than 10^4. A plasmid carrying a defective *URA3* gene, a piece of rDNA coding sequence, and the *XPR2* gene was targeted either at the rDNA loci, or at *XPR2*. Amplification levels varied between 5 and more than 40, depending on how defective the marker used was, when the plasmid was targeted to the rDNA; several clusters of tandemly integrated plasmids were present in different rDNA units in some transformants. Targeting to the *XPR2* locus also yielded transformants, albeit at a lower frequency, but most of them eventually turned out to result from integration into the rDNA loci; a single integrant at the *XPR2* locus was observed which carried about 30 copies. Another plasmid carrying the defective *URA3* marker and a piece of Ylt1 was targeted to the transposon(s). Amplification levels of 5 to 15 were observed; here again, a given transformant may contain plasmids tandemly integrated in several Ylt1 transposons.

Stability of the integrated plasmids was high, except when *XPR2* transcription was turned on: in this case, high copy number transformants (more than 30 copies) appeared very sick, and rapidly threw off cells with less copies. No stability problem was observed with transformants carrying 10 copies or less of *XPR2*, but this may, of course, depend on the "toxicity" of the protein expressed.

5.2
Replicative Transformation

Replicative vectors carry both an origin of replication (*ORI*) and a centromeric region (see Sect. 4.2.3) and are thus usually relatively stable under nonselective conditions (about 1% loss per cell division) and present at 1–3 copies per cell.

Some strains, particularly W29 derivatives (like P01a or P01d, see below) seem to missegregate these plasmids at elevated frequency, thus yielding more unstable transformants (10–20% loss/cell division), and accordingly contain more plasmids per transformed cell (up to 13, Fournier, unpubl.).

Transformation of replicative vectors can be done either by the lithium acetate procedure or by electroporation. Both methods yield transformation frequencies in the range of 10^3–10^4 transformants/μg of DNA, or significantly less than integrative transformation.

Using the lithium acetate procedure, this frequency can be significantly increased by linearizing the plasmid, whereas this treatment nearly abolishes transformation by the electroporation method. The cut can be made either in a region of homology with the genome (and is then apparently cleanly repaired), or in regions without homology with the genome: no precise repair is then possible and a delection of variable extent is observed in surviving plasmids. This has been used to generate in vivo a set of nested deletions: a Bluescript based vector carrying *LEU2* and *ARS3018* was linearized within the polylinker, and deletions extending in the adjacent *LEU2* or *ARS* sequence were recovered from yeast transformants (B. Kudla, unpubl.).

Replicative plasmids carrying repeated sequences as short as 31 bp may suffer recombination between these sequences when transformed by the lithium acetate procedure, and thus excise the intervening sequences. Longer repeats have a higher propensity for recombination: up to 70% deletions were observed with 750-bp-long repeats. Recombination apparently takes place during or just after transformation, since undeleted plasmids are perfectly stable during subsequent mitotic divisions. This type of rearrangement was never seen with (cut) integrative plasmids using the lithium acetate method, or with uncut replicative plasmids transformed by electroporation. We thus advise the use of the lithium acetate method for integrative plasmids, and electroporation for uncut replicative plasmids.

Replicative plasmids provide a limited amplification and/or another genomic context which may result in strong phenotypic effects: the *scr2-II13* allele is lethal as a single copy integrated in the genome, but viable, although thermosensitive when plasmid-borne. Several multicopy suppressor genes could be cloned by using a replicative library to complement chromosomal mutations as diverse as morphogenesis mutations (Dominguez, unpubl.), regulatory mutations of *XPR2* (M. Lambert, unpubl.), suppressor mutations of a 7S RNA defect (Choukri, unpubl.), or mutations showing colethality with a 7S RNA defect (Boisramé, to be published).

Replicative plasmids are also useful for the study of essential genes by "plasmid shuffling" (He et al. 1992). In *Y. lipolytica*, this procedure is plagued by the frequent occurence of recombination events between the incoming and the resident plasmid, resulting in stable cointegrates (Fournier et al. 1993). In order to reduce the level of recombination, we developed *ARS* plasmids based on pACYC184; whereas cointegrates were present in 1–5% of the transformants when pBR322-based *ARS* plasmids were used, this frequency dropped to less than 0.05% when plasmids based on pINA443 and pINA894 were used (A. Boisramé, to be published).

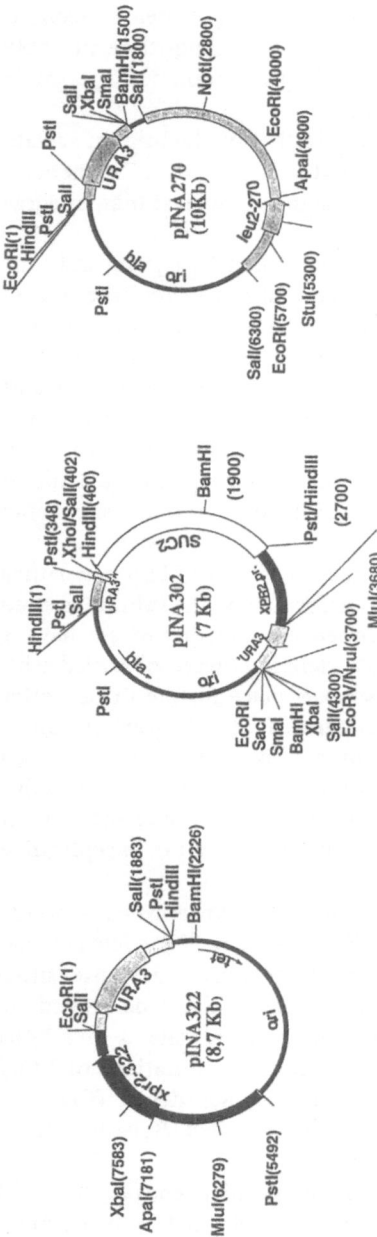

Fig. 3. Disruption and deletion vectors for creating nonrevertible mutations in the *LEU2, URA3,* and *XPR2* gene

5.3
Creation of a Set of Nonreverting Markers

In order to create in any genetic background a set of markers suitable for selecting transformants and devoid of the major secreted protease (AEP), a set of plasmids has been developed (Fig. 3).

The procedure outlined below has been tested on an industrial strain of *Y. lipolytica* (W29, ATCC24060) and on inbred derivatives (Kurischko et al. 1992). A disruption cassette was first used to replace the wild-type *URA3* gene by an *XPR2-SUC2* gene fusion (*Sal*I fragment of the disruption plasmid pINA302; Fig. 3); transformants were selected for sucrose utilization (Suc⁺ phenotype, Nicaud et al. 1989b). All of the wild-type *Y. lipolytica* strains so far tested are Suc⁻. The Suc⁺ transformants (1000–2000/µg of fragment) were then screened for the Ura⁻ phenotype; 10 to 20% were indeed Ura⁻, and the presence of the disrupted *URA3* copy (*ura3-302*) was checked by Southern hybridization.

In order to generate further derivatives of the Ura⁻Suc⁺ strains, pINA270 and pINA322 were constructed: they carry *URA3* and either a deleted *LEU2* or a deleted *XPR2* gene (Fig. 3). The deletions were successively substituted to the resident wild-type copies of *LEU2* or *XPR2* by the following two-step procedure. Ura⁺ transformants (10^5/mg of plasmid) were obtained with pINA270 cleaved at the unique *Not*I site located next to the *LEU2* gene. Five transformants were plated (10^6 cells per plate) on YNB containing 15 mg/l uracil, 100 mg/l leucine, and 1.25 g/l fluoroorotic acid (5-FOA) to counterselect Ura⁺ cells. 5-FOAᴿ clones appeared at a frequency of 10^{-4}/cell plated and were shown by Southern analysis to result from reexcision of the integrated plasmid; 2% of the 5-FOAᴿ clones were simultaneously Leu⁻ and carried the deleted *LEU2* gene (Southern analysis). These strains (Ura⁻, Suc⁺, Leu⁻) were again transformed to Ura⁺ using pINA322 cleaved at the unique *Mlu*I site. Reexcision of the plasmid was selected as above, and AEP⁻ derivatives (33 to 42% by plate assay) were isolated and checked for both *LEU2* and *XPR2* deletions (Southern analysis).

Using the same procedure, we derived isogenic strains from the industrial strain W29: P01a (*MatA, ura3.302, leu2.270*) and P01d (*MatA, ura3.302, leu2.270, xpr2.322*). This strain has superior growth capacities and gave better yields of heterelogous proteins (Nicaud and Tharaud, unpubl.).

A similar procedure can obviously be developed for the creation of other markers.

5.4
Marker Genes and Vectors

5.4.1
Biosynthetic Genes

Cloning of the *LEU2* gene of *Y. lipolytica* in pINA62 has been reported previously (Gaillardin and Ribet 1987). pINA62 can be targeted to the genome of a *Y. lipolytica leu2* recipient strain at a very high frequency (>10^5 transformant/µg of

plasmid DNA with a lithium acetate procedure, see Xuan et al. 1988) after linearization by either *ApaI*, *BglII* or *NotI* which are unique in the vector (Fig. 1A).
URA3-based vectors were derived from pLD55 (Davidow et al. 1987; see pINA300
in Fig. 1A). The *LYS5* and *ADE1* genes/have been occasionally used to disrupt
cloned genes (He et al. 1990; Nicaud et al. 1989a).

A genomic library has been constructed in pINA62 by inserting a *Sau*3A partial
digest into the *Bam*HI site of pINA62 (Xuan et al. 1988). Linearization of the library
by *NotI* avoids cutting in the inserts and should generally facilitate use of this
library. Several genes could be isolated from this library, including *LYS5* (Xuan et
al. 1988), *LYS1* (Dominguez, unpubl.), *LYC1* (Beckerich et al. 1994), and mating
type (Kurischko et al. 1992), and XPR6 (Enderlin and Ogrydziak 1994).

Smaller derivatives of pINA62 carry the 2.2-kb *Eco*RI fragment encompassing
the whole *LEU2* gene (Davidow et al. 1987) and preserve ability to confer resistance
to both ampicillin and tetracyclin; these vectors have been modified to include a
NotI-SfiI linker at the *Pvu*II site of pBR322 (see pINA214 in Fig. 1A). This facilitates
targeting to a "landing platform" consisting of a pBR322 plasmid inserted into the
genome of the recipient strain (Davidow et al. 1987; Blanchin-Roland et al. 1994).

Following identification of *ARS* sequences in *Y. lipolytica* (Fournier et al. 1991),
vectors carrying *ARS3018* or *ARS1068* together with *LEU2* or *URA3* were developed
(Fig. 1B). A gene library in pINA240 was constructed by inserting a *Sau*3A partial
digest of genomic DNA (size-selected in the 5-kb range) in the *Bam*HI site (Chasles
and Fournier unpubl.). Similar libraries have been built by G. Barth and R.
Rachubinski. Reisolation of complementing plasmids from the yeast transformants can be simply achieved by shuttling back to *E. coli* (Fournier et al. 1991).
This library is best transformed by electroporation, to avoid deletion in the
plasmids by recombination between direct repeats.

5.4.2
Heterologous Markers and Reporter Genes

Expression of the bleomycin resistance gene from Tn5 under the control of the
LEU2 promoter has been reported (Gaillardin and Ribet 1987): direct selection of
resistant clones was possible on YPD +15 mg/l phleomycin, although only half of
the resistant clones resulted from integration of the vector (remaining clones were
spontaneous resistants). The gene encoding hygromycin resistance gene from
Streptomyces hygroscopus was fused downstream from the *XPR2* promoter on both
integrative and replicative plasmids (Fig. 4; Cordero-Otero, unpubl.).
Untransformed control cells yielded less than one hygromycin resistant clone for
10^8 plated cells (on YPD pH 6.8 containing 100 mg/l hygromycin). Following either
integrative or replicative transformation, resistant clones appeared at a frequency
of 50–100/μg of input DNA; all expressed a second control marker present on the
plasmid (to be published).

The invertase-encoding gene of *S. cerevisiae* (*SUC2*) has been used as a selective
marker (Nicaud et al. 1989b): the SUC2 gene was put under the control of the XPR2
promoter, and directed periplasmic accumulation of invertase, thus allowing *Y.
lipolytica* to grow on sucrose. Although this selection could be used to select gene

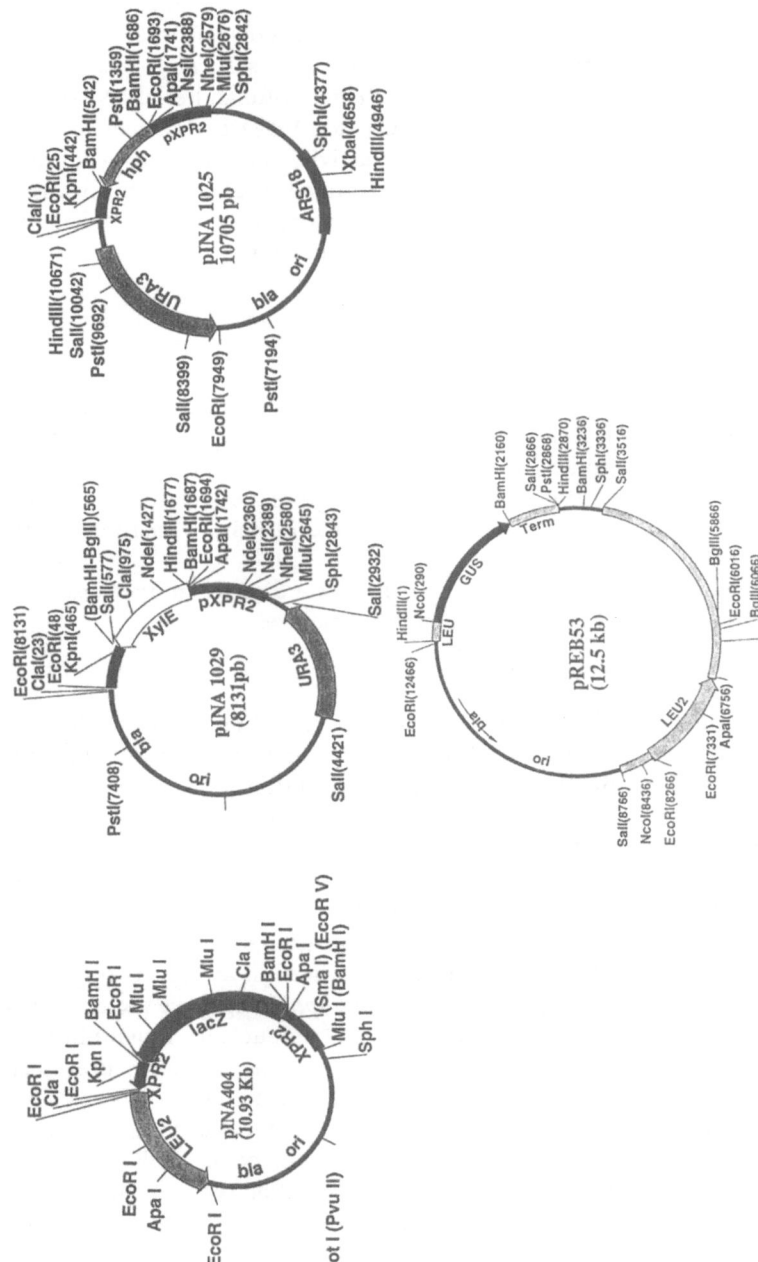

Fig. 4. Heterologous marker genes and expression vectors based on either the *XPR2* promoter (*above*), or the *LEU2* promoter (*below*)

disruptions in several wild-type contexts (see Sect. 5.1.3), a frequent problem is that wild-type strains grow slowly on (impurities of) sucrose plates.

Several bacterial genes have been used as reporter genes: *E. coli* β-galactosidase (Gaillardin and Ribet 1987; Xuan et al. 1990; Blanchin Roland et al. 1994) or β-glucuronidase (Bauer et al. 1993; Hamsa and Chattoo 1994), and catechol oxidase (R. Coredero-Otero, unpubl.). No interfering activity was detected in untransformed *Y. lipolytica* extracts, and plate tests exist in all cases to detect expression at the colony level. Very low levels of the *gusA* gene product can be detected by a fluorescence assay, making it suitable for most applications.

5.4.3
Expression and Secretion Vectors

The *XPR2* promoter and signal sequence have been used to express and secrete several proteins (Buckholz and Gleeson 1991). AEP is initially synthesized as a precursor with a 157 amino acid preproregion, ending with dibasic cleavage site (Lys$_{156}$Arg$_{157}$; Matoba et al. 1988). Two types of expression vectors were constructed, permitting either expression of a cytosolic protein or secretion to the outside of the cell.

An *Eco*RI and a *Bam*HI site were created by site-directed mutagenesis just downstream from the ATG thus generating the structure:

....ATG.GAA.TTC. AAG.GAT. CCT.....

A *Bam*HI deletion was made to remove most of *XPR2*-coding sequence. The final expression cassette consists of the *XPR2* upstream promoter region from the *Mlu*I site down to the *Bam*HI site, thus conserving transcription and translation initiation sites, and a 3' region from the BamHI site down to the *Eco*RI site, thus encompassing transcription termination signals. This cassette was inserted into a pBR322 vector carrying a *Not*I-*Sfi*I linker destroying the *Pvu*II site, a *Mlu*I linker destroying the *Bam*HI site, and the *LEU2* marker (see pINA404 in Fig. 4 and Blanchin-Roland et al. 1994). Derivatives of this construct carry the *URA3* gene or an *ARS* sequence (Fig. 4). Foreign genes can be spliced in as *Bam*HI fragments. Genes succesfully expressed include *xylE*, *lacZ*, and *hph*. Integrative vectors can be targeted either to the *XPR2*, *LEU2*, or *URA3* loci if unique sites remain in these sequences, or by *Not*I/*Sfi*I digestion to a pBR322 landing platform previously inserted into the recipient strain (Blanchin-Roland et al. 1994).

In order to direct secretion of foreign proteins, vectors carrying convenient restriction sites downstream from the prepro-sequence of the *XPR2* gene were constructed (Tharaud et al. 1992; Fig. 5). The *XPR2* coding sequence was deleted between the *Ava*I site (overlapping the Pro codon, 11 amino acids before the pro-mature junction) and the *Kpn*I site (overlapping the Gly80 codon of the mature enzyme) and replaced by the following oligonucleotide:

(C)CC.GAGATT.CCg.GCc.TCT.TCg.gcc.GCC.AAG.CGA/.cccgggtggacgtctagaGG-TAC(C).

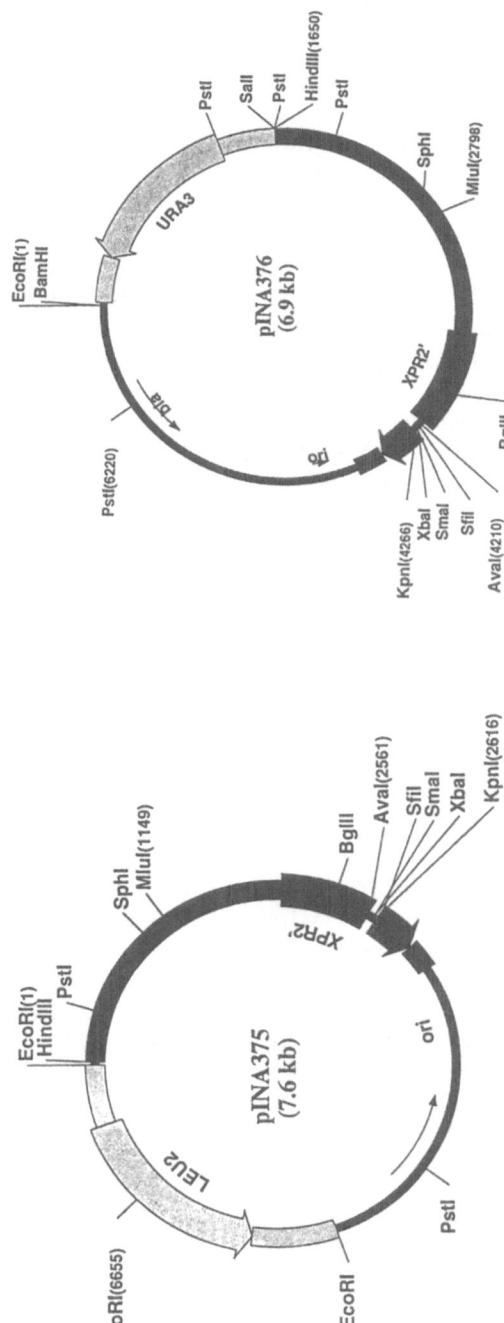

Fig. 5. Vectors for the expression of secreted proteins. The *XPR2* promotor and terminator sequences are depicted by *black boxes*. The region encoding the AEP preprosequences and C-terminal amibnoacids is indicated by a *thick black arrow*, interrupted by the polylinker sequence located at the Xpr6p cleavage site (see text)

The polylinker introduces *Ava*I, *Sfi*I, *Sma*I, *Xba*I, and *Kpn*I sites (underlined). In spite of the numerous changes made (indicated by lower letters in the above sequence), the pro-sequence is conserved up to the Lys-Arg cleavage site (indicated by A/in the above sequence), but for one substitution (Asn to Ala). A minimal terminator sequence was synthesized by PCR from the *XPR2* 3'-flanking region: this synthetic fragment carried a *Kpn*I site just in front of the *XPR2* stopcodon and the first 100 nucleotides of the *XPR2* terminator sequence, followed by a *Sal*I site. In the final vector (see pINA375 and pINA 376), foreign coding sequences can be spliced in between the *Sma*I and *Xba*I or *Kpn*I sites, although this will add one or more extra amino acids after the dibasic cleavage site. Preferably, target sequences can be spliced into the *Sfi*I site using an adaptor sequence with a (*Sfi*I compatible) 3' AAG overhang and restoring the correct AEP prosequence up to the dibasic cleavage site: (TCT.AAT.GCC.AAG.CGA/etc). Integrative plasmids can be targeted to the host strain *XPR2* locus by *Mlu*I digestion. The heterologous products should at best be secreted as mature proteins, after proteolytic removal of the AEP prodomain by the host dibasic endoproteinase Xpr6p. These constructions have been used to drive secretion of porcine α-interferon (Nicaud et al. 1991) and of human coagulation factor XIIIa (Tharaud et al. 1992). Similar constructs were used to express bovine prochymosin and human tissue plasminogen activator (Buckholz and Gleeson 1991). Except for the case of human coagulation factor XIIIa (a protein which is not normally secreted by a signal peptide-dependent pathway and which was retained intracellularly and degraded), all other proteins were correctly matured and efficiently secreted. Hepatitis B virus middle surface antigen has been expressed using a fusion to the AEP signal peptide and part of pro-sequence (Hamsa and Chattoo 1994): expression was growth-regulated and directed intracellular accumulation (presumably in the lumen of the endoplasmic reticulum) of preS2-HBsAg assembled into Dane particles (up to 2.35% of total soluble intracellular protein).

A *LEU2*-based expression vector has been constructed by Bauer et al. (1993). An A to C mutation was created at position −1 from the ATG to create an *Nco*I site (the effect of this mutation has not been evaluated, but see Sect. 4.2.6) and the *Eco*RI site upstream from the *LEU2* promoter was replaced by a *Hin*dIII site, so as to create a portable *LEU2* promoter on a *Hin*dIII-*Nco*I fragment. Similarly, the *Apa*I site downstream from the coding sequence was replaced by a *Bam*HI site, an *Nco*I site normally present within the 3' polyadenylation signal was destroyed, and the *Eco*RI site located further downstream was replaced by a *Hin*dIII site, so as to yield a portable *LEU2* terminator on a *Bam*HI-*Hin*dIII fragment (Bauer et al. 1993; Fig. 4).

5.4.4
Vectors for Multicopy Integration

Plasmids like pINA774 or pINA970 (Fig. 6) carry a defective *URA3* marker from which the entire promoter sequence has been deleted except for 6 bp upstream from the ATG. Transcription is driven from unknown adjacent sequences, and is

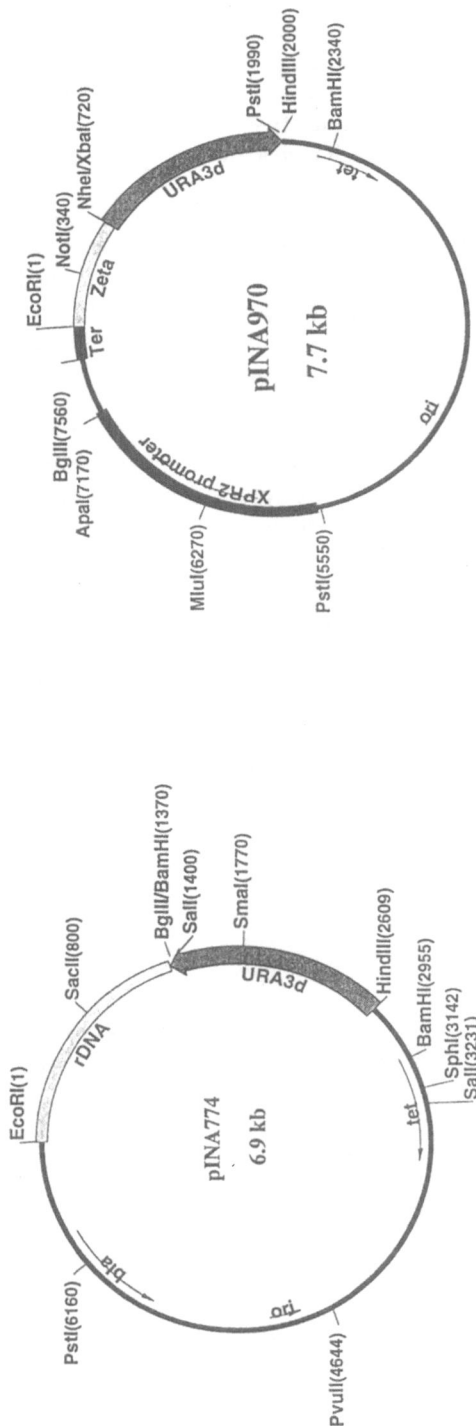

Fig. 6. Vectors for multicopy integration. They carry a defective *URA3* marker and a fragment of repeated DNA, originating either from rDNA (*left*) or from the Ylt1 transponson (*right*)

probably very dependent on the vector context: whereas single-copy Ura⁺ transformants have never been observed with either vector, copy numbers observed are in the range of 40–60 with pINA774 (a pINA773 derivative carrying the *XPR2* gene) and of 5–15 with pINA970 (Le Dall et al. 1994, and unpubl. results). Less defective *URA3* promoters are available, which give a lower amplification (Le Dall et al. 1994).

A nonleaky, nonreverting *ura3* strain is required, such as strains from the E129 series or similarly derived strains (see Sect. 5.3). Vectors like pINA774 have to be linearized by *Sac*II to target the plasmids to the rDNA units. Transformants appear after 5 to 20 days at a frequency of 10–50/μg of input DNA (vs. 10^4 for a nondefective plasmid). Vectors like pINA970 can be linearized by *Not*I to be targeted to zeta sequences. Transformants appear at a frequency of 50–100 after 5 to 15 days in this case. There is apparently no correlation between the delay of transformant appearance and copy number.

Stability of the transformants is likely to depend on the nature of genes inserted into these vectors: introduction in rDNA-based vectors of an expression cassette under the control of the *XPR2* promoter rapidly destabilized high copy number transformants under induction conditions (Le Dall et al. 1994). It may be advisable to use rDNA-based vectors for amplification of poorly expressed gens, and zeta-based vectors (or rDNA-based vectors with less defective *URA3* promoters) for highly expressed genes.

The structure of the integrants and their copy number are determined by PCR or Southern hybridization (Le Dall et al. 1994).

6
Media and Culture Conditions

6.1
Media

6.1.1
Complete Media

Yeast Extract – Peptone – Glucose – Medium (YPD)

Yeast extrace	10 g/l
Bacto peptone	20 g/l
Glucose	20 g/l

Yeast Extract – Malt Extract – Peptone – Glucose – Medium (YMG)

Yeast extract	3 g/l
Malt extract	3 g/l
Bacto peptone	5 g/l
Glucose	10 g/l

6.1.2
Synthetic Minimal Media

Yeast – Nitrogen – Base – Medium (YNB) (DIFCO)

YNB   6.7 g/l

Different concentrations of carbon sources are recommended:

Glucose	10–20 g/l (up to 100 g/l can be used)
Sodium acetate	4 g/l (higher concentrations reduce growth, toxic above 10 g/l), addition of 0.05 M phosphate buffer of pH 5 is recommended
Ethanol	up to 3% (higher concentrations are toxic)

n-Alkanes, fats, and fatty acids have no toxic effects and can therefore be also used at high concentrations. Emulsifiers like Brij 58 or Tween can be added but are not necessary because *Y. lipolytica* secretes very efficient bioemulsifiers.

Minimal – Medium with Thiamine (MMT)

Some strains of *Y. lipolytica* grow slowly in YNB, therefore this medium is recommended to obtain/higher growth rates.

$NH_4H_2PO_4$	5 g/l
KH_2PO_4	2.5 g/l
$MgSO_4 \times 7H_2O$	1 g/l
Thiamine $\times$ HCl	0.3 mg/l
Solution of trace elemets	1 ml

Solution of trace elements (mg in 100 ml):

$Ca(NO_3)_2 \times 4H_2O$	2000
$FeCl_3 \times 6H_2O$	200
H_3BO_3	50
$CuSO_4 \times 5H_2O$	10
$MnSO_4 \times 4H_2O$	40
$ZnSO_4 \times 7H_2O$	40
Na_2MoO_4	20
$CoCl_2$	10
KI	10

Addition of carbon sources as recommended for YNB.

The pH value should be adjusted at 6 for most carbon sources, but at pH 4 for acetate or fatty acids (or strongly buffered at pH 6 with phosphate buffer).

Sometimes, problems occur with strains harboring multiple amino acid auxotrophies: use of 0.1% glutamine as N-source is recommended in such cases.

6.1.3
Conjugation Media

Highest conjugation frequencies are obtained in liquid or on solid YM without glucose (Barth and Weber 1984). Addition of 0.05% sodium citrate increases slightly the conjugation frequency (Weber et al. 1988).

Yeast extract – malt medium (YM)

Bacto-peptone	5 g/l
Bacto-yeast extract	3 g/l
Malt extract	3 g/l
Bacto-agar	20 g/l

6.1.4
Sporulation Media

Sporulation can be induced in a variety of liquid and solid media. However, the medium of choice for induction of high frequency of sporulation depends on the strains used.

Sporulation of diploid cells can be induced on commercially available V8 medium or on YM (Gaillardin et al. 1973).

High sporulation frequency is induced by YNB or MMT containing 1.5% sodium citrate (citrate sporulation medium = CSM). Lower or higher concentrations of sodium citrate reduce the sporulation frequency dramatically (Barth and Weber 1985).

Reduced aeration or temperatures above 30 °C decrease sporulation frequency.

6.1.5
Special Media

Liquid Medium for Protease Induction

Strains are grown in late exponential phase with good aeration at 23 °C on YPDm:

Proteose peptone	50 g/l
Yeast extract	10 g/l
Glucose	1 g/l
Na citrate buffer 0.2 M	pH 4.0 for acid protease
Na_2HPO_4, KH_2PO_4 50 mM	pH 6.8 for alkaline protease

Induction starts around 2–3 OD 600 nm. Maximum levels are reached after 24 h (around 18 OD 600 nm).

Skim-Milk Complete Medium (Alkaline Protease)

Dissolve 10 g Difco skim milk in 60 ml Na phosphate buffer 0.1 M pH 6.8. Heat for 10 min in a boiling water bath and repeat after waiting 24 h at room temperature.

Dilute this solution ten times in the following sterile medium kept at 45 °C: to 480 ml of YNB solution (1.7 g/l YNB without amino acids and without ammonium sulfate, 20 g/l agar, amino acids as required for auxotrophies), add solution A and 60 ml 0.5 M phosphate buffer pH 6.8.

Positive clones form clear halos on a white background.

This medium can also be used to quantitate AEP activity from supernatants: omit YNB in this case, add 2 mM final sodium azide and pour into petri dishes. Take care to have all plates contain the same volume of medium (15 ml).

Punch holes in the agar either with a Pasteur pipette (large end) or with a 200 μl micropipette tip. Make 20 holes in a 90 mm dish.

Put 10–20 μl of (diluted) supernatant in the holes. Examine after 12 and 24 h.

Bovine Serum Albumin Medium (Acid Protease)

Same as skim milk, except for buffer (use sodium citrate pH 4.0 in this case and for protein source use 2% BSA).

α-Naphthyl Phosphate Medium (Acid Phosphatase)

Use phosphate-depleted YPD medium (Rubin 1974) or YNB-glucose with 30 mg (LP_{30}) or 10 mg (LP_{10}) KH_2PO_4/l. The major acid phosphatase is fully derepressed on LP_{10}, whereas Pho2p is significantly derepressed on LP_{30}.

Colonies grown on these media can be stained for APase activity as follows. Make patches on LP_{10}; when grown up, gently pour 5 ml of an overlay of soft agar (1%) containing (per ml) 0.2 mg of α-naphtylphosphate and 2 mg of Fast Garnett GBC salts (Sigma) in 0.05 M acetate buffer pH 4.0.

Colonies stain yellowish to dark brown, depending on the activity.

RNase Medium

Plate tests:

Glucose	10 g/l
Difco proteose peptone	4 g/l
Sigma type V yeast RNA	2 g/l
KH_2PO_4	145 mg/l
$MgSO_4$, $7H_2O$	400 mg/l
$CaCl_2$, $2H_2O$	150 mg/l
NaCl	100 mg/l
Thiamine	1 mg/l
Agar	15 g/l
Citrate-phosphate buffer pH 5.0	50 mM

RNA is solubilized by adding 1 M NaOH until the pH is between 5 and 6. Plates are inoculated with a toothpick and incubated 48 h at 25 °C. They are then overlaid with 0.1% toluidine blue. RNase-positive colonies are surrounded by a pink halo on a blue background.

Liquid medium for RNase production:
Same as for protease induction (see Cheng and Ogrydziak 1986)

Lipase Induction Media

Lipase-positive clones are able to grow and form clear halos on YNB medium with
tributyrin (10 ml/l) as sole carbon source (Nga et al. 1988). Faster growth is observed if 10 g/l glucose is present, lipase-positive clones are distinguished from
negative ones by the presence of halos.

Liquid media containing olive oil as inducer have been used for production and
purification of lipase (Gomi et al. 1986).

Mycelial Induction Medium

Cells of *Y. lipolytica* are harvested by centrifugation from a glucose containing
medium (YNB or MMT + 1% glucose) during the logarithmic phase of growth
at 28 °C (about 10 to 14 h of growth), resuspended in water and cooled down to
4 °C for at least 15 min. The cells are then heat-shocked by inoculating 1–5 ×
10^6 cells/ml into prewarmed (30 °C) YNB containing 50 mM sodium citrate pH 6.0
as buffer and 1% N-acetyl-glucosamine as carbon source. The culture is then
incubated at 30 °C.

5-Fluoroorotic Acid-Containing Medium (5FOA Medium)

Ura-clones are positively selected on YNB medium containing

Uracil	15 mg/l
5-Fluoroorotic acid	1.25 g/l

Leaky Ura⁻ clones are killed on such a medium. For an unknown reason, direct
selection of spontaneous *ura3* mutants could not be achieved on this medium
from various wild-type strains, but it worked well for selecting *ura3* disruption
or deletions in the same strains following transformation (see Sect. 9.1). A maximum of $10^6 - 10^7$ cells have to be plated on 9 cm 5-FOA plates, higher densities
result in high background growth. If background growth occurs, replica plating
on a fresh 5-FOA plate can be done to rescue the Ura⁻ clones. The selection does
not work well at temperatures below 20 °C, although raising the pH to 6.8 helps
somewhat.

6.2
Temperature

Most strains of *Y. lipolytica* grow only up to 34 °C, but there are some strains which
are adapted to higher temperatures (Blagodatskaya and Kockova-Kratochvilova
1973; Barnett et al. 1990). The recommended temperature for growth is 25 to 30 °C.
Induction of sporulation is highest at 23 °C. Conjugation and sporulation frequencies are highest at 23 to 28 °C, but decrease strongly above 30 °C.

AEP productivity is highest at 23 °C and decreases when temperature is increased (Dedeoglu and Ogrydziak, unpubl.).

6.3
pH Values of the Growth Media

Most strains of *Y. lipolytica* tolerate well low pH values down to pH 3, but growth is reduced above pH 7 and stops above pH 8. Therefore, addition of buffer with a pH lower than pH 7 or low pH values (pH 3.5 to 4) of the medium at the beginning of the cultivation are recommended for utilization of carbon sources such as acetate.

6.4
Aeration

Y. lipolytica is an obligatory aerobe and cannot grow or ferment without oxygen. Low oxygen pressure reduces the growth rate strongly. Therefore vigorous shaking of batch cultures or high oxygen supply in fermentors is recommended. However, development of foam may occur, especially during cultivation in complete media like YEPD. This has to be inhibited, because cells of *Y. lipolytica* migrate rapidly into the foam, thus reducing the yield. Addition of small amounts of oil efficiently reduces the formation of foam.

6.5
Conservation of Strains

Cells of *Y. lipolytica* die without oxygen. Therefore strains of this fungus cannot be stored for a longer period under paraffin-closed tubes, or on plates which are closed by parafilm, or in closed plastic bags. However, strains of *Y. lipolytica* can be sent out in sealed tubes because they survive for short periods (up to 1 month) quite well.

6.5.1
Conservation with Glycerol

A culture of *Y. lipolytica* cells from the mid logarithmic growth phase is mixed with sterilized glycerol at a final concentration of 25 to 50% in a plastic tube. Tubes are stored at −70 to −80 °C.

6.5.2
Liquid Nitrogen and Freeze-Drying Preservation

Culture. After subcloning on solid YPD, one colony is pregrown in 2 ml of YM liquid medium for 24 h at 28 °C under aeration (C1 culture); 25 ml of fresh YM medium are inoculated with 0.5 ml of C1 culture, and grown at 28 °C under aeration for 24 h. Cells from this culture (C2) will be preserved by freezing and freeze-drying.

Freezing (following Daggett and Simione 1987). Ten ml of the C2 culture are mixed with 2.5 ml of fresh YM medium containing 50% of glycerol (final glycerol concentration 10%).

The mixture is dispensed in ten cryotubes and chilled at 4 °C for 1 h. The tubes are frozen in the Nalgene cryobox at −80 °C in order to obtain a temperature decrease of −1 °C per min. The tubes are then preserved in liquid nitrogen at −150 °C.

Freeze-drying (following Berny and Hennebert 1991). Eight ml of the C2 culture are centrifuged at 4000 rpm for 5 min and cells are suspended in 2 ml of the following mixture: 10% skim milk, 5% honey, and monosodium glutamate 5%.

The suspension is chilled at 4 °C for 1 h and 0.1 to 0.15 ml of cell suspension are dispensed in glass tubes.

After being capped with cotton, the tubes are frozen at −80 °C for 1 h. Freeze-drying is performed in a Virtis 12 EL freeze-dryer for 5 h. The temperature at the begining is about −40 °C and rises to 20 °C at the end. Tubes are sealed under vacuum and stored at 4 °C in the dark.

7
Genetic Techniques

7.1
Mutagenesis

7.1.1
UV-Light Mutagenesis

Cells are grown for 22 h in YPD, 1 ml harvested by centrifugation (6000 g, 5 min), washed once with water, and resuspended in water at a cell density of about 5×10^3 cells/ml.

Ten ml of this suspension are given in a Petri dish and are exposed to UV light in the dark. Normally, UV light of 80 to 120 J/m² inactivate about 50% of the cells of a haploid strain; 30 to 40% inactivation is recommended for efficient selection of single-site mutants.

After irradiation, cells are plated onto YEPD plates (100 and 10 μl per plate) and incubated for 2 days at 28 °C in the dark. Colonies should be not too big before replica plating otherwise it is more difficult to transfer them.

YEPD plates are replica-plated onto selective media and incubated at 28 °C or 32–34 °C (for selection of its mutants).

7.1.2
N-Methyl-N′-Nitro-N-Nitroso-Guanidine (MNNG) Mutagenesis

Cells of *Y. lipolytica* are harvested by centrifugation after cultivation in YPD overnight, washed once with water, and resuspended in 0.9% sodium chloride soultion at about 2×10^7 cells/ml.

MNNG is added at a final concentration of 125 μg/ml.

The suspension is shaken at 28 °C for 40 min.

Cells are harvested by centrifugation, washed once with water, resuspended in the same volume of water, and diluted 10 and 100 times.

One hundred μl of the diluted suspensions are spread on YPD plates. Plates are replica-plated onto selective media after 2 days of incubation at 28 °C. The frequency of survival cells is about 1–5%.

7.1.3
Enrichment of Mutants by Nystatin

This method can be used for enrichment of spontaneous mutants or concentration of mutants after mutagenesis.

Cells are cultivated overnight in YPD (for 16 h), harvested by centrifugation, washed, and transferred into nitrogen-free MMT + 1% glucose. For enrichment of mutants defective in utilization of certain carbon sources, glucose has to be replaced in the minimal medium by the corresponding carbon source.

The culture is shaken at 28 °C for 8 h to deplete metabolic pools. After 8 h, a sterile solution of ammonium sulfate is added at a final concentration of 5 g/l. Cultivation is continued until growth has resumed (about one doubling), which normally takes 4 to 10 h. Alternatively, we found it convenient to let the culture sit on the bench at room temperature overnight and to shake it for 2 h at 28 °C before nystatin addition.

Nystatin is added to a final concentration of 15 μg/ml from a stock solution (1 mg/ml in ethanol).

The length of the nystatin treatment (usually 1.5 h on MMT-type media) has to be adjusted to the growth rate of the strain on the selective medium: optimal results are obtained if the treatment is carried out for about one doubling of the culture. Survival should be lower than 10^{-4} after this step.

If required, the whole treatment can be repeated to further enrich for mutants.

Serial dilutions of the washed suspensions are made in 0.8% NaCl-20% glycerol and plated on YPD plates. Dilutions can be stored at −80 °C until colonies on the YPD plates can be counted; additional samples of the frozen suspension can then be plated (we routinely observe about 90% survival in the frozen samples).

In order to select independent mutants, we usually treat 10–20 cultures in parallel. Five ml of cells can be conveniently grown in 25 ml Corex tubes, slanted to facilitate washing, which are rotated at 28 °C.

7.2
Mating and Sporulation

7.2.1
Conjugation on Solid Medium

Strains of opposite mating type and complementary auxotrophic requirements are streaked onto YPD plates. These plates will be incubated for 1 day at 28 °C.

A loop of cells is transferred onto a fresh YPD plate, so as to form a line across the plate. Strains of identical mating type are streaked on parallel lines across the same plate. A maximum of five strains can be put on one plate. The plates are incubated for 18 to 21 h (important for high conjugation frequency) at 28 °C.

Two YPD plates containing strains of opposite mating type are replica-plated crosswise onto one YM (without glucose) plate. This YM plates do not longer incubate than 24 h, as zygotes of some strains die fast after 24 h of incubation on YM. The YM plate is replica-plated onto a plate of selective minimal medium. Diploid colonies can be harvested from the cross points of the strains from these plates after 3 days of incubation at 28 °C.

7.2.2
Conjugation in Liquid Medium

Strains of opposite mating type are separately inoculated into 10 ml of YPD and shaken at 28 °C for 1 day.

Of these cultures 0.1 ml is transferred into 10 ml of YPD and shaken at 28 °C for 18 to 21 h. This time is strain-dependent and has to be checked carefully to obtain high frequency of conjugation.

One ml of each culture is taken, centrifuged (6000 g, 3 min) and resuspended in 1 ml YM without glucose (about $1-2 \times 10^7$ cells/ml). However, weakly conjugating strains should be mixed at higher cell densities (up to 1×10^8 cells/ml). One ml of each strain of opposite mating type is mixed together with 8 ml YM without glucose in a flask and shaken at 28 °C.

Zygotes can be selected after about 16 h. High numbers of diploid cells can be isolated by plating the conjugation mixture onto selective medium until 24 h after mixing the strains. After this time, the number of colony forming zygotes decreases in some strains.

7.2.3
Sporulation

Sporulation of many diploid strains can be efficiently induced by transferring the cells onto solid CSM or into liquid CSM. The percentage is higher and sporulation occurs more rapidly at 23 °C than at 28 °C. Cells can be precultivated in YPD or in synthetic media until the beginning of the stationary growth phase. Cells grow only slowly in CSM, therefore inoculation of about 2 to 5×10^7 cells/ml is recommended. Well-sporulated cultures can be easily detected by a brownish color, which is different from the white or cream-like color of unsporulated cultures. Asci containing up to four spores occur after 3 days. After 4 days the cultures should be stored at 4 °C, otherwise the asci start to disrupt and the spores are relased spontaneously.

For an unknown reason, some diploids fail to sporulate on CSM; sporulation in this case can be obtained on V8 or (to a lesser extent) on YM. Asci take more time to develop on V8 (usually 1 week or more).

7.3
Isolation of Ascospores

In the case where analysis of meiotic recombination is the aim, it is best to induce sporulation immediately after isolating the diploid, because some diploid strains are unstable during storage.

7.3.1
Random Spore Selection Using Nystatin

Sporulated strains are transferred into 10 ml of fresh YPD medium. In the case of cells taken from solid sporulation medium, sterile glass beads should be added to disperse cell aggregates, or the suspension may be briefly sonicated. The YPD culture is shaken for 2 h at 28 °C to activate vegetative cells without inducing germination of spores. Some drops of 0.1 M HCl are added to adjust the pH to 4.5 to 5.0. 0.25 ml of the nystatin stock solution (1 mg/1 ml ethanol) is added and the incubation at 28 °C is continued for another 1.5 h.

Cells and spores are harvested by centrifugation (6000 g, 5 min), washed twice with water and resuspended in the Helicase solution (200 mg/7 ml water). This suspension is shaken for 1 h at 28 °C. Cells and spores are collected by centrifugation (6000 g, 5 min), washed twice with water, and resuspended in 5 ml MMT or YNB. The suspension is sonicated (4×10 s) and then plated onto YPD plates (dilutions from 10^{-1} to 10^{-4}, depending on the sporulation frequency and germination of spores).

After 2 days' incubation at 28 °C, the YPD plates are replica-plated onto selective medium to check the genotype of the spores. Surviving diploid cells can be detected by replica plating of the YPD plates onto CSM plates. Diploid cells with form brownish colonies after 3 days at 23 °C.

7.3.2
Micromanipulation

Dissection of ascospores can be done as described for *S. cerevisiae*. Pretreatment of the sporulated culture with glusulase or zymolyase 100T (Miles) to break up the walls of the asci is strongly recommended. Note that some strains (E129 series) do not require enzymatic pretreatment (which actually rapidly kills cells and spores), since asic break up easily. For some strains, asci are easier to dissect after storage of the sporulated cultures in the cold for one day at least. Some strains form spherical spores and are easier to dissect than helmet-shaped spores (Barth and Weber 1985).

Spore viability in highly inbred lines is in the range of 75–80% (Ogrydziak 1988a). For a still unknown reason, maybe the presence of several clusters of ribosomal RNA genes (see above), this frequency could not be increased until now.

7.4
Use of Dimethylformamide to Induce Chromosome Loss

(This procedure was kindly communicated by Dr. J. De Zeeuw.) Initial attempts were made to adapt to *Y. lipolytica* the benzimidazole-2yl-carbamate (MBC) procedure used to induce chromosome loss in *S. cerevisiae*. It became apparent that dimethylsulfoxide (DMSO), used for solubilization of MBC, itself induced chromosome loss. Since *Y. lipolytica* metabolizes DMSO and generates a bad odor, other solvents were surveyed. Dimethylformamide (DMF) which is not metabolized, has a high boiling point, and is freely miscible with water, was selected.

A single colony of the diploid is inoculated in 10 ml of YPD medium and grown overnight. The cell pellet is resuspended in YPD at a density of 10 Klett Units. Ten ml of this suspension is shaken until the Klett reading reaches 50 (about 5 h), then 250 ml of DMF are added. Shaking is resumed for an additional 20 h. About 50 to 95% of the cells die during the treatment. Cells are centrifuged, washed, resuspended in 10 ml of fresh YPD, and allowed to grow for 24 h.

Suitable dilutions are plated on minimal supplemented medium or on YPD containing phloxine B (25 mg/l) and incubated for 3–5 days. Strains having suffered chromosome loss (10–60% depending on the diploid and markers) yield small, smooth, deep red colonies among a background of rough, large, and pink-white colonies of the parent. Candidates are tested for uncovering of parental markers.

Mitotic conversions (1–3%) are eliminated by determining the mating-type status of the putative segregants.

8
Methods of Cell Biology (Structural Studies)

8.1
Available Antibodies

Below follows a list of antibodies which have been successfully used to detect proteins of *Y. lipolytica* on Western blots or to label cellular structures. Antibodies are listed in Table 4 according to the main cellular compartment where their target protein is (supposedly) localized.

Table 4. Antibodies used in Western blots or the label cellular structures of *Y. lipolytica*

Compartment of the cell	Antigen	Origin of antigen	Reference
Cytoplasm	Phosphoglycerate kinase	*S. cerevisiae*	Unpubl.
Peroxisomes	Isocitrate lyase	*Y. lipolytica*	Barth, unpubl.
	Thiolase	*S. cerevisiae*	Berninger et al. (1993); Barth, unpubl.
	Pay4p	*Y. lipolytica*	Nuttley et al. (1993)
	anti-SKL	Synthetic	Nuttley et al. (1993)
Mitochondria	Glycerolphosphate acyltransferase		Unpubl.
Endoplasmic reticulum	226p	*Y. lipolytica*	Boisramé, unpubl.
	Kar2p	*S. cerevisiae*	Unpubl.
	AEP	*Y. lipolytica*	Matoba et al. (1988); Fabre et al. (1991)
Golgi	Sec14p	*S. cerevisiae*	Lopez et al. (1994)
	Kex2p	*S. cerevisiae*	Lopez et al. (1994)
Cell wall	AEP	*Y. lipolytica*	See above
	Acid phosphatase	*Y. lipolytica*	Lopez et al. (1988)
Nucleus	Asr1p-lacZ	Hybrid prot.	Barth, unpubl.

8.2
Immunofluorescence

1. Grow cells in 20 ml medium until a density of $1-2 \times 10^7$ cells/ml.

2. Add 2.5 ml of 1 M phosphate buffer (pH 6.5) and 2.5 ml formaldehyde (37%). Incubate at room temperature for 2 h (shake gently).

3. Cetrifuge at 3000 g (5 min). Wash pellet 3× with 10 ml SP. Resuspend in 1 ml SPβ.

4. Add 25 µl of freshly dissolved Zymolyase 20T (15 mg/ml SPβ) and 50 µl Glusulase. Incubate at 37 °C for 1 to 1.5 h (shake gently).

5. Harvest cells by centrifugation (2000 g, 10 min). Wash pellet with 3 ml SP and resuspend in 0.5 ml SP.

6. Transfer 10 µl of fixed cells onto glass slides (multiwell slides are recommended). Slides have to be covered before with poly-L-lysine in the following way. Wash slides with absolute ethanol. Add 10 ml poly-L-lysine (1% in water) to each well, wait 10 s, aspirate, and let dry. Wash 3× with water, let dry completely.

7. Let cells settle down and adhere for 15–30 min. Slides should be kept in a humid chamber at room temperature during this time and for following steps.

8. Rinse 3× with PBS (do not let the cells dry).
   Cover cells with 10 ml PBT and incubate for 5 min at room temperature.

9. Remove PBT and add 10 μl primary antibody diluted in PBT (check suitable dilution). Incubate for 30–60 min at room temperature.
   Rinse 10× with PBT.

10. Cover cells with 10 μl fluorochrome-conjugated secondary antibody diluted in PBT [a 1:200 dilution of rhodamine conjugated F(ab)2 fragment goat anti-mouse IgG is often used]. Incubate for 30–60 min in the dark.

11. Aspirate the secondary antiserum and rinse 10× with PBS.

12. Place 5 μl mounting solution (1% n-propyl gallate) protect against photobleaching and cover with coverslip. Seal with clear nail polish. Slides can now be stored at −20 °C in the dark for months.
    View with fluorescence microscope equipped with proper filters.

Solutions

1 M phosphate buffer	46.6 ml 1 M K_2HPO_4
	53.4 ml 1 M KH_2PO_4
SP	1.2 M sorbitol
	0.1 M phosphate buffer
SPβ	SP + 20 mM β-mercaptoethanol
PBS	53 mM Na_2HPO_4
	13 mM NaH_2PO_4
	75 mM NaCl
PBT	PBS
	1% bovine serum albumine
	0.1% Triton X-100
	0.1% NaN_3

8.3
Embedding, Thin Sectioning, and Immunogold Labeling

We have obtained best results for immunogold labeling by using LR white resin with the method of R. Wright (Univ. of California, Berkeley).

Embedding

1. Transfer cells into microfuge tubes. Resuspend for fixation in PBS containing formaldehyde (3%) and glutaraldehyde (0.2%). Incubate 2 h at RT or overnight at 4 °C.

2. Wash 3× with 1 ml PBS at RT. Resuspend in 1 ml PBS containing 1% sodium metaperiodate. Incubate 30 min at RT.

3. Wash 3× with PBS. Resuspend cells in 1 ml PBS with 50 mM NH$_4$Cl.

4. Wash 3× with PBS.

5. Embed in 3% low gelling agarose + 100 μl/ml beads. Mix well with a tooth pick. Let solidify on ice and cut in small pieces.

6. Transfer pieces into a glass vial, add 50% ethanol, cap vial, and place it on a rotating drum for 30 min. Continue dehydration as follows.

7. Replace liquid by 75% ethanol for 60 min.

8. Replace by 90% ethanol for 60 min.

9. Replace by 100% ethanol for 60 min two times.

10. Replace by a 2:1 mixture of ethanol/LR white resin for 60 min.

11. Replace by a 1:1 mixture of ethanol/LR white resin for 60 min.

12. Replace by a 1:1 mixture of ethanol/LR white resin. Remove the cap of the vial and leave it overnight on the rotating drum to evaporate the ethanol.

13. Replace by LR white resin for 60 min.

14. Replace by fresh resin and let stand under vacuum for 15 min and without vacuum for 60 min.

15. Replace the resin and let stand for 1–2 h under vacuum.

16. Transfer the pieces into gelatine capsules containing some resin. Let stand under vacuum for 15 min. Cover samples with an aluminum foil and let polymeri for 2 days at 45–50 °C.
Preparation of grids (nickel grids are recommended), sectioning, and section mounting are done with standard procedures.

Immunolabeling should be done with affinity-purified antibodies. Depletion of the serum with the cell extract of a "null"-mutant sometimes gives satisfactory results; however, very often causes severe problems in interpretation. Background staining of the cell wall, vacuole, and nucleus are especially problematic. Therefore, special care has to be taken to include proper controls (labeling sections in the absence of primary antibody; labeling of a "null"-mutant; using preimmune serum in the primary incubations; reproducibility of results).

Use a petri dish of sufficient size to contain all the grids and all the solution droplets. Place a padding of paper towels into the bottom of the dish and thoroughly wet to maintain humidity throughout the labeling steps. The surface should be fairly flat.

Onto the wet pad, a length of Parafilm is positioned. Across the top of the Parafilm, four sticky dots are also positioned. The first and third represent the position where droplets of blocker will be placed. The second is the position is the position of the primary antibody (1°) and the fourth is the position of the gold-conjugated secondary antiserum (2°). It is good practice to do duplicates of each incubation, so that a backup is available if technical problems occur.

Solutions

PBST: 140 mM NaCl, 3 mM KCl, 8 mM Na_2HPO_4, 1.5 mM KH_2PO_4, and 0.05% Tween-20.

Blocker. PBST containing 2% ovalbumin. Solution can be prepared, filtered, aliquoted, and stored frozen at −20 °C. If the blocking solution has been frozen, it has to be refiltered before use.

Glass-distilled water of high purity has to be used and solutions are filtered through a 0.22 μm filter before use. Both 1° and 2° are diluted into blocker. The dilution factor for the 1° must be empirically determined, using a dilution series. The gold-conjugated 2° should be adjusted to $A_{525} = 0.3$ for 15 nm particles and to $A_{525} = 0.13$ for 5–10 nm particles.

Labeling

Blocking. Place 20 μl droplets of 2% ovalbumin in PBST in the appropriate position on the parafilm sheet. Submerge the appropriate grid into the solution and allow to incubate for 15 min at RT. Submerging is preferable to floating since it will allow labeling of exposed antigen on both sides of the section.

Incubation in primary antiserum. Remove the grid from the blocker, touched to a Kimwipe to removed excess fluid. For this and subsequent blottings, the forceps tip and grid should be held sideways on the tissue surface, so that fluid between the forceps tips is also removed. This step should be done rapidly, not allowing the sections to dry.

Submerge the grid in a 20-μl droplet of diluted 1° antiserum. Length of incubation is probably a matter of convenience; 2-h incubations at RT or overnight incubations at RT or 4 °C have been successful.

Washes. Perform in the wells of porcelain or glass spot plates. The wells are marked and filled with PBST, and the spot plate is placed in an orbital shaker.

Submerge the grid in the appropriate will. The shaker is adjusted so that the solution moves as rapidly as possible without spilling out of the well (5 min).

Remove the grid, blot, and transfer to the next well. A total of 3 washes (each 5 min) are performed and the grid is blotted and transferred to a second droplet of blocker (15 min).

Incubation in secondary (gold-conjugated) antiserum (2°). Blot the grid and transfer to the diluted 2°. Incubate the grid for 1 h at RT and then wash as above.

After the final PBST wash, wash the grid in distilled water by dipping 10 times with rapid up and down motion in a 5-ml beaker of water. This wash removes salts that would crystallize on the section. The grid is blotted on a Kimwipe, transferred to a labeled silicon mat, and allowed to air-dry.

View in the electron microscope.

9
Methods of Molecular Biology/Gene Technology

9.1
Transformation Systems

The protoplast method, initially used for transformation (Gaillardin et al. 1985) has been superseded by the following methods.

9.1.1
LiAc/LiCl Method

Competent Cells

1. Inoculate in the morning one loop of cells in 5 ml of YPD pH 4 in a 100-ml flask (YPD pH 4.0 is 1% yeast extract, 1% bactopeptone, 1% glucose, 50 mM citric acid-Na citrate pH 4.0). Shake at 28 °C. Count the preculture in the evening and inoculate 10 ml of culture on YPD pH 4.0 at 10^5 cells/ml from the preculture in a 250-ml flask. Grow overnight at 28 °C.

2. Harvest the culture between 9×10^7 and 1.5×10^8 cells/ml. This density is critical! When proceeding with a new strain, it is best to inoculate three overnight cultures at 5×10^4, 1×10^5, and 2×10^5 cells/ml, and to take the best one in the morning.

3. Spin and wash twice the cells in 10 ml TE. Resuspend the cells at 5×10^7 cells/ml in 0.1 M Li-acetate pH 6.0 (adjusted with acetic acid) and incubate 1 h with gentle shaking at 28 °C. Spin and resuspend in one tenth of the volume or around 5×10^8 cells/ml in LiAc buffer.

Storing of Competent Cells

Competent cells can be stored for a week at 4 °C, spin and resuspend them before use. Long-term conservation of competent cells is achieved by adding 25% glycerol (final conc.) to cells in LiAc buffer and freezing at −80 °C. Thaw on ice when needed, and wash once in LiAc before use.

Transformation

On ice:

1. Put 5 µl of carrier DNA at the bottom of a 5-ml test tube (carrier DNA is salmon sperm DNA, 5 mg/ml in 50 mM Tris, 5 mM EDTA, pH 8.0, sonicated in the 500-bp range). Put 200 ng of transforming DNA in 1–5 µl (restriction enzyme buffer does not hurt). Add 100 µl of competent cells, gently mix cells and DNA, and incubate at 28 °C for 15 min.

2. Add 0.7 ml of 40% PEG4000 (solution done in 0.1 M lithium acetate, adjusted to pH 6.0 with acetic acid).

3. Incubate for 1 h at 28 °C, with shaking (we use a rotary shaker set at 250 rpm).

4. Heat shock the mix at 39 °C for 10 min.

5. Add 1.2 ml of 0.1 M lithium acetate buffer pH 6.0.

6. Platce very gently 200 μl on selective medium.

7. Transformants appear after 2 days and can be counted after 4 days.

Comments

Effect of the pH of the culture. We have tested from pH 6.8 to 3.9 and found increasing values of transformation frequencies with decreasing pH.

Transformation frequency is ten times higher with a cell density of 5×10^7 cells/ml during treatment with LiAc than at 5×10^8 cells/ml. The number of transformants increases in a linear way according to the number of cells mixed with the transforming DNA, up to 10^8 cells, where an abrupt drop is observed. We therefore chose to mix 5×10^7 cells with the DNA.

Effect of carrier DNA. Transformation frequency increases with time of sonication, showing a maximum for a DNA size below 1 kb. The amount of carrier should be at least 25 μg, but it can be increased without inconvenience. PEG should be at a concentration of 40%, as 10, 20, or 30% give poor results. Agitation during the PEG treatment produces a two- to fourfold increase of transformation frequency. Heat shock is critical, as absence of heat shock results in at least a 50-fold decrease in efficiency. Temperature above 37 °C (10 min at 39 °C or 5 min at 42 °C) are preferable, and 10 min at 39 °C corresponds to a maximal efficiency.

9.1.2
Electroporation

Competent Cells

1. Pregrow the strain for about 5 h at 28 °C in YPD. Inoculate in the evening 200 ml of the same medium in a 1-l flask with about 10^3–10^4 cells/ml.

2. Grow overnight at 28 °C with shaking. Density in the morning should be around 1×10^7 cells. Centrifuge the culture at $6,000 \times g$ for 2 min at 4 °C.

3. Wash cells once in one culture volume of ice cold water, spin at $6,000 \times g$ for 2 min at 4 °C. Repeat washings in half this volume.

4. Resuspend in 1/25 vol of 1 M sorbitol at 4 °C (8 ml for 200 ml); spin as before. Resuspend the pellet in 1/1000 volume of sorbitol 1 M at 4 °C (200 μl). Starting from 200 ml, this should make about 400 μl, or enough for 8 transformations.

Keeping Competent Cells

1. Add 7% DMSO (vv).

2. Distribute 40 μl of this suspension in Eppendorf tubes.

3. Place for a few min in a mixture of dry ice and ethanol.

4. Store the cells at −80 °C.

5. Before use, thaw one tube at 4 °C.

6. Spin at 4 °C for 1 min.

7. Resuspend the pellet gently in 40 μl of cold 1 M sorbitol (spinning down and washing of the cells can be avoided).

Electroporation and Plating

1. Mix 40 μl of competent cells with 10–40 ng of DNA in a maximum volume of 5 μl, keeping everything on ice.

2. Place the mix between the electrodes (1.5 mm spacing, Jouan electroporator GHT 1287), apply one pulse of 900 V for 15 ms.

3. Place the mix back into an Eppendorf tube and dilute immediately with 1 ml of 1 M sorbitol at 4 °C.

4. Place one or more samples of 50 to 400 μl on selective medium.

5. Place gently, or try to spread the cells by tilting the plate.

Comments

The amount of DNA used in the transformation is rapidly inhibitory: a two times effect is observed with 25 ng versus 10 ng. Presence of carrier DNA (5 μg) gives a three- to nine-times inhibitory effect. Presence of sorbitol in the plates has been tested and does not give better results. A time pulse of 15 s is better than 10 or 5 ms at all voltages tested, and a lower voltage than 900 V (at 15 ms) results in a twofold decrease in the efficiency. A slight increase in transformation frequency seems to be obtained if buffering the sorbitol with 10 mM Tris at pH 8, but a detailed study of conditions has not been performed. We have also observed a slight increase in the number of transformants when plating the cells on the so-called T medium (E. Meilhoc, pers. comm.) prepared as follows: Autoclave 1 l of the basal mixture containing: 20 g glucose, 2 g Na glutamate, 9 g lysine-HCl, 3 g KH_2HPO_4, 3 g $Na_2HPO_4 12H_2O$. Add 1 ml of the following (filter sterilized) solutions: $CaCl_2$ (0.25 g/ ml), $MgSO_4$ (0.25 g/ml), $FeSO_4$(1.5 mg/ml), $(NH_4)_2SO_4H_2O$ (1.5 mg/ml), $ZnSO_4$ (5 mg/ml), biotine (2 mg in 100 ml of 50% ethanol), Ca pantothenate (1 mg/ml), inositol (5 mg/ml), nicotinic acid (5 mg/ml), p-aminobenzoate (0.3 mg/ml), pyridoxine-HCl (1 mg/ml), riboflavin (0.2 mg/ml), and thiamine-HCl (1 mg/ml). Vitamins solutions can be prepared separately, then mixed, aliquoted (8 ml for 1 l medium) and kept frozen.

9.1.3
Single Colony Method

This method is useful if many strains have to be transformed simultaneously, e.g., when one looks for highly transformable strains among the progeny of a cross.

1. Streak strains overnight on YPD plates in the morning.

2. In the evening make 1-cm² patches on YMC. YMC is YM medium buffered with 50 mM citric acid-Na citrate, pH 4.0.

3. Resuspend cells in the morning in 1 ml TE (one loop).

4. Spin and resuspend in 600 μl LiAc buffer 0.1 M pH 6.0 (about 5×10^7 cells/ml). Try to adjust turbidity of all samples by visual inspection.

5. Take 600 μl of this suspension into a new Eppendorf tube. Incubate for 1 h at 28 °C (no shaking necessary).

6. Spin and resuspend in one tenth of the volume (60 μl) of LiAc buffer pH 6.0.

7. For reversion controls, take 20 μl to a new tube. For transformation, take 40 μl of cells, add 2.5 μl of carrier DNA (see Sect. 4.1.1.), and 200 ng of digested DNA (in 1.5 μl).

8. Incubate 15 min at 28 °C.

9. Add 350 μl of PEG 4000 (see Sect. 4.1.1.).

10. Incubate 1 h at 28 °C.

11. Heat shock 10 min at 39 °C.

12. Add 600 μl of LiAc buffer 0.1 M pH 6.0.

13. Plate 200 μl on selective medium.

14. Treat the control identically but do not add DNA.

This procedure gives between 100 and 1000 transformants per plate; although not overefficient, it permits comparison of transformability of different strains.

9.2
Preparation of Protoplasts

The formation of protoplasts depends mainly on the age of the cells and the kind of enzymes used for digestion of the cell wall. The rule is that the younger the cells, the better the yield of protoplasts. In our hands, a combination of Glusulase and Zymolyase works most efficiently for digestion of cell walls of *Y. lipolytica*.

1. Collect the cells of 200 ml culture by centrifugation at 6000 g (5 min).

2. Wash the cells once with 100 ml sterile water and centrifuge again.

3. Resuspend the pellet in 20 ml SPβ (1.2 M sorbitol, 0.1 M KPO$_4$, pH 6.5, 20 mM β-mercaptoethanol).

4. Add 0.5 ml Zymolyase 100T (Miles) or 20T (Seikagaku Kogyo) (stock solution of 15 mg/ml) and 0.2 ml Glusulase (Dupont Comp.).

5. Incubate at 37 °C, protoplasts should occur after 30 min, more than 90% of the cells should be converted to protoplasts after 60 min.

Formation of protoplasts can be checked by microcopy or by lysis test ($100\,\mu$l suspension $+\ 800\,\mu$l water $+\ 100\,\mu$l 10% SDS solution – turbid solution should become clear when protoplast formation has taken place).

Protoplasts can be collected by centrifugation at 2,000$\,g$ (10 min) and resuspended in a buffer of choice, depending on the following procedure.

9.3
Isolation of Genomic DNA

9.3.1
Minipreparation

This method is derived from Hoffman and Winston (1987).

1. Grow the strain in 10 ml YPD or YNB until stationary phase.

2. Spin and resuspend the pellet in 0.5 ml of water.

3. Transfer to an Eppendorf tube.

4. Spin 5 s and pour off supernatant.

5. Spin 1 s and resuspend the pellet in whatever supernatant is left.

6. Add $200\,\mu$l of the following mix:
   2% Triton X-100
   1% SDS
   100 mM NaCl
   10 mM Tris-HCl pH 8.0
   1 mM EDTA

7. Add $200\,\mu$l of phenol-chloroform and 0.3 g of glass beads (acid-washed, $45\,\mu$m).

8. Vortex 3–4 min, then add $200\,\mu$l TE.

9. Spin 5 min, and save the aqueous phase in a new tube.

10. Add 1 ml 100% ethanol, mix by inverting the tube, spin 2 min, discard the supernatant. Do not dry at this step.

11. Resuspend the pellet in $400\,\mu$l of TE with $30\,\mu$g of RNase I (boiled).

12. Incubate for 5 min at 37 °C.

13. Add $10\,\mu$l of 4 M ammonium acetate, then 1 ml of 100% ethanol. Mix by inverting the tube, spin 2 min, discard the supernatant.

14. Dry under vacuum, then resuspend in $50\,\mu$l of TE.
    Use 5–10 μl for one digestion.

One should obtain about 2 μg of DNA; this is the method of choice for analyzing transformants by Southern hybridization. Bands up to 12 kb are detectable, higher ones might be sheared. Probably because of the presence of impurities, this method does not reproducibly yield transformants in *E. coli*.

9.3.2
Maxipreparation

Several methods have been used. Here is one which has been found suitable for library preparation.

1. Pregrow the strain on 100 ml YPD for 24 h.

2. Grow the strain for 24 h in 2 l of YPD (1 l in 5-l jars, or strong aeration).

3. Harvest the cell by centrifugation, wash twice in 100 ml TE/l.

4. Resuspend in 100 ml of PTP buffer (100 mM Tris-HCl pH 8.0, 10 mM EDTA, 1 M sorbitol).

5. Add DTT (20 mM final concentration), incubate 10 min at 30 °C.

6. Centrifuge the cells and wash in 100 ml PAPS buffer (10 mM AP adjusted to pH 6.8 with dry PIPES, 1 M sorbitol; AP is 2-amino 2-methyl 1,3-propanediol).

7. Resuspend in 100 ml PAPS buffer.

8. Add 100 ml of PAPS buffer containing 20 mg of Zymolyase 100T (Miles).

9. Incubate at 37 °C and check protoplasting (under the microscope or following the turbidity of 1/1000 dilution in water): it should take 20–30 min before osmosensitive cells are obtained.

10. Spin at low speed 10 min in a preweighed tube, pour off the supernatant gently, the pellet may be loose.

11. Weigh the protoplasts, add 1.5 ml PAPS buffer/g of protoplasts.

12. Slowly add the lysis buffer (3.5 ml/g of cells). We usually put the cells in a siliconized beaker with low magnetic stirring. Lysis buffer is 3% Sarkosyl, 0.5 M Tris pH 9.0, 200 mM EDTA.

13. Incubate 15 min at 65 °C.

14. Add one volume of phenol-chloroform, mix gently, and spin 10 min at 5,000 × g. Repeat phenol extraction twice, then extract with chloroform, let cool on ice.

15. Add ice-cold 3 M sodium acetate pH 5.0 up to 0.3 M final concentration.

16. Add cold isopropanol, 3/4 of the former volume.

17. A clump of DNA should form; spool it on a glass rod, dip it into 70% ethanol to remove the salts, dry it partially (do not overdry) and resuspend in TE (5 ml).

RNase treatments are optional.

9.3.3
Isolation of Yeast Plasmid DNA for *E. coli* Transformation

This procedure is derived from Ausubel et al. (1989).

1. Grow the yeast transformant overnight in 5 ml liquid YPD.

2. Spin and resuspend in 500 μl of sorbitol solution (0.9 M sorbitol, 0.1 M Tris-HCl pH 8.0, 0.1 M EDTA) Add 20 μl of 3 mg/ml Zymolyase 100T in sorbitol solution and 50 μl of 0.28 M β-mercaptoethanol. Incubate for 1 h at 37 °C.

3. Spin and resuspend the cells in 500 μl of Tris-EDTA solution (50 mM Tris, 20 mM EDTA, pH 8.0) by pipetting up and down repeatedly.

4. Add 50 μl of 10% SDS at 65 °C and invert the tube several times to mix. The solution should be very viscous at this point.

5. Add 200 volumes of 5 M potassium acetate. Mix well. Let stand on ice for at least 30 min. The content may be semisolid at this stage.

6. Centrifuge 3 min at room temperature in a table top centrifuge. Pour the supernatant into a new Eppendorf tube and fill the tube with 100% ethanol (about 1 ml). Centrifuge 10 s at room temperature. Discard the supernatant and partially dry the pellet under vacuum. Add 300 μl TE. It may take overnight to dissolve, heating at 65 °C will speed it up, or gentle vortexing.

7. Add 5 μl of 10 mg/ml RNaseI. Incubate 1 h at 37 °C.

8. Add 500 μl of 100% isopropanol. Mix content gently until the DNA precipitates in a single clump. Recover this clump with a 200-μl pipettor tip, squeeze dry against the wall of the tube, and transfer to a new tube containing 125 μl of TE. If DNA does not make a single clump, spin for 10 s, wash the pellet with 70% ethanol, dry and resuspend in 125 μl of TE. Transform highly competent *E. coli* cells (10^7 transformants/μg of control DNA). In our hands, 1 to 4 μl of this DNA solution will give 10 to 60 *E. coli* transformants.

9.4
Separation of Chromosomes (see also Chap. 3, this Vol.)

9.4.1
Plug Preparation

1. 50 ml of the following medium are inoculated and incubated overnight, to obtain about 10^7 cells/ml in the morning (see Sect. 4.1.2.):

   3% glucose
   0.5% yeast extract

0.5% peptone
0.0745% L-Methionine
0.0675% DL-Homocysteine thiolactone
0.1% DL-Methionine methyl sulfonium chloride

(the addition of the last three compounds is optional).

2. Spin and wash in 100 mM Tris-HCl pH 7.5, 100 mM EDTA.

3. Resuspend at a cell density of 2×10^9 cells/ml in TESorb buffer (0.9 M sorbitol, 0.1 M Tris-HCl pH 8.0, 0.1 M EDTA). Add 1/10 of the volume of a stock solution of Zymolyase 100T at 3 mg/ml and β-mercaptoethanol to a final concentration of 0.02 M. Incubate 5–10 min at 37 °C.

4. Prepare 1% low melting agarose in 0.25 M EDTA pH 8.0.

5. Mix one volume of cells and one volume of molten agarose.

6. Distribute the mix in the molds (or in a small petri dish), let it solidify at 3 °C for at least 20 min.

7. Take out the plugs (or cut to the size of the wells) and incubate them in 2 ml of TE-Sorb buffer containing Zymolyase and β-mercaptoethanol for 2 h at 37 °C (as above).

8. Rinse the plugs in 0.5 M EDTA, and incubate in lysis buffer (0.5 M EDTA, 0.01 M Tris-HCl pH 8.0, 1% Sarkosyl, 1 mg/ml proteinase K) at 50 °C for 24 h; change lysis buffer and repeat (addition of RNaseI is optional at this step). Use 5 vol lysis buffer for 1 volume plug. Plugs should become clearer upon this treatment.

9. Store the plugs in 0.5 M EDTA pH 8.0 at 4 °C (storage for several months is possible).

Before use, dialyze them extensively against TE (change bath 3 times).

9.4.2
Chromosome Separation

Using Beckman's Geneline. Gel is 0.8% agarose in 1× TAFE buffer (20× TAFE is 2.42% Trisbase, 0.29% EDTA, 0.5% glacial acetic acid). Plugs are sealed with agarose in the gel slots. Run conditions are: 20 mA, 70 min pulse time, for 90 h; then 25 mA, 50 min pulse time for 75 h, at 10 °C.

Using Biorad's CHEF DRII. Gel is 1% agarose in 0.5× TAE buffer. Plugs are sealed with agarose in the gel slots. Run conditions are: 40 V, 55 min pulse time for 47.7 h; then 43 V, 50 min pulse time for 70 h; and finally 50 V, 40 min pulse time for 48 h, at 12 °C.

9.5
Isolation of RNA

It is not easy to obtain high-quality preparations of RNA from cells of *Y. lipolytica*. Several methods were checked for isolation of RNA. Best results were obtained with the methods described by Chomczynski and Sacchi (1987), Domdey et al. (1984), or the commercially available RNA extraction kit of Pharmacia LKB Biotechnology.

9.5.1
The Procedure Described by Chomczynski and Sacchi (1987)

1. One hundred ml of an overnight culture are centrifuged for a few min at 5,000 × *g*, washed with 10 ml of 1 M sorbitol, and resuspended in 10 ml of 1 M sorbitol.

2. Add 200 µl of Zymolyase 100T (stock solution at 50 mg/ml) and incubate 1 h at 37 °C.

3. Centrifuge 2 min at 3,000 × *g* and resuspend in 0.5 ml sorbitol (1 M).

4. Add this protoplast suspension to 10 ml of denaturating mix, which is prepared as follows:
   To avoid manipulation of toxic products, we directly dissolve in the supplier flask at 65 °C the 250 g of guanidium thiocyanate in 293 ml water + 17.6 ml 0.75 M Na-citrate pH 7.0 + 26.4 ml Sarkosyl 10%. This solution can be stored for 3 months at room temperature. To 50 ml of solution, add before use 0.36 ml of β-mercaptoethanol. This mix can be stored for 1 month at 4 °C.

5. Add 20 µl DEPC, vortex and add 1 ml 2 M Na-acetate, pH 4.0.

6. Vortex and add 10 ml phenol (saturated with water).

7. Vortex and add 2 ml chloroform-isoamyl-alcohol. Vortex and put on ice for 15 min.

8. Centrifuge 20 min at 9,000 × *g* and dispense the aqueous phase in Eppendorf tubes with 750 µl per tube. Add 750 µl isopropanol and store 1 h at −20 °C.

9. Centrifuge 20 min in the cold in a table top centrifuge at maximum speed, and resuspend the pellets in 0.5 ml of the same guanidium-β-mercaptoethanol-DEPC solution as above.

10. Add 0.5 ml isopropanol and store overnight at −20 °C (or at least 1 h).

11. Centrifuge 10 min at 4 °C, wash pellets in 600 µl 75% ethanol (prepared with DEPC-treated water), centrifuge again, dry the pellet, and resuspend in 50 µl of DEPC-treated water. Vortex and do not heat to dissolve. Store the RNA prep at −80 °C.

9.5.2
The Procedure Described by Domdey et al. (1984)

1. Twenty five ml of the culture (5–8×10^7 cells/ml; if less concentrated increase the amount of culture) are centrifuged or filtrated over a Millipore filter ($1.2\,\mu$m pore size). Cells are resuspended in 1.5 ml of MMT or YNB without carbon source and transferred to a 2.2-ml microfuge tube. After centrifugation the supernatant is carefully removed and the cells are frozen in a dry ice-ethanol bath. The tubes can be stored at −80 °C for about 2 months. The quality of RNA decreases when cells are stored longer.

2. Add 0.7 ml AE buffer, 100 ml 10% SDS solution and 1 ml fresh phenol (equilibrated with AE buffer) to the frozen cells. Vortex vigorously until the pellet is dissolved completely.

3. Incubate at 65 °C for 4 min.

4. Rapidly chill the tubes in a dry ice-ethanol bath until phenol crystals appear.

5. Centrifuge the frozen mixture in a microfuge at maximum speed for 10 min.

6. Remove the phenol phase but leave the interphase intact intact!

7. Add a second aliquot of phenol and repeat the extraction step.

8. Transfer the aqueous phase into a new tube (without the interphase!).

9. Add an equal volume of phenol:chloroform (1:1 mixture), mix, and centrifuge again. Transfer the aqueous phase into a new tube.

10. Adjust the salt concentration to 0.4 M LiCl and add 2.5 vol ethanol. This preparation can be stored at −80 °C. At this step a small aliquot could be pelleted separately and the quality and amount of RNA could be checked on an EtBr-containing agarose gel.

11. To pellet the RNA, the tubes are centrifuged in a microfuge (at highest speed) for 10 min. Wash the pellet once with 75% ethanol.

12. Dry the pellet at room temperature, resuspend in $50\,\mu$l sterile water (pretreated with DEPC).
    Materials

AE buffer	50 mM sodium acetate (pH 5.3)
	10 mM EDTA
LiCl stock solution	4 M, pH 5.3

9.6
Available Gene Libraries

Genomic and cDNA libraries of *Y. lipolytica* are described in Table 5.

Table 5

Recommended hosts for screening	Vector	Source of inserts and library construction			Nb. of clones	Features of the library			Reference
		Strain	Type of insert	Insert size		% hybrids	Representation[a]	Notes	
E. coli	pBR322	H222	EcoRI or BamHI	3–7 kb	20,000	80%	0.83		Barth and Scheuber (1993)
E. coli	pHC79	15901-4	Sau3AI partial	40 kb	5,000	75%	0.999	1	C. Gaillardin (unpubl.)
E. coli	λ Charon 4A	CX-161-1B	HaeIII + AluI partial	15–20 kb	65,000	89%	0.999		Matoba et al. (1988)
Y. lipolytica leu2	pINA62	W29	Sau3AI partial	7–15 kb	6,500	89%	0.99	2	Xuan et al. (1988)
Y. lipolytica leu2	pINA240	W29	Sau3AI partial	3–6 kb	21,000	80%	0.83	3	M. Chasles and P. Fournier (unpubl.)
Y. lipolytica leu2	pINA237	B204-12C-112	Sau3AI partial	4–8 kb	50,000	82%	0.999		Schmid-Berger et al. (1994)
Y. lipolytica leu2	pINA445	E122	Sau3AI partial	5–7 kb	23,000	83%	0.999		Nuttley et al. (1993)
S. cerevisiae ura3	pFL61	W29	cDNA	0.8–4 kb	18,000	85%	YPD grown cells	4	Lopez et al. (1994)
S. cerevisiae trp1	YRp7	W29	Sau3AI partial	12–18 kb	3,780	100%	0.98	5	Tréton et al. (1992)

[a] Probability of covering one genome equivalent, evaluated by the Clarke and Carbon formula (see Xuan et al. 1988); except for the cDNA library where the growth conditions used for mRNA preparation are indicated.

1 Cosmids available as phage particles.
2 Available as 5 pools of 1,500 clones.
3 Available as 5 pools of 4,200 clones.
4 Made on 2 μ plasmid, cDNA are expressed under PGK promoter.
5 Available as 14 pools of 270 clones.

Acknowledgments. The authors wish to thank colleagues from the *Y. lipolytica* community for their friendly sharing of unpublished results and procedures. Particular thanks are due to D.M. Ogrydziak, P. Fournier, J.M. Nicaud, and M.T. Le Dall for reviewing all or part of the manuscript. We are grateful to D.M. Ogrydziak for critical comments on the manuscript. Part of the work presented here was supported by a grant from the European Community (BIOT-CT91-0267DSCN).

References

Abou-Zeid AA, Khoja SM (1993) Utilization of dates in the fermentative formation of citric acid by *Yarrowia lipolytica*. Zentralbl Mikrobiol 148: 213–221

Akiyama SI, Suzuki T, Sumino Y, Fukada H (1973a) Induction and citric acid productivity of fluoroactetate-sensitive mutant strains of *Candida lipolytica*. Agric Biol Chem 37: 879–884

Akiyama SI, Suzuki T, Sumino Y, Nakao Y, Fukada H (1973b) Relationship between aconitate hydratase activity and citric acid productivity in fluoroacetate-sensitive mutant strains of *Y. lipolytica*. Agric Biol Chem 37: 885–888

Atchinson JD, Szilard RK, Nutley WM, Rachubinski RA (1992) Antibodies directed against a yeast carboxyl terminal peroxisomal targeting signal specifically recognize peroxisomal proteins from various yeasts. Yeast 8: 721–734

Ausubel FM, Brent R, Kingston RE, Moore DD, Seidman JG, Smith JA, Struhl K (1989) In: Current protocols in molecular biology, vol 2. chpt. 13.11.3. J Wiley, New York

Barnett JA, Payne RW, Yarrow D (1990) Yeasts: characteristics and identification. Cambridge University Press, Cambridge

Barns SM, Lane DJ, Sogin ML, Bibeau C, Weisburg WG (1991) Evolutionary relationships among pathogenic *Candida* species and relatives. J Bacteriol 173: 2250–2255

Barth G (1985) Genetic regulation of isocitrate lyase in the yeast *Yarrowia lipolytica*. Curr Genet 10: 119–124

Barth G, Kunkel W (1979) Alcohol dehydrogenases (ADH) in yeast. II. NAD⁺- and NADP⁺-dependent alcohol dehydrogenases in *Saccharomycopsis lipolytica*. Z Allg Mikrobiol 19: 381–390

Barth G, Scheuber T (1993) Cloning of the isocitrate lyase gene (*ICL1*) from *Yarrowia lipolytica* and characterization of the deduced protein. Mol Gen Genet 241: 422–430

Barth, G, Weber H (1983) Genetic studies on the yeast *Saccharomycopsis lipolytica*. Inactivation and mutagenesis. Z Allg Mikrobiol 23: 147–157

Barth G, Weber H (1984) Improved conditions for mating of the yeast *Saccharomycopsis lipolytica*. Z Allg Mikrobiol 24: 403–405

Barth G, Weber H (1985) Improvement of sporulation in the yeast *Yarrowia lipolytica*. Antonie Leeuwenhoek J Microbiol 51: 167–177

Barth G, Weber H (1987) Genetic analysis of the gene *ICL1* of the yeast *Yarrowia lipolytica*. Yeast 3: 255–262

Bassel J, Mortimer RK (1973) Genetic analysis of mating-type and alkane utilization in *Saccharomycopsis lipolytica*. J Bacteriol 114: 894–896

Bassel J, Warfel J, Mortimer RK (1971) Complementation and genetic recombination in *Candida lipolytica*. J Bacteriol 108: 609–611

Bassel J, Hambright P, Mortimer RK, Bearden AJ (1975) Mutants of the yeast *Saccharomycopsis lipolytica* that accumulates and excretes protoporphyrin IX. J Bacteriol 123: 118–122

Bassel JB, Mortimer RK (1982) Genetic and biochemical studies on n-alkane non-utilizing mutants of *Saccharomycopsis lipolytica*. Curr Genet 5: 77–88

Bassel JB, Mortimer RK (1985) Identification of mutations preventing n-hexadecande uptake among 26 n-alkane non-utilizing mutants of *Yarrowia* (*Saccharomycopsis*) *lipolytica*. Curr Genet 9: 579–586

Bauer R, Paltauf F, Kohlwein SD (1993) Functional expression of bacterial β-glucuronidase and its use as a reporter system in the yeast *Yarrowia lipolytica*. Yeast 9: 71–75

Beckerich JM, Heslot H (1978) Physiology of lysine permease in *Saccharomycopsis lipolytica*. J Bacteriol 133: 492–498

Beckerich JM, Lambert M, Heslot H (1981) Mutations affecting the lysine and polyphosphate pools in the yeast *Saccharomycopsis lipolytica*. Biochem Biophys Res Commun 100: 1292–1298

Beckerich JM, Colonna Ceccaldi B, Lambert M, Heslot H (1984) Evidence for the control of a mutation in lysine catabolism by the mating-type in *Yarrowia lipolytica*. Curr Genet 8: 531–536

Beckerich JM, Pommies E, Faivre C, Lambert M, Heslot H (1986) Estimation of compartmentation of lysine inside the cells of *Yarrowia lipolytica*. Biochimie 68: 517–529

Beckerich JM, Lambert M, Gaillardin C (1994) LYC1 is the structural gene of N-acetyl-lysine transferase in yeast. Curr Genet 25: 24–29

Berninger G, Schmidtchen R, Casel G, Knörr A, Rautenstrauss K, Kunau W-H, Schweizer E (1993) Structure and metabolic control of the *Yarrowia lipolytica* peroxisomal 3-oxoacyl-CoA-thiolase gene. Eur J Biochem 216: 607–613

Berny J-F, Hennebert GL (1991) Viability and stability of yeast cells ands filamentous fungus spores during freeze-drying; effect of protectant and cooling rates. Mycologica 83: 805–815

Blagodatskaya VM, Kockova-Kratochvilova A (1973) The heterogeneity of the species *Candida lipolytica*, *Candida pseudolipolytica* n. sp. and *Candida lipolytica* var. *thermotolerans* n. var. Biologia (Bratislava) 28: 709–716

Blanchin-Roland S, Cordero Otero R, Gaillardin C (1994) Two upstream UAS control expression of the *XPR2* gene encoding an extracellular alkaline protease in the yeast *Yarrowia lipolytica*. Mol Cell Biol 14: 327–338

Bojnanska A (1977) Determination of mating-types and frequency of zygotes in strains of the species *Candida lipolytica*. Acta Facultatis Rerum Naturalium Universitalis Conenianae. Genetica VIII: 55–65

Booth JL, Vishniac HS (1987) Urease testing and yeast taxonomy. Can J Microbiol 33: 396–404

Bruyn J (1954) An intermediate product in the oxidation of hexadecene-1 by *Candida lipolytica*. K Ned Acad Wet Proc Ser C 57: 41–44

Buckholz RG, Gleeson MAG (1991) Yeast systems for the commercial production of heterologous proteins. Bio/technology 9: 1067–1072

Campbell I (1975) Numerical analysis and computerized identification of the yeast genera *Candida* and *Torulopsis*. J Gen Microbiol 90: 125–132

Cardillo R, Fronza G, Fuganti C, Grasseli P, Mele A, Pizzi D (1991) Stereochemistry of the microbial generation of delta-decanolide, gamma-dodecanolide and gamma-nonanolide from C18 13-hydroxy, C18 10-hydroxy, and C19 14-hydroxy unsaturated fatty acids. J Org Chem 56: 5237–5239

Cheng SC, Ogrydziak DM (1986) Extracellular RNase produced by *Yarrowia lipolytica*. J Bacteriol 168: 581–589

Cheng SC, Ogrydziak DM (1987) Processing and secretion of the *Yarrowia lipolytica* RNase. J Bacteriol 169: 1433–1440

Chomczynski P, Sacchi N (1987) Single-step method of RNA isolation by acid guanidium thiocyanate-phenol-chloroform extraction. Anal Biochem 162: 156–159

Cirigliano MC, Carman GM (1984) Isolation of a bioemulsifier from *Candida lipolytica*. Appl Environ Microbiol 48: 1154–1155

Cirigliano MC, Carman GM (1985) Purification and characterization of liposan, a bioemulsifier from *Candida lipolytica*. Appl Environ Microbiol 50: 846–850

Clare JJ, Davidow LS, Gardner DCJ, Oliver SG (1986) Cloning and characterization of the ribosomal RNA genes of the dimorphic yeast, *Yarrowia lipolytica*. Curr Genet 10: 449–452

Dagett P-M, Simione FP (1987) Cryopreservation manual. ATCC & Nalge Company, Rochester

Davidow LS, Apostolakos D, O'Donnell MM, Proctor AR, Ogrydziak DM, Wing RA, Stasko I, De Zeeuw JR (1985) Integrative transformation of the yeast *Yarrowia lipolytica*. Curr Genet 10: 39–48

Davidow LS, O'Donnell MM, Kaczmarek FS, Pereira DA, De Zeeuw JR, Franke AE (1987) Cloning and sequencing of the alkaline extracellular protease gene of *Yarrowia lipolytica*. J Bacteriol 169: 4621–4629

Delaisse JM, Martin P, Verheyen-Bouvy MF, Nyn EJ (1981) Subcellular distribution of enzymes in the yeast *Saccharomycopsis lipolytica*, grown on n-hexadecane, with special reference to the omega-hydroxylase. Biochim Biophys Acta 676: 77–90

Dell'Angelica EC, Stella CA, Ermacora MR, Ramos EH, Santone JA (1992) Study of fatty acid binding by proteins in yeast. Dissimilar results in *Saccharomyces cerevisiae* and *Yarrowia lipolytica*. Comp Biochem Physiol 102B: 261–265

De Zeeuw J, Stasko I (1983) Fermentation process for the production of alpha-isopropyl malic acid. US patent US4407853

Domdey H, Apostol B, Lin R-J, Newman A, Brody E, Abelson J (1984) Lariat structures are in vivo intermediates in yeast pre-mRNA splicing. Cell 39: 611–621

El-Sherbeini M, Bostian KA, Levitre J, Mitchell DJ (1987) Gene-protein assignments within the yeast *Yarrowia lipolytica* dsRNA viral genome. Curr Genet 11: 483–490

Enderlin CS, Ogrydziak DM (1994) Cloning, nucleotide sequence and functions of *XPR6*, which codes for a dibasic processing endoprotease from the yeast *Yarrowia lipolytica*. Yeast 10: 67–79

Ercoli B, Fuganti C, Graselli P, Servi S, Allegrano G, Barbeni M, Pisciotta A (1992) Stereochemistry of the biogeneration of C-10 and C-12 gamma-lactones in *Yarrowia lipolytica* and *Pichia ohmeri*. Biotechnol Lett 14: 665–668

Esser K, Stahl U (1976) Cytological and genetic studies of the life cycle of *Saccharomycopsis lipolytica*. Mol Gen Genet 146: 101–106

Fabre E, Nicaud JM, Lopez MC, Gaillardin C (1991) Role of the proregion in the production and secretion of the *Yarrowia lipolytica* alkaline extracellular protease. J Biol Chem 266: 3782–3790

Fabre E, Tharaud C, Gaillardin C (1992) Intracellular transit of a yeast protease is rescued by *trans*-complementation with its prodomain. J Biol Chem 267: 15049–15055

Finogenova TV, Glazunova LM (1976) Activity of enzymes of the citrate and glyoxylate cycles in the synthesis of citric and isocitric by various strains of *Candida lipolytica*. Mikrobiologiya 51: 21–30

Finogenova TV, Shishkanova IT, Kataeva IA (1982) Properties of *Candida lipolytica* mutants with the modified glyoxylate cycle and their ability to produce citric and isocitric acid. Appl Microbiol Biotechnol 23: 378–383

Fournier P, Gaillardin C, Persuy MA, Klootwijk J, van Heerikuizen H (1986) Heterogeneity in the ribosomal family of the yeast *Yarrowia lipolytica*: genomic organization and segregation studies. Gene 42: 273–282

Fournier P, Guyaneux L, Chasles M, Gaillardin C (1991) Scarcity of *ars* sequences isolated in a morphogenesis mutant of the yeast *Yarrowia lipolytica*. Yeast 7: 25–36

Fournier P, Abbas A, Chasles M, Kudla B, Ogrydziak DM, Yaver D, Xuan J-W, Peito A, Ribet A-M, Feynerol C, He F, Gaillardin C (1993) Colocalization of centromeric and replicative functions on autonomously replicating sequences isolated from the yeast *Yarrowia lipolytica*. Proc Natl Acad Sci USA 90: 4912–4916

Gaillardin C, Heslot H (1971) Physiological and genetical studies on yeasts of the genus *Candida*. in Radiation and radioisotopes for industrial microorganisms. International Atomic Energy Agency, Vienna, pp 93–111

Gaillardin C, Heslot H (1979) Evidence for mutations in the structural gene for homocitrate synthetase in *Saccharomycopsis lipolyica*. Mol Gen Genet 172: 185–192

Gaillardin C, Heslot H (1988) Genetic engineering in *Yarrowia lipolytica*. J Basic Microbiol 28: 161–174

Gaillardin C, Ribet AM (1987) *LEU2* directed expression of β-galactosidase activity and phleomycin resistance in *Yarrowia lipolytica*. Curr Genet 11: 369–375

Gaillardin C, Charoy V, Heslot H (1973) A study of copulation, sporulation and meiotic segregation in *Candida lipolytica*. Arch Microbiol 92: 69–83

Gaillardin C, Poirier L, Heslot H (1975) Studies on an unstable phenotype induced by UV irradiation: the lysine-excreting (lex⁻) phenotype of the yeast *Saccharomycopsis lipolityca*. Arch Mikrobiol 104: 89–94

Gaillardin C, Poirier L, Heslot H (1976a) A kinetic study of homocitrate synthase activity in the yeast *Saccharomycopsis lipolytica*. Biochim Biophys Acta 422: 340–406

Gaillardin C, Fournier P, Sylvestre G, Heslot H (1976b) Mutants of *Saccharomycopsis lipolytica* defective in lysine catabolism. J Bacteriol 125: 48–57

Gaillardin C, Poirier L, Ribet AM, Heslot H (1979) General and lysine specific control of saccharopine dehydrogenase levels in the yeast *Saccharomycopsis lipolytica*. Biochimie 61: 173–182

Gaillardin C, Ribet AM, Heslot H (1982) Wild-type and mutant forms of homoisocitric dehydrogenase in the yeast *Saccharomycopsis lipolytica*. Eur J Biochem 128: 489–494

Gaillardin C, Ribet AM, Heslot H (1985) Integrative transformation of the yeast *Yarrowia lipolytica*. Curr Genet 10: 49–58

Gomi K, Ota Y, Minoda Y (1986) Role of lipase activators produced by *Saccharomycopsis lipolytica* and calcium ions in its lipase reaction. Agric Biol Chem 50: 2531–2536

Groves P, Clare JJ, Oliver SG (1983) Isolation and characterization of a double stranded RNA virus-like particle from the yeast *Yarrowia lipolytica*. Curr Genet 7: 185–190

Guevara-Olivera L, Calco-Mendez C, Ruiz-Herrera J (1993) The role of polyamine metabolism in dimorphism of *Yarrowia lipolytica*. J Gen Microbiol 193: 485–493

Hadeball W (1991) Production of lipase by *Yarrowia lipolytica*. Acta Biotechnol 11: 159–167

Hagihara T, Mishina M, Tanaka A, Fukui S (1977) Utilization of tristane by the yeast *Candida lipolytica*. Fatty acid composition of pristane-grown cells. Agric Biol Chem 41: 1745–1748

Hamsa PH, Chatto BB (1994) Cloning and growth-regulated expression of the gene encoding the hepatitis B virus middle surface antigen in *Yarrowia lipolytica*. Gene 143: 165–170

Hann BC, Walter P (1991) The signal recognition particle in *S. cerevisiae*. Cell 67: 131–144

He F, Beckerich JM, Ribes V, Tollervey D, Gaillardin C (1989) Two genes encode 7SL RNAs in the yeast *Yarrowia lipolytica*. Curr Genet 16: 347–350

He F, Yaver D, Beckerich JM, Ogrydziak D, Gaillardin C (1990) The yeast *Y. lipolytica* has two, functional signal recognition particle 7S RNA genes. Curr Genet 17: 289–292

He F, Beckerich JM, Gaillardin C (1992) A mutant of 7SL RNA in *Yarrowia lipolytica* affecting the synthesis of a secreted protein. J Biol Chem 267: 1932–1937

Herman AI (1971) Mating responses in *Candida lipolytica*. J Bacteriol 107: 371

Heslot H (1990) Genetics and genetic engineering of the industrial yeast *Yarrowia lipolytica*. Adv Biochem Eng Biotechnol 43: 43–73

Heslot H, Gaillardin C, Beckerich JM, Fournier P (1979) Control of lysine metabolism in the petroleum yeast *Saccharomycopsis lipolytica*. In: Sebek O, Laskin A (eds) Genetics of industrial microorganisms. American Society for Microbiology, Washington DC, pp 54–60

Hoenes I, Simon M, Weber H (1991) Characterization of isocitrate lyase from the yeast *Yarrowia lipolytica*. J Basic Microbiol 31: 251–258

Hoffmann CS, Winston F (1987) A ten-minute DNA preparation from yeast efficiently releases autonomous plasmids for transformation of *Escherichia coli*. Gene 57: 267–272

Holzschu DL, Chandler FW, Ajello L, Ahearn, DG (1979) Evaluation of industrial yeasts for pathogenicity. Sabouraudia 17: 71–78

Ilchenko AP, Mauersberger S, Matyashova RN, Lozinov AB (1980) Induction of cytochrome P-450 in the course of yeast growth on different substrates. Mikrobiologija 49: 452–458

Ishikura T, Forster JW (1961) Incorporation of molecular oxygen during microbial utilization of olefins. Nature 192: 892–893

Kalle GP, Gadkari SV, Deshpande Y (1972) Inducibility of lipase in *Candida lipolytica*. Indian J Biochem Biophys 9: 171–175

Kamiryo T, Nishikawa Y, Mishina M, Terao M, Numa S (1979) Involvement of long-chain acyl coenzyme A for lipid synthesis in repression of acettylcoenzyme A carboxylase in *Candida lipolytica*. Proc Natl Acd Sci USA 76: 4390–4394

Kautola H, Rymowicz W, Linko YY, Linko P (1991) Production of citric acid with immobilized *Yarrowia lipolytica*. Appl Microbiol Biotechnol 35: 447–449

Kemp GD, Dickinson FM, Ratledge C (1990)Light sensitivity of the n-alkane-induced fatty alcohol oxidase from *Candida tropicalis* and *Yarrowia lipolytica*. Appl Microbiol Biotechnol 32: 461–464

Klug MJ, Markovetz AJ (1967) Degradation of hydrocarbons by members of the genus Candida. II. Oxidation of n-alkanes and l-alkenes by *Candida lipolytica*. J Bacteriol 93: 1847–1852

Kohlwein SD, Paltauf F (1983) Uptake of fatty acids by the yeasts, *Saccharomyces uvarum* and *Saccharomycopsis lipolytica*. Biochim Biophys Acta 792: 310–317

Köttig H, Rottner G, Beck KF, Schweizer M, Schweizer E (1991) The pentafunctional *FAS1* genes of *Saccharomyces cerevisiae* and *Yarrowia lipolytica* are co-linear considerably longer than previously estimated. Mol Gen Genet 226: 310–314

Kreger-van Rij NJW (1984) The yeasts, a taxonomic study. Elsevier Amsterdam

Kreger-van Rij NJW, Veenhuis M (1973) Electron microscopy of septa in ascomycetous yeasts. Mol Gen Genet 146: 101–106

Kück V, Stahl U, Lhermitte A, Esser K (1980) Isolation and characterization of mitochondrial DAN from the alkane yeast *Yarrowia lipolytica*. Curr Genet 2: 97–101

Kujau M, Weber H, Barth G (1992) Characterization of mutants of the yeast *Yarrowia lipolytica* defective in acetyl-Coenzyme A synthetase. Yeast 8: 193–203

Kurischko C (1984) Analysis of genetic markers in new breeding stocks of the yeast *Saccharomycopsis lipolytica*. Z Allg Mikrobiol 24: 545–550

Kurischko C (1986) Parasexual process in the yeast *Yarrowia lipolytica*. J Basic Microbiol 26: 33–41

Kurischko C, Weber H (1986) Temporal relationship of diploidization and haploidization in the yeast *Yarrowia lipolytica*. J Basic Microbiol 3: 137–144

Kurischko C, Inge-Vechtonov SG, Weber H (1983) Development of breeding stocks of the yeast *Saccharomycopsis lipolytica* by methods of moderate inbreeding. Z Allg Mikrobiol 23: 513–515

Kurischko C, Fournier P, Chasles M, Weber H, Gailardin C (1992) Cloning of the mating-type gene *MatA* of the yeast *Yarrowia lipolytica*. Mol Gen Genet 232: 423–426

Kurtzman CP, Phaff HJ (1987) Molecular taxonomy. In: Rose AH, Harrisson JS (eds) The Yeasts, vol 1. Biology of yeasts. Academic Press, London, pp 63–94

Le Dall M-T, Nicaud J-M, Gaillardin C (1994) Multiple-copy integration in the yeast *Yarrowia lipolytica*. Curr Genet 26: 38–44

Lopez MC, Dominguez A (1988) Purification and properties of a glycoprotein acid phosphatase from the yeast form of *Yarrowia lipolytica*. J Basic Microbiol 28: 249–263

Lopez MC, Nicaud JM, Skinner H, Vergnolles C, Kader JC, Bankaitis V, Gaillardin C (1994) A phosphatidylinositol/phosphatdylcholine transfer protein is required for differentiation of the dimorphic yeast *Yarrowia lipolytica* from the yeast to the mycelial form. J Cell Biol 124: 113–127

Maldonado P, and Gaillardin C (1972) Procédé d'obtention de souches diploides de *Candida lipolytica* et utilisation de ces souches dans un procédé de Préparation d'acide alpha cétoglutarique. French patent 72/41913

Matoba S, Ogrydziak DM (1989) A novel location for dipeptidyl aminopeptidase processing sites in the alkaline extracellular protease of *Yarrowia lipolytica*. J Biol Chem 264: 6037–6043

Matoba S, Fukuyama J, Wing RA, Ogrydziak DM (1988) Intracellular precursors and secretion of alkaline extracellular protease of *Yarrowia lipolytica*. Mol Cell Biol 8: 4904–4916

Matsuoka M, Ueda Y, Aiba S (1980) Role and control of isocitrate lyase in *Candida lipolytica*. J Bacteriol 144: 692–697

Matsuoka M, Uchida K, Aiba S (1982) Cytoplasmic transfer of oligenycin resistance during protoplast fusion of *Saccharomycopsis lipolytica*. J Bacteriol 152: 530–533

Matsuoka M, Matsubara M, Daidoh H, Imanaka T, Uchida K, Aiba S (1993) Analysis of regions essential for the function of chromosomal replicator sequences from *Yarrowia lipolytica*. Mol Gen Genet 237: 327–333

Matsuoka M, Himoeno T, Aiba S (1984) Characterization of *Saccharomycopsis lipolytica* mutants that express temperature-sensitive synthesis of isocitrate lyase. J Bacteriol 157: 899–908

Mattey M (1992) The production of organic acids. Crit Rev Biotechnol 12: 87–132

Mattey M, Adoga G (1991) Low molecular weight thermostable enzymes. Enzyme Microb Technol 13: 525

Mauersberger S (1991) Mutants of alkane oxidation in the yeasts *Yarrowia lipolytica* and *Candida maltosa*. In: Finogenova TV, Sharyshev AA (eds) Alkane metabolism and oversynthesis of metabolites by microorganisms. Centre for Biological Research USSR Academy of Sciences, Pushchino

Mauersberger S, Matyashova RN (1980) The content of cytochrome P-450 in yeast cells growing on hexadecane. Mikrobiologija 49: 571–577

Mauersberger S, Boehmer A, Schunck W-H, Muller H-G (1991) Cytochrome P-450 of the yeast *Yarrowia lipolytica*. Abstr Int Symp on Cytochrome P-450 of Microorganisms, Berlin, p 63

Mauersberger S, Drechsler H, Oehme G, Muller H-G (1992) Substrate specificity and stereoselectivity of fatty alcohol oxidase from the yeast *Candida maltosa*. Appl Microbiol Biotechnol 37: 66–73

May R, Barth G (1977) Tubulare Einschlüsse in Microbodies von. *Saccharomycopsis (Candida) lipolytica* Protoplasten. Protoplasma 91: 83–91

McKay IA (1992) Growth of fermentative and non fermentative yeasts in natural yoghurt, stored in polystyrene cartons. Int J Food Microbiol 15: 383–388

McKay IA, Maddox IS, Brooks JB (1990) Citrate production by yeast. Ferment Technol Ind Appl 285–291

Moat AG, Peters N, Srb AM (1959) Selection and isolation of auxotrophic yeast mutants with the aid of antibiotics. J Bacteriol 77: 673–677

Moran A, Burguillo FJ, Lopez MC, Dominguez A (1989) Kinetic properties of derepressible acid phosphatase from the yeast form of *Yarrowia lipolytica*. Biochim Biophys Acta 990: 288–296

Morzycka E, Sawnor-Korszynska D, Pasjewski A, Grabsky J, Raczynska-Bojanowska K (1976) Methionine overproduction in *Saccharomycopsis lipolytica*. Appl Environ Microbiol 2: 125–130

Mortimer RK, Bassel BB (1973) Genetic studies of *Saccharomycopsis lipolytica*. Proc of the conference on genetics of industrial microorganisms, USSR. 1–6

Murphy GL, Perry JJ (1984) Assimilation of chlorinated alkanes by hydrocarbon-utilizing fungi. J Bacteriol 160: 1171–1174

Nakase T, Komagata K (1971) Signification of DNA base composition in the classification of yeast genes *Candida*. J Gen Appl Microbiol Tokyo 17: 259–279

Naumova E, Naumov G, Fournier P, Nguyen HV, Gaillardin C (1993) Chromosomal polymorphism of the yeast *Yarrowia lipolytica* and related species: electrophoretic karyotyping and hybridization with cloned genes. Curr Genet 23: 450–454

Nga B-H, Heslot H, Gailardin CM, Fournier P, Chan K, Chan YN, Lim EW, Nai PC (1988) Use of nystatin for selection of tributyrin non-utilizing mutants in *Yarrowia lipolytica*. J Biotechnol 7: 83–86

Nga B-H, Gaillardin GM, Fournier P, Heslot H (1989) Genetic analysis of lipase low-producing mutants of *Yarrowia lipolytica*. J Gen Microbiol 135: 2439–2443

Nicaud JM, Fabre E, Beckerich JM, Fournier P, Gaillardin C (1989a) Cloning, sequencing and amplification of the alkaline extracellular protease gene of *Yarrowia lipolytica* J Biotechnol 12: 285–298

Nicaud J-M, Fabre E, Gailardin C (1989b) Expression of invertase activity in *Yarrowia lipolytica* and its use as a selective marker. Curr Genet 16: 253–260

Nicaud JM, Fournier P, La Bonnardière C, Chasles M, Gaillardin C (1991) Use of *ARS18* based vectors to increase protein production in *Yarrowia lipolytica*. J Biotechnol 19: 259–270

Nuttley WM, Brade AM, Gaillardin C, Eitzen GA, Glover JR, Aitchinson JD, Rachubinski RA (1993) Rapid identification and characterization of peroxisomal assembly mutants in *Yarrowia lipolytica*. Yeast 9: 507–517

Nuttley WM, Brade AM, Eitzen GA, Veenhuis M, Aitchinson JD, Szilard RK, Glover JR, Rachubinski RA (1994) *PAY4*, a gene required for peroxysome assembly in the yeast *Yarrowia lipolytica*, encodes a novel member of a family of putative ATPases. J Biol Chem 269: 556–566

Nyns EJ, Auquiere JP, Chiang N, Wiaux AL (1967) Comparative growth of *Candida lipolytica* on glucose and n-hexadecane. Nature 215: 177–178

Ogrydziak DM (1988a) Development of genetic maps of non-conventional yeasts. J Basic Microbiol 28: 185–196

Ogrydziak DM (1988b) Production of alkaline extracellular protease by *Yarrowia lipolytica*. CRC Crit Rev Biotechnol 8: 177–187

Ogrydziak DM (1993) Yeast extracellular proteases. Crit Rev Biotechnol 13: 1–55

Ogrydziak DM, Mortimer RK (1977) Genetics of extracellular protease production in *Saccharomycopsis lipolytica*. Genetics 87: 621–632

Ogrydziak DM, Demain AI, Tannenbaum SR (1977) Regulation of extracellular protease production in *Candida lipolytica*. Biochim Biophys Acta 497: 525–538

Ogrydziak D, Bassel J, Contopoulou R, Mortimer RK (1978) Development of genetic techniques and the genetic map of the yeast *Saccharomycopsis lipolytica*. Mol Gen Genet 163: 229–239

Ogrydziak DM, Cheng S-C, Scharf SJ (1982) Characterization of *Saccharomycopsis lipolytica* mutants producing lowered levels of alkaline extracellular protease. J Gen Microbiol 128: 2271–2280

Okuma M, Hwang CW, Masuda Y, Nishida H, Sugiyama J, Ohta A, Takagi M (1993) Evolutionary position of n-alkane-assimilating yeast Candida maltosa shown by nucleotide sequence of small-subunit ribosomal RNA gene. Biosci Biotechnol Biochem 57: 1793–1794

Oogaki M, Nakahara T, Uchiyama H, Tabuchi T (1983) Extracellular production of D-(+)-2-hydroxyglutaric acid by *Yarrowia lipolytica* from glucose under aerobic thiamine-deficient conditions. Agric Biol Chem 47: 2619–2624

Ota Y, Morimoto Y, Sugiura T, Minoda Y (1987) Soybean fraction increasing the extracellular lipase production by *Saccharomycopsis lipolytica*. Agric Biol Chem 42: 1937–1938

Ota Y, Gomi TK, Kato S, Sugiura T, Mindoa Y (1982) Purification and some properties of cell-bound lipase from *Saccharromycopsis lipolytica*. Agric Biol Chem 46: 2885–2893

Ota Y, Oikawa S, Morimoto Y, Minida Y (1984) Nutritional factors causing mycelial development of *Saccharomycopsis lipolytica*. Agric Biol Chem 48: 1933–1939

Ozegowski R, Kunath A, Schick H (1993) Tetrahedron-Asymmetry 4: 695–698

Pertuiset B, Beckerich Gaillardin C (1994) Molecular cloning of Rab-related genes in the yeast *Yarrowia lipolytica*. Analysis of *RYL1*, an essential gene encoding a *SEC4* homologue. Curr Genet 27: 123–130

Peters II, Nelson FE (1948a) Factors influencing the production of lipase by *Mycotorula lipolytica*. J Bacteriol 55: 581–591

Peters II, Nelson FE (1948b) Preliminary characterization of the lipase of *Mycotorula lipolytica*. J Bacteriol 55: 593–600

Poncet S, Arpin M (1965) Les *Candida* sans pouvoir fermentaire (Cryptococcacées). Antonie Leeuwenhoek J Microbiol 31: 433–464

Poritz MA, Siegel V, Hansen W, Walter P (1988) Small ribonucleoproteins in *Schizosaccharomyces pombe* and *Yarrowia lipolytica* homologous to signal recognition particle. Proc Natl Acad Sci USA 85: 4315–4319

Prodromou C, Wright IP, Evans IH, Bevan EA (1991) Cloning of the HIS3 gene of *Yarrowia lipolytica*. Antonie Leeuwenhoek 60: 95–99

Rodriguez C, Dominguez A (1984) The growth characteristics of *Saccharomycopsis lipolytica*: morphology and induction of mycelium formation. Can J Microbiol 30: 605–612

Rodriguez C, Lopez MC, Dominguez A (1990) Macromolecular synthesis during the yeast-mycelium transition in *Yarrowia lipolytica*. Exp Mycol 14: 310–321

Roiha H, Shuster EO, Brow DA, Guthrie C (1989) sn RNAs from budding yeasts: phylogenetic comparisons reveal extensive size variation. Gene 82: 113–124

Rothstein M, Cooksey KE, Greenberg DM (1962) Metabolic conversion of pipecolic acid into α-aminoadipic acid. J Biol Chem 237: 2828–2830

Rubin GM (1974) Three forms of the 5.8-S ribosomal RNA species in *Saccharomyces cerevisiae*. Eur J Biochem 41: 197–202

Sawnor-Korszynska D, Morzycka E, Zaborowska-Bojanowska K (1977) Compartmentation of the amino acid pool in *Saccharomycopsis lipolytica*. Acta Biochim Pol 24: 75–85

Schmid-Berger N, Schmid B, Barth G (1994) Ylt1, a highly repetitive retrotransposon in the genome of the dimorphic fungus *Yarrowia lipolytica*. J Bacteriol 176: 2477–2482

Schweizer E, Kötting H, Regler R, Rottner G (1988) Genetic control of *Yarrowia lipolytica* fatty acid synthetase synthesis and function. J Basic Microbiol 28: 283–292

Sekula BC (1991) Wax esters production by yeast. Biotechnol. Plant Fats Oils: 162–176

Sen K, Komagata K (1979) Distribution of urease and extracellular DNase in yeast species. J Gen Appl Microbiol 25: 127–135

Shah DN, Chattoo BB (1993) Starch hydrolyzate, an optimal and economical source of carbon for the secretion of citric acid by *Yarrowia lipolytica*. Starch 45: 104–109

Shah DN, Purohit AP, Sriprakash KS (1982) Preliminary genetic studies on a citric acid-producing strain of *Saccharomycopsis lipolytica*. Enzyme Microb Technol 4: 116–117

Simms PC, Ogrydziak DM (1981) Structural gene for the alkaline extracellular protease of *Saccharomycopsis lipolytica*. J Bacteriol 145: 404–409

Snow R (1966) An enrichment method for auxotrophic yeast mutants using the antibiotic nystatin. Nature 211: 206–207

Stahl U (1978) Zygote formation and recombination between like mating types in the yeast *Saccharomyces lipolytica* by protoplast fusion. Mol Gen Genet 160: 111–113

Strick CA, James LC, O'Donnell MM, Gollaher MG, Franke AE (1992) The isolation and characterization of the pyruvate encoding gene from the yeast *Yarrowia lipolytica*. Gene 118: 65–72

Sugiura-M, Isobe-M, Oikawa-T, Oono-H (1976) Sterol ester hydrolytic activity of lipoprotein lipase from *Pseudomonas fluorescens*. Chem Pharm Bull (Tokyo) 24: 1202–1208

Tharaud C, Ribet A-M, Costes C, Gaillardin C (1992) Secretion of human blood coagulation factor XIIIa by the yeast *Yarrowia lipolytica*. Gene 121: 111–119

Theil F, Schick H, Winter G (1991) Transesterification of meso-cyclopentane diols. Tetrahedron 47: 7569–7582

Tréton BY, Heslot H (1978) Etude de quelques propriétés de l'aconitase de la levure *Saccharomycopsis lipolytica*. Agric Biol Chem 42: 1201–1206

Tréton BY, Le Dall MT, Heslot H (1985) Virus like particles from the yeast *Yarrowia lipolytica*. Curr Genet 9: 279–284

Tréton BY, Le Dall MT, Heslot H (1987) UV-induced curing of the double-stranded RNA virus of the yeast *Yarrowia lipolytica*. Curr Genet 12: 37–39

Tréton BY, Le Dall MT, Gaillardin CM (1992) Complementation of *Saccharomyces cerevisiae* acid phosphatase mutation by a genomic sequence from the yeast *Yarrowia lipolytica* identifies a new sequence. Curr Genet 22: 345–355

Tsugawa R, Nakase T, Koyabashi T, Yamashita K, Okumura S (1969) Fermentation of *n*-paraffins by yeast. Part III. alpha-ketoglutarate productivity of various yeast. Agric Biol Chem 33: 929–938

van der Walt JP, von Arx JA (1980) The yeast *Yarrowia* gen. nov. Antonie Leeuwenhoek J Microbiol 46: 517–521

van Heerikhuizen H, Ykema A, Klootwijk J, Gaillardin C, Ballas C, Fournier P (1985) Heterogeneity in the ribosomal RNA genes of the yeast *Yarrowia lipolytica*; cloning and analysis of two size classes of repeats. Gene 39: 213–222

Vega R, Dominguez A (1986) Cell wall composition of the yeast and mycelial forms of *Yarrowia lipolytica*. Arch Mikrobiol 144: 124–130

Vega R, Dominguez A (1988) Partial characterization of alpha-mannosidase from *Yarrowia lipolytica*. J Basic Microbiol 28: 371–379

Weber H (1979) Substructural studies on sporulation of *Saccharomycopsis lipolytica*. Z Allg Mikrobiol 19: 283–297

Weber H, Barth G (1988) Nonconventional yeasts: their genetics and biotechnological applications. CRC Crit Rev Biotechnol 7: 281–337

Weber H, Forster W, Jacob HE, Berg H (1980) Enhancement of yeast protoplast fusion by electric field effects. Adv in Biotech: Proceedings of the 6th Int Symp/5th Int Symp on Yeasts, London, Canada, pp 219–224

Weber H, Kurischko C, Barth G (1988) Mating in the alkane-utilizing yeast *Yarrowia lipolytica*. J Basic Microbiol 28: 229–240

Weissbrodt E, Gey M, Barth G, Weber H, Stottmeister U, Duresch R, Richter H-P (1988) Verfahren zur Herstellung von 2-Oxoglutarsäure durch Hefen. DD Patent no. 267 999 A1.

Wésolowski M, Algeri A, Fukuhara H (1981) Gene organization of the mitochondrial DNA of yeasts: *Kluyveromyces lactis* and *Yarrowia lipolytica*. Curr Genet 3: 157–162

Wickerham LJ, Kurtzman CP, Herman AI (1970) Sexual reproduction in *Candida lipolytica*. Science 167: 1141

Wing RA, Ogrydziak DM (1985) Development of the genetics of the dimorphic yeast *Yarrowia lipolytica*. In: Timberlake WE (ed) Molecular genetics of filamentous fungi. A.R. Liss New York, pp 367–381

Wojatatowicz M, Rymowicz W, Kautola H (1991) Comparison of different strains of *Yarrowia lipolytica* for citric acid production from glucose hydrol. Appl Microbiol Biotechnol 31: 165–174

Xuan JW, Fournier P, Gaillardin C (1988) Cloning of the *LYS5* gene encoding saccharopine dehydrogenase from the yeast *Yarrowia lipolytica* by target integration. Curr Genet 14: 15–21

Xuan JW, Fournier P, Declerck N, Chasles M, Gaillardin C (1990) Overlapping reading frames at the *LYS5* locus in the yeast *Yarrowia lipolytica*. Mol Cell Biol 10: 4795–4805

Yamada T, Ogrydziak DM (1983) Extracellular acid proteases produced by *Saccharomycopsis lipolytica*. J Bacteriol 154: 23–31

Yamada Y, Nojiri M, Matsuyama M, Kondo K (1976) Goenzyme Q systems in the classification of the ascosporogenous yeast genera *Debaryomyces*, *Saccharomyces*, *Kluyveromyces* and *Endomycopsis*. J Gen Appl Microbiol Tokyo 22: 325–337

Yarrow D (1972) Four new combinations in yeasts. Antonie Leeuwenhoek J Microbiol 38: 357–360

Yaver DS, Matoba S, Ogrydziak DM (1992) A mutation in the signal recognition particle 7SRNA of the yeast *Yarrowia lipolytica* preferentially affects synthesis of the alkaline extracellular protease: in vivo evidence for translational arrest. J Cell Biol 116: 605–616

Zviagintzeva IS, Dmitriev VV, Ripan EL, Fichte BA (1980) Localization of an extracellular lipase of the yeast *Candida paralipolytica*. Microbiology (USSR) 49: 417–420 (in Russian)

CHAPTER 11

Arxula adeninivorans

Gotthard Kunze and Irene Kunze

1
History of *Arxula adeninivorans* Research

In 1984, Middelhoven et al. described a yeast species which was isolated from soil by the enrichment culture method. The yeast was known as *Trichosporon adeninovorans* at that time. This strain (CBS 8244T) displays unusual biochemical activities. It was shown to assimilate adenine and several other purine compounds as sole source of carbon and energy. Moreover, it grows at the expense of several amines.

A second strain of this species, Ls3 (PAR-4), originated from Siberia (Kapultsevich, Institute of Genetics and Selection of Industrial Microorganisms, Moscow, Russia). It was isolated during production of single-cell protein from wood hydrolysates.

In 1990, three further *Tr. adeninovorans* strains were isolated from chopped maize herbage ensiled at 25 or 30 °C in The Netherlands. Four strains of the same species were also found in humus-rich soil in south Africa (Van der Walt et al. 1990). A new genus, *Arxula* Van der Walt, M.T. Smith & Yamada (Candidaceae) was proposed for the classification of all these strains, which are xerotolerant, ascomycetous, anamorphic, arthroconidial, and nitrate-positive yeasts (Van der Walt et al. 1990). The genus *Arxula* accommodates two species, viz. *Arxula terrestre* (Van der Walt and Johanssen) Van der Walt, M.T. Smith & Yamada, nov. comb., which is the type species of the genus, and *Arxula adeninivorans* (Middelhoven, Hoogkamerte Niet and Kregervan Rij) Van der Walt, M.T. Smith and Yamada, nov. comb.

2
Physiology and Biochemical Procedures

2.1
Physiology

A. adeninivorans and *A. terrestre* are conspicuous because of the utilization of nitrate. They can grow on adenine (very slowly), uric acid, butylamine,

Institut für Pflanzengenetik und Kulturpflanzenforschung, Corrensstr. 3, 06466 Gatersleben, Germany pentylamine, and putrescine as sole source of carbon and nitrogen. However, *A. terrestre*, in contrast to *A. adeninivorans*, has no fermentative ability and does not assimilate soluble starch, melibiose and melizitose, propylamine and hexylamine (Middelhoven et al. 1984).

Middelhoven et al. (1991) published a detailed description of the yeast species *A. adeninivorans*. It assimilates all the sugars, polyalcohols, and organic acids used in the conventional carbon compound assimilation test rapidly, except for L-rhamnose, inulin, lactose, lactate, and methanol. All conventionally used compounds except creatine and creatinine were suitable as nitrogen sources. Several nitrogenous compounds, e.g., amino acids, purine derivatives, served as sole

Table 1. Properties of secretory enzymes of the strain *A. adeninivorans* Ls3 (Büttner et al. 1987, 1988, 1989a, 1990b,c, 1991a,b, 1992a,b; Büttner and Bode 1992)

Enzyme	Optimum			Molecular weight
	Temperature	pH	K_m value	
Glucoamylase (1.4-α-D-glucan glucohydrolase, EC 3.2.1.3)	60 to 70 °C	4.0–5.0	1.2 g/l for starch 11.1 mM for maltose	225 000
Acid phosphatases I and II (orthophosphoric-monoester phosphohydrolase, EC 3.1.3.2)				
I	50 to 55 °C	5.2–5.5	3.5 mM for p-nitro-phenylphosphate	320 000
II	"	"	"	250 000
Trehalase (α,α-Trehalose-Glucohydrolase, EC 3.2.1.28)	45 to 55 °C	4.5–4.9	0.8–1.0 mM for trehalose	250 000
Cellobiase I and II (β-D-glucosidase, EC 3.2.1.21)				
I	60 to 63 °C	4.5	4.1 mM for cellobiose	570 000
II	"	"	3.0 mM for cellobiose	525 000
Invertase (β-D-fructofuranoside fructohydrolase, EC 3.2.1.26)	60 to 70 °C	5.0	71–83 mM for sucrose	650 000
β-D-xylosidase (1,4-β-D-xylan xylohydrolase, EC 3.2.1.37)	60 °C	5.0	0.23–0.33 mM for p-nitrophenyl-β-xylopyranoside	60 000

source of carbon, nitrogen, and energy. This was also true of many primary n-alkylamines and terminal diamines, but of nitrogen-less analogous compounds such as alcohols, dialcohols, carboxylic acids, and dicarboxylic acids only intermediates of general metabolism were assimilated (Middelhoven et al. 1991).

The growth on some industrially available substrates is combined with a relatively high thermotolerance. Surprisingly, the strain Ls3 can grow at temperatures higher than 45 °C without an adaptation phase (Böttcher et al. 1988).

Arxula adeninivorans can not only grow on many different carbon sources but produces and secretes also several extracellular enzymes into the culture medium during growth. Besides RNase and some proteases, various glucosidases, such as glucoamylase, β-glucosidases, pectinases, xylosidase, and invertase could be detected. Table 1 summarizes important secretory enzymes and some of their properties, analyzed from *A. adeninivorans* Ls3.

Kunze and Kunze (1993b) compared the properties of secretory invertase and glucoamylase of the strain Ls3 with those of six further *A. adeninivorans* strains (CSIR 1136, CSIR 1138, CSIR 1147, CSIR 1148, CSIR 1149, CBS 8244T). All strains show similar activities within a pH and temperature range of 3.5 to 6.5 and 20 to 80 °C, respectively.

Further investigations were done concerning the secretion of invertase and glucoamylase during the cultivation of the seven *A. adeninivorans* strains (Kunze and Kunze 1993b). With the exception of CSIR 1147, which secretes invertase activities up to 5 nkat/ml, the secreted invertase of all the other strains varied between 1 and 2 nkat/ml. The highest activity of the constitutive expressed invertase could be found after about 40 h of cultivation in minimal medium, with 1% maltose as sole source of carbon.

For glucoamylase secretion it could be observed that especially the strain CSIR 1138 secretes the highest enzyme activities (4.5 nkat/ml). In the culture medium of the other *A. adeninivorans* strains, 1–2 nkat/ml activity of glucoamylase could be estimated. The highest activity of the inducible expressed glucoamylase was reached after about 60 h of cultivation in minimal medium with 1% maltose (Wartmann et al. 1995).

2.2
Biochemical Procedures

2.2.1
Cell Mass Determination

The yeast cell density can be determined by absorbance in a spectrophotometer. An absorbance value of 1.2 at 600 nm corresponds to a cell density of about 1×10^8 cells/ml. However, the conditions for cultivation can influence the morphology of the strains. Some of them tend to form mycelia (for example CSIR 1138), which affects this method of cell mass determination immensely.

2.2.2
Preparation of DNA

Chromosomal DNA

The chromosomal DNA is prepared from *A. adeninivorans* with a modified method described by Ledeboer et al. (1985).

Cells cultivated in 200–400 ml YE medium or SD medium at 37 °C are harvested, washed, disrupted mechanically by liquid nitrogen, and extracted with about 10 ml 2 mM Tris/pH 8.0; 50 mM EDTA, 2.5% SDS.

The suspensions are centrifuged at 15 000 g for 20 min. Ethanol (1.5 vol) is added to the supernatant, followed by incubation for 30 min at −20 °C.

Subsequently, the pellet is collected by centrifugation and resuspended in 4 ml TE buffer (10 mM Tris/pH 8.0; 1 mM EDTA).

Three hundred μg RNase A is added and the suspension is incubated for 30 min at 37 °C.

Then 5 mg pronase N (Serva) is mixed into the suspension, which is incubated for 30 min at 30 °C.

The DNA is treated with phenol, phenol/chloroform, and chloroform, and subsequently the DNA is precipitated, centrifuged, and suspended in a small volume of TE buffer.

Chromosomal DNA is prepared from the other nucleic acids by means of density gradient centrifugation described by Ledeboer et al. (1985).

Mitochondrial DNA

The method for isolation of mitochondrial DNA is described by Pich and Kunze (1992).

2.2.3
Preparation of Probes for Enzyme Activity

Assay for Extracellular Enzymes

The activity of secretory enzymes can determined directly from the culture medium after induction of the respective enzyme during growth of the strains in the presence of the appropriate substrate. For most of the secretory enzymes like glucoamylase, invertase, trehalase, and β-glucosidase, the release of glucose from the carbon source used can be determined with glucoseoxydase/ peroxydase reagent.

Methods for preparation of secretory enzymes were described by Büttner et al. for glucoamylase (1987), invertase (1990b), trehalase (1992a), β-glucosidase (1991a), and acid phosphatase (1991b).

Cell Disruption for Assays of Intracellular Enzymes

The activity of most of the intracellular enzymes can be determined after disruption of the cells by vortexing with glass beads, by X-pressure cell, or by permeabilization of the cells by means of 0.1% Triton ×100 in 100 mM Tris/HCl, pH 8.4, combined with freezing.

3
Growth Media

For routine laboratory studies, the basic culture conditions are similar to those generally used for *S. cerevisiae*. The YE medium contains 0.5% yeast extract, 0.5% peptone, and 1% of the carbon source. The minimal medium (SD medium) is made as described by Tanaka et al. (1967) supplemented with appropriate carbon sources to 1% and a vitamin mix containing pantothenic acid, thiamine, pyridoxine, nicotinic acid, and biotin.

4
Available Strains and Preservation Methods

4.1
Strains

Strains of the genus *Arxula* have been isolated from different continents of the earth. A general view of available wild-type strains is given in Table 2.

Most of the available *Arxula adeninivorans* strains were compared concerning their colony morphology on YE medium after growth on SD medium with 1% maltose as sole source for carbon (Kunze and Kunze 1993a). Figure 1 shows typical single colonies of the strains, which differ in size and surface structure.

Diverse auxotrophic mutants of the strain *A. adeninivorans* Ls3 were induced and biochemically characterized by Samsonova et al. (1989). After N-methyl-N′-

Table 2. Available wild-type strains of *Arxula adeninivorans*

Strain		Source
Ls3	= PAR-4	Wood hydrolysates (Siberia)
CBS 8244T	= CSIR 577	Soil (The Netherlands)
CBS 7370	= CSIR 1117	Soil (South Africa)
CBS 7377	= CSIR 1118	Soil (South Africa)
CSIR 1136		Maize silage (The Netherlands)
CSIR 1138		Maize silage (The Netherlands)
CSIR 1147		Soil (South Africa)
CSIR 1148		Soil (South Africa)
CSIR 1149		Soil (South Africa)

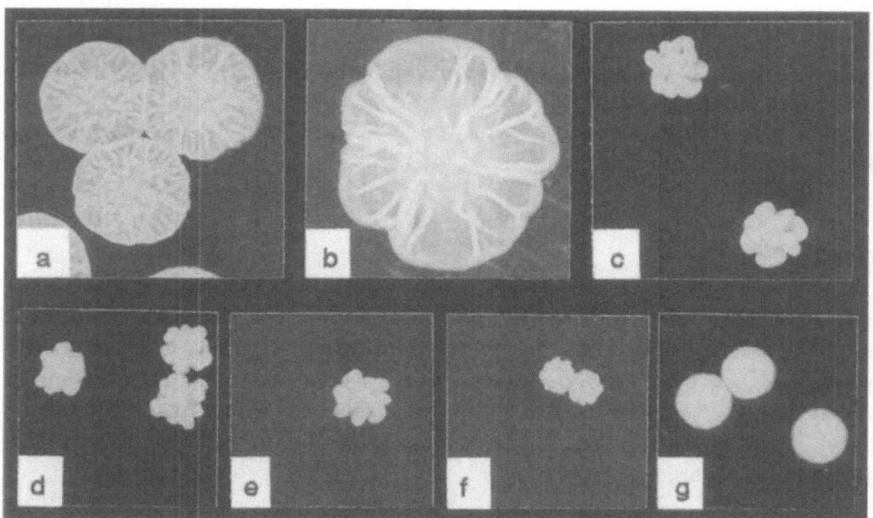

Fig. 1a–g. Single colonies of *A. adeninivorans* CSIR 1136 (**a**), CSIR 1138 (**b**), CSIR 1147 (**c**), CSIR 1148 (**d**), CSIR 1149 (**e**), CBS 8244T (**f**), Ls3 (**g**) cultivated on YE medium at 30 °C for 2 days. 1.5×

nitro-N-nitrosoguanidine (NG) and UV mutagenesis 3–4% auxotrophic mutants, could be found at a survival rate of 10^{-3}.

The markers are especially requirements for amino acids, inositol, nicotinic acid, and pyridoxine. They provide suitable chromosomal markers, an essential basis for further genetic study of this yeast.

Büttner et al. (1989b) isolated and characterized carbon catabolite derepression mutants from Ls3 using 2-deoxy-D-glucose. These mutants are especially characterized by a high synthesis of some extracellular enzymes in the presence of different carbon sources.

Auxothrophic mutants could also be selected from the other CBS and CSIR wild-type strains. After NG and UV mutagenesis they were induced by similar frequencies compared with Ls3 (Samsonova, unpubl.).

4.2
Preservation Methods

All strains of *A. adeninivorans* can be preserved by freezing or on YE agar (Kirsop 1988). To cryopreserve, the strains are cultivated in YE medium with 1% glucose and frozen with 10% glycerol. Before the yeasts are frozen to −80 °C, the cells are cooled to −20°C for 2 h. Similar cryocultures have been stored in our laboratory since 1986. By the second method all strains are transferred at 2-yearly intervals on YE agar. Therefore, cultures must be covered by a layer of paraffin oil and stored at 4 °C.

5
Parasexual Genetics

5.1
Protoplast Fusion (see also Chap. 2, this Vol.)

A useful technique for the asexual and haploid yeast species *A. adeninivorans* is the polyethyleneglycol (PEG)-induced fusion of protoplasts followed by mitotic haploidization. According to a method developed by Klinner and Böttcher (1984) protoplasts can be stabilized by 1 M sorbitol and regeneration of the fusion products takes place on the surface of agar. Up to now, many fusions between most of the wild-type strains, listed in Table 2, could be achieved (Samsonova, unpubl.).

Spontaneous or induced haploidization has been developed as a gene mapping tool also for other yeast species, such as *Schizosaccharomyces pombe* (Gutz 1966), *Rhodosporidium toruloides* (Samsonova and Böttcher 1978), *Pichia guilliermondii* (Samsonova and Böttcher 1978), *Saccharomyces cerevisiae* (Wood 1982), *Pichia pinus* (Tolstorukov et al. 1983), and *Yarrowia lipolytica* (Kurischko 1986).

5.2
Mitotic Haploidization

Mitotic haploidization can be achieved by benomyl. Benomyl is known to induce haploidization in yeast species without mitotic recombination processes (Wood 1982; Böttcher and Samsonova 1983). Therefore this technique is useful for the location of a given marker. A method suitable for the analysis of *A. adeninivorans* was developed by Samsonova and Böttcher (1980). They used 5 mg/l benomyl and cultivated the strains at 17–20 °C. By using benomyl-induced haploidization, four linkage groups could be found for the *Arxula* strain Ls3 (Büttner et al. 1990b; Samsonova, unpubl.). These findings agree with physical analyses using pulsed field gel electrophoresis (Sect. 6). By both methods 26 genes could be mapped. The gene of the extracellular glucoamylase from Ls3 could be located on linkage group II (Büttner et al. 1990). They assume that the organization of the linkage group in CBS 8244T differs from that in Ls3.

6
Chromosomal DNA

DNA reassociation, pulse field gel electrophoresis (PFG) and DNA fingerprinting have been chosen to characterize the genomic structure of some *A. adeninivorans* strains (Gienow et al. 1990; Kunze and Kunze 1993a).

6.1
DNA Reassociation

With the help of DNA reassociation, genome complexities have been analyzed. The *Arxula* strain Ls3 possesses genome complexity of 16.1×10^9 Da and strain CBS

8244T of 16.9×10^9 Da. In comparison to the complexity of *S. cerevisiae* (9.2×10^9 Da) these are relatively high values. Both tested *Arxula* strains could be identified as haploid strains after the total DNA concentration per cell and the mitochondrial DNA contents were measured (Table 3).

6.2
Pulsed Field Gel Electrophoresis

Chromosomes of seven *A. adeninivorans* strains (CSIR 1136, CSIR 1138, CSIR 1147, CSIR 1148, CSIR 1149, CBS 8244T, and Ls3) were separated by PFG electrophoresis and compared with those of the laboratory strains *S. cerevisiae* S288C and *Schizosaccharomyces pombe* 296R (Fig. 2; Table 4). Four chromosomes were identified in each *Arxula* strain. The molecular weights of these chromosomes were estimated between 4.6 Mbp (chromosome 1) and 1.6 Mbp (chromosome 4).

Based on these chromosomal patterns, the tested strains can be divided into four groups. The Arxula strain CSIR 1136 and CSIR 1138 (lanes 2 and 3) were classified into the first group. Both strains show clear differences in the chromosome pattern in comparison to the *Arxula* strains CSIR 1147 and Ls3 belonging to the second and the third group, respectively (lanes 4 and 8). Chromosomes 2 and 3 of both strains, respectively, differ in their molecular weight from the corresponding chromosomes of the strains belonging to the first group. In contrast to the strains of the first three groups, strains in the fourth group, including CSIR

Table 3. The degree of ploidy of *Arxula adeninivorans* Ls3

	S. cerevisiae D10	*A. adeninivorans* CBS 8244T	*A. adeninivorans* Ls3
Haploid genome size of the chromosomal DNA (g)	1.6×10^{-14}	2.7×10^{-14}	2.5×10^{-14}
Total DNA concentration per cell (g)	4.1×10^{-14}	2.6×10^{-14}	2.9×10^{-14}
Portion of mtDNA of the total DNA (%)	8.0	4.3	8.0
Portion of pDNA on the total DNA (%)	3.0	0	0
Final concentration of chromosomal DNA (g)	3.6×10^{-14}	2.5×10^{-14}	2.7×10^{-14}
Ploidy	2n	1n	1n

Table 4. PFG electrophoresis running conditions. Using 0.8% agarose, 11 °C, modified $0.25 \times$ TBE (0.025 M Tris/HCl; 0.025 M boric acid; 0.05 mM Na_2EDTA) and 46 V

Run time (h)	Pulse time (s)	Electrode configuration
72	3000–1000 (linear)	120° (constant)
24	1000 (constant)	120° (constant)

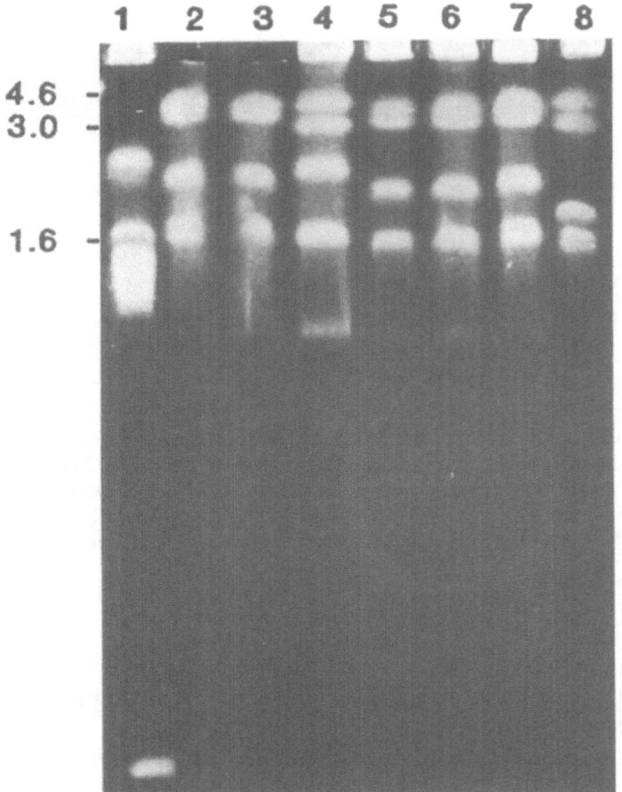

Fig. 2. PFG electrophoretic separation of chromosomes from seven *A. adeninivorans* strains. 1 *S. cerevisiae* S288C; 2 *A. adeninivorans* CSIR 1136; 3 *A. adeninivorans* CSIR 1138; 4 *A. adeninivorans* CSIR 1147; 5 *A. adeninivorans* CSIR 1148; 6 *A. adeninivorans* CSIR 1149; 7 *A. adeninivorans* CBS 8244T; 8 *A. adeninivorans* Ls3. Molecular weight markers (Mbp) are shown on the *left*

1148, CSIR 1149, and CBS 8244T (lanes 5–7), are very similar in their chromosomal patterns.

All tested *Arxula* strains (lanes 2–8) differ in their chromosome patterns in comparison to *S. cerevisiae* strain S288C (lane 1). Only the latter yeast strain contains the additional 2-μm plasmid DNA band. Naturally existing DNA and RNA plasmids could not be identified in any of the described *A. adeninivorans* strains.

6.3
DNA Fingerprinting

Different combinations of DNA probes and restriction enzymes were tested to obtain fingerprints of DNA from several *A. adeninivorans* strains.

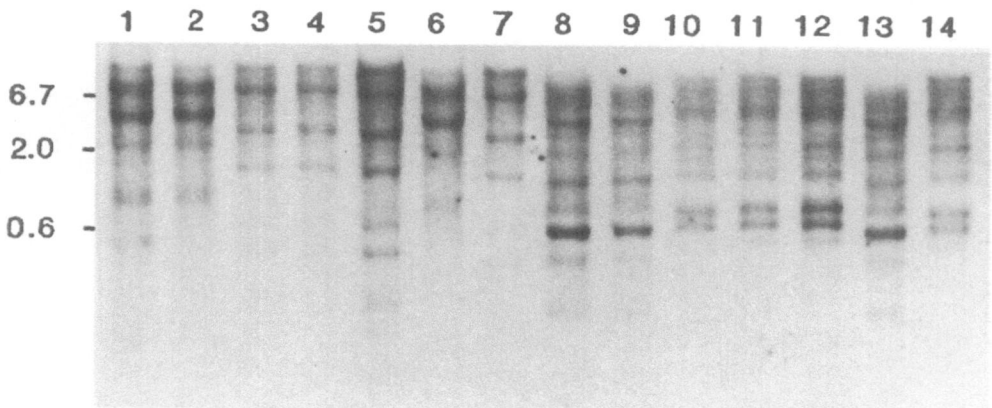

Fig. 3. Hybridization of *Eco*RI (lanes *1–7*) and BamHI (lanes *8–14*) digested chromosomal DNA of several *Arxula* strains with $(GT)_{10}$. **1** and **8** *A. adeninivorans* CSIR 1136; **2** and **9** *A. adeninivorans* CSIR 1138; **3** and **10** *A. adeninivorans* CSIR 1147; **4** and **11** *A. adeninivorans* CSIR 1148; **5** and **12** *A. adeninivorans* CSIR 1149; **6** and **13** *A. adeninivorans* CBS 8244T; **7** and **14** *A. adeninivorans* Ls3. Molecular weight markers (kb) are shown on the *left*

Oligonucleotides $(CT)_{10}$, $(GACA)_4$, and $(CT)_{10}$ were found to be suitable probes for fingerprints. *Eco*RI and *Bam*HI-digested DNA gives rise to a more appropriate restriction pattern than HinfI digests.

Figure 3 shows results of the analyses after EcoRI and BamHI digestions and of $(GT)_{10}$ hybridization. On the basis of these results, *Arxula* strains can be classified into two distinguishable groups. The first group contains the strains CSIR 1136, CSIR 1138, and CBS 8244T (lanes 1, 2, 6, and lanes 8, 9, 14). Within this group, only very small differences were visible in DNA pattern after digestions with *Eco*RI and *Bam*HI. The second group comprises the *Arxula* strains CSIR 1147, CSIR 1148, CSIR 1149, and Ls3 (lanes 3–5, 6, and lanes 10–12 and 14). Also these four strains show very similar hybridization patterns.

Analyses of *Eco*RI and *Bam*HI digested chromosomal DNA fragments, which were hybridized with the oligonucleotides $(GACA)_4$ and $(CT)_{10}$ (data not shown), give the same classification which is in accordance to that obtained by $(GT)_{10}$ hybridization experiments.

7
Mitochondrial DNA

Mitochondrial DNA from the strain *A. adeninivorans* Ls3 was characterized (Pich and Kunze 1992). It is circular and may amount to some 8% of the total cellular DNA. The average copy number is about 50 per cell, and the GC content was determined as 30.3 mol%. It seems to be a particularly GC-rich genome compared to the mtDNA of *S. cerevisiae* (20.2 mol%).

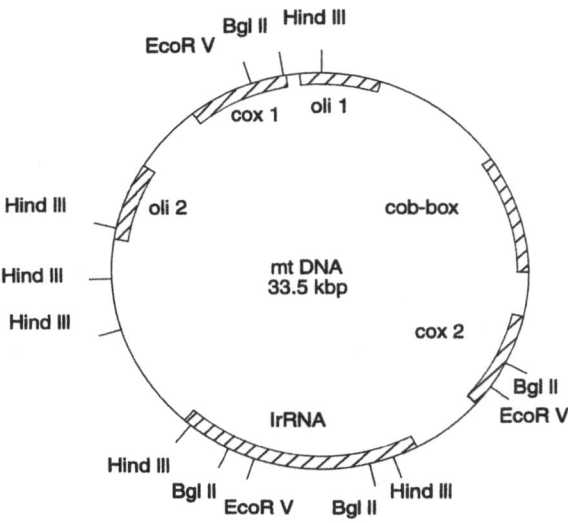

Fig. 4. Circular restriction and genetic map of mtDNA from *A. adeninivorans*. Abbreviations: *cob* apocytochrome b; *oli1* ATPase subunit 9; *oli2* ATPase subunit 6; *cox1* cytochrome c oxidase subunit 1; *cox2* cytochrome c oxidase subunit 2; *1rRNA* 21S rRNA

The size, as estimated by restriction analysis performed with nine endonucleases, is 35.5 kbp. The map comprises all restriction sites of *Hin*d III, *Bgl*II, and *Eco*RV (Fig. 4). Compared to other yeasts, *Arxula* harbors a relatively small mitochondrial genome, which is similar to those of *Candida rhagii* (30.0 kbp, Kovac et al. 1984) and *Kluyveromyces lactis* (37.0 kbp, Wilson et al. 1989).

Using mt gene probes from *S. cerevisiae*, six structural genes (*cob*, *cox1*, *cox2*, *oli1*, *oli2*, and 21S rRNA) were located on the mitochondrial genome of *A. adeninivorans*. The genetic map shows differences in the genome organization to other yeast mtDNAs.

The restriction patterns of mtDNAs of other strains differ to the data obtained with the strain Ls3. This points to polymorphic rearrangments inside the *Arxula* strains.

8
Transformation System

8.1
Genetic Markers and Isolation of Genes

Mutants of *A. adeninivorans* can be isolated at relatively high frequency after NG and UV mutagenesis. About 600 auxotrophic mutants (see below), such as *lys2* mutants, are available for the strain *A. adeninivorans* Ls3.

Kunze et al. (1990) describe the isolation of genes from *A. adeninivorans* by means of DNA hybridization and the complementation test of the corresponding *E. coli* mutations.

Thus, for example, the restricted *Arxula* DNA hybridize with the *S. cerevisiae* DNA fragments containing the *LYS2*, *AROM*, *CUP1*, *STE6*, *TEF1* genes and the regions of *CEN5*, *CEN6* (Table 5). With the help of radiolabeled DNA probes from the DNA of the *S. cerevisiae LYS2* gene the *Arxula LYS2* gene could be isolated from a Charon 4A gene library inserted 10–15 kbp DNA fragments of the chromosomal DNA from *Arxula* (Fig. 5).

Another screening test was the complementation of the respective *E. coli* mutants. Thus the *ILV1* gene was isolated by means of complementation of the *E. coli ilvA* mutation. Therefore, a plasmid gene library of the *S. cerevisiae* Yep13 plasmid was used which contains 5–8 kbp DNA fragments from *Arxula* (Fig. 6).

With the help of pulsed field gelelectrophoresis the *LYS2* gene and the *ILV1* gene were identified on the largest chromosome (chromosome 1) and on the chomosome with the lowest molecular weight (chromosome 4), respectively (Fig. 7A,B).

Table 5. Hybridization of different genes, centromere regions and promoter-terminator sequences from various organisms

Organism	Gene	Hybridization
S. cerevisiae	*LYS2*	+
	LEU2	−
	ARG4	−
	URA3	−
	TRP1	−
	HIS3	−
	AROM	+
	CEN6	+
	CEN5	+
	CUP1	+
	SUP4	−
	STE6	+
	TEF1	+
	TY96	−
	GAL1 promoter	−
	GAP promoter	−
	ADH1 promoter/terminator	−
	PHO5 promoter/terminator	−
	CYC1 promoter/terminator	−
N. crassa	*Bm1⁻*	−
	HIS4	+
	TUB2	+
Bac. macerans	*bg1M*	−
E. coli	*aroA*	+

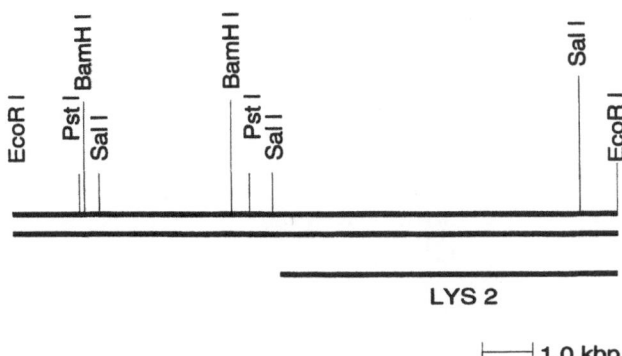

Fig. 5. Restriction map of the *LYS2* gene from *A. adeninivorans*

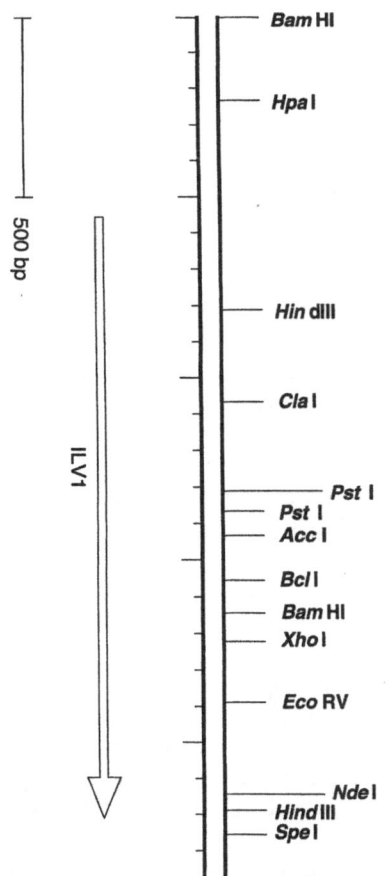

Fig. 6. Restriction map of the *ILV1* gene from *A. adeninivorans*

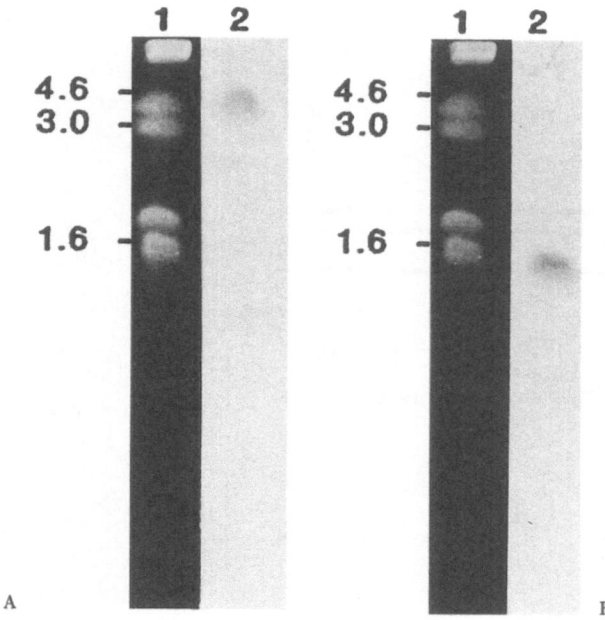

A   B

Fig. 7A,B. Pulsed field gel electrophoresis of *A. adeninivorans* Ls3 (A/*1* and B/*1*) and hybridization with the *LYS2* gene (A/*2*) and the *ILV1* gene (B/*2*) from *Arxula*

8.2
Transformation Markers

The *S. cerevisiae* marker *LYS2* has been conveniently used in *A. adeninivorans* Ls3, because the corresponding mutation in *A. adeninivorans* can be complemented by these genes under their own promoter (Kunze et al. 1990). Additionally, the corresponding *Arxula* gene was isolated from a genomic library and used as a transformation marker. The product of this gene – the 2-aminoadipate reductase – complements the *lys2* mutation from *Arxula* strain Ls3 as well as from *S. cerevisiae*.

8.3
Various Methods of Transformation

As a rule, our laboratory transforms plasmids containing the *LYS2* gene from *S. cerevisiae* or from *A. adeninivorans* as selective markers into the corresponding yeast strains with a *lys2* mutation, for example the strain G704 (*lys2 cys*; Fig. 8). DNA fragments or plasmids could not be transformed into other *Arxula* wild-type strains or mutants from these strains (CBS 8244T, CSIR 1136, CSIR 1138, CSIR 1147, CSIR 1148, CSIR 1149).

Three methods (lithium salt treatment, frozen competent cells, or electroporation) can be used to transform DNA into *A. adeninivorans* Ls3. These

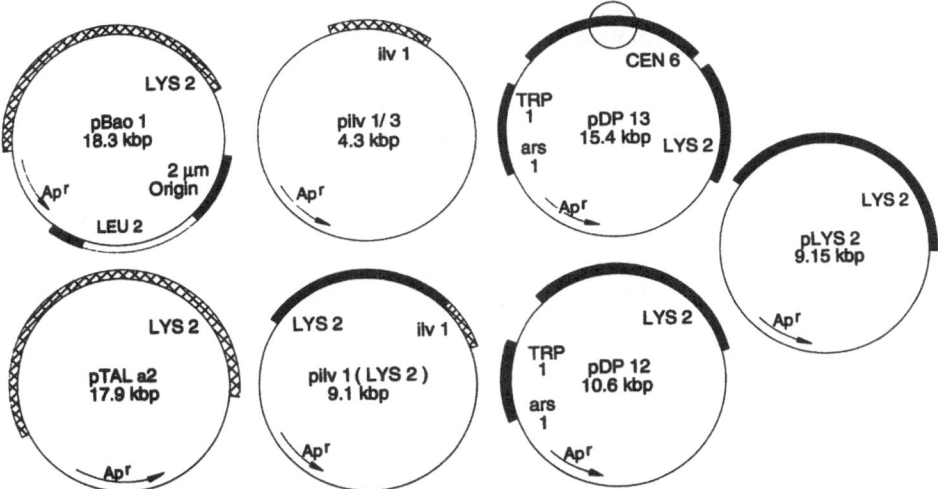

Fig. 8. Plasmids used to transform *A. adeninivorans*. The plasmids contain sequences from following sources: pBluescript *bla* and *ori* region *thin line*; *S. cerevisiae* sequences *thick line and open box*; *A. adeninivorans* sequences *dark box*

methods were also described for the transformation of *S. cerevisiae* and can also be applied with minor modifications for *Arxula*. The efficiencies of these methods are lower in comparison to the transformation frequencies obtained for *S. cerevisiae* (Table 6).

The highest frequencies of transformants were obtained using electroporation. Linearized DNA, as well as circularized DNA, can be integrated into the *Arxula* chromosome 1 containing the *LYS2* gene. Differences in the integration mechanisms using either of these DNA types cannot be determined. Most of the yeast transformants contain additional modified plasmid DNA of the originally transformed DNA plasmid or fragment.

For routine transformations, it is possible to use the lithium salt treatment (Ito et al. 1983). The obtained transformation frequencies are low and depend on the strains and the plasmids used. Lower concentrations of PEG 4000 (50%) advanced the efficiency of this method in some cases.

Recently, we have optimized the procedure described by Dohmen et al. (1991) for using frozen competent yeast cells. By using this method, plasmid DNA and DNA fragments were transformed and integrated into the genome. The transformation frequencies are higher compared with those of the lithium salt method, but cannot achieve the frequencies obtained by electroporation. However, a relatively high number of revertants were selected by this transformation procedure.

The transformed DNA plasmids or DNA fragments were integrated into chromosome 1 as well as replicated autonomously as plasmid with an altered restriction pattern. All transformants are mitotically stable during cultivation under selective and nonselective conditions. The presence of the plasmid DNA was de-

Table 6. Transformation frequencies of *A. adeninivorans*
G704 (lys2 cys) with different plasmids

Plasmid	Transformation frequency	Number of transformants per μg DNA
pDP12 (1)	1.8×10^{-6}	18
pDP13 (1)	7.0×10^{-7}	7
pLYS2 (1)	2.0×10^{-7}	2
pTALa2 (1)	9.0×10^{-7}	9
pBaol (1)	2.0×10^{-7}	2
pilv1 (*LYS2*) (1)	3.0×10^{-7}	3
pilv1/3 (1)	0	0
pLYS2 (2)		497
pDP12 (2)		875
pTALa2 (2)		547

The yeast cells were transformed by lithium salt treatment (1),
and by electroporation (2)

tected by hybridization of the transformant DNA preparations to fragments of the
original plasmid DNA. After transformation of these DNAs in *E. coli*, we were able
to select ampicillin-resistant colonies. The isolated plasmids of these colonies
show differences in the restriction pattern to the original plasmids. On the basis of
these results, we suggest that transformation proceeds integratively.

8.4
Transformation Protocols

8.4.1
Transformation by Lithium Salt Treatment

This procedure is a modification of the method developed by Ito et al. (1983).
All steps described are carried out at room temperature unless indicated
otherwise.

1. *Arxula* cells are grown in 100 ml YE-medium at 37 °C for about 18 h to a density
   of 10^8 cells/ml.

2. 10^8 cells/ml are harvested by centrifugation of 8000 g at 20 °C, and washed in TE
   buffer (10 mM Tris/pH 8.0; 1 mM EDTA).

3. After repeated centrifugation, the cells are suspended in 1.0 ml TE buffer, 0.1 M
   LiCl, and incubated for 1 h at 30 °C.

4. 100–μl portions of this suspension containing 10^7 competent cells are pipetted
   into sterile Eppendorf tubes containing 1–3 μg plasmid DNA. Each tube is
   mixed carefully and incubated for 30 min at 30 °C.

5. After addition of 0.1 ml 70% PEG 4000 to the suspension, the cells are incubated for 1 h at 30 °C and for 5 min at 37 °C.

6. The yeast cells are washed twice in water and plated on SD medium with the required amino acids.

7. The transformed cells can be selected 2–5 days later.

The *A. adeninivorans* Ls3 strain can be transformed with a frequency of 2–20 transformants/μg DNA by this method. Although the obtained transformation frequencies are relatively small, the method is easy, applicable for all known *Arxula* strains containing the equivalent mutation and reproducible. The application of other lithium salts reduces the transformation frequency.

8.4.2
Transformation of Frozen Competent Cells

This procedure developed by Dohmen et al. (1991) has been modified.

1. Cells are cultivated in YE medium at 37 °C for about 18 h to a density of 10^8 cells/ml.

2. After harvesting and washing with water, the yeast cells are suspended in a half-volume of Bicin buffer (1 M sorbitol; 10 mM Bicin-NaOH/pH 8.35; 3% PEG 1000; 5% DMSO), centrifuged, and suspended in 1/50 volume of the same buffer; 200-μl portions of this suspension containing competent cells are transferred into sterile Eppendorf tubes and frozen at −80 °C. At this temperature the cells can be stored for about 6 months.

3. One to three-μg plasmid DNA (about 5 μl) are applied on top of the frozen competent cells, which are thawed with vigorous agitation at 37 °C for 5 min.

4. After addition of 1 ml 40% PEG 1000–0.2 M Bicin-NaOH/pH 8.35 and careful mixing, the cells are incubated at 37 °C for 1 h.

5. The cells are centrifuged at 3000 g at 20 °C for 5 min, washed with 1.5 ml 0.15 M NaCl–10 mM Bicin-NaOH/pH 8.35 and suspended in a small volume (about 100 μl) of the same buffer.

6. All the yeast cells are plated on SD medium containing the required amino acids and cultivated at 37 °C for 3–5 days.

By this method, about 10 to 100 transformants per μg DNA are obtained. This frequency depends on the DNA plasmid and the yeast strain used.

8.4.3
Transformation by Electroporation

The Gene Pulser system from BioRad/FRG is used for the transformation in *A. adeninivorans* Ls3. The following protocol is based on the procedure described by Delmore (1989).

1. Cells are cultivated in YE medium from 1/50 volume of fresh overnight culture at 37 °C. The cells are harvested at the late exponential phase of the overnight culture (about 10^8 cells/ml) by centrifugation of 8000 g at 4 °C.

2. Cells are washed twice with distilled water, and subsequently twice with 1 M sorbitol at 4 °C.

3. Cells are suspended in 1 M sorbitol (about 10^9 cells/ml) at 4 °C. The following steps are carried out at 4 °C.

4. 40 µl of suspension are placed in a sterile Eppendorf tube, mixed with 2–5 µl DNA (about 500 ng), and incubated for 5 min.

5. The suspension is placed in sterile Gene Pulser Cuvettes (0.2-cm electrode gap).

6. The following parameters of the poration apparature are used:

    25 µF, 200 Ω, 1.8 kV for about 4.5 ms.

7. Subsequently 50–100 µl of the porated suspension is plated on appropriate selective media, containing 1 M sorbitol, and incubated at 37 °C for 3–5 days.

For most *Arxula* strains, a transformation frequency of about 5×10^2 per µg DNA can be reached. Particular strains, however, can not be transformed by electroporation.

9
Expression of Heterologous Genes in *A. adeninivorans*

To express heterologous genes in *Arxula*, the transcription promoter of the *ILV1* gene was identified, isolated, and characterized (Fig. 6). After construction of an

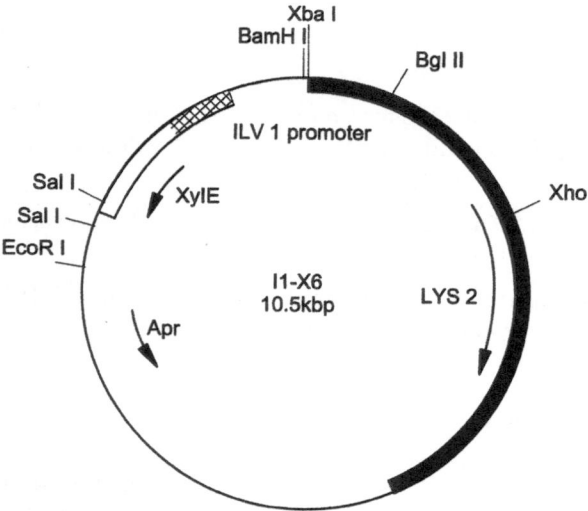

Fig. 9. Restriction map of the plasmid I1-X6

Table 7. Specific activity (pkat per mg protein) of catechol 2,3-dioxygenase in mutants and transformants of *S. cerevisiae* and *A. adeninivorans*

Strain	Specific activity	
	Intracellular	Extracellular
S. cerevisiae SHY2	0	0
S. cerevisiae SHY2/Bsc-T-X-Io8	0	0
S. cerevisiae SHY2/Bsc-T-X-Io12	0	0
S. cerevisiae SHY2/Bsc-T-X-Iu6	0.15	0.46
S. cerevisiae SHY2/Bsc-T-X-Iu12	0	0
A. adeninivorans G704	0	0
A. adeninivorans G704/pLYS2	0	0
A. adeninivorans G704/I1-X6	0.38	0

expression cassette containing the *ILV1* promoter and *XylE* gene from *Pseudomonas putida* as a heterologous gene encoding catechol-2,3-dioxygenase, this cassette was cloned into a plasmid containing the *LYS2* gene as a selectable marker. This plasmid I1-X6 (Fig. 9) was transformed into *A. adeninivorans* G704 (*lys2 cys*) and *S. cerevisiae* SHY2-lys2 (*ura3 trp1 his3 leu2 lys2*). Characterization of obtained *Arxula* transformants shows that the plasmid I1-X6 is integrated within the chromosomal DNA. In all transformants, an intracellular catechol-2,3-dioxygenase activity expressed by the *XylE* gene from *Pseudomonas putida* has been detected. In contrast to these results, in the *S. cerevisiae* transformants the catechol-2,3-dioxygenase activity has been localized intracellularly as well as extracellularly.

Acknowledgments. We are grateful to Prof. R. Bode, Prof. F. Böttcher, and Dr. I.A. Samsonova for helpful discussions, and to Dr. J. Phillipps and Prof. K. Müntz for critical reading of the manuscript. We also thank Dr. J.P. van der Walt and Prof. Kapultsevich for the generous gift of the yeast strains. The experimental work described in this chapter was supported by the Bundesminister für Forschung und Technologie, Bonn, FRG, Grant No. 0319691A, Grant No. 0310135A and the Minister für Wissenschaft und Forschung – Sachsen/Anhalt, Magdeburg, FRG, Grant No. 236A0731.

References

Böttcher F, Samsonova IA (1983) Zur fungiziden und genetischen Wirkung von Benomyl. In: Lyr H, Polter C (eds) Systemische Fungizide und antifungale Verbindungen. Akademie-Verlag, Berlin, pp 255-258

Böttcher F, Klinner U, Köhler M, Samsonova IA, Kapultsevich J, Bliznik X (1988) Verfahren zur Futterhefeproduktion in zuckerhaltigen Medien. DD-Patentschrift 278 354 A1

Büttner R, Bode R (1992) Purification and characterization of β-xylosidase activities from the yeast *Arxula adeninivorans*. J Basic Microbiol 32: 159–166

Büttner R, Bode R, Birnbaum D (1987) Purification and characterization of the extracellular glucoamylase from the yeast *Trichosporon adeninovorans*. J Basic Microbiol 27: 299–308

Büttner R, Bode R, Scheidt A, Birnbaum D (1988) Isolation and some properties of two extracellular β-glucosidases from *Trichosporon adeninovorans*. Acta Biotechnol 8: 517–525

Büttner R, Bode R, Birnbaum D (1989a) Purification and characterization of the extracellular glucoamylase from the yeast *Trichosporon adeninovorans*. J Basic Microbiol 27: 299–308

Büttner R, Scheit A, Bode R, Birnbaum D (1989b) Isolation and characterization of mutants of *Trichosporon adeninovorans* resistant to 2-deoxy-D-glucose. J Basic Microbiol 29: 67–72

Büttner R, Bode R, Samsonova IA, Birnbaum D (1990a) Mapping of the glucoamylase gene of *Trichosporon adeninovorans* by mitotic haploidization using hybrids from protoplast fusions. J Basic Microbiol 30: 227–231

Büttner R, Schubert U, Bode R, Birnbaum D (1990b) Purification and partial characterization of external and internal invertases from two strains of *Trichosporon adeninovorans*. Acta Biotechnol 10: 361–370

Büttner R, Bode R, Birnbaum D (1990c) Extracellular enzymes from the yeast *Trichosporon adeninovorans*. Wiss Z Ernst-Moritz-Arndt-Univ Greifswald, Math Naturwiss Reihe 39: 20–22

Büttner R, Bode R, Birnbaum D (1991a) Comparative study of external and internal β-glucosidases and glucoamylase of *Arxula adeninivorans*. J Basic Microbiol 31: 423–428

Büttner R, Bode R, Birnbaum D (1991b) Characterization of extracellular acid phosphatases from the yeast *Arxula adeninivorans*. Zentralbl Mikrobiol 146: 399–406

Büttner R, Bode R, Birnbaum D (1992a) Purification and characterization of trehalase from the yeast *Arxula adeninivorans*. Zentralbl Mikrobiol 147: 291–296

Büttner R, Bode R, Birnbaum D (1992b) Alcoholic fermentation of starch by *Arxula adeninivorans*. Zentralbl Mikrobiol 147: 237–242

Delmore E (1989) Transformation of *S. cerevisiae* by electroporation. Appl Environ Microbiol 55: 2242–2246

Dohmen RJ, Strasser AM, Höner CB, Hollenberg CP (1991) An efficient transformation procedure enabling long-term storage of competent cells of various yeast genera. Yeast 7: 691–692

Gienow U, Kunze G, Schauer F, Bode R, Hofemeister J (1990) The yeast genus *Trichosporon* spec. Ls3; molecular characterization of genomic complexity. Zentralbl Mikrobiol 145: 3–12

Gutz H (1966) Induction of mitotic segregation with p-fluorophenylalanine in *Schizosaccharomyces pombe*. J Bacteriol 92: 1567–1568

Ito U, Fukada Y, Murata K, Kimura A (1983) Transformation of intact yeast cells treated with alkali cations. J Bacteriol 153: 163–168

Kirsop BE (1988) Culture and preservation. In: Kirsop BE, Kurtzman CP (eds) Yeasts Cambridge University Press, Cambridge, pp 74–98

Klinner U, Böttcher F (1984) Hybridization of yeast by protoplast fusion: ploidy level of hybrids resulting from fusions in haploid strains of *Pichia guilliermondii*. Z Allg Mikrobiol 8: 533–537

Kovac L, Lazowska J, Slonimski PP (1984) A yeast with linear molecules of mitochondrial DNA. Mol Gen Genet 197: 420–424

Kunze G, Kunze I (1993a) Characterization of *Arxula adeninivorans* strains originated from different sources. Antonie Leeuwenhoek J Microbiol 65: 29–34

Kunze I, Kunze G (1993b) Comparative study of *Arxula adeninivorans* strains concerning morphological characteristics and activities of secretory invertase. J Eur Microbiol 212: 24–28

Kunze G, Pich U, Lietz K, Barner A, Büttner R, Bode R, Conrad U, Samsonova I, Schmidt H (1990) Wirts-Vektor-System und Verfahren zu seiner Herstellung. DD-Patentschrift 298 821 A5

Kurischko C (1986) Spontaneous haploidization in early zygote progeny and its use for mapping in the yeast *Yarrowia lipolytica*. Curr Genet 10: 709–711

Ledeboer AM, Edens L, Maat J, Visser C, Bos JW, Verrips CT, Janowicz Z, Eckart M, Roggenkamp R, Hollenberg CP (1985) Molecular cloning and characterization for methanol oxidase in *Hansenula polymorpha*. Nucleic Acids Res 13: 3063–3082

Middelhoven WJ, Hoogkamer-Te Niet MC, Kreger van Rij NJW (1984) *Trichosporon adeninovorans* sp. nov., a yeast species utilizing adenine, xanthine, uric acids, putrescine and primary n-alkylamines as the sole source of carbon, nitrogen and energy. Antonie Leeuwenhoek J Microbiol 50: 369–378

Middelhoven WJ, de Jong IM, de Winter (1991) *Arxula adeninivorans*, a yeast assimilating many nitrogenous and aromatic compounds. Antonie Leeuwenhoek J Microbiol 59: 129–137

Middelhoven WJ, Coenen A, Kraakman B, Gelpke MDS (1992) Degradation of some phenols and hydroxybenzoates by the imperfect ascomycetous yeast *Candida parapsilosis* and *Arxula adeninivorans*: evidence for an operative gentisate pathway. Antonie Leeuwenhoek J Microbiol 62: 181–187

Pich U, Kunze G (1992) Genome organization of mitochondrial DNA from the non-saccharomycete yeast *Arxula adeninivorans* LS3. Curr Genet 22: 505–506

Samsonova IA, Böttcher F (1978) Acriflavine-induced mitotic segregation in *Rhodosporidium toruloides* and *Candida quilliermondii*. Abstr XIVth Int Congr Genetics, Moscow, p 89

Samsonova IA, Böttcher F (1980) Mitotische Segregation durch antimikrotubuläre Agenzien bei *Pichia guilliermondii* und *Rhodosporidium toruloides*. Wiss Z Ernst-Moritz-Arndt-Univ Greifswald, Math Naturwiss Reihe 29: 71–72

Samsonova IA, Böttcher F, Werner C, Bode R (1989) Auxotrophic mutants of the yeast *Trichosporon adeninovorans*. J Basic Microbiol 29: 675–683

Tanaka A, Ohishi N, Fukui S (1967) Studies on the formation of vitamins and their function in hydrocarbon fermentation. Production of vitamin B6 by *Candida albicans* in hydrocarbon medium. J Ferment Technol 45: 617–623

Tolstorukov II, Efremov BD, Bliznik KM (1983) Genetic mapping of the yeast *Pichia pinus*. I. Identification of linkage groups using induced mitotic haploidization. Genetika 19: 897–902

Van der Walt JP, Smith MT, Yamada Y (1990) *Arxula* gen. nov. (Candidaceae), a new anamorphic, arthroconidial yeast genus. Antonie Leeuwenhoek J Microbiol 57: 59–61

Wartmann T, Kunze I, Bui MD, Manteuffel R, Kunze G (1995) Comparative biochemical, genetical and immunological studies of glucoamylase producing Arxula adeninivovans yeast strains. Microbiol Res 150: 113–120

Wilson C, Raguini A, Fukuhara H (1989) Analysis of the regions coding for transfer RNAs in *Kluyveromyces lactis* mitochondrial DNA. Nucleic Acids Res 17: 4485–4491

Wood JS (1982) Genetic effects of methyl benzimidazole-2-yl-carbamate on *Saccharomyces cerevisiae*. Mol Cell Biol 2: 1064–1079

CHAPTER 12

Candida maltosa

Stephan Mauersberger[1], Moriya Ohkuma[2,3], Wolf-Hagen Schunck[1], and Masamichi Takagi[2]

More than 30 years have passed since the first description of the yeast species *Candida* (*C.*) *maltosa* by Komagata et al. (1964a,b). Since then, *C. maltosa* has become of considerable academic and commercial interest. Now, together with some related *Candida* species and *Yarrowia* (*Y.*) *lipolytica* (cf. Barth and Gaillardin, Chap. 10, this Vol.), it is best known for its ability to grow on a wide variety of substrates including n-alkanes, fatty acids, or carbohydrates, and is therefore intensively investigated in its physiology, biochemistry, and molecular genetics. More recent investigations also use these yeast species for the study of fundamental cellular processes such as protein targeting, organelle biosynthesis, and drug resistance.

In this chapter we intend to summarize the present knowledge on the alkane-assimilating yeast *C. maltosa* according to the following main topics:

- History of research on *C. maltosa* and its taxonomic position
- Physiology, biochemistry, and cytology with special emphasis to the analysis of metabolic pathways like hydrocarbon and phenol degradation
- Cytochrome P450 systems and their regulation
- Methods of genetic analysis and available mutants
- Cloning and characterization of genes
- Development of host-vector systems and expression of foreign genes in *C. maltosa*
- Potential practical application of *C. maltosa*

Other reviews on these topics have been published on *Candida* yeasts, containing information mainly on *C. albicans*, and including also some data on *C. tropicalis*, *C. utilis*, and *C. maltosa* (Magee et al. 1988; Tanaka and Fukui 1989; Kirsch et al. 1990; Rachubinski 1990; Reiser et al. 1990; Takagi 1992; Müller et al. 1991a; Su and Meyer 1991; Romanos et al. 1992; Sudbery 1994).

Each section is supplied with detailed descriptions of selected methods currently used in the authors' laboratories to investigate the biology of *C. maltosa*.

[1] Laboratory of Membrane Proteins, Cell Biology Department, Max-Delbrück-Center for Molecular Medicine, Robert-Rössle-Str. 10, 13125 Berlin-Buch, Germany
[2] Cellular Genetics Laboratory, Department of Biotechnology, The University of Tokyo, Bunkyo-ku, Tokyo 113, Japan
[3] Present address: Laboratory of Microbiology, Institute of Physical and Chemical Research (RIKEN), Wako-shi, Saitama 351-01, Japan

Generally, most methods developed and described in the literature for other yeasts, especially for *Saccharomyces* (*S.*) *cerevisiae*, can also be applied to *C. maltosa* and other yeasts (cf. other chapters, this Vol.). Therefore the reader is referred also to the recent literature on methods for biochemical, genetic, and molecular biological investigations in yeast, mostly developed for *S. cerevisiae* (Campbell and Duffus 1988; Sambrook et al. 1989; Goeddel 1990; Rose et al. 1990; Guthrie and Fink 1991; Ausubel et al. 1994).

Abbreviations

aa	Amino acid(s)
AASA	α-aminoadipate-δ-semialdehyde
ACO	Acyl-CoA oxidase
alk	Alkane nonutilizing mutants
APA	Auxotrophy-prototrophy-auxotrophy alteration (APA change)
ARS	Autonomously replicating sequence
bp	base pair(s)
C.	*Candida*
CEN	Centromere region
CDOG	Catechol-1,2-dioxygenase or pyrocatechase
CHEF	Electrophoresis with a contour-clamped homogeneous electric field
Cm1	P450Cm1 = P450 52A3, as for P450Cm2 = P450 52A4
CoA	Coenzyme A
DAPA	Sodium-p-dimethylaminobenzene-diazosulfonate
DCA	Dicarboxylic (fatty) acid
EMS	Ethylmethanesulfonic acid
ER	Endoplasmic reticulum
FA	Fatty acids
FADH	Fatty alcohol dehydrogenase
FALDH	Fatty aldehyde dehydrogenase
FAOD	Fatty alcohol oxidase
5FOA	5-fluoro-orotic acid
HAP	6-N-hydroxy-aminopurine
HMG-CoA	3-hydroxy-3-methyglutaryl-CoA
ICR-170	2-methyl-6-chloro-[3-(ethyl-2-chloroethyl)amino]acridine hydrochlorid
IPM	Isopropylmalate
kb	Kilo base pairs
MNNG	N-methyl-N′-nitro-N-nitrosoguanidine
mtDNA	Mitochondrial DNA
NA	Nitrous acid
NCCR	NADPH-P450 (cytochrome c) reductase
NQO	4N-nitroquinoline oxide
OFAGE	Orthogonal field alternation gel electrophoresis
ORF	Open reading frame
P450	Cytochrome P450
P450$_{14DM}$	14α-demethylase P450 (P450 51A1)
PAH	Polycyclic aromatic hydrocarbons
Paprin	Paraffin-based single-cell protein from *C. maltosa*
PEG	Polyethylene glycol
PHM	Pseudohyphal morphology (mutant)
PPM	Proteophosphomannan complexes

SCP	Single-cell protein
SDS-PAGE	Polyacrylamide gelelectrophoresis with sodium dodecyl sulfate
UV	Ultraviolet light
UFA	Uneven fatty acids
Y.	*Yarrowia*
TRA	Transformation ability region

For further abbreviations used see Figs. 3 and 4, for abbreviations used for culture collections and strain designations see Table 1, for proteins and enzymes see Tables 4 and 5, for designation of genes see Table 11.

1
History and Taxonomy of *Candida maltosa*

1.1
History of Research on Candida maltosa *and on Its Taxonomic Position*

The development of research on most nonconventional yeasts was closely connected with increasing interest in the biotechnological application of these yeasts. This is particularly obvious for research on alkane-assimilating yeasts like *C. tropicalis*, *C. maltosa*, *Pichia guilliermondii*, and *Yarrowia lipolytica* (cf. other chapters, this Vol.), especially stimulated due to the fact that in the early 1960s hydrocarbon biochemistry became a theme of industrial research, when mainly petroleum companies became interested in the application of microorganisms for the production of single-cell protein (SCP) as foodstuff (animal fodder) on the basis of n-alkanes as well as for biochemical synthesis of amino acids, fatty acids, sterols, vitamins, and other substances of commercial interest (for reviews see Shennan and Levi 1974; Levi et al. 1979; Fukui and Tanaka 1981a,b; Rehm and Reiff 1981; Einsele 1983; Boulton and Ratledge 1984; Bühler and Schindler 1984; Shennan 1984; Tanaka and Fukui 1989, cf. Sect. 6).

Thus, the asporogenous yeast species *Candida maltosa* Komagata, Nakase et Katsuya was first isolated and described in 1964 in a study on hydrocarbon-assimilating yeasts in connection with its ability to grow on n-alkanes as the only source of carbon and energy, made by the Research Laboratories of the Ajinomoto Co. at Kawasaki, Japan (Komagata et al. 1964a,b). The name was originally chosen because this yeast strongly fermented maltose, although in later studies this was not confirmed (Meyer et al. 1975). This first isolate of *C. maltosa* was obtained from adhesives of neutralizing tanks in the monosodium glutamate manufacturing process. The first data on *C. maltosa* were published together with the description of a nearly related new species named *Candida cloacae* Komagata, Nakase et Katsuya, isolated from mud as a kerosene-enrichment culture (Komagata et al. 1964a,b; Table 1). These two strains, *C. maltosa* (later becoming the type strain CBS5611) and *C. cloacae* (type strain CBS5612), together with *C. tropicalis*, belong to the best hydrocarbon-assimilating yeasts (Bos and de Bruyn 1973) and have similar fermentation and assimilation patterns (van Uden and Buckley 1970). The yeasts *C. maltosa* and *C. cloacae* were first considered to be separate species by

Table 1. Wild-type strains of *Candida maltosa* available from culture collections and used in different laboratories and institutions. Other species names formerly used in the literature for different *C. maltosa* strains are given with the respective references. Additionally, in some cases publications are given since the strains were renamed into *C. maltosa*

Candida maltosa strains in Culture Collections or Institutions	Formerly designated strains of (isolated from, by)	Reference
Candida maltosa CBS 5611 (Type strain Cm) = ATCC 28140 = AJ 4718 = IAM 12247 = JCM 1504 = CCY 29-88-1 = VKM 1506 = FERM P733 = SBUG 542 = DBVPG 6143 = IFO 1977	*Candida maltosa* Y-2-3 (industrial equipment) *Candida sake* *Candida maltosa*	Komagata et al. (1964a,b, 1979) Shiio and Uchio (1971) Celma Calamita et al. (1971) Van Uden and Buckley (1970) Meyer et al. (1975) Kaneko et al. (1977)
Candida maltosa CBS 5612 (Type strain Cc) = ATCC 20184 = AJ 4719 = IAM 12248 = JCM 1505 = CCY 29-88-2 = VKM 1441 = DBVPG 6136 = IFO 1978	*Candida cloacae* YO-140 (original isolate from mud by Komagata) *Candida sake* *Candida maltosa*	Komagata et al. (1964a,b, 1979) Scheda (1966) Van Uden and Buckley (1970) Nakase and Komagata (1971) Meyer et al. (1975) Kaneko et al. (1977)
Candida maltosa 310 AJ 5463 = FERM P736, AJ 5341 = FERM P410 Ajinomoto Co., Ltd, Central Research Laboratory, Kawasaki, Japan	*Candida cloacae* 310 (from oil-soaked soil by T. Nakase) *Candida maltosa*	Shiio and Uchio (1971) Uchio and Shiio (1972a,b,c) Uchio and Shiio (1974) Uchio (1978) Fukui and Tanaka (1981b) Casey et al. (1990) Kaneko et al. (1977)
Candida maltosa CBS 6465 (Type strain Cst) = ATCC 28241 = IFO 1975 = AJ 4476 = JCM 1511	*Candida subtropicalis* (from air, by Nakase strain YO-146) *Candida maltosa*	Nakase et al. (1972) Fukuzawa et al. (1975) Hirata and Ishitani (1978) Meyer et al. (1975) Smith et al. (1976) Kaneko et al. (1977)
Candida maltosa CBS 6680 = AJ 4480 = IFO 1976 = ATCC 28284	*Candida subtropicalis* (from soil, by Nakase)	Nakase et al. (1972) Meyer et al. (1975) Su and Meyer (1991) Umemura et al. (1992)
Candida maltosa GSU 42 Department Biology Georgia State University Atlanta, USA	*Candida subtropicalis* (from oil slick)	Meyer et al. (1975) Smith et al. (1976) Crow et al. (1980) Cerniglia and Crow (1981)
Candida maltosa CBS 6658 = ATCC 20275 = FERM 705 = DBVPG 60211 Kanegafuchi Chemical Ind., Co., Ltd., Hyogo, Japan	*Candida* sp. 36 *Candida novellus* 36 (from soil) *Candida maltosa* 36	Watanabe et al. (1973a,b,c, 1975) Watanabe (1974) Shennan and Levi (1974) Meyer et al. (1975) Kaneko et al. (1977) Komagata (1979)
Candida maltosa EH15 D Institute of Technical Chemistry, Leipzig, later the Institute Biotechnology,	*Candida guilliermondii* EH 15 *Candida* sp. EH 15	Schunck et al. (1978a,b) Müller et al. (1979) Riege et al. (1980) Bauch et al. (1978) Müller et al. (1980) Bode et al. (1983)

Table 1. (*Contd.*)

Candida maltosa strains in Culture Collections or Institutions	Formerly designated strains of (isolated from, by)	Reference
Leipzig, Germany and Institute of Molecular Biology, Berlin-Buch, Germany and Petrochemical Enterprise (PCK) Schwedt, Germany	*Lodderomyces elongisporus* EH 15 D (or IMET H128)	Bode and Casper (1983) Kunze et al. (1984a) Bley et al. (1980) Heinritz et al. (1981, 1983a,b) Müller and Voigt (1981, 1984) Schneider and Triems (1981) Stichel et al. (1981, 1982) Sattler and Wünsche (1981, 1983) Wünsche et al. (1981) Mauersberger et al. (1981, 1984) Riege et al. (1981) Glombitza (1982) Gradova et al. (1983) Schneider et al. (1983) Honeck et al. (1982, 1985) Müller et al. (1982, 1983a) Brendler et al. (1983) Schunck et al. (1983a,b) Voigt et al. (1984a,b, 1985) Blasig et al. (1984) Bayer et al. (1985) Heinritz et al. (1985)
	Candida maltosa	Bode and Birnbaum (1984) Golubev et al. (1986)
Candida maltosa H or H62 University Jena Biology section and Institute of Biotechnology Leipzig, Germany	*Candida guilliermondii* H *Candida* spec. H (or H62)	Metz and Reuter (1977) Rademacher and Reuter (1978) Heinritz and Bley (1979) Röber and Reuter (1979) Popov et al. (1980) Triebel et al. (1980) Grimmecke and Reuter (1980) Brückner and Tröger (1981a,b) Kölblin and Birkenbeil (1981) Grimmecke and Reuter (1981a,b,c,d) Grimmecke et al. (1981) Kölblin and Tröger (1982) Fischer and Reuter (1982) Fischer et al. (1982) Nüske et al. (1982) Röber and Reuter (1982, 1984c)
	Candida maltosa H	Röber and Reuter (1984a)
Candida maltosa L4 = SBUG 700 = G217 University Greifswald, Section of Biology	*Candida maltosa* L4 (closely related with *Cm* EH15, from IBT Leipzig)	Kunze et al. (1984b) Kunze et al. (1985a) Hofmann and Schauer (1988) Schauer (1988)
Candida maltosa SBUG 420 = CBS 7327 University Greifswald, Section of Biology	*Candida maltosa* T420 (from waste water of industrial plant)	Kunze et al. (1984b) Hofmann and Schauer (1988)
Candida maltosa VSB NP4 and other strains: 1, NP2, NP4/1113, 143, 243, N540, 542B, 569,	*Candida guilliermondii* VSB NP4 *Candida maltosa* NP4	Maximova et al. (1972) Kasanzev et al. (1975) Ilchenko et al. (1980) Mauersberger et al. (1980) Mauersberger and Matyashova (1980) Krauzova et al. (1986)

Table 1. (*Contd.*)

Candida maltosa strains in Culture Collections or Institutions	Formerly designated strains of (isolated from, by)	Reference
Candida maltosa VSB 779	*Candida salmonicola*	Golubev et al. (1986) Ioffe et al. (1990)
Candida maltosa VSB 777 and 778	*Candida sake*	Golubev et al. (1986)
Candida maltosa VSB 899, and other strains NP4-111, 542 900, 906, 907, 908	*Candida parapsilosis*	Golubev et al. (1986) Sasnauskas et al. (1991, 1992) Jomantiene et al. (1991)
Candida maltosa ATCC 38040, 38041 ATCC 38042	*Saccharomyces cerevisiae* mutants, *C. maltosa*	Soom (1973) Rabinovich et al. (1974) Bassel et al. (1978)
Candida maltosa CYY 29-88-3, 29-88-4	*Candida maltosa* M6.D7, M18.D7 (from denitrifying sludge of laboratory unit)	Slavikova and Grabinska-Loniewska (1990)
Candida maltosa JCM 1504 Tanabe Seiyaku Co., Ltd. Osaka, Japan		Umemura et al. (1992)

Candida maltosa strains from Culture Collections mentioned in the literature. In parentheses the former designation of the strains in the collections are shown as Cc – *C.*
cloacae, Cst – *C. subtropicalis*, Cn – *C. novellus*, Cs – *C. sake*, Csa – *C. salmonicola*, Cp – *C. parapsilosis*, Le – *L. elongisporus*; nothing shown when initially described as *C. maltosa*.

AJ	4476, 4480, 4718, 4719 (Cc) from 1983, 5341 (Cc), 5463 (Cc)
ATCC	20184 (Cc = CBS5612), 20275 (Cn = CBS6658), 28140 (= CBS5611), 28241 (= CBS6465), 28284 (= CBS6680), 38040, 38041, 38042; compare Meyer et al. (1975) and Yoshida and Hashimoto (1986a,b)
CBS	5611, 5612 (Cc), 6465 (Cst), 6658, 6680 (Cst), 7327
CCY	29-88-1, 29-88-2 (Cc or Cs), 29-88-3, 29-88-4
DBVPG	6136, 6143, 60211
FERM	P-410 (Cc = AJ 5341), 705 (Cn), P-733, P-736 (Cc = AJ 5463)
IFO	1975, 1976, 1978
JCM	1504 (Umemura et al. 1992), 1505 (Cc)
SBUG	388, 420, 434, 443, 700 (L4), 701 (Hofmann and Schauer 1988)
VKM	1441 (Cc), 1506 (Cm) (1976), Y2359 (1986)
VSB	1, NP2, NP4, NP4/1113, 143, 249, N-540, 542B, 569, 640, 774 (formerly, all Cg); EH15 (Cg-Cp-Le); 777 and 778 (formerly Cs), 779 (Csa); 875 (Cm), NP4-111, 542, 899, 900, 906, 907, 908 (formerly all Cp), (Data taken from Golubev et al. 1986).

Abbreviations used for Culture Collections, Institutes or Laboratories.

AJ	Central Research Laboratory, Ajinomoto Co., Inc., Kawasaki, Japan
ATCC	American Type Culture Collection, Rockville, Maryland, USA
CBS˙	Centralbureau voor Schimmelcultures, Delft and Baarn, The Netherlands
CYY	Czechoslovak Collection of Yeasts, Institute of Chemistry, Bratislava, Slovak Republic
GSU	Georgia State University, USA
FERM	Fermentation Research Institute, Chiba, Japan
IAM	Institute of Applied Microbiology, University of Tokyo, Japan
IBT	Institute of Biotechnology (formerly Institute of Technical Chemistry), Leipzig, Germany
IBPM	Institute of Biochemistry and Physiology of Microorganisms, Pushchino, Russia
IFO	Institute of Fermentation, Osaka, Japan
JCM	Japan Collection of Microorganisms, Wako, Japan
SBUG	Section of Biology, University Greifswald, Germany
VSB	Allunion Institute of Protein Biosynthesis, Moscow, Russia
VKM	Allunion Culture Collection, Moscow, Russia.

Nakase and Komagata (1971) and Bos and de Bruyn (1973) because they grow at higher maximal temperature (40–42 °C) than *C. sake* (32–34 °C), in contrast to the view of van Uden and Buckley (1970), who discussed these yeast strains as members of the heterogeneous species *C. sake* (Saito et Ota) van Uden et Buckley.

Independently, two other *C. cloacae* strains were isolated in the Ajinomoto Central Research Laboratory (Shiio and Uchio 1971; Table 1). With mutants of the strain *C. cloacae* 310, a technology for production of dicarboxylic acids from n-alkanes was developed (see Sect. 6).

At the same time, the closely related alkane-assimilating strains of *Candida subtropicalis* Nagase, Fukazawa et Tsuichiya (Nakase et al. 1972; Hirata and Ishitani 1978) and *Candida* sp 36, named *Candida novellus* nom. nud. (Watanabe et al. 1973) were described (Table 1), and processes for industrial application of the latter strain for SCP production were developed (Watanabe et al. 1973a,b; Watanabe 1974a,b,c; Kaneko et al. 1977).

The taxonomic status of *C. maltosa*, *C. cloacae*, *C. novellus*, and *C. subtropicalis* was thus not very clear for many years. Their almost identical biochemical characteristics, the very narrow range of distribution (35.6–36.6%) of the GC content, high degree of DNA relatedness (at least 86%, Table 2) concluded from DNA hybridization experiments (Meyer et al. 1975), and the same agglutination factors and PMR spectra of cell wall polysaccharides (Fukazawa et al. 1975) suggested a very close relationship among these species. Fukazawa et al. (1975) and Meyer et al. (1975) thus concluded from their studies that these strains are very similar to each other and should form an independent species different from *C. sake*. Later, Kaneko et al. (1977) confirmed these results and added the strain *C. novellus* 36 to this group of yeasts. All these authors proposed that *C. cloacae*, *C. subtropicalis*, and *C. novellus* are synonyms for *Candida maltosa*, and chose the latter to name this species, which is now mostly accepted in the literature (cf. Table 1). These cultures fermented maltose weakly, failed to utilize soluble starch, and had maximal growth temperatures of 40–42 °C (Table 3). These results were later supported by Montrocher (1980), who performed serological analysis of these strains using the immunoprecipitation method in addition to morphological and physiological observations.

Today, *C. maltosa* is recognized as a species separate from *C. sake* van Uden et Buckley, *C. tropicalis* (Castellani) Berkhout, *C. albicans*, and *C. parapsilosis* (Ashford) Langeron et Talice on the basis of their physiological, morphological, and immunological properties and their DNA relatedness (Meyer et al. 1975; Kaneko et al. 1977; Kunze et al. 1984a,b; Table 2), although their near relationship is obvious according to most physiological parameters (cf. Sect. 1.2).

Additionally, the literature on *C. maltosa* was in the past even more confused by the fact that, especially until the middle of the 1980s, results on different strains later identified as *C. maltosa* were published, in addition to the taxa mentioned above (Table 1A), as belonging to other *Candida* species, such as *C. guilliermondii*, *C. parapsilosis*, *C. salmonicola*, or *Candida* sp., as well as to *L. elongisporus* or even as mutants of *Saccharomyces cerevisiae*. These investigations, after their relation to *C. maltosa* had been clearly shown later, have been therefore included in Table

Table 2. DNA relatedness of *Candida maltosa* with other yeasts

Source of unlabeled DNA	Homology as relative binding (%) of DNA from (source of labeled DNA)				
	C. maltosa [a]ATCC28241 ATCC20275 [b]L4, T420 CBS5611	*S. cerevisiae* D10	*L. elongisporus* CBS2605	*C. sake* ATCC14478	*C. tropicalis* ATCC750 CCY29-7-23
Yeast DNA					
C. maltosa	[c]86–100	–	–	11–17	11–32
C. maltosa	[d]87–100	–	53	–	15–24
C. sake	[c]7–20	–	–	99–100	12–21
C. tropicalis	[c]21–26	–	–	14	88–100
C. tropicalis	[d]14–33	–	–	–	100
P. guilliermondii	[d]34	18	22	–	–
L. elongisporus	[d]54	–	100	–	–
S. cerevisiae	[d]60	100	50	–	–
Nonspecific controls					
E. coli	[c]3–7	–	–	6	5
Calf thymus	[d]1–3	–	–	–	2

Abbreviations: – not determined; *C. maltosa* – *Candida maltosa*; *C. tropicalis* – *Candida tropicalis*; *P. guilliermondii* – *Pichia guilliermondii*; *L. elongisporus* – *Lodderomyces elongisporus*; *S. cerevisiae* – *Saccharomyces cerevisiae*. Homology between chromosomal DNA (unlabeled) from different sources and the ^{32}P-labeled (Kunze) or ^{3}H-labeled (Meyer) DNA from different yeast species: determined from DNA reassociation experiments using a C_0t-value of 100 and a reassociation temperature of 60 °C or 65 °C in 0.12 M sodium phosphate buffer pH 6.8 (Kunze) or 2 × SSC (Meyer), respectively.
[a,c] Data taken from Meyer et al. (1975): *C. maltosa* strains used – ATCC 28241, 28284, 20184, 20275, 28140, GSU42; *C. sake* – ATCC 14478, 28141, 28722, 28138, 28136, 28723, 28137, 28139, 28721; *C. tropicalis* – ATCC 750, 188807, 28142.
[b,d] Data taken from Kunze et al. (1984b): *C. maltosa* strains used – L4 (SBUG700), T420 (SBUG420 = CBS7327), CBS5611 = ATCC28140; *C. tropicalis* CCY29-7-23; *L. elongisporus* CBS2605; *Pichia guilliermondii* fp1-61; *S. cerevisiae* D10.

1B/C and the following sections, as they substantially extended the knowledge of *C. maltosa*.

Detailed reexaminations of the yeast species used for the production of single-cell protein (SCP) on n-alkanes in the former Soviet Union and GDR (cf. Sect. 6), made on the basis of progress and changes in yeast taxonomy, revealed that most of these strains belong to the species *C. maltosa* and *C. tropicalis* (Kunze et al. 1984a,b; Golubev et al. 1986; Table 1B/C). Golubev et al. (1986) found that more than 90% of the strains used for the large-scale production of alkane-based fodder yeast (Paprin, more than 1 million t per year in 1985) in the former USSR were reclassified as belonging to *C. maltosa*. Therefore it might be supposed that most data published earlier in these countries on alkane-utilizing *C. guilliermondii*, *C. parapsilosis*, *C. salmonicola*, and *C. sake* strains (without reference to strain numbers or culture collections, for example Gradova and Kovalsky 1978; Davidov et al.

1980, 1981a,b, 1982) should also be regarded as data on *C. maltosa*. For example, the strain *C. maltosa* VSB NP4 was formerly designated *C. guilliermondii* VSB NP4 (Maximova et al. 1972; Kasanzev et al. 1975; Table 1). Later, it was reclassified as belonging to *C. maltosa* (Golubev et al. 1986; Krauzova et al. 1986; Sharyshev and Krauzova 1988; Table 1B/C).

The yeast *C. maltosa* EH15 was the production strain for the SCP process developed in the former GDR in cooperation with the USSR, and applied at the Petrochemical Enterprise (PCK) Schwedt, GDR (Bauch et al. 1978; Brendler et al. 1983; cf. Sect. 6). The strain, selected at the Institute of Technical Chemistry (ITC) in Leipzig, was first designated *C. guilliermondii* (Michaleva et al. 1973a,b) or *C. parapsilosis* according to Gradova et al. (1983), and later *Candida* sp. EH15 (Table 1). In 1981, this strain was renamed as belonging to the species *Lodderomyces elongisporus* (Recca et Mrak) Van der Walt, due to ascospores allegedly detected (Wünsche et al. 1981), which was not confirmed by the results of Kunze et al. (1984a) and Golubev et al. (1986). Therefore this strain was later again reclassified as a strain of *C. maltosa* (Golubev et al. 1986).

This could be stated also for *C. maltosa* H, another strain intensively investigated at the Schiller University Jena, Germany (Table 1), formerly designated *C. guilliermondii* H or *Candida* sp. H62. In 1982, it was reclassified by D. Yarrow, CBS Delft, as *C. maltosa* (Röber et al. 1984a; Röber and Reuter 1985).

Kunze et al. (1984b) concluded from DNA reassociation experiments and from morphological-physiological criteria of classic yeast taxonomy (Barnett et al. 1979) made in comparison with the type strain *C. maltosa* CBS 5611 that the two alkane-assimilating yeast strains L4 (SBUG 700) and T420 (SBUG 420), investigated by some groups at the University of Greifswald, belong to the species *C. maltosa* and not to *L. elongisporus*, *P. guilliermondii*, or *C. tropicalis*. It should be noted that the *C. maltosa* strains L4 and EH15 are probably closely related, if not identical, although this was not shown in the literature.

In the early 1970s, the group of Inge-Vechtomov at the Leningrad State University reported the isolation of mutants of *Saccharomyces cerevisiae* that were capable of utilizing hydrocarbons as growth substrate (Soom 1973; Rabinovich et al. 1974). Later, Bassel et al. (1978) were unable to confirm these results and identified these strains as members of *C. maltosa* (Table 1).

There is no evidence for the formation of ascospores in *C. maltosa* and therefore this yeast has been considered to be a strictly imperfect one up till now. Nevertheless Viljoen et al. (1988) first proposed a possible relationship between *C. maltosa* (CBS 5611) as the anamorphic form and *Pichia etchelsii* (CBS 2011T) as the teleomorphic form of this yeast, as also as for *C. cacaoi* and *P. farinosa*, respectively, on the basis of the long-chain fatty acid compositions compared with other phenotypic criteria (i.e., assimilation of carbon sources, coenzyme Q type, G + C content, and proton magnetic resonance spectra). Although their G + C content was discussed by these authors as being similar, the mentioned values of 36.6 mol% for *C. maltosa* and of 38.5 mol% for *P. etchelsii* are not very convincing in comparison with other yeasts with known anamorph/teleomorph relationships like *C. shehatae* and *P. stipitis* or *P. guilliermondii* and *C. guilliermondii*.

1.2
Phylogenetic Relation of Candida maltosa *to Other Yeasts*

The progress made in molecular cloning and sequencing of comparable genes from different yeast species is the basis for a better understanding of their natural relationships. This is especially important for the genus *Candida*, a heterologous assemblage unified mainly by the absence of any sexual state (teleomorph). The phenotypic character-based classification has created a confusing taxonomy and a great heterogeneity within this genus.

The evolutionary relationships among several *Candida* species and their relatives have been described recently on the basis of small-subunit ribosomal RNA (srRNA) sequences (Barns et al. 1991; Hendriks et al. 1991). The srRNA sequences are now commonly used for phylogenetic analysis of yeast and fungi (for reviews see Kurtzman 1992, 1994; Barns et al. 1991; Bruns et al. 1992).

Recently, Ohkuma et al. (1993b) investigated the evolutionary position of the nonpathogenic n-alkane-assimilating yeast *C. maltosa* after cloning and sequencing the small-subunit (18S) ribosomal RNA gene (*SSRR*, cf. Table 11) The phylogenetic tree (Fig. 1) constructed by the neighbor-joining method shows that

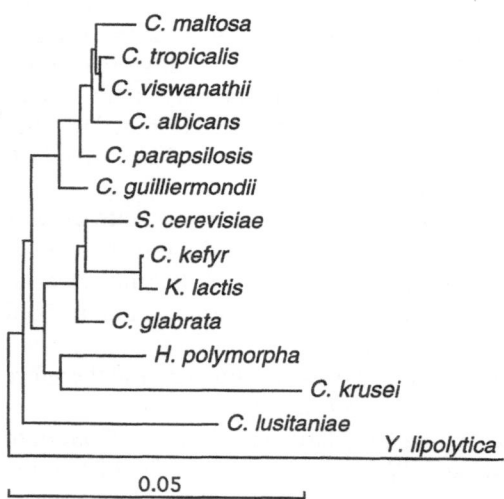

Fig. 1. Evolutionary tree based on the srRNA nucleotide sequences of different yeasts. Evolutionary position of *Candida maltosa* among different yeast species. The sequences for the srRNA genes were taken from the DDBJ, EMBL, and GenBank Nucleotide Sequence Databases for *C. maltosa* (Ohkuma et al. 1993b), *C. tropicalis*, *C. viswanathii*, *C. albicans*, *C. parapsilosis*, *C. guilliermondii*, *C. kefyr*, *C. glabrata*, *C. krusei*, *C. lusitaniae*, *Hansenula polymorpha*, *Kluyveromyces lactis*, *Saccharomyces cerevisiae*, and *Yarrowia lipolytica*. Bar under the tree indicates the distance corresponding to five changes per 100 nucleotides.

C. maltosa forms a subgroup lineage with *C. tropicalis*, *C. viswanathii*, *C. albicans*, *C. parapsilosis*, and *C. guilliermondii* in this genus, a cluster of relatively closely related organisms. *Saccharomyces cerevisiae*, *C. kefyr*, *Kluyveromyces lactis*, *C. glabrata*, *Hansenula polymorpha*, and *C. krusei* form another subgroup, and being more distantly related to *C. maltosa*, *C. lusitaniae*, and *Yarrowia (C.) lipolytica*, are excluded from these two subgroups. Consistent with the results of 5S RNA comparison, *Yarrowia (C.) lipolytica* was among the more divergent species.

The results are in accordance with the conclusions made from taxonomic studies using other parameters. Thus, *C. maltosa* was originally identified as a species closely related to *C. parapsilosis* and *C. tropicalis*, showing the same morphological characteristics and sharing many biochemical properties.

Candida maltosa was distinguished from *C. sake* (11–17%) and from *C. tropicalis* (27–32%) by insignificant DNA reassociation (Meyer et al. 1975; Table 2). In addition, *C. maltosa* was distinguished from *C. sake* by its growth on octane, its higher maximal growth temperature (40–42° to 30–34 °C) and lower GC content of its DNA (35.6–36.8% to 37.8–41%) and from *C. tropicalis* by its failure to utilize soluble starch for growth and its resistance to cycloheximide (Meyer et al. 1975). It should be mentioned that, according to DNA reassociation data (Meyer et al. 1975) and serological characteristics (Montrocher 1980), *C. maltosa* is obviously more closely related to *C. tropicalis* (DNA relatedness of 27–32% or 21–26%) than to *C. sake* (11–17% or 7–20%; Table 2). *Candida maltosa* is also distinguished from *C. tropicalis*, other *Candida* species, and *Lodderomyces elongisporus* by its high molecular weight of mitochondrial DNA (52 kb) and the low GC content (21–22%) of the mtDNA in comparison with data (25–30 kb and 27–37 GC%) obtained for these yeasts (Kunze et al. 1984a, 1986b; Su and Meyer 1991; cf. Sect. 4.3.3).

The close relationship between *C. maltosa* and members of its subgroup is also supported by the comparison of other macromolecular sequence data. A comparison of the sequences of L41 ribosomal protein (gene *RIM-C* or *L41Q*; cf. Table 11) shows an amino acid identity of 99.1% between *C. maltosa* and *C. tropicalis*. Those between *C. maltosa* and *C. guilliermondii* or *S. cerevisiae* are 85.4 and 84.9%, respectively.

A comparison of the amino acid sequences of the orotidine 5′-phosphate decarboxylase (encoded by the *URA3* gene) also shows good correlation with the srRNA tree. Amino acid identities between *C. maltosa* and *C. albicans*, *K. lactis*, or *S. cerevisiae* are 84, 77, and 72%, respectively (Ohkuma et al. 1993a).

The distribution of the hydrocarbon assimilation ability (Barnett et al. 1983; Schauer and Schauer 1986) is well correlated with the phylogenetic tree. Members of the *C. maltosa*-containing subfamily can assimilate hydrocarbons, except for *C. viswanathii*. *Candida lusitaniae* and *Y. lipolytica* can also assimilate hydrocarbons, but members of the subfamily belonging to *S. cerevisiae* cannot. Probably, a deleterious event for the assimilation of hydrocarbon occurred in the subgroup belonging to *S. cerevisiae* when the two subgroups were separated from each other. The different yeast subgroups probably also differ in the codon usage (cf. Sect. 4.4.3).

1.3
Handling of Candida maltosa *Strains*

Available Wild-Type Strains and Mutant Collections

The wild-type strains of *C. maltosa* used in different laboratories are shown in Table 1. Characterized mutants obtained from these strains are summarized in Table 6.

Conservation of Strains

Handling of *C. maltosa* strains is comparable with that for other yeasts. Storing and conservation is possible as

- slant cultures at 4 °C with YPD or minimal medium agar,

- deep freeze cultures with 20–30% glycerol in YPD or minimal medium in liquid nitrogen or at −70 to −80 °C for a longer time, or as

- freeze dried cells as well.

2
Physiology and Biochemistry of *Candida maltosa*

Among alkane-utilizing yeasts, *Candida maltosa*, with *Candida tropicalis* and *Yarrowia lipolytica*, stands out as the organism which has been most thoroughly studied concerning its physiology, biochemistry, and molecular biology.

2.1
Occurrence in Nature

Strains of *C. maltosa* have been isolated from air, soil, and water, particularly if enriched with hydrocarbons, but normally not from humans (Meyer et al. 1975; Golubev et al. 1986; Table 1).

Seeding of hydrocarbonoclastic yeasts into estuarine environment has also been tested (Cook et al. 1973). Out of six or eight yeasts used in seeding oil in estuarine or in freshwater environment, two hydrocarbon-assimilating yeasts persisted for the longest period of time. *Yarrowia (C.) lipolytica* disappeared 3 to 5 months after seeding, while *C. (subtropicalis) maltosa* persisted for over 1 year in freshwater and 7 months in estuarine environments. No adverse ecological side effects were observed as a result of seeding the nonpathogenic *Candida* yeasts. Berner et al. (1975) also investigated the survival of *C. maltosa* (*C. subtropicalis*) in freshwater and estuarine habitats. They found that this species maintained itself in oil-enriched sites, but did not spread to adjacent hydrocarbon-free areas (Ahearn et al. 1976).

More recently, Gradova et al. (1991) investigated the interactions of the industrial produced alkane-grown *C. maltosa* as a technogenic factor of pollution with the surrounding ecosystems (air, plants, soil) of the SCP plants in field tests and in model test systems under laboratory conditions. These data showed a 100- to 1000-fold decrease in *C. maltosa* content in soil during the first 100 days of the experiments, then reaching the normal yeast content in soil of 10^3–10^4 cells per g. A high self-purifying ability of soil leading to homeostasis stability in the natural ecosystems was observed in these experiments. The authors concluded that the pollution of the environment with *C. maltosa* produced in SCP plant equipped with special cleaning systems of the air exhaust may be regarded as a nonhazardous burden for the ecological balance.

2.2
The Problem of Pathogenicity and Toxicity for Candida maltosa *and Its SCP*

Due to the development of the SCP technologies based on alkanes or alcohols in the 1970s (Tannenbaum and Wang 1975; Shennan 1984; cf. Sect. 6), different *Candida, Pichia,* and *Yarrowia* species had to be evaluated for safety. The properties of food yeasts most often examined were nutritive value and toxicity (Smith and Palmer 1976). Since strains of unknown disease potential were candidates for SCP, they had also to be studied for pathogenicity.

Since some strains of the genus *Candida* are known to be opportunistic pathogens (Odds 1987), it has been pointed out that the possible pathogenicity of yeast used for the production of single-cell protein (SCP) should be carefully checked beforehand. SCP as a feedstuff is properly heat-treated and not living, so that the product has a priori no pathogenicity, regardless of the pathogenic ability of the living cells.

The possible pollution of the environment around a factory producing SCP through leakage of the living cells from the factory or through infection of employees of the factory has been discussed. Therefore, animal models and procedures for evaluating the potential for human pathogenicity had to be developed to compare the pathogenic potential of medically important yeasts, e.g., *C. albicans* and *Cryptococcus neoformans*, with industrially important yeasts. Although industrial use of fungi has not been documented as a health hazard, certain species of commercially used genera such as *Candida* and *Saccharomyces, Aspergillus,* and *Fusarium* have been implicated in human disease. Since members of such complex genera as *Candida* may vary in their pathogenic potential, new yeast strains being considered for industrial use should be tested individually, as has repeatedly been pointed out (Holzschu et al. 1979; Yoshida and Hashimoto 1986a,b).

Except for the opportunistic pathogen *C. tropicalis,* the particularly industrially important yeasts *C. maltosa, C. guilliermondii, C. utilis,* and *Y. (C.) lipolytica,* used mostly for the production of fodder yeast (SCP) on different substrates (hydrocarbons or carbohydrates) or of metabolites of commercial interest (citrate), are regarded as nonpathogenic organisms, which has been shown repeatedly by several authors. In contrast to *C. albicans* and *C. tropicalis,* these yeast strains showed no pathogenic evolution in human cell cultures or in tested laboratory animals (mice, guinea pigs, and rabbits) for every manner of injection (Gargani et al. 1978,

1979; Gargani 1979; Ahearn et al. 1979; Holzschu et al. 1979; Faggi and Mennini 1985; Yoshida and Hashimoto 1986a,b). Therefore most of these strains have been approved in several countries for use in industrial processes.

In Japan, a long-term research project was performed in the Ministry of Agriculture, Forestry, and Fisheries to establish a procedure for assessing the safety of single-cell protein, including a procedure for assessing the potential pathogenicity of yeasts (expressed as scores for the remaining living cells in organs of tested mice and histopathological examinations), not previously known to be an opportunistic pathogen, and expected to be useful for production of single-cell protein (Yoshida and Hashimoto 1986a,b). Interestingly, among the 42 yeast strains tested in these studies, most of which were *Candida* species, the *C. maltosa* strain IAM12248, repeatedly revealed as belonging to *C. maltosa*, was found to show a high possibility of being pathogenic. This strain belonged to the Group I, which contained almost all the strains known to be opportunistic pathogens (*C. albicans* and *C. tropicalis*). Therefore *C. maltosa* IAM12248 was strongly suspected to exhibit potential pathogenicity, although its pathogenicity was not known and not expected, and no information was available indicating that this strain causes disease in certain host animals. In contrast, none of the strains known to exhibit pathogenicity is included in the Group II containing all other additionally tested strains of *C. maltosa* (nine including IAM12247 and eight ATCC strains; cf. Table 1) together with *C. utilis*, *C. rugosa*, *C. lipolytica*, and *C. guilliermondii*, showing a low possibility of being potential pathogens. It should be mentioned that not all strains of *C. maltosa* exhibit pathogenicity, only strain IAM12248 as one of ten strains tested. Pathogenicity is therefore specific to a strain, not to a species (Yoshida and Hashimoto 1986b). These authors discriminate therefore between strains which could be used and handled as usual in the traditional fermentation industry and others which should not be used, or should at least be handled very carefully, or should be carefully reexamined before use (Yoshida and Hashimoto 1986a).

It should also be mentioned that in some recent publications the occurrence of *C. maltosa* in immunocompromised patients or animals, in addition to other pathogenic yeasts like *C. albicans*, *C. tropicalis*, *C. glabrata*, *C. parapsilosis*, and *Trichosporon cutaneum*, was demonstrated (Kitamura et al. 1990; Silva et al. 1990). Whether the action of *C. maltosa* itself would be pathogenic for these organisms remains to be clarified.

Hypersensitivity (Hypersensitive Reactions). The large-scale production of SCP in the industrial biotechnology plants can, however give rise to some serious problems for employees and for the population in the environment, as was reported in the past few years by authors from the former Soviet Union, where the production of alkane-based SCP with *C. maltosa* and other strains is being continued (for ref. see Gradova et al. 1991; cf. Sects. 1 and 6). Thus, Litovskaya (1988) reported hypersensitive reactions among workers in plants producing SCP using *C. guilliermondii*, *C. sake*, or *C. maltosa*. Immediate and cellular hypersensitive reactions, with disturbances in T-dependent and humoral immunity, were frequent. Workers who had been in contact with living cells, rather than the protein concentrate (Paprin), tended to show allergic symptoms of the skin, or bronchial asthma. These phenomena were studied by several authors using laboratory animals (Ermolaev et al. 1987, 1991; Ilyina et al. 1988; Pekelis et al. 1989; Spivak et al. 1988, 1989; Gukasyan et al. 1990; Litovskaya and Mokeeva 1990; Dalin et al. 1991; Kravtsov et al. 1991; Artamonova and Svitina 1991; Artamonova et al. 1993).

Furthermore, Pogorelskaia et al. (1991) reported that the appearance of fungal contaminations of the upper respiratory ways due to the production strains in workers and residents in the near environment of the SCP plants is very dependent on the use of stable, satisfactory air exhaust purification equipment (Sect. 2.1). Therefore, it seems to be that health or ecological hazard arising from SCP production plants is more a problem of technology, and thus of economics, than of the pathogenic potential, or toxicological or ecological action of the yeast strains used, especially for *C. maltosa*.

Toxicity of SCP. Because the yeasts grown on paraffins represent a suitable food for farm animals, in view of the unusual nature of this material, it appeared worthwhile to determine the effects of long-term feeding to animals with a diet containing varying amounts of single-cell protein (SCP) produced by growing *C. maltosa* or other yeasts on alkanes. These protein concentrates produced in several countries were named Liquipron (Italy), Paprin (ex-USSR), Fermosin (ex-GDR) when obtained from *C. maltosa*, or Toprina (Italy, France) when from *Y. lipolytica*.

Experimental data on the effect of long-term feeding of laboratory animals (rats, mice, dogs) and domestic animals (chicken, swine) with diets containing SCP obtained from alkane-grown yeast (*C. maltosa*, *Y. lipolytica*, or other yeasts) demonstrated no significant adverse effects of the protein concentrate diets on the physiological and biochemical parameters of the animals, on pathological patterns, on genetic material, or on reproductive function (Bizzi et al. 1980; Perri et al. 1981; Muramatsu et al. 1982; Nunziata et al. 1982; Lusky et al. 1988, 1989; Heinz et al. 1989; Ioffe et al. 1990). Especially the protein concentrates Liquipron, Toprina, and Fermosin obtained from alkane-grown yeast biomass were investigated to ascertain the biological significance and possible toxicological implications of their high content of uneven fatty acids (UFA). It was confirmed that the extent to which UFA (especially C17-fatty acids) accumulate in fat and tissues of the animals tested (rats, swine) fed with SCP products only partially reflects their UFA contents. The presence of UFA in rat or pig tissues does not appear to alter intermediate metabolism. In nature, odd-numbered fatty acids are ubiquitous. Uptake and storage of the UFA suggest that the organism reacts to an increased offer of C17-fatty acids with an accelerated decomposition of these fatty acids and/or with inhibition of their endogenous synthesis (Bizzi et al. 1980; Lusky et al. 1988, 1989). A study on a denucleinized protein product obtained from Paprin suggested that the product can be used as a fodder protein and even as a protein source for the food industry (Ioffe et al. 1990).

2.3
Physiology of Growth

Due to its application in biotechnology (cf. Sects. 1, 2, and 6) *C. maltosa* has been intensively studied concerning its growth physiology and biomass composition. The following physiological parameters were included in these investigations: temperature and pH for growth, growth rate (μ) and yield (biological carbon source conversion into microbial biomass), thermodynamic considerations including the auxiliary substrate concept, and biomass composition.

2.3.1
Temperature and pH

Candida maltosa belongs to the mesophilic yeasts having an optimal temperature for growth of about 32–34 °C, but it still grows at 37°C, and weak growth is observed up to 40–41 °C (Nakase et al. 1972; Meyer et al. 1975; Golubev et al. 1986). With increasing temperature, the yield and protein content decrease, but the degree of decrease is strain-dependent (Gradova et al. 1983). In contrast to *C. tropicalis* and other thermotolerant yeast (*C. albicans, C. rugosa, C. blankii*), it cannot grow at 43 °C and higher temperatures (Gradova et al. 1983, 1990; Golubev et al. 1986). In comparison with the psychrotolerant yeast *Y. (C.) lipolytica*, it is less active at low temperatures of 5–10 °C (Crow et al. 1980). The yeast *C. maltosa* can grow in a broad range of pH from 2.5 to 6.5. The best pH range for growth is between 3.5 and 5.5 (Gradova and Kovalsky 1978; Gradova et al. 1983; Zentgraf 1991b). For cultivation on alkanes in fermenters, the optimal pH is about 4.2–4.6.

For optimal growth in minimal media *C. maltosa* needs vitamins, biotin in particular. The biotin dependence is higher for growth on hydrocarbons than on carbohydrates; it is especially important for the utilization of short-chain n-alkanes (Gradova et al. 1983; Schauer 1988; cf. Sect. 2.4.2).

2.3.2
Growth Rate and Yield Coefficients

In mineral salt media *C. maltosa* NP4 is able to grow under optimal conditions without any limitation in batch cultures in the laboratory with comparable maximal specific growth rates (μ) of 0.55 to 0.6 h^{-1} on glucose (as the only source of carbon and energy) and 0.45 to 0.5 h^{-1} on hexadecane, respectively (Mauersberger et al. 1980; Mauersberger 1985). A similar relation, but with lower μ values, was reported by Brückner and Tröger (1981b) using glucose and an alkane mixture (Parex) for the *C. maltosa* strain H. Under conditions of decreased oxygen content in the medium, μ (or corresponding growth parameters) declined strongly, but the degree of oxygen limitation is very dependent on the substrate used (Mauersberger 1985; Schunck et al. 1987a).

In continuous culture, *C. maltosa* strains can be cultivated on carbohydrates or alkanes at dilution rates D from 0.05 to 0.5 h^{-1} (Gradova et al. 1983; Büttner et al. 1985). The optimal dilution rates for obtaining high yield in biomass production are between 0.2 and 0.3 h^{-1}. These values were reached only in high-performance industrial reactors that guarantee high mass and heat transfer, whereas in other industrial fermenters D is normally below these values. The best mass transfer conditions are obtained with recycle reactors, where the culture broth is continuously repumped in a range of one to several minutes, giving rise to areas of high and low turbulence. The produced oxygen-, pH-, and nutrient gradients affect microbial growth and cause an increase in specific material consumption coefficients.

With the aim of better understanding and improving the economy of the biotechnological process of biomass production, the dependence of physiological parameters (yield coefficients Y or substrate conversion rates) on milieu conditions (concentrations of carbon source and oxygen) and cell states has been investigated with *C. maltosa* (formerly designated *C. guilliermondii* or *L. elongisporus*; Table 1) cultures aerobically growing on carbohydrates, n-alkanes, or crude oil fractions. These studies for the analysis of the named physiological parameters were performed including calorimetric measurements and thermodynamic considerations:

- under normal conditions in discontinuous (batch) cultures on carbohydrates or alkanes or in continuous (chemostat) cultures (Glombitza and Heinritz 1979; Bley et al. 1980; Stichel et al. 1982; Heinritz et al. 1983a; Büttner et al. 1985; Minkevich et al. 1988; Zentgraf 1991a,b,c);

- under the influence of perturbations, like substrate or oxygen limitations, which occur in industrial recirculation reactors used for SCP production on alkanes, as well as under the influence of alternating milieu changes as substrate (glucose, sucrose, or n-alkanes) and oxygen supply during cultivation in laboratory or industrial fermenters (Glombitza and Heinritz 1979; Heinritz and Bley 1979; Heinritz et al. 1981, 1985; Glombitza 1982; Riege et al. 1989; Zentgraf 1991a,b,c, 1993);

- for the phenomenological description of the microbial substrate conversion by connecting energy and material balance equations (Heinritz et al. 1982, 1983a) proving the auxiliary substrate concept of Babel (Babel 1979, 1980, 1986; Müller and Babel 1988, 1989) for mixed substrate utilization;

- by using the dynamic process control conception during continuous cultivation (Heinritz et al. 1983b, 1985).

The results of the influence of alternating milieu changes, like the oxygen concentrations in the medium, on physiological parameters of *C. maltosa* cultures, are of importance for applying microorganisms in high-performance fermenters with changing gradients of dissolved oxygen tension in particular. It has been demonstrated that nonhomogeneity of the oxygen concentration in the culture medium during continuous cultivation on crude oil fractions or n-alkanes diminished the biomass production efficiency of *C. maltosa* (Heinritz et al. 1985; Riege et al. 1989). Strong oxygen limitation decreased protein formation and carbon incorporation into the biomass with a simultaneous increase in CO_2 formation, whereas periodic changes of oxygen supply caused a decrease only in carbon incorporation into the biomass and an increase in CO_2 formation, but not the incorporation into the nitrogen-containing biomass (protein), probably due to the presence of different regulation sites in the cell. For the industrial SCP production from crude oil fractions in recycle reactors, these results show that periodic oxygen gradients can cause lower yield coefficients (g biomass/g substrate) and higher CO_2 formation, whereas the protein formation per g utilized substrate might remain constant (Riege et al. 1989).

Included in these investigations for optimizing biomass production with *C. maltosa* was the concept of using mixed substrates or auxiliary substrate concept (Babel 1979, 1980). According to this concept, the right combinations of energy-deficient (glucose, sucrose, acetate, formate) and energy-rich substrates (n-alkanes, fatty acids, ethanol) should lead to an improvement of specific yield coefficients linked to a reduction of the specific heat production, and an increase in the growth rate and velocity of the substrate utilization for biomass production by microorganisms able to consume simultaneously both types of substrates without inhibition or (catabolite) repression effects. It was demonstrated that the auxiliary substrate concept is appropriate to improve the yield both for SCP- and product synthesis with different microorganisms and substrates (Babel 1986). Thus, this concept was proved also with *C. maltosa* (*L. elongisporus*) utilizing simultaneously sucrose and hexadecane (using a mixing ratio of 85% w/w sucrose/ 15% w/w paraffin) as substrates in batch cultures (Heinritz et al. 1982) and hexadecane and formate, hexadecane and isopropanol, or ethanol and formate in chemostat cultures as well (Müller and Babel 1988, 1989).

Besides the energetic efficiency, the binding state of the substrate is important in the search for optimal substrate mixing ratios. Thus, the fermentation systems sucrose/yeast, paraffin/yeast, and sucrose/paraffin/yeast were studied using synchronous populations produced by the method of phased cultures. The theoretically optimal mixing conditions (85% w/w sucrose/15% w/w paraffin) for optimal growth rate, carbon and oxygen utilization coefficients, and specific heat formation were proved by experimental data. Under other mixing conditions, simultaneous utilization of both substrates was not observed (Heinritz et al. 1982).

Additionally, several studies were also performed with *C. maltosa* at the biochemical level, showing regulation by the oxygen supply not only on the growth rate but also on biomass yield, biomass composition, and the content of enzymes involved in alkane degradation (Davidov and Gololobov 1980a,b; Mauersberger et al. 1980, 1984; Mauersberger 1985; Wiedmann et al. 1987, 1988a; Schunck et al. 1987a,b; Riege et al. 1989; Ilchenko et al. 1989; Shilova et al. 1989; cf. also Sects. 2.3.3. and 2.5).

2.3.3
Biomass Composition

The biomass composition of *C. maltosa* varies depending on the growth conditions used. Normally, different strains contain about 50–60% crude protein (total nitrogen $N \times 6.25$) or 40–45% of true protein according to Lowry on a dry weight basis

(Brückner and Tröger 1981a,b; Kölblin and Birkenbeil 1981; Kölblin and Tröger 1982; Gradova et al. 1983, 1990, our own data). This biomass of alkane-grown cells contains further about 25–30% total carbohydrates (polysaccharides as 8–10% high molecular mannan, 3–4% low molecular mannan, 11–15% glucan, mainly localized in the cell walls, 1.5–2% glucogen, and 0.7% trehalose), 6–7% total nucleic acids and 15–18% lipids. In contrast, in glucose-grown cells, the total carbohydrate content is significantly higher (43–46%, including 11% mannan, 12–13% glucan, 3% trehalose, and 4–6% glucogen), whereas the lipid content (13–14%) and the protein content (35–37%) are lower, without changed nucleic acid content (Brückner and Tröger 1981b; Blasig et al. 1984, 1989).

After selection of the best hydrocarbon utilizing strains, the crude protein content was in the range of 55–62%. Interestingly, this value did not vary significantly when using glucose or alkanes as carbon sources (Gradova et al. 1983, 1990). Under nitrogen limitation conditions, the crude protein content (up to 17–23%) and the nucleic acid (3–4%) content of alkane-grown cells decreased strongly, connected with an increase of both lipid (28–30%) and total carbohydrate (38–44%) contents (Brückner and Tröger 1981a,b; Gradova et al. 1983). Comparable results were obtained to some extent also under oxygen limitation conditions (Riege et al. 1989). These data for *C. maltosa* are in agreement with results obtained with other hydrocarbon-utilizing *Candida* yeasts (Nabeshima et al. 1970; Käppeli et al. 1975; see Brückner and Tröger 1981b) and with the protein content known for *S. cerevisiae* (Gradova et al. 1983). The biomass of *C. maltosa* contains about 4–5% lysine, 0.8–1.5% sulfur-containing amino acids, and a total of 23.5–26.0% of essential amino acids (Gradova et al. 1983).

The content of neutral lipids varied between 12 and 32% of the total lipid content without a recognizable relationship to the n-alkane chain length or other experimental conditions. The polar lipid fractions contained phosphatidylcholine and phosphatidylinositol as the major components (20% of total phosphate content, each), phosphatidylethanolamine (15%), phosphatidylserine and cardiolipin (6% each), and some further unidentified phosphate-containing lipids (Blasig et al. 1989). The fatty acid content in glycerol-grown cells was lower than in alkane-grown cells, and the total lipid content in glucose cells was about 50% (6–9%) of that (12–17%) of alkane-grown cells (Blasig et al. 1984, 1989).

2.3.4
Media

Complete (Complex) Medium: YPD (YEPD) composition as used for other yeasts (Rose et al. 1990), 1% yeast extract, 2% Bacto peptone, 2% Dextrose (glucose).

Synthetic Mineral Salt (Minimal) Media: YNB – Yeast nitrogen base without amino acids (Difco): *Candida maltosa* grows well in YNB with addition of a carbon source (1–2%) as glucose, glycerol, ethanol, or n-alkanes. YNB is often designated SD (0.67% YNB and 2% Dextrose (glucose) or SG (0.67% YNB and 2% Galactose).

Other Mineral Salt Media: Two mineral salt media were mainly used for investigation of *C. maltosa* EH15 in the authors' laboratory in Berlin-Buch (Mauersberger et al. 1981, 1984; Huth et al. 1990a). Medium 1 is composed according to Reader (Biochem J 21, 1927), whereas medium 2 is modified mainly for the production of higher biomass concentration on alkanes. These two media contain, in 1000 ml bidistilled water:

Components	Medium 1 (g/l)	Medium 2 (g/l)
$(NH_4)_2SO_4$	3.0	12.75
KH_2PO_4	1.0	1.56
$K_2HPO_4 \times 3H_2O$	0.16	0.33
$MgSO_4 \times 7H_2O$	0.70	0.41
$Ca(NO_3)_2 \times 4H_2O$	0.40	0.40
NaCl	0.50	0.50
KCl		0.08
$FeCl_3 \times 6H_2O$		0.01
Trace element solution	0.2 ml	0.45 ml

The trace element solution containes in 1000 ml aqua bidest.:
10.0 g $FeSO_4 \times 7H_2O$, 80.0 g $ZnSO_4 \times 7H_2O$, 64.0 g $MnSO_4 \times 4H_2O$, 8.0 g $CuSO_4 \times 5H_2O$, 10 ml conc. HCl

Vitamins: Only biotin (10 µg/l or up to 1 mg/l) is necessary for growth of *C. maltosa*, especially when grown on short-chain n-alkanes (Schauer 1988; cf. Sect. 2.4). Bacto Yeast extract (Difco) in a concentration of 0.1% is often used as complex vitamin and amino acid source.

Good growth of *C. maltosa* is also observed in the minimal salt medium according to Tanaka et al. (1967), supplemented with biotin (1 mg/l) and with the required vitamins, amino acids, and organic bases, when using auxotrophic mutants (Bode et al. 1983; Bode and Casper 1983; Schmidt et al. 1985, 1989a,b, Becher et al. 1991).

2.3.5
Cultivation Conditions

- pH: stop of growth near pH 8, tolerates low pH values to 2.5.

- Temperature: up to 40–41 °C, not at 43 °C (mesophilic yeast).

- Cultivation in shaking flasks at 100–240 rpm using rotary shakers (100 ml medium in 500-ml shaking flasks) or in fermenters.

- Ventilation (oxygen consumption): high oxygen supply necessary for good growth on hydrocarbon substrates.

- Method for monitoring growth and substrate utilization (see Sect. 2.4.3).

2.4
Substrate Utilization Spectrum of Candida maltosa

Due to its ability to grow on a variety of carbon sources including n-alkanes and fatty acids, the yeast *C. maltosa* has been studied since its discovery as an organism useful for industrial purposes, mainly for the production of single-cell protein (SCP) on alkanes and for obtaining commercially interesting oxidation products of these substrates (Levi et al. 1979; Einsele 1983; Shennan 1984; Bühler and Schindler 1984; cf. also Sects. 1 and 6). Therefore, mainly the alkane and fatty acid metabolic pathways have been intensively studied in this yeast.

Besides their application in taxonomic studies, the metabolism of carbohydrates was mainly studied in comparison with the n-alkanes as carbon sources, and in connection with cell wall biosynthesis. Recently, the metabolism of ethanol was also studied (Sect. 2.4.2).

Data obtained for these main studied pathways in *C. maltosa* are mostly comparable with the results obtained for the other thoroughly studied alkane-assimilating yeasts such as *C. tropicalis*, *P. guilliermondii*, and *Y. lipolytica* (Barth and Gaillardin, Chap. 10, this Vol.).

2.4.1
Nitrogen Sources and Amino Acid Catabolism

In spite of the fact that the *Candida* yeasts have been widely used for a long period of time for obtaining SCP from different kinds of raw material, the nitrogen metabolism of the yeast involved has been insufficiently studied until recently. The yeast *C. maltosa* is not able to utilize nitrate (NO_3^-) or nitrite (NO_2^-), but it can use ammonium ions (NH_4^+), urea, or amino acids as sole source of nitrogen for growth (Metz and Reuter 1977; Casper et al. 1985a,b; Guselnikova et al. 1989, 1991; Gradova et al. 1990; Huth et al. 1990a,b,c; Table 3).

Ammonium and Urea Assimilation

Ammonium assimilation. The ammonium assimilation and its regulation in *C. maltosa* was studied by Popov et al. (1980), Casper et al. (1985a), Huth et al. (1990a,b,c), Gradova et al. (1990), and Guselnikova et al. (1991). Data on the kinetics of ammonium transport into the cell using methylamine as an analogous material nonmetabolizable by the cells were published by Gradova et al. (1990) and Guselnikova et al. (1991). After cultivation on ammonium-containing medium, *C. maltosa* possesses a constitutive and active system for methylamine transport having a K_m of 2×10^{-4} M (low affinity permease) not significantly influenced by the carbon source (glucose or octadecane). Cultivation of the yeast under conditions of ammonium deficiency ($100\,\mu$M) resulted in expression of a second active system for methylamine transport, having a K_m of 2×10^{-5} M. This high affinity permease has no analogy with the two or three permeases of *S. cerevisiae* (Magasanik 1992), and occurs during cultivation with amino acids or yeast autolysate as the sole nitrogen source. At physiological ammonium concentration in medium (10 100 mM) the high affinity methylamine transport system was irreversibly inhibited.

Ammonium assimilation in *C. maltosa* is mainly connected with the glutamate/glutamine formation as reported for other yeast and fungi (Magasanik 1992). The synthesis of these two primary products involves three enzymes, the glutamate dehydrogenase (GDH NADP and GDH NAD), glutamine synthetase (GS), and glutamate synthase (GOGAT NAD/NADP). In *C. maltosa* the regulation of these enzymes is somewhat different from that known for bakers yeast (Casper et al. 1985a). The GDH exists in *C. maltosa* in two forms, the catabolic GDH (NAD) and the biosynthetic GDH (NADP). The GS is derepressed during growth on low ammonium or on a variety of alternative mitrogen sources, whereas the catabolic GDH (NAD) is repressed under these conditions. Due to its low K_m value for ammonium, the GS is obviously the most active enzyme during growth on low ammonium concentrations, whereas ammonium limitation did not significantly affect the biosynthetic GDH (NADP) or the GOGAT. The latter is present in amounts higher in *C. maltosa* than are normally found in *S. cerevisiae*. Casper et al. (1985a) concluded that under N-limitation ammonium assimilation is achieved via GS/GOGAT and under N-excess via GS/GDH(NADP) in *C. maltosa*. The characterization of a GOGAT mutant (*glu1*) supported the role of this enzyme in ammonium assimilation (Casper et al. 1985b).

During assimilation of ammonium as the only nitrogen source, yeasts like *C. maltosa* exhibit a strong correlated proton extrusion leading to an acidification of the medium. The proton extrusion of growing yeast cultures, measured from the alkali consumption to maintain a constant pH of the culture, was shown to be an exact and reliable on-line parameter for the description of biomass production and consumption of the nitrogen (with an NH_4^+/H^+ exchange ratio of exactly 1.0) and carbon sources quantitatively in batch and fed-batch fermentation processes. The biochemical basis of this proton extrusion was shown to be connected with the nitrogen metabolism of growing yeasts (Schunck et al. 1987a; Riege et al. 1989; Huth et al. 1990a,b,c, see Sect. 2.4.3).

Regulation of Ammonium Catabolism by Nitrogen Catabolite Repression. Changes in ammonium concentration in the medium influence both the ammonium transport system and the entry of other N-containing components into the cells and vice versa. These regulation phenomena, named nitrogen catabolite repression in yeast (Cooper 1982; Wiame et al. 1985), were also detected in *C. maltosa*. Thus, ammonium assimilation by *C. maltosa* is repressed by the presence of high amounts (>2 mg/ml) of asparagine (Popov et al. 1980). Otherwise, mechanisms for regulating nitrogen metabolism probably exist in this yeast which are different from the mechanism of nitrogen catabolite repression, as concluded from experimental data on peptide and amino acid utilization from yeast autolysate by *C. maltosa* (Gradova et al. 1990).

Urea assimilation. During growth of *C. maltosa* on glucose or n-alkane with urea as nitrogen source, no urease activity (urease negative yeast) and no accumulation of ammonia in the medium were detectable (Metz and Reuter 1977). Urea was obviously utilized via a urea-amydolase (UALase, forming CO_2 and NH_3, consisting of urea carboxylase, EC 6.3.4.6, and allophanate hydrolase, EC 3.4.1.13) activity, detected in cell-free extracts in the presence of ATP, biotin, and cations. This enzyme complex ATP: urea-amidolyase (ADP) is induced by urea and arginine and repressed by catabolite ammonia, as reported for other yeasts, including *S. cerevisiae* (Metz and Reuter 1977).

Amino Acid and Peptide Catabolism

The catabolism of amino acids in *C. maltosa* has been studied mainly in the laboratory of Dr. R. Bode at the University of Greifswald, Germany, including their enzymology and the use of respective mutants, and in connection with the anabolic pathways.

Several amino acids (glutamate, aspartate, asparagine, proline, arginine, ornithine, lysine, acetyl lysine, serine, leucine, isoleucine, valine, alanine) are used by *C. maltosa* as the only nitrogen source (Casper et al. 1985b; Umemura et al. 1992). Among several compounds tested, *C. maltosa* utilizes also phosphonoalanine as the sole nitrogen source (Bode and Birnbaum 1989). Most of the amino acids are also used as carbon sources, except the aromatic amino acids, leucine, isoleucine, valine, and glutamine, the latter because *C. maltosa* has no glutaminase activity (Casper et al. 1985b). Obviously, *C. maltosa* (strain JCM1504) assimilates the L-isomers of racemic amino acids such as alanine, arginine, asparagine, glutamate, proline, and serine with high stereoselectivity as the only source of carbon and nitrogen (Umemura et al. 1990, 1992; cf. Sect. 6).

Amino acids are transported into *C. maltosa* cells by a suggested nonspecific or general amino acid permease, which is able to catalyze the transport of several amino acids and probably other compounds like glyphosate (Bode et al. 1985b).

The main amino acid degradation pathways studied for *C. maltosa* are:

Glutamate, glutamine, aspartate, and asparagine degradation is included in their general metabolism due to their important role in nitrogen metabolism in *C. maltosa*, as discussed above (Casper et al. 1985a,b). No glutaminase was detected in this yeast. Asparagine is degraded via aspartate (asparaginase) to fumarate (aspartate lyase with catabolic function only), and aspartate is also converted by its aminotransferase to glutamate (Casper et al. 1985b).

Lysine catabolism occurs in *C. maltosa* via N-acetylated intermediates. The first two enzymes of this pathway, the novel acetyl-CoA:L-lysine N-acetyltransferase (EC 2.3.1, Schmidt et al. 1988; Hammer et al. 1991) and the N^6-acetyl-L-lysine:2-oxoglutarate aminotransferase (AcL-AT, Schmidt and Bode 1992) were first detected in yeast in *C. maltosa*. The two enzymes were characterized after partial and total purification, respectively. Further degradation of the 2-keto-acetamido-caproate and 5-aminovalerate to glutarate has been described in this yeast as in *Y. lipolytica*, whereas *P. guilliermondii* and *C. albicans* use the oxidative transamination as the first step (Hammer et al. 1991).

Aromatic amino acids are used as sole nitrogen source but not as carbon source by *C. maltosa* (Casper et al. 1985b). This yeast possesses three L-aromatic aminotransferase (EC 2.6.1.57) activities (ArAT I-III), one of which was inducible by aromatic amino acids (Bode and Birnbaum 1984, 1987). This manner of degradation of L-aromatic amino acids results in phenylpyruvate, p-hydroxy-phenylpyruvate, and indolepyruvate, which are converted into the corresponding aromatic acetates, catalyzed by aromatic lactate dehydrogenases (ArLDH), present in *C. maltosa* in high activities. This enzyme (EC 1.1.1.110) was first reported in yeast and characterized after purification (Lippoldt et al. 1986; Bode et al. 1986a). D-aromatic amino acids are degraded in an initial deamination step by a D-amino oxidase (DAO, EC 1.4.3.2), widely distributed in yeasts, although to a lower extent in *C. maltosa* (Lippoldt et al. 1986). The possibility of biotechnological application of these enzyme activities from *C. maltosa* has been tested (Bode and Birnbaum 1987, 1991a; see Sect. 6).

Alanine catabolism in *C. maltosa* JCM1504 is initiated by the enzyme alanine aminotransferase (AlaAT, EC 2.6.1.2) intensively studied by Umemura et al. (1991, 1994) after purification (Table 4). The AlaAT isolated from cells grown on L-alanine as the only source of carbon and nitrogen is stereospecific towards L-alanine, leaving the D-alanine untouched. This enantioselectivity of the AlaAT of *C. maltosa*, connected with a very low D-amino acid oxidase activity under the conditions used, was applied to produce D-alanine from a racemic D,L-alanine substrate mixture with *C. maltosa* cells (Umemura et al. 1990, 1992; cf. Sect. 6).

Leucine, isoleucine, valine are used not as carbon, but as nitrogen sources. The first step of degradation is probably catalyzed by two branched aminotransferases AT-I and AT-II (EC 2.6.1.42), detected in cytosol only and purified from *C. maltosa* (Bode and Birnbaum 1988).

Threonine is further used in isoleucine biosynthesis, whose first specific enzyme threonine dehydratase catalyzing its deamination was studied (Bode et al. 1986b; Bode and Birnbaum 1988).

Arginine is used as nitrogen source via arginase (endogenous urea formation from arginine during arginine catabolism) as in *S. cerevisiae*. The synthesis of arginase is regulated by arginine and ammonium as described for other yeasts. From the lower level of arginase in relation to ornithine carbamyl-transferase it was concluded that, especially in alkane-growing *C. maltosa*, the arginine catabolism is not very intensive (Metz and Reuter 1977).

Peptide utilization from yeast autolysate by *C. maltosa* was studied by Belov and Guselnikova (1988), Guselnikova et al. (1989), Gradova et al. (1990), and Belov et al. (1991). The peptide transport is independent of that of amino acids. Autolysate addition to the growth medium of *C. maltosa* decreases the intracellular pool of free amino acids and accelerates phosphatidyl inositol metabolism. The peptide components of the autolysate are presumed to act as biologically active compounds by exerting an essential effect on the phosphatidyl inositol system of *C. maltosa*.

2.4.2
Carbon Sources

The yeast *C. maltosa* is characterized by its ability to grow on a wide range of substrates as the only source of carbon and energy including long-chain alkanes and fatty acids, ethanol, acetate, and carbohydrates (Table 3).

Table 3. Substrate utilization spectrum and other physiological properties of *Candida maltosa*

Substrates and properties	Utilization as carbon and energy source or nitrogen source (or cooxidation demonstrated)
Fermentation of	
Glucose, galactose, maltose, sucrose, trehalose	+
Lactose, raffinose, melibiose, cellobiose	−
Soluble starch, α-methyl-glucoside	−
Assimilation of	
Carbohydrates	
Glucose, galactose, maltose, sucrose, trehalose	+
L-Sorbose, cellobiose, melezitose, D-xylose	+
Lactose, melibiose, raffinose, soluble starch	−
Inulin, arabinose, D-ribose, L-rhamnose	−
Alcohols and organic acids	
Ribitol, sorbitol (D-glucitol), D-mannitol	+
Erythritol, arabitol, galactitol (dulcitol)	−
Ethanol, glycerol, acetate, lactate	+
Succinate, pyruvate, citrate, 2-ketogluconate	+
Methanol	−
Other substrates	
Salicin, α-methyl-glucoside (glycosides)	+
Glucosamine, amino acids	+
Hydrocarbons (hc)	
n-Alkanes (C_6–C_{40}), alkenes, fatty alcohols, fatty acids	+
Single branched hc (e.g., 2-methyl-pentadecane)	+ (+)
Multiple branched hc (e.g., pristane)	− (+)
Cycloalkanes	− (+)
Aromatic hc (biphenyl, naphthalene, benzo(a)pyrene)	− (+)
Selected steroids	− (+)
Phenol, catechol, cresol	+
Nitrogen sources	
Ammonia, urea, amino acids	+
Nitrate, nitrite	−
Other properties	
Growth at 30–40 °C	+
at 43 °C	−
Acid formation	weak
Cycloheximide or formaldehyde resistance	+
Fat splitting	+
Urease	−
Vitamin (biotin) dependent	+
Growth on 50% glucose	−
Sporulation	−
Pseudomycel formation	+
Coenzyme Q type	9
G + C content	36–37%

Data taken mostly from Barnett et al. (1983) and Kreger-van Rij (1984), and from other references mentioned in the text (see Sect. 2.4); (−) negative; (+) positive.

Utilization of Carbohydrates

C. *maltosa* assimilates a broad range of carbohydrates which are used for taxonomic classification (Barnett et al. 1979, 1983; Kreger van Rij 1984; cf. Table 3). Carbohydrate transport and metabolism in C. *maltosa* were studied also for comparison (growth kinetics, thermodynamics, biomass composition; see Sect. 2.3) with other noncarbohydrate substrates, and to investigate the biosynthesis of polysaccharides and the regulation of catabolic and anabolic pathways (see Sects. 2.5 and 3.2).

Glucose Transport. Two different glucose transport systems were detected in C. *maltosa* which differed only three-fold in the magnitude of their affinity constants (K_m of 0.13 and 0.35 mM) measured with labeled glucose. The capacity of the carriers (V_{max}) was higher in the presence of glucose in the medium. The expression of further uptake systems under other environmental conditions was not excluded (Hofmann and Polnisch 1990c).

Metabolism of Glucose and Other Hexoses. Carbohydrate metabolism in C. *maltosa* was intensively studied in the 1980s in the laboratory of Dr. G. Reuter at the Schiller University Jena, Germany, in the course of investigations made on structure and regulation of biosynthesis of the polysaccharides, especially of the cell wall of this yeast (see also Sects. 2.5 and 3.2). The yeast C. *maltosa* exhibits some peculiarities in the metabolism of carbohydrates compared with the thoroughly studied S. *cerevisiae.* When oxygen is not limiting and glucose is in excess, glucose uptake is obviously controlled and no ethanol production was observed in C. *maltosa* during monophasic growth on glucose. Under oxygen limitation, however, mostly occurring at the end of the exponential growth in shake flasks, ethanol formation from the excess of glucose or other hexoses was observed (Röber and Reuter 1979, 1984a; Popov et al. 1980; Röber 1985; Schauer 1988). In contrast to S. *cerevisiae* and other true fermenting yeasts, C. *maltosa* showed no true Crabtree effect, e.g., no glucose fermentation to ethanol at high glucose concentrations of 1–10% under aerobic conditions was observed (Schauer 1988), although under oxygen limitation the formation of ethanol occurred. For resting cells, glucose utilization was accompanied by formation of small amounts of glycerol (Schauer 1988). In this respect, C. *maltosa* is obviously similar to C. *tropicalis* classified as glucose-insensitive and O_2-sensitive yeast. In contrast, obligatory oxidative yeasts like Y. *lipolytica* and *Trichosporon cutaneum* produce no or low amounts of ethanol during growth on glucose (Fiechter et al. 1987; Fiechter and Gmünder 1989).

The oxidation of glucose and other hexoses (mannose, fructose) by C. *maltosa* under aerobic conditions via tricarboxylic acid cycle is relatively low, suggesting its repression. Glucose degradation occurs mainly via the pentose phosphate pathway (PPP, oxidative and recyclic), some steps of glycolysis, and simultaneously by operation of the tricarboxylic acid cycle and pyruvate decarboxylation, as concluded from experiments with specifically labeled hexoses (Röber and Reuter 1979, 1984a; Röber 1985). In agreement with these results, the specific activity of the phosphofructokinase (PFK) was very low. The glyceroaldehyde-3-phosphate arising from the PPP is catabolized via the last steps of glycolysis into pyruvate, further decarboxylized to acetyl-CoA to yield energy, or transformed partially into ethanol. These steps of glycolysis were found to be increased in activity during oxygen limitation of the culture growing on hexoses, reflected in a higher glucose consumption rate under these conditions (Röber and Reuter 1984a; Röber 1985).

Glucose (hexose) degradation is initiated by formation of glucose-6-phosphate in the hexokinase reaction. The two hexokinase isoenzymes purified from *C. maltosa* (Table 4) are a target for effective control of glucose-6-phosphate formation by energy charge of the cell or by the intracellular level of ATP, ADP, and glucose instead of allosteric interconversion of the enzymes (Röber et al. 1984a). The glucose-6-phosphate dehydrogenase (G6PDH), initiating and controlling the pentose phosphate cycle by its metabolic regulation, was also purified and characterized in its kinetic regulation (Röber et al. 1984b).

The functioning of the pentose phosphate cycle is in glucose-growing *C. maltosa* the prerequisite for the synthesis of the cell wall polysaccharides (glucan and mannan) and of other energy reserve compounds (glycogen and trehalose), all having fructose-6-phosphate as common precursor intermediate for their biosynthesis (Röber and Reuter 1979, 1984b,c,d, 1985). The hexose utilization and these biosynthetic pathways in *C. maltosa* were studied in comparison with two mutants (H3 and H5) partially desensibilized in the catabolite repression by glucose (Röber and Reuter 1982, 1984a,b; Röber 1985).

The presence of glucose in the medium generally generates in *C. maltosa* a *catabolite repression* and/or *catabolite inactivation* of different metabolic pathways like the catabolism of hydrocarbons (alkanes, phenol – Mauersberger et al. 1980, 1981, 1984; Mauersberger 1991; Hofmann and Krüger 1985; Hofmann and Vogt 1987, 1988; see Sects. 2.4.2, 2.5), and the catabolism of ethanol or acetate, including the glyoxylate cycle and gluconeogenesis (Polnish and Hofmann 1989; Hofmann and Polnisch 1990a,b,c). Aerobic growth on glucose resulted also in repression of the tricarboxylic acid cycle. The degree of catabolite repression was shown to depend on the carbon sources (nature of hexoses) used and on the oxygen concentration in the medium (Röber and Reuter 1984a, 1985).

The utilization of carbohydrates mixtures as occurring in wood hydrolysates (hemicellulose beech and others) by *Candida* yeast including *C. maltosa* was studied by Manakov and Prishepov (1986) and Kostov et al. (1991).

Hydrocarbon and Phenol Assimilation

The yeast *C. maltosa* belongs, in addition to *C. tropicalis, C. intermedia, Pichia(C.) guilliermondii*, and *Y. (C.) lipolytica*, to the best-investigated alkane-assimilating yeasts due to its application in the industrial production of alkane-based SCP and oxidation products (see Sects. 1 and 6). The knowledge of alkane-assimilating yeasts was repeatedly reviewed, especially in the golden age of petroleum microbiology and biochemistry until the end of the 1970s (for earlier references see Levi et al. 1979; Fukui and Tanaka 1981a,b; Rehm and Reiff 1981, 1982; Bühler and Schindler 1984; Tanaka and Fukui 1989).

Table 4. Proteins purified and characterized from the yeast *Candida maltosa*

Protein	Strain and carbon source	Function	Gene isolated	Antibodies obtained	Reference
AlaAT	JCM1504	Alanine catabolism	–	–	Umemura et al. (1991, 1994)
ArLDH	L4	Catabolism of aromatic aa	–	–	Bode et al. (1986a)
ASG[c]	L4	Tryptophan biosynthesis	–	–	Bode et al. (1985a)
ASN[c]	L4	Tryptophan biosynthesis	–	–	Bode et al. (1985a)
BAAT	L4	Branched aa metabolism	–	–	Bode and Birnbaum (1988)
CDO	ATCC20184	Catechol degradation	–	–	Gomi and Horiguchi (1988)
Cell wall glycoprotein		Alkane uptake	–	+	Belov et al. (1983)
CM[c]	L4	Aromatic aa biosynthesis	–	–	Bode et al. (1985d)
Co-binding protein			–	–	Belov and Toneva-Davidova (1983)
Cyt b$_5$	VSB779		–	–	Avetisova (1991) Avetisova et al. (1993)
Cytochromes P450					
P450Cm1 (52A3)	EH15[a] n-alkanes	Alkane hydroxylation	+	+	Riege et al. (1981) Schunck et al. (1983a,b)
P450alk (52A3)	IAM2247 n-alkanes		+	–	Takagi et al. (1989)
	VSB779 Decane		(+)[d]	–	Mauersberger et al. (1992a)
P450Cm2 (52A4)	[b]	Fatty acid hydroxylation	(+)[d]	+	Scheller et al. (1992, 1996)

Protein	Strain	Function			Reference
P450Cm3 (52A5)	VSB779, Decane	Alkane and fatty acid hydroxylation	(+)[d]	—	Mauersberger et al. (1992a)
NADPH-P450 reductase	EH15, a	Reduction of P450	+	+	Honeck et al. (1982); Vogel et al. (1992); Kärgel et al. (1996); Bode et al. (1984a, 1985c); Bode et al. (1984d)
DAHP-S[c]	L4	Shikimate pathway	—	—	Sokolov et al. (1991, unpubl.)
EPSP-S[c]	L4		—	—	
FAOD	VSB779, Decane	Fatty alcohol oxidation	—	—	Mauersberger et al. (1992b); Hofmann and Polnisch (1990b)
F1,6BPase[c]	L4		—	—	Röber et al. (1984b)
G6PDH	H(H62)		—	—	Röber et al. (1984a)
HK I/II[c]	H		—	—	
α-IPM-S[c]	L4	Leucine biosynthesis	—	—	Bode and Birnbaum (1991c); Becher et al. (1991)
IPMDHT[c]	L4	Leucine biosynthesis	—	—	Bode and Birnbaum (1991c); Becher et al. (1991)
β-IPMDH[c]	L4	Leucine biosynthesis	+	—	Bode (1991); Bode and Birnbaum (1991c); Becher et al. (1991)
LAcT[c]	L4	Lysine metabolism	—	—	Schmidt et al. (1988); Schmidt and Bode (1992)
P-DHT[c]	L4	Aromatic aa biosynthesis	—	—	Bode et al. (1985d)
P-DH[c]	L4	Aromatic aa biosynthesis	—	—	Bode et al. (1985d)
PRAI[c]	L4	Tryptophan biosynthesis	—	—	Bode et al. (1985a)
PRT[c]	L4	Tryptophan biosynthesis	—	—	Bode et al. (1985a)
SOD	VSB779		—	—	Avetisova (1991)

Table 4. (*Contd.*)

Protein	Strain and carbon source	Function	Gene isolated	Antibodies obtained	Reference
ThrDHT	L4	Isoleucine biosynthesis	–	–	Bode et al. (1986b)
TS[c]	L4	Tryptophan biosynthesis	–	–	Bode et al. (1985a)

[a,b] Proteins were additional [a] or only [b] purified after heterologous expression in the yeast S. *cerevisiae* GRF18 (Schunck et al. 1991; Scheller et al. 1992, 1994, 1996; Kärgel et al. 1996).
[c] Partially purified proteins.
[d] Genes isolated from other C. *maltosa* strains.

Abbreviations.

ArL-DH   – D-aromatic lactate dehydrogenase (EC 1.1.1.110)
Ala-AT   – Alanine aminotransferase (EC 2.6.1.2)
ASG   – Glutamine-dependent anthranilate synthase (EC 4.1.3.27) in complex with the indole-3-glycerol-phosphate synthase (InGPS, EC 4.1.1.48)
ASN   – Ammonia dependent anthranilate synthase (EC 4.1.3.27)
BAAT   – Branched amino acid aminotransferase (EC 2.6.1.42) forms I and II, for leu, ile, val
CDO   – Catechol-1,2-dioxygenase (pyrocatechase)
CM   – Chorismate mutase (EC 5.4.99.5)
DAHP-S   – 3-Deoxy-D-arabinoheptulosonic acid 7- phosphate (DAHP) synthase, two isoenzymes
EPSP-S   – 5-Enolpyruvylshikimate 3-phosphate (EPSP) synthase (EC 2.5.1.19)
F1,6BPase – Fructose-1,6-bisphosphatase
G6PDH   – Glucose-6-phosphate dehydrogenase (EC 1.1.1.49)
HK I/II – Hexokinase I and II (EC 2.7.1.1)
LAcT   – Lysine acetyltransferase [N6-acetyl-L-lysine: 2-oxoglutarate aminotransferase] (EC 2.3.1.–)
α-IPM-S – α-Isopropylmalate (IMP) synthase (EC 4.1.3.12, coded by LEU4 gene)
IPMDHT   – IPM-dehydratase (EC 4.2.1.33, LEU1 gene)
β-IPMDH – β-Isopropylmalate dehydrogenase (EC 1.1.1.85, LEU2 gene)
P-DHT   – Prephenate dehydratase (EC 4.2.1.51)
P-DH   – Prephenate dehydrogenase (EC 1.3.1.12)
PRAI   – Phosphoribosylanthranilate isomerase (EC 5.3.1.16)
PRT   – Anthranilate phosphoribosyltransferase (EC 2.4.2.18)
SOD   – Superoxide dismutase
ThrDHT   – Threonine dehydratase
TS   – Tryptophan synthase (EC 5.3.1.16).

The Metabolism of n-Alkanes and Alkenes. Approximately 20% of the nearly 500 yeast species, mainly belonging to the genera *Candida*, *Pichia*, and *Yarrowia*, are able to grow alternatively either on carbohydrates or on middle- or long-chain n-alkanes as the only source of carbon and energy (Schauer and Schauer 1986; Müller et al. 1991a,b). Frequently, comparable growth rates can be achieved with both substrates (cf. Sect. 2.3.2). The different properties of both groups of compounds – the former rich in oxygen and, therefore, hydrophilic, the latter, without oxygen and strongly apolar – require different pathways for uptake and metabolism. Additionally to the data given here on the catabolic pathways of hydrocarbons in *C. maltosa*, an overview on the enzymology of the monoterminal and diterminal alkane oxidation pathway, including the role of cytochromes P450 and its regulation, will be given in Sect. 2.5. The aspects of the intracellular localization and the subcellular organization of this pathway will be discussed below in Section 3 (cf. Fig. 3). For the other hydrocarbons utilized by *C. maltosa*, the available data concerning enzymes participating in their oxidation will be discussed here.

In comparison with other alkane-assimilating yeasts *C. maltosa* shows a broad substrate specificity towards hydrocarbons. It utilizes n-alkanes of a broad chain-length spectrum from short-chain liquid n-alkanes (C_6 to C_9), via middle- or long-chain liquid alkanes (C_{10} to C_{18}) to long-chain solid alkanes (C_{19} to C_{32}), or even up to 40 carbon atoms (Celma Calamita et al. 1971; Michaleva et al. 1973a; Demanova et al. 1980a,b,c; Mauersberger et al. 1981; Davidov et al. 1980, 1982; Gradova et al. 1983; Schunck et al. 1987a; Huth 1987; Schauer 1988; Blasig et al. 1989; Huth et al. 1990a,b,c).

The short-chain alkanes ($<C_9$) are not assimilated by yeasts when they are added as a liquid to the mineral medium. However for *C. maltosa*, *C. tropicalis*, *C. parapsilosis*, and *C. sake* it was found that these yeast species can assimilate these volatile hydrocarbons when they are supplied in vapor phase to the submerse or emerse culture, when the substrate concentration is diminished to a nontoxic level (Celma Calamita et al. 1971; Meyer et al. 1975; Huth 1987; Hofmann and Schauer 1988; Schauer 1988; Huth et al. 1990b; Mauersberger 1991). In contrast, *C. rugosa*, *C. albicans*, *P. guilliermondii*, *L. elongisporus*, or *Y. lipolytica* are not able to assimilate these short alkane chain lengths, although they are able to oxidize these substrates (Schauer 1988). When supplied in small amounts in the presence of an inert hydrocarbon phase (1.2% of pristane), utilization of these short-chain n-alkanes was demonstrated also in a fermenter culture of *C. maltosa* (Schunck et al. 1987a; Huth et al. 1990a,b). Interestingly, there is an obvious correlation between alkane (especially short-chain) and phenol assimilation phenotypes among yeasts (Hofmann and Schauer 1988). Among ascomycetous and imperfect yeasts with ascomycetous cell wall structure (DBB−), all hydrocarbon utilizers have the coenzyme Q_9 type (Bos and deBruyn 1973; Hofmann and Schauer 1988).

For optimal growth on n-alkanes, especially on shorter-chain n-alkanes ($<C_{10}$), *C. maltosa* needs the vitamin biotin. This stimulating effect of biotin on utilization of short-chain alkanes is connected with the presence of carbon dioxide in the medium (Gradova et al. 1983; Schauer 1988), reflecting the activity of the acetyl-CoA carboxylase, the key enzyme in fatty acid de novo biosynthesis. Mutants with

biotin auxotrophic growth were obtained (Gradova et al. 1976, 1983). When *C. maltosa* was cultivated on long-chain alkanes ($>C_{14}$) the biotin dependence was not very strong, which is obviously connected with the direct incorporation of the fatty acids, derived from the alkane, into cell lipids, therefore omitting biotin dependent fatty acid synthesis (Gradova et al. 1983; Schauer 1988). Substrates containing mixtures of n-alkanes (crude oil fractions, diesel oil, petroleum, kerosene, alkane mixtures – Parex, Mepasin) are good carbon sources for *C. maltosa*.

The growth of *C. maltosa* on hexadecane is not very strongly influenced or inhibited by the presence of the chlorinated hydrocarbon pesticide heptachlor (velsicol) dissolved in the alkane phase, which itself is probably degraded by the yeast to form 1-hydroxychlordene (Smith et al. 1976).

The substrate n-alkanes generate an excess in energy and reduction equivalents, if the β-oxidation results in generation of NADPH with ammonium as the nitrogen source and comparable P/O-quotients of >1.85. Theoretically, a yield for hexadecane of about 1.26 g/g should be obtained. In experiments during growth on hexadecane/ammonium, yields of only 0.94 g/g could be reached; but in the presence of formate (or isopropanol) an improvement up to 1.26 g/g (or 1.13) was possible, which corresponds to the maximum carbon conversion efficiency (Babel 1986; Müller and Babel 1988, 1989). The behavior of *C. maltosa* on both alkane and ethanol as substrates shows that the energy conversion must be the limiting factor because, due to the growth yields on both substrates, the P/O-quotient is approximately 1.4. This means that with this species an excess energy situation is not established (Müller and Babel 1988), although for other substrates and organisms this could be clearly demonstrated (Müller et al. 1983b; for references see Babel 1986).

For *C. maltosa*, as for most other alkane-assimilating yeasts (Rehm and Reiff 1981), mainly the *monoterminal oxidation pathway* of n-alkanes was observed. With intact cells the first detectable intermediates were fatty acids of the same chain length as the alkane substrate (Blasig et al. 1984, 1988, 1989; Schauer 1988; Fig. 3, for alkane oxidizing enzymes see Sect. 2.5). The formed fatty acids are after activation to acyl-CoA degraded in the β-oxidation yielding acetyl-CoA or propionyl-CoA used in the intermediate metabolism for energy production and synthesis of all cellular components. The *fatty acid composition* of *C. maltosa* cells grown on different chain lengths of alkanes was found to be changed in comparison with the normal fatty acid composition of glucose- or glycerol-grown cells, where predominantly $C_{16:0}$ (30–40%), $C_{18:1}$ (35–55%), and to a lesser extent $C_{16:1}$, $C_{18:0}$ and $C_{18:2}$ fatty acids were present (Brückner and Tröger 1981b; Gradova et al. 1983; Schauer 1988; Blasig et al. 1984, 1989). If n-alkanes up to dodecane were used as growth substrates, the main portion of fatty acids of substrate chain length was degraded via β-oxidation, and the fatty acids required for lipids were synthesized de novo, thus having the same fatty acid composition as glucose- or glycerol-grown cells. With increasing chain lengths of the n-alkane substrate (from C_{14} up to C_{18}), the proportion of de novo synthesis of fatty acids decreased while the incorporation of fatty acids of substrate chain length into lipids directly, or after desaturation, chain elongation or chain shortening, gained more importance. Thus, cells precultured on C_{16}- or C_{17}-alkanes contained mostly (96–99%) even-chain and odd-chain fatty acids (saturated and desaturated), respectively (Blasig et al. 1984, 1989). Using solid n-alkanes up to C_{28} as substrates, fatty acids in the range C_{16} to C_{18} were predominant, independent of the n-alkane chain length (Blasig et al. 1989). In this case, after monoterminal oxidation the resulting long-chain fatty acids with substrate chain length were obviously chain-shortened by C_2 units down to an optimal range of chain length from C_{16} to C_{18} and again incorporated into cellular lipids directly or after desaturation. The de novo fatty acid synthesis was negligible under these conditions.

Additionally to monoterminal, *diterminal alkane oxidation* was also demonstrated to occur in yeast leading to the formation of dicarboxylic acids, further degraded in the β-oxidation to succinate. The formed dicarboxylic acids of shorter chain lengths are partially excreted into the medium (Schauer 1988; Blasig et al. 1984, 1988).

In contrast to various filamentous fungi, no *subterminal alkane oxidation* was observed in *C. maltosa* cells cultivated on n-alkanes (Schauer 1988; Blasig et al. 1984, 1988, 1989). However, 1-alkenes are degraded by *C. maltosa* via secondary alcohols, methylketones, and, resulting in a two-carbon atoms shortened primary alcohol, oxidized further by the terminal oxidation pathway. Therefore, after cultivation on 1-alkenes, the capacity of cells to oxidize secondary alcohols and ketones is enhanced. Moreover, these subterminal oxidation products could be identified in the cultivation medium (Schauer et al. 1986; Schauer 1988). Zinchenko et al. (1990) reported the occurrence of ω-1 fatty acids derived from n-alkenes incorporated in all lipid classes. These fatty acids are preferably incorporated in phospholipids in contrast to fatty acids derived from n-alkanes, which seems to be associated with higher affinity of the phospholipid acyltransferases to unsaturated fatty acids.

Normally, *C. maltosa* also grows well on the terminal and diterminal oxidation products of alkanes, such as 1-alkanols (fatty alcohols), fatty aldehydes, fatty acids, ω-hydroxy fatty acids, and dicarboxylic acids of the chain length spectrum such as that of the hydrocarbons (Mauersberger et al. 1981; Schunck et al. 1987a,b; Schauer 1988; Casey et al. 1990; Mauersberger 1991). Shorter chain lengths ($<C_{10}$) of these substrates are found to be toxic for yeast cells, or showing inhibitory effects for growth on alkanes.

Utilization of Branched Hydrocarbons. Davidov et al. (1981a,b, 1982) investigated the simultaneous utilization of single-branched alkanes, 2- or 3-(^{14}C)methyl hexadecane, together with octadecane as carbon sources by *C. (guilliermondii) maltosa*, when both substrates were dissolved in pristane (2,6,10,14-tetramethylpentadecane). The branched alkanes were partially incorporated into lipids as 15- or 14-methyl hexadecanoic acids, converted into water-soluble parts of the cells (proteins, amino acids) and other intermediate metabolites (DCA with methyl groups in β-position) and partially oxidized to CO_2. Multiple-branched hydrocarbons like pristane are not used as sole carbon and energy source by *C. maltosa*, in contrast to some *Y. (C.) lipolytica* strains growing well on these substrates (Hagihara et al. 1977; Crow et al. 1979, 1980). However, *C. maltosa* was shown also to be capable of uptake of the branched-chain hydrocarbons from a broth containing growth-supporting n-alkanes (Crow et al. 1979; Schunck et al. 1987a,b). Furthermore, slow oxidation of pristane to pristanoic acid by *C. maltosa* was observed (Blasig et al. 1989), and pristane induced cytochrome P450 in *C. maltosa* (Mauersberger et al. 1981; Mauersberger 1985; Schunck et al. 1987b). It is suggested that the same enzymes as known for n-alkanes (see Sect. 2.5) are participating in the oxidation of branched alkanes.

Cooxidation of Cycloalkanes. Cyclopentane to cyclododecane and derivatives are not utilized as growth substrates by *C. maltosa*, but partial oxidation of these compounds to cycloalkanols and cycloalkanones by alkane- or glucose-grown cells was observed, when supplied in nontoxic concentrations (Mauersberger et al. 1981; Schauer 1988). A further oxidation of the cycloalkanols and cycloalkanones formed by unknown mechanisms of cooxidation to CO_2 in the presence of glucose was observed (Schauer 1988). However the oxidation rates of cycloalkanes or their first oxidation products are only 5–10% that of n-alkanes. The enzymes for cycloalkane oxidation in *C. maltosa* were not characterized, cytochrome P450 was not induced by the cycloalkanes tested, and the cycloalkane oxidation was not enhanced in alkane-grown cells (Mauersberger 1985; Schauer 1988).

Utilization of Phenylalkanes. Phenylalkanes (alkylbenzenes) with n-alkyl chains longer than nine carbon atoms were utilized as carbon sources for growth by *C. (guilliermondii) maltosa*, although also shorter alkyl chains were oxidized rapidly, as checked by growth-dependent proton extrusion (Huth 1987; Huth et al. 1990b), by growth experiments on plates, by oxygen consumption measurements (Schauer 1988), or by measuring CO_2 generation using labeled substrates (Sokolov et al. 1981; Davidov et al. 1982). The oxidation of phenylalkanes occurred mainly via phenylalkanols and phenylalkanoic acids, and after β-oxidation finally resulting in phenylacetic acid for even-numbered and benzoic acids for odd-numbered alkyl chains, respectively. No products of α-oxidation (phenylpropionic acid or cinnamic acid) were detected (Schauer 1988). These compounds were accumulated and excreted into the medium, although their slow further oxidation by alkane-grown cells was observed when supplied at low concentrations (<0.06%). Additionally, they showed inhibitory effects on growth and respiration of *C. maltosa* (Schauer 1988; Huth et al. 1990b).

The primary oxidation of the alkyl chain is presumed to occur by the same enzyme systems as shown for alkane or fatty acids (see Sect. 2.5). This is supported by the facts that alkanes and phenylalkanes were utilized by *C. maltosa* simultaneously, that phenylalkanes, especially octyl-benzene, were strong inducers for P450 in *C. maltosa* (Mauersberger 1985, and unpubl. results), and their utilization was inhibited by low concentration of CO in the medium (Schunck et al. 1987a). Additionally, some short-chain 1-phenylalkan-1-ols were good substrates for the alkane-induced fatty alcohol oxidase activity in *C. maltosa* (Mauersberger et al. 1992b).

Assimilation of Phenols. *Candida maltosa*, like several other n-alkane assimilating yeasts (*C. albicans, C. tropicalis, P. guilliermondii, Y. lipolytica*, for review see

Neujahr 1990), is capable of using phenol as a source of carbon and energy (Hofmann and Krüger 1985; Hofmann and Vogt 1987, 1988; Gomi and Horiguchi 1988; Hofmann and Schauer 1988; Polnisch et al. 1992). Interestingly, there is good correlation between the alkane and phenol utilization phenotypes among a group of yeasts with ascomycetous cell wall structure. Especially all n-octane-utilizing yeasts of this group also assimilate phenol. Furthermore, it could be shown that the phenotypes hydrocarbon and phenol utilization are strongly correlated with the coenzyme Q_9 structure (CoQ_9) of these yeasts (Hofmann and Schauer 1988).

Yeasts assimilate phenol and other phenolic compounds (resorcinol, catechol, quinol, hydroxyquinol) via the β-ketoadipate pathway. As studied in detail for *Trichosporon cutaneum* and *C. tropicalis* (Krug et al. 1985; Krug and Straube 1986; Neujahr 1990; Kalin et al. 1992), the initial step in the assimilation of phenol is catalyzed by a phenol hydroxylase (EC 1.14.13.7) leading to catechol. Catechol is then further converted by an intradiol cleavage (*ortho*-fission) to *cis,cis*-muconate using the enzyme catechol-1,2-dioxygenase (or pyrocatechase, EC 1.13.1.1). The muconic acid is converted by means of lactonizing, isomerizing, and hydrolytic enzymes to β-ketoadipate. A coenzyme A derivative of β-ketoadipate undergoes thiolytic cleavage to acetyl-CoA and succinate, used in the central metabolism of the cell.

The inducible enzymes of the β-ketoadipate pathway have been found also in *C. maltosa* when growing on phenol (up to 0.1%) or catechol. Whereas the phenol hydroxylase of *C. maltosa* (strain L4) has been characterized only in cell-free extracts so far (Hofmann and Krüger 1985; Polnisch et al. 1992), the catechol-1,2-dioxygenase was purified and characterized (dimer of 33 kDa subunits) from catechol-grown *C. maltosa* ATCC 20184 cells, a strain useful for muconic acid production from catechol in a bioprocess (Gomi and Horiguchi 1988; Table 4). These first enzymes of the pathway, phenol hydroxylase and catechol-1,2-dioxygenase, are strongly regulated by the carbon source present in the medium. No activites were found in cells grown on glucose, glycerol, succinate, or hexadecane. Glucose repressed the induction of both enzymes by phenol, suggesting that the phenol metabolism in *C. maltosa* is controlled by induction-repression of these enzymes rather than by derepression-repression (Hofmann and Krüger 1985; Hofmann and Vogt 1987, 1988; Gomi and Horiguchi 1988). The phenol hydroxylase in preinduced *C. maltosa* cells is obviously regulated by a strong catabolite inactivation under conditions of carbon and energy excess (Hofmann and Krüger 1985). In carbon-limited (chemostat) cultures, *C. maltosa* assimilated phenol besides glucose (Hofmann and Vogt 1987), as was shown also for hexadecane in combination with sucrose, formate, or isopropanol (Heinritz et al. 1982; Müller and Babel 1988, 1989). Interestingly, Hofmann and Vogt (1988) later described a delayed degradation of phenol by *C. maltosa* L4 in comparison with *C. tropicalis* or *Trichosporon cutaneum* in the presence of hexadecane, obviously due to repression of the synthesis of phenol hydroxylase and catechol-1,2-dioxygenase. This is the first report on a metabolic repression by n-alkanes, which were degraded before phenol. The delayed phenol degradation could be connected with changes of the phenol uptake caused by the adaptation of the yeast cell to n-alkane as carbon source.

For the yeasts *C. tropicalis* and *C. maltosa*, the observed oxidation rates of phenol (0.5 g/l in 5 to 10 h) in fermenter systems are under the same conditions significantly higher than observed for the *T. cutaneum* (80 to 85 h, Hofmann and Vogt 1988), a strain mostly discussed besides bacteria for the application in biological phenol degradation processes (Präve et al. 1982).

Phenol-assimilating *C. maltosa* cells are able to degrade 2-, 3- or 4-monochlorophenols (Polnisch et al. 1992), which have been classified as priority pollutants, but cannot grow on these substrates as sole carbon source, as was earlier demonstrated for phenol-grown *Rhodotorula glutinis*, *C. tropicalis*, and *T. cutaneum* (Hasegawa et al. 1990; Neujahr 1990). Especially 3- and 4-chlorophenols were broken down very rapidly by phenol-grown *C. maltosa* cells under formation of several oxidation products (of the *ortho*-fission pathway of phenol degradation, chlorocatechols and chloropyrogallos, and 4-carboxymethylenebut-2-en-4-olide) with concomitant release of chloride, providing the first evidence of dehalogenation activity in yeasts. However 2-chlorophenol was only partially converted into *cis*, *cis*-2-chloromuconic acid (Polnisch et al. 1992).

Cooxidation of Polycyclic Aromatic Hydrocarbons. The fate of polycyclic aromatic hydrocarbons (PAH) has received attention since benzo(a)pyrene and similar compounds were shown to be toxic, carcinogenic, and/or mutagenic. The yeast *C. maltosa*, like other alkane-assimilating species (*C. tropicalis*, *Y. lipolytica*, *C. guilliermondii*), is not able to utilize most PAH as the sole sources of carbon and energy, but is able to cooxidize such PAH as naphthalene, biphenyl, and benzo(a)pyrene by means of the monooxygenase pathway (Crow et al. 1980; Cerniglia and Crow 1981). This pathway is characteristic for eukaryotic microorganisms (fungi and yeast) and higher eukaryotes, and results in the transformation of aromatic hydrocarbons (benzene) and PAH into hydrophobic (primary hydroxylated products) and more hydrophilic compounds (formed after glucuronide and sulfate conjugation at the primary appearing hydroxy groups), contrary to the dioxygenase pathway present in several genera of bacteria such as *Pseudomonas*, where the PAH are oxidized into *cis*-dihydrodiol and catechol forms, which are further metabolized to ring cleavage products, and utilized as carbon sources for growth of the bacteria (Cerniglia 1981). In the eukaryotic monooxygenase pathway, the aromatic hydrocarbons are oxidized to reactive arene oxides that can isomerize to phenols or undergo enzymatic hydration to yield *trans*-dihydrodiols. In the case of naphthalene as substrate, 1-naphthol (the major metabolite), 2-naphthol, 4-hydroxy-1-tetralone, and *trans*-1,2-dihydroxy-1,2 dihydronaphthalene were detected as products with the yeasts *C. maltosa* and *C. lipolytica* (Cerniglia and Crow 1981).

The cooxidation pathway of PAH in yeasts is obviously not directly connected with the alkane-induced cytochrome P450-dependent monooxygenase systems of these yeasts, because *C. utilis* and *S. cerevisiae*, yeasts not able to grow on alkanes, also transform naphthalene into 1-naphthol and other oxidation products

(Hofmann 1986a,b). Therefore, fungi and yeast oxidize PAH via cytochrome P450 monooxygenase systems, which are expressed even in the absence of alkanes, and are different from their alkane-inducible hydroxylase systems.

Fatty Acid, Acetate, and Ethanol Metabolism

The metabolism of fatty acids, dicarboxylic acids, acetate, and ethanol in *C. maltosa* results in the formation of acetyl-CoA, further utilized in the glyoxylate cycle, tricarboxylicacid cycle, and gluconeogenesis. The metabolism of these substrates can be regarded mainly as a part of the alkane metabolism (Fig. 3, cf. Sect. 2.5). Some aspects of the ethanol metabolism in *C. maltosa* were reported in connection with the thermodynamics of growth (Minkevich et al. 1988; Müller and Babel 1988), and with the regulation of glyoxylate cycle and gluconeogenesis (Polnisch and Hofmann 1989; Hofmann and Polnisch 1990a; see Sect. 2.5).

Ethanol is metabolized without fermentative metabolism in yeasts. It is an energy excess substrate with ammonium as nitrogen source if a P/O > 1.85 is realized. Theoretically, a yield of about 0.77 g/g can be obtained. In chemostat cultivation experiments with *C. maltosa* on ethanol/ammonium, submaximal growth yields of only 0.61 g/g were attained. Addition of formate improved this value to 0.76 g/g, according to the auxiliary substrate concept (Müller and Babel 1988). In the case of acetate, no excess in energy or reduction equivalents can be reached, and only if a P/O = 3 is taken into account is a balanced carbon/energy ratio approximated for nitrogen source ammonium. Theoretically a growth yield of about 0.49 g/g can be obtained with ammonium as nitrogen source (Müller and Babel 1988).

2.4.3
Miniaturized Fermenter System for Physiological and Biochemical Studies

Note on Method. The proton extrusion of *C. maltosa* and other yeast cultures growing on different nitrogen-free carbon sources in a medium containing ammonium salts as the only nitrogen source was shown to be an exact and reliable on-line parameter for the description of biomass production, consumption of the nitrogen (with an NH_4^+/H^+ exchange ratio of exactly 1.0) and carbon sources quantitatively in fermentation processes (Schunck et al. 1987a,b; Riege et al. 1989; Huth et al. 1990a,b,c; cf. Sect. 2.4.1). This proton extrusion of a growing yeast in batch cultures, leading to an acidification of the medium and measured as alkali consumption required to maintain a constant pH value, was used to develop a miniaturized fermentation system to describe the growth of yeasts, and the growth-related substrate consumption (different carbon sources and NH_4 as N source) in batch or fed-batch cultures on-line, and more exactly than by determining these values with usual off-line methods (dry weight, optical density, and protein content). The miniaturized fermentation system (working volume about 90 ml) was applied to study various physiological regulation phenomena in yeasts like the regulation of growth rate and of cytochrome P450 and other enzyme contents in *C. maltosa* and other yeasts (*Y. lipolytica*, *Debaryomyces formicarius*, *S. cerevisiae*) by oxygen concentrations in the medium (Schunck et al. 1987a,b; Mauersberger 1985, and unpubl. results), and to test the potential for utilization of various substrates as carbon sources for growth of yeasts (Schunck et al. 1987a; Huth et al. 1990b).

The proton extrusion can be registered in a simple way and with high accuracy as an on-line parameter in the miniaturized fermentation system (Fig. 2). The small working volume (50–100 ml) permits the description of growth from substrate amounts of 5 to 50 mg quantitatively in fed-batch fermentation processes (Huth et al. 1990b). Application of the practically inert hydrocarbon phase pristane (1–2%) was used for testing the growth of yeasts on substrates (short-chain alkanes, 1-alcohols, aldehydes, fatty acids, and others) exhibiting potential toxic or inhibitory effects at the yeast culture, or to dissolve solid hydrocarbon substrates (long-chain alkanes, fatty alcohols, fatty acids, and others) in the pristane phase.

Procedure. The miniaturized fermentation apparatus (shown in principle in Fig. 2) was especially developed for these experiments and was in the main composed of simple laboratory equipment. A mantled cylindrical glass vessel (total volume 200 ml) served as the fermenter. An oxygen electrode (Clark type) was incorporated into this vessel from the side 2 cm above the bottom. The plastic cover contained drill holes for the pH electrode, the supply of the culture with air (and other gases like nitrogen or carbon monoxide), NaOH, and HCl to maintain a constant pH, for addition of substrates as well as for taking samples manually. This instrumentation was anchored in a piatherm plate, located on the cover, and the whole structure was fixed to the fermenter vessel by steel springs. A larger drillhole in the center of the cover served as an entry for the stirrer, as well as an outlet for the air stream. The experiments were performed under protected but not sterile conditions.

Yeast biomass was produced in shaking flasks or in a fermenter before using the substrates n-alkanes, glycerol, or glucose (Huth et al. 1990a). The yeast biomass can be used directly from the preculture after dilution with fresh medium (without yeast extract) or after harvesting the preculture by centrifugation. Approximately 1 to 3 g wet weight biomass was used for short-term substrate utilization experiments, whereas for longer-term cultivation experiments, the starting biomass was lower. The mineral salt media 1 and 2 (see Sect. 2.3.4) were used in these experiments and the cultivation conditions were as described (Huth et al. 1990a) below.

The fermentation conditions were as follows:

- Working volume 90 ml at the beginning of the experiments containing maximally about 3 g yeast wet weight biomass.
- Temperature 32 °C maintained by an ultrathermostat.
- Agitation 1800 rpm, six-blade-stirrer, driven by a laboratory stirrer.
- Aeration through a gas nozzle with 0.2–4 vvm (maximally 25 l/h), or streaming with nitrogen to regulate the pO_2 of the medium, or with CO to make inhibition studies for cytochromes P450 or a_3.

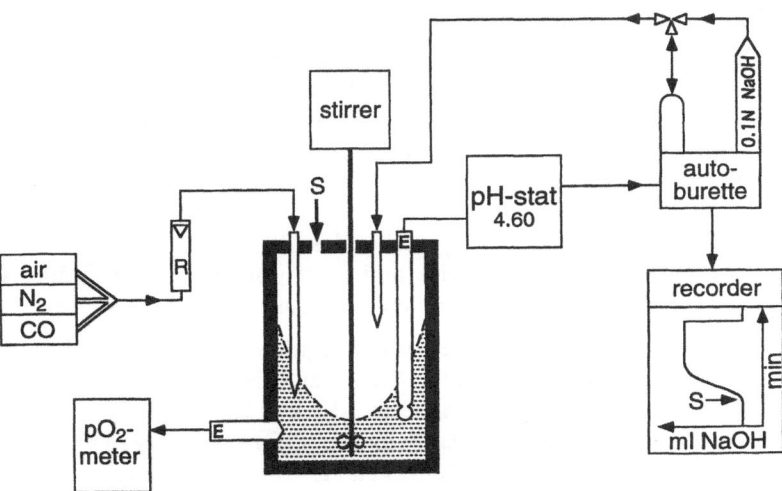

Fig. 2. Schematic representation of the miniaturized fermentation system used for testing substrate utilization, inhibition, and regulation experiments with *Candida maltosa* and other yeasts. *E* pH-electrode, and pO₂-electrode; *S* Manual substrate supply to and withdrawal of samples from the fermenter

- pH maintenance up scale at pH 4.6 with 0.1 N NaOH (for substrate utilization test in short time experiments) or with 0.5 N NaOH (for longer-term experiments) using an autotitrator (Radiometer, Copenhagen, Denmark) which registers alkali consumption over time.
- Substrate supply manually, after reaching carbon limitation, or after supplying a distinct amount of alkali.

Experimental data obtained with this miniaturized fermentation equipment using *C. maltosa* or other yeast strains and hydrocarbons, their primary oxidation products (fatty alcohols, fatty acids), glycerol or carbohydrates as carbon sources have been published (Mauersberger 1985; Schunck et al. 1987a,b; Blasig et al. 1989; Riege et al. 1989; Huth et al. 1990a,b,c).

2.5
The Enzymology of the Alkane Catabolic Pathway and Its Regulation in Candida maltosa

An important characteristic of alkane assimilation by yeasts is the flow of carbon from alkane substrates to syntheses of all cellular carbohydrates via fatty acids, which is quite different from the case of conventional substrates like carbohydrates. The alkane assimilation by yeasts was found to occur mainly via the monoterminal and diterminal oxidation pathways (cf. Sect. 2.4.2). It can be divided into several steps (cf. Fig. 3):

1. Transport of alkanes into the cell (Sect. 2.5.2).
2. Primary oxidation of alkanes to corresponding fatty acids (Sect. 2.5.3).
3. Activation of fatty acids to their CoA esters and subsequent metabolism of fatty acyl-CoA, degradation to acetyl-CoA and propionyl-CoA (in β-oxidation) or incorporation of fatty-acyl moieties into cellular lipids (Sect. 2.5.4).
4. Synthesis of tricarboxylic acid cycle intermediates from acetyl-CoA via the glyoxylate cycle (Sect. 2.5.5).
5. Gluconeogenesis and synthesis of polysaccharides (Sects. 2.5.5, 2.6).
6. Synthesis of amino acids, nucleic acids and other constituents, including a portion of the cellular fatty acids and lipids (Sect. 2.6).

Steps 1 and 2 are the most specific processes of the microbial alkane utilization. Steps 3 and 4 are as in fatty acid metabolism. Steps 4 and 5 are as in gluconeogenic substrates such as ethanol and acetate (cf. Sects. 2.5.5 and 2.6), and step 6 is as in various other carbon sources including carbohydrates (for reviews see Fukui and Tanaka 1981a; Tanaka and Fukui 1989; Müller et al. 1991a). This section deals with an overview on the enzymology and its regulation of the alkane catabolic pathway in the yeast *C. maltosa*, with special emphasis on the metabolic function of cytochrome P450 systems and the further enzymes involved in the primary oxidation of the alkanes to fatty acids and dicarboxylic acids. Some aspects of the intracellular localization and the subcellular organization of these pathways will be discussed below in Sect. 3. Especially the further metabolic pathways and the function of peroxisomes in them have been very intensively investigated in *C. tropicalis* and *Y. lipolytica* (for reviews see Tanaka and Fukui 1989; Barth and Gaillardin, Chap. 10, this Vol.). Therefore some results obtained with these yeast species will also be discussed shortly here.

2.5.1
Alteration in Yeast Cells During Growth on n-Alkanes

Genetic analysis indicates that the function of more than 80 genes may be required to bring about the phenotype of alkane assimilation. Among them, at least 26 genes are linked to alkane uptake and oxidation to fatty acids, as concluded from studies with *Y. lipolytica* (Bassel and Mortimer 1982, 1985). The development of the alkane utilization phenotype during transition from glucose to alkane as carbon source produces alterations and characteristic modifications in the yeast cell both at the biochemical (this section) and morphological levels (cf. Sect. 3):

- Chemical and structural alterations at the cell surface which are related to hydrocarbon transport (Osumi et al. 1975a; Meissel et al. 1976; Käppeli and Fiechter 1976, 1977; Käppeli et al. 1978, 1984; Dmitriev et al. 1980; Fischer et al. 1982; Belov et al. 1983; Bode and Köhler 1984 unpubl., Egorenkova and Belov 1984; Röber and Reuter 1984a,b,c,d; Schauer 1988).

- Induction of alkane-hydroxylating cytochrome P450(s) and NADPH-cytochrome c (P450) reductase (Mauersberger et al. 1980, 1984; Loper et al. 1985; Schunck et al. 1987a,b).

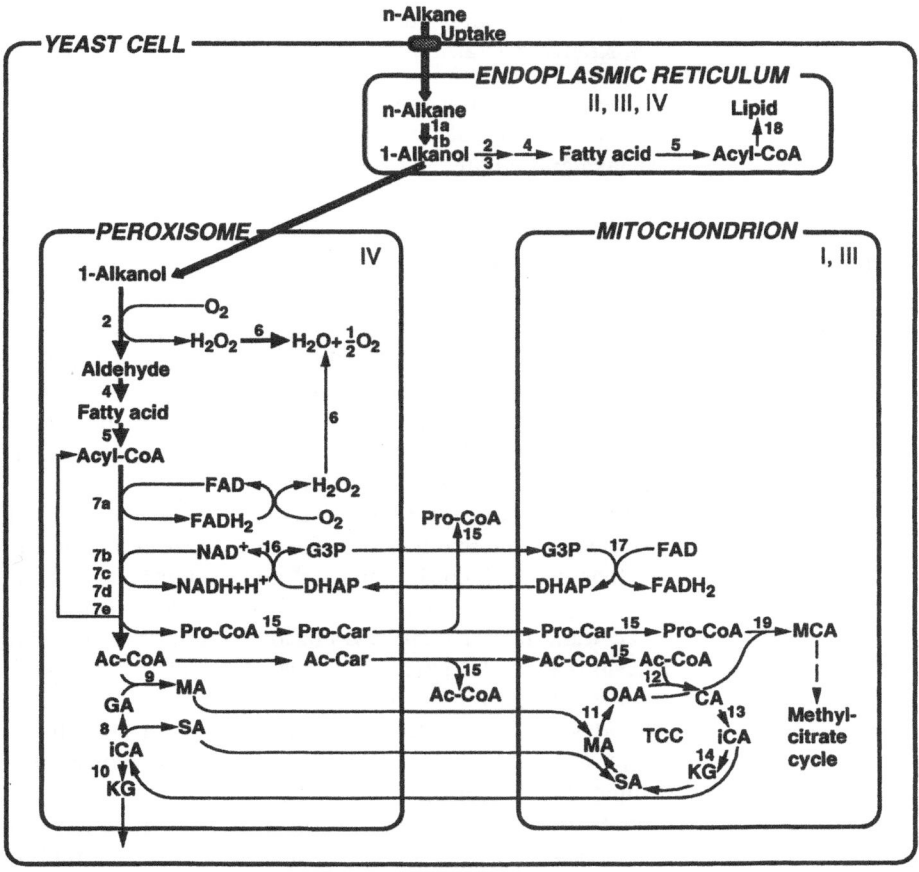

Fig. 3. The monoterminal n-alkane oxidation and its subcellular localization in yeasts. *Enzymes*: *1a* cytochrome P450 alkane monooxygenase, with *1b* NADPH-P450 reductase; *2* fatty alcohol oxidase (FAOD); *3* fatty alcohol dehydrogenase (NAD-dependent, FADH); *4* fatty aldehyde dehydrogenase (NAD-dependent, FALDH); *5* acyl-CoA synthetases (*5a* ACSI in ER; *5b* ACSII peroxisomal); *6* catalase; *7* β-oxidation system; *7a* fatty acyl-CoA oxidase (AOX); *7b* enoyl-CoA hydratase; *7c* 3-hydroxyacyl-CoA dehydrogenase; *7d* hydroxyacyl epimerase; *7e* 3-oxoacyl-CoA thiolase; *8* isocitrate lyase (ICL); *9* malate synthase (MS); *10* NADP-isocitrate dehydrogenase; *11* malate dehydrogenase; *12* citrate synthase; *13* aconitase; *14* NAD-isocitrate dehydrogenase; *15* carnitine acetyltransferase (CAT); *16* NAD-glycerol-3-phosphate dehydrogenase; *17* FAD-glycerol-3-phosphate dehydrogenase; *18* glycerol-phosphate acyl transferase; *19* methylcitrate synthase. *Abbreviations and substrates*: *n-Alkanes, 1-alkanol* fatty alcohol; *Aldehyde* fatty aldehyde; *Fatty acid, Acyl-CoA* fatty acyl-CoA; *Ac-CoA* acetyl-CoA; *Pro-CoA* propionyl-CoA; *Ac-Car* acetyl carnitine; *Pro-Car* propionyl carnitine; *Car* carnitine; *GA* glyoxylate; *iCA* isocitrate; *CA* citrate; *MA* malate; *OAA* oxalacetate; *SA* succinate; *KG* α-ketoglutarate; *G3P* glycerol-3-phosphate; *DHAP* dihydroxyacetone phosphate; *MCA* methylcitrate; *TCC* tricarbxylic acid cycle. The *Roman numbers* symbolize other transformations of fatty acids as key intermediates; *I* elongation of fatty acids; *II* desaturation of fatty acids (ER); *III* lipid biosynthesis (ER or mitochondria); *IV* ω-hydroxylation (P450-dependent), and further diterminal oxidation of the ω-hydroxy fatty acids by FAOD, FADH, and FALDH to dicarboxylic acids, which, after activation by ACSII, are degraded in the β-oxidation (*7*)

• Induction of fatty alcohol and fatty aldehyde-oxidizing enzymes (Mauersberger et al. 1984, 1987, 1992b; Krauzova et al. 1986).

• Increased formation of peroxisomes (Osumi et al. 1975b) and induction of peroxisomal β-oxidation, of the enzymes of glyoxylate cycle, and of gluconeogenesis (Tanaka et al. 1982; Tanaka and Fukui 1989).

2.5.2
Uptake of n-Alkanes

Microbial n-alkane utilization starts with the uptake of the apolar substrate by the cell. Despite many efforts, the basic principle of this process has not yet been identified. Most findings agree with the suggestion that the uptake of n-alkanes by the yeast cell is a passive process, which is facilitated by special hydrophobic properties and structures of the yeast cell (for details see Davidova and Rachinskii 1979; Bühler and Schindler 1984; Tanaka and Fukui 1989).

The chemical composition, structure, and biosynthesis of C. maltosa cell wall components mannan (proteophosphomannan – outer layer) and glucan fibrils (forming the inner layer together with chitin, mannan, and protein) were intensively studied by several authors (Bos 1975; Hirata and Ishitani 1978; Rademacher and Reuter 1978; Grimmecke and Reuter 1980, 1981a,b,c,d; Grimmecke et al. 1981; Fischer and Reuter 1982, Nüske et al. 1982; Belov et al. 1983; Egorenkova and Belov 1984; Röber and Reuter 1984a,b,c,d; Bovina et al. 1986, 1988). The structure of these cell wall polysaccharides in yeasts is strain- and species-specific, and therefore used as taxonomic characteristics using the proton magnetic resonance spectroscopy (Hirata and Ishitani 1978; cf. Sect. 1.2).

Significant differences in the chemical composition of the cell wall between glucose- and alkane-grown cells of C. maltosa were established. The cell wall of alkane-grown yeast cells is more hydrophobic compared with glucose-grown cells (Röber and Reuter 1984d; cf. Sects. 2.3.3 and 3). The lipid content in the cell wall of alkane-grown cells increased twofold (Bos 1975).

The lipopolysaccharides of the cell wall, obviously responsible for adhering microemulsions of alkanes (as reported by Käppeli and Fiechter 1976, 1977; Käppeli et al. 1978, 1984 for C. tropicalis), whose formation is induced by alkanes, have been isolated and characterized as proteophosphomannan complexes (PPM) for C. maltosa (for C. tropicalis as fatty acid-proteomannan complexes) containing covalently linked lipids in form of threonine-bound phosphoacyl-glycerides (Grimmecke and Reuter 1980; Lerche and Kretzschmar 1980, 1986; Röber and Reuter 1984d for C. maltosa; Käppeli et al. 1978).

Belov et al. (1983) isolated from the culture liquid of alkane-grown C. (guilliermondii) maltosa two glycoproteins of 95 and 105 kDa molecular weight and a fraction of proteomannan fragments. Immunocytochemically, with ferritin-labeled antibodies against the two fractions, both proteins were shown to be localized in the outer part of the cell wall channels only with alkane-grown yeast cells, whereas the mannan was found distributed diffusely over the cell wall surface of alkane- and glucose-grown cells. With these antibodies obtained, Davidov and

Belov (pers. comm.) were able to inhibit alkane utilization by this yeast. Egorenkova and Belov (1984) reported a looser cell wall structure of alkane cells compared with glucose cells, containing more high molecular mannan.

The regulation phenomena of the biosynthesis of the cell wall components were studied by Röber and Reuter (1984a,b,c, 1985) and Röber et al. (1984a,b) in the strain *C. maltosa* H. Cell wall hydrophobicity is obviously regulated by changes in the proteophosphomannan complexes, especially by a decrease in the saccharide-phosphates (Grimmecke and Reuter 1981a,b,c; Grimmecke et al. 1981) and an increase in the threonyl-phosphoglyceride components (Grimmecke and Reuter 1980) in alkane-, ethanol- or acetate-grown cells (Röber and Reuter 1984d). Interestingly, the hydrophobicity of the *C. maltosa* cell wall is already increased after consumption of the glucose or in resting cells, probably connected with the derepression of the gluconeogenesis and the glyoxylate cycle (Röber and Reuter 1984d; Schauer 1988).

These chemical alterations of the cell wall are obviously related to hydrocarbon transport, and closely connected with peculiarities and rearrangements of the cell wall structures (such as formation of special channels, accompanied by slime-like outgrowths or also called protrusions on the cell surface, increased plasma membrane invaginations and membrane vesicles) discussed in detail in Sect. 3. The slime-like outgrowths reach the cell membrane through the electron-dense channels, and endoplasmic reticulum (ER) is arranged regularly beneath each channel. This observation has led to the hypothesis that alkanes attached to the protrusions or hydrophobic outgrowths may migrate through the channels via plasma membrane to the ER, the site of alkane hydroxylation by cytochrome P450 monooxygenase systems (Tanaka and Fukui 1989; cf. Sects. 2.5.3 and 3.4).

The role of mainly neutral lipids containing membrane vesicles (lipid granules, lipid vacuoles, or lipid inclusions) in the alkane transport from the plasma membrane to the ER or in the intracellular lipid transport was often discussed (Kasanzev et al. 1975; Belov et al. 1976; Davidova et al. 1977a,b; Davidova and Rachinskii 1979; Davidova et al. 1979), but there is still a lack of clear evidence for their transport function of n-alkanes.

2.5.3
The Enzymes of Primary Alkane Oxidation to Fatty Acids and Their Regulation in *Candida maltosa*

The enzymology of the alkane degradation pathway in yeasts and its regulation by external factors like carbon source and oxygen concentration in the medium were intensively studied in connection with the application of *Candida* and *Yarrowia lipolytica* strains for production of SCP and of useful oxidation products (cf. Sects. 1, 2.4, and 6).

The Role of Cytochrome P450 and Fatty Alcohol Oxidase in Alkane Oxidation

The enzymes catalyzing the first steps of the n-alkane assimilation in *C. maltosa* up to the fatty acid have been identified and characterized.

Cytochrome P450 Dependent Alkane Monooxygenase Systems. As already mentioned, the first enzymatic step of hydrocarbon assimilation by yeasts is the terminal hydroxylation of the n-alkane by a cytochrome P450 enzyme system (Fig. 3) leading to the fatty alcohol as the first intermediate of alkane catabolism. The enzyme system is ER membrane-bound and consists of a NADPH-cytochrome P450 reductase transferring electrons, and a cytochrome P450 acting as hydroxylase in a typical monooxygenase reaction. Both enzymes (P450Cm1 in this case) were first obtained in a highly purified state from alkane-grown *C. maltosa* EH15 (Riege et al. 1981; Honeck et al. 1982; Schunck et al. 1983a,b), and later from other *C. maltosa* strains (Avetisova et al. 1985; Sokolov et al. 1986a,b; Takagi et al. 1989; Mauersberger et al. 1992a; cf. Table 4). Successful reconstitution experiments with the highly purified enzymes (Honeck et al. 1982; Schunck et al. 1983a,b; Mauersberger et al. 1992a) and the selective inhibition by carbon monoxide of the first step of alkane degradation under in vivo conditions (Schunck et al. 1987a,b) provided strong evidence that the initiating reaction is catalyzed by cytochrome P450. A second P450 system is involved in the ω-hydroxylation of fatty acids as the first step of the diterminal oxidation pathway occurring in yeast during alkane degradation (Fig. 3). This P450Cm2 protein has been characterized only after cDNA cloning and heterologous expression in *S. cerevisiae* (Schunck et al. 1991), which is the general approach for the characterization of individual P450 forms from *C. maltosa* (Schunck et al. 1991; Scheller et al. 1992, 1994, 1996; Zimmer et al. 1995). Additionally, Mauersberger et al. (1992a) purified a third P450Cm3, additionally to the P450Cm1, from decane-grown *C. maltosa* VSB779 (Table 4) with activities towards both substrate classes (alkanes and fatty acids). Today, eight alkane-inducible cytochrome P450 genes are known in *C. maltosa* (Ohkuma et al. 1995a). Their functions are currently under investigation (Zimmer and Schunck 1995; Zimmer et al. 1995; see also Sect. 4.4.2 for their characteristics and molecular biology data).

Fatty Alcohol Oxidase (FAOD) and Fatty Aldehyde Dehydrogenase (FADH). The next step in the alkane catabolic pathway, the oxidation of the n-alkan-1-ol to the corresponding fatty aldehyde (Fig. 3), is catalyzed mainly by a FAOD enzyme instead of a NAD(P)-dependent dehydrogenase (FADH) activity, as assumed earlier in the literature (see Mauersberger et al. 1987, 1992b). Since the middle of the 1980s, a growing body of evidence has indicated that molecular oxygen instead of NAD is the essential electron acceptor of fatty alcohol oxidation and that simultaneously H_2O_2 arises. Consequently, the enzyme responsible for fatty alcohol oxidation is a H_2O_2-forming oxidase (FAOD). Ilchenko and Tsfasman (1988) have isolated this enzyme from alkane-grown *C. famata* (formerly *Torulopsis candida*). Studies with *C. maltosa* (Krauzova et al. 1986; Mauersberger et al. 1987, 1992b; Blasig et al. 1988; Shilova et al. 1989), *P. guilliermondii*, *C. tropicalis*, and *Y. lipolytica* (for references see Mauersberger et al. 1992b) support the assumption of a wide distribution of FAODs in alkane-assimilating yeasts. The alkane-induced FAOD of *C. maltosa*, obviously localized in peroxisomes (see Sect. 3.4), was characterized in its broad substrate specificity and stereoselectivity (oxidation of primary and secondary alcohols, 1,ω-diols, ω-hydroxy fatty acids, phenylalkanols, and terpene alcohols, see also below) in the membrane-bound state (Mauersberger et al. 1992b) and after purification (Sokolov and Mauersberger, unpubl. results; Table 4).

Fatty Aldehyde Dehydrogenase (FALDH). The subsequent step, the oxidation of fatty aldehydes into fatty acids, was clearly demonstrated as a membrane-bound, NAD-dependent dehydrogenase (FALDH) reaction by its detection in *C. maltosa* (Mauersberger et al. 1984, 1987; Krauzova et al. 1986) and other yeasts, and finally by its purficiation from *C. (T. candida) famata* (Ilchenko and Tsfasman 1987).

Besides the inducible membrane-bound FAOD and FALDH activities, there exist also cytosolic NAD- or NADP-dependent alcohol and aldehyde dehydrogenases (FADH and FALDH) in alkane-assimilating yeasts. For *C. maltosa*, two such soluble FADH activities have been demonstrated (Sharyshev 1989, unpubl. results; Mauersberger et al. 1992b). Therefore, the conclusion that only oxidase-type enzymes are involved in fatty alcohol oxidation in yeast should be considered with caution. Additionally, there are hints for the presence of different FAOD activities in *Y. lipolytica* (Ilchenko et al. 1994).

Regulation of Enzymes Oxidizing Alkanes to Fatty Acids

The formation of the alkane-induced P450 system, FAOD and FALDH in *C. maltosa* cells was investigated under the influence of such extracellular factors as carbon source and oxygen due to their key position as the rate-limiting step in the alkane oxidation by yeast cells (Mauersberger et al. 1980; Schunck et al. 1987a,b), especially under the oxygen limitation conditions often occurring in the industrial SCP fermentation processes (Riege et al. 1989).

Regulation of Cytochrome P450 Biosynthesis in C. maltosa. As outlined above and in Sect. 4.2.2. (see also Fig. 6), multiple alkane-inducible P450 forms may occur in *C. maltosa* which differ in physiological functions and regulation of expression. Therefore, the unequivocal identification of the induced P450 form(s) and the availability of specific probes have become the most crucial points of regulation studies. Based on recent progress in DNA cloning, first details have been reported concerning regulation mechanisms of alkane-inducible P450 forms of *C. maltosa* (Ohkuma et al. 1995a; cf. Sect. 4.4.2), *C. tropicalis* (Seghezzi et al. 1991, 1992) and of the lanosterol 14α-demethylating P450 in yeast (see Müller et al. 1991a).

On the other hand, a number of earlier studies cannot be clearly interpreted with respect to the P450 form(s) actually occurring under the applied cultivation conditions. However, they will be included in this chapter because they might stimulate future experiments to elucidate the regulation of individual P450 forms in yeasts.

Induction by Long-Chain n-Alkanes. Following the initial reports in the 1970s for *C. tropicalis*, induction of P450 by long-chain n-alkanes has been reported for a large number of yeast species being able to utilize these unconventional substrates (reviewed by Müller et al. 1991a). Various compounds having chemical structures

Table 5. Regulation of the enzyme content in *Candida maltosa* by the carbon source and by the oxygen concentration in the medium. Content of some constituents in the cell free extract (S_3) obtained from spheroplasts of *C. maltosa* cells harvested from the exponential growth phase of batch cultures on the carbon sources glucose (1.5%, in shaking flasks), glycerol (3%), and n-alkanes (Parex 3%, in a fermenter) under oxygen non-limited (1.2 vvm air supply) and oxygen limited conditions (0.2 vvm air supply, for details see Mauersberger et al. 1984; Mauersberger 1985)

Constituent content or activity	Glucose (n = 2)	Glycerol (n = 3)	Alkanes, oxygen not limited (n = 6)	Alkanes, oxygen limited (n = 3)
Cytochrome P450	0[a]	10	62	156[b]
NADPH-cyt c reductase	61	67	213	239
AHA	10	30	95[c]	52[c]
Cytochrome b_5	46	101	108	141
NADH-cyt c reductase	52	81	61	78
FAOD	3	20	98	96
FADH	38	56	268	307
FALDH	46	167	290	256
Catalase	76	280	597	1201
Isocitrate lyase	10	n.d.[d]	190	n.d.
Cytochrome (a + a_3)	109	48	58	146
Cytochrome oxidase	261	105	135	374
Protoheme	905	439	594	1127

Values are means of indicated (n) preparations; cytochrome and heme content: pmol/mg protein; enzyme activities: nmol/min × mg protein (mU/mg), excepting catalase activity − mU × 10^{-3} and AHA with hexadecane as substrate − pmol/min × mg protein.
Abbreviations: AHA − Alkane hydroxylase activity determined with ^{14}C-C_{16}
FAOD − Fatty alcohol oxidase, determined with 1-decanol
FADH − Fatty alcohol dehydrogenase, NAD$^+$ dependent, with 1-decanol
FALDH − Fatty aldehyde dehydrogenase, NAD$^+$ dependent, with tridecanol.
[a] Small amounts of P450 were detected in glucose-grown cells by RIA (3 pmol/mg microsomal protein), Kärgel et al. (1984, 1985), Schunck et al. (1987b).
[b] Values up to 250 pmol/mg protein during long-term oxygen limitation on alkanes (hexadecane or tetradecane), Schunck et al. (1987b).
[c] Values are not representative due to different residual alkane content in the cell extracts.
[d] n.d. − not determined.

related to n-alkanes were found to induce the formation of P450 and its corresponding NADPH-P450 reductase in *C. maltosa* (Table 5; Mauersberger and Matyashova 1980; Ilchenko et al. 1980; Mauersberger et al. 1980, 1981; Mauersberger 1985; Schunck et al. 1987b). Among these, long-chain n-alkanes (C_{12}–C_{14}), alkenes, and some phenyl alkanes turned out to be the most potent inducers. A remarkable induction effect has been also described for long-chain secondary alcohols. A three to four times lower P450 content was detected in fatty alcohol- and fatty acid-grown cells when compared with the values reached during cultivation on n-alkanes, which has now been verified by Northern blot analysis for the expression of the individual P450 genes *ALK1* to *ALK7* (Ohkuma et al. 1995a). The expression of these individual P450 genes of the *CYP52* family is differently induced using the substrates alkanes, alkanols, alkanals, or fatty acids. Nevertheless, the spectral P450 content of alkane-grown cells can be almost completely attributed to P450 forms involved in terminal alkane and fatty acid hydroxylation. However, the possibility that several other alkane-inducible P450 forms with unknown substrate specificities exist has to be considered (see Sect. 4).

Tetradecane- and hexadecane-mediated P450 induction can be almost completely repressed by the simultaneous addition of glucose (above 0.5%) to the culture medium (Takagi et al. 1980a,b; Mauersberger et al. 1981). Interestingly, induction of P450 in *C. maltosa* and *Y. lipolytica* by hexadecane was more sensitive to glucose repression than the induction by decane, which could be a result of the presence of differently regulated P450 genes in these yeasts (Mauersberger 1991). Glycerol and galactose were found to be no or less repressive substrates (Table 5; Mauersberger et al. 1981; Mauersberger 1985).

Alkane-mediated P450 induction requires protein and lipid de novo synthesis, as suggested by inhibition with cycloheximide and cerulenin (Takagi et al. 1980; Mauersberger et al. 1981). Using a radioimmunoassay based on antibodies against the highly purified alkane hydroxylating P450Cm1, Kärgel et al. (1984) demonstrated that microsomes of alkane-grown *C. maltosa* cells have a 100–300-fold higher specific content of the corresponding protein than detectable after cultivation on glucose (Table 5). The relative amounts of P450-specific mRNA were found to parallel the changes in the spectrally and immunologically determined P450 content (Wiedmann et al. 1986). These results indicate that the formation of the alkane-inducible P450 forms is mainly regulated at the transcriptional level. Cells cultivated on glucose and n-alkanes can be considered to represent the repressed and induced state, respectively. Glycerol-grown cells may represent a derepressed state. Microsomes isolated from glycerol-grown *C. maltosa* cells are active in alkane hydroxylation and contain significant amounts of P450 (Mauersberger et al. 1984), which is immunologically indistinguishable from that of alkane-grown cells (Kärgel et al. 1984; Schunck et al. 1987b). Moreover, the occurrence of P450 during cultivation on glucose upon transition to the stationary phase may indicate derepression of alkane-inducible P450 forms and/or induction by fatty acids due to activation of lipid turnover (Ilchenko et al. 1980).

During batch cultivation of *C. maltosa* on n-hexadecane, P450 formation and growth curve correlate in a characteristic manner (Mauersberger and Matyashova 1980). The cellular P450 content is strongly increased in the lag phase of growth upon transition from glucose to n-alkane utilization, reaches a nearly constant level in the exponential phase, and declines rapidly after consumption of the carbon source in the stationary phase. The time course of P450 induction in *C. maltosa* growing on n-decane revealed two steps, a first maximum in the middle of the exponential phase of growth, and the higher second maximum in the stationary phase (Avetisova et al. 1985, 1990; Sokolov et al. 1986a,b). It was supposed that the high P450 content found in the stationary phase on decane may be a result of the induction of more than one P450 form. This was demonstrated finally by purification of the P450Cm3 besides P450Cm1 from decane-grown cells (Mauersberger et al. 1992a).

Influence of Oxygen on Alkane-Mediated P450 Expression. In particular, under oxygen limitation conditions, P450-catalyzed alkane hydroxylation appears to represent the rate-limiting step of the whole pathway (Schunck et al. 1987a). In fact, significant differences were found in the pO_2 dependence of utilization rates of n-alkanes, fatty alcohols, fatty acids, and glucose. It was concluded that the P450 system is distinguished by the lowest oxygen affinity among the O_2 activating enzymes involved in alkane assimilation (Schunck et al. 1987a). As studied in detail with *C. maltosa* (Table 5; Mauersberger et al. 1980, 1984; Schunck et al. 1987b; Ilchenko et al. 1989, 1991) and with other yeasts (*C. tropicalis*, *Debaryomyces formicarius*, *Y. lipolytica*, Mauersberger et al. 1991; see also Müller et al. 1991a), the cellular P450 content is strongly increased by transition to oxygen-limited growth during cultivation on long-chain n-alkanes. Enhanced P450 formation was also observed maintaining oxygen saturation but inhibiting more selectively the alkane hydroxylation reaction by introducing low concentrations of carbon monoxide into the culture medium (Mauersberger 1985; Schunck et al. 1987a,b). However, this effect did not occur when, instead of n-alkanes, fatty alcohols were used as the carbon source. In conclusion, enhanced P450 formation appears to depend on two factors, a decrease in the enzymatic activity of the alkane-hydroxylating P450 system (O_2-deficiency or CO inhibition) and the presence of n-alkanes, which act both as substrate and as inducer of P450. The most simply interpretation of these findings may be that the P450 formation is mainly regulated by the intracellular inducer concentration which depends on the relative rates of alkane transport into the cell and the actual alkane hydroxylating activity of the enzyme system (Schunck et al. 1987b). According to this, oxygen limitation results in enhanced substrate induction by inducer accumulation. Immunological studies revealed that it is indeed the alkane hydroxylating P450 which increases during oxygen limitation (Schunck et al. 1987b). In fact, Wiedmann et al. (1988a) found a parallel increase of the P450 encoding mRNA.

However, to qualify the simple scheme of inducer accumulation, it must be considered that oxygen limitation does not only affect the synthesis of the alkane-inducible P450. Moreover, in different degrees, catalase, cytochrome oxidase, cytochrome c peroxidase, and other hemoproteins respond to oxygen limitation (Table 5; Mauersberger et al. 1984; Ilchenko et al. 1989, 1991). Therefore, it may be assumed that derepression of heme biosynthesis plays an important role. It remains for further studies on promoter functions to distinguish between enhanced substrate induction, heme control, and the not completely excluded possibility of a more direct oxygen effect, shown for other yeast proteins like catalase, as discussed by Müller et al. (1991a,b).

All our results indicate that the main conclusions about the physiological function and the regulation of alkane-induced P450 forms of *C. maltosa* also apply to other yeast species like *C. tropicalis*, *Debaryomyces formicarius*, and *Y. lipolytica* (Mauersberger et al. 1991).

Regulation of the FAOD and FALDH in C. maltosa. Like the alkane-hydroxylating P450 system, the two other enzymes, membrane-bound FAOD and FALDH, involved in the oxidation of hydrocarbons to the corresponding fatty acids were also induced by growth of *C. maltosa* on n-alkanes in comparison with the carbon source glycerol (Table 5; Mauersberger et al. 1984, 1992a; Shilova et al. 1989). All these enzymes were found to be repressed significantly using glucose as growth substrate, although to a different degree. The regulation of the cytosolic FADH or ADH activities present in *C. maltosa* (Sharyshev 1991, unpubl.; Mauersberger et al. 1992b) remains to be clarified.

In contrast to the alkane-induced P450, the content of these two enzymes was not significantly influenced by oxygen limitation of the cells (Table 5). A further increase in FAOD activity during prolonged strict oxygen-limited growth on alkanes was also observed, although this oxygen-mediated effect was not so strong as seen in P450 (Shilova et al. 1989; Mauersberger et al. 1992b).

Spectral Determination of P450 in Whole Yeast Cells

Spectral determination of P450 using the CO-difference spectrum can be performed using whole cells. This method has been widely used to characterize P450 formation in yeast cells under the influence of such extracellular factors as carbon source and oxygen. Dithionite-reduced CO-difference spectra yield a reliable quantitation of the total P450 content, provided that there is no severe spectral interference with mitochondrial cytochrome a_3. To avoid this problem, a simple method was found to improve the estimation of low P450 content in whole cells (Sharyshev et al. 1983; Schunck et al. 1987b; Ilchenko et al. 1989). It involves omitting dithionite and adding antimycin A, potassium cyanide (1–2 mM in the assay), or oxidizing the cells in advance by streaming with air, to avoid interference with cytochrome a_3 by blocking its reduction. This modified method was successfully applied to detect small amounts of P450 in whole cells of *C. maltosa* (Schunck et al. 1987b) and other yeasts like *S. cerevisiae* (Yekhvalova et al. 1989) and *Y. lipolytica*, where the interference problem is more striking due to higher cytochrome oxidase content (Mauersberger 1991; Mauersberger et al. 1991). The cytochrome P450 content can be calculated from CO-difference spectra using an extinction coefficient of $91 \, mM^{-1} \times cm^{-1}$. It should be mentioned that by this method it is not possible to discriminate between different forms of cytochrome P450 with different substrate specificities or between enzymatically active and inactive, but spectral detectable forms of cytochromes P450.

Induction of P450 in yeasts by Different Hydrophobic Compounds

The induction experiments can be performed as described by Mauersberger et al. (1981) in shaking flasks or in a (miniaturized) fermenter by incubating glucose-precultured yeast cells (0.5–1 g wet weight cells) for at least 4 h at 30 °C in 100 ml mineral salt medium containing 0.05% (w/v) yeast extract and the hydrophobic compound to be tested (like 0.5% v/v decane or hexadecane) or no carbon source as control (Mauersberger and Matyashova 1980; Mauersberger et al. 1980, 1981;

Takagi et al. 1980a; Mauersberger 1991). Glycerol and galactose as nonrepressive substrates are suitable substrates when the induction potential of nondegradable hydrocarbons is to be tested for P450 induction (Mauersberger et al. 1981; Mauersberger 1985).

Heterologous Expression of P450 Genes of *Candida maltosa* in *Saccharomyces cerevisiae*

To test the function of individual alkane-inducible P450 forms of *C. maltosa*, their heterologous expression in *S. cerevisiae* under the control of the *GAL10* promoter using the vector YEp51 (Schunck et al. 1991; Scheller et al. 1994; Zimmer and Schunck 1995; Zimmer et al. 1995), or in *C. maltosa* under control of the homologous *GAL10* promoter (Ohkuma et al. 1995b) was performed. For expression of the authentic P450 forms in *S. cerevisiae*, recombinant PCR was used to exchange the CTG codons in the genes or cDNAs of cytochrome P450s by the triplet TCT (Zimmer and Schunck 1995, see also Sect. 4.4.3). Both the unchanged and modified cDNAs were ligated into the *SalI/Bam*HI sites of the expression vector YEp51. The recipient strain *S. cerevisiae* GRF18 (*leu2 his3*) was transformed according to Keszenman-Pereyra and Hieda (1988) and cultivated in yeast minimal medium (0.67% YNB without amino acids), 50 mg/l L-histidine, and 2% raffinose as the carbon source to a cell density of 0.7–1.0 × 10^8 cells/ml. P450 induction was performed by adding 2% galactose to the cultivation medium. The amount of P450 expressed was determined by CO-difference spectra with whole cells or with cell-free extracts. The P450 expression is maximal after 8 to 15 h cultivation in the presence of galactose.

Purification of Proteins from *Candida maltosa*

A number of proteins from *C. maltosa* have been purified and characterized (see Table 4). Among them, interest was mostly concentrated on enzymes systems (P450 systems and FAOD) involved in the oxidation of n-alkanes to the corresponding primary oxidation products fatty alcohols and fatty acids, obtained from alkane-grown *C. maltosa* (Riege et al. 1981; Honeck et al. 1982; Schunck et al. 1983a,b; Sokolov et al. 1986a,b; Takagi et al. 1989; Mauersberger et al. 1992a,b). Furthermore, a complex method for the simultaneous large-scale purification of P450s, NADPH-P450 reductase, cytochrome b_5, and FAOD from decane-grown *C. maltosa* VSB779 by ultrafiltration, phase partition, and one-step chromatography was developed, and superoxide dismutase, epoxide hydrase and catalase were purified from this strain (Avetisova 1991; Avetisova and Davidov 1993; Avetisova et al. 1993).

Additionally, P450Cm1 (P450 52A3, or its modified variant), P450Cm2 (P450 52A4), and the NADPH-P450 reductase of *C. maltosa* have also been purified after high-level heterologous expression of their genes in *S. cerevisiae* (Scheller et al. 1992, 1994, 1996; Kärgel et al. 1996; Schunck et al. 1996).

Purification of Cytochromes P450 from C. maltosa *Grown on n-Alkanes.* Isolation of P450 was mostly started from cells cultivated on n-alkanes under strong oxygen limitation, thus strongly increasing the specific content of alkane-induced P450 in the biomass by the factor 4 to 6 (Mauersberger et al. 1980, 1981, 1984; Schunck et al. 1987a,b). Up to now, only the P450Cm1 (P450 52A3) and Cm3 (P450 52A5) have been purified from different strains of *C. maltosa*. P450Cm1 was first isolated and characterized from *C. maltosa* EH15 grown on an n-alkane mixture from C_{11} to C_{19} (Parex) under oxygen limitation (Riege et al. 1981; Schunck et al. 1983a,b). Similar results were obtained later by Takagi et al. (1989) for the purification of the P450Alk1 (P450 52A3) protein from the strain IAM12247 grown on tetradecane. From *C. maltosa* VSB779 grown on decane under strong oxygen limitation, both P450Cm1 and P450Cm3 were isolated by a modified method (Avetisova 1991; Mauersberger et al. 1992a) using chromatography on octyl-Sepharose.

Purification of NADPH-Cytochrome P-450 Reductase. A modified method according to Honeck et al. (1982) is described, which can be applied for the purification of reductases from *C. maltosa* and also after its heterologous expression in *S. cerevisiae* (Kärgel et al. 1996; Scheller et al. 1996). The reductase purified by this method is active in reconstitution experiments with homologous yeast and heterologous mammalian P450s (Shkumatov 1993; Scheller et al. 1994, 1996), respectively.

Procedure

1. Microsomal membrane fractions containing the NADPH-cytochrome P450 reductase were prepared from the yeast *C. maltosa* EH15 as described (Riege et al. 1981; Schunck et al. 1978, 1983a,b).

2. Microsomes (4–5 g of protein) were diluted to a final concentration of 10 mg/ml in 100 mM Tris/HCl buffer, pH 7.7, containing 30% glycerol, 1 mM EDTA, 1 mM DTT, 0.5 mM PMSF, and 5 μM FAD/FMN. A solution of 10% (w/v) CHAPS was added to give a final concentration of 1.5% (w/v).

3. After stirring for 30 min at 4°C, the solubilized microsomes were centrifuged at 75 000 g for 60 min.

4. The supernatant was diluted to a protein concentration of 3 mg/ml with 25 mM potassium phosphate buffer, pH 7.7, containing 0.3% Präwozell WON-100, 0.1% sodium cholate, 0.1 mM EDTA, 0.1 mM DTT, 0.5 mM PMSF, 1 μM FAD/FMN (buffer B), and applied to a DEAE-Toyopearl 650 column (1 × 10 cm) previously equilibrated with buffer B.

5. Ion-exchange chromatography and the following purification step on hydroxylapatite were carried out as described by Honeck et al. (1982), with the exception that Präwozell WON-100 was omitted from the equilibration and elution buffers for chromatography on hydroxylapatite.

6. Finally, the NADPH-cytochrome P450 reductase was purified by affinity chromatography on 2,5-ADP-Sepharose. Conditions used for loading, washing, and elution of the affinity resin were identical to those described by Ardies et al.

(1987), with the exception that CHAPS, Lubrol PX and BHT were omitted from all buffers.

The estimated NADPH-cytochrome P450 reductase activity of the preparation was 68.2 μmol cytochrome c reduced/min × mg protein when assayed at 25 °C in 50 mM Tris/HCl buffer, pH 7.7, 0.05 mM cyt c, 3.3 mM KCN, 0.1 mM NADPH, using an extinction coefficient of 21 mM^{-1} × cm^{-1} at 550 nm. The reductase prepared by this method was free of other constituents of the microsomal electron transfer systems, e.g., P450, cytochrome b$_5$, and NADH-cytochrome b$_5$ reductase. A single band in the SDS-PAGE shows a homogenous protein with an apparent molecular weight of 79 kDa (Honeck et al. 1982).

Fatty Alcohol Oxidase and Its Stereoselectivity in 2-Alkanol Oxidation

Note on Method. Fatty alcohol oxidase (FAOD) of *C. maltosa* can be used to perform stereoselective oxidation of secondary alcohols to obtain enantiomeric pure S(t) secondary alcohols. The enzyme is present in a crude membrane fraction obtained from cells grown on alkanes (Mauersberger et al. 1992b, cf. Sect. 6).

Preparation of the Crude Membrane Fraction. Candida maltosa cells of the late (linear) growth phase were harvested by centrifugation and the preparation of the crude membrane fraction after mechanical cell disintegration with glass beads in a Dyno mill was performed according to earlier publications using the method of calcium precipitation of membranes at 6000 g or high-speed centrifugation (100 000 g, 1 h) from the 6000 g supernatant (Riege et al. 1981; Mauersberger et al. 1987).

Enzyme Assays. Fatty alcohol oxidase (FAOD) activity was assayed measuring the substrate-dependent oxygen consumption or hydrogen peroxide formation in presence of sodium azide to inhibit catalase activity as described earlier in detail (Mauersberger et al. 1987; Blasig et al. 1988).

The enzyme activity was determined at 30 °C under magnetic stirring in 2.5 or 3 ml 100 mM Tris-HCl buffer, pH 8.8, in an open chamber equipped with a hydrogen peroxide electrode or in a closed chamber using an oxygen electrode.

After addition of the FAOD-containing protein sample, the reaction was started with substrates dissolved in up to 150 μl acetone. The amount of protein sample added was varied from 0.1 to 1 mg protein for determination of initial rates to up to 5 mg for determination of total amounts of the cosubstrate oxygen used or hydrogen peroxide formed during the reaction per substrate concentration added (measurements of stoichiometry). Saturation of the buffer with air resulted in an O$_2$ concentration of 240 μM under the conditions used. For H$_2$O$_2$ measurement, the electrode was calibrated with hydrogen peroxide solutions defined by spectral determination at 240 nm.

Product Identification. To identify the products of the alkanol oxidation, the volume of the above-described buffered reaction mixture was increased up to 10 ml, containing 25 to 50 mg protein of the crude membrane fraction.

The reaction was started by adding about $10\,\mu$mol substrate (1-alkanols or 2-alkanols) dissolved in 250–500 μl acetone. The incubations were continued until no further O_2 consumption in a parallel 3-ml analytical assay was observed, and stopped with 5 ml 8% H_2SO_4. Control assays were stopped before addition of substrates.

Both the stopped reaction and control mixtures were extracted according to the procedure of Dole and Meinerts (1960). The heptane extracts were analyzed by TLC and GLC as described by Blasig et al. (1988).

Further analysis of enantiomeric 2-alkanols was carried out by GLC (Hewlett Packard 5890; isotherm at 125 or 150 °C) after derivatization with iso-propylisocyanate on a 30-m fused silica capillary coated with an XE-60 bonded tertiary butylvalinamide phase (König 1987). Peaks of the racemic mixture could be resolved to about 80%. However, the reproducibility was measured better than ±2% and the accuracy was estimated better than ±3% in the range of small amounts of the R(–)-enantiomers.

2.5.4
Fatty Acid Oxidation

The first three steps of n-alkane oxidation up to the fatty acid outlined above (Fig. 3; Sect. 2.5.3) proceed without variation in the chain length. The arising fatty acids are intermediates with a central position in the metabolism which can be used directly for the biosynthesis of lipids or for obtaining acetyl-CoA. For lipid biosynthesis, desaturation and elongation or chain shortening of the fatty acids originating from n-alkanes are necessary. However, de novo synthesis of fatty acids is without importance in cells of *C. maltosa* growing on long-chain n-alkanes (Blasig et al. 1984, 1989, cf. Sect. 2.4.2).

The enzymology of the further oxidation of the fatty acid was not investigated in detail for *C. maltosa*, as had been done for *C. tropicalis* and *Y. lipolytica* (for a review see Tanaka and Fukui 1989). There are some results (cloned gene *POX* for the acyl-CoA oxidase, ACO, see Sect. 4.4.2; induced ACO activity in oleic acid-grown cells, Kamiryo et al. 1989) that allow the assumption that the enzymology could be comparable with the results obtained for these two yeasts. As shown for *Y. lipolytica*, two different acyl-CoA-synthetases (ACSI and II) introduce the fatty acids into the lipid biosynthesis (normally the minor part) or into the β-oxidation (Mishina et al. 1978). In alkane-utilizing yeasts, β-oxidation is inducible and exclusively localized in peroxisomes, in contrast to in animal cells, where both mitochondrial and peroxisomal β-oxidation systems exist. In most yeasts and fungi, β-oxidation is introduced by a hydrogen peroxide-generating fatty acyl-CoA oxidase converting acyl-CoA to the 2-enoyl-CoA. The following reactions are catalyzed by a trifunctional protein displaying the activities of enoyl-CoA hydratase and 3-hydroxyacyl-CoA dehydrogenase and of 3-hydroxyacyl-CoA epimerase. This three-functional protein has been found in all fungi, including *S. cerevisiae* (Kunau et al. 1988; Kunau and Hartig 1992). In the final reaction, 3-oxoacyl-CoA is cleaved by 3-oxoacyl-CoA thiolase to acetyl-CoA and a saturated acyl-CoA having two carbons less than the original fatty acyl-CoA (for references see Fukui and Tanaka 1981a,b; Tanaka and Fukui 1989; Müller et al. 1991a).

Additionally, fatty acids are ω-hydroxylated by alkane-inducible P450 systems for initiating diterminal oxidation (Fig. 3; Schunck et al. 1991), although this pathway is not achieved to any remarkable extent in growing cells (Blasig et al. 1984, 1988; Schauer 1988). However, in mutants, diterminal oxidation can become the main pathway allowing the industrial production of long-chain dicarboxylic acids (Uemura et al. 1988; Casey et al. 1990; cf. Sects. 4.2,6). At least the P450Cm2, characterized after cloning and heterologous expression in S. cerevisiae (Schunck et al. 1991; Zimmer et al. 1995), is involved in the ω-hydroxylation of fatty acids (see Sect. 2.4.2). Furthermore, Mauersberger et al. (1992a) purified a third P450Cm3, also active in ω-hydroxylation from decane-grown C. maltosa VSB779 (Table 4). Today, eight alkane-induced cytochrome P450 genes are known in C. maltosa. Their characteristics and molecular biology data will be discussed in Section 4.

The cellular fatty acid composition of alkane-grown yeasts is quite stable. For C. maltosa it was shown that 96 to 98% of the fatty acids have a chain length of 16–18C atoms and the relation saturated/unsaturated is about 35/65 (Blasig et al. 1984, 1989), independent of the chain length of the n-alkane utilized. The content of odd-numbered fatty acids, mainly C_{17}, is increased in cells grown on odd-numbered n-alkanes or a distillation fraction of long-chain n-alkanes (e.g., C_{14}–C_{18}), demonstrating the direct incorporation of fatty acids arising from alkane oxidation into lipids. As experiments with odd-numbered n-alkanes as carbon source show, even a high percentage of fatty acids with this unusual chain length does not impair the yeast cell (Blasig et al. 1989, cf. Sect. 2.4.2).

2.5.5
Intermediate Metabolism and Gluconeogenesis

In the catabolism of alkanes and fatty acids, the final product of the β-oxidation acetyl-CoA is used in the intermediate metabolism in the same way as during utilization of ethanol or acetate. Thus, the prevailing acetyl-CoA is the starting point for gaining energy and for all anabolic pathways needed in the alkane-utilizing yeast cell. The acetyl-CoA is consumed in the inducible glyoxylate cycle to produce gluconeogenic intermediates and TCA cycle intermediates used in biosynthetic processes. Catabolizing an odd-numbered n-alkane, the concluding cycle of β-oxidation supplies propionyl-CoA, which is further metabolized via the methyl citrate cycle in yeasts (Fig. 3; Tanaka et al. 1982; Tanaka and Fukui 1989; Müller et al. 1991a). Thus, alkane or fatty acid utilization does not supply C_3-units which are needed, e.g., for the gluconeogenesis and the biosynthesis of the pyruvate family of amino acids. This bottleneck is overcome by the glyoxylate cycle. After the condensation of oxaloacetate with acetyl-CoA to citric acid and its conversion to isocitric acid, the latter is cleaved by the inducible isocitrate lyase (ICL), forming succinate and glyoxylate (Table 5; Davidov and Gololobov 1980a,b; Krauzova and Sharyshev 1987). Malate synthetase catalyzes the condensation of glyoxylate and acetyl-CoA to malate which is dehydrogenated to oxaloacetate. With this, the acceptor compound for acetyl-CoA is recycled. Malate and oxaloacetate are sources for pyruvate and phosphoenolpyruvate, the C_3 key intermediates (Tanaka and Fukui 1989; Fig. 3; cf. also Barth and Gaillardin, Chap. 10, this Vol.).

Catabolite Repression and Catabolite Inactivation by Glucose. During growth on n-alkanes, phenol, fatty acids, acetate, or ethanol, gluconeogenesis is a necessary anabolic sequence for cell substance synthesis on these substrates. In the presence of glucose and other substrates metabolized via glycolytic sequences, the so-called glucose effect leads in yeast to a strong reduction of the activities of intermediate metabolic sequences such as citrate and glyoxylate cycles, respiration, and gluconeogenesis by the catabolite repression and/or catabolite inactivation mechanisms.

During discontinuous growth of *C. maltosa* on ethanol or glucose, the differences in the activities of cytoplasmic and mitochondrial malate dehydrogenases (cMDH and mMDH), phosphoenolpyruvate carboxykinase (PEPCK), and fructose-1,6-bisphosphatase (F-1,6-bPase) were found to be much smaller (factor 2 to 5) than in *S. cerevisiae* and other yeasts (factor 15 to 50) under similar conditions. Therefore, it was concluded that catabolite repression does not play an essential role in the control of gluconeogenesis in *C. maltosa* (Hofmann and Polnisch 1990a,b).

On the other hand, these and other gluconeogenic enzymes (isocitrate lyase, ICL) underlie a fast catabolite inactivation in the presence of glucose. Addition of glucose to derepressed (or starved) cells of *C. maltosa* causes a rapid increase within 30 s of the concentration of cyclic AMP and fructose-2,6-bisphosphate (F-2,6-bP) followed by phosphorylation of at least 12 different proteins and inactivation of gluconeogenic enzymes as the ICL, cMDH, PEPCK, and F-1,6-bPase activities (Polnisch and Hofmann 1989; Hofmann and Polnisch 1990a,b,c). The F-1,6-bPase was demonstrated to be phosphorylated in the course of inactivation; no phosphorylation of the cMDH could be detected, showing that cAMP-dependent phosphorylation plays an important role in the catabolite inactivation of F-1,6-bPase in *C. maltosa*, as described for *S. cerevisiae* and other yeasts (Hofmann and Polnisch 1990b).

A spontaneous mutant (PHM, stands for pseudohyphal morphology) of *C. maltosa* was isolated, being defective in catabolite inactivation of gluconeogenic enzymes (Hofmann and Polnisch 1990c). This mutant was selected due to its different colony and cell (pseudohyphal) morphology under all growth conditions tested (see Sect. 3.1), although the mutation was without significant influence on growth velocity on both glucose or hexadecane. The cAMP-linked signaling pathway is therefore connected with the morphological transition from mycelium to yeast in *C. maltosa*, as was previously demonstrated for other yeasts and fungi (for literature see Hofmann and Polnisch 1990c).

The weak reaction of the gluconeogenic enzymes to the glucose effect is obviously connected with the mentioned metabolic peculiarities of carbohydrate utilization in *C. maltosa*, as the increased part of the pentose phosphate pathway compared with the glycolysis (Röber and Reuter 1979, cf. Sect. 2.4.2). This might be connected also with the fact that *C. maltosa* strains occur mainly in hydrocarbon-containing habitats, where the functioning of gluconeogenic metabolic sequences is almost essential (cf. Sect. 2.1).

The enzymes of primary alkane oxidation in *C. maltosa* (P450, FAOD, FALDH) are repressed in the presence of glucose (Mauersberger et al. 1980, 1981, 1984), although a derepression of these enzymes occurred already in the beginning of the stationary phase of growth on glucose (Mauersberger et al. 1984; Kärgel et al. 1985; Schauer 1988). In this respect, it was of interest that P450 induction in *C. maltosa* (and *Y. lipolytica*) by hexadecane was more sensitive to glucose repression than induction by decane, which could be a result of the presence of differently regulated P450 genes in this yeast (Mauersberger 1991).

2.6
Biosynthetic Pathways

2.6.1
Amino Acid Biosynthesis

The anabolism (this Sect.) and catabolism (see Sect. 2.4.1) of amino acids in *C. maltosa* have been studied mostly in the laboratories of yeast genetics and molecular biology at the Biology Section of the University of Greifswald, Germany, including its enzymology, the characterization of several auxotrophic mutants (for genes and their functions as auxotrophic markers, cf. Sect. 4.4.2), and its connection with the respective catabolic pathways (see Sect. 2.4.1). Amino acid biosynthesis in *C. maltosa* was found to be mostly comparable with the main results obtained for the intensively studied yeast *S. cerevisiae*, and the alkane-utilizing yeasts *C. tropicalis* and *Y. lipolytica* in particular. Therefore, mainly the peculiarities found for *C. maltosa* will be discussed here in detail.

The isopropylmalate (IPM) pathway for *L-leucine biosynthesis* in *C. maltosa* shows general similarity to that of other lower eukaryotes, but there are individual differences in the numbers of genes responsible for single enzymatic steps and in the properties of the enzyme proteins itself. The four enzymatic steps in the conversion of α-ketoisovalerate to leucine, including 2-IPM synthase (EC 4.1.3.12, coded by the *LEU4* gene), IPM dehydratase (EC 4.2.1.33, *LEU1* gene), 3-IPM dehydrogenase (EC 1.1.1.85, *LEU2* gene) and L-leucine (branched-chain) aminotransferase (EC 2.6.1.42), were examined in the wild-type and leucine auxotrophic strains of *C. maltosa* (Takagi et al. 1986a; Bode and Birnbaum 1988, 1991c; Wedler et al. 1990; Bode 1991; Bode et al. 1991; Becher et al. 1991; see also Table 6 and Sects. 4.2 and 4.4). The enzymes could be purified at least partially (Table 4) and their catalytic properties were determined, showing some differences to the enzymes of other yeast and fungi (Bode and Birnbaum 1988, 1991c; Becher et al. 1991; Bode 1991). The pathway is metabolically controlled mainly by feedback inhibition of the first enzyme 2-IPM synthase by leucine. An additional novel control step exists, the inhibition of 3-IPM dehydrogenase by L-valine (Bode 1991; Bode and

Birnbaum 1991c). The enzyme biosynthesis is regulated by at least three different control mechanisms of gene regulation – leucine repression, valine induced derepression, and general control of amino acid biosynthesis (Becher et al. 1991; Bode 1991; Bode et al. 1991). The isolated *leu2* mutants were used to establish host-vector systems for *C. maltosa* using the cloned *LEU2* gene (Takagi et al. 1988a; Wedler et al. 1990; Becher et al. 1991; cf. Tables 14 and 15). The potential application of *leu1* and *leu2* mutants of *C. maltosa* for the production of 2-IPM and 3-IPM, respectively, was recently reported (Bode et al. 1991; see Sect. 6).

Threonine dehydratase (ThrDHT – EC 4.2.1.16), the first specific enzyme of *isoleucine biosynthesis*, was purified from *C. maltosa* (Table 4) and studied in its metabolic regulation (Bode and Birnbaum 1986; Bode et al. 1986b). Its activity is inhibited by isoleucine, stimulated by valine, and strongly increased, probably allosterically, by the presence of phosphate ions, as is the enzyme from *Y. lipolytica* and *P. guilliermondii*, but not from *S. cerevisiae*, *C. utilis* and *Hansenula* spp. (Bode and Birnbaum 1986). The biosynthesis of the ThrDHT is regulated under the general control of amino acid biosynthesis (Bode 1991).

Lysine is synthesized via the α-aminoadipate pathway in *C. maltosa* as in other yeasts and fungi (Schmidt et al. 1985, 1989a,b; Kunze et al. 1987a,b). Four enzymes of the lysine biosynthetic pathway of *C. maltosa*, homocitrate synthase (EC 4.1.3.21, first enzyme in the pathway, having low feedback inhibition), α-aminoadipate reductase (AAAR, EC 1.2.1.31, M_r 160 000, *LYS2* and *LYS5* genes), saccharopine reductase (SaR, EC 1.5.1.9, M_r 90 000, *LYS9* gene), and saccharopine dehydrogenase (SaDH, EC 1.5.1.7, M_r 45 000, *LYS1* gene) were investigated in their metabolic and biosynthetic regulatory pattern. The formation of saccharopine reductase was constitutive, whereas the synthesis of homocitrate synthase, α-aminoadipate reductase, and saccharopine dehydrogenase were regulated by the general control of amino acid biosynthesis. Contrary to in *S. cerevisiae*, in *C. maltosa* the four enzymes were not repressed by the presence of lysine in the medium (Schmidt et al. 1985). Mutants of *C. maltosa* lacking saccharopine reductase (*lys9*) and saccharopine dehydrogenase (*lys1*, see Table 6; Sect. 4.2) were found to accumulate and excrete α-aminoadipate-δ-semialdehyde (AASA, 80–90 mg/l within 48 h using the *lys1* mutant G285) into the medium when growing on glucose and supplemented with 50–100 mg/l L-lysine (Schmidt et al. 1989a,b), being the first report of lysine-requiring yeast mutants that accumulate and excrete AASA. The AASA accumulation by *C. maltosa* mutants may be explained by the low feedback regulation of their homocitrate synthase and the equilibrium of the enzyme reactions involved in the lysine biosynthesis (Schmidt et al. 1989a,b).

The *glutamate/aspartate anabolism* is coupled with the role of these amino acids in ammonium metabolism (see Sect. 2.4.1). From the analysis of mutants (*asn1*, *asp1*) it was concluded that asparate is synthesized in *C. maltosa* only from glutamate via transamination (aspartate aminotransferase) as in *S. cerevisiae*, while the synthesis of asparagine from aspartate is catalyzed by one asparagine synthetase, and not by isoenzymes as in *S. cerevisiae* (Casper et al. 1985b).

Arginine is synthesized from ornithine and carbamylphosphate via citrullin, as in most other yeasts.

Biosynthesis of aromatic amino acids (Bode and Birnbaum 1984, 1991c; Bode et al. 1984a,c, 1985a,c,d; Lippoldt et al. 1986). *Candida maltosa* synthesizes phenylalanine and tyrosine only via phenylpyruvate and p-hydroxyphenylpyruvate, respectively, using the enzymes chorismate mutase (EC 5.4.99.5), prephenate dehydratase (EC 4.2.1.51), and prephenate dehydrogenase (EC 1.3.1.12 or 13), whereas the arogenate pathway is not active in this yeast (Bode and Birnbaum 1984; Bode et al. 1984c, 1985d). In *C. maltosa*, a new type of regulation of phenylalanine and tyrosine biosynthetic enzymes was discovered. The biosynthetic pathway is inhibited by the end product tyrosine at the level of chorismate mutase, which converts chorismate (the main branch point in the biosynthesis of all three amino acids) into prephenate, the common intermediate of the phenylalanine-tyrosine biosynthesis. Additionally, tryptophan is absolutely necessary for the enzymatic reaction of the cytosolic enzymes chorismate mutase and prephenate dehydrogenase, activating these enzymes (Bode et al. 1984c, 1985d). Amont various different yeasts studied, chorismate mutase of *C. maltosa* showed the highest degree of activation by tryptophan (Bode and Birnbaum 1991b). The amount of aromatic amino acids in the free amino acid pool of *C. maltosa* cells (Bode and Casper 1983) indicates that the biosynthesis of phenylalanine and tyrosine can be regulated only by tryptophan. The regulation of tryptophan concentration is affected by the inhibition of anthranilate synthase by tryptophan, whereas the common branch of the biosynthesis of aromatic compounds is controlled by two isozymic tyrosine-sensitive and a phenylalanine-sensitive 3-deoxy-D-arabinoheptulosonic acid 7-phosphate (DAHP) synthases, constitutive enzymes catalyzing the first step in the biosynthesis of aromatic amino acids (Bode et al. 1984a, 1985c).

The enzymes of tryptophan biosynthesis (ammonia-dependent anthranilate synthase – ASN, anthranilate phosphoribosyltransferase – PRT, phosphoribosyl-anthranilate isomerase – PRAI, tryptophan synthase – TS, and the complex of glutamine-dependent anthranilate synthase with indole-3-glycerol-phosphate synthase – ASG-InGPS) were partially purified from *C. maltosa* and the regulation patterns were established. The formation of three enzymes ASN, PRT, ASG-InGPS is regulated by the general control of amino acid biosynthesis, whereas synthesis of PRAI and TS is constitutive (Bode et al. 1985a).

Three differently regulated aromatic aminotransferases (ArAT I–III) have been found and characterized in *C. maltosa*, leading to the conclusion that only the ArAT II has a biosynthetic function (Bode and Birnbaum 1984).

Glyphosate Action in C. maltosa (Bode et al. 1984a,b,d, 1985b; Bode and Birnbaum 1989). The broad-spectrum herbicide glyphosate (N [posphono-methyl]-glycine) was studied as an inhibitor of enzymes involved in the shikimate pathway, representing the first steps, finally leading to chorismate, in biosynthesis of aromatic amino acids in *C. maltosa*. These studies, being the first reports on glyphosate action in yeasts, showed the inhibition of only one isoenzyme DAHP synthase, namely the tyrosine-sensitive one, of the dehydroquinate synthase (EC 4.6.1.3), and of the 5-enolpyruvyl-shikimate 3-phosphate (EPSP) synthase (EC

2.5.1.19), the sixth step in the shikimate pathway being the main target of glyphosate action in this yeast (Bode et al. 1984a,b,d; Bode and Birnbaum 1989). Glyphosate is obviously transported into *C. maltosa* cells by a non-specific or general amino acid permease (Bode et al. 1985b). The glyphosate inhibition of growth causes accumulation of shikimic acid and shikimate-3-phosphate and leads to increased synthesis of several enzymes of unrelated pathways, which are regulated by the *general control of amino acid biosynthesis*, which is also known in other fungi. An identical derepression of enzymes could be obtained in *C. maltosa* by use of other amino acid antimetabolites (amitrol), amino acid mutants strains, or under conditions of external amino acid imbalance (Bode et al. 1983, 1990; Bode and Casper 1983). The general control of amino acid biosynthesis is characterized in *C. maltosa* as in other yeasts by a noncoordinated, parallel increase in the level of enzymes of several unrelated amino acid pathways in response to starvation for one of these amino acids. Thus, the presence of amitrol (3-amino-1,2,4-triazol), as a very efficient histidine antimetabolite leading to histidine depletion, leads to a two- to threefold derepression of the effected biosynthetic enzymes of lysine (Schmidt et al. 1985), leucine, isoleucine and valine (Bode 1991), and several other enzymes involved in different pathways (Bode et al. 1983). The activities of three amino acid biosynthetic enzymes, threonine dehydratase, tyrosine aminotransferase, and saccharopine dehydrogenase, increased about two- to fourfold (in response to conditions of histidine, tryptophan, or lysine limitation) as a result of action of the general control of amino acid biosynthesis in several yeasts, including *C. maltosa*, *Hansenula polymorpha*, *S. cerevisiae*, and *Y. lipolytica*, whereas no evidence for the existence of this general control was found in *C. brumptii*, *C. utilis*, *Pichia guilliermondii*, *Trichosporon adeninovorans*, and other yeasts (Bode et al. 1990).

Adenine and uracil biosynthesis of C. maltosa was investigated in connection with the isolation of respective mutants and genes (see Sects. 4.2.2 and 4.4.2).

2.6.2
Biosynthesis of Lipids

The lipid biosynthesis of *C. maltosa* was investigated by Dr. A. Belov and coworkers in Moscow under several aspects, including the role of lipid granules as the compartment of lipid synthesis in the yeast cell, phosphatidylinositol metabolism (Belov and Guselnikova 1988), the substrate specificity of acyltransferases from lipid granules (Davidova et al. 1989), the effects of n-alkenes and the derived ω-1 fatty acids on the lipid metabolism (Zinchenko et al. 1990), and the localization and specificity of enzymes of acylglycerol biosynthesis in lipid granules and microsomes (Zinchenko and Belov (1990). Additionally to the data of Kasanzev et al. (1975) and Maksimova et al. (1988) on the dynamics of lipid inclusions during cultivation on paraffins, some aspects of fatty acid and lipid biosynthesis in *C. maltosa* were discussed already in Sections 2.4, 2.5, and will be discussed for the ergosterol biosynthesis in Section 4.2.2. Like in other yeasts, the presence of a cytochrome P450 system involved in 14α-demethylation of lanosterol in *C. maltosa* was demonstrated by immunological methods (Kärgel et al. 1990).

2.6.3
Biosynthesis of Polysaccharides

Some aspects of polysaccharide biosynthesis in *C. maltosa* were discussed in Sects. 2.4.2, 2.5.2 and 3.2, and in connection with the regulation of gluconeogenesis (see Sect. 2.5.5). The structure (see also Sect. 2.5.2), biosynthesis, and its regulation of polysaccharides used as cell wall material (mannan and glucan, chitin) and as energy reserve compounds (glycogen and trehalose) in *C. maltosa* cells were intensively studied in vivo and in vitro, as reported in a set of publications from the University of Jena (Rademacher and Reuter 1978; Röber and Reuter 1979, 1982, 1984a,b,c,d, 1985; Grimmecke and Reuter 1980, 1981a,b,c,d; Lerche and Kretzschmar 1980, 1986; Popov et al. 1980; Triebel et al. 1980; Brückner and Tröger 1981b; Grimmecke et al. 1981; Nüske et al. 1982). The functioning of the pentose phosphate cycle in glucose-growing *C. maltosa* is the prerequisite for the synthesis of the cell wall polysaccharides glucan and mannan and of the energy reserve compounds glycogen and trehalose, all having fructose-6-phosphate as common precursor intermediate for their biosynthesis (Röber and Reuter 1979, 1984b,c,d, 1985). The hexose utilization and polysaccharide biosynthetic pathways in *C. maltosa* were studied in comparison with two mutants (H3 and H5) partially desensibilized in the catabolite repression by glucose (Röber and Reuter 1982, 1984a,b; Röber 1985).

2.7
Protein Transport and In Vitro Translation System

2.7.1
Protein Transport Studies with *Candida maltosa*

The yeast *C. maltosa* was used to study protein targeting to and translocation across the endoplasmic reticulum (ER) in lower eukaryotes in comparison with the mammalian system (Wiedmann et al. 1988b; Müsch 1993). Using this yeast, an in vitro translation/translocation system was established. Microsomal membranes of *C. maltosa* were isolated active in the translocation and core glycosylation of proteins synthesized in cell-free systems derived from *C. maltosa* (for details see Sect. 2.7.2), *S. cerevisiae*, or wheat germ.

As is well documented for *S. cerevisiae* (Hann and Walter 1991; Larriba 1993), both co- and posttranslational protein transport into the ER was observed also in *C. maltosa*. Translocation and core glycosylation of a heterologous secretory protein (prepro-α-factor of *S. cerevisiae*) were observed with *C. maltosa* microsomes added during or after translation. The signal peptide is cleaved off. The homologous integral ER-protein cytochrome P450Cm1 is also inserted both co- and posttranslationally into the microsomal membranes from *C. maltosa*. In contrast, its insertion into canine microsomes occurs efficiently only in a cotranslational SRP-dependent manner (Wiedmann et al. 1988b), showing the generality of the posttranslational mode of protein translocation in yeast in addition to the cotranslational mode.

Investigating the relationship of the eukaryotic translocation apparatus of the ER with cross-linking experiments during transport of the model secretory protein prepro-α-factor into yeast (*C. maltosa* and *S. cerevisiae*) and dog pancreas microsomes, Müsch (1993) identified the Sec61p and αSSR (34kDa) proteins as cross-linked in *S. cerevisiae* as well as in the mammalian system, respectively. The participation of homologous proteins in translocation machinery of *C. maltosa* was demonstrated.

Recent in vivo ER targeting studies, using the cloned *C. maltosa* ER proteins cytochrome P450Cm1 (P450 52A3), P450Cm2 (P450 52A4), and NADPH-P450 reductase were performed in a model system druing their high-level heterologous expression in *S. cerevisiae* under the control of the *GAL10* promoter (Menzel et al. 1994, 1996). In these experiments, different parts of the membrane anchor region of the P450 proteins (hydrophobic segments HS1 and HS2 at the N terminus) were used to construct several fusions with the cytoplasmic form of yeast invertase (*SUC2* gene) as reporter protein, to analyze the topogenic sequences in the P450 primary structure. It was shown that the same sequences, that confer the correct targeting to and retention within the ER membrane of the P450 and the fusion proteins, were found to induce also ER proliferation (cf. Sect. 3.2). The first hydrophobic segment HS1 proved to be sufficient for ER targeting and insertion, whereas the HS1 and the following hydrophilic sequence (first 44 amino acids) of P450 52A3 protein are required for a correct retention in the ER.

Kamiryo et al. (1989) demonstrated the in vivo transport of the subunits PXP-2 and PXP-4 (genes *POX2* and *POX4*) of peroxisomal fatty acyl-coenzyme A oxidase of *C. tropicalis* into peroxisomes of *C. maltosa*. Additionally to the results mentioned here, *C. maltosa* can be used to study ultrastructural changes such as the proliferation of ER and peroxisomes during adaptation of the yeast cell to assimilation of n-alkanes (cf. Sect. 3).

2.7.2
In Vitro Translation System Using Cell-Free Extracts Isolated from *Candida maltosa*

Preparation of Cell Lysates from *Candida maltosa* Active in In Vitro Translation

Note on Method. Active in vitro protein translating cell-free extracts can be obtained from *C. maltosa* as described in detail by Wiedmann (1987), Wiedmann et al. (1988b), and Zimmer and Schunck (1995). The method is adapted to that described by Erickson and Blobel (1983) and Waters and Blobel (1986) for *S. cerevisiae*. A detailed methodology for bakers yeast is described also by Feinberg and McLaughlin (1988).

Procedure

1. Grow *C. maltosa* EH15 cells in YPD (up to 3 1) in shaking flasks at 30 °C and 240 rpm to an optical density (OD_{600}) of about 5 to 10. Harvest the cells by centrifugation and wash with 1/20 volume of fresh YPD medium at room temperature.

   Comment   Cells should be taken from the exponential growth phase. Lower OD_{600} of about 1 give more active extracts but lower yield.

2. Resuspend the cells (20–50 g wet weight) to a final concentration of 0.3 g/ml in YPD medium containing 1.2 M sorbitol (YPDS) and 13 mM DTE, adjust the pH to 7.0 with 10 N NaOH, and incubate 10 min at 30 °C under shaking at 80 rpm. Add Zymolyase 100 T (Seikagaku, Kugyo Co., Ltd, Japan) to a final concentration of 1 mg/g wet weight cells and incubate the mixture with shaking at 80 rpm for up to 60 min at 30 °C to form spheroplasts. Control spheroplasting microscopically.

3. Sediment the spheroplasts at 2000 rpm for 10 min, wash the spheroplasts in YPDS and incubate in 60 ml YPDS per g wet weight cells for further 30 min at 30 °C under slightly shaking at 80 rpm for reactivation of the cells.

All following steps are performed at 4 °C.

4. Sediment at 2000 rpm (3000 g) for 10 min, wash the spheroplasts with 80 ml 1.2 M sorbitol, resuspend them in 0.5 ml buffer A (see below) per g cells and homogenize in a tightly fitting Potter homogenizer for 20 strokes.

5. Centrifuge the homogenate at 15 000 g for 25 min. Remove the supernatant thoroughly, and centrifuge for 30 min at 100 000 g.

6. Gently remove the clear supernatant and pass it through a Sephadex G-25 (or G50, fine) column of 2×50 cm equilibrated with buffer A. The column size should be adapted to the cell amount to be extracted.

7. Collect the eluates with OD_{260} higher than 60 (maximally OD_{260} is approx. 150) and mix these fractions. Freeze obtained lysate dropwise in liquid nitrogen and store until used for in vitro translation experiments. This material is referred to as the cell-free extract or cell lysate (S100).

8. Continue with steps 8 to 10 for preparation of a microsomal membrane fraction active in translocation (see below).

In Vitro Translation Experiments Using Cell Lysate from *Candida maltosa*

Note on Method. Preparation of the *C. maltosa* lysate was carried out as described above. In vitro translation experiments are performed according to Wiedmann et al. (1988b) by a modified method used for *S. cerevisiae* by Waters and Blobel (1986). The reader is also referred to the detailed discussion of the in vitro translation methodology given by Feinberg and McLaughlin (1988).

Procedure

1. Treat 100 µl cell-free extract with 1 µl micrococcal nuclease (0.4 U/µl, Boehringer Mannheim) for 6 min at 21 °C in the presence of 1.0 mM $CaCl_2$ followed by addition of EGTA to 2.0 mM to stop the nuclease.

   Comment The yeast lysates contain high amounts of endogenous RNA (interfering with the translation of exogenous RNA) to be eliminated by nuclease treatment. This step should be optimized for each lysate obtained. Normally, the conditions described are enough to reduce the endogenous synthesis and to promote the initiation of protein synthesis stimulated by heterologous mRNA.

2. Perform translation experiments in 0.5-ml reaction tubes with a 12.5 µl (or 25 µl) mixture containing 10 mM Hepes/KOH, pH 7.4, 1.2 mM ATP, 0.1 mM GTP, 2.0 mM dithiothreitol, 9.5 mM creatine phosphate, 0.9 µg creatine phosphokinase (Boehringer Mannheim), 30 µM L-amino acids expect the (^{3}H)-labeled compounds (Amersham), 5 µl cell-free extract (S100) and 1–2 µl of (^{3}H)leucine (1.33 TBq/mmol, 1 µCi/µl; or (^{3}H)serine, 1.33 TBq/mmol, 1 µCi/µl; or (^{3}H)isoleucine, 3.66 TBq/mmol, 1 µCi/µl; or (^{35}S) methionine), $Mg(OAc)_2$ (2–3 mM), KOAc (130–160 mM) and the mRNA sample (approx. 10 µg, the effective RNA amount should be tested).

   Comments: In principle, *C. maltosa* lysates function with noncapped RNA preparations, but usually capping increases the efficiency of translation. Therefore, in vitro transcriptions were performed using 0.5 mM of cap analog P1-5′-(7-methyl)-guanosine-P3-guanosine triphosphate (m7GpppG, Boehringer Mannheim; Zimmer and Schunck 1995). For each of the RNAs and lysates used the optimal concentrations of $Mg(OAc)_2$ and KOAc should be adjusted as shown by Erickson and Blobel (1983) or Wiedmann (1987) using an active standard mRNA (globin RNA, TMV RNA or total *C. maltosa* RNA). The concentration of K^+ and Mg^{2+} should be optimized for the RNA preparation to be tested.

   Pipette all components on ice before starting the reaction at 21 °C.

3. Incubate the reaction tubes at 21 °C for 30–90 min (*C. maltosa* system) or at 30 °C for 60 min (wheat germ system).

4. Finally, take samples:
   - for the determination of total radioactivity (2 µl) in the protein by counting the hot trichloroacetic acid-insoluble material in a scintillation counter;
   - for SDS-PAGE of the total proteins (5 µl); and
   - for the immunoprecipitation (rest of the mixture, see below).

The translation activity of the *C. maltosa* extract is mostly comparable with a wheat germ extract purchased or prepared according to Wiedmann (1987), but shows up to seven times higher activity when compared with a *S. cerevisiae* lysate (Wiedmann 1987; Feinberg and McLaughlin 1988).

In contrast to the system obtained from *S. cerevisiae*, active for globin-, phaseolin- and pre-prolactin-mRNAs, the *C. maltosa* system is more selective towards the mRNA, where mostly yeast mRNA species (α-factor transcripts, total mRNA from *C. maltosa* including P450-specific mRNA) were effectively translated (Wiedmann 1987). The described method was recently successfully applied to investigate codon usage in *C. maltosa* (Zimmer and Schunck 1995). In these experiments artificial mRNAs were translated.

Immunoisolation of In Vitro Translation Products

Immunoprecipitation can be performed according to Wiedmann et al. (1986, 1988b) using rabbit antisera obtained after immunization of rabbits with purified proteins, or using purified polyclonal antibodies against cytochrome P450 (Kärgel et al. 1984, 1985) or other proteins to be analyzed. Affinity purification of antibodies was performed according to Pringle et al. (1991) using purified P450 protein bound to nitrocellulose.

Preparation of Yeast Microsomal Membranes Active in Protein Transport

Procedure

1. to 7. These steps were the same as for obtaining cell-free extracts (shown above).

8. The pellet of the last centrifugation step consists of a tightly packed ribosome sediment and a flocculent membrane sediment above it. Remove the latter and resuspend it with a Dounce homogenizer (five strokes) in buffer B.

9. Centrifuge the membranes through a 14% glycerol cushion in a Ti50.2 Beckman rotor for 60 min at 41 000 rpm.

10. Resuspend the membranes finally in buffer B at a OD_{260} of about 100 and freeze in liquid nitrogen.

To perform in vitro translation/translocation experiments with *C. maltosa* membranes, add the obtained microsomal preparation to the in vitro translation mixture at a final concentration of 0.4 OD_{260} units per 25 μl assay (for details see Wiedmann 1987; Wiedmann et al. 1988b).

Buffers
Buffer A: 20 mM HEPES-KOH, pH 7.4, 100 mM KOAc, 2 mM Mg(OAc)$_2$, 2 mM DTT (column buffer)
Buffer B: 50 mM HEPES-KOH, pH 7.4, 1 mM DTT

2.8
Some Peculiarities of Candida maltosa

For the yeast *C. maltosa*, some additional peculiarities have been demonstrated and studied at biochemical and molecular biological levels. These are the appearance of

- Cycloheximide resistance (Takagi et al. 1986b; Sasnauskas et al. 1992c; see Sect. 4.4.2).
- Formaldehyde resistance genes, applicable as marker genes for the transformation of other yeasts (Sasnauskas et al. 1992a; for details see Sects. 4.4.2 and 5.2).
- Cyanide-insensitive respiration (Schauer 1988; Ilchenko et al. 1991).
- Growth at high cobalt concentrations in connection with cobalt-binding proteins and cobalt accumulation (Belov and Toneva-Davidova 1983; Belov et al. 1985).
- Intracellular and extracellular proteases (Dolgikh et al. 1990; Sinanyan et al. 1990).
- Thermal shock proteins in *C. maltosa* and investigation of its thermotolerant mutants (Sinanyan et al. 1989, 1990).

As described for many eukaryotic microorganisms, *C. maltosa* has an alternative respiratory pathway which is not inhibited by cyanide. The biochemical nature of this pathway and of the alternative oxidases in particular are poorly understood at present (for literature see Claisse et al. 1991; Sakajo et al. 1993). The development of probably two cyanide-insensitive respiration pathways in *C. maltosa* was induced by the presence of respiratory inhibitors like KCN or in dependence on the physiological state of glucose- and alkane-growing cells (Greiner 1985 and Schubert 1987, both cited by Schauer 1988; Ilchenko et al. 1991).

The accumulation of proteinases in the culture media of industrial *C. maltosa*, *C. rugosa*, and *C. tropicalis* strains was investigated by Dolgikh et al. (1990). It was found that these yeasts exhibit proteolytic activity, and therefore were able to use proteins as the only source of nitrogen. The regulation of the proteolytic activity of *C. maltosa* and its thermotolerant mutants was studied by Sinanyan et al. (1990).

Growth of *C. maltosa* cells at rather high concentrations of cobalt in the medium was demonstrated. Belov and Toneva-Davidova (1983) isolated a Co-binding protein, mostly located in the vacuoles, with a molecular mass of 5000 Da containing 4 g-atoms of cobalt/mole protein from the yeast *C. maltosa* grown on glucose in the presence of 3 mg/l Co^{2+}. This protein was supposed to be involved in reducing the intracellular cobalt concentration. Later, Belov et al. (1985) reported the distribution of different chemical forms of intracellular Co^{2+} between the vacuolar and cytoplasmic compartments studied in cells of *C. maltosa*.

Additionally to the properties shown in Table 3, some other peculiarities of *C. maltosa* discussed in other Sections should be mentioned here. This yeast is regarded as:

- a dimorphic yeast showing pseudomycelium formation under certain circumstances (see Sect. 3.1);
- a petite-negative yeast, showing no ability to form respiratory-deficient mutants on treatment with acriflavine (Bos and deBoer 1968);
- an asporogenous (imperfect) yeast without sexuality (see Sect. 4).

3
Cytology and Morphology of *Candida maltosa*

3.1
Morphology

The imperfect yeast *C. maltosa* exists mostly in the true yeast form. Cells are normally round, nearly round to short oval cells of about $(2.5-7) \times (3-8) \mu m$ after growth in rich medium (malt extract), and they reproduce by multilateral budding on a narrow base. Sometimes, the cells are prolonged to 10 to $13 \mu m$. In slide culture on potato dextrose agar (Difco), corn meal agar, or under certain conditions of growth limitation in submerse cultures, the occurrence of a pseudomycelial form (pseudohyphae or pseudomycelium) with blastospores, but no chlamydospores or true mycelium were observed (Komagata et al. 1964b; Kaneko et al. 1977; Kunze et al. 1984a,b; Golubev et al. 1986; Huth et al. 1990a), characterizing this yeast as a dimorphic one.

C. *maltosa* is characterized by the absence of true hyphae, chlamidospores, ballistospores, chlamydoconidia, and arthroconidia. Teleospores and ascospores were not observed in individual or mixed cultures (possible anamorph/ teleomorph relationship discussed in Sects. 1.1 and 1.2). A spontaneous mutant of *C. maltosa* SBUG700 was isolated showing pseudohyphal morphology (PHM) under all growth conditions tested. On solid malt extract medium the mutant formed dark yellow colonies and rough brim, in contrast to the round, cream or yellow-colored colonies with an entire margin and a smooth and shiny surface of the wild-type strain. This PHM mutant was shown to be defective in cAMP-dependent catabolite inactivation of gluconeogenic enzymes. The cAMP-linked signaling pathway is therefore connected with the morphological transition from mycelium to yeast (dimorphism) also in *C. maltosa*, as was previously demonstrated for other yeasts and fungi (for literature see Hofmann and Polnisch 1990c).

3.2
Ultrastructure of Glucose- and Alkane-Grown Candida maltosa *Cells*

During transition from growth on carbohydrates (glucose) to n-alkanes in cells of *C. maltosa*, as in other alkane-assimilating yeasts, in addition to the above-discussed physiological-biochemical changes (alterations in the biomass composition and enzyme induction, cf. Sects. 2.3 and 2.5), various ultrastructural alterations also occur, such as an increase in intracellular membranes (organelles) like the ER and peroxisomes.

Adaptation to alkane utilization also leads to several morphological and chemical alterations of the cell wall and of the plasma membrane of *C. maltosa* strains. The cell wall of hydrocarbon-grown *C. (cloacae) maltosa* cells has a lipophilic (lipoid surface layer) character, which is reflected in transmission electron micrographs of sections of glutaraldehyde-OsO_4-fixed whole cells as a dark edge, missing in glucose-grown cells (Bos and deBoer 1968; Bos 1975). In scanning electron micrographs, the surface of glucose-grown *C. maltosa* cells appeared smooth, but rough on alkane-grown cells (Schauer 1988). In transmission electron micrographs the changes in the cell wall appeared as hydrocarbon deposits, and formation of special ultramicroscopic channels, accompanied by slime-like outgrowths or protrusions (up to 100 per cell) on the cell surface (Fischer et al. 1982; Belov et al. 1983; Egorenkova and Belov 1984; Bode and Köhler 1984, unpubl. results cited by Schauer 1988), as was repeatedly described for other alkane-assimilating yeasts (Ludvik et al. 1968; Rylkiñ et al. 1974; Osumi et al. 1974, 1975b; Meissel et al. 1976; Käppeli et al. 1978, 1984; Dmitriev et al. 1980; for further references see Fischer et al. 1982; Müller et al. 1991a). These structural peculiarities are closely connected with the alterations in the chemical composition of the cell wall, directed to support the transport of the hydrophobic substrates into the cell (Fischer et al. 1982; Röber and Reuter 1984c; Egorenkova and Belov 1984; cf. Sect. 2.5.3).

The basic morphological structures of the yeast cell, such as the nucleus, vacuoles, mitochondria, plasma membrane, and cell wall, are distinctly visible in ultrathin sections of resin (Epon) embedded (Mauersberger et al. 1987) or in cryosections (Vogel et al. 1991, 1992) of alkane- and glucose-grown *C. maltosa* cells (Fig. 4). Whereas glycogen granules (not contrasted while solubilized under these conditions) were visible only in glucose-grown cells, alkane-grown cells contained in addition to vacuoles more lipid granules (spherosomes), surrounded by a single membrane as reported previously for *C. maltosa* (Davidova et al. 1989; Zinchenko et al. 1990) and other alkane-utilizing yeasts (Ludvik et al. 1968; Kasanzev et al. 1975; Meissel et al. 1976; Davidova et al. 1977a,b, 1979; Davidova and Rachinskii 1981; Belov and Davidova 1982).

A very striking feature of *C. maltosa* cells is a strong proliferation of peroxisomes (microbodies) and of the endoplasmic reticulum (ER) observed during transition from growth on glucose to n-alkanes, as has been described repeatedly for other alkane-assimilating *Candida* yeasts and *Yarrowia lipolytica* (Ludvik et al. 1968; Osumi et al. 1974, 1975a; Meissel et al. 1976; Fukui and Tanaka 1981a,b; Stepanjuk 1981). Proliferation of ER resulted in the formation of largely extended membrane tubules surrounding the increased amount of peroxisomes, in addition to ER membranes being in close vicinity of the plasma membrane (Mauersberger et al. 1987; Vogel et al. 1991, 1992; Fig. 4). The ER became more extensive, especially under oxygen limitation of the alkane-assimilating cells, when the P450 content increased strongly (Mauersberger et al. 1980, 1984; Schunck et al. 1987b). In the ultrathin sections of *C. maltosa* cells it has not yet been possible to distinguish between smooth and rough ER membranes on a morphological basis only. Additionally, no direct transitions can be found between the ER membranes, where the P450 system is located (Mauersberger et al. 1987; Vogel et al. 1992; cf. Sect. 3.4), and the plasma membrane or peroxisomes themselves. Furthermore, the number of plasma membrane invaginations is increased.

A strong proliferation of ER was also observed in *C. maltosa* cells growing on galactose instead of on alkanes, when its P450 *ALK* genes were overexpressed under the control of host-own *GAL1* promoter from the multicopy vector pNGH2 (Ohkuma et al. 1995b; see Sect. 5.3). This phenomenon of increased formation of

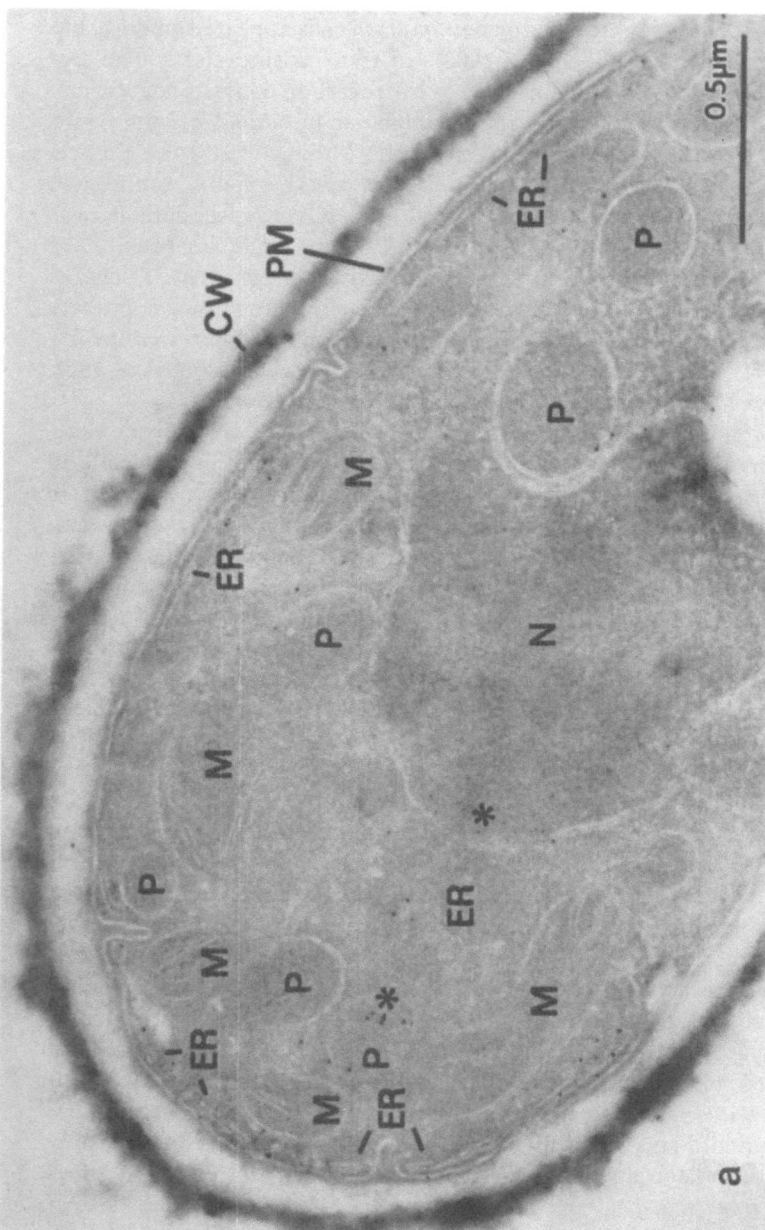

Fig. 4a-e. Immunocytochemical localization of cytochrome P450 52A3 and its NADPH-dependent P450 reductase in cryosections n-alkane-grown *Candida maltosa* cells (Vogel et al. 1992). Immunolabeling with two distinct particle sizes of colloidal gold for the P450 52A3 protein (5 nm) and the NADPH-P450 reductase protein (9 nm) after maximum P450 induction by n-alkanes and additional oxygen limitation. The specific labeling for both proteins over ER in the whole cells (a), in particular over ER forming the nuclear envelope (b), ER associated with peroxisomes (c) and with the plasma membrane (d) was achieved. Control section where the specific antibodies were omitted exhibit no immunolabel (e). Abbreviations: *CW* cell wall; *PM* plasma membrane; *N* nucleus; *M* mitochondrion; *V* vacuole; *Gly* glycogen; *P* peroxisome; *ER* endoplasmic reticulum. *Bars* 0.5 µm (a) and 0.1 µm (b-e)

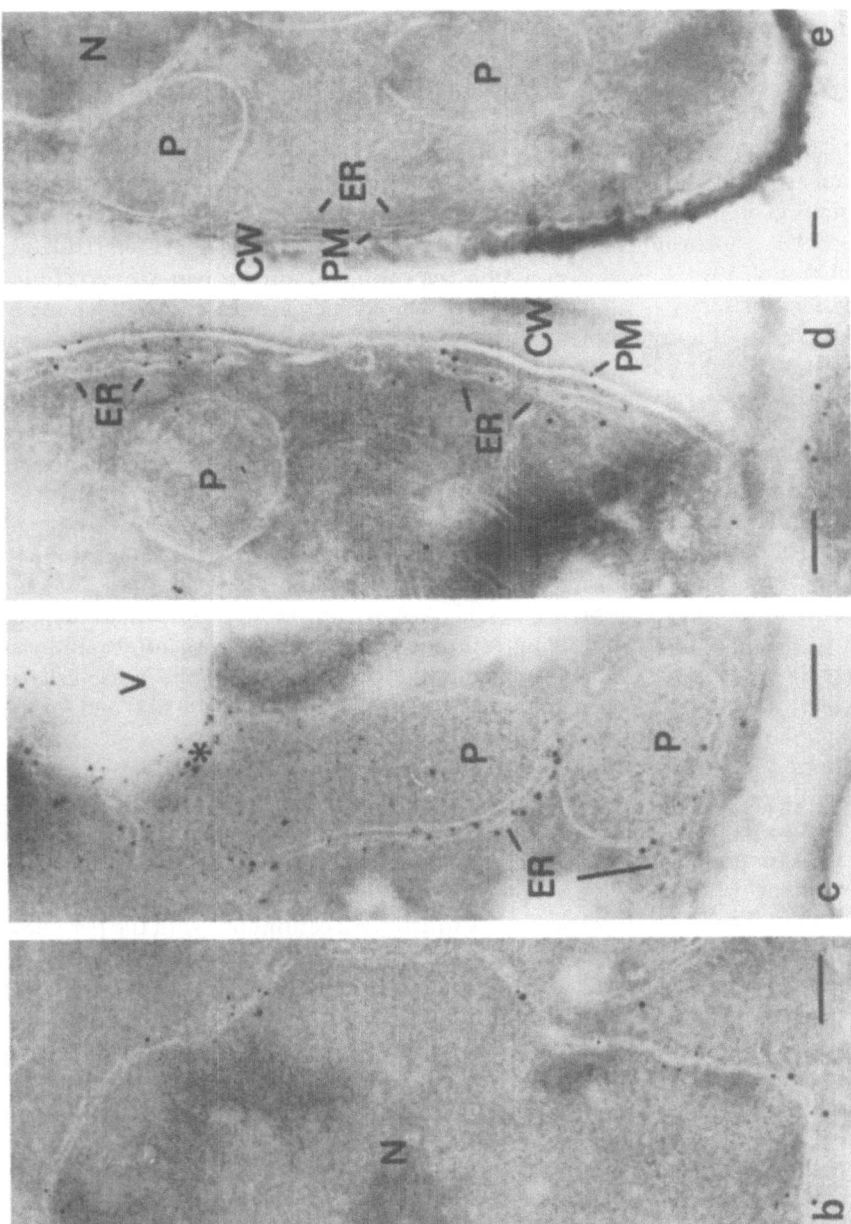

Fig. 4b–e

ER membranes was already observed in *S. cerevisiae* in response to high-level homologous or heterologous expression of certain ER-resident membrane proteins such as the 3-hydroxy-3-methylglutaryl-CoA (HMG-CoA) reductase (Wright et al. 1988, 1990), cytochromes P450Cm1 (P450 52A3), P450Cm2 (P450 52A4), P450$_{14DM}$ (P450 51A1) or chimeric proteins consisting of an N-terminal membrane anchor sequence of P450Cm1 or P450Cm2 and the reporter protein invertase (Schunck et al. 1991; Wiedmann et al. 1993; Menzel et al. 1992, 1994, 1996; Zimmer et al. 1995), cytochrome b$_5$ (Vergeres et al. 1993) and the NADPH-P450 reductase (Menzel et al. 1994; Kärgel et al. 1996). Depending on the protein expressed in *S. cerevisiae*, distinctly organized stacks of paired membranes and tubular membranes appeared and occupied considerable areas of the cytoplasm (for a recent review see Wright 1993). Overexpression of the NADPH-P450 reductase of *C. maltosa* in *S. cerevisiae* caused a proliferation of "karmellae"-like structures (Menzel et al. 1994; Kärgel et al. 1996), as initially reported for the HMG-CoA reductase (Wright et al. 1988). The cytochrome P450-mediated ER membrane proliferation in *S. cerevisiae* is accompanied by coinduction of the ER proteins KAR2p and SEC61p and accumulation of precursor forms of proteins (KAR2p, α factor) that have to translate across the ER membrane. Cytosolic proteins (SSA1p and 2p) and mitochondrial proteins (CYTc1p and F$_1\beta$p) are not affected (Wiedmann et al. 1993; Menzel et al. 1994). Interestingly, the heterologous overexpression of other (not ER) membrane proteins in *S. cerevisiae* also induces strong ER-membrane or nuclear-membrane proliferation in the yeast cell, as shown for the plant plasma membrane H$^+$-ATPase (Villalba et al. 1992) and for the *Drosophila copia* gag precursor (Yoshioka et al. 1992).

Another system of proliferating membranes in yeasts was described for *Hansenula polymorpha* as a response during adaptation of the cells to oleic acid-containing media (Veenhuis et al. 1990). Although this organism cannot grow on oleic acid, induction of peroxisomal β-oxidation enzymes and increased phospholipid content connected with a rapid proliferation of membranes, consisting of a variable number of membranous layers which were continuous with the peroxisomal membrane, was observed during incubation with this fatty acid.

The discussed phenomena of inducible membranes (reviewed by Wright 1993) may provide a suitable model to study the so far largely unknown mechanisms coordinating the synthesis and assembly of protein and lipid components during ER biogenesis. For this purpose, a system would be of particular interest that allows a direct comparison between physiological forms of ER proliferation and those elicited by experimental overexpression of individual membrane proteins. The alkane-assimilating yeast *C. maltosa* may offer such a possibility.

3.3
Electron Microscopy Methods

For electron microscopic examination of whole *C. maltosa* cells and organelle preparations, conventional methods using resin embedding, thin sectioning and staining of specimens (Mauersberger et al. 1987; Sharyshev and Krauzova 1988), as

well as an especially for yeast cells developed method of immunoelectron microscopy with high resolution of membrane structures (Vogel et al. 1991, 1992;
Ohkuma et al. 1995b) have been successfully used.

3.3.1
Electron Microscopy of Cells and Cell Fractions Using Resin Embedding

A method for obtaining electron micrographs of *C. maltosa* cells with good preserved intracellular membrane structures (Mauersberger et al. 1987) is given.

1. Fix spheroplasts of *C. maltosa* cells grown on n-alkanes or glucose, prepared as
   described by Mauersberger et al. (1984), in 3% glutaraldehyde in 50 mM sodium
   potassium phosphate buffer, pH 7.0, containing 0.6 M sorbitol for 30 min at
   room temperature.

2. Continue with staining in a buffered solution of 1% osmium tetroxide and 0.8%
   ferricyanide for 2 h according to Hepler (1981) for preservation of membrane
   staining.

3. Contrast in 0.5% uranylacetate for 18 h and dehydrate in an acetone series.

4. Embed the cells in Epon 812 (Serva, Heidelberg, FRG).

5. Grid stain the thin sections obtained with an Ultrotome II (LKB, Sweden) with
   1% uranyl acetate for 15 min and with Reynolds lead citrate.

6. View all specimens on an electron microscope at 80 kV.

Treatment of the particulate cell fractions for electron microscopy was performed
as described by Mauersberger et al. (1987). A comparable method was used by
Sharyshev and Krauzova (1988).

1. Fix the particulate fractions immediately after pelleting for 30 min with 3%
   glutaraldehyde in a 50 mM potassium phosphate buffer, pH 7.0, containing 5%
   sucrose and 10 mM $MgCl_2$ and wash with buffer.

2. Postfix the pellets in 1% buffered osmium tetroxide for 2 h and wash with
   buffer.

3. Embed in 2% agar. Dehydrate these agar blocks in an acetone series.

4. Embed the specimens in Mikropal (Ferak, Berlin, FRG).

5. Proceed with 5 and 6 as shown above.

3.3.2
Immunoelectron Microscopy

Note on Method. Because it has been very difficult to preserve both antigenicity
and ultrastructure of the yeast cell embedded even in Lowicryl K4M and related
resins at low temperature, a new strategy for cryosectioning of yeast cells (*C.
maltosa* and *S. cerevisiae*) and subsequent immunolabeling was introduced in Dr.
Frank Vogel's Laboratory of Electron Microscopy at the Max-Delbrück-Center for

Molecular Medicine (MDC) in Berlin-Buch. Based on the cryosection technique of Tokuyasu, some modifications were introduced to produce suitable croysections of yeast cells. The main modifications are the fixation of the yeast cells at the temperature of 30 °C and the pH of 4.7–4.9 used for growth, as well as the sucrose infiltration performed at 30 °C for 2 h. The frozen-thawed sections obtained by this new strategy for the preparation of yeast cells exhibit both an excellently preserved and highly resolved ultrastructure and a conserved antigenicity of membrane proteins to be localized in the yeast cell (Vogel et al. 1991, 1992) as well (Fig. 4). All membranes of the well-known yeast compartments (envelop of the nucleus, mitochondria, ER, vacuole and plasma membrane, peroxisomes) are clearly visible. The method makes it possible to examine the biogenesis of ER membranes and to assess the subcellular distribution of cytochrome P450, the NADPH-P450 reductase, and other antigens in the yeast cells.

Procedure. Fixation and processing for electron microscopy of *C. maltosa* cells are done by a modified cryosectioning method as descried by Vogel et al. (1992) and Ohkuma et al. (1995b):

1. Harvest the yeast cells cultivated in a fermenter under defined conditions and wash in 0.1 M Na-citrate buffer, pH 4.7, at 30 °C. Subsequently fix the cells in a mixture of 0.5% glutaraldehyde and freshly prepared 4% paraformaldehyde in 0.1 M Na-citrate buffer, pH 4.7, for 1 h at 30 °C, to preserve the growth conditions during fixation.

2. Wash the cells three times with phosphate-buffered saline (PBS: 0.1 M phosphate buffer, 150 mM NaCl), add 1% sodium metaperiodate (Sigma-Aldrich, Deisenhofen, Germany) to promote the cryoprotectants sucrose and PVP to penetrate the cell wall (Van Tuinen and Riezman 1987) and incubate for 1 h at 4 °C.

3. Wash the cells with PBS and immerse the specimens in a mixture of 25% polyvinylpyrrolidone (PVP) K15 (MW 10 000, Fluka, Buchs, Switzerland) and 1.6 M sucrose (Analytical grade, Serva, Heidelberg, FRG) for 2 h at 30 °C according to Tokuyasu (1989) and freeze in liquid nitrogen.

4. Mount the samples on specimen holders, freeze in liquid nitrogen, and prepare ultrathin thawed-cryosections (approximately 50–100 nm) according to Tokuyasu (1986, 1989) at −115 °C with glass knives using an ultracryotome (Leica, Vienna, Austria). Place the thawed cryosections on Formvar/carbon-coated copper grids (200 mesh, hexagonal).

5. Label the specimens with primary antibodies and protein A-gold (5 or 10 nm) complexes according to Griffiths et al. (1984) and Griffiths and Hoppeler (1986) using IgG fractions (20 μg/ml) containing antibodies against P450, P450 reductase, invertase, or other antigens.

6. Stain (contrast) and stabilize finally the frozen-thawed sections using a mixture of 0.3% uranyl acetate (A grade, Serva, Heidelberg, FRG) and 2% methyl cellu-

lose (25 cps, Sigma-Aldrich, Deisenhofen, FRG) as described by Griffiths et al. (1982).

7. View all specimens on an electron microscope at 80 kV.

As controls, the sections were incubated directly with protein A-gold omitting reaction with primary antibodies, or cells not expressing the antigen were used to check the specificity of primary antibody reaction. Polyclonal antibodies against SDS-denatured P450Cm1 and P450Cm2 were raised in rabbits. The IgG fraction was prepared as described by Kärgel et al. (1984, 1985). All polyclonal antisera used were incubated with washed *C. maltosa* cells to remove antibodies binding to cell wall carbohydrates (Brada and Schekman 1988) for 1 h at room temperature. The cells were removed by centrifugation and the procedure was repeated with fresh cells, if necessary. Affinity purification of antibodies was performed according to Pringle et al. (1991) using purified P450 protein bound to nitrocellulose.

3.4
Subcellular Organization of the Alkane Metabolism

The subcelluar distribution of enzymes involved in the primary oxidation steps of alkanes and fatty acids (Fig. 3, reviewed for yeasts by Tanaka and Fukui 1989; Müller et al. 1991a) in *C. maltosa* cells has been studied by differential and sucrose and Percoll gradient centrifugation (Mauersberger et al. 1984, 1987; Krauzova and Sharyshev 1987; Sharyshev and Krauzova 1988; Kamiryo et al. 1989).

Already in the early 1970s, first studies on alkane-hydroxylating cytochrome P450 of yeasts suggested its occurrence in the endoplasmic reticulum by comparison of the yeast with the well-investigated mammalian liver cell. In both cases, after cell disruption, cytochrome P450 sediments with the microsomal fraction. The components of alkane monooxygenase (cytochrome P450 and NADPH-cytochrome c reductase) have been shown to be located mainly in the light membrane fractions (microsomes) of yeast cells. This yeast microsomal fraction (when obtained after enzymatic cell wall lysis, followed by careful osmotic cell disruption and differential centrifugation) consists mainly of vesicles of the plasma membrane and of intracellular membranes. Furthermore, fragments of mitochondria and peroxisomes occur in small quantities. Separation of the microsomal fraction by sucrose gradient centrifugation was performed, resulting in an increased content of P450 and its reductase in the purified ER fractions, thus separated from the impurities of plasma membrane fragments present (Mauersberger et al. 1987). More recently, an immunocytochemical method was applied to localize the P450 and the corresponding NADPH-cytochrome P-450 reductase in the ER of *C. maltosa* cells grown on n-alkanes (Vogel et al. 1991, 1992) or on galactose during expression under the control of the *GAL1* promoter (Ohkuma et al. 1995b). For this purpose, a special immunogold method on cryosections of yeast cells was developed for conserving both the high resolution in ultrastructure and the antigenicity of membrane proteins to localize in the yeast cell (Sect. 3.3.2). Using this technique, P450 was found to be concentrated in the nuclear envelope during the early phase of the induction process. After maximal induction, the highest labeling was observed in membranes of the ER closely associated with peroxisomes and the plasma membrane. Double-labeling experiments revealed that P450 and its NADPH-dependent reductase were distributed in the same regions of the ER (Fig. 4; Vogel et al. 1992; cf. Mauersberger et al. 1987).

The enzymatic systems for the oxidation of higher alcohols and aldehydes (membrane-bound FAOD, FADH, and FALDH) occur predominantly in peroxisomes, accompanied, e.g., by catalase and isocitrate lyase, but not in the mitochondrial fraction, although their (or of isoenzymes) activity in the ER membranes could not be excluded (Fig. 3; Krauzova and Sharyshev 1987; Mauersberger et al. 1987). As known from the studies of Tanaka's group, mainly on alkane-grown *C. tropicalis* and *Y. lipolytica* (latest review Tanaka and Fukui 1989), this organelle is characterized by a high level of various oxidases, including the β-oxidation enzyme acyl CoA oxidase and catalase, and the glyoxylate cycle enzymes isocitrate lyase and malate synthase, which is suggested to be the case also in *C. maltosa* (Fig. 3; Krauzova and Sharyshev 1987; Mauersberger et al. 1987; Hill et al. 1988; Kamiryo et al. 1989).

Considering the different intracellular localization of the reactions, alkane hydroxylation in ER and alkanol and aldehyde oxidation in peroxisomes, the fatty alcohol must be an essential transport form for long alkyl chains in the alkane-assimilating yeast cell (Fig. 3), as proposed by Fukui and Tanaka (1981a,b) for other alkane-assimilating yeasts. Due to the fact that fatty acids are also substrates for the ER resident P450s forms active in ω-hydroxylation, fatty acids derived from fatty alcohols in peroxisomes should also be at least retransported into ER from peroxisomes, to form dicarboxylic acids (DCA). This suggests a direct functional importance of the proliferation of ER membranes and of their tight association with peroxisomes during growth on n-alkanes (Fig. 4), which was supported by first in vitro studies with organelle preparations (Mauersberger et al. 1987).

3.5
Cell Fractionation and Preparation of Organelles

Cell fractionation protocols applied for alkane-grown *C. maltosa* are comparable to those described for other yeasts. Several methods for the preparation of subcellular organelles from *C. maltosa* cells were described in the literature. Only references for *C. maltosa* are given here, although methods described for other yeasts such as *C. tropicalis*, *Y. lipolytica*, or *S. cerevisiae* should also be applicable (for further references see Davidova et al. 1975; Krauzova and Sharyshev 1987; Mauersberger et al. 1987; Sharyshev and Krauzova 1988; Tanaka and Fukui 1989).

Cell walls – from alkane-grown cells (Maximova et al. 1972; Fischer and Reuter 1982; Nüske et al. 1982; Belov et al. 1985; Maksimova et al. 1988); and cell wall components (Röber and Reuter 1984a,b,c,d): mannan (Rademacher and Reuter 1978; Bovina et al. 1986, 1988), proteophosphomannan (PPM, Grimmecke and

Reuter 1980, 1981a,b,c; Grimmecke et al. 1981), glucan (Grimmecke et al. 1981d; Nüske et al. 1982).

Plasma membranes – from alkane-grown cells after gradient centrifugation of the microsomal fraction (Mauersberger et al. 1987).

Vacuoles – from glucose-grown cells by Belov et al. (1985) according to Belov et al. (1976) using a technique of flotation in a Ficol gradient applied to *C. tropicalis.*

Peroxisomes – from alkane-grown (Krauzova and Sharyshev 1987; Sharyshev and Krauzova 1988) or oleic acid-grown cells (Kamiryo et al. 1989).

Mitochondria – from alkane-grown cells (Krauzova and Sharyshev 1987; Sharyshev and Krauzova 1988).

Microsomes (ER) – from glucose-, glycerol-, or alkane-grown cells by differential centrifugation after mechanical cell disintegration (resulting in crude membrane preparations) or after spheroplast lysis (Mauersberger et al. 1981, 1984; Röber and Reuter 1984c; Mauersberger 1985; Krauzova and Sharyshev 1987; Sharyshev and Krauzova 1988) or by sucrose gradient centrifugation of the microsomal fraction (Mauersberger et al. 1987) or Percoll gradient centrifugation of cell-free homogenates (Krauzova and Sharyshev 1987; Sharyshev and Krauzova 1988).

Lipid vesicles (inclusions or granules) – from protoplasts by a technique of flotation in a Ficoll gradient developed for *C. tropicalis* by Belov et al. (1976), Davidova et al. (1977a,b, 1989), and Zinchenko et al. (1990).

For the preparation of pure fractions of organelles, like ER or peroxisomes, several subcellular fractionation steps followed by density gradient centrifugation using sucrose or Percoll as gradient material were applied (Krauzova and Sharyshev 1987; Mauersberger et al. 1987; Sharyshev and Krauzova 1988). A method for fractionating subcellular membrane structures, especially for separation of mitochondria and peroxisomes, of *C. maltosa* grown on n-alkane has been developed (Krauzova and Sharyshev 1987; Sharyshev and Krauzova 1988). It involves the consecutive two-stage centrifugation of membrane organelles under isoosmotic conditions in self-generating Percoll gradients. The method is applicable for a rapid and effective purification of mitochondria and peroxisomes which retain their biochemical and morphological integrity as shown by analysis of the distribution of marker enzyme activities in different subcellular membranes as well as by electron microscopy examination of Percoll gradient fractions.

Further separation of the microsomal (ER-containing) fraction by sucrose gradient centrifugation resulted in an increased content of P450 and its reductase in the purified ER fractions separated from impurities of plasma membrane fragments present in the microsomal preparations (Mauersberger et al. 1987).

4
Genetics and Molecular Biology of *Candida maltosa*

4.1
Strains Used in Different Laboratories

The wild-type strains of *C. maltosa* mainly used in the past and currently applied for genetic and molecular biological studies in different laboratories are the following (cf. Sect. 1, and Tables 1, 6, 8, 10, 11, and 14):

IAM12247 (=CBS5611) – Department of Biotechnology (Prof. M. Takagi) at the University of Tokyo, Japan.

SBUG700 (L4) – Department of Biology, Institutes of Genetics, Biochemistry, and Applied Microbiology, at the University of Greifswald, Germany.

EH15 – Department of Cell Biology, Laboratory of Membrane Proteins (Dr. W.-H. Schunck) at the Max-Delbrück-Center for Molecular Medicine (formerly at the Institute of Molecular Biology), Berlin-Buch, Germany.

VSB899 – Institute of Biotechnology, formerly Applied Enzymology (Dr. K. Sasnauskas), Vilnius, Lithuanian Republic.

From these strains, several gene libraries have been constructed (Table 10) and all *C. maltosa* genes have been cloned so far (Table 11), with the exception of the *POX4* gene cloned from *C. maltosa* ATCC20184 (Hill et al. 1988). The characterized mutants derived from these strains are listed in Table 6 and are described below in Sect. 4.2).

Recommended Strains and Plasmids of E. coli. The molecular cloning procedures and recombinant DNA methods applied for *C. maltosa* mostly followed standard protocols (Sambrook et al. 1989). Therefore, the experimental work with *C. maltosa* DNA does not require special plasmids and *E. coli* strains. Commonly used strains of *E. coli* (HB101, JM103, DH5αc, SURE, and others) and plasmids derived from pBR322, pUC118/119, or pUCBM20/21 were successfully applied to construct genomic libraries of *C. maltosa* DNA or cDNA (cf. Table 10).

4.2
Mutagenesis and Mutants

4.2.1
Mutagenesis of *Candida maltosa*

Mutagens. Different mutants of *C. maltosa* were isolated after physical (ultraviolet or gamma radiation) or chemical (MNNG, EMS, NA, and other classical mutagenic agents) mutagenesis in several laboratories (Yano et al. 1981; Bode and Casper 1983; Bode et al. 1983; Chang et al. 1984; Klinner et al. 1984; Durasova et al. 1986, 1989; Mikhailova et al. 1987, 1991; Samsonova et al. 1987; Schwarz et al. 1987; Casey et al. 1990; Becher et al. 1991; Jomantiene et al. 1991; Sasnauskas et al. 1992c; cf.

Table 6). Contrary to other organisms, UV-induced inactivation curves in all tested *C. maltosa* strains derived from L4 showed more or less linearity without a shoulder at low dose area, suggesting a one-hit kinetics, as known from mutants with a defective repair system (Samsonova et al. 1987). Among the chemical mutagens N-methyl-N'-nitro-N-nitrosoguanidine (MNNG), 6-N-hydroxy-aminopurine (HAP), 4-N-nitroquinoline oxide (NQO), ethylmethanesulfonic acid (EMS), nitrous acid (NA), sodium-p-dimethylaminobenzene-diazosulfonate (DAPA), and 2-methyl-6-chloro-[3-(ethyl-2-chloroethyl)amino]acridine hydro-chloride (ICR-170) were tested alone and in combination with UV light, and compared in their effectiveness to induce mutants in *C. maltosa* by Chang et al. (1984), Mikhailova et al. (1987), and Durasova et al. (1989).

Frequency of Mutants. Generally, the frequency of mutants is relatively low in *C. maltosa* due to its diploid or aneuploid nature (Chang et al. 1984; Kunze et al. 1984a,b), when compared with haploid strains of *Y. lipolytica, P. guilliermondii,* or *S. cerevisiae.* Under comparable experimental conditions, the total mutant frequency obtained after MNNG treatment in *C. maltosa* EH15 was about 0.2–0.3% (at 0.5–1% survivings from 10^7 cells/ml with 500 μg MNNG/ml), being considerably lower than for the haploid *Y. lipolytica* strains, when mutant frequencies of up to 8–12% (at 0.5–5% survivings from 10^7 cells/ml with 250 μg MNNG/ml) were reached (Mauersberger and Barth 1985, unpubl. results; Mauersberger 1991). In the Laboratory of Yeast Genetics at Greifswald University numerous auxotrophic, amino acid catabolic and alkane nonutilizing (*alk*) mutants were obtained from the *C. maltosa* strains L4 or EH15 on exposure to UV or MNNG using the same conditions as described by Böttcher and Samsonova (1978) for *Rhodosporidium* Banno. The total mutant frequencies were 0.05 and 0.4% with UV or MNNG, respectively (Bode et al. 1983; Klinner et al. 1984; Samsonova et al. 1987; Schwarz et al. 1987). The highest frequencies of mutants resistant to nystatin were obtained with MNNG and NQO (about 0.04% at 0.1–1% surviving cells). Interestingly, other mutagens active in *S. cerevisiae,* including diepoxioctane (DEO), showed no mutagenic activity in *C. maltosa* VSB589 in these experiments (Durasova et al. 1989).

Stability of C. maltosa *Mutants.* The mutants obtained in the primary screening from *C. maltosa* were sometimes very instable. However, selected stable auxotrophic or *alk* mutants showed reversion rates between 10^{-6} and 10^{-8} (Chang et al. 1984; Schwarz et al. 1987; Samsonova et al. 1987; Mauersberger, unpubl. results; Mauersberger 1991). A relatively high instability of some auxotrophic mutants of *C. maltosa* L4 was observed and analyzed in more detail (APA-change, Schult et al. 1987; Becher et al. 1994; see Sect. 4.2.2).

Enrichment Procedures of Mutants. Due to the relatively low mutant frequencies observed for the diploid or aneuploid *Candida* yeasts (*C. maltosa, C. tropicalis, C. albicans*), several authors tried to apply mutant enrichment techniques with polyene antibiotics (nystatin, amphotericin B, pyrrolnitrin) to enhance the frequency of mutants. Mainly the nystatin-treatment method according to Snow

(1966), widely used for isolation of auxotrophic mutants in yeasts, has been modified and applied to enrich mutants after MNNG or UV mutagenesis of *C. maltosa* and *C. tropicalis* deficient in assimilation of alkanes (Yano et al. 1981), and of *C. (cloacae) maltosa* more recently by Casey et al. (1990), as well as auxotrophic mutants of *C. maltosa* (Chang et al. 1984). It appeared that the growing *Candida* cells were not efficiently killed by nystatin. Only for the combination nystatin and pyrrolnitrin (an antifungal antibiotic like nystatin) were enrichment ratios after three cycles observed, resulting in increased *alk* mutant frequencies from 0.9 to 3% (Yano et al. 1981). Although the frequency of auxotrophs increased to 2–14% after nystatin treatment, they were limited only to *ade* and *his* (Chang et al. 1984). Casey et al. (1990) reported an optimized nystatin enrichment procedure for *alk* mutants of *C. (cloacae) maltosa* (strain P410), showing that besides other factors the key to antibiotic sensitivity of *Candida* yeast cells is the nutrient status of the cells, particularly the nitrogen level. The experiments showed that the essence of achieving a high differential killing ratio was to starve the cells for nitrogen (the cells became insensitive to nystatin) and then resuspend them in a nitrogen-containing medium with alkane as carbon source, where only wild-type cells became nystatin-sensitive due to reestablished cellular nitrogen levels. Using this procedure, the frequency of *alk* mutants after MNNG mutagenesis increased to 1.15% in only one round of enrichment, thus minimizing the opportunity for mutant cloning.

Positive Selection of Mutants. The alternative to the mainly used negative selection approach followed by biochemical characterization for isolation of specific (auxotrophic or carbon assimilation) mutants is the use of methods which permit the positive (direct) selection of mutants with particular enzyme lesions. Positive selection can be achieved by using substrate analogs (antimetabolites with toxic or growth inhibiting effects when converted) or antibiotics for selection of resistant mutants, or after gene disruption experiments with an established host/vector system. A combination of these two approaches was recently applied for the isolation of *ura3* mutants in *C. maltosa* resistant to the substrate analog 5-fluoro-orotic acid (5FOA) by Ohkuma et al. (1993a) after cloning the *URA3* gene of *C. maltosa* by complementation of the *ura3* mutation of *S. cerevisiae*. To construct a useful host for genetic engineering of *C. maltosa* using *URA3* as a marker, one allele of *URA3* in a double auxotroph (*his5 ade1*) was disrupted by the homologous *ADE1* gene, and subsequently two kinds of *ura3* mutants were isolated by selecting for spontaneous 5FOA resistance. One of the mutants was homozygous for the disruption (*ura3::ADE1/ura3::ADE1*); the other was heterozygous (*ura3::ADE1/ura3*). The *ura3::ADE1* allele in the latter strain was resubstituted by *URA3* to rescue the adenine auxotroph (*his5 ade1 URA3/ura3*). Finally, by selecting a 5FOA-resistant mutant, a triple auxotroph (*his5 ade1 ura3/ura3*) was isolated (Table 6).

Adenine auxotrophic mutations occur frequently in *C. maltosa* and are easy to screen for red mutants, which may correspond to *ade1* or *ade2* mutants, as in *S. cerevisiae* (Jomantiene et al. 1987; Kawai et al. 1991; Sasnauskas et al. 1992b). A direct selection procedure for increased polyene antibiotic (nystatin, levolin, amphothericin) resistance in *C. maltosa* was used for the isolation of mutants with defects in ergosterol biosynthesis (Durasova et al. 1989).

4.2.2
Mutant Phenotypes

The main phenotype classes of mutants obtained in *C. maltosa* are the following:

- Auxotrophic (*aux*) mutants, including amino acid, purine, pyrimidine, and vitamin auxotrophs (Chang et al. 1984; Klinner et al. 1984; Kunze et al. 1987a,b; Samsonova et al. 1987; Schwarz et al. 1987; Sasnauskas et al. 1992c; further references see below and Table 6).

- Mutants in amino acid degradation pathways (Casper et al. 1985b; Umemura et al. 1991).

- Alkane nonutilizing (*alk*) mutants (Yano et al. 1981; Samsonova et al. 1987; Schwarz et al. 1987; Sunairi et al. 1988; Casey et al. 1990; Mauersberger 1991; Mauersberger and Barth 1985, unpubl. results; see Table 7).

Table 6. Characterized mutants of *Candida maltosa* available from different laboratories

Strain	Genotype	Enzyme defect	Reference
Mutants obtained from the wild-type strain IAM 12247[a]			
J288	*leu2*, Alk leaky[b]	3-IPMDH	Chang et al. (1984)
			Takagi et al. (1986a)
CH1	*his5* Alk[+b]	HPAT	Hikiji et al. (1989)
CHA1	*his5, ade1*	HPAT, SAICAR-S	Kawai et al. (1991)
CHU1	*his5, ade1,*	HPAT, OPDCase	Ohkuma et al. (1993a)
	ura3::ADE1/		
	ura3::ADE1		
CHAU1	*his5, ade1,*	HPAT, SAICAR-S,	Ohkuma et al. (1993a)
	ura3/ura3	OPDCase	
CHS1	*his5, cys2*	HPAT, SerAT	Ohkuma (unpubl. results)
CHSA1	*his5, cys2, ade1*	HPAT, SerAT,	Ohkuma (unpubl. results)
		SAICAR-S	
G630	*leu4, arg*	2-IPMS	Becher et al. (1991)
			Bode et al. (1991)
N-07		AlaAT	Umemura et al. (1991)
Mutants obtained from wild-type strain L4[c]			
G374	*ade1*	SAICAR-S	Kasüske et al. (1992)
	asp1-1	APAT	Casper et al. (1985b)
	asn1-1	AS	Casper et al. (1985b)
G344	*arg4-18*	ASL	Kunze et al. (1985a,b, 1986a)
	glu1-1	GOGAT	Casper et al. (1985b)
	ilv1-1	TDA	Schult et al. (1987)
G381, G321	*leu1*	2-IPMDHT	Bode et al. (1991)
G367, G368	*leu1*	2-IPMDHT	Bode et al. (1991)
G383, G746	*leu1*	2-IPMDHT	Becher et al. (1991)
G727, G735	*leu1, thi*	2-IPMDHT	Becher et al. (1991)

Table 6. (*Contd.*)

Strain	Genotype	Enzyme defect	Reference
G773	*leu1 ade*	2-IPMDHT	Becher et al. (1991)
G587, G785	*leu2*	3-IPMDH	Becher et al. (1991, 1994)
G946	*leu2*	3-IPMDH	Bode et al. (1991)
G755	*leu2, ura*	3-IPMDH	Becher et al. (1991, 1994)
G285; G358	*lys1*-14; *lys1*-16	SAPDH	Schult et al. (1987)
			Schmidt et al. (1989a,b)
G457	*lys2*-21, *ino*-3	2-AAR	Schmidt et al. (1985)
			Kunze et al. (1987a,b)
G107; G112	*lys5*-5; *lys5*-1, *arg*-5	2-ARR	Schmidt et al. (1985)
G329; G376	*lys5*-19; *lys5*-20	2-AAR	Kunze et al. (1987a,b)
G108	*lys9*	SAPR	Schmidt et al. (1989a,b)

Mutant obtained from wild-type strain VSB899

ade1	*ade1*	SAICAR-S	Jomantiene et al. (1987)
			Sasnauskas et al. (1992b)

[a] Strains obtained mainly by UV mutagenesis, except J288 *leu2* obtained after a second MNNG mutagenesis, and the spontaneous *ura3* mutations in CHAU1. These strains are used for the host-vector system. Mutants G630 and N-07 were obtained from the wild-type strains CBS5611 and JCM1504, respectively, both strains are regarded as being the same as IAM12247 (see Table 1).

[b] J288 strain has serious growth defect on alkanes, whereas CH1 and the mutants derived from it show normal growth on alkanes (Hikiji et al. 1989).

[c] G – Strain collection at the Greifwald University, Biology Section (in some publications also Institute of Genetics [IGG] – Becher et al. 1991; Most of these strains are included in the SBUG strain collection). The wild-type strain SBUG700 (L4) is similar if not identical to the *C. maltosa* strain EH15.

Mutant designation. When shown as *lys2* – genetically defined mutant; when *lys-2* – allele of mutation phenotype, genotype not characterized; *lys2*-2 and *lys2*-5 are two alleles of mutations in the same gene.

Abbreviations for substrates and enzymes.

2-AA	– 2-Aminoadipate
2-AAR	– 2-Aminoadipate reductase (EC 1.2.1.31), *LYS2* and *LYS5* genes
AlaAT	– Alanine aminotransferase (EC 2.6.1.2)
AS	– Asparagine synthetase (EC 6.3.1.1), *ASN1* gene
ASL	– Argininosuccinate lyase, *ARG4* gene
APAT	– Aspartate aminotransferase (EC 2.6.1.1), *ASP1* gene
GOGAT	– Glutamate synthetase (EC 2.6.1.53), *GLU1* gene
HPAT	– Histidinol-phosphate aminotransferase (EC 2.6.1.9), *HIS5* gene
IPM	– Isopropylmalate
2-IPMDHT	– 2-IPM dehydratase (EC 4.2.1.33), *LEU1* gene
3-IPMDH	– 3-IPM dehydrogenase (EC 1.1.1.85), *LEU2* gene
2-IPMS	– 2-IPM synthase (EC 4.1.3.12), *LEU4* gene
OPDCase	– Orotidine-5′-phosphate decarboxylase (EC 4.1.1.23), *URA3* gene
SAICAR-S	– Phosphoribosylamidoimidazol succinocarboxamide synthetase or SAICAR synthetase (EC 6.3.2.6), *ADE1* gene
SAPDH	– Saccharopine dehydrogenase (EC 1.5.1.7), *LYS1* gene
SAPR	– Saccharopine reductase (EC 1.5.1.9), *LYS9* gene
TDA	– Threonine deaminase, (*EC4.2.1.16*), *ILV1* gene
SerAT	– Serine O-acetyltransferase (EC 2.3.1.30), *CYS2*, *CYS1* genes

- Polyene antibiotic-resistant mutants, mostly having defects in ergosterol biosynthesis (Durasova et al. 1986, 1989, 1991; Mikhailova et al. 1987, 1991; Volchek et al. 1988; Ogorodnikova et al. 1991).

- Catabolite repression mutants H3, H5, H6 (Röber and Reuter 1982, 1984a,d, 1985), and the spontaneous catabolite inactivation mutant PHM (Hofmann and Polnisch 1990c).

- Morphology mutants obtained from *C. maltosa* H, having differences in mannan content (UV and MNNG mutants, Brückner and Tröger 1981a; Kölblin and Birkenbeil 1981; Kölblin and Tröger 1982).

Several of these mutants obtained from *C. maltosa* were characterized in detail at biochemical and molecular genetic levels, as shown below and in the Table 6.

Auxotrophic Mutants. Numerous *aux* mutants of *C. maltosa* have been isolated mainly from the two wild-type strains, IAM12247 (CBS5611) and SBUG700 (L4), in the Department of Agricultural Chemistry at the University of Tokyo (Chang et al. 1984; Hikiji et al. 1989; Kawai et al. 1991; Ohkuma et al. 1993a) and in the Laboratory of Yeast Genetics at the Greifswald University since the beginning of the 1980s (Bode and Casper 1983; Bode et al. 1983; Klinner et al. 1984; Schwarz et al. 1987; Becher et al. 1991, 1994; cf. Table 6).

Among the approximately 300 auxotrophic mutants obtained in *C. maltosa* L4, some double auxotrophs were isolated (*ade leu1, ade-26 pro-1, arg leu4, arg-2 lys5-1, arg-5 lys5-1, arg-2 his-4 lys5-1, his-11 met-6, his-23 met-6, leu2 ura, leu1 thi, lys2-21 ino-3, met-13 ino-3*), as partially shown in Table 6 (Klinner et al. 1984; Schmidt et al. 1985; Kunze et al. 1987a,b; Samsonova et al. 1987; Schwarz et al. 1987; Becher et al. 1991). Some of the auxotrophic mutants were characterized biochemically in detail in their genetic defects (Table 6) and used in host-vector systems (Table 14). The most frequent and easiest available auxotrophic mutants in *C. maltosa* are the *ade* (14%) and *his* (18%) phenotypes, which is in accordance with the above mentioned data obtained by Chang et al. (1984).

Mutants of the *ADE1* or *ADE2* genes in *C. maltosa*, as in the majority of yeast species, accumulate a red or pink intracellular pigment (amino-imidazole ribotide, or carboxyamino-imidazole ribotide, which is then converted by oxidation into a polymeric derivative which is colored and relatively nontoxic to the cells in which it accumulates) when grown on limiting adenine. These stable red mutants of *C. maltosa* showing low (<1%) activity of phosphoribosyl-aminoimidazole-succinocarboxamide-synthetase (SAICAR-synthetase, EC 6.3.2.6), coded by the *ADE1* gene (Jomantiene et al. 1987; Kawai et al. 1991; Sasnauskas et al. 1992b), were easily selectable and therefore included in the construction lines of multiple auxotrophic recipient strains derived from the isolated *his5* mutant CH1 (CHA1 from CH1, CHSA1 from CHS1) used in host-vector systems (Hikiji et al. 1989; Kawai et al. 1991; Ohkuma, unpubl. results; Tables 6, 14, cf. Sect. 5). Ohkuma et al. (1993a) constructed two further recipient strains, CHU1 and CHUA1, in this line using gene disruption technique to create an *ura3* defect. More recently, the mutants CHS1 and CHSA1 were obtained by a second mutagenesis of the strains CH1 and

CHA1. The *cys2* mutants were hard to obtain from the wild-type strain, but easier from the CH1 mutant, which may have already some defect (Ohkuma, unpubl. results).

Leucine auxotrophs (4.5% *leu* from L4) occurred with relatively low frequency, which might be related with the presence of silent gene copies in *C. maltosa*, as demonstrated recently for the *LEU2* gene (Becher et al. 1991, 1994). The first *C. maltosa leu2* host, J288, later used for transformation (see Sect. 5.3), was one of three *leu* mutants (J183, J288, J316) isolated after a second round of mutagenesis of the only *leu* mutant found in 30 000 survivors after the first round of UV mutagenesis. The original *leu* mutant was very unstable and reverted at high frequency (Chang et al. 1984). More recently, the four enzymatic steps in the conversion of 2-ketoisovalerate to leucine were examined in the wild-type and in 13 leucine-auxotrophic strains of *C. maltosa* L4 (Wedler et al. 1990; Becher et al. 1991; cf. Sect. 2.6.1). The genetic lesions in the auxotrophs involve at least five different loci (complementation groups) and are correlated with three enzymatic steps (coded by the *LEU1*, *LEU2*, and *LEU4* genes), which was confirmed by gene cloning, protoplast fusion, and enzyme assays. In analogy to *S. cerevisiae*, the corresponding *leu* mutants of *C. maltosa* were designated *leu1*, *leu2*, and *leu4*. In 4 out of 13 leucine-auxotrophic mutants, a defect in the 3-isopropylmalate dehydrogenase (coded by the *LEU2* gene) was shown (Table 6). The potential application of *leu1* and *leu2* mutants of *C. maltosa* for the production of 2-IPM and 3-IPM, respectively, was reported (Bode et al. 1991; see Sect. 6).

Five selected *C. maltosa lys* mutants (8% *lys* from L4) showed two complementation groups on the basis of complementation analysis using protoplast fusion, which might represent the two genes *LYS2* and *LYS5* coding for two different subunits of the 2-aminoadipate reductase, in analogy to the results known for *S. cerevisiae*. Only one of these two groups of mutants (G457 *lys2*) could be transformed by plasmids containing the *LYS2* gene of *S. cerevisiae* (Schmidt et al. 1985; Kunze et al. 1987a,b; Table 16). Further *lys* mutants defective in the saccharopine dehydrogenase (*lys1*) and in the saccharopine reductase (*lys9*) have been characterized by Schmidt et al. (1989a,b; cf. Table 6) and tested for their potential application in production of useful compounds, such as α-aminoadipate-δ-semialdehyde (AASA; cf. Sect. 6).

The mutant *C. maltosa arg4-18* (G344, 9% are *arg* mutants) was characterized by its transformation with the *ARG4* gene coding for the argininosuccinate lyase of *S. cerevisiae* (Kunze et al. 1985a,b, 1986a, 1987a,b).

It is remarkable and still without explanation that no *tyr*, *phe* or *trp* auxotrophs have been found in *C. maltosa* (Kunze et al. 1987a,b; Schwarz et al. 1987), a result similar to that obtained for *C. albicans* (Poulter 1990).

Several of the auxotrophic mutants of *C. maltosa* induced by UV or MNNG show an unusual behavior (Schult et al. 1987; Becher et al. 1994). After reversion to prototrophy (nearly one revertant per 10^8 cells) they mutated to the original auxotrophy, and these auxotrophy-prototrophy-auxotrophy alterations (APA change) took place successively many times in a cell line. The genetic stability of clones originating from APA mutations varied considerably. Spontaneous mutation frequencies of different clones covered the range between 10^{-9} and 10^{-1}.

APA mutations are not restricted to particular types of auxotrophic mutants, since they were observed in mutants with a blocked biosynthesis of arginine, isoleucine, leucine, aspartic acid, asparagine, histidine, lysine, methionine, adenine, or uracil.

A lineage of at least four types of spontaneous APA changes starting with the *leu2* mutant G587 derived from the wild-type *C. maltosa* strain L4, including even segregation of an *arg* marker, was recently demonstrated by Becher et al. (1994). Data obtained from investigations of the enzymes encoded by the unstable *LYS1* and *ILV1* genes (saccharophine dehydrogenase and threonine deaminase) indicate that the APA mutations are allelic and create novel enzymes species with altered catalytic and regulatory properties. Therefore a mechanism involving repeated site-specific insertions and excisions of DNA elements (insertion mutagenesis), probably by the action of mobile (transposon-like) genetic elements, was proposed by Böttcher and Samsonova (Schult 1987; Schult et al. 1987). An interpretation of the APA change based at experimental data on the occurrence of probably silent copies of *LEU2* genes in the genome of the *C. maltosa leu2* mutant G587 was recently presented by Becher et al. (1994; cf. Sect. 4.4.2). These authors propose further that the whole chromosome on which these alleles reside might be silent, controlled by at least two chromosome-silencing points on each chromosome arm. Assuming such a genome organization, the APA changes (see Sect. 4.2.2) can be easily interpreted as mitotic cross-over events between active and inactive wild-type and mutant alleles. The activation of the silent wild-type alleles requires their transposition into an active chromosome, which can be accomplished by a double cross-over event. As for the high frequency morphology switching in *C. albicans* (Soll 1990), the possible molecular basis of these phenomena in *Candida* yeast remains elusive, although several switching systems in other organisms have been elucidated at the molecular level and therefore provide models for experiments on *Candida* systems.

Alkane Nonutilizing (alk) Mutants. Mutants deficient in the alkane degradation pathway (*alk*), as well as being of scientific importance in studies of this pathway, are also of commercial interest because several of the degradative intermediates (fatty alcohols or dicarboxylic acids) are of value to the chemical industry. Therefore, these mutants are of interest for possible application to produce useful oxidation products directly derived from the n-alkane substrate (cf. Sect. 6). Thus, first *alk* mutants (M-1 and M-12) in *C. maltosa* were already obtained in the beginning of the 1970s from the strain *C. cloacae* 310 and applied for the production of dicarboxylic acids (Uchio and Shiio 1972b,c, for details see Sect. 6).

In several laboratories, *alk* mutants were obtained from different *C. maltosa* strains (Yano et al. 1981; Samsonova et al. 1987; Schwarz et al. 1987; Böttcher 1987; Sunairi et al. 1988; Casey et al. 1990; Mauersberger 1991; see Table 7). For *C. maltosa*, the frequency of *alk* mutants was about the same as that of auxotrophic mutants, being maximally approximately 0.1–0.2% after MNNG mutagenesis without special enrichment techniques for each type of mutants (Schwarz et al. 1987; Mauersberger, unpubl. results). Recently, techniques have been described which allow mutated populations of *C. maltosa* to be enriched efficiently for the isolation

Table 7. Phenotypes of *Candida maltosa* valk mutants

Strain	Alk phenotypes[a]						Reference Survivals (mutagens used) Frequency of alk mutants Comments
	A[c]	B	C	D	E	Others[b]	
C. maltosa IAM12247	119	2	8 (C/D)[a]		n.d.	n.d.	Yano et al. (1981) 14900 (EMS) 129 *alk* (0.86%) nystatin enrichment
C. maltosa L4 Several auxotrophic mutants[d]	32 (Aa 9/Ab 23/Ac 0)[c]	29 (B/C)[a]		24	59	13[b]	Schwarz et al. (1987) 100000 (MNNG) 157 *alk* (0.157%) without enrichment
C. maltosa J288 (*leu2, alk* leaky)		64 (A/B/C/D)[a]			9	2 *glu* (154 aux)	Sunairi et al. (1988) 10000 (UV) 73 *alk* (0.73%)
C. cloacae P410 (*C. maltosa*)	80 (A/B/C)[a]			84	124E, *glu* or respiratory deficient[b]		Casey et al. (1990) 25000 (MNNG) 288 *alk* (1.15%) nystatin enrichment

[a] For several *alk* mutants no differentiation into the *alkA* to *alkD* phenotypes was given by the authors. Mostly, no clear *alkB* phenotype occurred, mutants are alkane leaky or alcohol leaky and not very stable.
[b] No clear classification into *alkA* to *alkE* phenotypes possible, or other phenotypes observed like *glu*.
[c] *alkA* mutants show subtypes *alkAa*, *Ab* or *Ac* in alkane chain length utilization.
[d] Several auxotrophic mutants of *C. maltosa* L4 were used for obtaining *alk* mutants – G284 (*met2-6*), G347 (*met6-21*), G589 (*his-2*), L4 (*thi-2*).

of *alk* mutants, thus increasing their frequency up to 1.15% (Casey et al. 1990). After mutagenesis (UV-, EMS- or MNNG-treatment) alkane-nonutilizing *C. maltosa* mutants were isolated by negative selection due to their inability to grow on n-alkanes. On the basis of the main alkane oxidation pathways (Fig. 3), these *alk* mutants can be classified according to Bassel and Mortimer (1982, 1985) into the phenotypes *alkA* to *alkE* by substrate utilization tests on agar plates with different alkane oxidation intermediates (alkanes, fatty alcohols, fatty aldehydes, fatty acids), acetate or ethanol, and glucose as carbon sources (see Sect. 4.2.3 and Table 7). The phenotypes *alkA*, *D*, and *E* were found frequently, whereas the phenotypes *alkB* and *alkC* occurred with low frequency, depending on the number of genes involved in these metabolic steps (cf. Sect. 2.5 and Fig. 3). Comparable distribution patterns of the *alk* phenotype classes have been found for *Y. lipolytica* (Bassel and Mortimer 1982, 1985; Mauersberger 1991), *C. tropicalis* (Yano et al. 1981), and *Pichia guilliermondii* (Schauer et al. 1987) after chemical or UV-induced mutagenesis. According to Bassel and Mortimer (1982, 1985), about 84 genes are involved in the phenotype of alkane utilization in *Y. lipolytica*, as calculated from the frequencies of *alk* mutants in comparison with those of auxotrophic mutants. From genetic data these authors concluded that at least 26 genes control the substrate uptake and the oxidation of n-alkanes to the corresponding fatty acid. About 16 of these genes should be connected with the alkane uptake. A comparable genetic background should also apply for *C. maltosa*.

Mutants of Phenotype alkA. The *alkA* mutants are found with high frequency in *C. maltosa* (Table 7), because this phenotype occurs as a result of possible defects in genes connected with the n-alkane uptake, the activity of the cytochrome P450 alkane monooxygenase systems or with the transport to and from these enzyme systems to the fatty alcohol oxidizing enzyme(s), e.g., in the transport to and from endoplasmic reticulum to peroxisomes (Fig. 3). A striking feature is the occurrence of different subtypes of the phenotype *alkA*, which are based on changes in the utilization of different n-alkane chain lengths. Whereas Alk+ strains utilize middle- and long-chain n-alkanes as growth substrates, starting with the chain length C_6 for *C. maltosa*, one can divide the *alkA* mutants obtained into three subtypes (Schwarz et al. 1987; Mauersberger 1991; Table 7):

alkAa – no alkane utilization at all: C_8^-, C_{10}^-, C_{12}^-, C_{16}^-, C_{12}-ol+
alkAb – no utilization of shorter n-alkanes, but utilization of longer n-alkanes: C_8^-, C_{10}^-, $C_{12}^\pm$, C_{16}^+, C_{12}-ol+
alkAc – utilization of shorter n-alkanes, but no utilization of longer chain lengths: C_8^+, C_{10}^+, $C_{12}^\pm$, C_{16}^-, C_{12}-ol+

Interestingly, the phenotypes *alkAa* and *alkAb* were frequently obtained directly after mutagenesis of *C. maltosa*, whereas the phenotype *alkAc* was obtained with very low frequency and stability (Table 7; Schwarz et al. 1987; Mauersberger 1991). Different *alkA* mutants were found not only directly after mutagenesis but also as spontaneous or induced revertants from *alkE* mutants. Especially *alkAc* mutants have been described up to now only for *C. maltosa* as revertants from a distinct *alkE* mutant G988 (*alk2154 met*) or as recombinants (spontaneous or UV-induced mitotic segregation) of the fusion product fp306-14 for which *alkD* and *alkE* mutants (including G988 *alk2154*) were used as parent strains (Schwarz et al. 1987; Mauersberger 1991). Taken together, all these results were hints for genetically determined differences between the metabolism of short- and long-chain alkanes, including the presence of multiple alkane-induced P450 forms with different chain-length specificities for the substrate n-alkane (Mauersberger 1991). This is now studied in detail on the basis of cloned alkane-inducible P450 genes in the authors' laboratories (Ohkuma et al. 1991a,b, 1995a,d; Schunck et al. 1991; Scheller et al. 1992, 1996; Zimmer and Schunck 1995; Zimmer et al. 1995; Table 11).

Phenotypes alkB, alkD, *and* alkE. For some *C. maltosa* mutants the conditions for the accumulation of fatty alcohols (*alkB*) and dicarboxylic acids (*alkD*) from the substrate n-alkane were investigated (Uchio and Shiio 1972b,c; Casey et al. 1990; Mauersberger 1991; see Sect. 6). The low frequency *alkB* mutants could be useful as potential fatty alcohol- or diol-producing strains directly from n-alkane. Out of 80 *C. maltosa* mutants characterized by Casey et al. (1990) as *alkA* to *alkC* (Table 7), 16 produced small amounts of hexadecanol, as shown also for the first described *alkB* mutant of *Y. lipolytica* (Mauersberger 1991). Mutants of *alkD* phenotype are frequently found in agreement with the numerous genes involved in the activation and degradation of fatty acids to acetyl-CoA (Fig. 3, see Sect. 2.5), including the genes for the acyl-CoA synthetases I and II (ACSI/II) and for the β-oxidation enzymes. Yeast *alkD* mutants are of special interest for practical use to produce dicarboxylic acids (DCA), ω-hydroxy-fatty acids and fatty acids directly from n-alkanes (Uemura et al. 1988; Schindler et al. 1990; Atomi et al. 1994; cf. Sect. 6). The results of Casey et al. (1990) with *C. maltosa* mutants showed that of the 84 *alkD* (Alk⁻ PA⁻ Ace⁺) mutants, most could accumulate dicarboxylic acids from hexadecane and palmitic acid and at least one mutant also produced 3-hydroxy-hexadecanedioic acid. Mutants of the *alkE* phenotype were obtained with high frequency (Table 7), reflecting the considerable amount of possible defects in the intermediate metabolism leading to this phenotype. Among them there should be mutants with defective enzyme activities of the glyoxylate cycle, e.g., isocitrate lyase (ICL) or malate synthase (MS) or of gluconeogenesis. Probably defects in the peroxisome formation (Per⁻) can also result in this phenotype.

Catabolite Repression and Catabolite Inactivation Mutants. Mutants of *C. maltosa* H obtained after treatment with MNNG were selected in the presence of 2-deoxyglucose. These mutants, H3, H5, and H6, were characterized as partially desensibilized in catabolite repression. They showed several changes in carbohydrate metabolism, especially in the biosynthesis of polysaccharide components (mannan and glucan) of the cell walls (Röber and Reuter 1982, 1984a,d, 1985). More recently, a spontaneous mutant of *C. maltosa* SBUG700 was isolated showing pseudohyphal morphology under all growth conditions tested (see Sect. 3.1). This PHM (stands for pseudohyphal morphology) mutant was shown to be defective in catabolite inactivation of gluconeogenic enzymes (Hofmann and Polnisch 1990c).

Colony Morphology Mutants. Several mutants obtained by UV and MNNG mutagenesis from *C. maltosa* H were described as having differences in high molecular mannan content, but not in glucan or low molecular weight mannan content (Kölbin and Birkenbeil 1981; Kölblin and Tröger 1982). Another UV mutant H13 was isolated showing mainly filamentous growth (Brückner and Tröger 1981a).

Mutants in Ergosterol Biosynthesis. The biosynthesis of sterols in *Candida* yeasts has a number of distinctive features as compared with the biosynthesis of ergosterol in *S. cerevisiae*, well studied in this respect. This is apparently due to the physiological and metabolic peculiarities of *Candida* species. A simple procedure has been developed to isolate *C. maltosa* mutants defective in ergosterol biosynthesis. A series of mutants of the *C. maltosa* strain VSB569 were selected for their resistance to polyene antibiotic nystatin (>15 mg/ml) after UV or chemical (NA, MNNG, NQO, or HAP) mutagenesis (Durasova et al. 1986, 1989; Mikhailova et al. 1987; Volchek et al. 1988). The composition of sterol fractions of 192 nystatin-resistant mutants was studied by UV spectrometry, TLC, and GLC-MS. The data obtained suggest that resistance of *C. maltosa* mutants to polyene antibiotics like nystatin is associated mainly with changes in the composition of the membrane sterols due to defects in ergosterol biosynthesis. Comparative analysis of sterols from *C. maltosa* and *S. cerevisiae* mutants allows determining blocks at different stages of ergosterol biosynthesis in some *C. maltosa* mutants, C24-transmethylation, the formation of the double bond during C22 and $\Delta 8 \Rightarrow \Delta 7$ isomerization, 14α-demethylation, C-5(6)-dehydrogenation, or reduction of C-14(15) and C-24(28) double bonds (Volchek et al. 1988; Durasova et al. 1991; Mikhailova et al. 1991; Ogorodnikova et al. 1991).

4.2.3
Mutant Isolation in *Candida maltosa*

MNNG Mutagenesis

Note on Method. Methods for UV light or chemical mutagenesis of *C. maltosa* strains have been published (Chang et al. 1984; Klinner et al. 1984; Casey et al. 1990). The elements of various mutagenesis protocols are similar to those described in the laboratory manuals for *S. cerevisiae* (Campbell an Duffus 1988; Goeddel 1990; Rose et al. 1990; Guthrie and Fink 1991) and *Candida* yeasts (Kirsch et al. 1990). Although these protocols for obtaining yeast mutants are applicable, for the diploid/aneuploid yeast *C. maltosa* the total mutant yield (0.1–0.3% for UV or MNNG, respectively) is considerably lower than that observed for haploid yeasts like *Y. lipolytica* or *S. cerevisiae*, being in the range of 5–10% for MNNG. Therefore enrichment procedures were applied to increase the mutant frequency in *C. maltosa* (cf. Sect. 4.2.1).

Procedure for MNNG Mutagenesis
Take all necessary care for the work with the mutagen MNNG. Carry out inactivation of MNNG solutions in 10% H_2SO_4 for at least 24 h.

1. Grow *C. maltosa* EH15 cells in 10 ml YPD for 20–24 h to a density of 5×10^7 to 10^8 cells per ml (late exponential growth phase).

2. Wash with sterile 50 mM phosphate buffer, pH 7, count the cell number, and dilute with buffer to 4×10^7 cells/ml. Make samples of 0.5 ml of the diluted cell suspensions.

3. Add 0.5 ml of freshly prepared MNNG (125–500 μg/ml) solutions in phosphate buffer pH 7 to reach final concentrations of 62–250 μg MNNG/ml and starting cell concentration of 2×10^7/ml.

4. Incubate at 30 °C with slight shaking for 15 to 120 min, for determination of the inactivation curves, to reach at least 90 to 99.9% of inactivation (10 to 0.1% surviving cells).

5. Dilute the samples to the appropriate dilution with sterile 0.9% NaCl and plate onto YPD agar. The dilution rate should be between 10^{-1} and 10^{-5} to have an appropriate number (50 to 200) of surviving clones on one YPD plate.

6. Incubate 2–3 days at 30 °C and count the surviving cells.
   Select the desired mutants by replica plating onto minimal medium agar (see below).

Selection and Classification of *alk* Mutants

Note on Method. The *alk* mutants are differentiated from the auxotrophic mutants by testing their growth on minimal medium with n-alkanes (C_{10}, C_{12}, C_{16}) or glucose as carbon sources in comparison with the rich medium YPD using the replica-plating technique. On the basis of the main alkane oxidation pathways (Fig. 3) *alk* mutants can be classified according to Bassel and Mortimer (1982, 1985) into the phenotypes *alkA* to *alkE* by substrate utilization tests on agar plates with different alkane oxidation intermediates as carbon sources as follows:

Phenotype	Growth on carbon source (chain lengths)					
	Alkanes	Alcohols	Aldehydes	Fatty acids	Acetate or	Glucose
	(C_8, C_{10}) (C_{12}, C_{16})	(C_{12}-ol) (C_{16}-ol)	(C_{12}-al)	(C_{12}OOH) (C_{16}OOH)	ethanol	
Alk$^+$	+	+	+	+	+	+
alkA	−	+	+	+	+	+
alkB	−	−	+	+	+	+
alkC	−	−	−	+	+	+
alkD	−	−	−	−	+	+
alkE	−	−	−	−	−	+

Procedure

1. Prepare master plates of the cells after mutagenesis on YPD agar (100 clones per plate).

2. Use these plates as templates for velvet pad replications onto YNB plates containing as carbon sources either glucose (2%), octane, decane, dodecane, or hexadecane (supplied in the vapor phase by adding 150–300 μl alkane to a filter paper placed in the lid of the Petri dish), and onto YPD plates as the control for the first selection.

3. After 4–5 days growth on YNB and 2 days on YPD plates, identify the *alk* and the auxotrophic or *glu* mutants by comparison of the three plates.

4. Pick the mutants of with a needle and stab inoculate onto fresh YPD plates (50–100 per plate).

5. After overnight growth, use these new master plates for replica plating onto a range of YNB plates containing glucose, acetate, ethanol, fatty acids, fatty aldehydes, fatty alcohols, or n-alkanes (chain length see above) as carbon sources. For a sufficient and comparable growth on all substrates in the agar plate experiments at 30 °C, the alkanes, dodecane-1-ol and dodecanal are fed via the gas phase. For other water-insoluble substrates, such as hexadecane-1-ol and fatty acids, the growth is supported by adding 1% Brij 35 together with 0.1% substrate to the minimal medium agar.

6. After 3–5 days growth classify the mutants into the *alkA* to *alkE* phenotypes in this second selection. This selection should be repeated several times to isolate stable mutants.

The determination of the spectrum of auxotrophic mutants showed a distribution based on the length of biosynthetic pathways (but no *phe*, *tyr*, *trp* mutants occurred at all), and a low percentage of double mutations, indicating also mainly monogenic defects in the *alk* mutants. For more detailed characterization, *alkA* to *alkE* mutants were selected, which showed high stability of the *alk* phenotype (reversion rates to Alk⁺ smaller than 10^{-6} or 10^{-7}) and of the auxotrophic markers (Mauersberger 1991).

4.2.4
Classical Genetic Techniques for *Candida maltosa*

Because of the imperfect status of *C. maltosa*, the main classical genetic methods requiring sporulation and mating, and successfully used for the genetic investigation of other yeasts (*S. cerevisiae*, *Y. lipolytica*, or *P. guilliermondii*, cf. other chapters, this Vol.) are not applicable in *C. maltosa*. Thus, for *C. maltosa*, genetic results can be obtained only by means of parasexual procedures. The establishment of protoplast fusion technique for *C. maltosa* (Chang et al. 1984; Klinner et al. 1984; see Sect. 4.2.5) provided all prerequisites for a parasexual analysis. After protoplast fusion, the following events may occur: immediate karyogamy, long-lasting heterokaryosis, chromosomal rearrangements, enhanced mitotic recombination, or deletions. Mitotically stable fusion hybrids were obtained with frequencies between 5×10^{-5} and 10^{-7} (Klinner et al. 1984; Klinner and Böttcher 1985; Samsonova et al. 1987; Becher et al. 1991). Chang et al. (1984) reported even higher fusant frequencies of 5×10^{-3} between *ade* and *his* mutants, but these hybrids were very unstable. Complementation tests in *C. maltosa* were performed by protoplast fusion for two nonallelic *his* mutations (Klinner et al. 1984). Later, Kunze et al. (1987a,b) analyzed the complementation groups of 6 *lys* mutations in *C. maltosa* by means of protoplast fusion. More recently, this method was applied to classify the 13 leucine-auxotrophic mutants into five complementation groups (Becher et al. 1991).

The frequency of spontaneous mitotic segregants in stable hybrids was insufficient for genetic analysis. Substances that are known to induce frequent mitotic segregation in other yeast species such as benomyl (a benzimidazole derivative), p-fluorophenylalanine, and griseofulvin were ineffective in *C. maltosa*. Acriflavin enhanced the segregation frequency only slightly. Irradiation by UV induced mitotic segregation significantly up to 10%, mainly occurring as mitotic crossing over (without chromosomal losses) in *C. maltosa*. It is noteworthy that even the chemical mutagen MNNG induced mitotic segregation in *C. maltosa* L4 (Klinner et al. 1984; Böttcher 1987). The usefulness of hybridization by protoplast fusion and UV-induced mitotic segregation for the genetic analysis of *C. maltosa* was tested. Based on these methods, the construction of the linkage group I in *C. maltosa* with the sequence *CEN-ade-26-pro-1* was performed (Klinner et al. 1984; Samsonova et al. 1987).

Klinner and Böttcher (1985) demonstrated that in *C. maltosa* even chromosomal rearrangements can arise during hybrid formation after protoplast fusion, resulting in differences in the mitotic segregation pattern between fusion hybrids of one and the same fusion combination. The authors suggested that this process was caused by transposable DNA elements which had been mobilized by protoplast fusion.

Whereas intraspecies fusion hybrids could be obtained in any case, interspecies hybridization did not occur using *C. maltosa*, *C. utilis*, *C. tropicalis*, and *P. guilliermondii* (Klinner and Böttcher 1987). The intergeneric fusion hybrids between *C. maltosa* and *S. cerevisiae* reported by Chang et al. (1984) were unstable, probably existing only as an intermediate heterokaryonic state, as frequently observed for other yeasts (Weber and Barth 1988). Attempts to polyploidize *C. maltosa* or *P. guilliermondii* by protoplast fusion were not successful due to chromosome loss immediately after fusion (Klinner and Böttcher 1987).

4.2.5
Method of Protoplast Fusion

Note on Method. Due to the lack of a sexual cycle in *C. maltosa* this yeast can only be manipulated by parasexual processes. The method of protoplast fusion is a useful tool for genetic analysis in *C. maltosa*, and was performed for complementation analysis of auxotrophic mutants (Chang et al. 1984; Klinner et al. 1984; Kunze et al. 1987a; Becher et al. 1991; see also Chap. 2, this Vol.).

Procedure

1. Preparation of Protoplasts. Preparation of protoplasts of *C. maltosa* cells is easily performed by using snail gut juice from *Helix pomatia* according to a method described (Maraz et al. 1978). Sorbitol (1 M) is used as an osmotic stabilizer. Freeze-dried snail gut juice is added to a final concentration of 100 mg/ml per 10^9 cells. Rates of 95% conversion to protoplasts are usually obtained within 1 h of incubation at 30 °C (Becher et al. 1991).

2. Protoplast Fusion. Fusion experiments in PEG solutions can be performed as described by Becher and Böttcher (1983), Klinner et al. (1984), or Chang et al. (1984): 5×10^7 protoplasts of each of the corresponding mutants are used for the fusion experiments. The hybrid character of fusion products of several fusion combinations can be confirmed by UV irradiation-induced mitotic segregation according to Klinner et al. (1984).

4.3
Characterization of the Candida maltosa *Genome*

Compared to the situation of *S. cerevisiae*, the molecular biology of *Candida* yeast including *C. albicans*, *C. maltosa*, and *C. tropicalis* has not been so intensively investigated, although, especially during the past decade, several efforts have been made and new successful results have been obtained. These alkane-utilizing yeast species represent an interesting model for studies on imperfect yeasts (Kirsch et al. 1990).

4.3.1
Genome Characteristics and Ploidy

The first molecular biological characterization of the *C. maltosa* genome was done in the beginning of the 1980s (Meyer et al. 1975; Kunze 1982; Chang et al. 1984; Kunze et al. 1984a,b; Tables 8, 9). For the strain *C. maltosa* EH15, the molecular mass of the haploid genome was experimentally determined from DNA reassociation kinetics. The value for kinetic complexity of DNA was with approximately 8.1×10^9 Da relatively low in comparison with other yeasts. According to these data, the *C. maltosa* DNA contains a high degree (19.9%) of repetitive sequences. The determined total DNA amount per cell was approximately 52×10^{-15} g or 13×10^{-15} g per haploid nuclear genome (Kunze et al. 1984a; Table 8). Additionally, *C. maltosa* contains with 35% an extremely high amount of circular mitochondrial DNA (see Sect. 4.3.4). Its G + C content is with 22% lower than in several other yeasts, whereas for the chromosomal DNA it is in the range (36–37%) obtained for other *Candida* yeasts (*C. tropicalis* and *C. albicans*), but considerably lower than in *C. sake* strains with 38–40% (Meyer et al. 1975; Su and Meyer 1991). Recent data on

Table 8. Genome characteristics of *Candida maltosa*

Characteristic	DNA content per cell[a] (relative)	Haploid size of genome[b]	Reference
Total DNA	52 fg (100%)	8.1×10^9 Da (= 13 fg)	Kunze et al. (1984a)
Chromosomal DNA			
Content	34 fg (65%)	13 fg	Kunze et al. (1984a)
Size	14–20 Mb		Tanaka et al. (1987)
		15.5 fg	Becher et al. (1991, 1994)
Chromosome number	8–10 (11)		Ohkuma et al. (1993c, 1995a)
GC%	36–37%		Meyer et al. (1975) Su and Meyer (1991)
Mitochondrial DNA			
Content	18 fg (35%)	2.8×10^9 Da	Kunze et al. (1984a)
Size	51–52 kb	(total)	Kunze et al. (1986a)
	Molecular mass 3.5×10^7 Da		Su and Meyer (1991)
Copies	80 per cell		
GC%	21–22%		
Plasmid DNA	0 (0%)		Kunze et al. (1984a)
DNA concentration (fg) per haploid genome size of chromosomal DNA (fg)		2.6 (for nuclear DNA)	Kunze et al. (1984a)
Ploidy level		2n–3n, aneuploid	Kunze et al. (1984a) Becher et al. (1994)
Transposons	Not found, but suggested		Becher et al. (1994)
Cloned genes			
Sequenced	See Table 11		
Not sequenced	*ADE2, ARG4*		Jomantiene et al. (1987)
	ARS (two different fragments)		Polumienko and Grigorieva (1985)
	ARS1 and *ARS2*		Becher et al. (1991) Kasüske et al. (1992)
	CYS2 (SerAT)		Ohkuma (unpubl.)

Abbreviations: *ADE2* – Posphoribosylaminoimidazole carboxylase (EC 4.1.1.21)
            ARG4 – Argininosuccinate lyase (ASL)
            ARS – Autonomously replicating sequence
            CYS2 – Serine O-acetyltransferase (SerAT, EC 2.3.1.30), coded by *CYS1* and
                CYS2 genes.
            fg – femtogram (10^{-15} g).
Data obtained mainly for the *C. maltosa* strains EH15, L4, and IAM12247 (CBS5611)
[a] DNA content per cell determined experimentally.
[b] Haploid genome size determined as kinetical complexity of the chromosomal DNA per haploid genome from DNA reassociation kinetics (Kunze et al. 1984a).

Table 9. Electrophoretic karyotype and chromosomal localization of several cloned genes in *Candida maltosa* strain IAM12247 (CBS5611). Chromosomes were separated by pulsed field gel electrophoresis (CHEF or OFAGE). The molecular sizes were estimated using the chromosomes of *S. cerevisiae* as a standard, and genes were assigned by Southern blot analysis (see Sects. 4.3.2–3, 4.4.4, and Fig. 5)

Chromosome no.[a]	Size (Mbp)	Genes localized at the chromosome
VIII	2.20	*CYP52A3–5, 9–11 (ALK1-3, 5, 7–8); POX5*
VII	1.85	*POX2, POX4*
VI	1.47	
V	1.28	*HIS5, ARS*
IV	1.20	*CYP52C2 (ALK6)*[b], *LEU2*[b]
III	1.15	*LEU2*[b]
II	1.10	*CYP52C2 (ALK6)*[b], TRA (*ARS, CEN*)
I	0.88	*CYP52D1 (ALK4)*

For gene designation (see Table 11):
CYP52 (ALK):   Alkane-induced P450 genes, Ohkuma et al. (1993c, 1995a).
POX genes: Peroxisomal acyl-CoA oxidases, Masuda (unpubl. results).
HIS5, LEU2, TRA: Data of Tanaka et al. (1987).
[a] At least 8–10 chromosomal bands were separated in the *C. maltosa* strains L4 (Becher et al. 1991) and IAM12247 (Tanaka et al. 1987; Becher et al. 1994; Ohkuma et al. 1995a) using OFAGE and CHEF techniques, respectively.
[b] Becher et al. (1991) demonstrated that at least two *LEU2* genes situated on chromosomes of different size (probably bands III and IV) occur in the genome of *C. maltosa* CBS5611 (IAM12247), whereas *LEU2* was located only on band III (1.15 Mbp) in the strain L4. A comparable feature was observed for *CYP52C2 (ALK6)* located at the two chromosomes II and IV.

cloned genes from *C. maltosa* are in agreement with this. Other circular plasmids were not detected in *C. maltosa* (Kunze et al. 1984a, 1986b).

It is discussed that *C. maltosa* has a diploid or probably an aneuploid karyotype with the average ploidy being approximately 2. Several molecular data give evidence that this yeast has a highly aneuploid genome, suggesting that some chromosomes are present in one copy, whereas others are present in more than one copy, as proposed by different authors (Chang et al. 1984; Böttcher 1987; Samsonova et al. 1987; Tanaka et al. 1987; Becher et al. 1991, 1994; Kawai et al. 1991). Chang et al. (1984) concluded from measurements of nuclear DNA content by fluorescent microscope photometry and from UV-light inactivation curves that *C. maltosa* IAM12247 should contain the same ploidy level as a diploid strain of *S. cerevisiae*. The molecular data of Kunze et al. (1984a,b) argue more strongly for the aneuploid nature of *C. maltosa* EH15 than for a diploid state, for which a ploidy level of 2.6 was calculated. Additionally, the results of both mutagenesis and recombination suggest that *C. maltosa* is an extremely aneuploid yeast. Thus, the mutant frequency in *C. maltosa* (0.1% auxotrophic mutants after chemical mutagenesis; cf. Sect. 4.2) is generally less than expected for a haploid strain, but still too high to be explained by mitotic recombination (mitotic crossing-over, gene conversion, or nondisjunction and a corresponding segregational homozygotization) in a diploid strain (Böttcher 1987; Samsonova et al. 1987; Becher et al. 1991, 1994). The suggestion of an aneuploid genome for *C. maltosa* is in good agreement with the view of the genome provided by the electrophoretic karyotype analysis (Sect. 4.3.2). As discussed by Becher et al. (1994), the calculated haploid genome size of at least 14 Mbp (15.5×10^{-15} g) argues for a roughly triploid status, even taking into account the high proportion of mtDNA (Kunze et al. 1984a) in *C. maltosa*. The aneuploid genome structure is supported by recent sequence data for multiple alleles of the *LEU2* gene and *CYP52* genes and the respective chromosome assignments of these genes (Ohkuma et al. 1993c, 1995a; Becher et al. 1994; Table 9).

4.3.2
Electrophoretic Karyotype

Recently, alternating field electrophoresis using the methods of orthogonal field alternation gel electrophoresis (OFAGE, according to Carle and Olson 1984) and

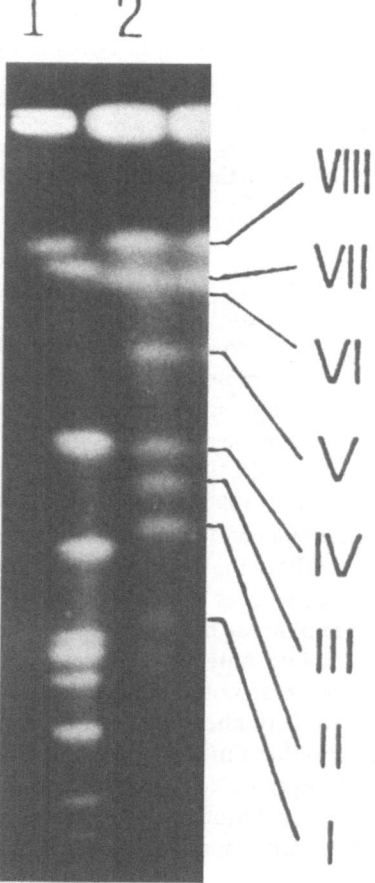

Fig. 5. Electrophoretic karyotype of *Candida maltosa*. Chromosomal patterns of *C. maltosa* IAM12247 (*lane 2*) and *S. cerevisiae* AB1380 (*lane 1*) analyzed by CHEF as described in Sect. 4.3.3. Chromosome bands *I* to *VIII* of *C. maltosa* are shown on the *right side* (see also Table 9)

electrophoresis with a contour-clamped homogeneous electric field (CHEF, according to Chu et al. 1986) have been applied for electrophoretic karyotyping of the *C. maltosa* strains SBUG700 (L4), IAM12247 (CBS5611) and VSB899 (Tanaka et al. 1987; Wedler et al. 1990; Becher et al. 1991, 1994, 1995; Ohkuma et al. 1993c, 1995a; see Sect. 4.3.3).

The genomic DNA of *C. maltosa* has been tentatively divided into 8 to 10 or 11 chromosomes (Table 9 and Fig. 5) using different conditions for the OFAGE and CHEF techniques applied in different laboratories, compared with 13–16 chromosomes in *S. cerevisiae*. Some chromosomes were larger than the largest *S. cerevisiae* chromosomes (1.6 Mbp). This is in agreement with data obtained for other *Candida* yeasts, where the average sizes of chromosomes are higher while the numbers are lower than in *S. cerevisiae* (Magee and Magee 1987; Kamiryo et al. 1991). The simple addition of differently sized chromosomal bands of *C. maltosa* allowed an estimated total haploid genome size of at least 14 Mbp to the determined, calculated according to size markers from *S. cerevisiae* chromosomes (Tanaka et al. 1987; Becher et al. 1994). Taking into account the differences in the intensity of band staining and the data mentioned above supporting aneuploidy, the total haploid genomic size has been calculated to be between 17 and 19.9 Mbp (Tanaka et al. 1987; Becher et al. 1994; Ohkuma et al. 1995a), which is approximately the same as the haploid genome sizes of *S. cerevisiae* or *C. tropicalis* (Kamiryo et al. 1991).

The three *C. maltosa* wild-type strains so far analyzed, CBS5611 (IAM12247), L4, and VSB899, can be differentiated by their individual electrophoretic karyotype pattern (Becher et al. 1991, 1994). Under the same experimental conditions, differences in size, number, and staining intensity of the chromosome bands were detected among these strains and the *leu2* mutant G587 derived from L4, indicating a high variability in the *C. maltosa* karyotype. Such polymorphism in chromosome band size was observed for *C. albicans* strains (Magee and Magee 1987). Certain chromosomes of the strain *C. maltosa* L4 undergo frequent mitotic recombinations, which lead to changes in the number and size of chromosomal bands in the electrophoretic karyotype (Becher et al. 1994).

4.3.3
Separation of Chromosomes by Pulsed Field Gel Electrophoresis

Note on Method. A method for separation of *C. maltosa* chromosomes by gel electrophoresis with a contour-clamped homogeneous electric field (CHEF, according to Chu et al. 1986) as applied in the authors' laboratory is given below (Ohkuma et al. 1995a; Fig. 5). Recently, chromosome separation of different *C. maltosa* strains by CHEF gel electrophoresis was also reported by Becher et al. (1994, 1995). Alternatively, the method of orthogonal field alternation gel electrophoresis (OFAGE) as described by Carle and Olson (1984) was applied for electrophoretic karyotyping of different *C. maltosa* strains (Wedler et al. 1990; Becher et al. 1991). Standard conditions in these reports were running times of 48 h at 200 V and pulse times of 180 s. Separation of smaller chromosomes was more efficient at a gel temperature of 13 °C, whereas the larger chromosomes were separated better at 18 °C (see also Chap. 3, this Vol.).

Procedure

1. Make a sample from a late exponential phase *C. maltosa* culture grown in YPD medium.

2. Harvest the cells of 1.5 ml culture in an microcentrifuge tube and wash with distilled water. Resuspend the cells in 210 μl SCE buffer.

3. Add to the cell suspension 150 μl of 25 mg/ml Zymolyase 20 T (Seikagaku Kogyo Co.) in 1 M sorbitol and 50 mM Tris-HCl (pH 8.0) solution and place at 42 °C.

4. Add 1% low-melting agarose (FMC BioProducts) kept at 42 °C and mix gently but thoroughly.

5. Put this mixture into the chambers of the Bio-Rad CHEF sample mould and allow to set at 4 °C.

6. Incubate the sample plugs in SCE buffer at 37 °C for 5 h.

7. Incubate the plugs in 0.45 M EDTA, 0.01 M Tris-HCl (pH 8.0), 1% sodium N-lauroyl sarcosine, 0.5 mg/ml proteinase K (Seikagaku Kogyo) at 50 °C overnight.

8. Finally wash the agar plugs in 0.05 M EDTA (pH 8.0), 0.01 M Tris-HCl (pH 8.0) for 1 h and store at 4 °C in this buffer.

9. Prepare CHEF gels from 1% agarose (SIGMA) in 0.5 × TBE buffer.

10. Load sample plugs into wells and seal with 0.5 × TBE solution containing 1% low-melting agarose and allow to set for 1 h in 0.5 × TBE buffer before electrophoresis.

11. Run the gel with a Bio-Rad CHEF-DRII Megabase DNA Electrophoresis System. Carry out the electrophoresis in two steps at 14 °C, with a 120-s pulse time at 150 V for 36 h and a 180-s pulse time at 150 V for 18 h.

12. At the end of the run, stain the gel with 0.5 μg/ml ethidium bromide in 0.5 × TBE solution for 30 min and destain for 1–3 h in distilled water.

13. After photographing, subject the gel to irradiation of a UV transilluminator (254 nm) for 15 min, then depurinate by soaking in 0.25 M HCl for 15 min at room temperture prior to a standard Southern transfer.

14. Carry out Southern blots of the CHEF gel using Hybond-N$^+$ membrane (Amersham) according to the instructions of supplier. The procedure of hybridization and DNA probes used were the same as in the case of Southern blots of genomic DNA of *C. maltosa*.

Buffers

SCE: 1 M sorbitol, 0.1 M sodium citrate, 0.06 M EDTA (pH 7.0),
             0.1 M β-mercaptoethanol
1 × TBE: 90 mM Tris, 90 mM boric acid, 2 mM EDTA

4.3.4
Mitochondrial DNA

Mitochondrial (mt) DNA of *C. maltosa* EH15 was first isolated and further characterized by Kunze et al. (1984a, 1986b), in order to develop an extrachromosomal genetic system for this yeast. The mtDNA is circular, and the size estimated from restriction analysis is approximately 52 kb or 3.5×10^7 Da. The total mass of mtDNA in a single cell is about 2.8×10^9 Da (or 18 fg), with 35% representing a high portion of total *C. maltosa* NDA. Thus, a copy number of about 80 per haploid genome was calculated (Kunze et al. 1984a, 1986b; Su and Meyer 1991; Table 8).

The mitochondrial genome organization (size, restriction patterns, and gene mapping) of several yeasts is different from that of *C. maltosa*. By its high molecular weight of mtDNA and its low GC content (21–22%), *C. maltosa* is distinguished from *C. tropicalis*, other *Candida* species, and *Lodderomyces elongisporus*, for which values of 25–30 kb and 27–37 GC% were determined (Kunze et al. 1984a, 1986b; Su and Meyer 1991; cf. Sect. 1.1). Two *C. maltosa* strains analyzed by Su and Meyer (1991) showed slight differences in restriction patterns of mtDNA. A restriction map of the mtDNA from the strain L4 was constructed. Six structural genes *LrRNA*, *CYTB* (*COB*), *COX1*, *COX2*, *ATP6* (*OLI2*), and *ATP9* (*OLI1*), encoding large rRNA, apocytochrome b, cytochrome c oxidase subunits I and II, mitochondrial ATPase subunits 6 and 9, respectively, were located on the *C. maltosa* chondriome by cross-hybridization experiments with the corresponding genes of *S. cerevisiae* (Kunze et al. 1986b). As known for most *Candida* species (Riggsby 1990), *C. maltosa* also is a petite-negative yeast, without the ability to form respiratory-deficient mutants on treatment with acriflavine (Bos and deBoer 1968).

4.4
Genes of Candida maltosa

4.4.1
Gene Cloning Strategies and Gene Libraries

Cloning Procedures. The strategies, their limitations, and problems arising in cloning *Candida* genes, as discussed in detail by Kurtz et al. (1990) for *C. albicans*, apply also to *C. maltosa*. The following main strategies were used for cloning of *C. maltosa* genes:

Cloning via Gene Expression. This approach includes the isolation of genes through the expression of a phenotype such as the complementation of a selectable auxotrophic marker or the selection of a dominant resistance marker. The complementation strategy can be realized by functional expression of the desired *C. maltosa* gene in corresponding mutants of *S. cerevisiae* or/and *E. coli*, as shown first for the *LEU2* gene (Kawamura et al. 1983; Takagi et al. 1987; Becher et al. 1991), and later for *ADE1* and *URA3* genes (Jomantiene et al. 1987, 1991; Sasnauskas et al. 1991; Ohkuma et al. 1993a, cf. Table 15). Alternatively, a selection of dominant selectable resistance marker genes (*RIM-C, CYH^R, FDH1*) in wild-type *Saccharomyces* strains was performed (Takagi et al. 1986b; Sasnauskas et al. 1992a,c). Additionally, gene expression cloning can be performed by detection of protein products with an antibody. In this way, cytochrome P450 (*CYP52*) or its reductase (*CPR*) cDNA clones were obtained by immunoscreening in *E. coli* with antibodies against purified proteins (Schunck et al. 1989a,b, 1991; Kärgel et al. 1996).

As was recently demonstrated, serious problems for cloning and functional expression of *C. maltosa* genes in heterologous systems may arise due to the observed deviation from the universal genetic code in some *Candida* species (Sugiyama et al. 1995; Zimmer and Schunck 1995; for details see Sect. 4.4.3). Now the available host-vector systems for *C. maltosa* (Table 14, cf. Sect. 5) enable self-cloning via gene expression in the homologous host and overcome the problems of codon usage. So far, this method has been successfully applied to isolate *HIS5* and *ADE1* genes of *C. maltosa* IAM12247 (Hikiji et al. 1989; Kawai et al. 1991) and is, on the other hand, limited by restricted availability of characterized mutants in *C. maltosa*. However, despite all uncertainties mentioned, there are compelling reasons to choose expression strategies in heterologous and homologous systems for cloning *C. maltosa* genes.

Cloning via Cross-Hybridization. As there are limitations to cloning genes via expression (lack of mutants, and certain genes being toxic in yeast, when overexpressed), this strategy becomes an important alternative. It includes screening of either plasmid or bacteriophage λ gene libraries with homologous or heterologous DNA (genomic DNA fragments, cDNA, or oligonucleotides) probes mainly by colony hybridization techniques. In this manner the *C. maltosa* genes *POX4, CYP52,* and *PGK1* were isolated (Hill et al. 1988; Takagi et al. 1989; Ohkuma et al. 1991a,b, 1993c, 1994a; Masuda et al. 1994). The nucleic acid hybridization strategies (including PCR-based methods) are probably the most commonly employed techniques for isolation of genes coding for structurally well-conserved protein or RNAs, or at least for enzymes, for which sufficient evolutionary conservation is expected. Nevertheless, this methodology in no way assures the experimentalist that the cloned sequence will encode the desired function. Pseudogenes and related sequences are readily isolated by this method in addition to the desired sequence.

Gene Libraries. For cloning of *C. maltosa* genes, several genomic DNA and cDNA libraries, listed in Table 10, have been established. These libraries were constructed mostly as partially *Sau*3AI digested genomic DNA fragments in both *E. coli* (pBR322, pUC) and *S. cerevisiae/E. coli* shuttle vectors (YEp13, YEp24, pL3, and others). These shuttle vectors, e.g., YEp13, contain origins of replication for both species, the *bla* (*Amp^R*) gene for selection in *E. coli* and the *LEU2* gene for selection in *S. cerevisiae*. Recently, gene banks were constructed in vectors (pTRA11, pBTH10B, pRA7) containing functional autonomously replicating sequences (TRA fragment with *ARS-CEN* or *ARS* only) and homologous auxotrophic marker

Table 10. Gene libraries of *Candida maltosa* obtained in different laboratories

Donor strain[a]	Vectors[b] (cloning site)	DNA (G or C): restricted by (c or p)[c]	No. of clones (average insert size) E. coli host	Reference
Genomic libraries				
IAM12247	pBR322 (*Bam*HI, *Hind*III, or *Sal*I)	G: *Bam*HI, *Hind*III, or *Sal*I (c)	6000–50 000 (>1 kb) in C600	Kawamura et al. (1983)
IAM12247	YEp13 (*Bam*HI)	G: *Sau*3AI (p)	20 000 (5–6 kb) in MC1061	Takagi et al. (1986a)
IAM12247	pTRA11 (*Bam*HI)	G: *Sau*3AI (p)	16 000 (5–10 kb) in JA221	Sunairi et al. (1988) Hikiji et al. (1989) Kawai et al. (1991)
IAM12247	pBTH10B (*Bgl*II)	G: *Sau*3AI (p)	60 000 (5–10 kb) in MC1061	Takagi et al. (1989) Hwang et al. (1991)
IAM12247	pBTH10B[d] (*Bgl*II)	G: *Sau*3AI (p)	2000 alkane induced clones from 60 000[d]	Takagi et al. (1989) Hwang et al. (1991)
IAM12247	λEMBL3 (*Bam*HI)	G: *Sau*3AI (p)	4 × 10⁵ plaques (9–20 kb) in LE392	Ohkuma et al. (1991b)
IAM12247	YEp13 (*Bam*HI)	G: *Sau*3AI (p)	20 000 (3–9 kb) in MC1061	Ohkuma et al. (1993a)
VSB899	pUC19 (*Bam*HI)	G: *Sau*3AI (p)	No data given (3.3–10 kb)	Jomantiene et al. (1987)
VSB899	pL3 (*Bam*HI)	G: *Sau*3AI (p)	No data given (4–10 kb)	Sasnauskas et al. (1991) Jomantiene et al. (1987, 1991)
VSB899	pRA7 (*Bam*HI)	G: *Sau*3AI (p)	No data given in HB101	Sasnauskas et al. (1992b)
L4	pUC9-*TRP1/ARS1* (*Eco*RI)	G: *Eco*RI (c)	No data given in HB101[e]	Wedler et al. (1990)
L4	pEMBLYe31 (*Bam*HI)	G: *Sau*3AI (p)	No data given in HB101[e]	Wedler et al. (1990)
L4	YEp24 (*Bam*HI)	G: *Bam*HI (c)	No data given in HB101[e]	Wedler et al. (1990)
L4	YEp24 (*Bam*HI)	G: *Sau*3AI (p)	11 000 clones (5 kb) in HB101	Becher et al. (1991)

Table 10. (*Contd.*)

Donor strain[a]	Vectors[b] (cloning site)	DNA (G or C): restricted by (c or p)[c]	No. of clones (average insert size) *E. coli* host	Reference
G587 (*leu2*)[f]	pUC19 (*Eco*RI)	G: *Eco*RI (c)	10 000 clones (5 kb) in HB101	Becher et al. (1991) Becher et al. (1994)
G587 (*leu2*)[f]	pUC18 (*Bam*HI)	G: *Sau*3AI (p)	20 000 (7 kb) in XL#1-Blue	Becher et al. (1994)
L4	CipL1 (*Eco*RI)	G: *Eco*RI (c)	500 clones in HB101	Becher et al. (1991)
cDNA library EH15	pUC119 (*Hind*II)	C: Alkane-grown cells	100 000 (0.3–3.0 kb) in DH5αc	Schunck et al. (1989a,b)

[a] For the donor (wild-type) strains cf. Table 1.

[b] Shuttle vectors for *C. maltosa* and *E. coli:* pTRA11, pBTH10B, pRA7, CipL1 (*Candida* integrative plasmid *LEU2*) CrLp1 (*Candida* replicative *LEU2* plasmid); for vectors see Table 14. Shuttle vectors for *S. cerevisiae* and *E. coli:* YEp13, YEp24, pL3 (high copy yeast shuttle vector), pEMBLYe31, pUC9-*TRP1/ARS1*.

[c] Abbreviations: G – Genomic (chromosomal) DNA of *C. maltosa*,
     C – cDNA obtained from mRNA of *C. maltosa*,
     c – complete digestion,
     p – partial digestion.

[d] Alkane-induced clones obtained after differential screening of the genomic library in pBTH10B with cDNA probes obtained from alkane-grown and glucose-grown cells, respectively (Takagi et al. 1989; Hwang et al. 1991).

[e] Two of three libraries constructed contained the whole *C. maltosa* genome with 99% confidence.

[f] Library constructed to clone the *leu2* mutant gene (Becher et al. 1991), and to clone silent allelic variants of *LEU2* (Becher et al. 1994).

genes (*CmADE1*, *CmHIS5*, *CmLEU2*, or *ScLEU2*) suitable for self-cloning in *C. maltosa* (Tables 10, 14). For cloning of alkane-induced genes, a cDNA library using mRNA from alkane-grown cells was constructed (Schunck et al. 1989a,b). Alternatively, such clones were isolated after differential screening of a genomic library constructed with pBTH10B with two kinds of cDNA population as probes, obtained from glucose- or tetradecane-grown cells (Takagi et al. 1989; Hwang et al. 1991). Some of the *C. maltosa* gene libraries shown in Table 10 were not described in detail, although the authors were able to isolate genes from them successfully. It is therefore possible that some of these libraries do not contain a complete representation of the *C. maltosa* genome.

4.4.2
Cloned Genes and Regulation of Their Expression

The isolated genes of *C. maltosa* are listed in Tables 8 and 11. For the designation of *C. maltosa* genes, the nomenclature rules developed for *S. cerevisiae* and applied to other *Candida* yeasts were followed. To discriminate between *Candida* (*maltosa*) and *Saccharomyces* (*cerevisiae*), the gene names in literature and data bases sometimes start with *Cm*, *C* or *C-*, e.g., *CmLEU2* or *C-LEU2* for the *LEU2* gene of *C. maltosa*, instead of *LEU2* for the *S. cerevisiae* gene. For most of these cloned genes (Table 11) the nucleotide and amino acid sequences are available from data bases (EMBL, Genbank, DDBJ Nucleotide Sequence Databases, or SwissProt), excluding the cloned genes *ADE2*, *ARG4* (Jomantiene et al. 1987), several *ARS* fragments (Polumienko and Grigorieva 1985; Becher et al. 1991; Kasüske et al. 1992), and *CYS2* (Ohkuma, unpubl. data), for which no sequences have been published yet (see Table 8).

Genes of Biosynthetic Enzymes

The auxotrophic marker genes *ADE1*, *ADE2*, *ARG4*, *CYS2*, *HIS5*, *LEU2* and *URA3* coding for biosynthetic enzymes in the intermediary metabolism have been cloned from *C. maltosa*, and are now available as markers in transformation (Tables 11, 15). Additionally, some heterologous marker genes from *S. cerevisiae* and *C. albicans* are functioning in *C. maltosa* and vice versa (Table 16, cf. Sect. 5).

Table 11. Genes isolated and sequenced from the yeast *Candida maltosa*. For sequences that have not been published their appearance in the data bases with the remark (unpublished) was used as reference.

Gene symbol[a]	Trivial name or function[a]	Accession no.[b] (length)	Isolated from strain (as G or C)[c]	Reference
ADE1	SAICAR synthetase	M58322 (2540 bp)	VSB899 (G)	Sasnauskas et al. (1991) Jomantiene et al. (1991)
		D00855 (1820 bp)	IAM12247 (G)	Kawai et al. (1991)
		X74785-88 (4 variants)	SBUG700 (L4) G374 (*ade1*)	Becher et al. (1995)
ALI1	Ubiquinone oxidoreductase subunit	M61102 (2689 bp) with *POX18*[d]	IAM12247 (G)	Hwang et al. (1991) Souza et al. (1993)
ARS	ARS (196 bp) fragment	D00136 (196 bp pt)	IAM12247 (G)	Kawai et al. (1987)
ARS	ARS (150 bp) and ORF[d]	M58330 (1331 bp)	VSB899 (G)	Sasnauskas et al. (1992b) Jomantiene et al. (1987)
CEN	Centromere (*ARS* and *CEN*)	D29758 (1489 bp)	IAM12247 (G)	Ohkuma (1994, unpubl.) Ohkuma et al. (1995e)
CHS1	Chitin synthase	D29760 (603 bp pt)	IAM12247 (G)	Ohkuma et al. (1994, unpubl.)
CHS2	Chitin synthase	D29761 (597 bp pt)	IAM12247 (G)	Ohkuma et al. (1994, unpubl.)

Table 11. (*Contd.*)

Gene symbol[a]	Trivial name or function[a]	Accession no.[b] (length)	Isolated from strain (as G or C)[e]	Reference
CHS3	Chitin synthase	D29762 (355 bp pt)	IAM12247(G)	Ohkuma et al. (1994, unpubl.)
CPR	NCCR	X76226 (2230 bp)	EH15 (C)	Kärgel et al. (1993, unpubl., 1996)
CPR	NCCR	D25327 (2827 bp)	IAM12247 (G)	Ohkuma et al. (1993, unpubl., 1995c)
CYH[R]	CYHR protein (deduced)	M64932 (2384 bp)	VSB899 (G)	Sasnauskas et al. (1992c)
CYP52A subfamily[e]	Alkane and fatty acid hydroxylation			
CYP52A3	P450Cm1	X51931 (1667 bp)	EH15 (C)	Schunck et al. (1989b, 1991)
	ALK1-A	D00481 (2803 bp)	IAM12247 (G)	Takagi et al. (1989)
	ALK1-B	D12475 (2047 bp)	IAM12247 (G)	Ohkuma et al. (1991a)
CYP52A4	P450Cm2	X51932 (1787 bp)	EH15 (C)	Schunck et al. (1991)
	ALK3-A	X55881	IAM12247 (G)	Ohkuma et al. (1991b)
	ALK3-B	D12715 (609 bp pt)	IAM12247 (G)	Ohkuma (1992, unpubl.) Ohkuma et al. (1995a)
CYP52A5	ALK2-A	X55881	IAM12247 (G)	Ohkuma et al. (1991b)
	ALK2-B	D12714 (342 bp pt)	IAM12247 (G)	Ohkuma (1992, unpubl.) Ohkuma et al. (1995a)
CYP52A9	ALK5-A	D12717 (2499 bp)	IAM12247 (G)	Ohkuma (1992, unpubl.) Ohkuma et al. (1995a)
	ALK5-B	D26160 (283 bp pt)	IAM12247 (G)	Ohkuma (1992, unpubl.) Ohkuma et al. (1995a)
CYP52A10	ALK7	D12719 (5525 bp)	IAM12247 (G)	Ohkuma (1992, unpubl.) Ohkuma et al. (1995a)
CYP52A11	ALK8-A	D12719 (5525 bp)	IAM12247 (G)	Ohkuma (1992, unpubl.) Ohkuma et al. (1995a)
	ALK8-B	D26159 (815 bp pt)	IAM12247 (G)	Ohkuma (1993, unpubl.) Ohkuma et al. (1995a)
CYP52C subfamily[e]	Unknown function			
CYP52C2	ALK6-A	D12718 (2196 bp)	IAM12247 (G)	Ohkuma (1992, unpubl.) Ohkuma et al. (1995a)
CYP52D subfamily[e]	Unknown function			
CYP52D1	ALK4	D12716 (2393 bp)	IAM12247 (G)	Ohkuma (1992, unpubl.) Ohkuma et al. (1995a)
FDH1	FAHR	M58332 (2258 bp)	VSB899 (G)	Sasnauskas et al. (1992a)
GAL1/GAL10	Galactokinase Gal epimerase (promoter region)	D29759 (2034 bp)	IAM12247 (G)	Park et al. (1994, unpubl.) Park et al. (1996)
HIS5	HPAT (second ORF)	X17310 (1923 bp)	IAM12247 (G)	Hikiji et al. (1989)
LEU2	3-IPMDH	X05459 (2193 bp)	IAM12247 (G)	Takagi et al. (1987)
		X72939 (2200 bp)	L4 (G)	Becher et al. (1991) Becher et al. (1994)
	4 allelic variants of LEU2 (one mutant)	X72937 (2225 bp)	G587 (G)	Becher et al. (1994)
		X72938 (2199 bp)	G587 (G) (mutant allele)	Becher et al. (1991) Becher et al. (1994)
		X72940 (2193 bp)	G587 (G)	Becher et al. (1994)

Table 11. (*Contd.*)

Gene symbol[a]	Trivial name or function[a]	Accession no.[b] (length)	Isolated from strain (as G or C)[e]	Reference
PGK1	PGK	D12474 (2594 bp)	IAM12247 (G)	Masuda et al. (1994)
POX2	Acyl-CoA oxidase	D21228 (2723 bp)	IAM12247 (G)	Masuda and Takagi (1993, unpubl.)
POX4	Acyl-CoA oxidase	X06721 (2975 bp)	ATCC20184 (G)	Hill et al. (1988)
POX18	Oleate inducible peroxisomal protein	M61102 (2689 bp) with *ALI1*[d]	IAM12247 (G)	Hwang et al. (1991)
RIM-C (*L41Q*)	Ribosomal protein L41 (L41Q-type)	D90488 (1141 bp) D43686	IAM12247 (G)	Takagi et al. (1986b) Kawai et al. (1992) Mutoh et al. (1995)
(*LEL41*) (*L41P*)	(L41P-type)	M62396 D43686/87	IAM12247 (G)	Kawai et al. (1992, unpubl.) Mutoh et al. (1995)
SSRR	Small subunit rRNA	D14593 (1784 bp)	IAM12247 (G)	Ohkuma et al. (1993b)
SETR	tRNA^{Ser}CAG gene	D26074 (312 bp)	IAM12247 (G)	Sugiyama et al. (1995)
URA3	OPDCase	D12720 (1274 bp)	IAM12247 (G)	Ohkuma et al. (1993a)

[a] Abbreviations for gene names in the date bases mostly start with *Cm*, *C* or *C*-, e.g., *CmLEU2* or *C-LEU2* for the *LEU2* gene of *C. maltosa*, to discriminate between *Candida* (*maltosa*) and *Saccharomyces* (*cerevisiae*) genes.

Abbreviations and designations for the genes and the corresponding gene products

ADE1	– Genes for the phosphoribosyl-aminoimidazole-succinocarboxamide-synthetase or SAICAR synthetase (EC 6.3.2.6), obviously existing in allelic variants (Becher et al. 1995)
ALI1	– Alkane inducible protein, with high homology to NADH ubiquinone oxidoreductase subunit (EC 1.6.5.3/ 1.6.99.3) precursor (30.4 kDa), Souza et al. (1993)
ALK	– Alkane-inducible genes for cytochrome P450
ARS	– Autonomously replicating sequence
CEN	– Centromere region: TRA *Pstl-Eco*RI fragment containing *ARS* and *CEN* (see Fig. 7)
CHS	– Chitin synthase, different genes *CHS1, CHS2, CHS3*
CPR	– Gene for the NADPH cytochrome P450 (cytochrome c) reductase (NCCR, EC 1.6.2.4)
CYH^R	– Gene for the cycloheximide resistance (CYHR) protein
FDH1	– Formaldehyde resistance (FADR) gene, coding for a glutathione dependent formaldehyde dehydrogenase (EC 1.2.1.1), with high homology to mammalian ADHIII class genes
HIS5	– Gene for histidinol-phosphate amino transferase (HPAT, EC 2.6.1.9)
LEU2	– Gene for 3-isopropylmalate dehydrogenase (3-IPMDH, EC 1.1.1.85)
PGK1	– Gene coding for phosphoglycerate kinase (PGK)
POX	– *POX2* and *POX4* genes coding for peroxisomal acyl-CoA oxidase (AOX, EC 1.3.99.3) forms (PXP2 and PXP4)
RIM-C	– Ribosome modification by cycloheximide, coding for the L41Q-type protein (resistant); in contrast to *LEL41* coding for the L41P-type protein (sensitive to cycloheximide)
SSRR	– Gene for small subunit ribosomal RNA (CmSSRR)
SETR	– tRNA^{Ser}CAG gene – responsible for translation of the CUG codon as serine
URA3	– Gene for the orotidine-5'-phosphate decarboxylase (OPDCase, EC 4.1.1.23).

[b] The nucleotide sequence data appeared in the EMBL, Genbank and DDBJ Nucleotide Sequence Databases; Sequence lengths given in bp, pt – partial sequence.

[c] Genes isolated from genomic DNA (G) or cDNA (C) libraries, respectively. For donor strains cf. Tables 1 and 6.

[d] *ALI1* and *POX18* are located together at the same cloned fragment (Hwang et al. 1991) The *ARS* sequence is located in the neighborhood with an ORF encoding a polypeptide with high homology (70%) to the small ribosomal subunit protein No. 15 (RS15) of *Brugia pagangi* and contains an intron with canonical sites for correct splicing (Sasnauskas et al. 1992b).

[e] *CYP52* family in the *CYP* gene superfamily: alkane-inducible cytochrome P450 genes (P450alk genes) of yeast. Gene symbols are given according to the recommended nomenclature system for P450 genes (Nelson et al. 1993). *A* and *B* represent the allelic variants found for the genes *ALK1, ALK2, ALK3, ALK5,* and *ALK8*, respectively. Other members of the subfamily *CYP52* have been isolated from *Candida tropicalis* (Seghezzi et al. 1991, 1992) and *Candida apicola* (Lottermoser et al. 1994).

LEU2 (3-Isopropylmalate Dehydrogenase). Several allelic variants of the *LEU2* gene coding for 3-IPM dehydrogenase (EC 1.1.1.85, ORF of 373 aa) have been isolated from wild-type and mutant strains of *C. maltosa* (Kawamura et al. 1983; Takagi et al. 1987; Becher et al. 1991, 1994; Table 11). The *LEU2* gene was first cloned form *C. maltosa* IAM12247 by complementation of the corresponding mutations *leuB* in *E. coli* (C600) and *leu2* in *S. cerevisiae* strains AH22 or SHY3 (Kawamura et al. 1983; Takagi et al. 1987; cf. Table 15). This and the *LEU2* gene (CipLA) recently cloned by expression in *E. coli* (Becher et al. 1991, 1994) from the strain L4 have 99.9% homology over 2213 bp. The clone CipLA has been successfully used for complementation of the *leu2* mutation in the strain G587, a *leu2* mutant derived from L4. Additionally, three different alleles of the *LEU2* gene were cloned by colony hybridization from *C. maltosa* G587, among them two functionally active *LEU2* genes (CipLEa and CipLS) and the *leu2* mutant allele CipLEb (Becher et al. 1991, 1994; Table 11). Sequence comparison of the four alleles revealed about 2–2.5% interallelic divergence in the coding and noncoding nucleotide sequences, leading also to ten amino acid changes in the individual ORF for 3-IPM dehydrogenase. Polymorphism of the *LEU2* genes with at least two alleles was demonstrated also for the wild-type strain L4. These data argue strongly for an aneuploid genome with multiple *LEU2* alleles. The cloning of two functionally intact wild-type *LEU2* genes from the *leu2* mutant strain G587 let the authors suggest the presence of silent gene copies in *C. maltosa* L4 (Becher et al. 1994), which might be one possibility to explain the occurrence of APA changes (see Sect. 4.2.2). The *LEU2* gene of *C. maltosa*, like that of *C. utilis* and *C. tropicalis*, is considerably identical at both the nucleotide (72%) and amino acid (76%) levels with the *LEU2* gene of *S. cerevisiae* (Takagi et al. 1987; Becher et al. 1991). Interestingly, this homology might not be enough to ensure cross-hybridization cloning using high stringent conditions, as demonstrated for the *C. albicans LEU2* gene, although it can complement the corresponding mutations of *S. cerevisiae* and *E. coli* (Kurtz et al. 1990).

ADE1 *(Phosphoribosyl-Aminoimidazole-Succinocarboxamide-Synthetase).* The genes *ADE1* encoding SAICAR synthetase (EC 6.3.2.6) and *ADE2* (phosphoribosyl-aminoimidazole-carboxylase, EC 4.1.1.21) of *C. maltosa* VSB899 were first cloned by complementation of the corresponding mutations in *S. cerevisiae* (Jomantiene et al. 1987), although sequence data were published only for *ADE1* (Jomantiene et al. 1991; Sasnauskas et al. 1991). At the same time the *ADE1* gene of the strain IAM12247 was cloned by complementation of *ade1* mutation (self-cloning) in *C. maltosa* (Kawai et al. 1991). These two sequences for the *ADE1* coding regions are 100% identical over 1668 bp but differ in the 5′-upstream regions. Recent sequencing data (Becher et al. 1995; Table 11) give evidence that the *ADE1* gene exists in allelic variants in *C. maltosa* L4, as previously reported for *LEU2*. Four different *ADE1* alleles (with up to 8% divergence) have been sequenced from one prototrophic strain, which show some differences to the mentioned identical genes at both nucleotide and deduced amino acid leves. The *C. maltosa ADE1* genes are used in the host-vector systems for this yeast in different laboratories (cf. Sect. 5 and Table 14).

HIS5 (Histidinol-Phosphate Amino Transferase). The *C-HIS5* (HPAT, EC 2.6.1.9) was the first gene obtained by self-cloning via complementation of the *his5* mutation in *C. maltosa* strain CH1 using a previously established host-vector system for this yeast, and was then used for its further improvement by constructing the pBTH vectors series (Hikiji et al. 1989; cf. Sect. 5 and Table 14). The *HIS5* gene contains downstream a second noncharacterized ORF, and it complements also the corresponding *his5* mutation in *S. cerevisiae*, although the amino acid homology (389 aa) is relatively low with 51% (cf. Table 15).

URA3 (Orotidine-5'-Phosphate Decarboxylase). The *C-URA3* gene coding for the OPDCase (EC 4.1.1.23) has been cloned by complementation in the corresponding *ura3* mutant of *S. cerevisiae* (Ohkuma et al. 1993a). The *URA3* or *C. maltosa* has the highest homology to the corresponding *C. albicans* gene, 76% for the nucleotide, and 84% for the deduced aa sequences. The *S. cerevisiae URA3* gene is with 65 and 72%, respectively, more distantly related. The isolated *URA3* gene was used to construct a triple auxotroph (*his5 adel ura3*) as a useful host for genetic engineering of *C. maltosa* (Ohkuma et al. 1993a; cf. Tables 6, 14).

CYS2 (Serine O-Acetyltransferase). More recently *CYS2*, one of the two genes *CYS1* and *CYS2* coding for the serine O-acetyltransferase (SerAT, EC 2.3.1.30), has been cloned and applied for the development of the host-vector system in *C. maltosa* (Ohkuma 1993, unpubl.; cf. Tables 6, 8, 14).

The auxotrophic marker gene *ARG4* of *C. maltosa* VSB899 was cloned by complementation of the corresponding mutations in *S. cerevisiae* (Jomantiene et al. 1987), but no sequence data have been published yet.

CHS (Chitin Synthase Genes). The genes *CHS1*, *CHS2*, and *CHS3* involved in the biosynthesis of the cell wall component chitin in yeast have been cloned recently by homology screening from *C. maltosa* IAM12247, and partial sequences were obtained (Ohkuma 1993, unpubl.). These three different *CHS* genes coding for chitin synthease 1–3 (chitin synthetase-chitin-UDP acetylglucosaminyl-transferase) in yeast, the enzymatic activity responsible for polymerizing N-acetylglucosamine into the structural cell wall polysaccharide, chitin, have been described already for *S. cerevisiae* and *C. albicans*. They probably play an essential role in the determination of yeast cell morphology, in particular in yeast-hyphal transition (Sudoh et al. 1993).

Genes of Carbon Source Catabolism: Alkane Metabolism

The Cytochrome P450 Gene Subfamily CYP52 and Its Reductase Gene CPR. The key role of P450s in the degradation pathway of alkanes and fatty acids has been discussed already in Sects. 2.5.3 and 3.4 (cf. Fig. 3). Alkane-inducible P450 genes (P450alk, *ALK*, or *CYP52*) in *C. maltosa* were first cloned from alkane-induced cDNA and genomic DNA libraries of the strains EH15 (Schunck et al. 1989a,b) and IAM12247 (Takagi et al. 1989) by immunoscreening and colony hybridization, respectively. These two nearly identical *CYP53A3* clones (first named P450Cm1 and *ALK1*, with 98% aa identity) exhibit high homology to the P450 *alk* genes of *C. tropicalis* (57% aa identity to P450 *alk1*; Sanglard and Loper 1989), but quite low homology to the *CYP51* (P450$_{14DM}$) of *S. cerevisiae* or *C. tropicalis* (17–18%).

First evidence for the presence of a multigene family (*CYP52*) of P450alk genes in yeasts has been provided by Sanglard and Fiecher (1989) for *C. tropicalis* (*CYP52A1*, *CYP52A2*) and in our laboratories for *C. maltosa* (Schunck et al. 1991) after the expression of two cloned P450alk genes (cDNAs named P450Cm1 and P450Cm2) in *S. cerevisiae* and after gene disruption of both alleles of the first cloned P450alk (*ALK1*) gene (Ohkuma et al. 1991a). Further cloning revealed the presence of additional alkane-inducible P450 genes in *C. maltosa* (Ohkuma et al. 1991b, 1993c, 1995a,b,d). Resembling the situation in *C. tropicalis* (Seghezzi et al. 1991, 1992), the *CYP52* family of *C. maltosa* consists of 14 P450alk-related sequences, representing eight genes and five of their alleles (Table 11, Fig. 6). According to the recommended nomenclature system, these P450 genes are members of the *CYP52* family in the *CYP* gene superfamily, containing more than 300 individual P450 genes in 36 families (Nelson et al. 1993). Trivial names introduced by us relate to the nomenclature system of the *CYP52* family as follows: P450Cm1 and *ALK1* (*CYP52A3*), P450Cm3 and *ALK2* (*CYP52A5*), P450Cm2 and *ALK3*

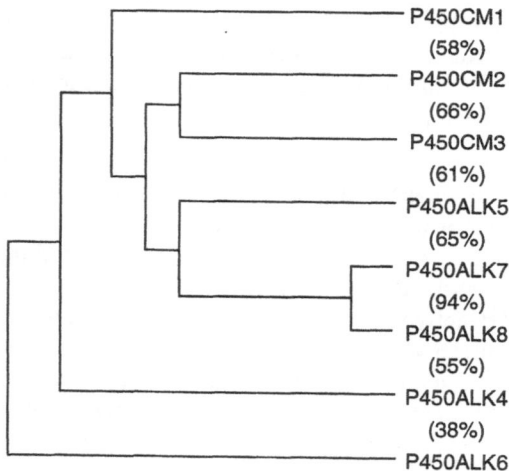

Fig. 6. Members of the cytochrome P450 family *CYP52* in *Candida maltosa*. Dendrogram of the alignment of the amino acid sequences derived from the cloned *CYP52* genes. Alignment performed with the PCgene program. The *values in parentheses* show the homology percentage of the both neighboring P450 proteins. Trivial gene or protein names relate to the recommended nomenclature system of the *CYP52* family (Nelson et al. 1993): P450CM1 and *ALK1* (*CYP52A3*), *ALK2* or P450CM3 (*CYP52A5*), P450CM2 and *ALK3* (*CYP52A4*), *ALK4* (*CYP52D1*), *ALK5* (*CYP52A9*), *ALK6* (*CYP52C2*), *ALK7* (*CYP52A10*), *ALK8* (*CYP52A11*)

(*CYP52A4*), *ALK4* (*CYP52D1*), *ALK5* (*CYP52A9*), *ALK6* (*CYP52C2*), *ALK7* (*CYP52A10*), *ALK8* (*CYP52A11*). Allelic variants (designated *ALK1-A* and *ALK1-B*) were found for the *CYP52* genes *ALK1*, *ALK2*, *ALK3*, *ALK5*, and *ALK8*, respectively (Ohkuma et al. 1991b, 1993c, 1995a). These allelic P450 protein sequences showed 94–98% homology. Interestingly, an additional cDNA clone encoding the same P450Cm1 protein with several nucleotide exchanges in the third codon position only was detected (Schunck et al., unpubl.). Two sets, each consisting of two genes (both alleles of *ALK2* and *3*, and *ALK7* and *8-A*), were tandemly arranged in the genome (cf. Sect. 4.4.4). Thus, the *CYP52* family probably reflects an adaptation to the utilization of different hydrocarbons and is assumed to be a result of gene duplications and divergent evolution from an ancestral gene. The homologies of the encoded nonallelic P450 poteins are mostly in the range of 55 to 65% (Fig. 6), except for the higher values with P450 pairs encoded by the tandemly arranged genes *ALK2/ALK3* (66%) and *ALK7/ALK8* (94%), and for the lower homology values of *ALK4* (48–55%) and *ALK6* (38–41%) being therefore classified to subfamilies *CYP52D* and *CYP52C*, respectively. The six genes (*ALK1-3* and *ALK5, 7, 8*) showing relatively higher similarities to each other than to *ALK4* and *ALK6* were thus classified in the subfamily *CYP52A*. All *CYP52A* genes were assigned to the same chromosome (cf. Table 9, and Sect. 4.4.4).

Other members of the family *CYP52* have been isolated from *C. tropicalis* (seven nonallelic genes, Seghezzi et al. 1992) and *C. apicola* (two *CYP52E* genes, Lottermoser et al. 1994). The presence of such an extended P450 family as disclosed in alkane-assimilating yeast species resembles the situation in higher eukaryotes but is a so far unique example among microorganisms. A distinct relationship of several members of *CYP52* with that of the *CYP4* and *CYP102* families (mammalian and bacterial fatty acid hydroxylating P450 forms) was observed (Ohkuma et al. 1991b; Schunck et al. 1991).

The enzymatic functions attributed so far to individual *C. maltosa* P450 forms are the primary hydroxylation of alkanes and the ω-hydroxylation of fatty acids (see Sect. 2.5). First attempts to investigate the substrate specificity of individual P450 forms have been done by means of their heterologous expression is *S. cerevisiae* (Schunck et al. 1991; Scheller et al. 1992, 1994). Recently, an efficient P450/reductase coexpression system could be established which allowed the high-level coproduction of authentic P450 forms and the homologous redox partner and led to the characterization of four *C. maltosa* P450 forms (Cm1 or *ALK1*, Cm2 or *ALK3*, *ALK2* or Cm3, and *ALK5*) with respect to their substrate specificities toward various alkanes and fatty acids. Each of these P450 forms seems to have its own individual substrate specificity with respect to the preferred type (n-alkane or fatty acid) and chain length of the substrate (Zimmer et al. 1995; Scheller et al. 1996; cf. also Sect. 2.5). A problem that complicates these studies, however, is the fact that *C. maltosa* exhibits a deviation from the universal genetic code: the codon CUG is read as serine instead of leucine (Sugiyama et al. 1995; Zimmer and Schunck 1995; see Sect. 4.4.3). Therefore, *C. maltosa* genes containing the corresponding triplet have to be corrected according to the universal genetic code in order to produce the authentic P450 proteins in *S. cerevisiae*.

The P450 forms mainly involved in alkane degradation are *ALK1-3* and *ALK5*, as demonstrated by their high alkane inducibility (Ohkuma et al. 1993c, 1995a) and by recent stepwise disruption of these genes and their alleles in the genome, an independent approach applied to reveal the in vivo function of *C. maltosa* P450 forms. A strain DA1235 in which all of these four genes were disrupted could not assimilate n-alkanes of any chain length (Ohkuma 1993; Ohkuma et al. 1993c, 1995d). This strain has been used recently to identify the role of individual P450 forms after their plasmid-based expression for growth of *C. maltosa* on different alkanes (Ohkuma and Zimmer, unpubl.). The function of the P450 genes *ALK4* (a probable pseudogene) and *ALK6* to *ALK8* is completely unknown yet. It may be speculated that these P450 forms are involved in the oxidation of other alkane-related compounds or even of aromatic substrates.

Although all P450alk genes, except for *ALK4*, are more or less induced by n-alkanes, but repressed by glucose and not by glycerol, their inducibilities by some other aliphatic carbon sources (fatty alcohols, aldehydes, and acids) showed variabilities. The main regulatory step is the transcriptional level (Schunck et al. 1987b; Ohkuma et al. 1991b, 1995a). Oxygen limitation during cultivation on alkane caused a superinduction of P450alk (cf. Sect. 2.5). Several *cis*-acting elements probably regulated by carbon sources were found in the promoter region of *ALK1* and *ALK3* genes. Two upstream activation sequences (UAS), responsible for induction by alkanes, and one upstream repression sequence (URS), responsible for repression by glucose, were defined (Muraoka et al. 1993; Takagi 1993).

CPR genes for the NADPH-P450 (cytochrome c) reductase (NCCR, EC 1.6.2.4) were recently cloned from alkane-induced cDNA and genomic DNA libraries of the *C. maltosa* strains EH15 (Kärgel et al. 1996) and IAM12247 (Ohkuma et al. 1995c), respectively. The two nearly identical clones (99% aa identity) exhibit high sequence homology to the *CPR* gene of *C. tropicalis* (86%), but quite low homology to the *CPR* of *S. cerevisiae* (49%). The *CPR*, and also the *POX* genes, are, as well as the P450 genes, inducible by growth of *C. maltosa* on alkanes.

Homologous or heterologous expression of individual *CYP52A* or *CPR* genes in *C. maltosa* or in *S. cerevisiae* under the control of the corresponding *GAL1-GAL10* promoters led to a dramatic proliferation of ER structures (Schunck et al. 1991; Wiedmann et al. 1993; Menzel et al. 1994, 1996; Kärgel et al. 1996; Ohkuma et al. 1995b). The molecular mechanisms of these membrane proliferation processes are very interesting for future study (see Sects. 3 and 5.3).

POX – Peroxisomal Protein (PXP) Genes. Hill et al. (1988) first cloned the *POX4* gene, encoding one form of alkane- or fatty acid-inducible fatty acyl-CoA oxidases (AOX, EC 1.3.99.3), the first enzyme in peroxisomal β-oxidation in yeast (see Fig. 3), from *C. maltosa* ATCC20184 by cross-hybridization with a cDNA probe coding for the C-terminus of the *C. tropicalis* AOX gene *Ct-POX4*. The deduced protein (PXP-4, 709 aa) exhibited 83 and 59% aa similarities to the *POX4* and *POX5* genes of *C. tropicalis*, respectively. Later, *POX2* coding for a different peroxisomal AOX form (protein PXP-2) were cloned and the presence of a third gene *POX5* was demonstrated by DNA hybridization techniques (Masuda and Takagi 1993, unpubl.; cf. Tables 9, 11), resembling the situation observed for *C. tropicalis* (Kamiryo et al. 1989; Rachubinski 1990).

Another peroxisomal gene *POX18* is located together with the *ALI1* gene (see below) at a fragment of an alkane-inducible, highly expressed clone TIG1, found in a genomic library by differential colony hybridization with cDNA (alkane/glucose) probes (Hwang et al. 1991). This *Cm-POX18* gene is transcribed in alkane-, fatty alcohol- and fatty acid-grown cells and encodes a deduced protein (127 aa) with very high homology of 94.5% to the small oleate-inducible peroxisomal protein PXP-18 of *C. tropicalis*, that was demonstrated to be a peroxisomal nonspecific lipid-transfer protein (Tan et al. 1990).

ALI1 (NADH Ubiquinone Oxidoreductase Subunit). The gene *ALI1* was first designated as coding for an alkane-inducible protein due to its tandemly location together with *POX18* (see above) at the same fragment of an alkane-inducible, highly expressed clone TIG1 (Hwang et al. 1991). From gene disruption experiments, it was concluded that *ALI1* was essential for alkane assimilation by *C. maltosa*, although it was not induced by alkanes. Recently, it was shown that the gene product of *ALI1* (276 aa) has extensive homology to a 30 kDa subunit precursor of respiratory complex I (NADH ubiquinone oxidoreductase, EC 1.6.5.3/ EC 1.6.99.3) known from other organisms (Souza et al. 1993). The deduced protein contains an N-terminal sequence that suggests a mitochondrial localization and indicates its role as being essential for mitochondrial respiration in alkane assimilation.

Genes of Carbon Source Catabolism: Carbohydrate Metabolism

GAL1 (Galactokinase)/GAL10 (UDP-Galactose Epimerase). These two highly inducible genes of the yeast galactose metabolism were recently cloned from *C. maltosa* in order to isolate their promoter region (Park et al. 1996). The galactose-inducible and glucose-repressible promoter region *GAL1-GAL10* was used to construct the high-copy expression vector pNGH2 for *C. maltosa* (Ohkuma et al. 1995b; Sugiyama et al. 1995; see Sect. 5.3), thus creating a plasmid which works in analogy to the highly efficient expression vector YEp51 for *S. cerevisiae* (Schunck et al. 1991; Scheller et al. 1994; Zimmer and Schunck 1995).

PGK1 (Phosphoglycerate Kinase). A *PGK1* gene coding for the glycolytic enzyme phosphoglycerate kinase (PGK, EC 2.7.2.3), a 47 kDa protein of 417 as a residues, was isolated form a *C. maltosa* genomic library by cross-hybridization with a *PGK* fragment of *S. cerevisiae* (Masuda et al. 1994). The gene exhibits the codon bias of highly expressed genes in yeast. Expression of this gene assayed by Northern blot analysis was significantly induced in cells grown on glucose but not in cells grown on alkane, alkanol, or fatty acid. The promoter region of *PGK1* was applied for construction of an expression vector (pMEA1) for *C. maltosa* (Masuda et al. 1994; cf. Sect. 5.3, Table 16).

Cycloheximide and Formaldehyde Resistance

The alkane-assimilating yeast *C. maltosa* possesses native resistance to different growth inhibitors (formaldehyde and other aldehydes) or antibiotics (cycloheximide).

RIM-C (L41Q) – Cycloheximide (CYH) Resistance. Whereas numerous yeasts, including *S. cerevisiae*, are highly sensitive to low concentrations (<0.5 μg/ml) of the antibiotic cycloheximide, *C. maltosa* and some other yeast species are resistant to this antibiotic at rather high concentrations (100 μg/ml or more). The resistance of *C. maltosa* to cycloheximide was demonstrated to be inducible (Takagi et al. 1985).

Two different genes related with this phenomenon have been cloned from *C. maltosa*, first the *RIM-C* (Ribosome modification by cycloheximide) gene (Takagi et al. 1986b), and later the *CYH^R* (Cycloheximide resistance) gene (Sasnauskas et al. 1992c; Table 11). Both genes were cloned in *S. cerevisiae* by selecting cyclohexim-ide-resistant clones using a cycloheximide sensitive wild-type strain as a host and a YEp-type plasmid as vector. It was proved that the *RIM-C* gene provides a convenient dominant vector marker for recombinant DNA technology in other yeast without genetic markers (Hino et al. 1992). Southern blot analysis with *RIM-C* as a probe indicated that there are at least three copies of *RIM-C*-related sequences in the genome of *C. maltosa*, in comparison with two copies in *S. cerevisiae* (Kawai et al. 1992). The nucleotide sequence of *RIM-C* gene contains a putative ORF (106 aa) interrupted by an intron (Table 12), coding for a protein homologous to the L41 ribosomal protein of the large subunit of *S. cerevisiae*. This protein obviously determines the sensitivity of the cells to cycloheximide. The 56th amino acid residue of the L41 protein in the sensitive ribosome is proline (L41P-type), while it is glutamine in the resistant ribosome (L41Q-type). The *RIM-C* (or *L41Q*, alleles 1a, 1b) genes of *C. maltosa* encoding L41Q-type protein are inducible in the presence of cycloheximide, while the L41P protein-encoding genes (*LEL41* or *L41P*, alleles 1a, 1b) are constitutively expressed. Further analysis revealed that the induction of *L41Q* was caused by inhibition of protein synthesis, a specific phenomenon of *L41Q*, not of other ribosomal protein genes examined Mutoh et al. (1995).

Similar genes with introns at the same position coding for P-type proteins were cloned from *C. tropicalis*, *Kluyveromyces fragilis*, *Pichia* sp. (M63395, Kawai, unpubl.) and *Schwanniomyces occidentalis*, whereas *S. cerevisiae* (L41a and L41b) or mammalia contain the Q-type proteins (Kawai et al. 1992, unpubl.; Delpozo et al. 1993). The highest amino acid homology was determined for the *L41Q* of *C. tropicalis* (99.1%), compared with 84–85% homology for the other yeasts (*Kf* and *Sc*), and 68–70% homology for the *L41P* proteins of rat and human.

CYH^R – Cycloheximide Resistance (CYHR) Protein. The second cloned *C. maltosa* gene, *CYH^R*, conferring resistance to cycloheximide in *S. cerevisiae*, has no homology to the previously cloned *RIM-C* gene and has no corresponding gene in the bakers yeast genome. The gene *CYH^R* is present in a single copy and the deduced gene product is a highly hydrophobic, probably membrane protein of 61 kDa without significant homology to any known proteins from databases (Jomantiene et al. 1992; Sasnauskas et al. 1992c). The gene *CYH^R* (as *RIM-C*) transforms *S. cerevisiae* at a frequency similar to auxotrophic markers (cf. Table 15) and can be used as dominant positive selectable marker for introducing

Table 12. Canonical intron splicing signals in yeasts

5'splice site	Distance (n)	Branch point	Distance (n)	3'splice site	Genes investigated Reference
Saccharomyces cerevisiae					
R/GTATGT	30–946	TACTAAC	5-77 (aver. 40)	YAG/N	54 different genes Rymond and Rosbach (1992)
Schizosaccharomyces pombe					
G/GTAWGT		TRCTAAC	9–16	TAG/N	55 introns from 28 genes Rymond and Rosbach (1992)
Candida maltosa					
A/GTATGT	214	TACTAAC	6	CAG/C	*FDH1* gene Sasnauskas et al. (1992a)
C/GTATGT	356	TACTAAC	17	TAG/G	SR15p gene (M58330) Sasnauskas et al. (1992b)
G/GTAYGT	35–277	TGCTAAC	7–16	TAG/T	*L41Q* and *L41P* genes Kawai et al. (1992) Mutoh et al. (1995)
Candida tropicalis					
G/GTATGT	35	TGCTAAC	7	TAG/T	*L41Q* gene Kawai et al. (1992)
Candida albicans					
N/GTATGT		TACTAAC		TAG/N	*ACT1*, *TUB2* Kurtz et al. (1990)
Yarrowia lipolytica					
N/GTGAGT	342–509	TACTAAC	1–2 (Y,A)	CAG/N	PYK: Strick et al (1992) SEC14: Lopez et al. (1994) ICL1: Mauersberger and Barth (1994, unpubl.)

N = A, C, G or T; R = A or G; W = A or T; Y = C or T.

recombinant plasmids into wild-type strains of *S. cerevisiae* and possibly for other eukaryotic organisms, as well as for gene disruption experiments. The first experiments with *Nicotiana tabacum* showed that the CYH^R gene can be used as a dominant selectable marker gene for introducing recombinant plasmids into plants (Jomantiene et al. 1992).

FDH1 – Formaldehyde Resistance. Additionally, *C. maltosa* is resistant to high concentrations of formaldehyde (8–10 mM MIC – minimum inhibitory concentration), in contrast to 1–4 mM MIC for other yeasts. Sasnauskas et al. (1992a) described a *C. maltosa* gene, *FDH1*, which confers resistance to formaldehyde in *S. cerevisiae*. The codon usage of the cloned gene corresponds to that of highly expressed yeast genes. The deduced 381 aa sequence (40 kDa) of the cloned gene *FDH1* showed high homology (66.5%) to mammalian alcohol dehydrogenases (ADH), especially to human ADH class III (gene *ADH5*, chi subunit) and only 24.9% similarity to *S. cerevisiae ADH1*. Recetly, it was shown that human and rat class III ADH and glutathione-dependent formaldehyde dehydrogenase (EC 1.2.1.1) are practically identical enzymes. Therefore, the predicted mechanism of resistance probably is glutathione-dependent oxidation of formaldehyde (Lebediene et al. 1992). The *FDH1* gene (at the plasmid pRAF1) was also successfully employed as a dominant selectable marker in transformation of wild-type *S. cerevisiae* (Sasnauskas et al. 1992a; cf. Table 15). The *FDH1* gene has a single intron in its coding sequence (see Table 12).

Further Ribosomal and Other Genes

SSRR Gene for Small Subunit Ribosomal RNA. This *C. maltosa* gene was detected in the highly expressed clone TIG2, isolated from a genomic library by differential screening with cDNA (alkane/glucose) probes in order to find alkane-inducible genes. In contrast to the clone TIG1 containing the genes *ALI1* and *POX18* (see above), TIG2 had homology to genomic DNA of *S. cerevisiae* (Hwang et al. 1991). The sequence of this clone disclosed high homology to srRNA from other organisms and allowed to establish the position of *C. maltosa* in the evolutionary tree of yeasts (Ohkuma et al. 1993b; cf. Sect. 1.2 and Fig. 1).

Ribosomal Protein RS15p Gene. Despite the above-mentioned ribosomal proteins L41, connected with the cycloheximide resistance in *C. maltosa*, an assumed ribosomal protein gene of *C. maltosa* was found in the neighborhood with an *ARS* sequence, indicated by one intron containing ORF and encoding a deduced polypeptide with high homology (70%) to the small ribosomal subunit protein No. 15 (RS15, 17.4 kDa) of *Brugia pagangi* (Sasnauskas et al. 1992b), suggesting that it belongs to the S15P family of ribosomal proteins.

SETR – tRNA^{Ser}CAG Gene. This tRNA gene, responsible for translation of the CUG codon as serine, was recently cloned using PCR and primers derived from the corresponding *C. albicans* gene, to prove the unusual amino acid assignment of the CUG codon *C. maltosa* (Sugiyama et al. 1995; cf. Sect. 4.4.3). This unusual tRNA shows high similarity in its sequence and secondary structure to the corresponding gene products of *C. albicans* and related *Candida* yeast (Yokogawa et al. 1992; Suzuki et al. 1993; Ohama et al. 1993).

   Transposons have not been detected in *C. maltosa* up till now, although the instability of auxotrophic mutations (APA changes, see Sect. 4.2.2) suggests that comparable mechanisms may also exist in this yeast. The structures of *ARS* and *CEN* sequences isolated from *C. maltosa* will be discussed below in Sect. 5.1 (see also Tables 8, 11, 15).

General Features of *Candida maltosa* Genes

The general structure of *C. maltosa* genes and the main motifs in 5'-upstream and 3'-downstream regions (TATA-box, consensus for translation initiation, polyadenylation signals, consensus for transcription termination) are comparable with those of *C. albicans* and *C. tropicalis* (Kurtz et al. 1990; Rachubinski 1990) or other yeast genes (Jones et al. 1991, 1992) and will not be discussed in detail here. The peculiarities of condon usage and condon bias in *C. maltosa* will be discussed below in Sect. 4.4.3.

Homology Between Structural Genes of Different Yeasts. Information on several common genes in *C. maltosa* and other yeasts including *Candida* and *Saccharomyces* is available. Comparison of the deduced amino acid sequences of several *C. maltosa* genes discussed above (*CYP52, CPR, URA3, POX, L41Q*) with those of other yeasts shows the closest taxonomic relationship between *C. maltosa* and the alkane-assimilating and pathogenic *C. tropicalis* and *C. albicans*. This close relationship is in agreement with the results of a comparison of the sequences of the 18S (*SSRR* genes) rDNA of *C. maltosa* and other *Candida* yeasts, as discussed in Sect. 1.2 (Barns et al. 1991; Ohkuma et al. 1993b).

The amino acid sequence of the *Cm-URA3* gene product shares 84 and 72% identity with that of the *URA3* gene product of *C. albicans* and *S. cerevisiae*, respectively. For the *Cm-CPR* the homology ranges from that of *C. tropicalis* (86%) to *S. cerevisiae* (49%), and rat (31%). This relation is obvious also for the *POX4* and *POX18* gene products showing 83 and 94.5% amino acid homology to the deduced *C. tropicalis* proteins. The deduced amino acid sequences of L41 ribosomal protein genes (*L41Q* or *L41P*) of *C. tropicalis, S. cerevisiae*, and rat share 99.1, 84.9, and 70.4% identities, respectively, indicating the expected high sequence conservation (Kawai et al. 1992). The sequence homology of *Cm-LEU2* and *Sc-LEU2* is 76% (Takagi et al. 1987), whereas the corresponding sequences of the *ADE1* (Kawai et al. 1991; Sasnauskas et al. 1991), *HIS5* (Hikiji et al. 1989), and *PGK1* (Masuda et al. 1994) genes share considerably lower homologies, with 66, 51, and 59% identitites, respectively. This may reflect the divergence in the evolutionary rates of each gene.

Interallelic Divergence Between Strains of the Same Species. The recent sequencing data for different *C. maltosa* genes gave evidence and hints that certain alleles of quite different genes concerned with nucleic acid biosynthesis (*ADE1*), amino acid biosynthesis (*LEU2*), and hydrocarbon utilization (*CYP52, POX*), are highly conserved in different strains of *C. maltosa*. On the other hand, an unexpected high sequence divergence or restriction fragment polymorphism between the multiple alleles of a single gene is tolerated within the genome of different strains. This may support the chromosome inactivation theory proposed recently by Becher et al. (1994, 1995; cf. Sect. 4.2.2). Whereas the nucleotide sequences of cloned *ADE1* and *LEU2* genes are 100 and 99.9% identical in different strains (VSB899, IAM12247, and L4, respectively), the *LEU2* genes cloned from one strain L4 are more divergent (2.5%) than the two copies from different strains. This is even more evident for P450 genes of the *CYP52A* subfamily, where 6.2% base deviations were detected in the coding region of two allelic variants of *CYP52A3* (*ALK1-A* and *ALK1-B*) cloned from strain IAM12247 (Ohkuma et al. 1991b). However, the cDNA sequence of the *CYP52A3* (P450Cml) from *C. maltosa* EH15 has only 2.2% base deviations from the *CYP52A3* (*ALK1-A*) allele from IAM12247 (Schunck et al. 1989a,b; Takagi et al. 1989). The inter-allelic divergence between the alleles of genes of even one *C.*

maltosa strain is therefore several times higher than the allelic differences between different *S. cerevisiae* strains, which were reported to reach 0.6% for chromosome III (Oliver et al. 1992).

Introns. First evidence on introns present in the assumed structural gene *FDH1*, in the ribosomal protein genes *L41Q* (*RIM-C*) or *L41P* and in the gene for the deduced ribosomal protein S15 of *C. maltosa* have been reported by Sasnauskas et al. (1992a,b) and Kawai et al. (1992). The canonical intron splicing signals were similar to that of *S. cerevisiae*, *S. pombe* (Rymond and Rosbach 1992), *Kluyveromyces lactis* (Deshler et al. 1989), *C. tropicalis* (Kawai et al. 1992), or *C. albicans* (Kurtz et al. 1990) genes, but different from that found in *Y. lipolytica* (Table 12; Barth and Gaillardin, Chap. 10, this Vol.).

A single intron was found in the coding sequence of *FDH1* (see above) with highly conserved sites (5'-GTATGT ... TACTAAC ... YAG-3') for correct splicing. It was shown that splicing of the *FDH1*-derived mRNA occurs identically in both *C. maltosa* and *S. cerevisiae*, demonstrating that *S. cerevisiae* accurately processes *C. maltosa* pre-mRNA (Sasnauskas et al. 1992a). The nucleotide sequences of the *L41Q* (*RIM-C*) and *L41P* (*LEL41*) genes of *C. maltosa*, coding for ribosomal proteins L41 of the large subunit, contain an intron at the same position of the putative ORF of 106 amino acids (Table 12, Mutoh et al. 1995). This intron position in *L41P* genes is highly conserved in the phylogenetic tree due to its presence in the different, not closely related, yeast species *C. maltosa*, *C. tropicalis*, *Kluyveromyces fragilis*, *Pichia* sp., *Schwanniomyces occidentalis*, and *S. cerevisiae* (Kawai et al. 1992, and unpubl. results; Delpozo et al. 1993). An intron was also found in a sequence which is located in the neighborhood of an *ARS*, and encodes an ORF for an assumed polypeptide with high homology (70%) to the small ribosomal subunit protein No. 15 (RS15) of *Brugia pagangi* (Sasnauskas et al. 1992b).

Sugiyama et al. (1995) recently reported an intron in the tRNA[Ser]CAG gene. Interestingly, these tRNA genes of *C. albicans*, *C. parapsilosis*, *C. rugosa*, and *C. zeylanoides*, which show relatively high sequence homology to that of *C. maltosa*, have no intron, but those of *C. maltosa*, *C. melibiosa*, and *C. cylindracea* have one intron (Ohama et al. 1993; Sugiyama et al. 1995). Up till now no introns were observed in numerous sequenced auxotrophic marker genes *ADE1* (6), *LEU2* (5), *HIS5*, and *URA3*, as well in the *CYP52* (15 P450 sequences including allelic variants), *CPR* (2) and *POX* (4) structural genes of *C. maltosa* (cf. Table 11).

4.4.3
Codon Usage

Codon usage in most genes of *C. maltosa* is comparable with that analyzed for *S. cerevisiae*, *C. tropicalis*, or *C. albicans* (Table 13; Rachubinski 1990; Kurtz et al. 1990). The most frequent codons calculated from several sequenced *C. maltosa* genes are in good agreement with the codon preference of *S. cerevisiae* (except lysine), according to Fröhlich et al. (1985). A total number of 60 different codons is found in these sequences. In most analyzed genes 63–85% of the amino acids were coded by only a restricted number of 24 codons (codon selectivity, % of most

Table 13. Codon usage of *Candida maltosa*. Preferred codons in *C. maltosa* are shown in **bold**. Preferred codons in *S. cerevisiae* according to Fröhlich et al. (1985) are indicated by a star (*)

Amino acid	Codon	Number[a]	Amino acid	Codon	Number[a]
Ala	**GCT***	245	Lys	**AAA**	312
	GCC*	73		AAG*	71
	GCA	35	Met	ATG	91
	GCG	11	Phe	**TTT**	124
Arg	CGT	18		**TTC***	141
	CGC	1	Pro	CCT	41
	CGA	2		CCC	0
	CGG	1		**CCA***	180
	AGA*	198		CCG	6
	AGG	5	Ser	**TCT***	114
Asn	AAT	113		**TCC***	64
	AAC*	116		TCA	38
Asp	**GAT***	201		TCG	12
	GAC*	81		AGT	38
Cys	**TGT***	51		AGC	6
	TGC	1		CTG[b]	5
Gly	**GGT***	275	Thr	**ACT***	210
	GGC	10		**ACC***	88
	GGA	31		ACA	32
	GGG	21		ACG	7
Gln	**CAA***	168	Tyr	**TAT***	79
	CAG	8		**TAC***	107
Glu	**GAA***	331	Trp	TGG	51
	GAG	3	Val	**GTT***	239
His	**CAT**	55		**GTC***	83
	CAC*	45		GTA	17
Ile	**ATT***	179		GTG	28
	ATC*	93			
	ATA	14	Sum aa		5064
Leu	CTT	42			
	CTC	11	Stop	TAA	8
	CTA	1		TAG	5
	TTA	175		TGA	0
	TTG*	270			

[a] Total number of times each codon is used in the *C. maltosa* genes: *LEU2, HIS5, URA3, ADE1, ALK1A, ALK1B, ALK2A, ALK3A, L41Q, ALI1, POX4, POX18, PGK* (for references see Table 11).
[b] Nonuniversal decoding of the codon CTG for serine instead of leucine in *C. maltosa* (Sugiyama et al. 1995; Zimmer and Schunck 1995).

frequent aa). The highest values (above 82%) were obtained for *ALK1-3, POX4*, and *PGK* genes (Ohkuma 1994, unpubl.), as for the highly expressed peroxisomal genes (*POX, CAT, HDE*) of *C. tropicalis* (Rachubinski 1990).

Codon Bias Index (CBI). In general, a codon bias index (CBI) correlates with the level of protein production, and highly expressed yeast genes have strong codon bias (on a scale from 0 to 1), as judged by the number of codons found compared to the 61 possible codons (Bennetzen and Hall 1982). Compared with other genes (Hwang et al. 1991) of *C. maltosa* the *PGK1* and *FDH1* genes were also found to have a strong codon bias, suggesting that they are highly expressed in *C. maltosa* (Sasnauskas et al. 1992a; Masuda et al. 1994), whereas *ADE1* is with a CBI of 0.68 assumed to be a moderately highly expressed gene (Sasnauskas et al. 1991). The *CYH*[R] (Sasnauskas et al. 1992c) is suggested to be a low level expressed gene.

On the other hand, it has been reported that the codon CUG, a universal leucine codon, is read as serine in some *Candida* species which are phylogenetically related to *C. maltosa*. This deviation from the universal genetic code was found in *C. cylindracea* (Kawaguchi et al. 1989; Yokogawa et al. 1992) and some related yeast species like *C. albicans* and *C. parapsilosis* (Ohama et al. 1993). Based on phylogenetic considerations, the same nonuniversal decoding of CUG was proposed to occur also in *C. maltosa* (Ohkuma et al. 1993b), because its evolutionary position is close to *C. albicans* and *C. parapsilosis* (cf. Sect. 1.2, Fig. 1).

Recently, in both our laboratories it was conclusively evidenced by in vitro translation experiments (Zimmer and Schunck 1995) and by in vivo studies (Sugiyama et al. 1995) that serine instead of leucine is specified by the CUG codon in the yeast *C. maltosa*. Depending on the cell-free system used, either serine, in the *C. maltosa*-derived in vitro translation system, or leucine, in the control with the conventional wheat germ system, was found to be incorporated in the translation products of artificial CUG-containing mRNAs. Moreover, it was possible to transfer the nonuniversal decoding of CUG to the wheat germ system by adding a tRNA fraction isolated from *C. maltosa*. This unusual amino acid assignment of the CUG codon read as serine, is obviously based on the presence of an unusual tRNA in *C. maltosa*, proved by cloning and sequencing of its gene (*SETR* – tRNA[Ser]CAG gene; Sugiyama et al. 1995) from this yeast with a structure almost identical to the corresponding gene of *C. albicans* and related *Candida* yeasts (Yokogawa et al. 1992; Ohama et al. 1993; Suzuki et al. 1993). Additionally, the comparison of the N-terminal amino acid sequences of the P450 52A3 protein overproduced in *C. maltosa* with a corresponding newly constructed mutant protein of P450 52A3, whose one triplet ATT (isoleucine) in this region of the *CYP52A3* gene has been changed to the CTG codon, demonstrates the use of CUG as a serine codon in vivo in *C. maltosa* (Sugiyama et al. 1995). This is supported by the result that the *S. cerevisiae URA3* gene, which has one CTG codon, could not complement *ura3* mutation of *C. maltosa*, but when its CTG codon was changed to another codon, CTC, the codon-changed *URA3* did complement the mutation in *C. maltosa* (Sugiyama et al. 1995). These results clearly indicate for the first time that the functional expression of a heterologous gene is possible only by changing the CTG codon to another leucine codon and that the deviation of the codon CUG from the universal genetic code is responsible for the defect of heterologous gene expression in *C. maltosa* (cf. Sect. 5.3). As a consequence of this alteration in the usage of the CUG codon in *C. maltosa*, heterologous expression in *S. cerevisiae* of some *C. maltosa* genes is expected to produce mutant proteins that have serine to leucine substitutions at each position coded by CUG. Therefore, an exchange of the CTG triplets of these structural genes by TCT encoding serine is required in order to produce the authentic proteins in *S. cerevisiae*. This was recently demonstrated for the heterologous expression in *S. cerevisiae* of the *C. maltosa* P450 genes *CYP52A4* and *CYP52A5* (Zimmer and Schunck 1995). In contrast, heterologous expression of the original *C. maltosa* P450 genes in *S. cerevisiae* resulted in the formation of still active but unstable enzymes probably subject to selective proteolysis in the endoplasmic reticulum of the host cells.

In *C. maltosa*, the codon CTG is a rare codon (Table 13) as it is in *C. albicans* (Brown et al. 1991). Except for the P450alk genes, the codon CTG can be found only in *HIS5* and *URA3* among *C. maltosa* genes sequenced so far. However, both genes can function in *S. cerevisiae* (Table 15; Sugiyama et al. 1995). In general, the frequency of codon usage positively correlates with the amount of isoacceptor tRNA present in the cells and an isoacceptor tRNA corresponding to the preferred codon is transcribed from multiple genes in the genome (Sharp et al. 1986; Lloyd and Sharp 1992). So it is reasonable that there are only one or two copies of the tRNA[SER]CAG gene in the genome of *C. maltosa*, as shown by genomic Southern blot analysis (Sugiyama et al. 1995).

4.4.4
Gene Mapping

Genetic Mapping in C. maltosa. Classical genetic mapping techniques developed mainly for *S. cerevisiae* and applied to other nonconventional yeasts (Ogrydziak 1988) are mostly not useful to study the genome organization of *C. maltosa* due to the lack of sexuality in it. Mapping of *C. maltosa* genes is only possible by parasexual methods (see Sect. 4.2.4). Such a parasexual cycle based on protoplast fusion and mitotic segregation demonstrated a linkage group with the sequence *CEN-ade*-26-*pro*-1 (Klinner et al. 1984). Some special organization features in the genome were elucidated after gene cloning and sequencing larger DNA fragments of *C. maltosa* (see below).

Physical Mapping in C. maltosa. The progress in recombinant DNA technology, an increasing number of cloned genes from *C. maltosa* (Table 11), and developed methods for electrophoretic separation of large DNA molecules (Pulsed or alternating field gel electrophoresis; see Sect. 4.3) made new approaches possible. Electrophoretic fractionation of intact yeast chromosomes represents an alternative to classical karyotyping, and largely overcomes the difficulties of cytogenetic analysis with imperfect yeast. Southern blots of chromosomes separated by these alternating field electrophoresis techniques (OFAGE or CHEF) are probed with labeled fragments of genes to determine to which of the chromosomes the gene hybridizes. In this way, the chromosomal assignments of several cloned *C. maltosa* genes were revealed (Tanaka et al. 1987; Becher et al. 1991, 1994; Ohkuma et al. 1993c, 1995a; Table 9, Fig. 5). All six genes of the *CYP52A* subfamily in *C. maltosa*, the alkane-inducible P450 genes *ALK1-3*, *ALK5*, *ALK7-8* (and their alleles), were assigned on the same chromosome VIII, which is the largest chromosome under the separation conditions applied. This is in contrast to results obtained for *C. tropicalis*, where the five members of the *CYP52A* subfamily are located on four different chromosomes (Seghezzi et al. 1992). Two other known *ALK* genes (*ALK4* = *CYP52D1* and *ALK6* = *CYP52C2*) are located on different chromosomes, indicating a close relationship between sequence homology and chromosomal localization of these members of the *CYP52* gene family. Interestingly, the *CYP52C2* was found on two different chromosomes II and IV (Ohkuma et al. 1993c, 1995a). The *POX2, 4*, and *5* genes coding for peroxisomal acyl-CoA oxidases were located at chromosomes VII and VIII. For the *ARS* sequence (TRA fragment) isolated from *C. maltosa* (Takagi et al. 1986a) the strongest signal of hybridization was observed for chromosomal band II, but there is another signal for band V, probably for a highly homologous sequence (Tanaka et al. 1987). The *HIS5* and *LEU2* genes were mainly localized at the chromosome bands V and III, respectively (Tanaka et al. 1987; Wedler et al. 1990; Becher et al. 1991, see Table 9). Additionally, for the *LEU2* gene, present in at least three allelic variants in *C. maltosa* (Table 11), a disomic state of the chromosomes carrying this and probably *LEU4* gene was elucidated (Becher et al. 1991, 1994). At least two *LEU2* genes situated on chromosomes of different size (probably bands III and IV) occur in the genome of *C. maltosa* CBS5611. Three differently sized chromosomes were found to carry *ADE1* genes in *C. maltosa* L4 (Becher et al. 1995). The diffuse nature of the chromosomal bands carrying *ADE1* specific sequences is probably associated with frequnent chromosomal rearrangements occurring in *C. maltosa* strains as discussed by Becher et al. (1995).

Cloning of *C. maltosa* genes revealed some close linkages of them in the genome (cf. Sect. 4.4.2). There are three pairs of tandem ($\Longrightarrow$ $\Longrightarrow$) arrangements of *CYP52A* (P450alk) genes, for both alleles of *ALK2-ALK3* and for *ALK7-ALK8-A* as well (Ohkuma et al. 1991b, 1993c, 1995a), as it is also found for the closely related *alk1* and *alk2* genes in *C. tropicalis* (Seghezzi et al. 1991, 1992), probably reflecting that the *CYP52* family arose by duplications of an ancestral gene. Similar tandem arrangements were found for the L41 ribosomal genes *L41P* and *L41Q1* (Kawai et al. 1992; Mutoh et al. 1995), and for the genes *POX18* and *ALI1* in the clone TIG1 (Hwang et al. 1991). The distances of the genes are 0.7 to 1.2 kb. There are two examples of nontandem clusters, for the peroxisomal acyl-CoA oxidase genes *POX2-POX4* ($\Longrightarrow$ $\Longleftarrow$) and the galactose utilization genes *GAL1-GAL10* ($\Longleftarrow$ $\Longrightarrow$). Analyzing the clustering of *CYP52A* genes by replacing each gene with the restriction site of *Not*I and pulsed-field gel electrophoresis, *ALK1, ALK5, ALK7-ALK8* were found to be clustered in a 160-kb region on chromosome VIII (Ohkuma 1993, unpubl.). In the TRA region the *ARS* sequence and the *CEN* (centromere) sequences are colocalized (cf. Sect. 5.1).

4.5
Preparation of DNA From Candida maltosa Cells

Note on Method. The methods described by Cryer et al. (1975) and Davis et al. (1980) for other species have been successfully applied to prepare total or genomic

DNA from *C. maltosa* cells in the laboratories of Takagi (Tokyo) and Sasnauskas (Vilnius), respectively. Here a method will be given for miniscale isolation of total DNA used in Takagi's laboratory. The method is valid for preparation of chromosomal and plasmid DNA of the yeast for restriction analysis, Southern blotting, and for *E. coli* transformation to rescue a plasmid from *C. maltosa* cells.

Procedure

1. Grow the cells in 10 ml YPD or YNB-glucose (or other carbon source) at 30 °C to $5 \times 10^7 - 1 \times 10^8$ cells/ml (late log phase). Harvest by centrifugation and resuspend the cells in 1 ml of ST buffer.

2. Spin down the cells in a microcentrifuge tube (1.5 ml) and resuspend in 0.5 ml of TS buffer.

3. Add 10 μl β-mercaptoethanol and mix.

4. Add 0.1 ml Zymolyase (5 mg/ml), incubate with occasional shaking at 30 °C for 40–60 min.

5. Harvest the cells by centrifugation (5000 rpm, 1 min) and resuspend completely in SE buffer.

6. Incubate for 15 min at 70 °C.

7. Cool down the tube to room temperature, add 30 μl proteinase K (0.5 mg/ml) and 10 μl RNase A (10 mg/ml), mix, and incubate 30–60 min at 37 °C.

8. Add 150 μl of 5 M potassium acetate solution, mix, and stand in ice for 30 min.

9. Centrifuge (8000 rpm, 1 min), transfer supernatant to a fresh microcentrifuge tube and add 0.5 ml phenol (TE buffer saturated) with vortexing.

10. Centrifuge (12 000 rpm, 5 min), and with the supernatant, perform three or more phenol-chloroform (1 : 1) extractions.

11. Transfer the final supernatant to a fresh microcentrifuge tube and add equal volume of 2-propanol with mixing.

12. Centrifuge for 12 000 rpm 5 min, drain off liquid, wash with 70% ethanol, and vacuum-dry briefly.

13. Dissolve the DNA in 100–150 μl TE buffer, mix, and resolve completely.

14. Add equal volume of 5 M ammonium acetate, vortex, and stand on ice for 30 min-1 h.

15. Centrifuge (12 000 rpm, 5 min), and transfer the supernatant to a fresh microcentrifuge tube and precipitate DNA with ethanol.

16. Centrifuge, wash with 70% ethanol, and dry. Dissolve the DNA in appropriate volume of TE buffer.

17. Prepared DNA will be sufficient for restriction analysis, for Southern blot analysis, and for transformation of *E. coli* (recovery of plasmids from *C. maltosa* cells).

Buffers

ST buffer: 1.2 M sorbitol, 50 mM Tris-HCl, pH 7.5
SE buffer: 0.2% SDS, 50 mM EDTA, pH 8.0
TE buffer: 10 mM Tris-HCl, pH 7.4, 1 mM EDTA

4.6
Preparation of RNA

The RNA preparation methods described for *S. cerevisiae* (in Campbell and Duffus 1988; Rose et al. 1990) can be generally applied also for *C. maltosa* (Ohkuma et al. 1995a). Essentially all special conditions commonly used for work with and preparation of RNA should be applied (DEPC treated RNase free solutions and equipment).

4.6.1
Isolation of Translatable mRNA

Note on Method. The method described here for *C. maltosa* was used for the isolation of mRNA highly active in subsequent in vitro translation experiments and immunoprecipitation with polyclonal antibodies against cytochrome P450Cm1 (Wiedmann et al. 1986, 1988a; cf. Sect. 2.7.2). Using these preparations, the mRNA coding for P450Cm1 in *C. maltosa* was quantified in dependence on growth conditions (regulation by carbon source and oxygen), and a cDNA library of n-alkane induced clones was constructed (Schunck et al. 1989a,b; see Table 10).

Procedure

1. Grow *C. maltosa* in a fermenter under defined conditions or in shaking flasks. Harvest about 200 ml culture containing approx. 5 g wet weight cells by centrifugation (6000 g for 10 min at 4 °C) and wash the cells with cold distilled water.

2. Freeze the cells in liquid nitrogen (and store at −70 °C before use) and disintegrate by grinding with a mortar in liquid nitrogen.

3. Isolate total RNA by the phenol-chloroform-method according to Palmiter (1974).

4. Prepare the poly(A) RNA from the total RNA fraction using poly(U) Sepharose chromatography according to Adesnik and Darnell (1972).

The eluted mRNA preparations can be used for in vitro translation experiments in the homologous *C. maltosa* system or in a wheat germ system (see Sect. 2.7.2), and as the template for cDNA synthesis.

4.6.2
Isolation of Total tRNA

Note on Method. Preparation of the total tRNA fraction from *C. maltosa* can be performed as described by Zimmer and Schunck (1995) on the basis of a standard method according to Holley (1967).

Procedure

1. Cultivate a 250-ml culture of *C. maltosa* at 30 °C in YPD to a density of 7×10^7 cells/ml and harvest by centrifugation ($3000\,g$ at 4 °C for 10 min).

2. After washing the cells with ice-cold distilled water, add 1.5 ml ice-cold extraction buffer (0.01 M Tris/HCl, pH 7.4, 0.1 M LiCl, 0.01 M EDTA and 0.2% SDS), 6 g of glass beads (0.45–0.50 mm), and 1.5 ml of phenol pre-equilibrated in extraction buffer to the cell pellet.

3. Vortex the mixture for 30 s and chill on ice for about 1 min. Repeat this step four times in order to break at least 90% of the cells.

4. Centrifuge the sample after adding 2.5 ml extraction buffer ($8000\,g$ at 4 °C for 5 min), and purify the supernatant by five phenol-chloroform ($1:1$) and two chloroform extractions followed by ethanol precipitation.

5. Load the resolved RNA onto a DEAE-cellulose (Pharmacia) column according to a standard method of total tRNA isolation (Holley 1967), wash with 0.1 M Tris-HCl, pH 7.5, and elute with the same buffer containing 1 M NaCl. Monitor the nucleic acid elution by A_{260} measurements, and precipitate the fractions containing total tRNA with ethanol.

6. Resolve the pellet in RNase-free water and store at −70 °C after freezing in liquid nitrogen.

Additionally, methods for RNA isolation from *C. maltosa* and its successful use in Northern blotting experiments and for cDNA synthesis have been described (Sunairi et al. 1984, 1988; Ohkuma et al. 1991b, 1995a; Sugiyama et al. 1995).

5
Host-Vector Systems for *Candida maltosa*

The development of convenient host-vector systems for *C. maltosa* by genetic engineering techniques represents a significant step towards its more widespread use in biotechnology, and will also increase the usefulness of this yeast for fundamental research. Because *C. maltosa* belongs to the fungi imperfecti and, therefore, conventional yeast genetic techniques are not applicable for this type of nonsexual yeast, recombinant DNA technology represents the only efficient way to analyze genomeits at the molecular level. On the other hand, due to its special properties as an alkane-utilizing yeast, *C. maltosa* might be a useful host for the functional expression of heterologous proteins and their application in biotransformation reactions (see Sects. 2.5 and 5.3).

For the development of host-vector systems for *C. maltosa*, all necessary steps of genetic engineering work were made during the past decade, mainly with the *C. maltosa* strain IAM12247 (at Tokyo University), as well as with strains L4 (at Greifswald University) and VSB788 (in Dr. Sasnauskas' group in Vilnius):

1. Autonomously replicating (*ARS*) and centromere (*CEN*) sequences have been isolated from the genome of this yeast and used for the construction of vectors

with independent replication within the transformed cells (YRp-type vectors; see Sect. 5.1).

2. Several auxotrophic mutants and the corresponding genes which complement these mutations have been isolated and used as markers for transformation of *C. maltosa* (Tables 6, 14; see Sect. 5.2).

3. Different expression vectors have been constructed using promoters of highly expressed homologous genes, such as the phosphoglycerate kinase (*PGK1*) gene (Masuda et al. 1994), the *GAL1* gene (Ohkuma et al. 1995b; Park et al. 1996), and the P450 gene *ALK1* (Ohkuma 1993; Muraoka et al. 1993; Ohkuma unpubl.). These vectors were applied for heterologous and homologous gene expression in *C. maltosa* (see Sect. 5.3).

5.1
ARS and CEN Regions of Candida maltosa

The construction of the host-vector system for *C. maltosa* by application of genetic engineering requires a vector to introduce any DNA sequences into the host cell. This vector should have at first a stable maintenance (integrated or autonomously replicating) in the host cell, should contain a selective marker (homologous or heterologous auxotrophic marker genes like *LEU2*) for transformation, and be available for recombinant DNA technology (appropriate cloning sites and amplification in *E. coli*, like the commonly used pBR or pUC plasmids).

Since *C. maltosa* does not have any plasmids like 2 μm DNA of *S. cerevisiae*, and neither the 2 μm ORI (YEp-type vectors containing this replication origin) nor the *ARS1* of *S. cerevisiae* are functional in *C. maltosa* (Takagi et al. 1986a), it was necessary to isolate an autonomously replicating sequence (*ARS*) for vector construction. Although integrative transformation is applicable for *C. maltosa*, it is at considerably low frequency. Because of the high transformation frequency in general, transformation with an *ARS* is favorable. Moreover, the transformed plasmid should exist extrachromosomally, and should thus be easily rescued in *E. coli*. In many fungal species, *ARS* elements have been identified (Oliver 1988). Vectors carrying such sequences will facilitate cloning of nuclear genes by mutant complementation.

Chromosomal DNA fragments containing functional *ARS* have been isolated and characterized from several *C. maltosa* strains and applied for the construction of host-vector systems independently in several laboratories (cf. Fig. 7, Tables 8, 11, 14, 15). First attempts to clone *ARS* sequences from *C. maltosa* by screening for replicative function in *S. cerevisiae* failed to isolate any sequences which were active in the donor organism (Kawamura et al. 1983). Moreover, plasmids containing *S. cerevisiae* 2 μ ORI or *ARS1* sequences failed to replicate in *C. maltosa* (Takagi et al. 1986a), as in *C. albicans* (Kurtz et al. 1990). Screening for *ARS* function in *C. maltosa* itself has been far more successful, and two groups (Takagi et al. 1986a; Jomantiene et al. 1987) have isolated sequences which promote high frequency transformation. The cloning procedure of the best characterized *ARS* from *C. maltosa* IAM12247 and its application for the construction of a host-vector system

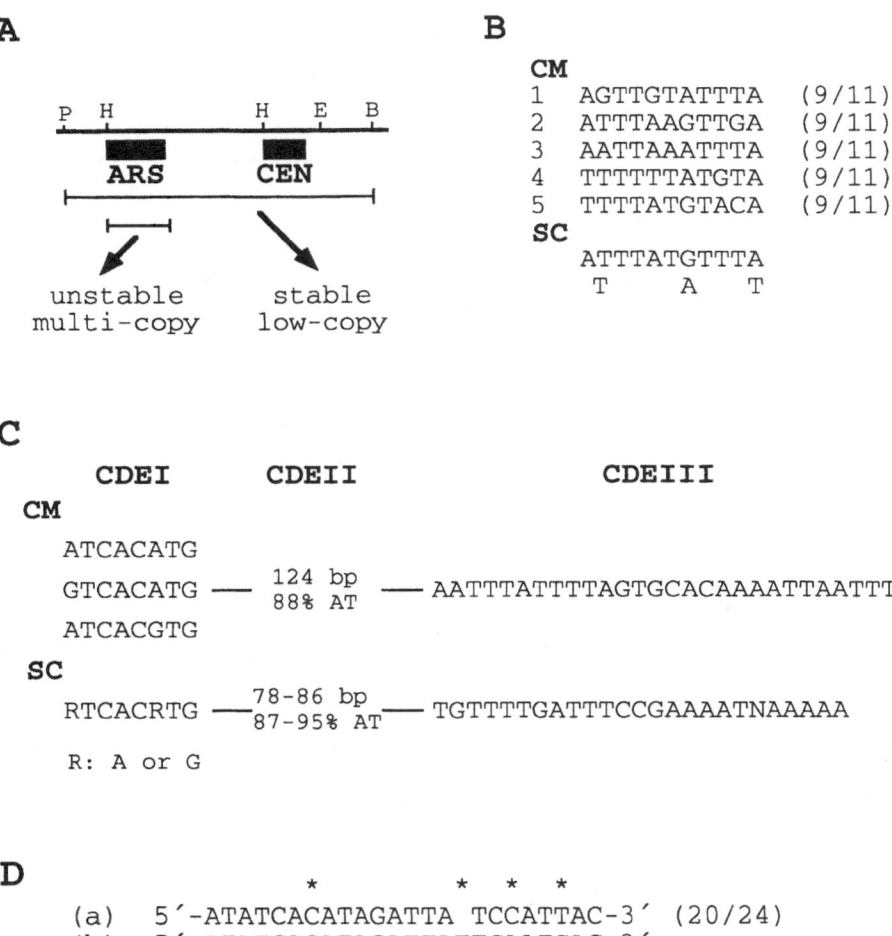

Fig. 7A–D. Autonomously replicating sequences (*ARS*) and centromere region (*CEN*) of *Candida maltosa*. **A** – Restriction map of the TRA (transforming ability) region. **B** – Consensus sequences found in *ARS* of *C. maltosa* (CM) and of *S. cerevisiae* (SC), taken from Kawai et al. (1987). **C** – Consensus sequences found in *CEN*. **D** – A further homology region in *ARS* sequences from two different *Candida maltosa* strains. (a) Strain IAM12247 (Kawai et al. 1987), **(b)** Strain VSB899 (Sasnauskas et al. 1992b). *Stars* indicate nonhomologous base pairs

has been reviewed recently (Takagi 1992, 1993; Ohkuma et al. 1994a). The *C. maltosa* mutant J288 (*leu2*) was transformed with the gene library of *C. maltosa* (20 000 clones) constructed in the plasmid YEp13 (*S. cerevisiae* vector, *LEU2* as selectable marker, see Table 10). From the resulting highly frequent Leu[+] transformants, a plasmid pCS1 was isolated, which carried an *ARS* fragment of *C. maltosa*, because of its ability for efficient transformation and its recovery in *E. coli*. The subcloned 3.8 kb *Bam*HI fragment which contained the *ARS* was designated as TRA (**TR**ansformation **A**bility) region. The *ARS* in the TRA region was finally subcloned into a 0.2 kb *Hind*III-*Sau*3AI fragment. This minimal 196 bp DNA fragment from *C. maltosa* IAM12247 was essential for the transformation ability. It exhibited *ARS* activity in both *C. maltosa* and *S. cerevisiae* (Takagi et al. 1986a; Kawai et al. 1987; cf. Table 15), as did the mentioned *ARS* isolated from *C. albicans* (Kurtz et al. 1990). The *C. maltosa ARS* contains five 11 bp sequences that were homologous to the *ARS* core consensus sequence of *S. cerevisiae* in 9 out of 11 nucleotides (Fig. 7; Marahrens and Stillman 1992). A similar consensus is also present in the *C. albicans ARS* (Kurtz et al. 1990).

Later, it was shown that the TRA region contained a centromere (*CEN*) of *C. maltosa* (Ohkuma et al. 1995e; Fig. 7, Tables 11, 14). Using the 3.8 kb TRA region, the stability of the plasmid under nonselective condition was very high (approximately 80% after ten generations) and the copy number of the plasmid is low (one or two copies per genome), sharing the properties of YCp-type vectors of *S. cerevisiae*. Using only the 0.2 kb *ARS* region, the stability was dramatically decreased (below 10% after ten generations) and the copy number is increased (more than 20), sharing the properties of YRp-type vectors of *S. cerevisiae*. The *CEN* region was subcloned into the 0.3 kb *Hind*III-*Spe*I fragment (Fig. 7). This *CEN* region was homologous to *CEN* of *S. cerevisiae*. The *CEN* of *S. cerevisiae* consists of three motifs, *CDEI*, *CDEII*, and *CDEIII*. The *CDEI* is an 8 bp consensus, the *CDEIII* is 25 bp consensus showing partial dyad symmetry. Located between *CDEI* and *CDEIII* is a 78–86 bp stretch containing more than 90% AT. In the case of *C. maltosa CEN*, there are three sequences completely conserved with *CDEI* of *S. cerevisiae*, and a 124 bp long stretch of 88% AT, a little longer than the *CDEII* of *S. cerevisiae*. Following *CDEII*, a partial dyad symmetry in the *CEN* of *C. maltosa* was found , but it had no sequence homology with *CDEIII* of *S. cerevisiae* (Fig. 7C; Ohkuma et al. 1995e). The isolated TRA (*CEN-ARS*) was used for the construction of different plasmids containing several auxotrophic marker genes (like pTRA1, pBTH10B, pUTA1, pUTC1) applied in the development of new host-vector systems of *C. maltosa* (cf. Sect. 5.2 and Table 14). Additionally, *ARS* elements of *C. maltosa*, isolated by comparable strategies, were described from the strains VSB747 (two fragments, no sequence data, Polumienko and Grigorieva 1985), VSB899 (one fragment, Jomantiene et al. 1987; Sasnauskas et al. 1992b), and L4 (*ARS1* and *ARS2*, no sequence data; Becher et al. 1991, 1994; Kasüske et al. 1992), which differ in their functioning in *S. cerevisiae* (Table 15). Interestingly, the *ARS1* and *ARS2* from *C. maltosa* strain L4 as the *ARS* from strain VSB899 (in the vector pRJ1 with *C-ADE1*, Sasnauskas et al. 1992b) showed no *ARS* activity in *S. cerevisiae*, although the *C-ADE1* gene of *C. maltosa* functions in *S. cerevisiae*, whereas the *ARS* isolated from strains IAM12247 and VSB747 were active. The noncentromeric plasmids containing homologous *ARS* (or replicators) from *C. maltosa* are maintained with the same frequency in both wild-type and mutant strains of *S. cerevisiae* (Karpova et al. 1987). On the other hand, there are several earlier reports indicating at *ARS1*, the 2μ *ORI*, and the *CEN6* regions of *S. cerevisiae* were not functionally active in *C. maltosa* (cf. Sect. 5.3, Table 16; Takagi et al. 1986a; Kunze et al. 1987a,b).

In general, the *C. maltosa ARS* elements (at least 150–200 bp) produce an increased transformation rate and mitotically unstable transformants, typical for autonomously replicating plasmids. The stability of transformants was, with about 60–65% after ten generations under selective growth conditions (Sasnauskas et al. 1992b), or with less than 10% under nonselective conditions (Kawai et al. 1987; Hikiji et al. 1989) strikingly lower than the stability of the above mentioned TRA (*CEN-ARS*) containing plasmids (about 80% after ten generations even under nonseletive conditions). The *ARS* without *CEN* containing plasmids showed a high copy number of the recombinant plasmids in transformants of about 10–20 copies per cell (Takagi et al. 1986a; Kawai et al. 1987; Hikiji et al. 1989; Sasnauskas et al. 1992b). Comparison of the nucleotide sequences of two sequenced *ARS* elements (Kawai et al. 1987; Sasnauskas et al. 1992b) revealed an additional 24 bp homologous sequence (20/24) specific for the *C. maltosa ARS* element (Fig. 7D). The function of these sequences still remains unclear. This consensus sequence of *C. maltosa ARS* has no similarity to the *S. cerevisiae ARS* consensus sequence. The *ARS* sequence cloned from *C. maltosa* VSB899 is located in the neighborhood with an ORF encoding a polypeptide with high homology (70%) to the small ribosomal subunit protein No. 15 (RS15) of *Brugia pagangi*, and containing an intron with canonical sites for correct splicing (Sasnauskas et al. 1992b).

5.2
Development of Host-Vector Systems

The progress made in transformation and the development of host-vector systems for *C. maltosa* during the past decade are partially described in some reviews on nonconventional yeast (Oliver 1988; Weber and Barth 1988; Rachubinski 1990; Reiser et al. 1990; Sudbery 1994) and were reviewed more recently by Takagi (1992, 1993) and Ohkuma et al. (1994b).

5.2.1
Transformation Systems, Marker Genes, and Vectors

Transformation systems of *C. maltosa* have been developed using homologous or heterologous auxotrophic (nutritional) markers selectable by complementation (Tables 11, 14, 16 and references therein). At first, plasmids that had been developed for *S. cerevisiae* were tested for transformation of *C. maltosa*, and then homologous selectable phenotypic genes and *ARS* elements were isolated in an attempt to obtain more efficient and more stable transformation of *C. maltosa* cells. Thus, first transformation systems for *C. maltosa* had been developed using *S. cerevisiae ARG4* (Kunze et al. 1985a,b), *LEU2* (Kawamura et al. 1983; Takagi et al. 1986a), and *LYS2* (Kunze et al. 1987a,b, obviously integrative transformation occurred) genes to complement the corresponding mutations in *C. maltosa* (Tables 6, 14, 16). Difficulties in isolating appropriate host mutations and in strain construction were the limiting factors in the development of host-vector systems for this organism (Chang et al. 1984; Hikiji et al. 1989; cf. Sect. 4.2, Table 6). Other marker genes (*HIS5*, *URA3*, *TRP1*) of *S. cerevisiae* were obviously not functionally active in *C. maltosa* (Table 16). Thereafter, the homologous genes *ADE1*, *ADE2*, and *ARG4* (Jomantiene et al. 1987, 1991; Kawai et al. 1991; Sasnauskas et al. 1991, 1992c) and *LEU2* (Takagi et al. 1987; Becher et al. 1991, 1994) have been cloned from *C. maltosa* and tested as transformation markers (Tables 11, 14) for host-vector systems. A breakthrough in the development of cloning systems for the alkane-utilizing yeast *C. maltosa* was the isolation of *ARS* and *CEN* fragments in TRA (Takagi et al. 1986a; cf. Sect. 5.1), stimulating the more recent isolation of additional homologous auxotrophic marker genes (*HIS5*, *URA3*, *CYS2*) and their application in transformation and gene disruption techniques (Tables 8, 11, 14, 15; see Sect. 4.4.2).

Thus, an improved host-vector system for *C. maltosa* using the homologous *HIS5* gene together with the TRA region was described. The *HIS5* gene was isolated from the genomic library of *C. maltosa* constructed in pTRA11 by self-cloning (Hikiji et al. 1989). The host-vector systems which were constructed before (Kawamura et al. 1983; Takagi et al. 1986a; Kawai et al. 1987), utilized *C. maltosa* J288 (*leu2*) as a host. As this recipient strain had a serious growth defect on n-alkanes, for the new host-vector system a newly isolated *C. maltosa* mutant, CH1 (*his5*, Alk⁺), was used as host, which showed growth on alkanes close to the wild-type strain (Hikiji et al. 1989; Table 6). Five vectors for *C. maltosa* CH1, each of which contains the 2.3 kb *HIS5* gene, the pBR322 sequence, and either the 3.8 kb TRA region (pBTH10B) or the 0.2 kb fragment (pBTH20A) of the TRA region were constructed (Table 14, for vector designation). Furthermore, to improve the host-vector systems for *C. maltosa*, a *his5 ade1* double-mutant (CHA1) was obtained after mutagenesis of the CH1 (*his5*) mutant, and the *ADE1* gene which complements the adenine auxotrophic mutation was cloned (Kawai et al. 1991). This host

Table 14. Available host-vector systems for *Candida maltosa*. For the vector composition genes were designated for example as *LEU2-Cm*, when the homologous *Candida maltosa* gene, and *LEU2-Sc* when genes from *Saccharomyces cerevisiae* were used for the vector construction.

Host strain	Vector	Composition of the vectors	Mode of maintenance i: integrative a: autonomously	Reference
J288[a]	pCMK3	*LEU2-Cm*, pBR322	i	Kawamura et al. (1983)
	pCMK31	*LEU2-Cm*, *TRP1-Sc*, pRC3	i	
	pCMK32	*LEU2-Cm*, *TRP1-Sc*, pRC3 (in both orientations)	i	
J288[a]	pTRA1	*ARS-Cm*, *CEN-Cm* (TRA 3.8 kb)	a	Takagi et al. (1986a)
	pTRA11	*LEU2-Sc*, pBR322 (in both orientations)	a	
J288[a]	pTRA2	*ARS-Cm* (TRA 0.2 kb)	a	Kawai et al. (1987)
	pTRA12	*LEU2-Sc*, pBR322	a	
CH1	pBTH10A	*ARS-Cm*, *CEN-Cm* (TRA 3.8 kb)	a	Hikiji et al. (1989)
CHA1	pBTH10B	*HIS5-Cm*, pBR322	a	
CHU1	pBTH110A	all plasmids	a	

Table 14. (*Contd.*)

Host strain	Vector	Composition of the vectors	Mode of maintenance i: integrative a: autonomously	Reference
CHAU1	pBTH110B		a	
	pBTH20A	*ARS-Cm*, (TRA 0.2 kb) *HIS5-Cm*, pBR322	a	Hikiji et al. (1989)
	pUTH18	*ARS-Cm*, *CEN-Cm* (TRA 3.8 kb) *HIS5-Cm*, pUC18	a	Okhuma (unpubl.) Sugiyama et al. (1995)
CHA1 CHAU1	pUTA1	*ARS-Cm*, *CEN-Cm* (TRA 1.8 kb) *ADE1-Cm*, pBR322	a	Kawai et al. (1991) Ohkuma (unpubl.)
CHU1 CHAU1	pTHU1	*ARS-Cm*, *CEN-Cm* (TRA 3.8 kb) *HIS5-Cm*, *URA3-Cm*, pUC18	a	Ohkuma et al. (1993a)
	pUTU1	*ARS-Cm*, *CEN-Cm* (TRA 1.8 kb) *URA3-Cm*, pUC19	a	Ohkuma (1994, unpubl.)
CHS1 CHSA1	pUTC1	*ARS-Cm*, *CEN-Cm* (TRA 3.8 kb) *CYS2-Cm*, pUC18	a	Ohkuma (1994, unpubl.)
G344 (*arg4*)	pYe(*ARG4*) 411	*ARG4-Sc*, pBR322 *ARS-Sc*	a	Kunze et al. (1985a,b)
G457 (*lys2*)	pDP12	*LYS2-Sc*, pBR322 *TRP1-Sc*, *ARS1-Sc* (not active)	i	Kunze et al. (1987a,b)
	pDP13	as pDP12, additional *CEN6-Sc* (not active)	i	
VSB899 (*ade1*)	pRA7-1 pRJ1	*ARS-Cm*, *ADE1-Cm*, pUC19 *ARS-Cm*, *ADE1-Cm* .pICEM-19H	a a	Jomantiene et al. (1987) Sasnauskas et al. (1992c)
G587 (*leu2*)	Not named	*ARS-Cm* (from VSB899)[b] *LEU2-Cm*	a	Wedler et al. (1990)
G587 (*leu2*)	CipL1 CrLp1	*LEU2-Cm*, pUC18 *ARS1-Cm* (1.8 kb) *LEU2-Cm*, pUC18	i a	Becher et al. (1991, 1994) Becher et al. (1991, 1994)
G374 (*ade1*)	CrAp2	*ARS1-Cm*, *ADE1-Cm* pBR325	a	Kasüske et al. (1992)
	CipA15	*ADE1-Cm*, pACYC177	i	Kasüske et al. (1992)
	pCCD5	*ARS2-Cm* (weak) *CYP53A3::ADE1-Cm* pACYC117	a/i	Kasüske et al. (1992)

[a] J288 strain has serious growth defect on alkanes, whereas CH1 and the mutants derived from it show normal growth on alkanes (Hikiji et al. 1989).
[b] *ARS* from VSB899 (Jomantiene et al. 1987).
Host strains used and their genotypes (cf. Table 6):
A) Derived from the wild-type strain *C. maltosa* IAM12247
    J288 – *leu2* (Takagi et al. 1986a); CH1 – *his5* (Hikiji et al. 1989)
    CHA1 – *his5*, *ade1* (Kawai et al. 1991)
    CHU1 – *his5*, *ura3::ADE1/ura3::ADE1*, and CHAU1 – *his5*, *ade1*, *ura3* (Ohkuma et al. 1993a)
    CHS1 – *his5*, *cys2*, and CHSA1 – *his5*, *cys2*, *ade1* (Ohkuma, unpubl.)
B) Derived from the wild-type strain *C. maltosa* L4
    G344 – *arg4*-18 (Kunze et al. 1985a,b); G457 – *lys2*-21, *ino*-3 (Kunze et al. 1987a,b)
    G587 – *leu2* (Becher et al. 1991, 1994; Kasüske et al. 1992)
    G374 – *ade1* (Kasüske et al. 1992).
Abbreviations: TRA – transforming ability region
               CipL1 – *Candida* integrative plasmid *LEU2*
               CrLp1 – *Candida* replicative *LEU2* plasmid.

is convenient, especially for gene disruption experiments. Using two kinds of marker genes, one can disrupt both alleles of distinct gene present in *C. maltosa*, as demonstrated for *ALI1* and *CYP53A3* genes (Hwang et al. 1991; Ohkuma et al. 1991a).

More recently, the triple auxotrophic mutants CHAU1 (*his5 ade1 ura3*) and CHSA1 (*his5 cys2 ade1*) were obtained (Ohkuma 1993; Ohkuma et al. 1993a; for details see Table 6 and Sect. 4.2.2) and the corresponding vectors, like pTHU1 and pUTC1, for transformation of these convenient recipient strains were constructed (Table 14). These hosts are more useful because, after disruption of both alleles of a gene, the remaining auxotrophic mutation is available for complementation with a next selective marker gene in a further transformation step (Ohkuma 1993).

Other host-vector systems for *C. maltosa* were developed in other laboratories based on independently isolated recipient strains (Table 6), the newly cloned *ARS* and marker genes *ADE1* (Sasnauskas et al. 1992c) and both *LEU2* and *ADE1* (Wedler et al. 1990; Becher et al. 1991, 1994; Kasüske et al. 1992) using autonomously replicating and integrative plasmids as well (Table 14).

Dominant (positive selectable) marker genes were up till now not used in transformation of *C. maltosa*. Contrary, several dominant marker genes isolated from *C. maltosa* (*RIM-C*, *CYH^R*, *FDH1*; see Sect. 4.4.2) are applicable for transformation of wild-type strains of *S. cerevisiae*, other yeasts and plants, if they

Table 15. Functional expression of genes isolated from *Candida maltosa* in the yeast *Saccharomyces cerevisiae* and in *E. coli*

Gene[a]	Complementation of mutation or functioning in	Reference
ARS two fragments[c] (strain VSB747)	Functions in *Sc*	Polumienko and Grigorieva (1985)
ARS (strain IAM12247)	Functions in *Sc* and *Cm*	Takagi et al. (1986a) Kawai et al. (1987)
ARS (strain VSB899)	Not functioning in *Sc*, functions in *Cm*	Jomantiene et al. (1987) Sasnauskas et al. (1992b)
ARS1 and *ARS2* (strain L4)	Not functioning in *Sc*, functions in *Cm*	Becher et al. (1991, 1994) Kasüsk et al. (1992)
Selectable auxotrophic markers[b]		
ADE1	*ade1* in *Sc*	Jomantiene et al. (1987, 1991) Sasnauskas et al. (1991) Kawai et al. (1991)
ADE2	*ade2* in *Sc*	Jomantiene et al. (1987)
ARG4	*arg4* in *Sc*	Jomantiene et al. (1987)
CYS2	*cys2* in *Sc*	Ohkuma (unpubl.)
HIS5	*his5* in *Sc*	Hikiji et al. (1989)
LEU2	*leuB* in *E. coli* *leu2* in *Sc*	Kawamura et al. (1983) Takagi et al. (1987) Wedler et al. (1990) Becher et al. (1991)
URA3	*ura3* in *Sc*	Ohkuma et al. (1993b)

Table 15. (*Contd.*)

Gene[a]	Complementation of mutation or functioning in	Reference
Functional expression of *other enzymes*[d]		
CYP52A3 (Cm1)	In *E. coli* (not active) in *Sc* GRF18	Schunck et al. (1989b) Schunck et al. (1991) Scheller et al. (1992, 1994, 1996)
CYP52A4 (Cm2)	In *Sc* GRF18	Schunck et al. (1991) Scheller et al. (1992, 1996) Zimmer and Schunck (1995)
CYP52A5 (*ALK2-A*)	In *Sc* GRF18	Zimmer (1995, unpubl.)
CYP52A9 (*ALK5-A*)	In *Sc* GRF18	Zimmer (1995, unpubl.)
CPR	In *E. coli* (not active) in *Sc* GRF18	Kärgel et al. (1996)
CYP52A3 to A5 and CPR	In *Sc* GRF18 (coexpression)	Zimmer (1995, unpubl.) Zimmer et al. (1995)
Dominant selectable markers		
RIM-C (*L41Q*)	Selectable in *Sc* and other yeasts for cycloheximide resistance	Takagi et al. (1986b) Kawai et al. (1992) Hino et al. (1992)
CYH[R]	Selectable in *Sc* for cycloheximide resistance	Sasnauskas et al. (1992c)
FDH1	Selectable in *Sc* for formaldehyde resistance	Sasnauskas et al. (1992a)

Abbreviations: Cm – Candida maltosa
      Sc – Saccharomyces cerevisiae.
[a] In the literature *Candida* (*maltosa, tropicalis,* or *albicans*) genes are mostly designated *C-ADE1* or *Ca-ADE1* to distinguish them from *S. cerevisiae* genes (*ADE1*).
[b] The genes *ADE1, ADE2, HIS3, LEU2, TRP1,* and *URA3* from *C. albicans* also complement the respective mutants in *S. cerevisiae* (Ohkuma 1994, unpubl.; for references see Kurtz et al. 1990).
[c] Two *ARS* fragments not sequenced. Their function in *C. maltosa* itself was not tested.
[d] Functional expression of P450 and NADPH-P450 reductase in the ER of *S. cerevisiae* GRF18 under control of the *GAL10* (*Sc*) promoter by using the vector YEp51. Nonfunctional expression of partial protein sequences in *E. coli* detected with antibodies.

are expressed in the recipient organism, as shown for the *RIM-C*-directed transformation of a freeze-tolerant yeast *Kluyveromyces thermotolerans* (Hino et al. 1992) and for the transformation of the plant *Nicotiana tabacum* using *CYH[R]* as a dominant selectable marker gene (Jomantiene et al. 1992). Unfortunately, many nonconventional yeasts are cycloheximide-resistant themselves.

5.2.2
Transformation Methods

Note on Methods. Although it was shown that *C. maltosa* could be transformed by the conventional spheroplast method with high frequency, this method is complicated and time-consuming. Therefore the more convenient lithium acetate and recently electroporation methods were applied for transformation of *C. maltosa* (Takagi et al. 1986a; Kasüske et al. 1992; Sasnauskas et al. 1992; Takagi 1992, 1993; Becher et al. 1994; Ohkuma et al. 1994a,b; Becher and Oeiver 1995). The procedures used in our laboratories are given below. For the quantitative determination of the transformation frequency, each plasmid multiplied in *E. coli* was gently purified before transformation into a *C. maltosa* recipient strain.

Transformation by the Spheroplast Method

Procedure

1. Grow the yeast culture with shaking in 100 ml YPD medium at 30 °C to 2–5 × 10^7 cells/ml.

2. Harvest by centrifugation and wash in distilled water.

3. Resuspend in 10 ml of 1.2 M sorbitol solution, add 25 µl of β-mercaptoethanol and incubate 10 min at 30 °C.

4. Centrifuge, wash in 10 ml of 1.2 M sorbitol, and resuspend in 1.2 M sorbitol.

5. Add 0.1 ml Zymolyase 60 000 (0.5 mg/ml), incubate at 30 °C with gentle or occasional shaking for 1 h.

6. Centrifuge at 2500 g for 3 min, wash twice in 10 ml 1.2 M sorbitol and wash once in SCT buffer (2 M sorbitol, 10 mM $CaCl_2$, 10 mM Tris-HCl, pH 7.5).

7. Resuspend in 5 ml SCT buffer, divide into 0.1 ml aliquots in sterile microcentrifuge tubes and add 0.1–10 µg DNA solution.

8. Place at room temperature for 10 min, add 1 ml of 10% PEG 4000 and mix gently.

9. Place at room temperature for 15 min.

10. Add transformation mixture to 7–10 ml of regeneration agar (1.2 M sorbitol, 0.67% YNB without amino acids, 2% glucose, 3% agar) in plastic tubes held at 45–50 °C in a water bath. Mix briefly and immediately pour over plates of selective media (the SD plates).

11. Incubate plates for 2–5 days.

Transformation by a Modified Lithium Acetate Method

Note on Method. Transformation of *C. maltosa* was performed by a slightly modified lithium acetate procedure according to Ito et al. (1983). For optimization of the frequency of transformation, the concentration of the lithium acetate and the pre-incubation time were changed and their effects on the frequency of transformation were assessed using the vector pTRA1 and strain J288 (*leu2*) as a host (see Tables 6 and 14). The frequency was raised as high as that of the spheroplast method, about 1800 colonies per μg DNA, when the cells were suspended in 0.05 M instead of 0.1 M lithium acetate and shaken at 30 °C for 90 min instead of for 60 min. Using 1 μg of the vector pBTH10B, a few hundred of transformants could be obtained by this method using the host strain *C. maltosa* CHA1 (*his5 ade1*).

Procedure

1. Inoculate a single colony from a YPD plate into 2 ml YPD. Grow the yeast culture with shaking overnight at 30 °C (preculture or seed culture).

2. Inoculate 10 ml of YPD medium with appropriate number of these freshly grown cells (0.2–0.5 ml of preculture), and incubate with shaking at 30 °C for 3 to 6 h.

3. Harvest the cells at 2–5 × 10⁷ cells/ml (late log phase) by centrifugation, and wash the cells twice in 5 ml of TE buffer (10 mM Tris-HCl, pH 8.0, 20 mM EDTA).

4. Resuspend the cells in 1 ml of the solution of 0.05 M lithium acetate in TE buffer, pH 8.0, and incubate at 30 °C for 1.5 h with gentle shaking.

5. Dispense 0.1 ml aliquots into sterile microcentrifuge tubes and add appropriate amount of plasmid DNA (0.1–10 μg), mix, and incubate 30 min at 30 °C.

6. Resuspend the cells completely (otherwise cells could not suspend in PEG solution), then add 0.1 ml 70% PEG 4000 solution in TE buffer, mix by vortexing tubes, and incubate at 30 °C for 1 h.

7. Transfer the tubes to a 42 °C waterbath for 5 min (heat shock).

8. Wash the cells with 1 ml sterile distilled water, harvest, and spread onto selective agar (YNB-glucose agar without the marker amino acid to select for, e.g., Leu⁻ plates for plasmid pTRA1).

9. Incubate the plates for 2–5 days at 30 °C.

Transformation by Electroporation

Note on Method. Electroporation has recently become the preferred method for gene transfer due to its ease and efficiency of operation in comparison to alternative methods. Although electroporation has been used successfully for other yeasts and fungi (for references see Kasüske et al. 1992, and other chapters, this Vol.) the optimal conditions vary significantly between different yeast strains. Therefore, it was necessary to optimize the conditions (cell growth and density, field strength, pulse duration, and DNA concentration) for the used *C. maltosa* strains (Kasüske et al. 1992; Becher and Oliver 1995; Ohkuma et al., unpubl.; Fig. 8). A maximum of 7000 transformants per 100 ng of plasmid DNA was reached under optimal conditions described by Kasüske et al. (1992), demonstrating that electropulse transformation is more efficient than other standard methodologies used with *C. maltosa* and perhaps other yeasts. It has the advantages of permitting very low concentrations of DNA to the used and of promoting the cotransformation events which may be essential to produce disruption or replacement mutants in this imperfect yeast.

Procedure

1. Grow the *C. maltosa* cells as described above in 10 ml YPD to a cell density of about 5×10^7 cells/ml.

2. Chill on ice and harvest by centrifugation at 0–4 °C.

3. Wash the cells with 5 ml sterile ice-cold 1 M sorbitol.

4. Resuspend the cells with 0.2 ml sterile ice-cold 1 M sorbitol (50-fold concentration of the cells).

5. Pipette 50-μl aliquots into a microcentrifuge tube on ice, add an appropriate amount of plasmid DNA (100–1000 ng), mix, and transfer into a prechilled sterile electroporation cuvette.

6. Perform electroporation using the following conditions:
   Equipment – Genepulser and Pulse Controller (BioRad)
   Cuvettes – 0.2 cm (0.4 cm also applicable, but increase voltage)
   Capacitor – 25 μF

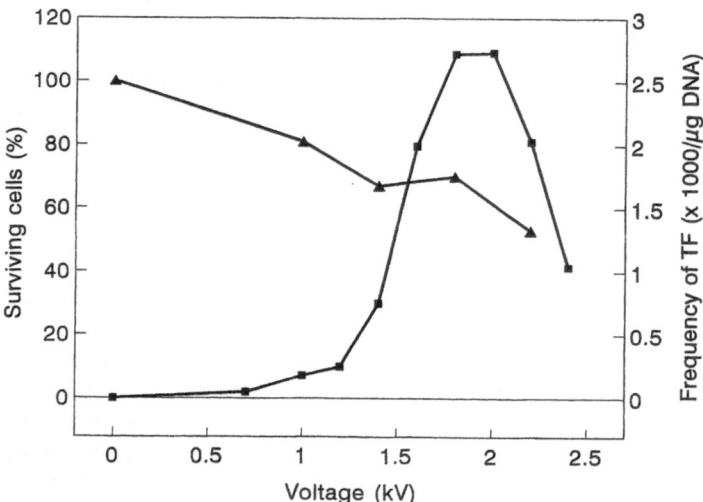

Fig. 8. Effect of initial voltage on frequency of *Candida maltosa* transformation. Host strain: *Candida maltosa* CHA1. Plasmid: pBTH10B (200 ng DNA). Conditions: Electroporation cuvettes 0.2 cm. Symbols: ■ Frequency of transformants per μg DNA (×1000); ▲ Surviving cells (%)

Parallel resistor – 200 ohm
Initial voltage – varied from 1.4 to 2.5 kV (optimal 1.8 to 2.0 kV)

7. Plate out onto selective agar (YNB-glucose), and incubate at 30 °C for 2–4 dyas.

5.3
Heterologous Gene Expression in Candida maltosa

The construction of expression vectors for *C. maltosa* is interesting for several reasons. First, by using an expression vector, it may be possible to induce a gene of interest under conditions where the gene is otherwise uninduced to analyze the function of the gene product. Second, there is interest in the expression of heterologous genes in the n-alkane assimilating yeast *C. maltosa*, because this yeast, especially when grown on alkanes or fatty acids, has the ability to develop intracellular membrane structures such as ER and peroxisomes (see Sect. 3.2). Consequently, *C. maltosa* may have an advantage when used as a host for the synthesis of heterologous proteins targeting to these organelles. Especially, the ER might be utilized as a residence for enzymes (like P450 systems), catalyzing the biotransformation of many kinds of mostly hydrophobic organic compounds. For this purpose, it should be advantageous to express heterologous genes in *C. maltosa* to confer new metabolic functions.

In this way, it will be possible to use the specific peculiarities of the alkane-assimilating yeast cell for optimization of special biotransformation reactions, as there are the facilitated transport of hydrophobic substrates into the cell, the high activity of cofactor supplying and electron transfer systems (NADPH-P450 reductase, cytochrome b_5 and its NADH-dependent reductase) supporting the extremely high in vivo turnover rate of the host's own alkane- and fatty acid-oxidizing P450 enzyme systems, and the optimized intracellular transport of hydrophobic sub-

Table 16. Expression of heterologous genes in *Candida maltosa*

Gene[a]	Complementation of a mutation or expression of a functional protein	Plasmid nature (i: integrative) (a: autonomously)	Reference
Heterologous auxotrophic markers genes			
ADE1 Ca[b]	*ade1*	(a)	Ohkuma et al. (1994,
ADE1 Sc	Not tested		unpubl.)
ARG4 Sc	*arg4*	(a)	Kunze et al. (1985a,b)
LEU2 Sc	*leu2*	(a)	Kawamura et al. (1983)
			Takagi et al. (1986a, 1988)
LYS2 Sc	*lys2* (weakly expressed)	(i)	Kunze et al. (1987a,b)
URA3 Sc	*ura3*[c]	(a)	Sugiyama et al. (1995)
URA3 Ca[b]	*ura3*	(a)	Ohkuma et al. (1994,
			unpubl.)
HIS5 Sc	Not functioning[c]	(a)	Sugiyama et al. (1995)
TRP1 Sc	Not expressed[e]	(i)	Kunze et al. (1987a)

Table 16. (*Contd.*)

Gene[a]	Complementation of a mutation or expression of a functional protein	Plasmid nature (i: integrative) (a: autonomously)	Reference
Dominant markers: not tested			
ARS/CEN function			
ARS Sc[d]	Functioning in *Cm*		Kunze et al. (1986a,b)
ARS1 Sc	Not functioning in *Cm*		Takagi et al. (1986a)
2μ ORI Sc	Not functioning in *Cm*		Takagi et al. (1986a)
CEN6 Sc	Not active in *Cm*		Kunze et al. (1987a,b)
Heterologous genes			
Amp^R Ec	β-Lactamase	(a)	Kunze et al. (1985a,b)
(*bla*)		(i)	Kunze et al. (1987a)
POX2 Ct	Subunits of peroxisomal	(a)	Kamiryo et al. (1989)
POX4 Ct	acyl-CoA oxidase		
LAC4 Kl	β-Galactosidase, as reporter gene for promoter testing of:		
	PGK1	(a)	Masuda et al. (1994)
	CYP52A3 or *CYP52A4*	(a)	Muraoka et al. (1993)
Several other genes: *lacZ Ec, gusA Ec*, P450 and others	No functional expression[c]		Sugiyama et al. (1995) Ohkuma et al. (unpubl.)

Abbreviations: *Cm – C. maltosa, Ca – C. albicans*
        Ct – C. tropicalis, Ke – Kluyveromyces lactis
        Sc – S. cerevisiae, Ec – E. coli.

[a] In the literature *Candida* (*albicans, maltosa,* or *tropicalis*) genes are mostly designated *C-ADE1* or *Ca-ADE1* to distinguish them from *S. cerevisiae* genes (*ADE1*).

[b] *Candida albicans* genes *ADE1* and *URA3* also complement the respective mutants in *S. cerevisiae* (Ohkuma et al. 1994, unpubl. results).

[c] *HIS5 Sc* – no functional expression in *Cm*. For the *URA3 Sc* firstly no expression in *Cm* was observed, which was probably due to the codon usage problems in this host. After changing its CTG codons in other leucine codons functioning in *Cm*, complementation of the corresponding *ura3* mutations in *Cm* by the modified genes of *Sc* was observed. The different codon usage might be the reason for the failure in functional expression of several other heterologous genes (*lacZ* and *gusA* genes of *E. coli* coding for β-galactosidase and β-glucuronidase, respectively, the 2,3-dihydroxybiphenyl dioxygenase gene of *Pseudomonas* sp., the catechol-2,3-dioxygenase gene of *P. putida*, mammalian P450 genes, e.g., P450C21) in *Cm* (Sugiyama et al. 1995, see Sects. 4.4.3 and 5.3).

[d] The vector pYe(ARG4)411 used in the experiments contained an *ARS* of *Sc* (Tschumper and Carbon 1982; Kunze et al. 1985a,b).

[e] Only insignificant increase in the activity of phosphoribosylanthranilate (PRA) isomerase (coded by *TRP1*) after transformation of *C. maltosa* G457 (*lys2*) mutant with the plasmid pDP13, containing additionally the *TRP1* gene of *S. cerevisiae*, due to lack of *trp1* mutants in *C. maltosa* (Kunze et al. 1987a).

strates and products due to the subcellular organization of the alkane oxidation pathway (cf. Sects. 2.5 and 3.4).

Based on recent progress in the development of host-vector systems in *C. maltosa*, first-expression vectors were constructed to test several isolated promoters of strongly expressed *C. maltosa* genes for heterologous and homologous protein expression as well (Table 16).

Using the promoter and terminator regions of the newly isolated regulated *PGK1* (phosphoglycerate kinase) gene of *C. maltosa* in the low-copy expression vector pMEA1 (*ARS-CEN, ADE1*), Masuda et al. (1994) demonstrated the functional expression of an endogenous P450 gene (*CYP52A3*) and a heterologous gene (*LAC4-Kl* encoding β-galactosidase of *Kluyveromyces lactis*) in glucose-grown cells, but not in cells grown on alkane, alkanol, or fatty acid (Table 16). Additionally, the *ALK1* (*CYP52A3*) and *ALK3* (*CYP52A4*) P450 promoters were fused with this reporter gene *Kl-LAC4* in the low-copy vector pPL1 (*ADE1, CEN-ARS*) and expressed in *C. maltosa*, as reported by Muraoka et al. (1993, 1994). Using this system, first n-alkane responsive regions in the promoter of P450alk genes were identified by deletion analysis. More recently, a similar host-vector system (vector pUTU1 – *URA3, CEN-ARS*) was used to express P450alk genes in *C. maltosa* DA1235 (*ura, alk*) and related strains (obtained from strain CHAU1 by stepwise gene disruption of the P450 genes *ALK1-3* and *5*) to study the function of individual alkane-inducible P450 forms in *C. maltosa* in vivo (Ohkuma et al. 1993c; Ohkuma et al. 1995d; and unpubl. results).

The *LAC4-Kl* gene is therefore useful as a reporter gene in *C. maltosa* and other *Candida* species, in which some other commonly used reporter genes for promoter testing in yeast, such as the *lacZ* or *gusA* genes of *E. coli*, could not be expressed, probably due to codon usage problems, as it was also reported for expression of these reporter genes in *C. albicans* and *C. tropicalis* (Leuker et al. 1992).

The strongly regulated, galactose-inducible and glucose-repressible *GAL1* (galactokinase) promoter in the high-copy (>20) vector pNGH2 (*ARS, HIS5*) was recently used to overexpress the homologous P450 genes *ALK1-A*, its mutant forms, and *ALK-3A* in *C. maltosa* CHA1 or CHU1 (Ohkuma et al. 1995b; Sugiyama et al. 1995). Immunoelectron microscopy revealed that, upon overexpression, a dramatic proliferation of ER occurred, in which the overproduced P450alk proteins accumulated. The artificially proliferated ER membranes were mainly tubular forms, and stacks of paired membranes were also observed after prolonged expression. The tubular forms were morphologically very similar to the physiologically proliferated forms of ER present in alkane-induced *C. maltosa* cells (cf. Sect. 3, Fig. 4). The observed proliferation of ER membranes by overproduction of homologous P450alk will provide a unique opportunity for investigating the mechanisms by which cells regulate ER biogenesis, in comparison with the intrinsic form of ER proliferation. Thus, using *C. maltosa* as a host, it is now possible to compare physiologically proliferated forms of ER with those induced artificially.

Additionally, the *POX2* and *POX4* genes encoding the peroxisomal acyl-CoA oxidase proteins PXP-2 and PXP-4, respectively, of the peroxisomal ACO of *C. tropicalis* were expressed from vectors based on pTRA3 (multicopy vector with

0.6 kb TRA, *Sc-LEU*), introduced into *C. maltosa* J288 (Kamiryo et al. 1989). The fact that *C. tropicalis* promoters were recognized and the gene products were targeted into the *C. maltosa* peroxisomes indicates that gene expression and targeting signals and mechanisms in these two yeasts are similar.

Functional expression of β-lactamase (*Amp^R* gene) of *E. coli* in addition to heterologous marker genes in transformed *C. maltosa* cells, using the expression vector pYe(ARG4)411, which carried the *ARG4* gene of *S. cerevisiae* (Kunze et al. 1985a,b), or using the vectors pDP12 or pDP13, which carried the *LYS2* gene of *S. cerevisiae* (Kunze et al. 1987a), was reported (Table 16). The expression of this prokaryotic gene in yeasts after previous selection using a selective marker is a potentially useful tool for finding transformants.

By taking advantage of the recently developed host-vector system for *C. maltosa*, it was possible to test the expression of several other heterologous genes, such as *lacZ* and *gusA* genes of *E. coli* coding for β-galactosidase and β-glucuronidase, respectively, the 2,3-dihydroxybiphenyl dioxygenase gene of *Pseudomonas* sp., the catechol-2,3-dioxygenase gene of *P. putida*, and the mammalian P450C21 gene in *C. maltosa* (Sugiyama et al. 1995; Ohkuma et al., unpubl.). The expression of these genes has failed, although in some cases synthesis of a transcript of the genes was confirmed, suggesting difficulties in expressing heterologous genes at the posttranscriptional level, which might be connected with the alteration in codon usage in this yeast, as discussed in Sect. 4.4.3.

As shown in Sect. 5.2.1, several auxotrophic marker genes of *S. cerevisiae* (*Sc*) and *C. albicans* (*Ca*) were tested to be expressed under the control of their own promoter in *C. maltosa* with different results (Table 16). Expression of *Ca-ADE1* and *Ca-URA3*, as well as expression of *Sc-ARG4*, *Sc-LEU2*, and *Sc-LYS2* was demonstrated due to complementation of the corresponding mutations in *C. maltosa*. In contrast, several other heterologous marker genes including *Sc-HIS5*, *Sc-URA3*, and probably *Sc-TRP1* cannot be functionally expressed in *C. maltosa* (Table 16). Only an insignificant increase in the activity of phosphoribosyl-anthranilate isomerase (coded by *TRP1*) after transformation of *C. maltosa* G457 (*lys2*) mutant with the plasmid pDP13, containing additionally to *Sc-LYS2* the *Sc-TRP1* gene, was observed (Kunze et al. 1987a).

As mentioned above, the CTG codon usage alteration may be a primary cause of occurring difficulties in expressing a heterologous gene in *C. maltosa* (cf. Sect. 4.4.3). This was clearly demonstrated recently for the *URA3* gene of *S. cerevisiae*. After changing the only one CTG codon of the *Sc-URA3* gene, which could not complement the *ura3* mutation of *C. maltosa*, by the CTC leucine codon, the modified *Sc-URA3* gene was functionally expressed in *C. maltosa* (Sugiyama et al. 1995). This result is the first example of succeeding in functional expression of a heterologous gene in *Candida* species having an altered codon usage by changing the CTG codon in the gene to another. On the other hand, *S. cerevisiae LEU2* (with its own promoter) and *K. lactis LAC4* (under control of the host's own promoters, Masuda et al. 1994; Muraoka et al. 1993, 1994; Table 16) genes were shown to be expressed functionally in *C. maltosa*, although they have one or two CTG codons, respectively. In these genes, the position of the CTG codon may not affect the function of the encoded proteins.

Therefore, it is possible to overcome problems occurring in heterologous gene expression by changing the codons CTG (used for serine) in the gene to express into another codon for leucine. However, it may be a laborious task to change CTG codons, because there are many CTG codons in other genes from bacteria and mammals. The construction of a suppressor tRNA which translates the CUG codon as leucine in *C. maltosa* should be tested to avoid the labor of changing all the CTG codons in any gene to be expressed. Thus, as a consequence of the altered codon usage (Sect. 4.4.3) in *C. maltosa*, the application of the developed host-vector systems for heterologous expression of different genes in *C. maltosa* is at present still limited, when the presence of CUG codons in the genes to be expressed has serious consequences for the functioning of the respective gene product.

On the other hand, the *C. maltosa* host-vector system should be very useful for the expression of genes isolated from related *Candida* yeasts such as *C. albicans*, *C. cylindracea*, *C. melibiosa*, *C. rugosa*, *C. zeylanoides*, and *C. parapsilosis* (Kawaguchi et al. 1989; Ohama et al. 1993; Ohkuma et al. 1993b), where the unusual CUG codon usage also occurs. This might be true also for *C. tropicalis* and other *Candida* yeast related to *C. maltosa*.

6
Potential Biotechnological Application of *Candida maltosa*

In several Sections of this Chapter the particular industrial importance of alkane-utilizing yeasts like *C. maltosa*, *C. tropicalis*, and *Y. lipolytica* has been mentioned already. As will be shown below, several possibilities exist for potential biotechnological application of *C. maltosa*, although there are only few examples for a current realization directly in industry, or the industrial production has been stopped, as for the most SCP processes. This seems, for different reasons, to be a general feature in the application of nonconventional yeast today.

In the context of data obtained for alkane-assimilating yeasts in general (reviewed by Levi et al. 1979; Rehm and Reiff 1981, 1992; Fukui and Tanaka 1981a,b; Bühler and Schindler 1984), and based on progress made in the past decade as given in this chapter, some of the potential practical applications of *C. maltosa* in particular can be discussed as follows:

- Production of alkane-based SCP (Liquipron, Paprin, Fermosin, see Shennan 1984; Senez 1986), carbohydrate-based SCP, and different by-products of these processes, like biolipid extract and some of its components (ergosterol, ubiquinon), glutamic acid, and RNA (see below).
- Production of some hydrophobic intermediates of the alkane oxidation pathway, like fatty alcohols, fatty acids, and especially dicarboxylic acids (DCA), of commercial interest with *C. maltosa alk* mutants (Shiio and Uchio 1971; Uchio and Shiio 1972a,b,c; Casey et al. 1990; see below).
- Use of *C. maltosa* enzymes as biocatalysts. The production of homologous enzyme proteins directly from *C. maltosa* grown on alkanes and other substrates (P450s, NADPH-P450 reductase, cytochrome b_5, FAOD, SOD, catalase; Avetisova 1991; Mauersberger et al. 1992b; Avetisova and Davidov 1993; Avetisova et al.

1993; cf. Fukui and Tanaka 1981b) or after heterologous expression of their genes in *S. cerevisiae* (P450alk genes, Schunck et al. 1991; Scheller et al. 1992, 1994, 1996; Kärgel et al. 1996) and application of these enzymes in biotransformation reactions (see Sect. 2.5.3) was tested. The alkane-induced FAOD of *C. maltosa* can be applied for stereoselective oxidation of secondary alcohols to gain enantiomeric pure *S*(t) secondary alkanols of potential interest by using a crude membrane fraction or the purified enzyme obtained from cells grown on alkanes (Mauersberger et al. 1992b; see Sect. 2.5.3). Recently, Zimmer et al. (1995) described an efficient biotransformation system of lauric acid to dodecanoic acid with intact recombinant cells of *S. cerevisiae*, coexpressing the NADPH-P450 reductase and *CYP52A4* P450 genes from *C. maltosa* (see below). Bode and Birnbaum (1991a) investigated the enzymatic production of indolepyruvate, and of analogs of indolepyruvate from tryptophan and methyl- and fluoro-derivatives of tryptophan using the tryptophan aminotransferase of *C. maltosa*.

- Application of *C. maltosa* as a potential host for the production of heterologous proteins. The potential application of alkane-utilizing yeast as a host for the functional expression of enzymes like P450 systems and the use of recombinant *C. maltosa* cells as biocatalysts in biotransformation reaction was discussed in Sect. 5.3.
- Production of muconic acid from catechol in a bioprocess using catechol-grown *C. maltosa* ATCC 20184 cells (Gomi and Horiguchi 1988).
- Enantioselective production of D-amino acids from racemic substrates DL-amino acids, demonstrated for alanine (Umemura et al. 1990, 1992, Tanabe Seiyaku Company, Ltd., Osaka, Japan, see below).
- Production of 3- and 2-isopropylmalate with *leu2* and *leu1* mutants of *C. maltosa* (Bode et al. 1991, see below).
- Production of α-aminoadipate-δ-semialdehyde (AASA) with *lys1* and *lys9* mutants of *C. maltosa* (Schmidt et al. 1989a,b, see below).
- Mannan isolated from *C. maltosa* and its sulfated derivatives introduced into the intercellular space of tobacco and cucumber leaves promoted an increase of their resistance to the viral (tobacco mosaic virus) and bacterial (pathogenic strain *Pseudomonas syringae* pv. *lachrimans*) infections, respectively (Kovalenko et al. 1992).

Single Cell Protein (SCP) Production. The nonpathogenic yeast *C. maltosa* and other alkane-utilizing yeasts, like *Y. (C.) lipolytica* and *C. tropicalis*, have in the last 30 years been of particular interest for industrial application mainly because of their ability to produce single-cell protein (SCP) based on alkanes on an industrial scale (reviewed by Levi et al. 1979; Einsele 1983; Shennan 1984; cf. Sects. 1 and 2). However due to the rise in price of petroleum derivatives and the competition with conventional feed protein, the development of SCP processes in Western industrial countries and in Japan was stopped for economical and political reasons in the mid 1970s. Nevertheless, the long-term perspectives of SCP are sometimes discussed more positively (Rehm 1986; Senez 1986).

Mainly *C. maltosa* and *C. tropicalis* have been used for the production of SCP as a by-product of the oil-refining process in the former Soviet Union, GDR, and in the East European countries Bulgaria and Romania (Shennan 1984; Senez 1986). In several plants in the former Soviet Union mainly n-alkane fractions (paraffins) were used to produce fodder yeast (Paprin – hydrocarbon-based single-cell protein) with *C. maltosa* strains (Golubev et al. 1986; cf. Sect. 1). In the mid 1980s, this production reached about 1 million tons per year, and is still being continued in the countries of the former Soviet Union (Russia, Belarus, Ukraina), although on a lower production level. The SCP products and production-related problems were intensively investigated for their biological action (Ioffe et al. 1990), their antigenicity (Gukasyan et al. 1990), and in ecological studies (Gradova et al. 1991; see Sect. 2.2). The SCP product Paprin was recommended and used as new fodder protein for animals (Ioffe et al. 1990). In the former GDR at the oil processing plant at Schwedt/Oder (PCK Schwedt), a technology of combined utilization of oil distillates (gas-oil) by microbial deparaffinization to fodder yeasts (Fermosin), deparaffinized oil distillate, and microbial lipids was elaborated using a *C. maltosa* strain (Bauch et al. 1978; Bohlmann et al. 1979, 1982; Wünsche et al. 1981; Brendler et al. 1983; Ringpfeil 1983; Kozlova and Meshchankin 1991; see Sects. 1.1 and 2.2). The production of this plant was stopped at the beginning of the 1990s.

The microbial lipid (biolipid) extract occurred as a special by-product of this particular fermentation process using fuel-oil distillates as substrate. This biolipid extract contained mainly hydrocarbons, fats, phospholipids, free fatty acids, glycerides, sterols, ubiquinones, and vitamins (Müller and Voigt 1981), and was investigated for production of ergosterol and ubiquinones and for its direct application (Voigt et al. 1979, 1984a,b, 1985; Müller and Voigt 1984; Vier and Voigt 1984; Bergmann et al. 1987; Schuster et al. 1990). In the 1980s and more recently, there were several attempts to develop more complex industrial technologies for the isolation of additional useful by-products from the biomass or the cultural fluid of industrial SCP processes using alkane-assimilating *C. maltosa* strains (for earlier reviews on yeasts see Levi et al. 1979; Fukui and Tanaka 1981b). Such products are glutamic acid, RNA, and derived components (ribonucleosides, ribonucleotides), and their production was scaled up to a laboratory-sized pilot plant in Schwedt/GDR (Eckart et al. 1988) or are directly produced on an industrial scale additionally to the alkane-based SCP Paprin in Belarus (Baraji et al. 1990; Truchatshova et al. 1991. Shkumatov 1993, pers. comm.). Like other yeasts, *C. maltosa* can be used to produce SCP biomass also on sugars (glucose, galactose, xylose, mannose, arabinose, rhamnose) of hemicellulose beech wood hydrolysates obtained by phosphoric acid treatment (Kostov et al. 1991).

Production of Fine Chemicals Derived From n-Alkanes. Although initial interest focused mainly on the production of SCP from n-alkanes, later, as a result of detailed investigations into the metabolism of alkane-utilizing yeasts, and as an answer to the problems arising from oil price increase, attention turned toward the production of primary oxidation products and other metabolites directly from alkanes (as reviewed for yeast by Rehm and Reiff 1981; Fukui and Tanaka 1981b;

Bühler and Schindler 1984; Schindler et al. 1990). The aim was to exploit mutants or even wild-type strains of alkane-assimilating yeast by causing accumulation of commercially valuable intermediates. Several primary oxidation products (fatty alcohols, fatty acids, hydroxy fatty acids, and especially dicarboxylic acids) of long- and middle-chain alkanes are of industrial importance in the production of surfactants, lubricants, detergents, and cosmetics (Uchio and Shiio 1974; Uemura et al. 1988; Schindler et al. 1990).

In this case, *C. maltosa* was one of the first alkane-utilizing yeast for which the production of such hydrophobic metabolic intermediates, like dicarboxylic acids (DCA), was investigated up to pilot plant scale in the Ajinomoto Company in Japan. Shiio and Uchio (1971) tried to produce longer-chain DCAs, such as DCA-16, a precursor of synthetic muscone, from hexadecane. They selected the strain *C. (maltosa) cloacae* 310 (cf. Sect. 1 and Table 1) as a potent DCA producer among other hydrocarbon-utilizing yeasts. While growing in a medium containing both yeast extract and n-alkane, it oxidized the alkane to mainly shorter-chain DCAs (about 0.3 g per liter) containing the same or lesser number of carbon atoms. To increase the productivity of DCA formation, a mutant M-1 from the strain 310 was obtained by MNNG treatment which could not assimilate DCAs (Uchio and Shiio 1972a,b). Resting cells grown on a medium containing acetate and n-alkane produced significant quantities of dicarboxylic acid homologous to the alkane cosubstrate. The highest yield of DCA produced was with n-hexadecane as substrate (29.3 g of hexadecanedioic acid per liter). Since strain M-1 did not grow well on n-alkanes compared with the parent strain 310; the addition of other carbon sources which supported the growth of strain M-1 was necessary for the conversion of n-alkane to DCA by the growing cells. Later, Uchio and Shiio (1972c) derived a mutant of *C. cloacae* strain M-1 which was completely unable to assimilate n-alkanes and accumulated the homologous DCA when an n-alkane was present as cosubstrate with a substrate that would support growth of the mutant. This mutant strain M-12 cooxidized significant quantities of n-alkane (C_{10} to C_{16}) to the homologous DCA (up to 61 g/l DCA-16 in shaking flasks and 100 g/l DCA-16 under optimized condition in a fermenter during 100 h cultivation) during growth on acetate. These results suggest that the alkane-oxidizing pathway up to the DCA is intact in these mutants, but the ability to utilize the oxygenated product in the β-oxidation is not. The scale-up of the process to a 300-l tank has been performed (Uchio and Shiio 1974; Uchio 1978). However, for unknown reasons, this process was not introduced into industry by the Ajinomoto Co. It might be that the productivity was not high enough due to the reduced alkane oxidation rate of 10% in the mutants compared with the wild-type strain (Schindler et al. 1990). More recently, accumulation of DCA-16 (up to 7 g per l) from hexadecane and palmitic acid was demonstrated in 75 out of 84 isolated *alkD* mutants of *C. (cloacae) maltosa* (Casey et al. 1990; Table 7). At least one mutant also produced 3-hydroxyhexadecanedioic acid (up to 2.5 g/l). Of the 80 mutants characterized as *alkA–C* (Alk⁻, PA⁺), 16 produced small amounts (82 mg/l) of hexadecanol (cf. Sect. 4.2.2). Additionally, *C. maltosa* was tested together with *C. albicans* to produce long-chain alcohols from glucose as carbon source (White et al. 1989). Obviously, these strains converted the excess fatty acyl residues (C_{14}–C_{18}) under certain conditions (aerobic glucose utilization under nitrogen limitation, or anaerobic glucose utilization), into low amounts of fatty alcohols (60 or 120 $\mu g/g$ dry weight).

Numerous investigations have been performed to produce DCA with other yeasts including *C. tropicalis* and *Y. lipolytica* (references see Rehm and Reiff 1981; Bühler and Schindler 1984; Schindler et al. 1990). The first industrial process for production of DCA from n-alkane was developed in the Nippon Mining Company, Japan, using mutants of *C. tropicalis* (Uemura 1985; Taoka 1986; Uemura et al. 1988). Since its introduction in 1987, this industrial-scale production line at Nippon Mining (now Japan Energy Co.) is able to produce approximately 150 t brassylic acid (DCA-13) per year from n-tridecane, which represents the world demand for this acid, used in the perfume industry to produce macrocyclic musk compounds (Okino et al. 1986; Uemura et al. 1988). A recent biochemical investigation of this very productive mutant M2030 revealed that it has mutations leading to the absence of the two acyl-CoA oxidases and 3-ketoacyl-CoA thiolase, the first and the fourth enzymes involved in the peroxisomal β-oxidation system (Atomi et al. 1994). Furthermore, it was confirmed that the activity of the second and third enzyme of the β-oxidation system, the bifunctional enzyme (enoyl-CoA hydratase/ 3-hydroxyacyl-CoA dehydrogenase), drastically decreased in the mutant cells. The site of mutation causing the absence of the β-oxidation enzymes in connection with a decrease in the amount of peroxisomes is not clear.

Recently, several attempts were made to improve the DCA production with *C. tropicalis* by gene disruption experiments of two (*POX4* and *POX5*) of the three genes (*POX2*, *POX4*, *POX5*) coding for the acyl CoA-oxidase or of thiolase genes and by amplification of genes coding for enzymes of rate-limiting steps (*CYP52* and *CPR* for the P450 enzyme system) of the alkane oxidation pathway in yeasts (Schindler et al. 1990; Picataggio et al. 1991, 1992).

Another example is the efficient biotransformation of lauric acid to ω-hydroxy lauric acid and dodecanoic acid with intact cells of *S. cerevisiae*, in which the NADPH-P450 reductase from *C. maltosa* was coexpressed at high level with the *CYP52A4* P450 gene of this yeast (Zimmer et al. 1995). For this purpose, a multicopy plasmid was constructed that contained two independent expression units controlled by the galactose-inducible *GAL10* promoter. The recombinant *S. cerevisiae* cells within 100 min converted the applied lauric acid nearly completely to dodecanedioic acid, which is found in the cell-free supernatant.

Production of D-Amino Acids. The strain *C. maltosa* JCM1504 was applied to produce D-amino acids from a racemic DL-amino acid substrate (Umemura et al. 1990, 1991, 1992, 1994). After screening 253 yeast strains for their stereospecific degradation activity against DL-alanine *C. maltosa* JCM1504 was selected as the most favorable strain. This strain assimilates the L-isomers of racemic amino acids such as alanine, arginine, asparagine, glutamate, proline, and serine with high selectivity as the only source of carbon and nitrogen (Umemura et al. 1990, 1992). This property of *C. maltosa* JCM 1504 is connected with the stereoselectivity of its respective amino acid aminotransferases (shown for the alanine aminotransferase AlaAT, EC 2.6.1.2) and its relatively low D-amino acid oxidase activity under the conditions used. The conditions for enantioselective L-alanine degradation, leaving D-alanine mostly untouched, were established with this *C. maltosa* strain. The L-isomer from 200 g/l DL-alanine was nearly completely degraded within 40 h, and 90 g D-alanine remained in the reaction mixture. The chemical and optical purity of the D-isomer product thus obtained was 99 and 99.9% enantiomeric excess, respectively (Umemura 1990, 1992).

Production of 3- and 2-Isopropylmalate (IPM). The potential application of *leu1* and *leu2* mutant strains of *C. maltosa* for the production of 2-IPM and 3-IPM was demonstrated by Bode et al. (1991). These mutant strains, isolated from *C. maltosa* L4 lacking 3-isopropylmalate (3-IPM) dehydrogenase (*leu2* mutant G587) or 2-IPM dehydratase (*leu1* mutant G368), were able to excrete 3-IPM and 2-IPM, respectively, into the culture medium. The optimal conditions for the production of 3-IPM with the mutant G587 were determined. In the presence of 20 g glucose/l, 5 g $NH_4H_2PO_4$/l and 75 mg L-leucine/l in a minimal salt medium, *C. maltosa* G587 produced about 900 mg 3-IPM/l during 72 h of growth. By a simple procedure, both chromatographically pure dicarboxylic acids, 90 mg of 2-IPM and 1.4 g 3-IPM with 50% yield, were isolated from 3 l medium. The 2-IPM and 3-IPM, which are not available commercially, are useful for assays for both IPM dehydratase and 3-IPM dehydrogenase in the laboratory. Generally, they may be prepared by microbial production (for literature see Bode et al. 1991). The yield of 3-IPM using the *C. maltosa leu2* mutant is more than ten times higher than the amounts obtained by the above-mentioned methods.

Production of α-Aminoadipate-δ-Semialdehyde (AASA). Mutants of *C. maltosa* lacking saccharopine reductase (*lys9*) and saccharopine dehydrogenase (*lys1*; Table 6, see Sect. 4.2.2) were found to accumulate and excrete AASA (80–90 mg/l within 48 h using the *lys1* mutant G285) into the medium when growing on glucose and supplemented with 50–100 mg/l L-lysine (Schmidt et al. 1989a,b), being the first report of lysine-requiring yeast mutants that accumulate and excrete AASA. In contrast, *P. guilliermondii lys9* mutants lacked this AASA overproduction. The AASA accumulation by *C. maltosa* mutants may be explained by the low feedback regulation of their homocitrate synthase and the equilibrium of the enzyme reactions involved in the lysine biosynthesis (Schmidt et al. 1989a,b).

Acknowledgments. Part of the authors' work with *Candida maltosa* was supported by grants from the Bundesministerium für Bildung und Forschung of the FRG (BMBF 0310257A) and by the Deutsche Forschungsgemeinschaft (DFG) for SM and WS, and by Grants-in-Aid for Scientific Research from the Ministry of Education, Science, and Culture of Japan for MO and MT. MO was aided by JSPS Fellowships for Japanese Junior Scientists. The authors would like to thank Martin Griebenow for English corrections and support in the final preparation of the manuscript.

References

Adesnik M, Darnell JE (1972) Biogenesis and characterization of histone messenger RNA in HeLa cells. J Mol Biol 67: 397–406

Ahearn DG, Crow SA, Berner NH, Meyers SP (1976) Microbiological cycling of oil in estuarine marshlands. In: Wiley M (ed) Estuarine processes, vol 1 USES, Stresses and adaptation to the estuary. Academic Press, New York, pp 483–492

Ahearn DG, Holzschu D, Crow SA, Ibrahim AN (1979) Comparative studies on the potential pathogenicity of *Candida tropicalis* and *Candida maltosa*. In: Garattini S, Paglialunga S, Scrimshaw NS (eds) Single cell protein: safety for animals and human feeding. Pergamon Press, Oxford, pp 44–46

Ardies CM, Lasker JM, Bloswick BP, Lieber CS (1987) Purification of NADPH: cytochrome c (cytochrome P-450) reductase from hamster liver microsomes by detergent extraction and affinitiy chromatography. Anal Biochem 162: 39–46

Artamonova VG, Svitina NN (1991) On the current issues of bronchopulmonary diseases prevention among the workers engaged in industrial biotechnology (Russ). Gig Tr Prof Zabol 3: 31–33

Artamonova VG, Kuznetsov NF, Shleikin AG (1993) Experimental justification of the approach of medical genetics to individual prophylaxis of occupational diseases of respiratory organs. Cent Eur J Public Health 1: 16–18

Atomi H, Yu C, Hara A, Matsui T, Naito N, Kamasawa N, Osumi M, Ueda M, Tanaka A (1994) Characterization of a dicarboxylic acid-producing mutant of the yeast *Candida tropicalis*. J Ferment Bioeng 77: 205–207

Asubel FM et al. (eds) (1994) Current protocols in molecular biology. Current protocols, Green Publishing, John Wiley, New York, vol 2, chap 13

Avetisova SM (1991) Foundation of the purification method and biochemical characterization of cytochrome P450 of alkane-oxidizing yeast *Candida maltosa*. Dissertation, Moscow

Avetisova SM, Davidov ER (1993) Yeast cytochrome P450 substrate specificity and conformation of alkanes with different structure. 2nd Int Symp Cytochrome P450 Microorganisms Plants, Tokyo, June 13–17, 1993, Abstr, p 17

Avetisova SM, Sokolov YI, Kozlov VI, Davydov RM, Davidov ER (1985) The induction of cytochrome P-450 two forms in *Candida* yeast by n-alkanes of different chain length. In: Vereczky L, Magyar K (eds) Cytochrome P-450 – biochemistry, biophysics and induction. Akademiai Kiado, Budapest, pp 455–458

Avetisova SM, Popova LA, Davidov ER (1990) Two step induction of cytochrome P-450 in *Candida maltosa* yeast. Biocatalysis 4: 61

Avetisova SM, Popova LA, Davydov RM, Davidov ER (1993) Cytochrome b_5 from *Candida maltosa*: physico-chemical properties. 2nd Int Symp Cytochrome P450 Microorganisms Plants, Tokyo, June 13–17, 1993, Abstr, p 40

Babel W (1979) Bewertung von Substraten für das mikrobielle Wachstum auf der Grundlage ihres Kohlenstoff/Energie-Verhältnisses. Z Allg Mikrobiol 19: 671–677

Babel W (1980) Mischsubstratfermentationen – ein energetisch begründetes Konzept. Acta Biotechnol 0: 61–64

Babel W (1986) Theoretische Grundlagen des Auxiliarsubstratkonzeptes und seine praktischen Konsequenzen in biotechnischen Prozessen. Acta Biotechnol 6: 313–323

Baraji VN, Logatshova IA, Tsvigun IV, Truchatshova TV, Zinchenko AI, Shkumatov VM (1990) Patent Application USSR N1708845

Barnett JA, Payne RW, Yarrow D (1979) A guide to identifying and classifying yeasts. Cambridge University Press, Cambridge

Barnett JA, Payne RW, Yarrow D (1983) Yeasts: characteristics and identification. Cambridge University Press, Cambridge

Barns SM, Lane DJ, Sogin ML, Bibeau C, Weisburg WG (1991) Evolutionary relationships among pathogenic *Candida* species and relatives. J Bacteriol 173: 2250–2255

Bassel J, Phaff HJ, Mortimer RK, Miranda M (1978) Examination of hydrocarbon utilizing mutants of *Saccharomyces cerevisiae*. Int J Syst Bacteriol 28: 427–432

Bassel JB, Mortimer RK (1982) Genetic and biochemical studies on n-alkane non-utilizing mutants of *Saccharomycopsis lipolytica*. Curr Genet 5: 77–88

Bassel JB, Mortimer RK (1985) Identification of mutations preventing n-hexadecane uptake among 26 n-alkane non-utilizing mutants of *Yarrowia* (*Saccharomycopsis*) *lipolytica*. Curr Genet 9: 579–586

Bauch J, Koslova LI, Sobek K, Triems K, Meschtschankin GI, Roschkova MI (1978) Verfahren zur Gewinnung von "fermosin"-Futterhefe aus Erdöldestillaten. Chem Techn 30: 284–287

Bayer C, Iske U, Glombitza F, Nagel B (1985) Spektralphotometrische in-situ-Messungen der diffusen Reflexion bei der mikrobiellen Kohlenwasserstoffwandlung. Acta Biotechnol 5: 197–202

Becher D, Böttcher F (1983) The cell type of *Rhodosporidium toruloides* after protoplast fusion between strains of identical and opposite mating type. Curr Microbiol 9: 297–300

Becher D, Oliver SG (1995) Transformation of *Candida maltosa* by electroporation. Methods Mol Biol 47: 291–302

Becher D, Wedler H, Schulze H, Bode R, Kasüske A, Samsonova I (1991) Correlation of biochemical blocks and genetic lesions in leucine auxotrophic strains of the imperfect yeast *Candida maltosa*. Mol Gen Genet 227: 361–368

Becher D, Schulze S, Kasüske A, Schulze H, Samsonova IA, Oliver SG (1994) Molecular analysis of a *leu2(–)* mutant of *Candida maltosa* demonstrates the presence of multiple alleles. Curr Genet 26: 208–216

Becher D, Schulze S, Kasüske A, Stoll R, Wedler H, Oliver SG (1995) Chromosome polymorphism close to the *Cm-ADE1* locus of *Candida maltosa*. Mol Gen Genet 247: 591–602

Belov AP, Davidova EG (1982) Lipid granules as a compartment of lipid synthesis in the yeast cell (Russ). Mikrobiologiya 51: 302–307

Belov AP, Guselnikova TV (1988) The effect of peptides on phosphatidylinositol metabolism in *Candida* (Russ). Mikrobiologiya 57: 1042–1043

Belov AP, Toneva-Davidova EG (1983) Co²⁺ accumulation and intracellular distribution during yeast growth. In: Environmental regulation of microbial metabolism. FEMS Symposium, Pushchino, USSR, 1983, Abstr, p 202

Belov AP, Davidova EG, Rachinskii VV (1976) Isolation of vacuoles from *Candida tropicalis*. Mikrobiologiya 45: 852–858

Belov AP, Loginova TM, Tyurin VS, Gololobov AD (1983) The composition and localization on the cell wall of substances secreted by the *Candida guilliermondii* yeast cultivated on a hydrocarbon-containing medium. Prikl Biokhim Mikrobiol 19: 98–103

Belov AP, Davidova EG, Rachinskii VV (1985) Cobalt distribution studied in the cells of *Candida maltosa*. Mikrobiologiya 54: 970–973

Belov AP, Guselnikova TV, Gradova NB (1991) Adaptive changes in the nitrogen metabolism of yeasts due to consumption of peptides of a yeast autolysate (in Russian with English Translation). Appl Biochem Microbiol (Prikl Biokhim Mikrobiol) 26: 560–565

Bennetzen JL, Hall BD (1982) Codon selection in yeast. J Biol Chem 257: 3026–3031

Bergmann H, Voigt B, Seidel H, Meisgeier G (1987) Einfluß von Lipidfraktionen mikrobieller Herkunft auf die Wasserausnutzung in der biologischen Stoffproduktion von Kulturpflanzen. Acta Biotechnol 7: 201–206

Berner NH, Ahearn DG, Cook WL (1975) Effects of hydrocarbonoclastic yeasts on pollutant oil and the environment. In: Bourquin AW, Ahearn DG, Meyers SP (eds) Ecol Res Ser EPA-660/3-75001. US Environmental Protection Agency, Corvallis, pp 199–219

Bizzi A, Veneroni E, Tacconi MT, Codegoni AM, Pagani R, Cini M, Garattini S (1980) Accumulation and metabolism of uneven fatty acids present in single cell protein. Toxicol Lett 5: 227–240

Blasig R, Schunck W-H, Jockisch W, Franke P, Müller H-G (1984) Degradation of long-chain n-alkanes by the yeast *Lodderomyces elongisporus* I. Products of alkane oxidation in whole cells. Appl Microbiol Biotechnol 19: 241–246

Blasig R, Mauersberger S, Riege P, Schunck W-H, Jockisch W, Franke P, Müller H-G (1988) Degradation of long-chain n-alkanes by the yeast *Candida maltosa*. II. In vitro oxidation of n-alkanes and intermediates using microsomal membrane fractions. J Appl Microbiol Biotechnol 28: 589–597

Blasig R, Huth J, Franke P, Borneleit P, Schunck W-H, Müller H-G (1989) Degradation of long-chain n-alkanes by the yeast *Candida maltosa* III. Effect of solid n-alkanes on cellular fatty acid composition. J Appl Microbiol Biotechnol 31: 571–576

Bley T, Heinritz B, Steudel A, Stichel E, Glombitza F, Babel W (1980) Yield coefficients in dependence on milieu conditions and cell states. I. Synchronized batch growth of a yeast. Z Allg Mikrobiol 20: 283–286

Bode R (1991) Valine inhibition of beta-isopropylmalate dehydrogenase takes part in the regulation of leucine biosynthesis in *Candida maltosa*. Antonie Leeuwenhoek J Microbiol 60: 125–130

Bode R, Birnbaum D (1981) Aggregation and separability of the shikimate pathway enzymes in yeasts (Germ). Z Allg Mikrobiol 21: 417–422

Bode R, Birnbaum D (1984) Characterization of three aromatic amino transferases from *Candida maltosa* (Germ). Z Allg Mikrobiol 24: 67–75

Bode R, Birnbaum D (1986) Threonine dehydratase activity from several yeast species is activated and affected by phosphate. FEMS Microbiol Lett 37: 369–377

Bode R, Birnbaum D (1987) D-amino acid oxidase, aromatic L-amino aminotransferase, and aromatic lactate dehydrogenase from several yeast species: comparison of enzyme activities and enzyme specificities. Acta Biotechnol 7: 221–225

Bode R, Birnbaum D (1988) Purification and properties of two branched-chain amino acid aminotransferases from the yeast *Candida maltosa*. Biochem Physiol Pflanz 183: 417–424

Bode R, Birnbaum D (1989) Specificity of glyphosate action in *Candida maltosa*. Biochem Physiol Pflanz 184: 163–170

Bode R, Birnbaum D (1991a) Enzymatic production of indolepyruvate and some of its methyl and fluoro-derivatives. Acta Biotechnol 11: 387–393

Bode R, Birnbaum D (1991b) Regulation of chorismate mutase activity of various yeast species by aromatic amino acids. Antonie Leeuwenhoek J Microbiol 59: 9–13

Bode R, Birnbaum D (1991c) Some properties of the leucine-biosynthesizing enzymes from *Candida maltosa*. J Basic Microbiol 31: 21–26

Bode R, Casper P (1983) Allgemeine Kontrolle der Aminosäurebiosynthese in Mutanten von *Candida* sp. EH15/D. Z Allg Mikrobiol 23: 419–427

Bode R, Casper P, Kunze G (1983) Auslösung einer allgemeinen Kontrolle der Aminosäurebiosynthese bei *Candida* sp. EH15/D durch Amitrol. Biochem Physiol Pflanz 178: 457–468

Bode R, Melo C, Birnbaum D (1984a) Inhibition of tyrosine-sensitive 3-deoxy-D-arabinose-heptulosonate 7-phosphate synthase by glyphosate in *Candida maltosa*. FEMS Microbiol Lett 23: 7–10

Bode R, Melo C, Birnbaum D (1984b) Enzymological basis for glyphosate action in *Candida maltosa*. Biochem Physiol Pflanz 179: 775–783

Bode R, Melo C, Birnbaum D (1984c) Absolute dependence of phenylalanine and tyrosine biosynthetic enzyme on tryptophan in *Candida maltosa*. Hoppe-Seyler's Z Physiol Chem 365: 799–803

Bode R, Melo C, Birnbaum D (1984d) Mode of action of glyphosate in *Candida maltosa*. Arch Microbiol 140: 83–85

Bode R, Melo C, Birnbaum D (1985a) Regulation of tryptophan biosynthesis in the n-alkane-utilizing yeast *Candida maltosa*. Biochem Physiol Pflanz 180: 301–308

Bode R, Kunze G, Birnbaum D (1985b) Reversal of glyphosate-induced growth inhibition of *Candida maltosa* by several amino acids and pyruvate. Biochem Physiol Pflanz 180: 613–619

Bode R, Melo C, Birnbaum D (1985c) Regulatory properties of 3-deoxy-D-arabinose-heptulosonate-7-phosphate synthase isoenzymes from *Candida maltosa*. J Basic Microbiol 25: 3–11

Bode R, Melo C, Birnbaum D (1985d) Regulation of chorismate mutase, prephenate de-hydrogenase and prephenate dehydratase of *Candida maltosa*. J Basic Microbiol 25: 291–298

Bode R, Lippoldt A, Birnbaum D (1986a) Purification and properties of D-aromatic lactate dehydrogenase, an enzyme involved in the catabolism of aromatic amino acids of *Candida maltosa*. Biochem Physiol Pflanz 181: 189–198

Bode R, Schult I, Birnbaum D (1986b) Purification and some properties of threonine dehydratase from *Candida maltosa*. J Basic Microbiol 26: 443–451

Bode R, Schüssler K, Schmidt H, Hammer T, Birnbaum D (1990) Occurrence of the general control of amino acid biosynthesis in yeasts. J Basic Microbiol 30: 31–35

Bode R, Samsonova IA, Birnbaum D (1991) Production of α- and β-isopropylmalate by a mutant from *Candida maltosa*. Zentralbl Mikrobiol 146: 35–39

Bohlmann D, Bauch J, Kozlova LI, Meshankin GI, Roshkova MI, Triens K, Ringpfeil M, Sobek K (1979) Process for the production of "Fermosin^R" — fodder yeast from petroleum distillates. In: Dechema Monogr, vol 83, Microbiology applied to biotechnology. Verlag Chemie, Weinheim, pp 147–156

Bohlmann D, Bauch J, Gentzsch H, Dzingel G, Katrusch R, Kozlowa L, Roshkova M, Meschankin G (1982) Biosynthese von Eiweißstoffen durch mikrobiologische Entparaffinierung und Qualität der erhaltenen Produkte. Abh Akad Wiss DDR Abt Math Naturwiss Tech 2: 323–329

Bos P (1975) Some aspects of hydrocarbon assimilation by yeasts. Dissertation, Technical High School, Delft

Bos P, de Boer WE (1968) Some aspects of the utilization of hydrocarbons by yeasts. Antonie Leeuwenhoek J Microbiol 34: 241–243

Bos P, de Bruyn JC (1973) The significance of hydrocarbon assimilation in yeast identification. Antonie Leeuwenhoek J Microbiol 39: 99–107

Böttcher F (1987) Genetics of imperfect yeasts. 12th Int Spec Symp Yeast, Genet of Nonconventional Yeasts, Weimar, 1987, Abstr, p 3

Böttcher F, Samsonova IA (1978) *Rhodosporidium* BANNO: Dosiseffektbeziehungen, Mutageneffektivität und Mutantenspektrum bei der Induktion Auxotrophieverursachender Mutationen durch ultraviolettes Licht und N-Methyl-N'-nitro-N-nitrosoguanidin. Z Allg Mikrobiol 18: 637–646

Boulton CA, Ratledge C (1984) The physiology of hydrocarbon-utilizing microorganisms. In: Wiseman A (ed) Introduction to topics in enzyme and fermentation biotechnology, vol 9, Ellis Horwood, Chichester, pp 11–77

Bovina EV, Deriabin VV, Lange AV, Yarotsky SV (1986) Structure of mannan from the yeast *Candida maltosa* (Russ). Prikl Biokhim Mikrobiol 22: 679–683

Bovina EV, Deriabin VV, Gagloev VN, Serebriakov NG (1988) Study of the structure of mannans from *Candida maltosa* and *Candida tropicalis* using ^13C-NMR spectroscopy (Russ). Prikl Biokhim Mikrobiol 24: 218–225

Brada D, Schekman R (1988) Coincident localization of secretory and plasma membrane proteins in organelles of the yeast secretory pathway. J Bacteriol 170: 2775–2783

Brendler W, Bauch J, Lübbert GA, Wünsche L, Hedlich R, Triems K, Shdannikowa EN (1983) Spezielle Aspekte der nichtsterilen Hefeproduktion auf der Basis von Kohlenwasserstoffen. Acta Biotechnol 3: 351–356

Brown AJP, Bertram G, Feldmann PJF, Peggie MW, Swoboda RK (1991) Codon utilization in the pathogenic yeast, *Candida albicans*. Nucleic Acids Res 19: 4298

Brückner B, Tröger R (1981a) Vergleichende physiologische Untersuchungen zwischen *Candida* sp. H und der Mutante H 13 unter Stickstoffmangelbedingungen. Z Allg Mikrobiol 21: 19–26

Brückner B, Tröger R (1981b) Einfluß der Kohlenstoffquelle auf die Reservestoffbildung von *Candida* sp. H. Z Allg Mikrobiol 21: 77–84

Bruns TD, Vilgalys R, Barns SM, Gonzalez D, Hibbett DS, Lane DJ, Simon L, Stickel S, Szaro TM, Weisburg WG (1992) Evolutionary relationship within the fungi: analyses of nuclear small subunit rRNA sequences. Mol Phylogenet Evol 1: 231–241

Bühler M, Schindler J (1984) Aliphatic hydrocarbons. In: Rehm HJ, Reed G (eds) Biotechnology, vol 6a, Biotransformations (Kieslich K, vol ed), Verlag Chemie, Weinheim, pp 329–385

Büttner R, Uebel B, Genz I-L, Köhler M (1985) Wachstumsuntersuchungen im substratlimitierten pH-Auxostaten I. Bistabiles Wachstumsverhalten unter kaliumlimitierten Bedingungen. J Basic Microbiol 25: 227–232

Campbell I, Duffus JH (eds) (1988) Yeast, a practical approach. IRL Press, Oxford

Carle GF, Olson MV (1984) Separation of chromosomal DNA molecules from yeast by orthogonal-field-alternation gel electrophoresis. Nucleic Acid Res 12: 5647–5664

Casey J, Dobb R, Mycock G (1990) An effective technique for enrichment and isolation of _Candida cloacae_ mutants defective in alkane catabolism. J Gen Microbiol 136: 1197–1202

Casper P, Bode R, Birnbaum D (1985a) Regulation of ammonia assimilation in _Candida maltosa_ (Germ). J Basic Microbiol 25: 95–101

Casper P, Bode R, Samsonova IA, Birnbaum D (1985b) Glutamate/aspartate metabolism of _Candida maltosa_ (Germ). J Basic Microbiol 25: 637–643

Celma Calamita E, Arntz P, Bos P (1971) Obtaining protein concentrates using _Candida maltosa_ cultivated in gaseous n-octane (Ital). An Inst Nac Invest Agrar Ser Gen N 1: 165–177

Cerniglia CE (1981) Aromatic hydrocarbons: metabolism by bacteria, fungi and algae. Rev Biochem Toxicol 3: 321–360

Cerniglia CE, Crow SA (1981) Metabolism of aromatic hydrocarbons by yeast. Arch Microbiol 129: 9–13

Chang MC, Jung HD, Suzuki T, Takagi M, Yano K (1984) Ploidy in the asporogenous yeast _Candida maltosa_, isolation of its auxotrophic mutants and their cell fusion. J Gen Appl Microbiol 30: 489–497

Chu G, Vollrath D, Davis RW (1986) Separation of large DNA molecules by contour-clamped homogeneous electric fields. Science 234: 1582–1585

Claisse ML, Boze H, Dubreucq E, Segueilha L, Moulin G, Galzy P (1991) Characterization of alternative respiratory pathways in the yeast _Schwanniomyces castellii_ by the study of mutants deficient in cytochromes $a + a_3$ and/or b. Acta Biochim Pol 38: 365–392

Cook WL, Massey JK, Ahearn DG (1973) The degradation of crude oil by yeasts and its effect on _Lesbistes reticulatus_. In: Ahearn DG, Meyer SP (eds) The microbial degradation of oil pollutants. Louisiana State University, Center for Wetland Resources, Baton Rouge, pp 279–283

Cooper TG (1982) In: Strathern JN, Jones EW, Broach JB (eds) The molecular biology of the yeast _Saccharomyces_: metabolism and gene expression, vol 2, Cold Spring Harbor Laboratory Press, Cold Spring Harbor, pp 39–99

Crow SA, Bell SL, Ahearn DG (1979) Uptake of aromatic and branched chain hydrocarbons by yeasts. Bot Mar 22: 406

Crow SA, Bell SL, Ahearn DG (1980) Uptake of aromatic and branched chain hydrocarbons by yeasts. Bot Mar 23: 117–120

Cryer DR, Eccleshall R, Marmur J (1975) Isolation of yeast DNA. Methods Cell Biol 12: 39–44

Dalin MV, Gukasian IA, Spivak SM, Fish NG, Kravtsov EG, Ermolaev AV (1991) Approaches to development of diagnostic allergens for observation of workers engaged in the production of microbial fodder biomass and population of development zones and regions of microbiological plants (Russ). Gig Tr Prof Zabol 5: 31–33

Davidov ER, Gololobov AD (1980a) Regulation of metabolism in yeast during growth on n-alkanes. Proc jt. US/USSR Conf Mech Kinet Growth on Various Substr, Contr Simul Optim Microbiol Proc, Proj I-II. PB81.219131, pp 46–65. Natl Sci Found Res Appl Natl Needs, [Rep] NSF/RA (US) 1980, NSF/RA-800527

Davidov ER, Gololobov AD (1980b) Effect of pO_2 on the regulation of metabolism in yeasts during cultivation on n-alkanes. Proc jt. US/USSR Conf Mech Kinet Growth on Various Substr, Contr Simul Optim Microbiol Proc, Proj I-II. PB81.219131, pp 94–103. Natl Sci Found Res Appl Natl Needs, [Rep] NSF/RA (US) 1980, NSF/RA-800527

Davidov ER, Demanova NF, Sokolov YI, Gololobov AD (1980) Oxidation of individual isoalkanes and alkylaromatic hydrocarbons by yeasts of the genus _Candida_ (Russ). Prikl Biokhim Mikrobiol 16: 775–781

Davidov ER, Sokolov YI, Demanova NF, Gololobov AD (1981a) Utilization of 2-methyl hexadecane by the yeast Candida guilliermondii (Russ). Prikl Biokhim Mikrobiol 17: 328–341

Davidov ER, Sokolov YI, Demanova NF, Gololobov AD (1981b) Utilization of 3-methyl hexadecane by the yeast Candida guilliermondii (Russ) Prikl Biokhim Mikrobiol 17: 523–532

Davidov ER, Demanova NF, Sokolov YI, Gololobov AD (1982) Kinetics of hydrocarbon assimilation by yeast of the genus Candida (Russ). Acta Biotechnol 2: 213–225

Davidova EG, Rachinskii VV (1979) Uptake of n-alkanes by yeast cells (Review, in Russian). Uspechi Sovremennoi Biologii 88: 198–209

Davidova EG, Rachinskii VV (1981) Transport of liquid n-alkanes into the yeast cell determined by gas-liquid radiogaschromatography (Russ). Mikrobiologiya 50: 349–352

Davidova EG, Demanova NF, Gololobov AD, Rachinskii VV (1975) Isolation and characterization of the cell structures of Candida tropicalis (Russ). Mikrobiologiya 44: 621–624

Davidova EG, Belov AP, Rachinskii VV (1977a) Study of the role of lipid granules of a yeast cell in the assimilation of n-alkanes (Russ). Dokl Akad Nauk SSSR 235: 1189–1192

Davidova EG, Belov AP, Rachinskii VV (1977b) Isolation and characteristics of lipid granules from Candida tropicalis (Russ). Mikrobiologiya 46: 1044–1049

Davidova EG, Belov AP, Rachinskii VV (1979) Electrophoretic characteristics of the protein from yeast lipid granules (Russ). Mikrobiologiya 48: 803–808

Davidova EG, Zinchenko GA, Belov AP (1989) Substrate specificity of acyltransferases from lipid granules of mesophilic yeasts (Russ). Biokhimiya 54: 587–592

Davis R, Thomas M, Cameron J, John TS, Scherer S, Padgeff R (1980) Rapid DNA isolation for enzymatic and hybridization analysis. Methods Enzymol 65: 404–411

Delpozo L, Abarca D, Hoenicka J, Jimenez A (1993) Two different genes from Schwanniomyces occidentalis determine ribosomal resistance to cycloheximide. Eur J Biochem 213: 849–857

Demanova NF, Davidov ER, Gololobov AD (1980a) Oxidation of n-alkanes with different carbon chain lengths (in the range of C_{11}–C_{25}) by the Candida yeast (Russ). Prikl Biokhim Mikrobiol 16: 5–12

Demanova NF, Davidov ER, Gololobov AD (1980b) Oxidation of n-alkanes with different carbon chain lengths by Candida yeast (Russ). Prikl Biokhim Mikrobiol 16: 149–155

Demanova NF, Davidov ER, Gololobov AD (1980c) Yeast growth on mixture of n-docosane and n-octadecane during continuous cultivation (Russ). Prikl Biokhim Mikrobiol 16: 883–889

Deshler JO, Larson GP, Rossi JJ (1989) Kluyveromyces lactis maintains Saccharomyces cerevisiae intron-encoded splicing signals. Mol Cell Biol 9: 2208–2213

Dmitriev VV, Tsiomenko AB, Kulaev IS, Fikhte BA (1980) A cytochemical study of the "canal" formation in the yeast cell wall. Eur J Appl Microbiol Biotechnol 9: 211–216

Dole VP, Meinertz H (1960) Microdetermination of long-chain fatty acids in plasma and tissues. J Biol Chem 235: 2595–2599

Dolgikh MS, Kravtsov EG, Ermolaev AV (1990) Proteolytic activity of protein-producing yeast-like Candida fungi. Mikol Fitopatol 24: 229–235

Durasova EN, Mikhailova NP, Vyunov KA, Bakulev VM, Sokolov VN, Makhrosenkova MO, Khromov-Borisov NN (1986) The resistance of Candida guilliermondii to polyene antibiotics (Russ). Mikrobiologiya 55: 607–611

Durasova EN, Mikhailova NP, Sorokoletova EF, Vyunov KA (1989) The use of various mutagens for the induction of nystatin-resistant mutants in Candida maltosa (Russ). Mikrobiologiya 58: 760–763, Microbiology (NY) 58: 610–614 (English Translation)

Durasova EN, Mikhailova NP, Zhakovskaya ZA, Vyunov KA (1991) Sterol content of Candida maltosa strains with high resistance to nystatin (Russ). Mikol Fitopatol 25: 487–492

Eckart V, Cech D, Kammel K, Bauch J (1988) Die Gewinnung von Labor-, Fein- und Biochemikalien im VEB Petrolchemisches Kombinat Schwedt. Teil III: Ribonucleinsäure und RNA-Bausteine. Chem Techn 40: 432–434

Egorenkova GN, Belov AP (1984) Structural organization of the cell walls in yeasts of the genus *Candida* (Russ). Mikrobiologiya 53: 300–304, Microbiology (NY) 53: 241–245 (English Translation)

Einsele A (1983) Biomass from higher n-alkanes. In: Rehm H-J, Reed G (eds) Biotechnology, vol 3. Verlag Chemie, Weinheim, pp 43–81

Erickson AH, Blobel G (1983) Cell-free translation of messenger RNA in a wheat germ system. Meth Enzymol 96: 38–50

Ermolaev AV, Gukasyan IA, Ogarkov VI (1987) Isolation of surface antigens of *Candida maltosa* responsible for *Candida* sensitization and their immunochemical characteristics (Russ). Vopr Med Khim 33: 42–46

Ermolaev AV, Gukasyan IA, Parfenova EV, Spivak SM, Kravtsov EG (1991) Obtaining the allergens from yeast-like fungi of the genus *Candida*, producers of fodder protein, for hygienic standardization of the strains. Gig Tr Prof Zabol 8: 19–20

Faggi E, Mennini S (1985) Comparative studies of the pathogenicity of *Candida albicans*, *Candida utilis* and *Candida maltosa* in laboratory animals. Ann Microbiol Enzimol 35: 111–122

Feinberg B, McLaughlin CS (1988) Isolation of yeast mRNA and in vitro translation in a yeast cell-free system. In: Campbell I, Duffus JH (eds) Yeast, a practical approach, IRL Press, Oxford, pp 147–162

Fiechter A, Gmünder FK (1989) Metabolic control of glucose degradation in yeast and tumor cells. In: Fiechter A (ed) Advances in biochemical engineering/biotechnology, vol 39, Springer-Verlag Berlin Heidelberg, pp 2–28

Fiechter A, Käppeli O, Meussdoerffer F (1987) Batch and continuous culture. In: Rose AH, Harrison JS (eds) The Yeasts, vol 2, 2nd edn, Yeasts and the environment. Academic Press, London, pp 99–129

Fischer W, Reuter G (1982) Mannan-Lokalisation durch Concanavalin A im Zusammenhang mit elektronenmikroskopischen und chemisch-analytischen Untersuchungen an unterschiedlich präparierten Zellwänden der Futtereiweißhefe *Candida* sp. H. Z Allg Mikrobiol 22: 29–40

Fischer W, Brückner B, Meyer HW (1982) Ultrastructural alterations at the cell wall and the plasma membrane of *Candida* sp. H induced by n-alkane. Z Allg Mikrobiol 22: 227–236

Fröhlich K-U, Entian K-D, Mecke D (1985) The primary structure of the yeast hexokinase PII (*HXK2*) which is responsible for glucose repression. Gene 36: 105–111

Fukazawa Y, Nakase T, Shinoda T, Nishikawa A, Kagaya K, Tsuchiya T (1975) Significance of cell wall structures on yeast classification: proton magnetic resonance and serological and deoxyribonucleic acid characterization of *Candida sake* and related species. Int J Syst Bacteriol 25: 304–314

Fukui S, Tanaka A (1981a) Metabolism of alkanes by yeasts. Adv Biochem Eng 19: 217–237

Fukui S, Tanaka A (1981b) Production of useful compounds from alkane media in Japan. In: Fiechter A (ed) Products from alkanes, celluloses and other feedstocks. Akademie Verlag, Berlin, pp 1–36 (Fukui S, Tanaka A (1980) Adv Biochem Eng 17: 1–35)

Gargani G (1979) Models of pathogenicity for yeasts of the genus *Candida*. In: Garattini S, Paglialunga S, Scrimshaw NS (eds) Single cell protein: safety for animals and human feeding, Pergamon Press, Oxford, (1980) pp 30–38

Gargani G, Campisi E, Faggi E (1978) The problem of *Candida* virulence (Ital). Riv Ital Ig 38: 266–285

Gargani G, Campisi E, Faggi E (1979) Cross reactions between several species of the genus *Candida* demonstrated by intradermal inoculation of the guinea pig. Bull Soc Fr Mycol Med 8: 17–20

Glombitza F (1982) Der Einfluß der Flockenbildung auf die Versorgung der Hefezellen mit Sauerstoff bei der Fermentation flüssiger Kohlenwasserstoffe. Acta Biotechnol 2: 43–50

Glombitza F, Heinritz B (1979) Thermodynamik mikrobieller Prozesse. Z Allg Mikrobiol 19: 171–179

Goeddel D (ed) (1990) Methods enzymology vol 185, Gene expression technology, Section IV. Expression in yeast, Academic Press, London, pp 230–482

Golubev VI, Naumov GI, Bibikova II, Blagodatskaya VM, Voustin MM, Nikitina TN, Buzurg-Zade DL, Gradova NB (1986) A novel species assignment of the hydrocarbon digesting strains of the *Candida* genus yeast (Russ). Biotekhnologiya 0(5): 17–21

Gomi K, Horiguchi S (1988) Purification and characterization of pyrocatechase from the catechol assimilating yeast *Candida maltosa*. Agric Biol Chem 52: 585–587

Gradova NB, Kovalsky YV (1978) Production of fodder yeast cultures on media containing hydrocarbons (Russ). Mikrobiologiya 47: 259–264

Gradova NB, Osipova VG, Davidova EG, Chunaev AS, Kvitko KV (1976) Populational and phenogenetic analysis of variability *of Candida* yeast for the character of protein content in biomass (Russ). Genetics (USSR) 12: 80–88

Gradova NB, Dikanskaya EM, Robysheva ZN, Rodionova GS, Butteyeva MB, Zaikina AI (1983) Characterization of hydrocarbon oxidizing yeasts. Peculiarities of their growth and biosynthetical processes (Russ). Acta Biotechnol 3: 241–249

Gradova NB, Belov AP, Guselnikova TV (1990) Some aspects of the regulation of nitrogen metabolism in the yeast genus *Candida*. Study of the kinetics of ammonium transport to the cells of hydrocarbon-oxidizing yeast of the genus *Candida* with a change in nitrogen nutrition. Acta Biotechnol 10: 169–177

Gradova NB, Zaitsev SA, Gadzhieva VI (1991) Ecological monitoring of hydrocarbon-oxidizing *Candida* yeasts as a technogenic factor (Russ). Biotekhnologiya (Moscow) 0(2): 57–60

Griffiths G, Hoppeler H (1986) Quantitation in immunocytochemistry: correlation of immunogold labeling to absolute number of membrane antigens. J Histochem Cytochem 34: 1389–1398

Griffiths G, Brands R, Burke B, Louvard D, Warren G (1982) Viral membrane proteins acquire galactose in trans Golgi cisternae during intracellular transport. J Cell Biol 95: 781–792

Griffiths G, McDowall A, Back R, Dubochet J (1984) On the preparation of cryosections for immunocytochemistry. J Ultrastruct Res 89: 65–78

Grimmecke HD, Reuter G (1980) Struktur der Zellwandpolysaccharide in der Futtereiweiß-Hefe *Candida* sp. H. 5. Die komplexe Struktur des Proteophosphomannans. Biochem Physiol Pflanz 175: 781–788

Grimmecke HD, Reuter G (1981a) Struktur der Zellwandpolysaccharide in der Futtereiweiß-Hefe *Candida* sp. H. 1. Struktur des alkalistabilen Mannan-Proteins. Z Allg Mikrobiol 21: 95–107

Grimmecke HD, Reuter G (1981b) Struktur der Zellwandpolysaccharide in der Futtereiweiß-Hefe *Candida* sp. H. 2. Charakterisierung der Bindung des Phosphats am Mannan-Protein-Phosphat-Komplex und Identifizierung der als Phosphodiester gebundenen Mono- und Oligosaccharide. Z Allg Mikrobiol 21: 109–116

Grimmecke HD, Reuter G (1981c) Struktur der Zellwandpolysaccharide in der Futtereiweiß-Hefe *Candida* sp. H. 4. Struktur der alkalilabilen Oligosaccharide im Mannan-Protein-Phosphat-Komplex. Z Allg Mikrobiol 21: 211–218

Grimmecke HD, Reuter G (1981d) Struktur der Zellwandpolysaccharide in der Futtereiweiß-Hefe *Candida* sp. H. 6. Isolierung und Strukturaufklärung der Glucane. Z Allg Mikrobiol 21: 643–650

Grimmecke HD, Meyer H, Scheller D, Reuter G (1981) Struktur der Zellwandpolysaccharide in der Futtereiweiß-Hefe *Candida* sp. H. 3. Charakterisierung unterschiedlicher Phosphatbindungen im Mannan-Protein-Phosphat-Komplex. Z Allg Mikrobiol 21: 201–210

Gukasyan IA, Ermolaev AV, Kravtsov EG, Kacharmina VA (1990) The level of antigenic relationship of surface conjugates of the hydrocarbon-assimilating strains of yeast-like *Candida* fungi used in fodder protein production. Mikol Fitopatol 24: 420–424

Guselnikova TV, Pavlov AA, Bezrukov MG, Gradova NB (1988) Effect of thermal treatment on the fraction composition of yeast proteins (Russ). Biotekhnologiya 4: 509–511

Guselnikova TV, Belov AP, Gradova NB (1989) The effect of yeast autolysates on the level and distribution of free amino acids in yeast cells of the genus *Candida* (Russ). Mikrobiologiya 58: 202–205

Guselnikova TV, Gromov YA, Belov AP (1991) Influence of trophic cultivation conditions on the kinetics of methylamine transport in the yeast *Candida maltosa* (Russ). Mikrobiologiya 60: 232–237

Guthrie C, Fink GR (eds) (1991) Methods Enzymology, vol 194, Guide to yeast genetics and molecular biology. Academic Press, New York

Hagihara R, Mishina M, Tanaka A, Fukui S (1977) Utilization of pristane by a yeast *Candida lipolytica*. Fatty acid composition of pristane-grown cells. Agr Biol Chem 41: 1745–1748

Hammer T, Bode R, Schmidt H, Birnbaum D (1991) Distribution of three lysine-catabolizing enzymes in various yeast species. J Basic Microbiol 31: 43–49

Hann BC, Walter P (1991) The signal recognition particle in *S. cerevisiae*. Cell 67: 131–144

Hasegawa Y, Okamoto T, Obata H, Tokuyama T (1990) Utilization of aromatic compounds by *Trichosporon cutaneum* KUY-6A. J Ferment Bioeng 69: 122–124

Heinritz B, Bley T (1979) Einfluß alternierender Störungen auf die Verbrauchskennziffern beim Wachstum von Mikroorganismen. Z Allg Mikrobiol 19: 247–252

Heinritz B, Stichel E, Bley T, Rogge G, Glombitza F (1981) Yield coefficients in dependence on milieu conditions and cell states. II. Influence of perturbations on continuous cultivation of the yeast *Lodderomyces elongisporus* on hydrocarbons. Z Allg Mikrobiol 21: 581–586

Heinritz B, Stichel E, Rogge G, Bley T, Glombitza F (1982) Theoretische Bestimmung energetischer Wirkungsgrade der mikrobiellen Kohlenstoffsubstratwandlung und Vergleich mit experimentellen Werten an Phasenkulturen. Z Allg Mikrobiol 22: 534–544

Heinritz B, Stoll P, Glombitza F (1983a) Heat flow measurements in aerobic microbial growth processes with a nonisothermal calorimeter operating directly in the fermenter. Acta Biotechnol 3: 83–87

Heinritz B, Rogge G, Stichel E, Bley T (1983b) Use of biorhythmic processes for increasing the efficiency of carbon-compound conversion by microorganisms. Acta Biotechnol 3: 125–131

Heinritz B, Bley T, Ringpfeil M (1985) Einsatz von Hochleistungsreaktoren zur mikrobiologischen Stoffwandlung. Chem Technol 37: 514–516

Heinz T, Henning U, Wünsche J, Henk G (1989) The prececal and total intestinal nutrient digestibility and amino acid absorption of food yeasts in swine (Germ). Arch Tierernaehr 39: 1007–1019

Hendriks L, Goris A, Van de Peer Y, Neefs JM, Vancanneyt M (1991) Phylogenetic analysis of five medically important *Candida* species as deduced on the basis of small ribosomal subunit RNA sequences. J Gen Microbiol 137: 1223–1230

Hepler PK (1981) The structure of endoplasmic reticulum revealed by osmium tetroxide-potassium ferricyanide staining. Eur J Cell Biol 26: 102–110

Hikiji T, Ohkuma M, Takagi M, Yano K (1989) An improved host-vector system for *Candida maltosa* using a gene isolated from its genome that complements the *his5* mutation of *Saccharomyces cerevisiae*. Curr Genet 16: 261–266

Hill DE, Boulay R, Rogers D (1988) Complete nucleotide sequence of the peroxisomal acyl CoA oxidase from the alkane-utilizing yeast *Candida maltosa*. Nucleic Acids Res 16: 365–366

Hino A, Wongkhalaung C, Kawai S, Murao S, Yano K, Takano H, Takagi M (1992) Construction of a transformation system for a freeze-tolerant yeast *Kluyveromyces thermotolerans*. Agric Biol Chem 56: 228–232

Hirata T, Ishitani T (1978) Studies on the discrimination of SCP-related yeast by proton magnetic resonance spectroscopy: structural changes in cell wall mannan of *Candida subtropicalis* grown in different media. Agric Biol Chem 42: 775–780

Hofmann KH (1986a) Microbial transformation of polycyclic aromatic hydrocarbons (Germ). Wiss Z EMA Univ Greifswald, Math Nat Reihe 35: 23–26

Hofmann KH (1986b) Oxidation of naphthalene by *Saccharomyces cerevisiae* and *Candida utilis*. J Basic Microbiol 26: 109–111

Hofmann KH, Krüger AK (1985) Induction and inactivation of phenol hydroxylase and catechol oxygenase in *Candida maltosa* L4 in dependence on the carbon source. J Basic Microbiol 25: 373–379

Hofmann KH, Polnisch E (1990a) Activities of gluconeogenic enzymes in the yeast *Candida maltosa* during growth on glucose or ethanol (Germ). J Basic Microbiol 30: 333–336

Hofmann KH, Polnisch E (1990b) Cyclic AMP-dependent phosphorylation of fructose-1,6-bisphosphate and other proteins in the yeast *Candida maltosa*. J Basic Microbiol 30: 555–559

Hofmann KH, Polnisch E (1990c) Characterization of a mutant of the yeast *Candida maltosa* defective in catabolite inactivation of gluconeogenic enzymes. Arch Microbiol 154: 514–517

Hofmann KH, Schauer F (1988) Utilization of phenol by hydrocarbon assimilating yeasts. Antonie Leeuwenhoek J Microbiol 54: 179–188

Hofmann KH, Vogt U (1987) Induction of phenol assimilation in chemostat cultures of *Candida maltosa* L4. J Basic Microbiol 27: 441–447

Hofmann KH, Vogt U (1988) Degradation of phenol by yeasts in the presence of n-hexadecane under growth conditions in a stirred reactor (Germ). Zentralbl Mikrobiol 143: 87–91

Holley RW (1967) Isolation of sRNA from intact yeast cells. Methods Enzymol 12: 596–598

Holzschu DL, Chandler FW, Ajello L, Ahearn DG (1979) Evaluation of industrial yeasts for pathogenicity. Sabouraudia 17: 71–78

Honeck H, Schunck W-H, Riege P, Müller H-G (1982) The cytochrome P-450 alkane monooxygenase system of the yeast *Lodderomyces elongisporus*: purification and some properties of the NADPH-cytochrome P-450 reductase. Biochem Biophys Res Commun 106: 1318–1324

Honeck H, Schunck W-H, Müller H-G (1985) The function of cytochrome P-450 in fungi and prospects of application. Pharmazie 40: 221–227

Huth J (1987) Über die Verwertung von n-Alkanen durch *Candida maltosa* EH15D unter besonderer Berücksichtigung der festen n-Alkane mit 20 und mehr C-Atomen. Dissertation, Akademie der Wissenschaften der DDR, Berlin

Huth J, Blasig R, Werner S, Müller H-G (1990a) The proton extrusion of growing yeast cultures as an on-line parameter in fermentation processes: determination of biomass production and substrate consumption in batch experiments with *Candida maltosa* EH15 D. J Basic Microbiol 30: 481–488

Huth J, Werner S, Müller H-G (1990b) The proton extrusion of growing yeast cultures as an on-line parameter in fermentation processes: quantitative determination of growth from milligram amounts of substrate in a minimized fed-batch fermentation apparatus. J Basic Microbiol 30: 489–497

Huth J, Werner S, Müller H-G (1990c) The proton extrusion of growing yeast cultures as an on-line parameter in fermentation processes: ammonia assimilation and proton extrusion are correlated by an 1:1 stoichiometry in nitrogen-limited fed-batch fermentations. J Basic Microbiol 30: 561–567

Hwang CW, Yano K, Takagi M (1991) Sequences of two tandem genes regulated by carbon sources, one being essential for n-alkane assimilation in *Candida maltosa*. Gene 106: 61–69

Ilchenko AP, Tsfasman IM (1987) Isolation and characterization of aldehyde dehydrogenase from *Torulopsis candida* yeast grown on hexadecane (Russ). Biokhimiya 52: 58–65

Ilchenko AP, Tsfasman IM (1988) Isolation and characterization of alcohol oxidase for higher alcohols of the yeast *Torulopsis candida* grown on hexadecane (Russ). Biokhimiya 53: 263–271

Ilchenko AP, Mauersberger S, Matyashova RN, Losinov AB (1980) Induction of cytochrome P-450 in the course of yeast growth on different substrates (Russ). Mikrobiologiya 49: 452–458

Ilchenko AP, Shilova NK, Matyashova RN, Galynkin VA (1989) Effect exerted by the concentration of oxygen dissolved in the medium on the biosynthesis of cytochromes by *Candida maltosa* cells in the course of their growth on paraffins (Russ). Mikrobiologiya 58: 716–722

Ilchenko AP, Vasilkova NN, Matyashova RN (1991) Changes in the activity of enzymes utilizing H_2O_2 under different conditions of yeast cultivation (Russ). Mikrobiologiya 60: 55–64

Ilchenko AP, Morgunov IG, Honeck H, Mauersberger S, Vasilkova NN, Müller H-G (1994) Purification and some properties of alcohol oxidase from the yeast *Yarrowia lipolytica* H222 (Russ). Biokhimiya (Moscow) 59: 1312-1319

Ilyina VI, Dalin MV, Gukasyan IA, Tikhomirov YG, Mokeeva NV (1988) Testing the usefulness of an erythrocyte immunoglobulin diagnostic agent for assessing the levels of the protein paprin in the air (Russ). Gig Sanit 3: 38–40

Ioffe ML, Maksimova GN, Tsygankova NV, Zhutchkov VN (1990) Investigation on biological action of denucleinized product obtained from paprin. Biotekhnologiya (Soviet Biotechnology) 0(6): 70–72

Ito H, Fukuda Y, Murata K, Kimura A (1983) Transformation of intact yeast cells treated with alkali cations. J Bacteriol 153: 163–168

Jomantiene R, Januska A, Sasnauskas K, Janulaitis A (1987) Cloning of *ADE1, ADE2, ARG4* genes and *ARS* sequence of the yeast *Candida maltosa*. 12th Int Spec Symp Genet of Non-conventional Yeasts, Weimar, GDR, 1987, Abstr, p 41

Jomantiene R, Geneviciute E, Januska A, Lebedys J, Sasnauskas K (1991) *ADE1* gene of the yeast *Candida maltosa*. Eksp Biol 0(3): 19–29 (published 1992)

Jomantiene R, Lebediene E, Proscevicius J, Meskauskiene R, Sasnauskas K (1992) Molecular analysis of the *Candida maltosa* gene, conferring resistance to cycloheximide in *Saccharomyces cerevisiae*. Int Spec Symp Genet Mol Biol Non-conventional Yeasts, Leuenberg near Basel, Switzerland 1992, Abstr

Jones EW, Pringle JR, Broach JR (eds) The molecular and cellular biology of the yeast *Saccharomyces*, vol 1 – Genome dynamics, protein synthesis, and energetics (1991), vol 2 – Gene expression (1992), vol 3 – Cell cycle and cell biology (1993), Cold Spring Harbor Laboratory Press, Cold Spring Harbor

Kalin M, Neujahr HY, Weissmahr RN, Sejlitz T, Johl R, Fiechter A, Reiser J (1992) Phenol hydroxylase from *Trichosporon cutaneum*: gene cloning, sequence analysis, and functional expression in *Escherichia coli*. J Bacteriol 174: 7112–7120

Kamiryo T, Sakasegawa Y, Tan H (1989) Expression and transport of *Candida tropicalis* peroxisomal acyl-coenzyme A oxidase in the yeast *Candida maltosa*. Agric Biol Chem 53: 179–186

Kamiryo T, Mito N, Nike T, Suzuki T (1991) Assignment of most genes encoding major peroxisomal polypeptides to chromosomal band V of the asporogenic yeast *Candida tropicalis*. Yeast 7: 503–511

Kaneko T, Ishii K, Kawaharada H, Kagotani K, Shimada Y, Watanabe K (1977) Taxonomic studies on a hydrocarbon-assimilating *Candida* strain. Agric Biol Chem 41: 2269–2276

Käppeli O, Fiechter A (1976) The mode of interaction between the substrate and cell surface of hydrocarbon-utilizing yeast *Candida tropicalis*. Biotechnol Bioeng 18: 967–974

Käppeli O, Fiechter A (1977) Component from the cell surface of the hydrocarbon-utilizing yeast *Candida tropicalis* with possible relation to hydrocarbon transport. J Bacteriol 131: 917–921

Käppeli O, Aeschbach H, Schneider AH, Fiechter A (1975) A comparative study of carbon energy reserve metabolism of *Candida tropicalis* growing on glucose and on hydrocarbons. Eur J Appl Microbiol 1: 199–211

Käppeli O, Müller M, Fiechter A (1978) Chemical and structural alterations at the cell surface of *Candida tropicalis*, induced by hydrocarbon substrate. J Bacteriol 133: 952–958

Käppeli O, Walther P, Müller M, Fiechter A (1984) Structure of the cell surface of the yeast *Candida tropicalis* and its relation to hydrocarbon transport. Arch Microbiol 138: 279–282

Kärgel E, Schmidt HE, Schunck W-H, Riege P, Mauersberger S, Müller H-G (1984) A solid-phase radioimmunoassay for yeast cytochrome P-450. Anal Lett 17 B18: 2011–2024

Kärgel E, Schunck W-H, Riege P, Honeck E, Claus R, Kleber H-P, Müller H-G (1985) A comparative immunological investigation of the alkane-hydroxylating cytochrome P-450 from the yeast *Candida maltosa*. Biochem Biophys Res Commun 128: 1261–1267

Kärgel E, Aoyama Y, Schunck W-H, Müller H-G, Yoshida Y (1990) Comparative study on cytochrome P-450 of yeasts using specific antibodies to cytochromes P-450alk and P-45014DM. Yeast 6: 61–67

Kärgel E, Menzel R, Honeck H, Vogel F, Böhmer A, Schunck W-H (1996) *Candida maltosa* NADPH-cytochrome P450 reductase: Cloning of a full-length cDNA, heterologous expression in *Saccharomyces cerevisiae* and function of the N-terminal region for membrane anchoring and proliferation of the endoplasmic reticulum. Yeast 12 (in press)

Karpova TS, Zhuravleva TS, Pashina OB, Nikolaishvili NT, Larionov VL (1987) Chromosome stability in *Saccharomyces* yeasts (Russ). Genetika 23: 2148–2156

Kasanzev EN, Maximova GN, Shekina EV, Vorobyeva GI (1975) Determination of the relative volumes of lipid inclusions of the yeast *Candida guilliermondii* NP4 grown on hydrocarbons (Russ). Appl Biokhim Mikrobiol 11: 640–648

Kasüske A, Wedler H, Schulze S, Becher D (1992) Efficient electropulse transformation of intact *Candida maltosa* cells by different homologous vector plasmids. Yeast 8: 691–697

Kawaguchi Y, Honda H, Taniguchi-Morimura H, Iwasaki S (1989) The codon CUG is read as serine in an asporogenic yeast *Candida cylindracea*. Nature 341: 164–166

Kawai S, Hwang CW, Sugimoto M, Takagi M, Yano K (1987) Subcloning and nucleotide sequencing of an *ARS* site of *Candida maltosa* which also functions in *Saccharomyces cerevisiae*. Agric Biol Chem 51: 1587–1591

Kawai S, Hikiji T, Murao S, Takagi M, Yano K (1991) Isolation and sequencing of a gene, *C-ADE1*, and its use for a host-vector system in *Candida maltosa* with two genetic markers. Agric Biol Chem 55: 59–66

Kawai S, Murao S, Mochizuki M, Shibuya I, Yano K, Takagi M (1992) Drastic alteration of cycloheximide sensitivity by substitution of one amino acid in the L41 ribosomal protein of yeasts. J Bacteriol 174: 254–262

Kawamura M, Takagi M, Yano K (1983) Cloning of a *LEU* gene and an *ARS* site of *Candida maltosa*. Gene 24: 157–162

Keszenman-Pereyra D, Hieda K (1988) A colony procedure for transformation of *Saccharomyces cerevisiae*. Curr Genet 13: 21–23

Kirsch DR, Kelly R, Kurtz MB (eds) (1990) The genetics of *Candida*. CRC Press, Boca Raton

Kitamura H, Anri A, Fuse K, Seo M, Itakura C (1990) Chronic mastitis caused by *Candida maltosa* in a cow. Vet Pathol 27: 465–466

Klinner U, Böttcher F (1985) Chromosomal rearrangements after protoplast fusion in the yeast *Candida maltosa*. Curr Genet 9: 619–621

Klinner U, Böttcher F (1987) Protoplast fusion as tool for genetic analysis and manipulation. 12th Int Spec Symp Genet of Non-conventional Yeasts, Weimar, 1987, Abstr, p 4

Klinner U, Samsonova IA, Böttcher F (1984) Genetic analysis of the yeast *Candida maltosa* by means of induced parasexual processes. Curr Microbiol 11: 241–246

Kölblin R, Birkenbeil S (1981) Zusammenhang zwischen Koloniemorphologie und Polysaccharidgehalt in zellwandmodifizierten Mutanten von *Candida* sp. "H". Z Allg Mikrobiol 21: 519–530

Kölblin R, Tröger R (1982) Zusammenhang zwischen Protein- und Polysaccharidgehalt in zellwandmodifizierten Mutanten von *Candida* sp. H. Z Allg Mikrobiol 22: 63–68

Komagata K (1979) Characteristics of *Candida maltosa*. In: Garattini S, Paglialunga S,

Scrimshaw NS (eds) Single-cell protein: safety for animal and human feeding. Proc protein-calorie advisory group of the United Nations System Symp, Milan, Italy, March 31-April 1, 1977. Pergamon Press, Oxford, pp 39–43

Komagata K, Nakase T, Katsuya N (1964a) Assimilation of hydrocarbons by yeast. I. Preliminary screening. J Gen Appl Microbiol 10: 313–321

Komagata K, Nakase T, Katsuya N (1964b) Assimilation of hydrocarbons by yeast. II. Determination of hydrocarbon-assimilating yeast. J Gen Appl Microbiol 10: 323–333

König WA (1987) The practice of enantiomer separation by capillary gas chromatography. Hüthig-Verlag, Heidelberg, p 42

Kostov V, Ratchev R, Lazarova G, Russeva L, Krasteva J, Ivanova V, Vassileva M, Sokolov T, Jelev S (1991) Yeast assimilation of sugars from hemicellulose beech wood hydrolysates. Acta Microbiol Bulg 28: 51–61

Kovalenko OG, Korobko OP, Korbelainen ES, Barkalova AO, Telegeeva TA, Papp VT (1992) Effect of mannan from *Candida maltosa* and its sulphated derivatives on plant susceptibility to viral and bacterial infections (Russ). Mikrobiol Zh (Kiev) 54: 63–69

Kozlova LI, Meshchankin GI (1991) Production technology of fodder yeasts on oil distillates. Biotekhnologiya 0(6): 60–63

Krauzova VI, Sharyshev AA (1987) Study on subcellular distribution of the enzymes of n-alkane oxidation primary steps in the yeast *Candida maltosa* (Russ). Biokhimiya 52: 599–606

Krauzova VI, Kuvichkina TN, Sharyshev AA, Romanova IB, Lozinov AB (1986) Lauric acid and NADH synthesis during dodecanol and dodecanal oxidation by membrane fractions of the yeast *Candida maltosa* grown on hexadecane (Russ). Biokhimiya 51: 23–27

Kravtsov EG, Gukasyan IA, Dolgikh MS, Ermolaev AV, Spivak SM (1991) Isolation of antigen-active biopolymers from *Candida maltosa* culture fluid for obtaining allergens of diagnostic value for examination of industrial microbiology workers (Russ). Gig Tr Prof Zabol 3: 33–34

Kreger-van Rij NJW (1984) The yeasts, a taxonomic study. Elsevier, Amsterdam

Krug M, Straube G (1986) Degradation of phenolic compounds by the yeast *Candida tropicalis* HP15 II. Some properties of the first two enzymes of the degradation pathway. J Basic Microbiol 26: 271–281

Krug M, Ziegler H, Straube G (1985) Degradation of phenolic compounds by the yeast *Candida tropicalis* HP15 I. Physiology of growth and substrate utilization. J Basic Microbiol 25: 103–110

Kunau W-H, Hartig A (1992) Peroxisome biogenesis in *Saccharomyces cerevisiae*. Antonie Leeuwenhoek J Microbiol 62: 63–78

Kunau W-H, Bühne S, Moreno de la Garza M, Kionka C, Mateblowski M, Schultz-Borchard U, Thieringer R (1988) Comparative enzymology of β-oxidation. Biochem Soc Trans 16: 418–420

Kunze G (1982) Molekularbiologisch/biochemische Charakterisierung der Genome von *Candida* sp. EH15, *Lodderomyces elongisporus* CBS 2605, *Saccharomyces cerevisiae* D10, *Pichia guilliermondii* S0809, *Pichia guilliermondii* S0799 und *Pichia guilliermondii fp* 1–61. Universität Greifswald, Math Nat Dissertation A, Greifswald

Kunze G, Hecker M, Birnbaum D (1984a) Molecularbiological characterization of genomes from *Candida* sp. EH 15, *Lodderomyces elongisporus* CBS 2605, *Pichia guilliermondii* SO809 and *Pichia guilliermondii* fp1–61 (Germ). Z Allg Mikrobiol 24: 33–40

Kunze G, Schauer F, Samsonova I, Klinner U, Bode R, Hecker M, Birnbaum D (1984b) Identifizierung zweier *Candida maltosa*-Stämme mittels DNA-Reassoziation. Z Allg Mikrobiol 24: 607–613

Kunze G, Petzoldt C, Bode R, Samsonova I, Hecker M, Birnbaum D (1985a) Transformation of *Candida maltosa* and *Pichia guilliermondii* by a plasmid containing *Saccharomyces cerevisiae* ARG4 DNA. Curr Genet 9: 205–209

Kunze G, Petzoldt C, Bode R, Samsonova IA, Böttcher F, Birnbaum D (1985b) Transforma-

566   S. Mauersberger et al.

tion of the industrially important yeasts *Candida maltosa* and *Pichia guilliermondii*. J Basic Microbiol 25: 141–144

Kunze G, Petzoldt G, Bode R, Samsonova JA, Hecker M, Birnbaum D (1986a) Transformations of the industrially important yeasts *Candida maltosa* and *Pichia guilliermondii*. Acta Biotechnol 6: 28

Kunze G, Bode R, Birnbaum D (1986b) Physical mapping and genome organization of mitochondrial DNA from *Candida maltosa*. Curr Genet 10: 527–530

Kunze G, Bode R, Schmidt H, Samsonova IA, Birnbaum D (1987a) Identification of a *lys2* mutant of *Candida maltosa* by means of transformation. Curr Genet 11: 385–391

Kunze G, Bode R, Schmidt H, Samsonova IA, Birnbaum D (1987b) Identification of a *lys2* mutant of *Candida maltosa* by means of tranformation. 12th Int Spec Symp Yeast Genet of Non-conventional Yeasts, Weimar, 1987, Abstr p 81

Kurtz MB, Kelly R, Kirsch DR (1990) The molecular genetics of *Candida albicans*. In: Kirsch DR, Kelly R, Kurtz MB (eds) The genetics of *Candida*. CRC Press, Boca Raton, pp 21–73

Kurtzman CP (1992) rRNA sequence comparison for assessing phylogenetic relationship among yeasts (Minireview). Int J Syst Bacteriol 42: 1–6

Kurtzman CP (1994) Molecular taxonomy of the yeasts. Yeast 10: 1727–1740

Larriba G (1993) Translocation of proteins across the membrane of the endoplasmic reticulum: a place for *Saccharomyces cerevisiae*. Yeast 9: 441–463

Lebediene E, Jomantiene R, Sasnauskas K (1992) Cloning and sequence analysis of a *Candida maltosa* gene conferring resistance to formaldehyde. Int Spec Symp Genet Mol Biol Non-conventional Yeasts, Leuenberg near Basel, 1992, Abstr

Lerche K-H, Kretzschmar G (1980) Zellelektrophoretische Charakterisierung der Oberfläche von *Candida guilliermondii*. Z Allg Mikrobiol 20: 641–652

Lerche K-H, Kretzschmar G (1986) Partikelelektrophoretische Charakterisierung der Oberflächeneigenschaften von alkanutilisierenden Hefezellen: chemische Zusammensetzung und Tensidadsorption. Acta Biotechnol 6: 221–231

Leuker CE, Hahn H-M, Ernst JF (1992) β-Galactosidase of *Kluyveromyces lactis* (Lac4p) as a reporter of the gene expression in *Candida albicans* and *Candida tropicalis*. Mol Gen Genet 235: 235–241

Levi JD, Shennan L, Ebbon GP (1979) Biomass from liquid n-alkanes. In: Rose AH (ed) Microbial biomass. Academic Press, London, pp 361–419

Lippoldt A, Bode R, Birnbaum D (1986) Degradation of aromatic amino acids in *Candida maltosa*. J Basic Microbiol 26: 145–154

Litovskaya AV (1988) Immune response of persons exposed to protein-synthesizing fungi (Russ). Z Mikrobiol Epidemiol Immunobiol 2: 71–75

Litovskaya AV, Mokeeva NV (1990) Comparative evaluation of efficiency of various immunologic reactions with *Candida* antigens in detecting immediate hypersensitivity (Russ). Z Mikrobiol Epidemiol Immunobiol 9: 89–93

Lloyd AT, Sharp PM (1992) Evolution of codon usage patterns: the extent and nature of divergence between *Candida albicans* and *Saccharomyces cerevisiae*. Nucleic Acids Res 20: 5289–5291

Loper JC, Chen C, Dey CR (1985) Gene engineering of yeast for biodegradation: immunological cross-reactivity among cytochrome P450 systems proteins of *Saccharomyces cerevisiae* and *Candida tropicalis*. Hazardous Waste Hazardous Mat 2: 131–141

Lopez MC, Nicaud JM, Skinner H, Vergnolles C, Kader JC, Bankaitis V, Gaillardin C (1994) A phosphatidylinositol/phosphatidylcholine transfer protein is required for differentiation of the dimorphic yeast *Yarrowia lipolytica* from the yeast to the mycelial form. J Cell Biol 124: 113–127

Lottermoser K, Asperger O, Schunck W-H (1994) Polymerase chain reaction mediated detection of cytochrome P450 gene in the yeast *Candida apicola*. In: Lechner MC (ed) Cytochrome P450. Biochemistry, biophysics and molecular biology. John Libbey Eurotext, Paris, pp 643–646

Ludvik J, Munk V, Dostalek M (1968) Ultrastructural changes in the yeast *Candida lipolytica* caused by penetration of hydrocarbons into the cell. Experentia 24: 1066–1068

Lusky K, Stoyke M, Gobel R, Busch A, Ackermann H (1988) The effect of microbial protein, obtained on a hydrocarbon base (fermosin), with a defined fatty acid composition on fat metabolism and fat composition in slaughter animals. 1. The effect of fermosin on the composition of broiler depot fat (Germ). Nahrung 32: 627–633

Lusky K, Stoyke M, Gobel R, Doberschütz KD, Macholz R (1989) The effect of microbial protein from a hydrocarbon base (fermosin) with a defined fatty acid composition on fat metabolism and fat composition in slaughter animals. 2. The effect of "fermosin" on the composition of back fat in hogs (Germ). Nahrung 33: 203–212

Magasanik B (1992) Regulation of nitrogen utilization. In: Jones EW, Pringle JR, Broach JB (eds) The molecular and cellular biology of the yeast *Saccharomyces*: gene expression, vol II. Cold Spring Harbor Laboratory Press, Cold Spring Harbor, pp 283–317

Magee BB, Magee PT (1987) Electrophoretic karyotypes and chromosome numbers in *Candida* species. J Gen Microbiol 133: 425–430

Magee PT, Rikkerink EH, Magee BB (1988) Methods for the genetics and molecular biology of *Candida albicans*. Anal Biochem 175: 361–372

Maksimova GN, Berestennikova ND, Antokhina VI, Levandovskaya YB, Pozmogova IN (1988) Content of lipid inclusions, free and bound lipids in the *Candida maltosa* cells at different paraffin concentrations in the culture medium. Appl Biochem Microbiol (Moscow) 24: 549–553

Manakov MN, Prishepov FA (1986) The kinetics of monosaccharide digestion by the yeast of the genus *Candida* (Russ). Biotekhnologiya 0(2): 13–18

Marahrens Y, Stillman B (1992) A yeast chromosomal origin of DNA replication defined by multiple functional elements. Science 255: 817–823

Maraz A, Kiss M, Ferency L (1978) Protoplast fusion in *Saccharomyces cerevisiae* strains of identical and opposite mating type. FEMS Microbiol Lett 3: 319–322

Masuda Y, Park SM, Ohkuma M, Ohta A, Takagi M (1994) Expression of an endogenous and a heterologous gene in *Candida maltosa* by using a promoter of a newly isolated phosphoglycerate kinase (PGK) gene. Curr Genet 25: 412–417

Mauersberger S (1985) Regulation und subzelluläre Verteilung des Cytochrom P-450 Monooxygenasesystems und anderer am Alkanmetabolismus beteiligter Enzyme in *Candida* Hefen. Dissertation, Akademie der Wissenschaften der DDR, Berlin

Mauersberger S (1991) Mutants of alkane oxidation in the yeasts *Yarrowia lipolytica* and *Candida maltosa*. In: Finogenova TV, Sharyshev AA (eds) Alkane metabolism and oversynthesis of metabolites by microorganisms. Center for Biological Research USSR Academy of Sciences, Pushchino, pp 59–78

Mauersberger S, Matyashova RN (1980) The content of cytochrome P-450 in yeast cells growing on hexadecane (Russ). Mikrobiologiya 49: 571–577

Mauersberger S, Matyashova RN, Müller H-G, Losinov AB (1980) Influence of the growth substrate and the oxygen concentration in the medium on the cytochrome P-450 content in *Candida guilliermondii*. Eur J Appl Microbiol Biotechnol 9: 285–294

Mauersberger S, Schunck W-H, Müller H-G (1981) The induction of cytochrome P-450 in *Lodderomyces elongisporus*. Z Allg Mikrobiol 21: 313–321

Mauersberger S, Schunck W-H, Müller H-G (1984) The induction of cytochrome P-450 in the alkane-utilizing yeast *Lodderomyces elongisporus*: alterations in the microsomal membrane fraction. Appl Microbiol Biotechnol 19: 29–35

Mauersberger S, Kärgel E, Matyashova RN, Müller H-G (1987) Subcellular organization of alkane oxidation in the yeast *Candida maltosa*. J Basic Microbiol 27: 565–582

Mauersberger S, Böhmer A, Schunck W-H, Müller H-G (1991) Cytochrome P-450 of the yeast *Yarrowia lipolytica*. Int Conf Biochemistry, Biophysics of Cytochrome P-450: Structure, Function, Biotechnological and Ecological Aspects, Moscow 1991, Abstr

Mauersberger S, Persiyanova TB, Avetisova SM, Sokolov YI, Kärgel E, Kraft R, Schunck W-H, Davidov ER, Müller H-G (1992a) Characterization of two cytochrome P-450 forms purified from the yeast *Candida maltosa*. In: Archakov AI, Bachmanova GI (eds) Cytochrome P-450: biochemistry and biophsics. INCO – TNC, Joint Stock Company Moscow, pp 651–653

Mauersberger S, Drechsler H, Oehme G, Müller H-G (1992b) Substrate specificity and stereoselectivity of the fatty alcohol oxidase from the yeast *Candida maltosa*. Appl Microbiol Biotechnol 37: 66–73

Maximova GN, Vorobyova GI, Grigoryeva SP (1972) The question of hydrocarbon localization in the yeast cells of *Candida guilliermondii* NP-4 grown in media with paraffins (Russ). Prikl Biokhim Mikrobiol 8: 197–206

Meissel MN, Medvedeva GA, Kozlova TM (1976) Cytological mechanisms of alkane assimilation by yeast (Russ). Mikrobiologiya 45: 844–851

Menzel R, Scheller U, Schunck W-H, Müller H-G (1992) Inducible high-level expression of cytochromes P-450 *CYP52A3* and *CYP52A4* from *Candida maltosa* in *Saccharomyces cerevisiae*. In: Archakov AI, Bachmanova GI (eds) Cytochrome P-450: biochemistry and biophysics, INCO – TNC, Joint Stock Company, Moscow, pp 654–656

Menzel R, Kärgel E, Wolff C, Vogel F, Schunck W-H (1994) High level expression of integral membrane proteins induces proliferation of the endoplasmic reticulum. In: Lechner MC (ed) Cytochrome P450. Biochemistry, biophysics and molecular biology, John Libbey Eurotext, Paris, pp 307–310

Menzel R, Kärgel E, Vogel F, Böttcher C, Schunck W-H (1996) Membrane integration of cytochrome P450 and ER-proliferation are related processes. (submitted)

Metz W, Reuter G (1977) Anabole und katabole Enzyme des Harnstoffmetabolismus in einem kohlenwasserstoffverwertenden Stamm von *Candida guilliermondii*. Z Allg Mikrobiol 17: 599–610

Meyer SA, Anderson K, Brown RE, Smith MT, Yarrow D, Mitchell G, Ahearn DG (1975) Physiological and DNA characterization of *Candida maltosa*, a hydrocarbon-utilizing yeast. Arch Microbiol 104: 225–231

Michaleva VV, Garbalinsky VA, Botnikova TA, Karnoz GV, Melnik RA (1973a) Utilization of n-paraffins of different molecular weight by *Candida guilliermondii* (Russ). Prikl Biokhim Mikrobiol 10: 35–41

Michaleva VV, Gradova NB, Koslova LJ, Roschkova MI, Shdannikova JN, Welikoslavinskaja OI, Triems K, Pohland D, Glombitza F, Wünsche L, Kersten D-C, Schneider J (1973b) Verfahren zur Gewinnung von Futterhefe. GDR patent, DD WP 105.825

Mikhailova NP, Durasova EN, Vyunov KA (1987) Analysis of sterol mutants of *Candida maltosa*: genetic and biochemical aspects. 12th Int Spec Symp on Yeast, Weimar, Abstr, p 79

Mikhailova NP, Sorokoletova EF, Durasova EN, Vyunov KA, Shapovalov OI (1991) Sterol composition of nystatin-resistant *Candida maltosa* mutants. Folia Microbiol 36: 148–152

Minkevich IG, Baumann F, Rogge G, Heinritz B (1988) Ratio of heat production to oxygen consumption during the cell cycle of *Candida maltosa* EH 15 grown on ethanol. Acta Biotechnol 8: 435–444

Mishina M, Kamiryo T, Tashiro S, Hagihara T, Tanaka A, Fukui S, Osumi M, Numa S (1978) Subcellular localization of two long-chain acyl-coenzyme-A-synthetases in *Candida lipolytica*. Eur J Biochem 89: 321–328

Montrocher R (1980) Significance of immunoprecipitation in yeast taxonomy: Antigenic analyses of some species within the genus *Candida*. Cell Mol Biol 26: 293–302

Müller H, Voigt B (1981) Untersuchungen zur chemischen Zusammensetzung der Lipidfraktion von *Lodderomyces elongisporus* EH 15. Acta Biotechnol 1: 279–284

Müller H, Voigt B (1984) Bestimmung von freiem und gebundenem Ergosterol in Mikroorganismen. Z Allg Mikrobiol 24: 61–64

Müller H-G, Schunck W-H, Riege P, Honeck H (1979) The alkane-hydroxylating enzyme system of the yeast *Candida guilliermondii*. Acta Biol Med Ger 38: 345–349

Müller H-G, Mauersberger S, Schunck W-H, Riege P, Honeck H, Huth J (1980) The alkane-hydroxylating cytochrome P-450 system of yeast: regulation in vivo and progress in isolation. In: Gustafsson J-A et al. (eds) Biochemistry, biophysics and regulation of cytochrome P-450. Elsevier, Amsterdam, pp 251–254

Müller H-G, Schunck W-H, Riege P, Honeck H (1982) The alkane monooxygenase system of the yeast *Lodderomyces elongisporus*: Purification of the cytochrome P-450 and the NADPH-cytochrome P-450 reductase and reconstitution experiments. In: Hietanen E et al. (eds) Cytochrome P-450 – biochemistry, biophysics and environmental implications. Elsevier, Amsterdam, pp 445–448

Müller H-G, Mauersberger S, Schunck W-H, Wiedmann B (1983a) Enzyminduktion in der Hefe *Lodderomyces elongisporus* in Gegenwart von n-Alkanen. Z Allg Mikrobiol 23: 589–593

Müller H-G, Schunck W-H, Kärgel E (1991a) Cytochromes P-450 of alkane-utilizing yeasts (Review). In: Ruckpaul K, Rein H (eds) Frontiers in biotransformation vol 4. Akademie Verlag, Berlin, pp 87–126

Müller H-G, Kärgel E, Mauersberger S, Schunck W-H, Wiedmann B (1991b) Alkane catabolism in yeast – new results of the 1980s. In: Finogenova TV, Sharyshev AA (eds) Alkane metabolism and oversynthesis of metabolites by microorganisms. Center for Biological Research USSR Academy of Sciences, Pushchino, pp 3–16

Müller R, Markuske KD, Babel W (1983b) Verbesserung der Y-Werte bei Wachstum von *Hansenula polymorpha* auf Methanol durch simultane Verwertung von Glucose. Z Allg Mikrobiol 23: 375–384

Müller RH, Babel W (1988) Energy and reducing equivalent potential of C_2-compounds for microbial growth. Acta Biotechnol 8: 249–258

Müller RH, Babel W (1989) Kontinuierliche nicht-fermentative Synthese von Aceton. Wiss Z Karl Marx Univ Leipz Math-Naturwiss Reihe 38: 269–302

Muramatsu S, Hanada H, Nirasawa K, Yoshida M (1982) Mutagenicity tests for mice bred under the condition of long-continued feeding of single-cell protein diets. Bull Natl Inst Anim Ind 38: 23–32

Muraoka S, Ohkuma M, Ohta A, Takagi M (1993) Regulation of gene expression on n-alkane-inducible cytochrome P450s in *Candida maltosa*. 2nd Int Symp Cytochrome P450 Microorganisms Plants, Tokyo, June 13–17, 1993, Abstr, p 2

Muraoka S, Ohkuma M, Takagi M (1994) Recent advances on regulation of gene expression by hydrophobic compounds using yeast systems (Japanese). Tanpakushitsu-Kakusan-Koso 39: 521–529

Müsch A (1993) Die molekulare Umgebung einer naszierenden Polypeptid-Kette während ihrer Translokation in Hefe-Mikrosomen. Dissertation, Humboldt Universität, Berlin

Mutoh H, Mochizuki M, Ohta A, Takagi M (1995) Inducible expression of a gene encoding a L41 ribosomal protein responsible for the cycloheximide-resistant phenotype in the yeast *Candida maltosa*. J Bacteriol 177: 5383–5386

Nabeshima S, Tanaka S, Fukui S (1970) Studies on the hydrocarbon utilization by microorganisms XII. Comparison of the polysaccharide contents of yeast cells grown on hydrocarbons and glucose. J Ferment Technol 4: 556–562

Nakase T, Komagata K (1971) Significance of DNA base composition in the classification of the genus *Candida*. J Gen Appl Microbiol 17: 259–279

Nakase T, Fukazawa Y, Tsuchiya T (1972) A comparative study on two forms of *Candida tropicalis* (Cast.) Berkhout. J Gen Appl Microbiol 18: 349–363

Nelson DR, Kamataki T, Waxman DJ, Guengerich FP, Estabrook RW, Feyereisen R, Gonzalez FJ, Coon MJ, Gunsalus IC, Gotoh O, Okuda K, Nebert D (1993) The P450 superfamily: update on new sequences, gene mapping, accession numbers, early trivial names of enzymes, and nomenclature. DNA Cell Biol 12: 1–51

Neujahr HY (1990) Yeast in biodegradation and biodeterioration processes. In: Verachtert H, De Mot R (eds) Yeast biotechnology and biocatalysis. Marcel Dekker, New York, pp 321–348

Nunziata A, Argentino-Storino A, Mercatelli P, Salerno RO (1982) Two year toxicity in beagle dogs fed a new protein source. Arch Toxicol Suppl 5: 378–381

Nüske J, Grimmecke HD, Reuter G (1982) Polysaccharid-Strukturen von Zellwand-Präparaten aus der Futtereiweiß-Hefe *Candida* sp. H. Z Allg Mikrobiol 22: 477–486

Odds FC (1987) *Candida* infections: an overview. CRC Crit Rev Microbiol 15: 1–5

Ogorodnikova TE, Durasova EN, Sinitskaya NA, Orlov AI, Mikhailova NP, Vyunov KA (1991) Biochemical basis of different nystatin resistance of *Saccharomyces cerevisiae* and *Candida maltosa* yeast mutants (Russ). Mikrobiologiya 60: 26–33 (Microbiology, New York 60: 680–686, English Translation 1992)

Ogrydziak DM (1988) Development of genetic maps of nonconventional yeasts. J Basic Microbiol 28: 185–196

Ohama T, Suzuki T, Mori M, Osawa S, Ueda T, Watanabe K, Nakase T (1993) Non-universal decoding of the leucine codon CUG in several *Candida* species. Nucleic Acids Res 21: 4039–4045

Ohkuma M (1993) Study on n-alkane-inducible cytochrome P-450 gene family in *Candida maltosa* (Japanese). PhD Thesis, Tokyo University

Ohkuma M, Hikiji T, Tanimoto T, Schunck W-H, Müller H-G, Yano K, Takagi M (1991a) Evidence that more than one gene encodes n-alkane-inducible cytochrome P-450s in *Candida maltosa*, found by two-step gene disruption. Agric Biol Chem 55: 1757–1764

Ohkuma M, Tanimoto T, Yano K, Takagi M (1991b) *CYP52* (cytochrome P450alk) multigene family in *Candida maltosa*: molecular cloning and nucleotide sequence of the two tandemly arranged genes. DNA Cell Biol 10: 271–82

Ohkuma M, Muraoka S, Hwang CW, Ohta A, Takagi M (1993a) Cloning of the *C-URA3* gene and construction of a triple auxotroph (*his5, ade1, ura3*) as a useful host for the genetic engineering of *Candida maltosa*. Curr Genet 23: 205–210

Ohkuma M, Hwang CW, Masuda Y, Nishida H, Sugiyama J, Ohta A, Takagi M (1993b) Evolutionary position of n-alkane-assimilating yeast *Candida maltosa* shown by nucleotide sequence of small-subunit ribosomal-RNA gene. Biosci Biotech Biochem 57: 1793–1794

Ohkuma M, Muraoka S, Ohta A, Takagi M (1993c) A cytochrome P450alk gene family in *Candida maltosa*: chromosomal mapping and gene disruption. 2nd Int Symp Cytochrome P450 Microorganisms Plants, Tokyo, June 13–17, 1993, Abstr, p 3

Ohkuma M, Kawai S, Takagi M (1994a) Subject 1. Isolation and characterization of an *ARS* for *Candida maltosa*. In: Maresca B, Kobayashi GS (eds) Molecular biology of pathogenic fungi, a laboratory manual. Telos Press, New York, pp 213–220

Ohkuma M, Muraoka S, Takagi M (1994b) Subject 2. Construction of host-vector systems in *Candida maltosa*. In: Maresca B, Kobayashi GS (eds) Molecular biology of pathogenic fungi, a laboratory manual. Telos Press, New York, pp 221–226

Ohkuma M, Muraoka S, Tanimoto T, Fujii M, Ohta A, Takagi M (1995a) *CYP52* (cytochrome P450alk) multigene family in *Candida maltosa*: identification and characterization of eight members. DNA Cell Biol 14: 163–173

Ohkuma M, Park S-M, Zimmer T, Menzel R, Vogel F, Schunck W-H, Ohta A, Takagi M (1995b) Proliferation of intracellular membrane structures upon homologous overproduction of cytochrome P-450 in *Candida maltosa*. Biochim Biophys Acta (Biomembranes) 1236: 163–169

Ohkuma M, Masuda Y, Park S-M, Ohtomo R, Ohta A, Takagi M (1995c) Evidence that the expression of the gene for NADPH-cytochrome P-450 reductase is n-alkane-inducible in *Candida maltosa*. Biosci Biotech Biochem 59: 1328–1330

Ohkuma M, Zimmer T, Iida T, Schunck W-H, Ohta A, Takagi M (1995d) Isozyme function of n-alkane-inducible cytochrome P450 in *Candida maltosa* by sequential gene disruption (in preparation)

Ohkuma M, Kobayashi K, Kawai S, Hwang CW, Ohta A, Takagi M (1995e) Identification of a centromeric activity in the autonomously replicating TRA region allows improvement of the host-vector system of *Candida maltosa*. Mol Gen Genet 249: 447–455

Okino H, Taoka A, Uemura N (1986) Production of macrocyclic musk compounds, via alkanedioic acids produced from n-alkanes. In: Lawrence BM, Mookherjee BD, Willis BJ (eds) Flavors and fragrances: a world perspective. Elsevier, Amsterdam, pp 753–760

Oliver SG (1988) Replication and recombination in gene establishment in non-*Saccharomyces* yeasts. J Basic Microbiol 28: 197–208

Oliver SG et al. (1992) The complete DNA sequence of yeast chromosome III. Nature 357: 38–46

Osumi M, Miwa N, Teranishi Y, Tanaka A, Fukui S (1974) Ultrastructure of *Candida* yeasts grown on n-alkanes. Appearance of microbodies and its relationship to high catalase activity. Arch Microbiol 99: 181–201

Osumi M, Fukuzumi F, Teranishi Y, Tanaka A, Fukui S (1975a) Development of microbodies in *Candida tropicalis* during incubation in a n-alkane medium. Arch Microbiol 103: 1–11

Osumi M, Fukuzumi F, Yamada N, Nagatani T, Teranishi Y, Tanaka A, Fukui S (1975b) Surface structure of some *Candida* yeast cells grown on n-alkanes. J Ferment Technol 53: 244–248

Palmiter RD (1974) Magnesium precipitation of ribonucleoprotein complexes. Expedient techniques for the isolation of undegraded polysomes and messenger ribonucleic acid. Biochem 13: 3606–3614

Park SM, Ohkuma M, Masuda Y, Ohta A, Takagi M (1996) Galactose-inducible expression systems in *Candida maltosa* using promoters of newly-isolated *GAL1* and *GAL10* genes. (submitted)

Pekelis MV, Ermolaev AV, Ushomirskaya MS, Orlova LM, Gukasyan IA (1989) Immunochemical study of surface glycoconjugates of yeast-like fungi of the *Candida* genus (Russ). Prikl Biokhim Mikrobiol 25: 390–396

Perri GC, Nunziata A, Argentino-Storino A, Salerno RO, Mercatelli P (1981) Long-term toxicity and carcinogenicity of a new protein source in rats. Toxicol Eur Res 3: 305–310

Picataggio S, Deanda K, Mielenz J (1991) Determination of *Candida tropicalis* acyl coenzyme A oxidase isozyme function by sequential gene disruption. Mol Cell Biol 11: 4333–4339

Picataggio S, Rohrer T, Deanda K, Lanning D, Reynolds R, Mielenz J, Eirich LD (1992) Metabolic engineering of *Candida tropicalis* for the production of long-chain dicarboxylic acids. Bio/Technology 10: 849–898

Pogorelskaia SA, Mokeeva NV, Makarova IB (1991) The rate of isolation of fungi in the genus *Candida* from the nasopharyngal mucosa of those in contact with the products from microbial protein manufacture (Russ). Zh Mikrobiol Epidemiol Immunobiol 3: 24–26

Polnisch E, Hofmann KH (1989) Cyclic AMP, fructose-2,6-bisphosphate and catabolite inactivation of enzymes in the hydrocarbon-assimilating yeast *Candida maltosa*. Arch Microbiol 152: 269–272

Polnisch E, Kneifel H, Franzke H, Hofmann KH (1992) Degradation and dehalogenation of monochlorophenols by the phenol-assimilating yeast *Candida maltosa*. Biodegradation 2: 193–199

Polumienko AL, Grigorieva SP (1985) New yeast vectors containing autonomously replicating sequences from *Candida maltosa* genome (Russ). Molek Genet Mikrobiol Vir 7: 26–31

Popov B, Reuter G, Meyer HW (1980) Cell wall regeneration of *Candida spec.* protoplasts. Z Allg Mikrobiol 20: 47–62

Poulter R (1990) Classical methods for the genetic analysis of *Candida albicans*. In: Kirsch DR, Kelly R, Kurtz MB (eds) The genetics of *Candida*. CRC Press, Boca Raton, pp 75–123

Präve P, Faust U, Sittig W, Sukatsch DA (1982) Handbuch der Biotechnologie. Akademische Verlagsgesellschaft Wiesbaden

Pringle JR, Adams AEM, Drubin DG, Haarer BK (1991) Immunofluorescence methods for yeast. Methods Enzymol 194: 565–602

Rabinovich EG, Yegorova VN, Smirnova OY, Inge-Vechtomov SG (1974) Hydrocarbon-utilizing mutants of *Saccharomyces cerevisiae*. Part II to VI. Suppression of sporulation, copulation, and mitotic recombination in Hyc° and Hyc+ mutants (Russ). Genetika 10: 93–99 and related papers of the series in this issue.

Rachubinski RA (1990) Genetic methods for and gene structure in other *Candida* species.

In: Kirsch DR, Kelly R, Kurtz MB (eds) The genetics of *Candida*. CRC Press, Boca Raton, pp 177–186

Rademacher K-H, Reuter G (1978) Zur Struktur des Mannans von *Candida guilliermondii* H. Z Allg Mikrobiol 18: 63–66

Rehm HJ (1986) Single cell protein production from petroleum derivatives and its utilization as food and feed. In: Alani DI, Moo-Young M (eds) Perspectives in biotechnology and applied microbiology. Elsevier, New York, pp 1–16

Rehm HJ, Reiff I (1981) Mechanisms and occurrence of microbial oxidation of long-chain alkanes. Adv Biochem Eng 19: 175–215

Rehm HJ, Reiff I (1982) Regulation der mikrobiellen Alkanoxidation mit Hinblick auf die Produktsynthese. Acta Biotechnol 2: 127–138

Reiser J, Glumoff V, Kälin M, Ochsner U (1990) Transfer and expression of heterologous genes in yeast other than *Saccharomyces cerevisiae*. In: Fiechter A (ed) Advances in biochemistry engineering/biotechnol vol 43. Springer, Berlin Heidelberg New York, pp 76–102

Riege P, Schunck W-H, Honeck H, Müller H-G (1980) Eigenschaften des Cytochrom P-450-abhängigen alkanhydroxylierenden Enzymsystems aus *Candida guilliermondii*. Wiss Z Ernst Moritz Arndt Univ Greifswald 29: 125–126

Riege P, Schunck W-H, Honeck H, Müller, H-G (1981) Cytochrome P-450 from *Lodderomyces elongisporus*: its purification and some properties of the highly purified protein. Biochem Biophys Res Commun 98: 527–534

Riege P, Blasig R, Müller H-G, Heidenreich G, Bauch J (1989) Influence of oxygen and substrate supply on the metabolism of *Candida maltosa* during cultivation on n-alkanes. Appl Microbiol Biotechnol 32: 101–107

Riggsby WS (1990) Physical characterization of the *Candida albicans* genome. In: Kirsch DR, Kelly R, Kurtz MB (eds) The genetics of *Candida*. CRC Press, Boca Raton, pp 125–145

Ringpfeil M (1983) SCP-Produktion auf der Basis von Kohlenwasserstoffen. Acta Biotechnol 3: 227–240

Röber B (1985) Katabole Repression bei aeroben und O_2-limitiertem Wachstum – energetische und stoffliche Bilanz des Kohlenhydratmetabolismus bei Hefen. J Basic Microbiol 25: 581–590

Röber B, Reuter G (1979) Biosynthese der Zellwand-Polysaccharide Mannan und Glucan als Spiegelbild unterschiedlicher Abbauwege der Glucose durch *Candida* sp. H. Z Allg Mikrobiol 19: 187–194

Röber B, Reuter G (1982) In vitro-Einbau von ^{14}C-Hexose-6-Phosphat in Mannan, β-Glucan und Glycogen bei *Candida* sp. H und ihren Mutanten. Z Allg Mikrobiol 22: 671–673

Röber B, Reuter G (1984a) Control of catabolic and anabolic sequences of carbohydrate utilization in the *scp* yeast *Candida maltosa* H and its mutants H3 and H5 (Germ). Z Allg Mikrobiol 24: 41–55

Röber B, Reuter G (1984b) Regulation of the glucopolysaccharide biosynthesis in the *scp* yeast *Candida* sp. H by precursor preparation (Germ). Z Allg Mikrobiol 24: 167–177

Röber B, Reuter G (1984c) Mannan-biosynthesis in microsome fractions from protoplast-lysates of *Candida maltosa* H (Germ). Z Allg Mikrobiol 24: 179–188

Röber B, Reuter G (1984d) Regulation of proteophosphomannan biosynthesis in the *scp* yeast *Candida maltosa* H by precursor preparation (Germ). Z Allg Mikrobiol 24: 317–328

Röber B, Reuter G (1985) Effector- and precursor-function of mannose-6-phosphate, mannose-1-phosphate, UDP-n-acetylglucosamine and dolichylphosphate in the proteophosphomannan biosynthesis of the *scp* yeast *Candida maltosa* H (Germ). J Basic Microbiol 25: 243–264

Röber B, Stolle J, Reuter G (1984a) Properties of hexokinase from *Candida maltosa* H, a SCP yeast (Germ). Z Allg Mikrobiol 24: 619–627

Röber B, Stolle J, Reuter G (1984b) Properties of the glucose-6-phosphate dehydrogenase from *Candida maltosa* H, a SCP yeast (Germ). Z Allg Mikrobiol 24: 629–636

Romanos MA, Scorer CA, Clare JJ (1992) Foreign gene expression in yeast: a review. Yeast 8: 423–488

Rose MD, Winston F, Hieter P (eds) (1990) Methods in yeast genetics. A laboratory course manual. Cold Spring Harbor Laboratory Press, Cold Spring Harbor

Rylkin SS, Berezov TB, Gurina LV, Belova LA, Shulga AV, Orlova VS, Saubenova MG (1974) Composition of cell wall of *Candida tropicalis* during growth on glucose and n-alkanes (Russ). Mikrobiologiya 43: 551–552

Rymond BC, Rosbach M (1992) Yeast pre-mRNA splicing. In: Jones EW, Pringle JR, Broach JR (eds) The molecular and cellular biology of the yeast *Saccharomyces*, vol 2, Gene expression. Cold Spring Harbor Laboratory Press, Cold Spring Harbor, pp 143–192

Sakajo S, Minagawa N, Yoshimoto A (1993) Characterization of the alternative oxidase protein in the yeast *Hansenula anomala*. FEBS Lett 8: 310–312

Sambrook J, Fritsch EF, Maniatis T (1989) Molecular cloning: a laboratory manual. Cold Spring Harbor Laboratory Press, Cold Spring Harbor

Samsonova I, Klinner U, Böttcher F (1987) Genetic studies on *Candida maltosa*. 12th Int Spec Symp Genet Non-conventional Yeasts, Weimar, 1987, Abstr, p21

Sanglard D, Fiechter A (1989) Heterogeneity within the alkane-inducible cytochrome P450 gene family of the yeast *Candida tropicalis*. FEBS Lett 256: 128–133

Sanglard D, Loper JC (1989) Characterization of the alkane-inducile cytochrome P450 (P450alk) gene from the yeast *Candida tropicalis*: identification of a new P450 gene family. Gene 76: 121–136

Sasnauskas K, Jomantiene R, Geneviciute E, Januska A, Lebedys J (1991) Molecular cloning of the *Candida maltosa ADE1* gene. Gene 107: 161–164

Sasnauskas K, Jomantiene R, Januska A, Lebediene E, Lebedys J, Janulaitis A (1992a) Cloning and analysis of a *Candida maltosa* gene which confers resistance to formaldehyde in *Saccharomyces cerevisiae*. Gene 122: 207–211

Sasnauskas K, Jomantiene R, Lebediene E, Lebedys J, Januska A, Janulaitis A (1992b) Molecular cloning and analysis of autonomous replicating sequence of *Candida maltosa*. Yeast 8: 253–259

Sasnauskas K, Jomantiene R, Lebediene E, Lebedys J, Januska A, Janulaitis A (1992c) Cloning and sequence analysis of a *Candida maltosa* gene which confers resistance to cycloheximide. Gene 116: 105–108

Sattler K, Wünsche L (1981) Aufnahme von Kohlenwasserstoffen durch Hefen (Teil I). Acta Biotechnol 0: 15–20

Sattler K, Wünsche L (1983) Möglichkeiten der Gewinnung von Koppelprodukten der mikrobiellen Eiweißsynthese auf der Basis von Kohlenwasserstoffen. Acta Biotechnol 3: 345–350

Schauer F (1988) Zur Physiologie des Kohlenwasserstoffabbaus in *Candida maltosa*. Universität Greifswald, Math-Nat, Dissertation B

Schauer F, Schauer M (1986) Alkanassimilierende Hefen. Systematische Stellung und Erfassung einiger Leistungsgrenzen. Wiss Z EMA Univ Greifswald Math Naturwiss Reihe 35: 14–23

Schauer F, Hofmann KH, Köhler M (1986) The subterminal oxidation of aliphatic hydrocarbons in *Candida maltosa*. In: Microbe 86, 14th Int Congr Microbiol, Manchester, Abstr, p 255

Schauer F, Lindow S, Schauer M, Samsonova I, Böttcher F (1987) Oxidation of n-alkanes by *Pichia guilliermondii* and induction of mutants. 12th Int Spec Symp Genet of Non-conventional Yeasts, Weimar, 1987, Abstr, p 57

Scheda R (1966) Kohlenwasserstoffe zehrende Hefen. Die Branntweinwirtschaft 106: 373–376

Scheller U, Schunck W-H, Müller H-G (1992) Characterization of two different alkane-inducible P-450 forms from *Candida maltosa* by means of heterologous expression in *Saccharomyces cerevisiae*. In: Archakov AI, Bachmanova GI (eds) Cytochrome P-450: biochemistry and biophysics. INCO – TNC, Joint Stock Company, Moscow, pp 662–664

Scheller U, Kraft R, Schröder K-L, Schunck W-H (1994) Generation of the soluble and functional cytosolic domain of microsomal cytochrome P450 52A3. J Biol Chem 269: 12779–12783

Scheller U, Zimmer T, Kärgel E, Schunck W-H (1996) Characterization of the n-alkane and fatty acid hydroxylating cytochrome P450 forms 52A3 and 52A4. Arch Biochem (in press)

Schindler J, Meusdoerffer F, Giesel-Bühler H (1990) Microbial production in industrial chemicals: Basic features of dicarboxylic acid production by yeasts (Germ). Forum Mikrobiol 5: 274–281

Schmidt H (1988) Lysinmetabolismus der Hefen Candida maltosa und Pichia guilliermondii. Univ Greifswald, Math-Nat Dissertation A

Schmidt H, Bode R (1992) Characterization of a novel enzyme, N^6-acetyl-L-lysine: 2-oxoglutarate aminotransferase, which catalyzes the second step of lysine catabolism in Candida maltosa. Antonie Leeuwenhoek J Microbiol 62: 285–290

Schmidt H, Bode R, Lindner M, Birnbaum D (1985) Lysine biosynthesis in the yeast Candida maltosa: properties of some enzymes and regulation of the biosynthetic pathway. J Basic Microbiol 25: 675–681

Schmidt H, Bode R, Birnbaum D (1988) Lysine degradation in Candida maltosa: occurrence of a novel enzyme, acetyl-CoA:L-lysine N-acetyltransferase. Arch Microbiol 150: 215–218

Schmidt H, Bode R, Samsonova IA, Birnbaum D (1989a) Isolation and characterization of alpha-aminoadipate-delta-semialdehyde overproducing mutants from yeasts. FEMS Microbiol Lett 60: 201–204

Schmidt H, Bode R, Samsonova IA, Birnbaum D (1989b) Production of alpha-aminoadipate-delta-semialdehyde by a mutant from Candida maltosa. Appl Microbiol Biotechnol 31: 463–466

Schneider JD, Triems K (1981) Einfluß extracellulärer Kaliumionenkonzentrationen auf die celluläre Natriumkonzentration von Lodderomyces elongisporus D. Acta Biotechnol 1: 197–199

Schneider JD, Hansel R, Hedlich R, Jechorek M (1983) Growth characteristics of a thermotolerant strain of Lodderomyces elongisporus grown on sucrose. Acta Biotechnol 3: 13–19

Schult I (1987) Enzymologische Studien über mutabile Gene bei Candida maltosa. Ernst-Moritz-Arndt-Universität Greifswald, DDR, Sektion Biologie, Dissertation A

Schult I, Samsonova I, Böttcher F (1987) Induction of unstable genes in Candida maltosa. 12th Int Spec Symp Non-conventional Yeast, Weimar, 1987, Abstr, p 22

Schunck W-H, Riege P, Blasig R, Honeck H, Müller H-G (1978a) Cytochrome P-450 and alkane hydroxylase in Candida guilliermondii. Acta Biol Med Ger 37: K3–K7

Schunck W-H, Riege P, Kuhl R (1978b) Cytochrome P-450 of eukaryotic microorganisms. Pharmazie 33: 410–415

Schunck W-H, Riege P, Müller H-G, Scheler W (1983a) Isolation and some molecular properties of cytochrome P-450 from the alkane assimilating yeast Lodderomyces elongisporus (Russ.) Biokhimiya (Moscow) 48: 518–526

Schunck W-H, Riege P, Honeck H, Müller H-G (1983b) Isolierung und Rekonstitution des Alkan-Monooxygenase-Systems der Hefe Lodderomyces elongisporus. Z Allg Mikrobiol 23: 653–660

Schunck W-H, Mauersberger S, Huth J, Riege P, Müller H-G (1987a) Function and regulation of cytochrome P-450 in alkane-assimilating yeast I. Selective inhibition with carbon monoxide in growing cells. Arch Micriol 147: 240–244

Schunck W-H, Mauersberger S, Kärgel E, Huth J, Müller H-G (1987b) Function and regulation of cytochrome P-450 in alkane-assimilating yeast II. Effect of oxygen-limitation. Arch Microbiol 147: 245–248

Schunck W-H, Kießling U, Strauss M, Kärgel E, Wiedmann B, Mauersberger S, Gaestel M, Gross B, Müller H-G (1989a) Cloning of a cDNA for the alkane hydroxylating P-450 from

Candida maltosa. In: Schuster I (ed) Biochemistry and biophysics of cytochrome P-450. Taylor & Francis, London, pp 656–659

Schunck W-H, Kärgel E, Gross B, Wiedmann B, Mauersberger S, Köpke K, Kießling U, Strauss M, Gaestel M, Müller H-G (1989b) Molecular cloning and characterization of the primary structure of the alkane hydroxylating cytochrome P-450 from the yeast *Candida maltosa*. Biochem Biophys Res Commun 181: 843–850

Schunck WH, Vogel F, Gross B, Kärgel E, Mauersberger S, Köpke K, Gengnagel C, Müller HG (1991) Comparison of two cytochromes P-450 from *Candida maltosa*: primary structures, substrate specificities and effects of their expression in *Saccharomyces cerevisiae* on the proliferation of the endoplasmic reticulum. Eur J Cell Biol 55: 336–345

Schunck W-H, Scheller U, Juretzek T (1996) Generation of the cytosolic domain of microsomal P450 52A3 after high-level expression in *Saccharomyces cerevisiae*. Methods Enzym (in press)

Schuster G, Voigt B, Müller H (1990) The influence of combined treatments with 2,4-dioxohexahydro-1,3,5-triazine (DHT) and lipophilic fractions from the yeast *Candida maltosa* IMET H128 on virus symptoms and tuber mass of identical potato eye cutting plants. Z Pflanzenkr Pflanzenschutz 97: 84–86

Schwarz E, Mülling K, Samsonova I, Schauer F, Böttcher F (1987) Genetic studies of n-alkane uitilization of *Candida maltosa*. 12th Int Spec Symp Genet of Non-conventional Yeasts, Weimar, 1987, Abstr, p 75

Seghezzi W, Sanglard D, Fiechter A (1991) Characterization of a second alkane-inducible cytochrome P450-encoding gene, *CYP52A2*, from *Candida tropicalis*. Gene 106: 51–60

Seghezzi W, Meili C, Ruffiner R, Kuenzi R, Sanglard D, Fiechter A (1992) Identification and characterization of additional members of the cytochrome P450 multigene family *CYP52* of *Candida tropicalis*. DNA Cell Biol 11: 767–780

Senez JC (1986) The economical aspects of single cell protein production from petroleum derivatives. In: Alani DI, Moo-Young M (eds) Perspectives in biotechnology and applied microbiology. Elsevier, New York, pp 33–48

Sharp PM, Touhy TM, Mosurski KR (1986) Codon usage in yeast: cluster analysis clearly differentiates highly and lowly expressed genes. Nucleic Acid Res 14: 5125–5143

Sharyshev AA, Krauzova VI (1988) Fractionation of subcellular membrane organelles of the yeast *Candida maltosa* in Percoll gradients (Russ). Biologicheskie Membrany (Biological Membranes) 5: 187–197

Sharyshev AA, Matyashova RN, Komarova GN (1983) Cytochrome P-450 assay in *Candida* and *Saccharomyces* cells under various growth conditions. Int Symp Environ Regulation Microbial Metabolism, Pushchino 1983, Abstr, pp 33–34

Shennan JL (1984) Hydrocarbons as substrates in industrial fermentations. In: Atlas PR (ed) Petroleum microbiology. Macmillan, New York, pp 643–683

Shennan JL, Levi JD (1974) The growth of yeasts on hydrocarbons. In: Hockenhull DJD (ed) Progress in industrial microbiology, vol 13. Churchill Livingstone Edinburgh, pp 1–57

Shiio I, Uchio R (1971) Microbial production of long-chain dicarboxylic acids from n-alkanes. Part I. Screening and properties of microorganisms producing dicarboxylic acids. Agric Biol Chem 35: 2033–2042

Shilova NK, Matyashova RN, Ilchenko AP (1989) The effect of aeration on the activity of alcohol oxidase and enzymes utilizing hydrogen peroxide in the course *Candida maltosa* growth on paraffin (Russ). Mikrobiologiya 58: 430–435

Shkumatov VM (1993) Heterologous reconstitution of monooxygenases. 2nd Int Symp Cytochrome P450 Microorganisms Plants, Tokyo, June 13–17, 1993, Abstr, p 28

Silva J, Laborda RR, Almendro G, Salim R (1990) Detection of opportunistic yeast pathogens in hospitalized immunocompromised patients. Rev Latinoam Microbiol 32: 261–264

Sinanyan ES, Davidova EG, Davtyan MA, Davidov ER (1989) Synthesis of thermal shock proteins in the mesophilic strain of the yeast *Candida maltosa* and in its thermotolerant mutant (Russ). Izv Timiryazev Skh Akad 0(2): 195–199

Sinanyan ES, Davtyan MA, Davidov ER (1990) Proteolytic activity of mesophilic and thermophilic yeast *Candida maltosa* in thermal and ethanol shock (Russ). Biol Zh Arm 43: 96–100

Slavikova E, Grabinska-Loniewska A (1990) Taxonomical study of yeasts and yeast-like microorganisms isolated from the denitrification unit Biocenosis. Acta Mycol (Warsaw) 23: 81–88 (1987, published 1990)

Smith NG, Bourquin AW, Crow SA, Ahearn DG (1976) Effect of heptachlor on hexadecane utilization by selected fungi. Dev Ind Microbiol 17: 331–336

Smith RH, Palmer R (1976). A chemical and nutritional evaluation of yeast and bacteria as dietary protein sources for rats and pigs. J Sci Food Agric 27: 763–770

Snow R (1966) An enrichment method for auxotrophic yeast mutants using the antibiotic nystatin. Nature 211: 206–207

Sokolov YI, Davidov ER, Demanova NF, Gololobov AD (1981) Utilization of alkyl aromatic hydrocarbons by the yeast *Candida guilliermondii* (Russ). Prikl Biokhim Mikrobiol 17: 660–668

Sokolov YI, Avetisova SM, Davidov ER (1986a) Isolation, purification, and porperties of cytochrome P-450 from yeast of the genus *Candida* grown on n-alkanes (Russ). Biokhimiya 51: 1649–1654

Sokolov YI, Avetisova SM, Davydov RM, Davidov ER (1986b) Detection of two cytochrome P450 forms participating in the alkane oxidation of *Candida* yeast (Russ). Dokl Akad Nauk SSSR 286: 1506–1511

Soll DR (1990) Dimorphism and high-frequency switching in *Candida albicans*. In: Kirsch DR, Kelly R, Kurtz MB (eds) The genetics of *Candida*. CRC Press, Boca Raton, pp 148–176

Soom YO (1973) Mutants of *Saccharomyces cerevisiae* utilizing n-alkanes. I. Isolation and characterization of mutants (Russ). Genetika 9: 95–101

Souza AE, Myler PJ, Stuart KD (1993) The alkane-inducible *Candida maltosa ALI1* gene product is an NADH:ubiquinone oxidoreductase subunit homologue. Gene 137: 349–350

Spivak SM, Gukasyan IA, Ogarkov VI (1988) Effect of non-pathogenic yeast-like fungi of the genus *Candida* on the process of forming immediate hypersensitivity to heterologous protein in guinea pigs (Russ). Gig Sanit 9: 75–76

Spivak SM, Gukasyan IA, Ermolaev AV, Ustinenko AN, Antonovicha LA (1989) Study of sensitizing properties of yeast-like fungi of the genus *Candida* in the production of dietary proteins (Russ). Gig Sanit 6: 77–79

Stepanjuk VV (1981) On the nuclear origin of peroxisomes as possible precursors of mitochondria in hydrocarbon-oxidizing yeasts of the genus *Candida* (Russ). Zitologiya 23: 369–377

Stichel E, Glombitza F, Iske U (1981) Paraffinübergang aus der Kohlenwasserstoffphase zur Hefezelle. Acta Biotechnol 1: 9–15

Stichel E, Rogge G, Bley T, Heinritz B (1982) Yield coefficients in dependence on milieu conditions and cell states. III. Induction of synchrony in continuous yeast cell cultivation by milieu changes (*Lodderomyces elongisporus*). Z Allg Mikrobiol 22: 717–722

Strick CA, James LC, O'Donnell MM, Gollaher MG, Franke AE (1992) The isolation and characterization of the pyruvate kinase encoding gene from the yeast *Yarrowia lipolytica*. Gene 118: 65–72 (and correction in Gene 140: 141–143)

Su CS, Meyer SA (1991) Characterization of mitochondrial DNA in various *Candida* species: isolation, restriction endonuclease analysis, size, and base composition. Int J Syst Bacteriol 41: 6–14

Sudbery PE (1994) The non-*Saccharomyces* yeasts. Yeast 10: 1707–1726

Sudoh M, Nagahashi S, Doi M, Ohta A, Takagi M, Arisawa M (1993) Cloning of the chitin synthase 3 gene from *Candida albicans* and its expression during yeast-hyphal transition. Mol Gen Genet 241: 351–358

Sugiyama H, Ohkuma M, Masuda Y, Park S-M, Ohta A, Takagi M (1995) In vivo evidence for non-universal usage of the codon CUG in *Candida maltosa*. Yeast 11: 43–52

Sunairi M, Watabe K, Takagi M, Yano K (1984) Increase of translatable mRNA for major microsomal proteins in n-alkane-grown *Candida maltosa*. J Bacteriol 160: 1037–1040

Sunairi M, Suzuki R, Takagi M, Yano K (1988) Self-cloning of genes for n-alkane assimilation from *Candida maltosa*. Agric Biol Chem 52: 577–579

Suzuki T, Ueda T, Ohama T, Osawa S. Watanabe K (1993) The gene for serine tRNA having anticodon sequence CAG in a pathogenic yeast, *Candida albicans*. Nucleic Acid Res 21: 356

Takagi M (1992) Host-vector system and reverse genetics in a non-conventional yeast, *Candida maltosa*. In: Mongkolsuk SP, Lovett PS, Trempy JE (eds) Biotechnology and environmental science: molecular approaches. Plenum Press, New York, pp 13–22

Takagi M (1993) Reverse genetics in a non-conventional yeast, *Candida maltosa*. In: Maresca E, Kobayashi GS, Yamaguchi H (eds), Molecular biology and its application ot medical mycology. NATO ASI Series, vol H 69. Springer, Berlin Heidelberg New York, pp 13–22

Takagi M, Moriya K, Yano K (1980a) Induction of cytochrome P450 in petroleum-assimilating yeast. I. Selection of a strain and basic characterization of cytochrome P450 induction in the strain. Cell Mol Biol 25: 363–369

Takagi M, Moriya K, Yano K (1980b) Induction of cytochrome P450 in petroleum-assimilating yeast. II. Comparison of protein synthesizing activity in cells grown on glucose and n-tetradecane. Cell Mol Biol 25: 371–375

Takagi M, Kawai S, Takata Y, Tanaka N, Sunairi M, Miyazaki M, Yano K (1985) Induction of cycloheximide resistance in *Candida maltosa* by modifying the ribosomes. J Gen Appl Microbiol 31: 267–275

Takagi M, Kawai S, Chang MC, Shibuya I, Yano K (1986a) Construction of a host-vector system in *Candida maltosa* by using an *ARS* site isolated from its genome. J Bacteriol 167: 551–555

Takagi M, Kawai S, Shibuya I, Miyazaki M, Yano K (1986b) Cloning in *Saccharomyces cerevisiae* of a cycloheximide resistance gene from the *Candida maltosa* genome which modifies ribosomes. J Bacteriol 168: 417–419

Takagi M, Kobayashi N, Sugimoto M, Fujii T, Watari J, Yano K (1987) Nucleotide sequencing analysis of a *LEU* gene of *Candida maltosa* which complements *leuB* mutation of *Escherichia coli* and *leu2* mutation of *Saccharomyces cerevisiae*. Curr Genet 11: 451–457

Takagi M, Uchino S, Sugimoto M, Kawai S, Hikiji T, Yano K (1988) Construction of promoter-probe vectors for *Candida maltosa*, a n-alkane-assimilating yeast, using the *LEU2* gene of *Saccharomyces cerevisiae*. J Basic Microbiol 28: 335–342

Takagi M, Ohkuma M, Kobayashi N, Watanabe M, Yano K (1989) Purification of cytochrome P-450alk from n-alkane-grown cells of *Candida maltosa*, and cloning and nucleotide sequencing of the encoding gene. Agric Biol Chem 53: 2217–2226

Tan H, Okazaki K, Kubota I, Kamiryo T, Utiyama H (1990) A novel peroxisomal nonspecific lipid-transfer protein from *Candida tropicalis*. Gene structure, purification and possible role in β-oxidation. Eur J Biochem 190: 107–112

Tanaka A, Fukui S (1989) Metabolism of n-alkanes In: Rose AH, Harrison JS (eds) The yeasts, vol 3, 2nd edn, Metabolism and physiology of yeasts. Academic Press, London, pp 261–287

Tanaka A, Ohishi N, Fukui S (1967) Studies on the formation of vitamins and their function in hydrocarbon fermentation. Production of vitamin B_6 by *Candida albicans* in hydrocarbon medium. J Ferment Technol 45: 617–623

Tanaka A, Osumi M, Fukui S (1982) Peroxisomes of alkane-grown yeast: fundamental and practical aspects. Ann NY Acad Sci 386: 183–199

Tanaka H, Takagi M, Yano K (1987) Separation of chromosomal DNA molecules of *Candida maltosa* on agarose gels using the OFAGE technique. Agric Biol Chem 51: 3161–3163

Tannenbaum SR, Wang DIC (1975) Single cell protein II. MIT Press, Cambridge

Tokuyasu KT (1986) Application of cryoultramicrotomy to immunocytochemistry. J Microsc 143: 139–149

Tokuyasu KT (1989) Use of poly(vinylpyrrolidone) and poly(vinyl alcohol) for cryoultramicrotomy. Histochem J 21: 163–171

Taoka A (1986) Production of brassylic acid by fermentation. BioIndustry 3: 867–874

Triebel H, Grimmecke HD, Kretzschmer K, Bär H (1980) Molecular weight determination on a mannan-protein-phosphate complex from the cell wall of the yeast *Candida* sp. H. Stud Biophys 82: 47–54

Truchatshova TV, Ermolenko TM, Gubina LP, Radyuk VG, Shkumatov VM (1991) Analysis, fractionation and industrial technology of hydrocarbon-assimilating *Candida maltosa*. 15th Int Spec Symp on Yeast, Riga, Latvia 1991, pp 180–181

Tschumper G, Carbon J (1982) Delta sequences and double symmetry in a yeast chromosomal replicator region. J Mol Biol 156: 239–307

Uchio R (1978) Microbial production of long-chain dicarboxylic acids from n-alkanes. Petrol Microorg 20: 13–16

Uchio R, Shiio I (1972a) Microbial production of long-chain dicarboxylic acids from n-alkanes Part II. Production by *Candida cloacae* mutant unable to assimilate dicarboxylic acid. Agric Biol Chem 36: 426–433

Uchio R, Shiio I (1972b) Production of dicarboxylic acids by *Candida cloacae* mutants unable to assimilate n-alkane. Agric Biol Chem 36: 1169–1175

Uchio R, Shiio I (1972c) Tetradecane-1,14-dicarboxylic acid production from n-hexadecane by *Candida cloacae*. Agric Biol Chem 36: 1389–1397

Uchio R, Shiio I (1974) Microbial production of long-chain dicarboxylic acids from n-alkanes. Petrol Microorg 11: 14–23

Uemura N (1985) Industrialization of the production of dibasic acids from paraffins using microorganisms (Japanese). Hakko to Kogyo 43: 436–441

Uemura N, Taoka A, Takagi M (1988) Production of dicarboxylic acids by fermentation. In: Applewhite TH (ed) World conference on biotechnology of fats and oil industry. American Oil Chemist's Society, pp 148–152

Umemura I, Yanagiya K, Komatsubara S, Sato T, Tosa T (1990) D-alanine production by using asymmetric degrading activity of *Candida maltosa*. Ann NY Acad Sci 613: 659–662

Umemura I, Yanagiya K, Komatsubara S, Sato T, Tosa T (1991) Characteristics of alanine aminotransferase from *Candida maltosa*. In: Fukui T, Kagamiyama K, Soda K, Wada H (eds) Enzymes dependent on pyridoxal phosphate and other carbonyl compounds as cofactors. Pergamon Press, Oxford, pp 229–231

Umemura I, Yanagiya K, Komatsubara S, Sato T, Tosa T (1992) D-alanine production from D,L-alanine by *Candida maltosa* with asymmetric degrading activity. Appl Microbiol Biotechnol 36: 722–726

Umemura I, Yanagiya K, Komatsubara S, Sato T, Tosa T (1994) Purification and some properties of alanine aminotransferase from *Candida maltosa*. Biosci Biotech Biochem 58: 283–287

Van Tuinen E, Riezman H (1987) Immunolocalization of glyceraldehyde-3-phosphate dehydrogenase, hexokinase, and carboxypeptidase Y in yeast cells at the ultrastructural level. J Histochem Cytochem 35: 327–333

Van Uden N, Buckley H (1970) Genus *Candida* Berkhout. In: Lodder J (ed) The yeasts, a taxonomic study. North-Holland Publ, Amsterdam, pp 893–1087

Veenhuis M, Kram AM, Kunau WH, Harder W (1990) Excessive membrane development following exposure of the methylothrophic yeast *Hansenula polymorpha* to oleic acid-containing media. Yeast 6: 511–519

Vergeres G, Yen TSB, Aggeler J, Lausier J, Waskell L (1993) A model system for studying membrane biogenesis. Overexpression of cytochrome b5 in yeast results in marked proliferation of the intracellular membrane. J Cell Sci 106: 249–259

Vier B, Voigt B (1984) Untersuchungen zur Anreicherung von Ergosterol und Ubichinon aus Lipid-Kohlenwasserstoff-Fraktionen. Acta Biotechnol 4: 377–379

Viljoen BC, Kock JLF, Britz TJ (1988) The significance of long-chain fatty acid composition and other phenotypic characteristics in determining relationships among some *Pichia* and *Candida* species. Gen Microbiol 134: 1893–1900

Villalba JM, Palmgren MG, Berberian GE, Ferguson C, Serrano R (1992) Functional expression of plant plasma membrane H⁺-ATPase in yeast endoplasmic reticulum. J Biol Chem 267: 12341–12349

Vogel F, Kärgel E, Schunck W-H (1991) In situ localization of cytochrome P-450, the first enzyme involved in aliphatic hydrocarbon degradation in the yeast *Candida maltosa*. Progr Histochem Cytochem 23: 383–389

Vogel F, Gengnagel C, Kärgel E, Müller H-G, Schunck W-H (1992) Immunocytochemical localization of alkane-inducible cytochrome P450 and its NADPH-dependent reductase in the yeast *Candida maltosa*. Eur J Cell Biol 57: 285–291

Voigt B, Seidel H, Müller H, Beck D, Ringpfeil M, Riedel M, Bauch J, Gentzsch H, Bohlmann D (1979) Biolipidextrakt – ein neuer Rohstoff aus der Produktion von "Fermosin"-Futterhefe auf Basis Erdöldestillat. Chem Techn 31: 409–411

Voigt B, Reutgen H, Worbs M, Sesser I (1984a) Untersuchungen zur Anreicherung von Ubichinon-9 aus Lipid-Kohlenwasserstoff-Fraktionen mittels Sephadex LH-20. Acta Biotechnol 4: 137–141

Voigt B, Müller H, Worbs M, Winkler F, Köhler U (1984b) Untersuchungen zur Anreicherung von Ubichinon-9 aus Lipid-Kohlenwasserstoff-Fraktionen mittels Kurzwegdestillation. Acta Biotechnol 4: 293–296

Voigt B, Müller H, Schuster G (1985) Antiphytovirale Aktivität von lipophilen Fraktionen aus der Hefe *Lodderomyces elongisporus* IMET H 128. Acta Biotechnol 5: 313–317

Volchek EA, Durasova EN, Mukhlenov AG, Mikhailova NP, Vyunov KA (1988) Sterol composition of nystatin-resistant *Candida maltosa* strains (Russ). Izv AN SSSR Ser Biol 0(6): 915–921

Watanabe K (1974) Production of SCP with hydrocarbon-assimilating yeasts (Japanese). J Ferm Assoc Japan (Hakko-Kyokai-Shi) 32: 239–248

Watanabe K, Shimada Y, Kawaharada K, Suzuki K, Tanaka F (1973a) Kanegafuchi Chemical Industry Co, Ltd, Japan, Japan Patent 48–43877

Watanabe K, Shimada Y, Kawaharada K, Suzuki K, Tanaka F (1973b) Kanegafuchi Chemical Industry Co, Ltd, Japan, US Patent 3725 200

Watanabe K, Shimada Y, Kawaharada K, Suzuki K, Tanaka F (1973c) Kanegafuchi Chemical Industry Co, Ltd, Japan, British Patent 1307 434

Watanabe K, Shimada Y, Kawaharada K, Suzuki K, Tanaka F (1975) Kanegafuchi Chemical Industry Co, Ltd, Japan, German Patent 2454 048

Waters MG, Blobel G (1986) Secretory protein translocation in a yeast cell-free system can occur posttranslationally and requires ATP hydrolysis. J Cell Biol 102: 1543

Weber H, Barth G (1988) Nonconventional yeasts: their genetics and biotechnological applications. CRC Crit Rev Biotechnol 7: 281–337

Wedler H, Schulze S, Budahn H, Januschka A, Sasnauskas K, Böttcher F, Becher D (1990) Gentechnische Bearbeitung von Hefearten mit biotechnologischer Bedeutung. Wiss Z Ernst Moritz Arndt Univ Greifswald, Math Naturwiss Reihe 39: 27–30

White MJ, Hodgson LF, Rose AH, Hammond RC (1989) Long-chain alcohol production by yeasts. Yeast Apr 5 Spec Issue S456–470

Wiame J-M, Grenson M, Arst HN (1985) Nitrogen catabolite repression in yeasts and filamentous fungi. Adv Microb Physiol 26: 1–88

Wiedmann B (1987) Untersuchungen zur Biosynthese des Cytochrom P450 aus *Candida maltosa*. Akademie der Wissenschaften der DDR, Dissertation, Berlin

Wiedmann B, Wiedmann M, Kärgel E, Schunck W-H, Müller H-G (1986) n-Alkanes induce the synthesis of cytochrome P-450 mRNA in *Candida maltosa*. Biochem Biophys Res Commun 36: 1148–1154

Wiedmann B, Wiedmann M, Schunck W-H, Mauersberger S, Kärgel E, Müller H-G (1987) Regulation of cytochrome P-450 biosynthesis in alkane assimilating yeasts. In: Zelinka J, Balan J (eds) Proc 6th Int Symp Metabol Enzymol Nucleic Acids Includ Gene Manipul, Bratislava 1987, pp 383–393

Wiedman B, Wiedmann M, Mauersberger S, Schunck W-H, Müller H-G (1988a) Oxygen limitation induced indirectly the synthesis of cytochrome P-450 mRNA in alkane-growing *Candida maltosa*. Biochem Biophys Res Commun 150: 859–865

Wiedmann M, Wiedmann B, Voigt S, Wachter E, Müller HG, Rapoport TA (1988b) Posttranslational transport of proteins into microsomal membranes of *Candida maltosa*. EMBO J 7: 1763–1768

Wiedmann B, Silver P, Schunck W-H, Wiedmann M (1993) Overexpression of the ER-membrane protein P-450 *CYP52A3* mimics *sec* mutant characteristics in *Saccharomyces cerevisiae*. Biochim Biophys Acta 1153: 267–276

Wright R (1993) Insights from inducible membranes. Curr Biol 3: 870–873

Wright R, Basson M, D'Ari L, Rine J (1988) Increased amounts of HMG-CoA reductase induce "Karmellae": A proliferation of stacked membrane pairs surrounding the nucleus. J Cell Biol 107: 101–114

Wright R, Keller G, Gould SJ, Subramani S, Rine J (1990) Cell-type control of membrane biogenesis induced by HMG-CoA reductase. New Biol 2: 915–921

Wünsche L, Sattler K, Gradova NB, Meinhold I, Hedlich R, Brendler W, Uhlig H, Rodionova GS, Saikina AI (1981) Composition of the microorganism population in an unprotected fermentation process (Germ). Z Allg Mikrobiol 21: 469–474

Yano K, Kanamuri M, Takagi M (1981) Enrichment of n-alkane assimilation-deficient mutants of *Candida* yeasts by synergistic effect of nystatin and pyrrolnitrin. Agric Biol Chem 45: 1017–1018

Yekhvalova TV, Sharyshev AA, Mikhailova NP, Vyunov KA (1989) Cytochrome P450 content in yeast *Saccharomyces cerevisiae* with alterations of different stages of sterol synthesis (Russ). Biokhimiya 54: 1344–1347

Yokogawa T, Suzuki T, Ueda T, Mori M, Ohama T, Kuchino Y, Yoshinari S, Motoki I, Nishikawa K, Osawa S, Watanabe K (1992) Serine tRNA complementary to the nonuniversal serine codon CUG in *Candida cylindracea*: evolutionary implications. Proc Natl Acad Sci USA 89: 7408–7411

Yoshida M, Hashimoto K (1986a) Assessment of the pathogenicity of yeast used in the production of single cell protein. Agric Biol Chem 50: 2117–2118

Yoshida M, Hashimoto K (1986b) Potential pathogenicity of *Candida maltosa* IAM 12248. Agric Biol Chem 50: 2119–2120

Yoshioka K, Fujita A, Kondo S, Miyake T, Sakaki Y, Shiba T (1992) Production of a unique multi-lamella structure in the nuclei of yeast expressing *Drosophila copia gag* precursor. FEBS Lett 302: 5–7

Zentgraf B (1991a) Microcalorimetric studies of aerobic growth of *Candida maltosa* I. Chemostat cultures. Thermochim Acta 187: 1–8

Zentgraf B (1991b) Microcalorimetric studies of aerobic growth of *Candida maltosa* II. Batch cultures. Thermochim Acta 187: 9–14

Zentgraf B (1991c) Bench-scale calorimetry in biotechnology. Thermochim Acta 193: 243–252

Zentgraf B (1993) Calorimetric studies for optimization of high-performance reactors. Pure Appl Chem 65: 1915–1920

Zimmer T, Schunck W-H (1995) A deviation from the universal genetic code in *Candida maltosa* and consequenses for heterologous expression of cytochromes P450 52A4 and 52A5 in *Saccharomyces cerevisiae*. Yeast 11: 33–41

Zimmer T, Kaminski K, Scheller U, Vogel F, Schunck W-H (1995) In vivo reconstitution of highly active *Candida maltosa* cytochrome P450 monooxygenase systems in inducible membranes of *Saccharomyces cerevisiae*. DNA Cell Biol 14: 619–628

Zinchenko GA, Belov AP (1990) Topography of enzymes of acylglycerol biosynthesis in yeast membranes (Russ). Izv Timiryazev Skh Akad 0(1) 1990: 123–129

Zinchenko GA, Belov AP, Balashova LD, Davidova EG (1990) Specific features of the lipid metabolism in *Candida* yeasts during assimilation of n-alkenes (Russ). Appl Biochem Microbiol (Moscow) 26: 237–241 (English Translation)

CHAPTER 13

Trichosporon

Jakob Reiser, Urs A. Ochsner, Markus Kälin, Virpi Glumoff, and Armin Fiechter

1
History of *Trichosporon* Research

The yeast *Trichosporon cutaneum* belongs to the genus *Trichosporon* Behrend, which was described as early as 1890 (Behrend 1890). This genus includes yeasts which are characterized by budding cells of various shapes, a more or less developed pseudomycelium, or a true mycelium and arthrospores (Fig. 1). *Trichosporon* yeasts may form asexual endospores, but sexual reproduction has not been demonstrated so far (Do Carmo-Sousa 1970). Biochemical characteristics such as hydrolysis of urea, utilization of mono-, di-, tri-, or polysaccharides, etc., as well as studies concerning DNA base composition and DNA relatedness, led Guého et al. (1984) to propose that *Trichosporon* yeasts should be classified into two separate groups. The first group, which appears to be related to the Ascomycetes, includes 13 species with a G+C content lower than 50% (34.7–48.8%) and lacks urease, with *T. margaritiferum* being an exception. The second group appears to be related to the Basidiomycetes and contains 15 species with a G+C content higher than 50% (57–64%) including *T. cutaneum*, *T. beigelii*, and *T. pullulans*, and has the ability to hydrolyze urea. The basidiomycetous nature of some of the *Trichosporon* yeasts is demonstrated by the lamellar structure of the cell walls (Kreger-Van Rij and Veenhuis 1971) and the presence of xylose (Weijman 1979). Furthermore, the diazonium blue B test (van der Walt and Hopsu-Havu 1976) has been applied to a number of *T. beigelii* strains (Kemker et al. 1991). The positive test results are in agreement with the described basidiomycetous affinity of these strains. The phylogenetic relationships of various basidiomycetous yeasts as deduced from ribosomal RNA sequences have been described (Guého et al. 1989; van de Peer et al. 1992). From these studies it appears that *T. beigelii* and *T. cutaneum* are related to the family Filobasidiaceae. Hara et al. (1989) have grouped 44 different strains of *T. beigelii* and related organisms based on differences in the ubiquinones and assimilation of melibiose and raffinose. These initial findings suggested that *T. beigelii* is a heterogeneous group of yeasts consisting of at least four different types of organisms.

Institut für Biotechnologie, ETH-Hönggerberg, 8093 Zürich, Switzerland

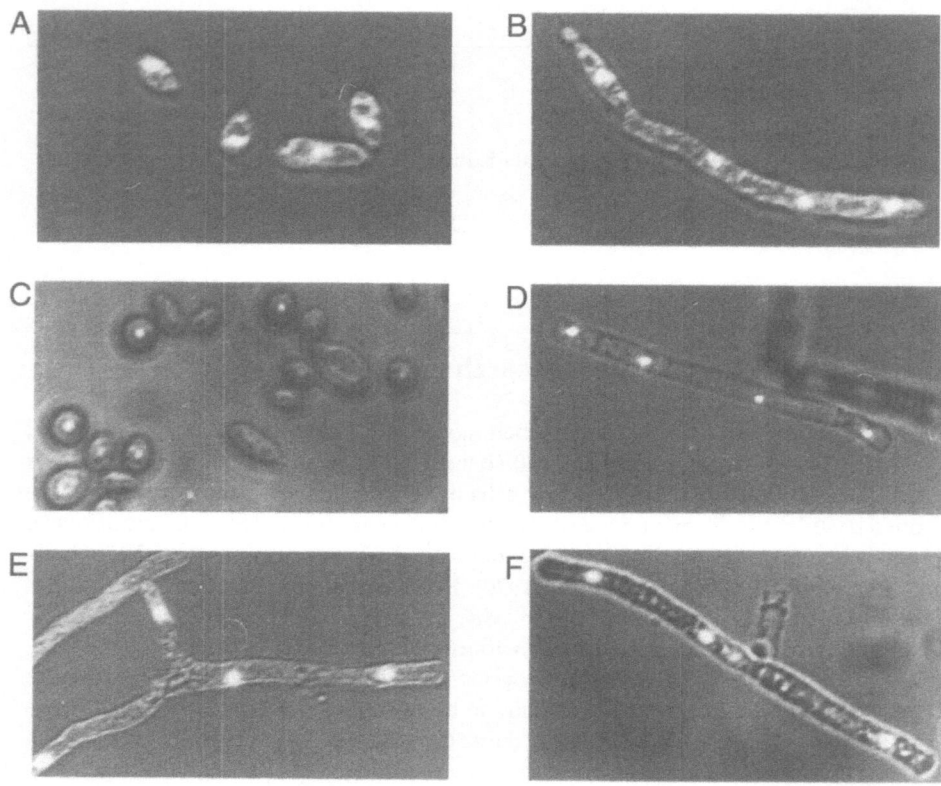

Fig. 1A–F. Cell morphologies of different *Trichosporon* strains. Nuclei were detected by fluorescent microscopy after staining the cells with mithramycin. **A** *T. cutaneum* DSM 70698. **B** *T. cutaneum* ATCC 58094. **C** and **D** *T. cutaneum* ATCC 46490. **E** *T. beigelii* CBS 5790. **F** *T. pullulans* ATCC 10677

T. cutaneum has been isolated from a number of sources including soil, industrial waste water, wood pulp, sludge, and clinical specimens. *T. beigelii* has been found to be the causative agent of white piedra, which is a relatively inconsequential infection of the hair. However, this organism does not appear to be part of the skin flora in healthy subjects (see, e.g., McBride et al. 1988). Certain strains have also been found as opportunistic pathogens causing deep-seated and disseminated infections in immunocompromised patients (reviewed in Hoy et al. 1986). Kemker et al. (1991) have analyzed 15 clinical and environmental strains of *T. beigelii* for similarities by using morphological features, biochemical profiles based on carbon compound assimilation and uric acid utilization, isoenzyme electrophoresis, and restriction fragment length polymorphisms in rRNA genes. The findings suggested that strains that cause invasive disease are distinct from the superficial and nonclinical isolates and that isolates from the skin and mucosae represent a number of different organisms, including some environmental forms. The study also revealed that *T. beigelii* is a complex of genetically distinct organisms and that more than one type is found in clinical samples. Only recently, the genus *Trichosporon* was revised once more based on a number of characteristics such as morphology, ultrastructure of septal pores, coenzyme Q system, G+C content of DNA, DNA/DNA reassociation, and 26S ribosomal RNA partial sequences (Guého et al. 1992a,b; Fig. 2). This work separates the *Trichosporon* isolates from soil into three species: *T. pullulans*, a psychrophilic species which is clearly different from all other *Trichosporon* species, *T. dulcitum*, a mesophilic species, and *T. moniliiforme*. *T. cutaneum* seems to be a rare species never isolated from soil (Guého et al. 1992b). A preliminary inspection of the *T. cutaneum* ATCC 46490 and DSM 70698 strains using culture studies and biochemical characteristics, classifies them into the *T. moniliiforme* group (Guého et al. 1992b; E. Guého, pers. comm.).

Recently, a thermotolerant *T. adeninovorans* strain was described (Middelhoven et al. 1984; Gienow et al. 1990). This yeast is able to grow at 45 °C and

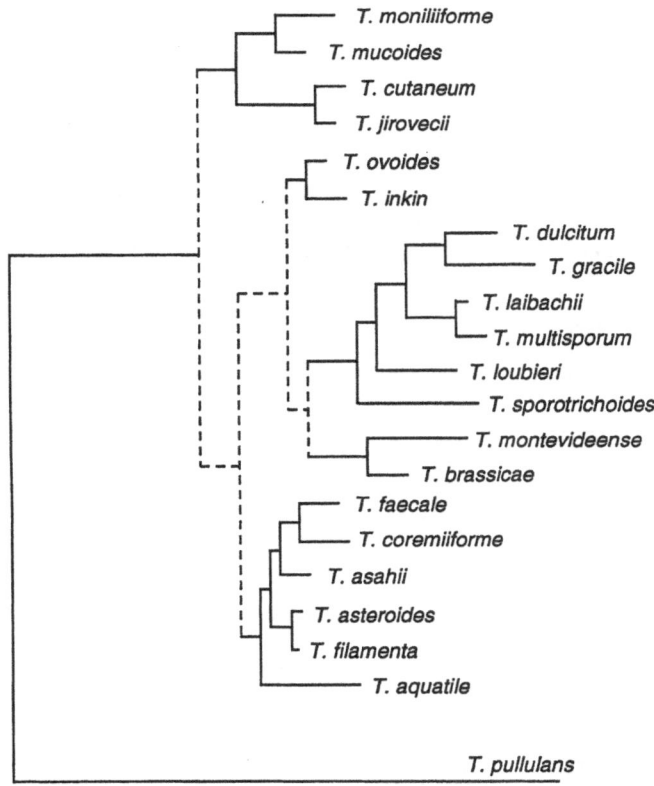

Fig. 2. Classification of *Trichosporon* yeasts based on 26S ribosomal RNA sequences. (After Guého et al. 1992b)

its cell wall structure was found to be ascomycete-like. *T. adeninovorans* is able to use a variety of carbon sources including mono- and disaccharides, starch and n-alkanes. This group of organisms has recently been reclassified and now belongs to the genus *Arxula* (Middelhoven et al. 1991).

Like *T. cutaneum*, *T. pullulans* has a wide substrate spectrum. The maximum temperature of growth for *T. pullulans* is around 23–27 °C (Do Carmo-Sousa 1970) but growth of *T. pullulans* was observed even at 5 °C (Berg et al. 1987).

2
Available Strains and Mutant Collections

A number of different *Trichosporon* strains have been used in diverse applications (Table 1). These strains are all available from the DSM, ATCC, or CBS collections, respectively. Ochsner et al. (1991) have prepared a collection of UV-induced auxotrophic mutants based on the *T. cutaneum* DSM 70698 strain as presented in Table 2. Auxotrophic mutants were also induced in other *T. cutaneum* strains by using nitrosoguanidine mutagenesis (Ykema et al. 1989; Sienko et al. 1992).

3
Media for Different Purposes

For maintenance and cultivation of *T. cutaneum*, a number of different media are being used. Yeast extract peptone dextrose medium (YEPD: 1% Bacto-yeast extract, 2% Bacto-peptone, 2% glucose); synthetic minimal medium (0.67% Bacto-yeast nitrogen base without amino acids (Difco), 2% glucose); minimal medium D

Table 1. *Trichosporon* strains and their applications

Strain	Collection number	Use	Reference
T. cutaneum	DSM 70698	Bioreactor studies Molecular genetics	Käppeli and Fiechter (1981)
T. cutaneum	ATCC 46490	Phenol degradation Phenol biosensor	Neujahr and Varga (1970)
T. cutaneum	ATCC 58094	Degradation of aromatic compounds	Sze and Dagley (1984)
T. cutaneum	ATCC 20509[a]	Lipid production from whey	West et al. (1990)
T. cutaneum	ATCC 62975	Peroxisome biogenesis	Veenhuis et al. (1986)
T. beigelii	CBS 5790	Production of polysaccharidases	Zimmermann and Emeis (1989)
T. pullulans	ATCC 10677	Production of amylases, cellulase, and xylanase	De Mot and Verachtert (1986) Stevens and Payne (1977)

[a] Originally designated *Apiotrichum curvatum*.

Table 2. Auxotrophic mutants of *T. cutanuem* DSM 70698. (After Ochsner et al. 1991)

Mutant number	Auxotrophic requirement	Mutant stability[a]	Remarks
8.1	Adenine	20	Brown colonies
13	Adenine	Leaky	White colonies
31	Adenine	13	Brown colonies
38	Adenine	2	Brown colonies
41[b]	Adenine	<1	Brown colonies
13.1	Arginine	18	
30.1	Arginine	Leaky	
30.2	Arginine	10	
33.1	Arginine	<10	
33.2	Arginine	<10	
70	Arginine	12	
71	Arginine	70	
72	Arginine	<10	
21.2	Histidine	<1	Sweet smell
21.5	Histidine	<1	Sweet smell
17	Histidine	>100	
18	Histidine	1.5	
37	Histidine	15	
29	(Iso)leucine[c]	70	
61	Isoleucine	8	
1.1	Leucine	15	
25.4	Leucine	Leaky	
15	Leucine	20	
16	Leucine	Leaky	
59	Leucine	45	
6	Lysine	<10	
24	Lysine	<1	
32	Lysine	12	
49	Lysine	5	
8	Methionine	2400	
22	Methionine	10	
1.2	Tryptophan	20	
30.4	Tryptophan	Leaky	
43.1	Tryptophan	Leaky	
57	Tryptophan	4	
62	Tryptophan	2	
4	Uracil	200	
7	Uracil	<1	Resistant to 5-fluoro orotate (5-FOA)
16.2	Uracil	<1	Resistant to 5-FOA
35.1	Uracil	Leaky	
46	Uracil	6	Resistant to 5-FOA

[a] The genetic stability of the mutants was determined by plating 10^8 mutant cells on minimal medium and counting the number of revertants after 5 days of incubation at 30 °C.
[b] This mutant was derived from strain ATCC 46490.
[c] This mutant was growing on minimal medium supplemented with either leucine or isoleucine.

(Table 3), and buffered minimal medium D (Table 4). Solid medium is made by adding 2% Bacto-agar (Difco). In supplemented minimal media, the appropriate amino acids are added at a final concentration of 5 to 50 mg ml^{-1}.

Table 3. Minimal medium D, g l^{-1}. (Hug et al. 1974)

Glucose	10
$(NH_4)_2SO_4$	2.0
$(NH_4)_2HPO_4$	0.64
KCl	0.29
$MgSO_4 \cdot 7H_2O$	0.15
$CaCl_2 \cdot 2H_2O$	0.09
$FeCl_3 \cdot 6H_2O$	4.8
$MnSO_4 \cdot H_2O$	3.5
$ZnSO_4 \cdot 7H_2O$	3.0
$CuSO_4 \cdot 5H_2O$	0.78
Biotin	0.01
m-Inositol	20.0
Ca-Pantothenat	10.0
Thiamine (Vitamin B$_1$)	2.0
Pyridoxine (Vitamin B$_6$)	0.5

The medium components are dissolved in tap water and the pH of the medium adjusted to around 5 to 5.5 using 2 M HCl before sterilization. Salts, trace elements, vitamins, and glucose are sterilized separately.

Table 4. Buffered minimal medium D. (Hug et al. 1974; Varga and Neujahr 1970)

Glucose	20 g l^{-1}
or phenol	0.5
NH_4NO_3	4.0
KH_2PO_4	1.5
$NaHPO_4$	1.5
$MgSO_4 \cdot 7H_2O$	200 mg l^{-1}
$CaCl_2 \cdot 2H_2O$	10
$FeSO_4 \cdot 7H_2O$	50.0
$MnSO_4 \cdot H_2O$	2.0
$ZnSO_4 \cdot 7H_2O$	3.5
$CuSO_4 \cdot 5H_2O$	0.78
Biotin	0.01
m-Inositol	20.0
Ca-Pantothenate	10.0
Thiamine (Vitamin B$_1$)	2.0
Pyridoxine (Vitamin B$_6$)	0.5

The medium components are dissolved in tap water and the pH of the medium adjusted to 6.7 before sterilization. Salts, trace elements, vitamins, and glucose are sterilized separately. Phenol is added as a unsterilized 6% solution and glucose as a 50% solution after sterilization.

4
Conservation of Strains

Stock cultures of the strains are maintained at −70 °C in YEPD medium containing 15% glycerol. Alternatively, they are lyophilized and kept at ambient temperature. Working cultures can be stored on YEPD plates or liquid medium at 4 °C. On plates, the colonies stay viable for 4 to 6 weeks, whereas in liquid medium the cells can be kept viable for at least 8 weeks.

5
Genetic Techniques

5.1
Mutant Induction

5.1.1
UV Mutagenesis

To obtain auxotrophic mutants of the *T. cutaneum* DSM 70698 strain, a UV mutagenesis protocol was used. The ATCC 46490 and ATCC 58094 strains turned out to be different in their UV sensitivity (Fig. 3) and the mutant yield was lower as far as the ATCC 46490 strain is concerned.

1. Grow overnight culture in YEPD medium. Remove a 0.1-ml aliquot and centrifuge briefly to pellet the cells. Resuspend cells in 1 ml of sterile water.

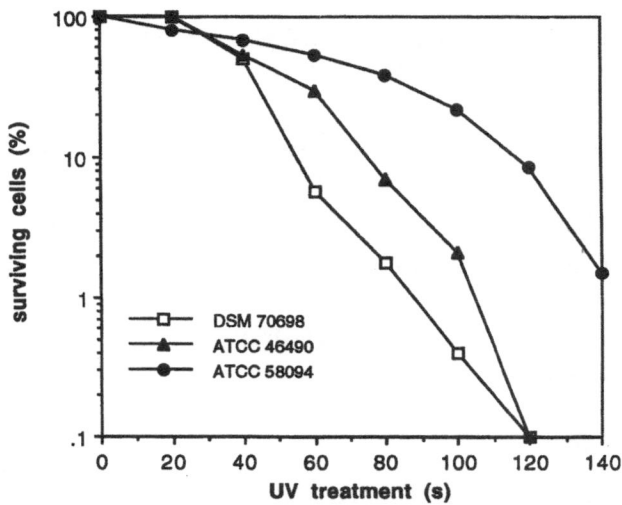

Fig. 3. UV sensitivity of the *T. cutaneum* DSM 70698, ATCC 46490, and ATCC 58094 strains. The curves show the survival of the cells as a function of the length of time of irradiation

2. Separate cell aggregates by an ultrasonic treatment (3 s) and determine the cell density by using a hemacytomoter.

3. Dilute cells into 5 ml sterile water in a sterile glass petri dish to give a final concentration of 10^5 cells ml^{-1}.

4. Rotate open petri dish with cell suspension at about 50 rpm min^{-1} using a laboratory shaker to mix the cells.

5. Turn on UV lamp (Philips-TUV lamp, 15 W) which is located 40 cm above the surface of the liquid. Keep dish rotating and remove 0.1-ml aliquots of the cell suspension every 20 to 30 s for 3 min. Dilute cell aliquots into 1.9 ml of sterile water and plate 0.1-ml samples onto YEPD agar plates. Incubate plates for 1 or 2 days at 30 °C, and count colonies.

For the isolation of mutants scale up the above procedure, choosing the conditions under which the percentage of survivors is between 10 to 20%. Auxotrophic mutants are identified by replica plating colonies from YEPD agar plate onto minimal medium D agar plates. About 100 to 200 colonies per petri dish can be conveniently screened in this way.

5.1.2
Nitrosoguanidine Mutagenesis

This procedure was initially applied for the isolation of fatty acid auxotrophic mutants of the oleaginous yeast *Apiotrichum curvatum* (Ykema et al. 1989). This strain has recently been reclassified as *T. cutaneum* ATCC 20509 (West et al. 1990). Thus, the same method is likely to be applicable to other *Trichosporon* strains as well.

1. Harvest cells from an overnight culture (exponential phase) in YEPD medium and resuspended pellet in 0.1 M phosphate buffer (pH 7.0) to give a final OD_{600} of 10.

2. Add N-methyl-N'-nitro-N'-nitrosoguanidine (MNNG) from a 20 mg ml^{-1} stock solution in acetone to give a final concentration of 0.5 to 1 mg ml^{-1}.

3. Incubate cells at 30 °C with shaking in the dark for various lengths of time.

4. Harvest cells and wash four times with 0.1 M phosphate buffer, pH 7.0. Resuspend cells in the same original volume of phosphate buffer.

5. Determine the percentage of survivors by plating small aliquots onto an appropriate medium (e.g., YEPD agar). The treated cells are viable at 4 °C for 1 to 2 days.

5.2
Preparation of Protoplasts and Protoplast Fusion

1. Grow mutant strains to be fused in 10 ml of YEPD medium to a density of 5×10^7 cells ml^{-1}.

2. Harvest cells and wash twice with sterile water and resuspend pellets in 2 ml of protoplasting solution containing 1% protoplast forming enzyme (Boehringer Mannheim, Germany), 2% Novozym 234 (Novo Industri A/S, Bagsvaerd, Denmark), and 0.6 M KCl.

3. Check kinetics of protoplast formation in a microscope and harvest protoplasts by centrifugation (3000 rpm, 5 min) as soon as 90% of the cells are protoplasted. Wash protoplasts twice with 2 ml of 0.6 M KCl and resuspend in 0.1 ml of 0.35 M CaCl₂.

4. Mix 25-μl aliquots of the protoplasts of the two strains to be fused, centrifuge as above, and resuspend in the liquid remaining in the tube. Add 1 ml of fusion solution (35% PEG 4000, 50 mM CaCl₂) and incubate mixture for 10 min at room temperature.

5. Mix 0.2-ml aliquots of the PEG/protoplast suspension with 7 ml of protoplast regeneration agar (minimal medium D containing 0.35 M CaCl₂ and 2% Bacto agar) at 48 °C and pour onto prewarmed protoplast regeneration agar plates. Incubate plates for 4 to 5 days at 30 °C.

The frequency of protoplast regeneration is measured by plating aliquots of the protoplast suspension onto protoplast regeneration agar plates supplemented with the appropriate amino acids.

6
Biochemical Techniques

6.1
Preparing Trichosporon *Chromosomal DNA*

6.1.1
Large-Scale Procedure 1

This is modification of a procedure which had initially been designed for the isolation of DNA from white-rot fungi (Raeder and Broda 1988).

1. Grow cells in YEPD or minimal medium D with shaking at 30 °C.

2. Harvest cells by centrifugation and freeze cell pellet at –20 °C.

3. Grind 10 g of the frozen cell pellet in liquid nitrogen using a mortar. Continue until the cell pellet is finely powdered.

4. Transfer ground material to a 50-ml centrifuge tube with a cap and resuspend in 20 ml of extraction buffer (200 mM Tris-HCl, pH 8.5, 250 mM NaCl, 25 mM EDTA, 0.5% SDS). Add 50 μl of RNase A (20 mg ml⁻¹) and incubate at 37 °C for 45 min.

5. Add an equal volume of a 1 : 1 mixture of salt-saturated phenol and chloroform. Close tube and seal well and mix gently overnight at room temperature using a rotating device.

6. Centrifuge tube at 10 000 rpm for 45 min (4 °C).

7. Transfer aqueous phase to a new tube and extract with an equal volume of chloroform. Centrifuge tube at 10 000 rpm for 30 min (4 °C).

8. Transfer aqueous phase to a 30-ml Corex tube. Add 0.54 vol isopropanol. Mix and store tube for 60 min at −20 °C.

9. Spin DNA at 10 000 rpm for 30 min at 4 °C.

10. Redissolve pellet in 1–2 ml TNE (10 mM Tris-HCl, pH. 8.0, 0.15 NaCl, 1 mM EDTA), and extract with an equal volume of chloroform. Spin, transfer aqueous phase, and add 2.5 vol ethanol.

11. Spool DNA using the tip of a Pasteur pipette and let dissolve in 1 to 2 ml TE buffer (10 mM Tris-HCl, pH. 8.0, 1 mM EDTA) overnight.

12. Read the OD_{260}.

This DNA may contain residual amounts of RNA. It can be used for restriction digestion and Southern blotting (see Fig. 4).

6.1.2
Large-Scale Procedure 2

This method is a modification of the one described by Rodriguez and Tait (1983) for the isolation of high molecular weight DNA from *Saccharomyces cerevisiae*.

1. Grow cells overnight in YEPD medium.

2. Harvest cells at 3000 rpm for 5 min and resuspend in sterile water. Spin as before.

3. Add 1 ml spheroplast buffer (SB buffer: 0.2 M Tris-HCl, pH 7.5, 1 M sorbitol, 0.1 M EDTA, 0.1 M β-mercaptoethanol) per g cell pellet, and resuspend cells.

4. Add 1 ml filter-sterilized Novozym 234 (12 mg ml^{-1} in spheroplast buffer) per g of cells. Mix well and incubate for 60 min at 30 °C. Shake tube gently at 60 to 100 rpm.

5. Centrifuge spheroplasts at 3000 rpm for 5 min at 4 °C and resuspend in 2 ml of SB buffer.

6. Add an equal volume of TEN buffer (10 mM Tris-HCl, pH 7.5, 1 mM EDTA, 10 mM NaCl) and SDS to a final concentration of 1%.

7. Add 0.5 ml RNase (10 mg ml^{-1} RNase A in 0.1 M Na acetate, pH 4.8, 0.3 mM EDTA), mix gently and incubate for 2 h at 37 °C.

8. Add 1 ml pronase (2 mg ml^{-1} in TEN buffer) and incubate for another 2 h at 37 °C. Swirl mixture periodically.

9. Heat mixture at 65 °C for 30 min and cool to room temperature.

1 2 3 4 5 6 7 8   1 2 3 4 5 6 7 8

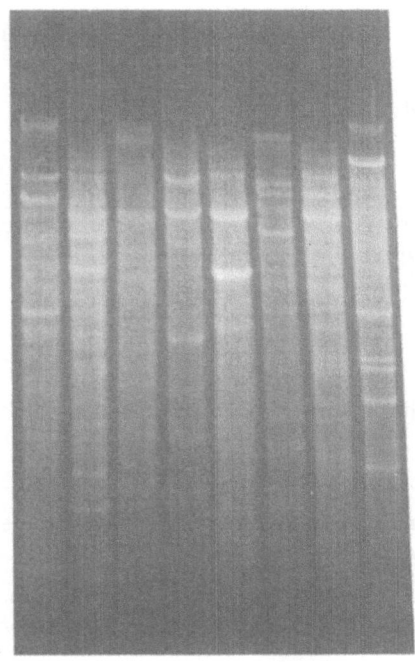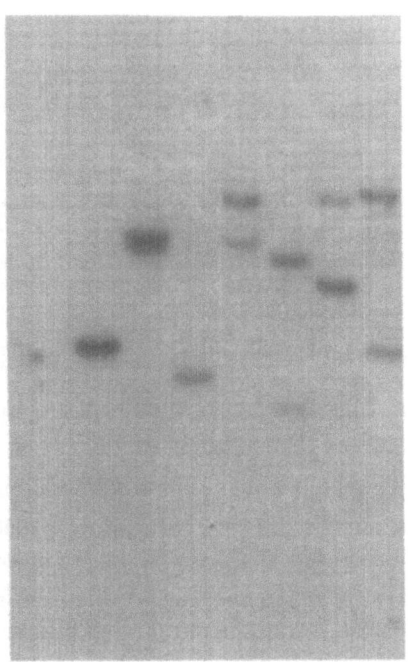

A   B

Fig. 4A,B. Southern blot analysis of *Trichosporon* DNAs. **A** Ethidium bromide-stained agarose gel. **B** Southern blot analysis using the *Neurospora crassa* β-tubulin gene (Orbach et al. 1986) as a probe. The DNAs from *T. cutaneum* ATCC 62975 (*lane 1*), ATCC 20509 (*lane 2*), DSM 70698 (*lane 3*), ATCC 58094 (*lane 6*), and ATCC 46490 (*lane 7*), and from *T. pullulans* ATCC 10677 (*lane 4*), *T. beigelii* CBS 5790 (*lane 5*), and *Cryptococcus elinovii* (*lane 8*) were cut with *Pvu*II. The fragments were separated in a 0.8% agarose gel, transferred to a Gene Screen Plus membrane (Du Pont), which was then hybridized with a ³²P-labeled, 3.1-kb long *Hind*III fragment containing the β-tubulin gene. Hybridization was performed in 50% formamide in the presence of 10% dextran sulfate at 42 °C overnight. The filters were washed in 2× SSC/0.1% SDS twice for 15 min at room temperature and then in 0.1× SSC/ 0.1% SDS for 30 min at 42 °C

10. Add an equal volume of phenol-chloroform-isoamyl alcohol (25:24:1, v/v/v) and mix gently. Centrifuge for 5 min at 3000 rpm (4 °C).

11. Transfer aqueous phase to a fresh tube using a wide-bore pipette.

12. Add an equal volume of chloroform-isoamyl alcohol (24:1, v/v), mix gently and centrifuge as in step 10.

13. Add 1/25 vol 5 M NaCl and 2 vol ethanol. Store for 15 min on ice.

14. Centrifuge for 15 min at 10 000 rpm (4 °C). Dry pellet and redissolve in 10 ml of TEN buffer.

15. Add 0.1 ml RNase A (10 mg ml^{-1} RNase A in 0.1 M Na-acetate, pH 4.8, 0.3 mM EDTA), mix gently and incubate for 1 h at 37 °C.

16. Repeat steps 10 to 13 and centrifuge for 30 min at 10 000 rpm (4 °C). Wash pellet with 80% ethanol and dry.

17. Dissolve pellet in 1 to 3 ml of TEN buffer.

18. Measure OD$_{260}$.

This DNA can be used for restriction digestion and Southern blotting and for the construction of genomic libraries (see Ochsner et al. 1991). RNase A and pronase should be pretreated as described by Maniatis et al. (1982).

6.1.3
Small-Scale Procedure

This procedure was originally designed for the isolation of DNA from *S. cerevisiae* (Hoffman and Winston 1987).

1. Grow a 10-ml culture in minimal medium D to stationary phase (1 to 2 days).

2. Collect cells by centrifugation at 4000 rpm for 5 min using a table top centrifuge and resuspend cell pellet in 0.5 ml of water.

3. Transfer cell suspension to an Eppendorf tube and spin for 5 s at 13 000 rpm. Decant supernatant and resuspend cells in the residual liquid by brief vortexing.

4. Add 0.2 ml of a buffer containing 2% Triton X-100, 1% SDS, 100 mM NaCl, 10 mM Tris-HCl, pH 8.0, 1 mM EDTA. Add 0.2 ml phenol/chloroform/isoamyl alcohol (25:24:1, v/v/v) and 0.3 g of acid-washed glass beads (0.45–0.5 mm diameter).

5. Vortex for 3 to 4 min at full speed. Add 0.2 ml TE buffer (10 mM Tris-HCl, pH 8.0, 1 mM EDTA).

6. Spin for 5 min at 13 000 rpm. Transfer aqueous layer to a new tube. Add 1 ml of ethanol. Invert the tube to mix.

7. Spin for 2 min at 13 000 rpm. Resuspend pellet in 0.4 ml TE buffer containing 50 μg RNase A. Incubate for 5 min at 37 °C. Add 10 μl 5 M potassium acetate and 1 ml of ethanol. Invert the tube to mix.

8. Spin for 2 min at 13 000 rpm. Dry DNA pellet under reduced pressure. Resuspend DNA in 50 μl of TE-buffer.

This DNA can be used for restriction digestion and Southern blotting.

6.2
Preparing Trichosporon *Total RNA*

This method was initially set up for the isolation of RNA from fungal mycelia (Teeri et al. 1987).

1. Grow cells in desired medium and collect by centrifugation at 5000 rpm for 5 min at 4 °C. Wash once in 4 ml of water and transfer to a plastic capped tube.

2. Spin down at 5000 rpm for 10 min at room temperature and thoroughly drain the pellet. Freeze tube in liquid nitrogen.

3. Grind frozen cell pellet to a fine powder using a mortar under liquid nitrogen.

4. Transfer frozen cell powder to a beaker containing 1 ml of a guanidine thiocyanate solution (4 M guanidine thiocyanate (Fluka), 0.5% sodium N-lauroylsarcosine (Fluka), 0.1 M β-mercaptoethanol, 25 mM Na-citrate, pH 7.0) per g of wet cells. Briefly warm beaker at 37 °C to thaw the cells and remove cell debris by centrifugation at 6000 rpm for 10 min at 4 °C.

5. To the supernatant add solid CsCl (1 g per 2.5 ml of supernatant) and mix to dissolve the CsCl.

6. To a Beckman SW41 polyallomer ultracentrifuge tube add 1.2 ml of a solution containing 5.7 M CsCl and 0.1 M EDTA, pH 7.0. Carefully layer supernatant on top and centrifuge tube using a Beckman SW41 ultracentrifuge rotor at 33 000 rpm for 20–24 h at 15 °C.

7. Remove liquid phase above the tightly packed interphase on top of the CsCl cushion using a Pasteur pipette. Carefully remove interphase and CsCl cushion and cut off the bottom of the tube using a scalpel. The RNA pellet should now be visible as a gelatinous, transparent layer at the bottom of the tube.

8. Rinse RNA pellet briefly with 50 μl of RNase-free water. Homogenize pellet in 50 to 500 μl of RNase-free water and transfer to a sterile Eppendorf microfuge tube, making sure that no RNA remains attached to the bottom of the ultracentrifuge tube.

9. Dissolve the RNA by mixing the contents of the tube every now and then on ice for at least 4 h. Spin tube at 13 000 rpm for 5 min at 4 °C. Transfer supernatant to a fresh tube and determine concentration by measuring the OD_{260}.

This RNA can be used for Northern blots and nuclease protection assays. For the subsequent isolation of poly(A)+ RNA the protocol of Maniatis et al. (1982) was followed. The procedure as described above failed when the isolation of RNA from cells grown on xylan was attempted. To circumvent this problem the cells were powdered as described in step 3 above and the RNA processed by using a Pharmacia QuickPrep Micro mRNA Purification kit (1 ml of extraction buffer per 2.5 g of powdered cells). This yielded poly(A)+ RNA which was useful for the synthesis of cDNA. The cell pellet as described in step 2 above can either be kept in this form at −70 °C or processed immediately. Pretreat CsCl solution, water, pipette tips and tubes with diethyl pyrocarbonate and autoclave (Maniatis et al. 1982). The polyallomer ultracentrifuge tubes are soaked for at least 30 min in 0.5% diethyl pyrocarbonate in ethanol and then rinsed thoroughly with autoclaved water.

6.3
Preparing Trichosporon *Protein Extracts*

6.3.1
Large-Scale Extracts

This procedure is used when it is necessary to make an extract from 100 g of cell paste or more.

1. Thaw frozen cell pellet in buffer P containing 50 mM potassium phosphate, pH 7.6, 1 mM β-mercaptoethanol, 0.1 mM EDTA, and 1 mM phenyl-methanesulfonyl fluoride (PMSF) to create a thick slurry.

2. Disrupt cells mechanically in a Dyno Mill (Bachofen AG, Basel, Switzerland) using glass beads (0.45 mm diameter).

3. Remove cell debris by centrifugation (25 000 g, 30 min, 4 °C).

Extracts of *T. cutaneum* cells have also been produced by sonication (see Sejlitz and Neujahr 1987).

6.3.2
Small-Scale Extracts

This method was first described by Lehmann et al. (1989) and Kemker et al. (1991).

1. Harvest 3 ml of a fresh culture by centrifuging and wash pellet once with ice-cold 50 mM Tris-HCl, pH 8.0.

2. Break cells by vortexing with 0.3 g of acid washed glass beads (diameter 0.4–0.5 mm) in the presence of 200 μl of yeast lysis buffer (50 mM Tris-HCl, pH 8.0, 0.1% Triton X-100, 0.5% SDS).

3. Centrifuge extract at 13 000 rpm for 2 min at room temperature.

For SDS PAGE analysis mix 30 μl of the supernatant with an equal volume of 2× concentrated SDS PAGE sample buffer (Laemmli 1970) and heat for 5 min at 100 °C. For the isolation of enzymatically active proteins, the same procedure is used, with the exception that the yeast lysis buffer is replaced with 100 mM Tris-HCl, pH 8.0. A protocol for measuring enzymatic activity in situ has been presented by Mörtberg and Neujahr (1985).

7
Molecular Techniques

7.1
Transformation Systems Based on Dominant Markers

1. One hundred ml of YEPD medium are inoculated with 5 ml of an overnight culture of *T. cutaneum* and the culture is grown at 30 °C (150 rpm) until the

OD_{600} is between 1.1 and 1.3, corresponding to approximately 2×10^7 cells ml^{-1}.

2. The cells are centrifuged for 5 min at 3000 rpm and the pellet washed twice with 5 ml of buffer I (0.65 M sorbitol, 0.1 mM EDTA, 0.1 mM DTT, 10 mM Tris-HCl, pH 7.5).

3. One ml of buffer II (2.0 M sorbitol, 0.1 mM EDTA, 0.1 mM DTT, 10 mM Tris-HCl, pH 7.5) is added per g of cells and the cells are resuspended.

4. To the cells, an equal volume of a sterilized solution of Novozym 234 (12 mg ml^{-1} in buffer II) is added and the cell suspension incubated at 30 °C without shaking. Spheroplast formation is tested by diluting a 50-μl cell aliquot into 1 ml of buffer II and by measuring the OD_{600}; 100 μl of 10% SDS are then added and the OD_{600} is determined again. After 15 to 20 min the ratio of the OD_{600} values of the SDS-treated sample and the untreated sample is usually less than 10%.

5. The suspension is centrifuged for 3 min at 3000 rpm and washed gently with 5 ml of buffer III (1.5 M sorbitol, 0.1 mM EDTA, 0.1 mM DTT, 10 mM Tris-HCl, pH 7.5) twice.

6. The washed spheroplasts are suspended in 10 ml SY (1.0 M sorbitol, 67% YEPD) and incubated for 1 h at 30 °C without shaking.

7. The spheroplasts are pelleted as decribed above and finally resuspended in 1 ml of SYTC (0.9 M sorbitol, 67% YEPD, 10 mM Tris-HCl, pH 7.5, 10 mM CaCl$_2$).

8. For transformation, 1 to 10 μg plasmid DNA (previously purified using a Quiagen column) are added to 0.1 ml of the spheroplast suspension in SYTC in a Falcon 2059 tube.

9. After a 15-min incubation at room temperature, 1 ml of a PEG solution (20% PEG 4000 (Merck), 10 mM Tris-HCl, pH 7.5, 10 mM CaCl$_2$) are added and the suspension is mixed gently and incubated at 30 °C for 20–30 min without shaking.

10. Five ml of top agar (YEPD, 0.9 M sorbitol, 2% agar) kept at 48–50 °C containing the desired antibiotic is added to 0.2–1.1 ml of the transformation suspension and the mixture is poured onto a regeneration agar plate (YEPD, 0.9 M sorbitol, 2% agar) containing the appropriate antibiotic (100 μg ml^{-1} of hygromycin B, Boehringer-Mannheim, Germany, or 10 μg ml^{-1} of phleomycin, Cayla, Toulouse, France).

All steps are carried out at room temperature unless noted otherwise. The transformants are counted after 3 to 6 days of incubation at 30 °C. For determining the regeneration frequencies, spheroplasts are treated as decribed above, except that no DNA is added, and then diluted with a buffer containing 0.9 M sorbitol and 10 mM Tris-HCl, pH 7.5, and plated with or without the addition of top agar onto regeneration agar plates. The Novozym 234 solution is sterilized by passing it through a 0.45-μ filter. The purity of the plasmid DNA is critical for high efficiency transformation, although less purified DNA preparations can also been used (see Ochsner et al. 1991).

The *Aspergillus nidulans* plasmid, pAN7-1 (Punt et al. 1987), which carries the *Escherichia coli* hygromycin B phosphotransferase (*hph*) gene (Gritz and Davies 1983) was used successfully for transforming the *T. cutaneum* DSM 70698, ATCC 46490 and ATCC 58094 strains, whereas the *A. nidulans* plasmid pAN8-1 (Mattern and Punt 1988), carrying the *Streptoalloteichus hindustanus* phleomycin resistance (*ble*) gene (Drocourt et al. 1990) was useful in the *T. cutaneum* DSM 70698 and the ATCC 58094 strains.

7.2
Transformation Systems Based on Cloned Biosynthetic Genes

In the transformation experiments dealing with auxotrophic mutants, the transformed spheroplasts are collected by centrifugation (4000 rpm, 3 min, room temperature) after the PEG treatment and resuspended in 2 ml of 0.9 M sorbitol, 10 mM Tris-HCl, pH 7.5, mixed with 2 ml of 2% alginate (Sigma, low viscosity) in 0.9 M sorbitol, 10 mM Tris-HCl, pH 7.5, and poured onto SRA plates (minimal medium D, 0.9 M sorbitol, 20 mM $CaCl_2$, 2% Bacto agar) and incubated at 30 °C. Note that the alginate procedure cannot be used in conjunction with hygromycin B or phleomycin selection.

7.3
Genes from Trichosporon cutaneum

With a view toward designing transformation and expression vectors for *Trichosporon* yeasts, a number of genes have been isolated. A cosmid clone capable of complementing the arginine auxotrophic arg70 mutant strain (Table 2) was isolated by using a sib-selection strategy (Ochsner et al. 1991). The cosmid clone was subsequently subcloned and the sequence of a complementing 4.5-kb DNA fragment determined (Glumoff 1992; Reiser et al. 1994). The DNA sequence revealed the presence of a gene encoding the large subunit of the mitochondrial carbamoyl phosphate synthetase (CPSA). The *T. cutaneum* CPSA gene has the capacity to encode a protein of 1170 amino acids, and its coding region is interrupted by a short intron. The derived amino acid sequence is 62% identical and 77% similar to the sequence of the corresponding protein of *Saccharomyces cerevisiae*. The codon usage in the CPSA gene (Table 5) was found to be similar to the one in filamentous fungi.

For the design of expression vectors, the strongly expressed phenol hydroxylase (PHY) gene has been isolated (Kälin et al. 1992; Kälin 1993). The comparison of the codon usage in the two genes reveals a high similarity, with the exception of the codons for arginine and glycine. In the case of arginine, the preferred codon in PHY is CGC (79%) compared to CGT (50%) in CPSA. In the case of CPSA, the

Table 5. Comparison of the codon usages in the phenol hydroxylase (PHY) and the carbamoyl phosphate synthetase A (CPSA) genes from *T. cutaneum*. (Glumoff 1992; Kälin et al. 1992)

		PHY	CPSA			PHY	CPSA			PHY	CPSA			PHY	CPSA
Phe	TTT	23	13	Ser	TCT	11	8	Tyr	TAT	8	8	Cys	TGT	11	18
	TTC	77	87		TCC	21	16		TAC	92	92		TGC	89	82
Leu	TTA	0	1		TCA	11	2	End	TAA	100	0	End	TGA	0	100
	TTG	2	4		TCG	44	66	End	TAG	0	0	Trp	TGG	100	100
	CTT	12	21	Pro	CCT	4	15	His	CAT	4	0	Arg	CGT	12	50
	CTC	81	70		CCC	68	67		CAC	96	100		CGC	79	32
	CTA	5	1		CCA	7	5	Gln	CAA	13	0		CGA	2	0
	CTG	0	3		CCG	21	13		CAG	87	100		CGG	5	5
Ile	ATT	29	30	Thr	ACT	20	29	Asn	AAT	15	0	Ser	AGT	4	0
	ATC	71	69		ACC	70	69		AAC	85	100		AGC	9	8
	ATA	0	1		ACA	0	1	Lys	AAA	2	0	Arg	AGA	0	3
Met	ATG	100	100		ACG	10	1		AAG	98	100		AGG	2	10
Val	GTT	18	19	Ala	GCT	17	22	Asp	GAT	10	0	Gly	GGT	23	46
	GTC	62	75		GCC	60	64		GAC	90	100		GGC	65	45
	GTA	2	1		GCA	17	2	Glu	GAA	2	0		GGA	7	6
	GTG	18	5		GCG	6	12		GAG	98	100		GGG	5	3

The numbers refer to the occurrence (in percent) of a given codon.

percentage of G and C in the third position of the glycine codon is unusually low (48%) compared to the other amino acids and to PHY (70%; Table 5). A cDNA encoding a *cis*, *cis*-muconate lactonizing enzyme from the *T. cutaneum* ATCC 58094 strain was isolated and sequenced. The deduced amino acid sequence exhibited moderate sequence similarity with 3-carboxy-*cis*, *cis* muconate lactonizing enzyme from *Neurospora crassa* (Mazur et al. 1994).

It has recently been found that the leucine codon CUG is read as leucine in *T. cutaneum* JCM 1533 (Ohama et al. 1993). This is in contrast to *Candida cylindraceae* and related species, in which CUG is read as serine.

8
Specific Biochemical Properties of *Trichosporon* Yeasts

8.1
Physiology of Trichosporon *Yeasts*

Trichosporon yeasts have the potential to use a very large variety of carbon sources (Laaser et al. 1989) and *T. cutaneum, T. beigelii*, and *T. pullulans* have been shown to grow on various monosaccharides including pentoses and hexoses, on disaccharides, such as cellobiose, maltose, lactose, sucrose, melibiose, and trehalose (Mörtberg and Neujahr 1986) and on polysaccharides including xylans (Hrmová et al. 1984), ball-milled filter paper (Stevens and Payne 1977), starch (De Mot and Verachtert 1986), and on pectic acid, carboxymethylcellulose, and locust-bean gum (Zimmermann and Emeis 1989). Other carbon sources include ethylamine (Veenhuis et al. 1986), uric acid (Middelhoven et al. 1983), propionate, D-alanine, D-methionine, and oleic acid (M. Veenhuis, pers. comm.), all of which lead to the formation of peroxisomes (Veenhuis et al. 1985, 1986). D-glucarate, galactarate and L-tartarate (Schneider et al. 1990), and cyclohexanecarboxylic acid (Hasegawa et al. 1982) have also been found to serve as carbon sources for certain strains of *Trichosporon*. Furthermore, *T. cutaneum* and *T. beigelii* are capable of using various aromatic compounds as sole carbon and energy sources. The biochemistry and physiology of phenol, cresol, salicylate, benzoate and anthranilate degradation and the metabolism of aromatic amino acids have been the subjects of extensive research (see Fig. 5; reviewed in Dagley 1985; Neujahr 1990). These properties indicate the extraordinary potential of *T. cutaneum* for the efficient conversion of various carbon sources into biomass. *T. cutaneum* has a purely oxidative metabolism and does not form ethanol even under oxygen limitation (Käppeli and Fiechter 1982) and it shows a high biomass yield of around 50 to 55% on glucose (Fiechter et al. 1987). Hess (1988) has worked out the conditions for high density cultivation of *T. cutaneum*. Biomass concentrations as high as 200 g (dry weight) per liter could be harvested in a continuous cultivation process using a cell recycling system. The highest productivity, 22 g per liter per h, was obtained at a cell density of 120 g per liter. Thus, the broad substrate range in conjunction with the bioreactor technology developed for *T. cutaneum* is an attractive feature in view of the synthesis of valuable foreign proteins.

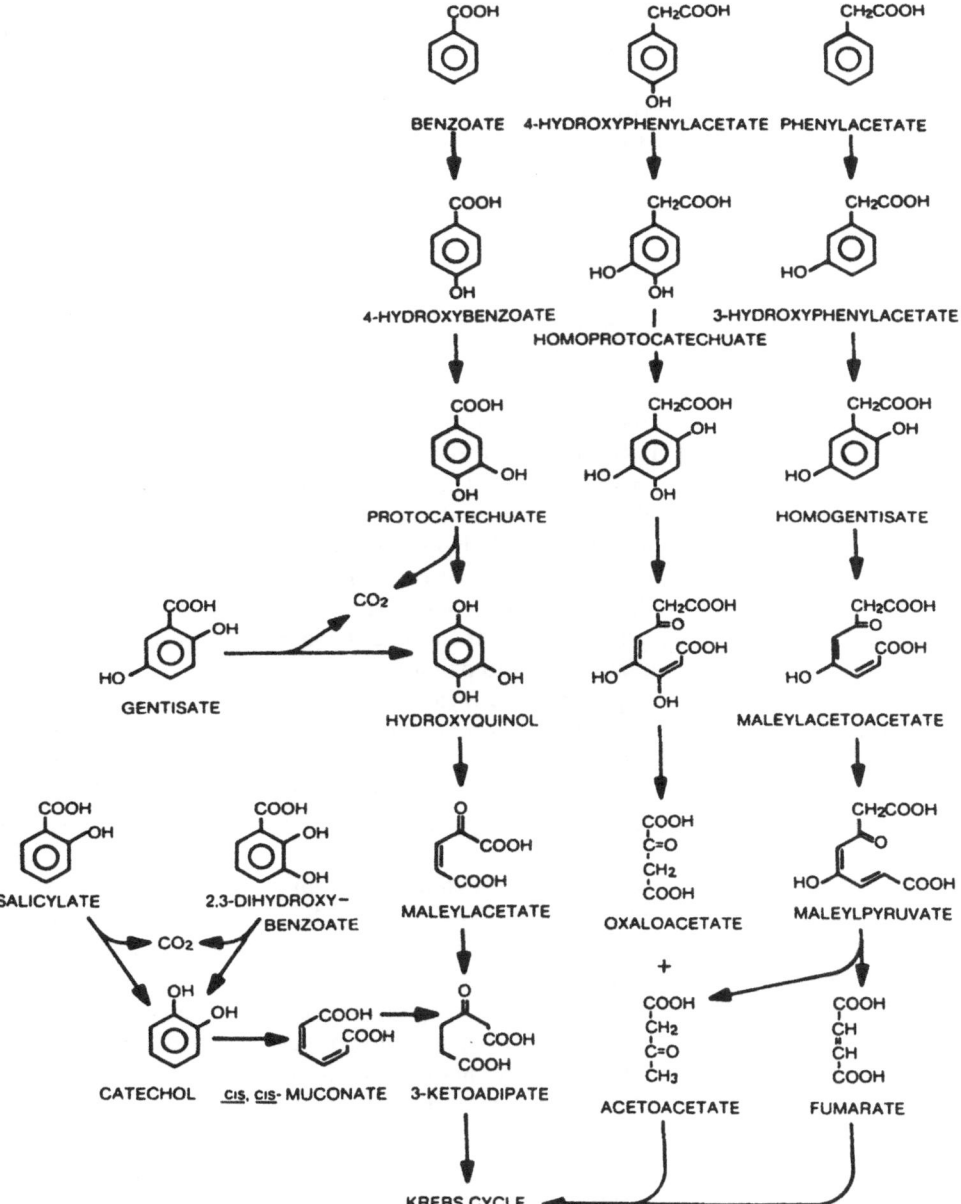

Fig. 5. Metabolism of aromatic compounds by *T. cutaneum.* (After Rochkind-Dubinsky et al. 1987)

Trichosporon yeasts have the capacity to accumulate lipids and thus they belong to the oleaginous yeasts (Ratledge 1988; West et al. 1990). The process of lipid accumulation occurs when an oleaginous yeast is grown in a medium with a high carbon to nitrogen ratio (usually about 30:1), so that the excess carbon is assimilated without conversion to protein or nucleic acids. *T. pullulans* has been found to accumulate more than 65% of its biomass as lipid. The order of abundance of the fatty acyl groups of the lipids was oleate > palmitate > linoleate > stearate (Ratledge 1988).

8.2
Biochemistry of Trichosporon *Yeasts*

T. cutaneum cells can be grown in the presence of 0.05% phenol as a carbon source and the growth rates were found to be equal on both phenol and glucose (Spånning and Neujahr 1987). The first three enzymes involved in phenol metabolism including phenol hydroxylase (Neujahr and Gaal 1973), catechol 1,2-oxygenase (Varga and Neujahr 1970) and *cis, cis*-muconate cyclase (Gaal and Neujahr 1980) have been isolated and characterized in detail. The levels of the enzymes involved in phenol degradation were found to be some 50–400 times higher in phenol-grown than in glucose-grown cells (Gaal and Neujahr 1981) and phenol hydroxylase comprises 2–5% of the total cell protein in fully induced cells (Neujahr and Gaal 1973). In addition to phenol, resorcinol, catechol, cresols, and fluorophenols can induce phenol hydroxylase (Gaal and Neujahr 1981). A full-length phenol hydroxylase cDNA has recently been sequenced and overexpressed in an enzymatically active form in *E. coli* (Kälin et al. 1992), and the recombinant enzyme was purified to homogeneity (Waters and Neujahr 1994). The *E. coli*-derived enzyme was crystallized and a preliminary X-ray analysis was reported (Enroth et al. 1994).

Similarly to *Kluyveromyces lactis*, *T. cutaneum* contains an inducible β-galactosidase (Mörtberg and Neujahr 1986; West et al. 1990). The *T. cutaneum* ATCC 46490 strain studied by Mörtberg and Neujahr revealed a cell-wall-bound extracellular enzyme, whereas the protein of the ATCC 20509 strain studied by West et al. (1990) was intracellular. A four- to sevenfold induction of β-galactosidase was detected in cultures grown on lactose, lactulose, or galactose and the enzyme-hydrolyzed lactose, lactulose, and nitrophenyl-β-D-galactoside (West et al. 1990). A second intracellular lactose hydrolase, a β-glycosidase, was described by West et al. (1990). It had a wider substrate spectrum and hydrolyzed lactose, nitrophenyl-β-D-galactosides, 4-nitrophenyl-β-D-glucoside, cellobiose, laminaribiose, laminaritriose, and sophorose efficiently. This enzyme was induced by lactose, lactulose, or galactose and also by cellobiose. Hrmová et al. (1984) had earlier detected an intracellular β-glucosidase capable of hydrolyzing cellobiose. The best inducer of this β-glucosidase was thiocellobiose.

Using ball-milled filter paper as a growth substrate, *T. cutaneum* and *T. pullulans* were found to produce appreciable amounts of cellulase activity (Stevens and Payne 1977). The main products of cellulose degradation were cellobiose and glucose. Xylanase activity was also present in the same culture filtrates.

A xylan- and xylose-inducible endo-1,4-β-D-xylanase has been purified from the culture supernatant of the *T. cutaneum* DSM 70698 strain grown on xylan (Stüttgen and Sahm 1982). This enzyme comprised about 27% of the secreted proteins at levels around 50 mg per liter. When the cells were grown on glucose or cellobiose, no β-xylanase activity could be detected. Hrmová et al. (1984) described a different strain of *T. cutaneum* which produced a xylan-inducible β-xylanase. The enzyme could not be induced by xylose, but the activity was increased 200-fold by xylan, about 700-fold by methyl-β-xyloside, but only slightly by cellobiose. Zimmermann and Emeis (1989) have isolated and characterized a number of polysaccharidases from the culture supernatant of *T. beigelii*, strain CBS 5790. An amylase was present when the cells were grown on starch, and pectinases were found in starch, carboxymethyl cellulose (CMC), xylan, or locust-bean gum-grown cells. The production of these enzymes was strongly repressed by glucose.

The *T. pullulans* ATCC 10677 strain produced α-amylase and glucoamylase activities in a medium containing corn steep liquor and corn starch or soluble starch (De Mot and Verachtert 1986). For both amylases, the maximum concentrations were found in stationary phase cultures. In addition, pullulanase activity was found in the glucoamylase fraction and cyclodextrinase activity was present in the α-amylase fraction. The electrophoretic analysis revealed that the α-amylase activity was due to a single protein. The glucoamylase, however, occurred in multiple forms. The four glucoamylases and the α-amylase were found to be glycosylated.

Büttner et al. (1987) have isolated an inducible amylase from *T. adeninovorans* which hydrolyzed starch by cleaving single glucose units, thus showing that it is a glucoamylase. Maltose was found to be a better inducer than soluble starch. To relieve carbon catabolite repression, mutants of *T. adeninovorans* resistant of 2-deoxy-D-glucose were selected (Büttner et al. 1989). Relative to wild-type cells, a 100-fold derepression of glucoamylase activity was detected in such mutant cells grown on glucose.

9
Trichosporon Cell Biology: Staining of Nuclei

Trichosporon cutaneum nuclei can be stained either using DAPI (4'-6-diamidino-2-phenylindole) as described by Williamson and Fennell (1975) or by using mithramycin A (Slater 1976; see Fig. 1).

1. Collect 1 ml of a cell culture and wash cells with water.

2. Sonicate cells briefly and adjust concentration to 10^8 cells ml^{-1}.

3. Mix 50 μl of the cell suspension with 50 μl of a 0.4 mg ml^{-1} solution of mithramycin in 50% aqueous ethanol, 30 mM MgCl$_2$. Alternatively use 50 μl of a 3 μg ml^{-1} solution of DAPI instead of mithramycin.

4. Examine cells under a fluorescence microscope using a 515-nm suppression filter for mithramycin stained nuclei and a 420-nm suppression filter for DAPI-stained nuclei.

10
Applications of *Trichosporon* Yeasts

Trichosporon yeasts are currently being used in a number of applications (Table 1). Due to its robustness, *T. cutaneum* has been used for many years as a model organism in different bioreactor systems (Käppeli and Fiechter 1981; Jaramillo 1985; Hess 1988; Küng and Moser 1986; Yonsel and Deckwer 1990). During the cultivation of *T. cutaneum*, no by-products like ethanol or acetate are detected and 96 to 100% of the carbon source used are recovered as biomass and CO_2 (Janshekar 1979).

    T. cutaneum has been used as a microbial sensor for determining the biological oxygen demand (BOD) in waste water (Riedel et al. 1990), for carbohydrate analysis (Riedel et al. 1990), for amperometric determination of ammonium ions in mixed cultures with *Bacillus subtilis* and *Pseudomonas aeruginosa* (Riedel et al. 1990) and for phenol determination (Neujahr and Kjellén 1979). For BOD determination, *T. cutaneum* E4 cells were immobilized by mixing them with 10% polyvinyl alcohol and pouring them on a glass plate to form a membrane. A modified oxygen electrode was coated with the immobilized cell membrane and a dialysis membrane. This biosensor had an operational stability of 48 days. Phenol hydroxylase and catechol-1,2-oxygenase isolated from *T. cutaneum* have also been used for the construction of biosensors (Kjellén and Neujahr 1979; Neujahr 1980) to detect phenol and catechol, respectively.

References

Behrend G (1890) Ueber *Trichomycosis nodosa*. Ber Klin Wochenschr 21: 464–467

Berg GR, Inniss WE, Heikkila JJ (1987) Stress proteins and thermotolerance in psychotrophic yeasts from arctic environments. Can J Microb 33: 383–389

Büttner R, Bode R, Birnbaum D (1987) Purification and characterization of extracellular glucoamylase from the yeast *Trichosporon adeninovorans*. J Basic Microbiol 27: 299–308

Büttner R, Scheit A, Bode R, Birnbaum D (1989) Isolation and characterization of mutants of *Trichosporon adeninovorans* resistant to 2-deoxy-D-glucose. J Basic Microbiol 29: 67–72

Dagley S (1985) Microbial metabolism of aromatic compounds. In: Moo-Young M, Bull AT, Dalton H (eds) Comprehensive biotechnology, vol 1. Pergamon Press, Oxford, pp 483–505

De Mot H, Verachtert H (1986) Secretion of α-amylase and multiple forms of glucoamylase by the yeast *Trichosporon pullulans*. Can J Microbiol 32: 47–51

Do Carmo-Sousa L (1970) *Trichosporon* Behrend. In: Lodder J (ed) The yeasts. A taxonomic study. 2nd edn North-Holland Publ, Amsterdam, pp 1309–1352

Drocourt D, Calmels T, Reynes J-P, Baron M, Tiraby G (1990) Cassettes of the *Streptoalloteichus hindustanus ble* gene for transformation of lower and higher eukaryotes to phleomycin resistance. Nucleic Acids Res 18: 4009

Enroth C, Huang W, Waters S, Neujahr H, Lindqvist Y, Schneider G (1994) Crystallization and preliminary X-ray analysis of phenol hydroxylase from *Trichosporon cutaneum*. J Mol Biol 238: 128–130

Fiechter A, Käppeli O, Meussdoerffer F (1987) Batch and continuous culture. In: Rose AH, Harrison JS (eds) The yeasts, vol 2. Academic Press, London, pp 99–129

Gaal A, Neujahr HY (1980) *cis, cis*-muconate cyclase from *Trichosporon cutaneum*. Biochem J 191: 37–43

Gaal A, Neujahr HY (1981) Induction of phenol-metabolizing enzymes in *Trichosporon cutaneum*. Arch Microbiol 130: 54–58

Gienow U, Kunze G, Schauer F, Bode R, Hofemeister J (1990) The yeast genus *Trichosporon* spec. LS3; Molecular characterization of genomic complexity. Zentralbl Mikrobiol 145: 3–12

Glumoff V (1992) Heterologous gene expression in the soil yeast *Trichosporon cutaneum*. Doct Thesis, Swiss Federal Institute of Technology, Zürich

Glumoff V, Käppeli O, Fiechter A, Reiser J (1989) Genetic transformation of the filamentous yeast, *Trichosporon cutaneum*, using dominant selection markers. Gene 84: 311–318

Gritz L, Davies J (1983) Plasmid-encoded hygromycin B resistance: the sequence of hygromycin B phosphotransferase gene and its expression in *Escherichia coli* and *Saccharomyces cerevisiae*. Gene 25: 179–188

Guého E, Tredick J, Phaff HJ (1984) DNA base composition and DNA relatedness among species of *Trichosporon* Behrend. Antonie Leeuwenhoek J Microbiol 50: 17–32

Guého E, Kurtzman CP, Peterson SW (1989) Evolutionary affinities of heterobasidiomycetous yeasts estimated from 18S and 25S ribosomal RNA sequence divergence. Syst Appl Microbiol 12: 230–236

Guého E, de Hoog GS, Smith MT (1992a) Neotypification of the genus *Trichosporon*. Antonie Leeuwenhoek J Microbiol 61: 285–288

Guého E, Smith MT, de Hoog GS, Billon-Grand G, Christen R, Batenburg-van der Vegte WH (1992b) Contributions to a revision of the genus *Trichosporon*. Antonie Leeuwenhoek J Microbiol 61: 289–316

Hara N, Tubota Y, Saito K, Suto T (1989) Biochemical characteristics and ubiquinone of 44 strains of *Trichosporon beigelii* and related organisms J Gen Appl Microbiol 35: 1–10

Hasegawa Y, Higuchi K, Obata H, Tokuyama T, Kaneda T (1982) A cyclohexanecarboxylic acid utilizing yeast: isolation, identification, and nutritional characteristics. Can J Microbiol 28: 942–944

Hess PN (1988) Untersuchungen zur Leistungsfähigkeit aerober Suspensionskulturen in gerührten Bioreaktoren. Doct Thesis, Swiss Federal Institute of Technology, Zürich

Hoffman CS, Winston F (1987) A ten-minute DNA preparation from yeast efficiently releases autonomous plasmids for transformation of *Escherichia coli*. Gene 57: 267–272

Hoy J, Hsu K-C, Rolston K, Hopfer RL, Luna M, Bodey GP (1986) *Trichosporon beigelii* infection: a review. Rev Infect Dis 8: 959–967

Hrmová M, Biely P, Vrsanska M, Petrakovà E (1984) Induction of cellulose- and xylan-degrading enzyme complex in the yeast *Trichosporon cutaneum*. Arch Microbiol 138: 371–376

Hug H, Blanch HW, Fiechter A (1974) The functional role of lipids in hydrocarbon assimilation. Biotechnol Bioeng 16: 965–985

Janshekar H (1979) Studies on continuous production of fodder yeast from molasses. Doct Thesis, Swiss Federal Institute of Technology, Zürich

Jaramillo A (1985) Fluid dynamics and oxygen transfer during the cultivation of *Trichosporon cutaneum* in jet loop bioreactors. Doct Thesis, Swiss Federal Institute of Technology, Zürich

Kälin M (1993) Isolation and characterization of the *Trichosporon cutaneum* phenol hydroxylase gene and its application for heterologous protein production. Doct Thesis, Swiss Federal Institute of Technology, Zürich

Kälin M, Neujahr HY, Weissmahr R, Sejlitz T, Jöhl R, Fiechter A, Reiser J (1992) Phenol hydroxylase from *Trichosporon cutaneum*. Gene cloning, sequence analysis and functional expression in *Escherichia coli*. J Bacterial 174: 7112–7120

Käppeli O, Fiechter A (1981) On the methodology of oxygen transfer coefficient measurements. Biotechnol Lett 3: 541–546

Käppeli O, Fiechter A (1982) Growth of *Trichosporon cutaneum* under oxygen limitation: kinetics of oxygen uptake. Biotechnol Bioeng 24: 2519–2526

Kemker BJ, Lehmann PF, Lee JW, Walsh TJ (1991) Distinction of deep versus superficial clinical and nonclinical isolates of *Trichosporon beigelii* by isoenzymes and restriction fragment length polymorphisms of rDNA generated by polymerase chain reaction. J Clin Microbiol 29: 1677–1683

Kjellén KG, Neujahr HY (1979) Immobilization of phenol hydroxylase. Biotechnol Bioeng 21: 715–719

Kreger-van Rij NJW, Veenhuis M (1971) A comparative study of the cell wall structure of basidiomycetous and related yeasts. J Gen Microbiol 68: 87–95

Küng W, Moser A (1986) Bioprocess engineering characteristics of the horizontal stirred tank. Bioproc Engin 1: 23–28

Laaser G, Möller E, Jahnke K-D, Bahnweg G, Prillinger H, Prell HH (1989) Ribosomal DNA restriction fragment analysis as a taxonomic tool in separating physiologically similar basidiomycetous yeasts. Syst Appl Microbiol 11: 170–175

Laemmli UK (1970) Cleavage of structural proteins during the assembly of the head of bacteriophage T4. Nature 227: 680–685

Lehman PF, Hsiao C-B, Salkin IF (1989) Protein and enzyme electrophoresis profiles of selected *Candida* species. J Clin Microbiol 27: 400–404

Maniatis T, Fritsch EF, Sambrook J (1982) Molecular cloning. A Laboratory Manual. Cold Spring Harbor Laboratory, Cold Spring Harbor

Mattern IE, Punt P (1988) A vector of *Aspergillus* transformation conferring phleomycin resistance. Fungal Genet Newslett 35: 25

Mazur P, Pieken WA, Budihas SR, Williams SE, Wong S, Kozarich JW (1994) *Cis-cis*-muconate lactonizing enzyme from *Trichosporon cutaneum*. Biochemistry 33: 1961–1970

McBride ME, Kalter DC, Wolf JE Jr (1988) Antifungal susceptibility testing of *Trichosporon beigelii* to imidazole compounds. Can J Microbiol 34: 850–854

Middelhoven WJ, van den Brink JA, Veenhuis M (1983) Growth of *Candida famata* and *Trichosporon cutaneum* on uric acid as the sole source of carbon and energy, a hitherto unknown property of yeasts. Antonie Leeuwenhoek J Microbiol 49: 361–368

Middelhoven WJ, Hoogkramer-Te Niet MC, Kreger-Van Rij NJW (1984) *Trichosporon adeninovorans* sp. nov., a yeast species utilizing adenine, xanthine, uric acid, putrescine and primary n-alkylamines as the sole source of carbon, nitrogen and energy. Antonie Leeuwenhoek J Microbiol 50: 369–378

Middelhoven WJ, de Jong IM, de Winter M (1991) *Arxula adeninivorans*, a yeast assimilating many nitrogenous and aromatic compounds. Antonie Leeuwenhoek J Microbiol 59: 129–137

Mörtberg M, Neujahr HY (1985) Uptake of phenol by *Trichosporon cutaneum*. J Bacteriol 161: 615–619

Mörtberg M, Neujahr HY (1986) Transport and hydrolysis of disaccharides by *Trichosporon cutaneum*. J Bacteriol 168: 734–738

Neujahr HY (1980) Enzyme probe for catechol. Biotechnol Bioeng 22: 913–918

Neujahr HY (1990) Yeasts in biodegradation and biodeterioration processes. In: Verachtert H, De Mot R (eds) Yeasts: biotechnology and biocatalysis. M Dekker, New York, pp 322–348

Neujahr HY, Gaal A (1973) Phenol hydroxylase from yeast. Purification and properties of the enzyme from *Trichosporon cutaneum*. Eur J Biochem 35: 386–400

Neujahr HY, Kjellén KG (1979) Bioprobe electrode for phenol. Biotechnol Bioeng 21: 671–678

Neujahr HY, Varga JM (1970) Degradation of phenols by intact cells and cell-free preparations of *Trichosporon cutaneum*. Eur J Biochem 13: 37–44

Ochsner UA, Glumoff V, Kälin M, Fiechter A, Reiser J (1991) Genetic transformation of the filamentous yeast *Trichosporon cutaneum* using homologous and heterologous marker genes. Yeast 7: 513–521

Ohama T, Suzuki T, Mori M, Osawa S, Ueda T, Watanabe K, Nakase T (1993) Non-universal decoding of the leucine codon CUG in several *Candida* species. Nucleic Acids Res 21: 4039–4045

Orbach MJ, Porro EB, Yanofsky C (1986) Cloning and characterization of the gene for β-tubulin from a benomyl-resistant mutant of *Neurospora crassa* and its use as a dominant selectable marker. Mol Cell Biol 6: 2452–2461

Punt PJ, Oliver RP, Dingemanse MA, Pouwels PH, van den Hondel CAMJJ (1987) Transformation of *Aspergillus* based on the hygromycin B resistance marker from *Escherichia coli*. Gene 56: 117–124

Raeder U, Broda P (1988) Preparation and characterization of DNA from lignin-degrading fungi. Methods Enzymol 161: 211–220

Reiser J, Glumoff V, Ochsner UO, Fiechter A (1994) Molecular analysis of the *Trichosporon cutaneum* DSM 70698 *argA* gene and its use for DNA-mediated transformations. J Bacteriol 176: 3021–3032

Ratledge C (1988) Yeasts for lipid production. Biochem Soc Trans 16: 1088–1091

Riedel K, Lange KP, Stein HJ, Huehn M, Ott P, Scheller F (1990) A microbial sensor for BOD – *Trichosporon cutaneum* microbial electrode used for waster-water BOD determination. Waste Res 24: 883–887

Rochkind-Dubinsky M, Sayler GS, Blackburn JW (1987) Metabolism of nonchlorinated aromatic compounds. In: Microbiological decomposition of chlorinated aromatic compounds. Marcel Dekker, New York, pp 48–72

Rodriguez RL, Tait RC (1983) Recombinant DNA techniques: an introduction. Addison-Wesley, Reading, pp 167–168

Schneider H, Biely P, Latta R, Lee H, Dorscheid D, Levy-Rick S (1990) Utilization by yeasts of D-glucarate, galactarate, and L-tartarate is uncommon and occurs in strains of *Cryptococcus* and *Trichosporon*. Can J Microbiol 36: 856–858

Sejlitz T, Neujahr HY (1987) Phenol hydroxylase from yeast. A model for phenol binding and an improved purification procedure. Eur J Biochem 170: 343–349

Sienko M, Stepien PP, Paszewski A (1992) Generation of genetic recombinants in *Trichosporon cutaneum* by spontaneous segregation of protoplast fusants. J Gen Microbiol 138: 1409–1412

Slater ML (1976) Rapid nuclear staining method for *Saccharomyces cerevisiae*. J Bacteriol 126: 1339–1341

Spånning Å, Neujahr HY (1987) Growth and enzyme synthesis during continuous culture of *Trichosporon cutaneum* on phenol. Biotechnol Bioeng 29: 464–468

Stevens BJH, Payne J (1977) Cellulase and xylanase production by yeasts of the genus *Trichosporon*. J Gen Microbiol 100: 381–393

Stüttgen E, Sahm H (1982) Purification and properties of endo-1,4-β-xylanase from *Trichosporon cutaneum*. Eur J Appl Microbiol Biotechnol 15: 93–99

Sze IS, Dagley S (1984) Properties of salicylate hydroxylase and hydroxyquinol 1,2-dioxygenase purified from *Trichosporon cutaneum*. J Bacteriol 159: 353–359

Teeri T, Kumar V, Lehtovaara P, Knowles JKC (1987) Construction of cDNA libraries by blut-end ligation: high frequency cloning of long cDNAs from filamentous fungi. Anal Biochem 164: 60–67

Van de Peer Y, Hendriks L, Goris A, Neefs J-M, Vancanneyt M, Kersters K, Berny J-F, Hennebert GL, de Wachter R (1992) Evolution of basidiomycetous yeasts as deduced from small ribosomal subunit RNA sequences. Syst Appl Microbiol 15: 250–258

Van der Walt JP, Hopsu-Havu VK (1976) A colour reaction for the differentiation of ascomycetous and hemibasidiomycetous yeasts. Antonie Leeuwenhoek J Microbiol 42: 157–163

Varga JM, Neujahr HY (1990) Purification and properties of catechol 1,2-oxygenase from *Trichosporon cutaneum*. Eur J Biochem 12: 427–434

Veenhuis M, Hoogkamer-Te Niet MC, Middelhoven WJ (1985) Biogenesis and metabolic significance of microbodies in urate-utilizing yeasts. Antonie Leeuwenhoek J Microbiol 51: 33–43

Veenhuis M, van der Klei IJ, Harder W (1986) Physiological role of microbodies in the yeast *Trichosporon cutaneum* during growth on ethylamine as the source of energy, carbon and nitrogen. Arch Microbiol 145: 39–50

Waters S, Neujahr HY (1994) A fermentor culture for production of recombinant phenol hydroxylase. Prot Express Purif 5: 534–540

Weijman ACM (1979) Carbohydrate composition and taxonomy of *Geotrichum, Trichosporon* and allied genera. Antonie Leeuwenhoek 45: 119–127

West M, Emerson GW, Sullivan PA (1990) Purification and properties of two lactose hydrolases from *Trichosporon cutaneum*. J Gen Microbiol 136: 1483–1490

Williamson DH, Fennell DJ (1975) The use of fluorescent DNA binding agents for detecting and separating yeast mitochondrial DNA. Methods Cell Biol 17: 335–351

Ykema A, Verbree EC, Nijkamp HJ, Smit H (1989) Isolation and characterization of fatty acid auxotrophs from the oleaginous yeast *Apiotrichum curvatum*. Appl Microbiol Biotechnol 32: 76–84

Yonsel S, Deckwer W-D (1990) Maßstabvergrößerung pneumatisch betriebener Reaktoren (Airlift/Blasensäule) für O$_2$-limitierte Fermentationen. BioEngineering 6: 12–23

Zimmermann M, Emeis CC (1989) Extracellular polysaccharidases from *Trichosporon beigelii*. In: Martini A, Vaughan Martini A (eds) Proc 7th Int Symp on Yeasts, Wiley, Chichester, S131

Subject Index

Springer-Verlag
and the Environment

We at Springer-Verlag firmly believe that an international science publisher has a special obligation to the environment, and our corporate policies consistently reflect this conviction.

We also expect our business partners – paper mills, printers, packaging manufacturers, etc. – to commit themselves to using environmentally friendly materials and production processes.

The paper in this book is made from low- or no-chlorine pulp and is acid free, in conformance with international standards for paper permanency.